RENEWALS 458-4574
DATE DUE

Polar Remote Sensing
Volume II: Ice Sheets

Robert Massom and Dan Lubin

Polar Remote Sensing

Volume II: Ice Sheets

Published in association with
Praxis Publishing
Chichester, UK

Dr. Robert Massom
Senior Research Scientist and Lecturer
Antarctic Climate and Ecosystems Cooperative Research Centre
University of Tasmania
Sandy Bay
Tasmania
Australia

Dr. Dan Lubin
Research Physicist and Senior Lecturer
Scripps Institution of Oceanography
University of California, San Diego
La Jolla
California
USA

SPRINGER–PRAXIS BOOKS IN GEOPHYSICAL SCIENCES
Published in association with Antarctic Climate and Ecosystems Cooperative Research Centre, Australia
SUBJECT *ADVISORY EDITOR*: Dr. Philippe Blondel, C.Geol., F.G.S., Ph.D., M.Sc., Senior Scientist, Department of Physics, University of Bath, Bath, UK

ISBN 3-540-26101-X Springer-Verlag Berlin Heidelberg New York

Springer is part of Springer-Science + Business Media (springeronline.com)

Bibliographic information published by Die Deutsche Bibliothek

Die Deutsche Bibliothek lists this publication in the Deutsche Nationalbibliografie; detailed bibliographic data are available from the Internet at http://dnb.ddb.de

Library of Congress Control Number: 20059276101

Apart from any fair dealing for the purposes of research or private study, or criticism or review, as permitted under the Copyright, Designs and Patents Act 1988, this publication may only be reproduced, stored or transmitted, in any form or by any means, with the prior permission in writing of the publishers, or in the case of reprographic reproduction in accordance with the terms of licences issued by the Copyright Licensing Agency. Enquiries concerning reproduction outside those terms should be sent to the publishers.

© Praxis Publishing Ltd, Chichester, UK, 2006
Printed in Germany

The use of general descriptive names, registered names, trademarks, etc. in this publication does not imply, even in the absence of a specific statement, that such names are exempt from the relevant protective laws and regulations and therefore free for general use.

Cover design: Jim Wilkie
Project management: Originator Publishing Services, Gt Yarmouth, Norfolk, UK

Printed on acid-free paper

Contents

Preface	ix
List of figures	xiii
List of tables	xxi
List of abbreviations	xxiii
About the author	xxix
Publisher credits	xxxi

1 Polar ice sheets: Introduction — 1
   1.1 The global importance of polar ice sheets — 1
   1.2 Ice-sheet mass balance—background information — 10
       1.2.1 Comparative characteristics of the Greenland and Antarctic Ice Sheets — 16
       1.2.2 Approaches to the measurement of ice-sheet mass balance — 16
   1.3 The recent revolution in satellite remote sensing — 21
   1.4 References — 24

2 Synthetic aperture radar interferometry and related techniques — 39
   2.1 Introduction and background — 39
   2.2 Underlying principles and terminology of InSAR — 41
       2.2.1 The basis of ice sheet surface elevation and motion measurement by satellite repeat-pass interferometry — 46
       2.2.2 Measurement of ice-surface displacement/motion — 50
       2.2.3 What is an interferogram? — 54
   2.3 Inherent constraints and sources of ambiguity and error — 55
       2.3.1 Thermal decorrelation — 56
       2.3.2 Temporal decorrelation — 57

		2.3.3	Decorrelation due to excessive target motion/rotation	59
		2.3.4	Atmospheric and ionospheric propagation delay	60
		2.3.5	Baseline (spatial) decorrelation	62
		2.3.6	Volume decorrelation, and ambiguities relating to radar penetration depth and volume scattering	63
		2.3.7	Image speckle	66
		2.3.8	Geometric distortions and steep terrain effects	67
		2.3.9	Estimation of phase decorrelation	67
		2.3.10	Processing errors and other sources of ambiguity	70
		2.3.11	Summary of baseline-related trade-offs	70
	2.4	Satellite-SAR interferometric data requirements and processing steps	71
		2.4.1	Selection of suitable SAR images	74
		2.4.2	Geometric co-registration	81
		2.4.3	Pre-processing and interferogram generation	82
		2.4.4	Phase unwrapping	84
		2.4.5	Phase-to-height conversion	89
		2.4.6	Interferogram geo-coding	89
	2.5	The measurement of ice-sheet elevation and surface velocity using differential InSAR techniques	90
		2.5.1	Ice-sheet topography from DInSAR	92
		2.5.2	Retrieval of ice-sheet surface motion	93
		2.5.3	Tidal effects on InSAR retrievals over floating ice	101
	2.6	Ice-sheet surface motion retrieval by speckle tracking	101
	2.7	Ice-sheet surface motion retrieval by coherence tracking	108
	2.8	Relative strengths and weaknesses of differential InSAR, coherence tracking, and intensity tracking	109
	2.9	Polarimetric Interferometric SAR (Pol-InSAR)	111
	2.10	Developments and outstanding issues	112
	2.11	References	116
3	**Satellite remote sensing of ice sheet parameters and processes**			137
	3.1	Introduction	137
	3.2	Improved detection and mapping of surface features	137
		3.2.1	High-resolution visible–thermal infrared methods	137
		3.2.2	Moderate-resolution visible/thermal infrared	145
		3.2.3	Synthetic aperture radar	151
		3.2.4	The mapping of ice-sheet facies	158
	3.3	Detection, mapping and monitoring of ice-sheet surface melt and refreezing	164
		3.3.1	Visible and thermal infrared techniques	164
		3.3.2	Active-microwave techniques	166
		3.3.3	Passive-microwave techniques	169
		3.3.4	Estimation of surface melt rates	174
	3.4	Detection and mapping of changes in ice-sheet margins	175
		3.4.1	The collapse of the Larsen Ice Shelf	177

	3.4.2	Examples of recent change in the configuration of other glacier systems	183
3.5	Icebergs		186
	3.5.1	Iceberg detection and size statistics	186
	3.5.2	Monitoring iceberg drift	199
	3.5.3	Measurement of iceberg thickness	206
3.6	Measurement of ice sheet topography/elevation and change in elevation		209
	3.6.1	Surface elevation from satellite radar altimetry	211
	3.6.2	Surface elevation from laser altimetry	229
	3.6.3	Improved digital elevation model construction using satellite image-based photoclinometry	234
	3.6.4	Surface elevation from SAR interferometry (InSAR)	237
3.7	Accumulation rate		240
3.8	Ice velocity, strain rate, and balance velocity/flux		247
	3.8.1	Feature tracking	250
	3.8.2	SAR interferometry and ice velocity	257
	3.8.3	Balance velocities and fluxes	265
	3.8.4	Strain rates	268
	3.8.5	Detection of variability in ice flow and surge behavior	271
	3.8.6	Ice shelf buttressing, and the impact of ice shelf removal on outlet glaciers	274
3.9	Tidal displacement of ice shelves and glacier tongues		276
3.10	Grounding-line (zone) detection and monitoring		279
3.11	Estimates of ice discharge flux and basal melt/freeze rates		281
3.12	The measurement of changes in ice mass by satellite gravity sensors		288
3.13	The radar sounding of ice sheets from space		289
3.14	Ice-sheet surface temperature		291
	3.14.1	Thermal infrared techniques	292
	3.14.2	Passive-microwave techniques	295
3.15	Ice sheet surface albedo		299
3.16	Grain size, impurity content, and surface to near-surface characteristics		309
	3.16.1	Snow-grain size and impurity content	309
	3.16.2	Changes in snow/firn characteristics inferred from passive-microwave data	312
	3.16.3	Ice sheet roughness characteristics and proxy wind measurements	315
	3.16.4	Active-microwave remote sensing of ice-sheet surface and near-surface characteristics	318
3.17	Conclusions		321
3.18	References		323
Appendix			387
Index			397

Preface

The objective of this two-volume book is to survey one of the most challenging and at the same time most necessary applications of satellite remote sensing: geophysical sciences in the Earth's polar regions. Much has happened in this field over the past decade. In addition to dedicated satellite instruments and programs that have monitored critical manifestations of climate and atmospheric change, such as the retreat of Arctic sea ice, the motion of the great Antarctic ice sheets, and the evolution of the ozone "hole" in both polar regions, many serendipitous applications of satellite remote sensing have arisen for polar research. In recent years, numerous imaginative researchers have adapted remote sensing instruments for high-latitude investigations in ways that were never envisioned by the sensors' designers. A survey of polar remote sensing accomplishments is particularly useful at this point in time, as the Earth Science community is experiencing a transition to a new generation of satellite remote sensing instruments with an order-of-magnitude greater capability than their predecessors.

This two-volume set is sub-divided into major geophysical disciplines related to climate change, and the intent is to introduce the physical principles involved with remote sensing derivation of important geophysical quantities in the polar regions. In this second volume, an introductory chapter on the key importance of ice sheets and their mass balance is followed bywith a chapter on Synthetic Aperture Radar interferometry or InSAR—an exciting new technique that is revolutionizing the measurement of vast and remote ice sheets from space. Examples are given of the extraordinary results that are emerging. The final chapter is devoted to a review of ice-sheet and glacier properties and parameters that can be measured by satellites, highlighting the strong interactions with atmosphere and ocean.

At the time of going to press, we were greatly saddened to learn of the loss of CryoSat during its launch (in October 2005). This innovative mission was to have made an immense contribution to reducing uncertainties in current estimates of ice mass balance (both ice-sheet and sea-ice). Our thoughts go out to those colleagues who have put years of energy (both physical and mental) into realizing this important mission, only to see it dematerialize at the 11th hour. We sincerely hope that a follow-on/replacement mission will be launched at some stage in the future. As such, we have retained sections on CryoSat in this volume. Please note that Volume 1 went to press prior to the launch of CryoSat.

Many individuals have played important roles in the recent growth and vitality of polar satellite remote sensing. In particular, we acknowledge Mr. Robert Whritner of the Scripps Institution of Oceanography. In the late 1980s, Bob recognized the need to archive the telemetry collected by polar orbiter tracking equipment then just installed at McMurdo and Palmer Stations by the U.S. National Science Foundation, for the benefit of the greater scientific community. Not only did he accomplish this, but in establishing the Arctic and Antarctic Research Center at Scripps, he also made his vast experience as a Navy meteorologist freely available to researchers interested in using the data. It is no exaggeration to state that Bob has helped hundreds of researchers worldwide, on an individual basis, to incorporate satellite remote sensing into their programs. Dr. Robert L. Bernstein, founder and now Chief Technical Officer of the SeaSpace Corporation, has also played a major role. SeaSpace, which in many respects sets the industry standard in Earth satellite tracking and scientific data processing, got its start with the NSF Antarctic contracts. The challenging Antarctic environment was a factor in the immediate technical excellence of the SeaSpace products, and these high standards in turn ensured that polar satellite data have been steadily available to a great many researchers and field expeditions, not only at fixed land-based stations but also from satellite tracking equipment aboard polar research vessels. We also acknowledge the extraordinary contribution of Dr. Joey Comiso of the NASA Goddard Space Flight Center, and of many other colleagues who are too numerous to mention here. Their immense contribution in shaping the development of polar remote sensing will become fully apparent in the text and references. Writing a book is a major responsibility, and we only hope that we have done justice to their important work.

On a personal as well as professional level, Rob would like to extend heartfelt thanks to a number of special colleagues who have given extraordinary help, encouragement and much-valued friendship during his quarter century in polar research. These are Ted Scambos (NSIDC, University of Colorado), Joey Comiso and Claire Parkinson (NASA Goddard Space Flight Center), Mark Drinkwater (European Space Agency), Mark Rosenberg (Antarctic Climate and Ecosystems [ACE] CRC), Mark Curran (ACE CRC and Australian Antarctic Division), Mike Pook (CSIRO Division of Marine and Atmospheric Research, Australia), Andrew Cowan (formerly Scott Polar Research Institute), Vernon Squire (University of Otago), Gareth Rees (Scott Polar Research Institute), Peter Wadhams (University of Cambridge), Matthew Sturm (U.S. Cold Regions Research and Engineering Laboratory) and Steve Ackley (Clarkson University) (CRREL).

Dan would like to extend a heartfelt thanks to colleagues and mentors who have given him so much encouragement, friendship, and so many outstanding opportunities as a research physicist. These are John Frederick and Doug MacAyeal (University of Chicago), Catherine Gautier (University of California Santa Barbara),

V. Ramanathan, Richard Somerville, Francisco P. J. Valero, Greg Mitchell, Osmund Holm-Hansen, Wolf Berger, Charles Kennel (Scripps Institution of Oceanography), Sally Ride, David Tytler, James Arnold (University of California San Diego), Knut Stamnes (Stevens Institute of Technology), David Bromwich (Byrd Polar Research Center), Charles R. Booth (Biospherical Instruments, Inc.), and Moshe J. Lubin (Laboratory for Laser Energetics, University of Rochester).

Many people have unselfishly and kindly given of their energy and precious time to contribute to this book by providing papers, expert comments and diagrams. These include Ted Scambos, Terry Haran, Florence Fetterer, Mark Serreze, Julienne Stroeve, Nancy Geiger Wooten, and Ken Knowles (all U.S. National Snow and Ice Data Center, University of Colorado); Helen Fricker (Scripps Institution of Oceanography); Ian Joughin (University of Washington [formerly of NASA Jet Propulsion Laboratory]); Mark Drinkwater and Evert Attema (European Space Agency); Joey Comiso, Claire Parkinson, Thorsten Markus, Bob Bindschadler, Dorothy Hall, Kimberley Casey, Nick DiGirolamo, Christopher Shuman, Jay Zwally, Rich McPeters, Ashwin Mahesh, and Patricia Vornberger (all NASA Goddard Space Flight Center); Henrik Steen Andersen (Danish Meteorological Institute); Mark Fahnestock (University of New Hampshire, U.S.A.); Klaus Grosfeld (University of Bremen, Germany); David Long (Brigham Young University, U.S.A.); Ian Boyd (University of St. Andrews, Scotland); Kevin Arrigo and Howard Zebker (Stanford University, U.S.A.); Frank Rau (University of Freiburg, Germany); Tony Liu (NASA Goddard Space Flight Center and U.S. Office of Naval Research, U.S.A.); Anne Nolin (Oregon State University, U.S.A.); Benoît Legrésy and Frédérique Rémy (LEGOS [CNES–CNRS–UPS], France); Eric Rignot, Ron Kwok, Ben Holt, and David Diner (NASA Jet Propulsion Laboratory, U.S.A.); Seelye Martin, Dale Winebrenner, Robert Drucker, Yanling Yu, and Ron Lindsay (all University of Washington); Bob Jacobel (St. Olaf's College, U.S.A.); Hermann Engelhardt (California Institute of Technology); Konrad Steffen, Russell Huff, and Chuck Fowler (University of Colorado); Leif Toudal Pedersen (Danish Centre for Remote Sensing); Wolfgang Rack (Alfred Wegener Institute, Germany); Helmut Rott (University of Innsbruck); Mike Manore, Dean Flett, and Katherine Wilson (Canadian Ice Service); Tavi Murray (University of Leeds, England); Cheryl Bertoia (COSPAS–SARSAT Secretariat, England); Yunhe Zhao (Caelum Research Corporation, U.S.A.); Harry Keys (Dept. of Conservation, New Zealand); David Vaughan (British Antarctic Survey); Jia Wang (IARC, University of Alaska Fairbanks); Jörg Haarpaintner (Norwegian Meteorological Institute); Massimo Frezzotti (ENEA CLIM–OSS, Italy); Baerbel Lucchitta (USGS, U.S.A.); Wolfgang Dierking (Alfred Wegener Institute, Germany); Christophe Genthon and Olivier Torinesi (LGGE CNRS, France); Miguel Angel Morales Maqueda (New York University); Kazutaka Tateyama (Kitami Inst. of Technology, Japan); Jon Bamber (University of Bristol, England); Pedro Skvarca (Instituto Antártico Argentino, Argentina); LeenKiat Soh (University of Nebraska, U.S.A.); Gennady Belchansky (IARC, University of Alaska Fairbanks); Bill Pichel (NOAA/NESDIS, U.S.A.); David Douglas (USGS Alaska Science Center, U.S.A.); Sheldon Drobot (National Academies, Washington D.C.); Jørgen Dall (Danish Centre for Remote Sensing); Laurence Gray (Canada Centre for Remote Sensing); Rick Forster

(University of Utah, U.S.A.); Johan Mohr (Technical University of Denmark); Bob Onstott (General Dynamics–Advanced Information Systems, U.S.A.); Birgitte Furevik (Nansen Environmental and Remote Sensing Center, Norway); Seymour Laxon (University College London, England); Hernán De Angelis (University of Stockholm, Sweden); Laurie Padman (Earth and Space Research, U.S.A.); Berndt Scheuchl (University of British Columbia, Canada); Sylviane Surdyk and Yan Ropert-Coudert (National Institute of Polar Research, Tokyo); Steve Rintoul (CSIRO Marine Research); Warren White (Scripps Institution of Oceanography, U.S.A.); Volkmar Wismann (IFARS, Germany); Roland Warner, Neal Young, and Barry Giles (Antarctic Climate and Ecosystems Cooperative Research Centre [ACE CRC] and Australian Antarctic Division); Susan Solomon (NOAA Aeronomy Laboratory); Mike Fromm (U.S. Naval Research Laboratory); Cora Randall (LASP, University of Colorado); Kathy Pagan (NASA Ames Research Center); Ray Smith (University of California, Santa Barbara); Sharon Stammerjohn (Lamont-Doherty Observatory, Columbia University); Stephen Warren, Thomas Grenfell, and Bonnie Light (University of Washington); Von Walden (University of Idaho); Don Perovich (U.S. Army Cold Regions Research and Engineering Laboratory); David Bromwich (Byrd Polar Research Center); Jennifer Francis (Rutgers University); Jeff Key (NOAA/NESDIS); Janet Intrieri (NOAA ETL); Matthew Lazzara and Shelly Knuth (University of Wisconsin); and Tom Charlock (NASA Langley Research Center). This book would not exist without the inestimable help and input of these people.

Rob extends his gratitude to Glenn Hyland (ACE CRC and Australian Antarctic Division) for reviewing Chapter 2, to Roland Warner, Vin Morgan, and Tas Van Ommen (ACE CRC and AAD) for their helpful comments regarding the ice-sheet mass balance section in Chapter 1, and to Mark Rosenberg (ACE CRC) for his inestimable help in editing references. He is also grateful to his many wonderful colleagues at the Antarctic Climate and Ecosystems Cooperative Research Centre (CRC) in Hobart, Tasmania (Australia), and the previous Antarctic CRC. For Rob, this work was supported by the Australian Government's Cooperative Research Centres Programme through the Antarctic Climate and Ecosystems Cooperative Research Centre (ACE CRC). He is very grateful for this support.

Dan is grateful to Gabrielle Ayres for invaluable editorial support; to Jo Griffith for excellent illustrations; and to Steve Hart and Rob Wittenmyer for assistance with satellite data processing.

Last but not least, we are extremely grateful to Philippe Blondel (University of Bath, England), who carried out the onerous task of scientifically reviewing the book with extreme diligence, thoughtfulness and good humour. All of his comments have been highly beneficial. We are equally grateful to Clive Horwood (Praxis Publishing) for inviting us to write this book, and for his patience and encouragement as this project evolved, and to Neil Shuttlewood and his team at Originator for outstanding editorial and production work.

This book is dedicated to Dan's wife Lorri, and Rob's wife Yuko and his beautiful daughter Adelle Yuki. This project would not have been possible without Yuko sour spouses' love and unstinting support throughout, and little Adelle was a wonderful bonus for Rob during the writing.

Figures

1.1	A simplified two-dimensional schematic cross-section of a coupled Antarctic ice shelf-ocean system.	3
1.2	A schematic diagram of an ice sheet	10
1.3	The location of automatic weather stations (AWSs) in Antarctica in March 2004, superimposed on a NOAA Advanced Very High Resolution Radiometer (AVHRR) image mosaic of the ice sheet	22
2.1	A schematic representation showing the relationship between the wavelength, amplitude, and phase characteristics of electromagnetic radiation	43
2.2	A schematic showing the basic principle of repeat-pass inteferometric phase measurement from space	44
2.3	Geometric configuration for satellite repeat-pass interferometry.	46
2.4	(Left) Interferogram of ice motion in the part of the Rutford Ice Stream, Antarctica. (Right) Location map of glaciological features shown	color
2.5	An example of "azimuth streaks" in an azimuth shift image for a Radarsat-1 Antarctic Mapping Mission (AMM) image pair.	61
2.6	Map of the C-band penetration depth of the Geikie Ice Cap (Greenland).	64
2.7	Effect of penetration depth d and effective baseline length on the interferometric phase	66
2.8	A simplified schematic illustration of forms of distortion in radar images of the Earth's surface	68
2.9	The typical chain of processing steps involved in the retrieval of a Digital Elevation Model (DEM) from a coherent interferometric SAR image pair	72
2.10	A series of images of Pine Island Glacier, West Antarctica, at ~75°S and 100°W, showing a sequence of processing steps to produce a InSAR-derived map of ice velocity	color
2.11	Map of ice motion of the lower reaches of Pine Island Glacier	color
2.12	(a) Map of Greenland showing ERS coverage obtained through the Greenmap Project. (b) Interferometric strip processing of ERS data	color
2.13	Results from the InSAR study of an ice stream in NE Greenland	color

xiv **Figures**

2.14	A 3-D shaded relief representation of the surface elevation of part of West Greenland, showing a swath derived from ERS InSAR data superimposed on a smoother ice sheet surface derived from pulse-limited radar altimetry	94
2.15	Schematic of the geometry for the measurement of ice sheet surface velocity from InSAR	96
2.16	Color-coded map of the InSAR-derived annual horizontal ice displacement in the region around Storstrømmen Glacier (NE Greenland)	color
2.17	Map of InSAR-derived ice flow vectors and elevations from ERS data, from the region around Storstrømmen Glacier (NE Greenland)	color
2.18	An important application of satellite radar interferometry is the systematic, precise and detailed detection of the grounding line of ice sheet outlet glaciers	color
2.19	The data-processing sequence to retrieve range and azimuth displacement data by speckle tracking from a coherent pair of SAR images	103
2.20	A schematic of (a) the satellite geometry on two passes separated by baseline B, and (b) the ground geometry for the calculation of ice sheet surface displacement in three dimensions using the SAR speckle-filtering technique	104
2.21	Schematic of the procedure for calculating ice surface velocity by speckle tracking	105
2.22	(a) Surface velocity derived by speckle tracking in three 1997 Radarsat-1 AMM repeat swaths on the Filchner Ice Shelf, Antarctica. (b) A detail of the Recovery Glacier, which enters the Filchner Ice Shelf, showing ice motion	color
2.23	Examples of interferometric satellite constellations investigated by CNES and DLR for a possible TerraSAR-L inteferometric cartwheel configuration.	115
3.1	(a) Landsat 2 MSS image of the Jutulstraumen ice stream, Queen Maud Land, East Antarctica, acquired on October 30, 1975. (b) Landsat 3 RBV image of the Rennick Glacier and the Oates Coast, northern Victoria Land, Antarctica, acquired on September 17, 1980	140
3.2	(a) A MODIS channel-1 image of the Larsen B Ice Shelf, Antarctic Peninsula from November 19, 2000. (b) Enlarged IKONOS panchromatic scene to show one of the dolines	142
3.3	An Ikonos image of part of the Bindschadler Ice Stream (formerly Ice Stream D), West Antarctica, on February 9, 2001, showing fields of sastrugi	144
3.4	The West Antarctic sector of the AVHRR image mosaic of Antarctica produced by Ferrigno et al. (2000), with major features marked	146
3.5	Comparison of (a) a Landsat TM scene, (b) an AVHRR image processed by the data-cumulation method, and (c) a single AVHRR scene	148
3.6	Data-cumulated (enhanced resolution) NOAA AVHRR image composite of the Ross Ice Shelf	149
3.7	Dynamic features derived from Figure 3.6: (a) flowstripes, (b) rifts and crevasse series, (c) provenance of the ice determined from tracing flow boundary tracks, and (d) flowlines and 100-year particle motion	150
3.8	An enhanced resolution MODIS multi-image composite of the southeastern section of the floating Ross Ice Shelf, created from about 10 images by the data cumulation technique	152
3.9	A mapped cloud-free MODIS band-2 image of the Amery Ice Shelf, at a full spatial resolution of 250 m and acquired on December 2, 2000	153
3.10	High-resolution Radarsat-1 image mosaic of Antarctica collected during AMM between September 19 and October 14, 1997	156

Figures xv

3.11	Two Radarsat-1 images derived from the Radarsat-1 AMM dataset shown in Figure 3.10, and homing in on the Lambert Glacier and Amery Ice Shelf region of East Antarctica to show the extraordinary and unprecedented level of detail obtainable from these data.	157
3.12	Daily image time series of satellite radar scatterometer A data of Greenland for four periods (days 209, 223, 237 and 251) in the boreal summer to autumn freeze-up period	161
3.13	Ice-sheet facies in Greenland observed in Seasat SASS data for the boreal autumn of 1978.	162
3.14	Comparison of A measurements (normalized to a 40° incidence angle) of the Greenland Ice Sheet made by the SASS with those of the NSCAT and ERS-2 scatterometer made after an 18-year time interval—i.e., 1978 to 1996.	color
3.15	(a) Radarsat-1 ScanSAR mosaic of the northern Antarctic Peninsula, December 12, 1998. (b) "Radar facies" observed during January 5, 1999.	color
3.16	Ice sheet facies and corresponding radar zones, with typical backscatter values for the Antarctic Peninsula region during the ablation season	163
3.17	A cloud-free Landsat TM image of an area of the west-central Greenland Ice Sheet, acquired on June 22, 1990	165
3.18	Plot of the change in σ_{40} as a function of volumetric snow moisture content for different thicknesses of a wet-snow layer	167
3.19	Maps of the areal extent of snowmelt in Greenland for the years 1992, 1995, 1996 and 1999, derived from ERS-1 and -2 radar scatterometer data.	color
3.20	Schematic summarizing results from an analysis of ERS-1 scatterometer σ_{40}^{o} data along the western slope of the Greenland Ice Sheet from 1992 to 1995, showing the different facies	color
3.21	A record of melt duration at the ETH/CU Camp station on the Greenland Ice Sheet from 1999 to 2003	169
3.22	Map of the mean annual duration of circum-Antarctic melt derived from analysis of Nimbus-7 SMMR and DMSP SSM/I data for 18 years of the period 1980–1999	color
3.23	A map of the maximum areal extent of surface melt on the Greenland Ice Sheet for (a) 1992 and (c) 2002, derived from analysis of Nimbus-7 SMMR and DMSP SSM/I data. (b) Time series of maximum annual melt extents for the Greenland Ice Sheet derived from the same data	171
3.24	Microwave emission at 19-GHz H-polarization measured by the DMSP SSM/I for two pixels on the Antarctic Peninsula, January 1, 1990 to December 31, 1994	173
3.25	Maps of melt season lengths on the Antarctic Peninsula.	color
3.26	(a) Photograph of a meltwater stream flowing into a large moulin in the ablation zone of the Greenland Ice Sheet in summer. (b) Schematic of glaciological features in the ablation and equilibrium zones of the West Greenland Ice Sheet	176
3.27	Cloud-free satellite images of the breakup and surface melt ponding on the Larsen B Ice Shelf.	178
3.28	An ERS SAR image composite of the northern Larsen Ice Shelf, from October 2000.	179
3.29	Ice front positions for the Larsen A Ice Shelf and Larsen Inlet, eastern Antarctic Peninsula for various dates.	181
3.30	(a) Trends of mean surface air temperature for December–February. (b) the retreat of ice shelves in the NE Antarctic Pensinsula	color

xvi **Figures**

3.31	(a) The RAMP image mosaic of Antarctica. (b) Plot of backscatter intensity .	182
3.32	An ASTER image of the Pine Island Glacier, West Antarctica, acquired on December 12, 2000	184
3.33	Images of the Pine Island Glacier acquired by the Terra MISR, showing the birth of a large tabular iceberg (42 × 17 km) some time between November 4 and 12. 2001	185
3.34	Map of ice-front positions for the Pine Island Glacier (West Antarctica), overlain on a sketchmap drawn from aerial photography collected in 1966	185
3.35	The location of the ice front of the Ross Ice Shelf, 1841–1997	187
3.36	A Radarsat-1 image of the Shirase Glacier (East Antarctica) from the 2000 AMM, with ice margin locations derived from other sources to show how the glacier retreated over the period 1962–2000	color
3.37	Front positions of the Amery Ice Shelf for ten epochs from 1936 to 2000	188
3.38	A true-color MODIS image (250-m resolution) of the Ross Sea and Ross Ice Shelf acquired on September 17, 2000, showing sections of the massive iceberg B15 that calved from the Ross Ice Shelf in March 2000	191
3.39	A MODIS image (250-m resolution) of large icebergs along the coast of George V Land	192
3.40	Radarsat-1 ScanSAR imagery of the Mertz Glacier region of East Antarctica from (a) November 3, 1997, and (b) October 1, 2002	194
3.41	A sequence of SAR amplitude images from the new 1-km-resolution GMM of Envisat for the period April through June 2004, showing the displacement over time in the image mosaic of two large tabular icebergs in the region of the Amery Ice Shelf, East Antarctica	196
3.42	An iceberg detection product derived operationally by the Danish Meteorological Institute	197
3.43	An enhanced resolution image from SeaWinds on QuikSCAT showing large named icebergs in the Weddell Sea (Antarctica) and adjacent to the Antarctic Peninsula on July 17, 1999	color
3.44	(a) Map showing the location of the Amery Ice Shelf rift system. (b) EOS Terra MISR false-color composite image	color
3.45	(a) Measured lengths of Amery Ice Shelf transverse rifts T1 and T2 (see Figure 3.44)	color
3.46	(a) Schematic map of the western George V Land coastline (East Antarctica) showing variations in the ice front and the movement of icebergs, derived from satellite images collected from 1963 to 1996 and listed in the legend. (b) A Soyuz Kosmos KATE 200 image of the same region from February 14, 1984	200
3.47	(a) ERS wind scatterometer coverage of the Antarctic region. (b) Drift tracks of eight large icebergs around Antarctica. (c) Map of seafloor bathymetry between 20 and 120°E, with the tracks of five icebergs marked. (d) Drift speeds of the five icebergs	color
3.48	An example of a DMSP OLS-based operational iceberg image product from the U.S. Navy/NOAA/Coastguard National Ice Center (NIC)	202
3.49	(a) The scatterometer-derived drift track of iceberg A22B overlain on a QuikSCAT SeaWinds image (polar stereographic projection) from 5 October 2000. (b) A compilation map of iceberg tracks derived from satellite radar scatterometer data for 1978 (Seasat) and 1992–2002 (ERS-1/-2, NSCAT and QuikSCAT SeaWinds)	204
3.50	A plot of the number of icebergs detected and tracked versus time	color

3.51	A sequence of cloud-free images (each ~130 × 145 km in size) from the EOS Terra MISR showing iceberg movements and resultant changes in sea-ice distribution in the Ross Sea (Antarctica) from December 11, 2000 to December 9, 2001	207
3.52	An along-track profile of the elevation (above sea level) measured by the Envisat RA-2 of a large tabular iceberg in the NW Weddell Sea	208
3.53	(a) A schematic of a radar altimeter idealized waveform with a Gaussian distribution of surface slopes and other effects	color
3.54	A schematic of the principle of pulse-limited altimetry	213
3.55	(a) A schematic of a typical radar altimeter waveform from a flat, uniformly rough surface—e.g., an ice sheet—illustrating the concept of the ice mode of tracking used by ESA in their processing of Envisat radar altimeter data acquired over ice-covered surfaces. (b) Schematic illustration of the concept of retracking radar altimeter data	215
3.56	The high-resolution RAMP DEM of Antarctica, which combines topographic data from a variety of sources to provide consistent coverage	220
3.57	(a) Surface elevation map of the PIG and surrounding ice sheet region derived from the Geodetic Mission of ERS-1. (b) Map of ice drainage basins in the same region and derived from the same dataset	221
3.58	Surface elevation of the Amery Ice Shelf, determined from ERS radar altimeter data	color
3.59	(a) A spatial plot of the trend in $\partial H/\partial t$ over the Antarctic Ice Sheet for the period June 1995 to April 2000. (b) The corresponding elevation-change time series for a circular region centered around the outlet of the Pine Island Glacier	color
3.60	(a). Elevation change of the Antarctic Ice Sheet from 1992 to 1996 for 63% of the grounded Antarctic Ice Sheet. (b) Map of elevation change of part of the West Antarctic Ice Sheet, derived from ERS-1 and -2 radar altimeter data (1992–1999)	color
3.61	High-resolution map of the surface slope of the Antarctic Ice Sheet	color
3.62	A schematic comparison of the illumination geometry (side view), footprint (plan view mapped to a flat-Earth surface) and impulse response for (a–c) a conventional pulse-limited radar altimeter, and (d–f) a delay-Doppler radar altimeter	228
3.63	An illustration of ice sheet elevation and cloud data from the GLAS onboard NASA's ICESat on its first day of operation, February 20, 2003	color
3.64	Surface elevation measurements of (a) Antarctica and (b) Greenland acquired by the ICESat GLAS during Mission 2c (May 18–June 21, 2004)	color
3.65	Two ICESat profiles across Byrd Glacier, which is the largest outlet glacier draining the East Antarctic Ice Sheet into the Ross Ice Shelf through the Transantarctic Mountains	231
3.66	An elevation profile obtained by ICESat GLAS on October 18, 2004 across rifts T2 and L1 on the Amery Ice Shelf, for the ground track shown in Figure 3.44c	233
3.67	Shaded-relief images of the Greenland Ice Sheet from (a) a satellite radar-altimeter-based DEM with a spatial resolution of ~5 km, and (b) the same DEM but photoclinometrically enhanced with data derived from NOAA AVHRR imagery and at a spatial resolution of ~1 km	236
3.68	Sub-scene from the enhanced MODIS-derived photoclinometric DEM of the area near the onset regions of the Ross ice streams	237

xviii **Figures**

3.69 A high-resolution DEM of Ryder Glacier, an outlet glacier at the northern edge of the Greenland Ice Sheet, derived from interferometric processing of ERS-1 SAR images 239

3.70 (a) Predicted snow accumulation in Greenland from NSCAT B values. (b) Comparison of NSCAT accumulation retrieval from days 261–266, 1966 along the transect in (a). . . . . color

3.71 A map of the accumulation-rate field in the dry-snow zone of Greenland. . . . color

3.72 Barrier motion is estimated as the ratio between mean annual ice shelf area change for a particular interval, and the length of the discharge periphery . . . 249

3.73 Surface ice velocity vectors superimposed on an ERS-1 image of the lower part of Pine Island Glacier, Antarctica. 251

3.74 (a) A Landsat TM image of the upstream portion of Bindschadler Ice Stream (D) from January 16, 1987. (b) Velocity field for the study region 253

3.75 Color-coded map of the ice surface velocity magnitude (speed) of Ice Streams Bindschadler (D) and E (West Antarctica), derived from eight pairs of Landsat TM images. . . . . color

3.76 An overview of the central Siple Coast region of Antarctica in an AVHRR image mosaic, with subtle morphological features revealed by enhanced image processing. 256

3.77 A shaded relief perspective view (looking from the east) of the ice-flow field and elevation of Storstrømmen, a large outlet glacier draining the northeastern part of the Greenland Ice Sheet. . . . . color

3.78 (a) Ice flow speed of West Antarctic ice streams determined from multiple swaths of Radarsat-1 AMM data co-registered with a mosiac of NOAA AVHRR data. (b) Ice speed superimposed on surface elevation. . . . . color

3.79 (a) An ice velocity map of the Northeast Greenland Ice Stream, derived from InSAR processing of ERS SAR data. (b) A shaded surface DEM of the same area . . . . color

3.80 (a) Flow velocity of the Amery Ice Shelf, Lambert Glacier, and several other outlet glaciers. (b) Map of the magnitude of the velocity error for (a). . . . . . . color

3.81 Spatial distribution of smoothed ice surface velocity magnitude over the Amery Ice Shelf and adjacent ice. 264

3.82 A computer model of the ice balance velocities of the Antarctic Ice Sheet, based on a digital elevation model that incorporates ERS radar altimeter elevation measurements . . . . color

3.83 (a) Balance velocities for part of the Siple Coast, West Antarctica, and (b) surface velocities derived from Landsat feature-tracking and from Radarsat-1 SAR data using speckle-tracking and InSAR techniques combined color

3.84 (a) Longitudinal strain rate for Ice Streams Bindschadler (D) and E (West Antarctica), calculated from Landsat TM-derived ice-surface velocity data collected over the period January 1987 to January 1992. (b) Spatial distribution of errors . . . . color

3.85 The spatial distribution of the transverse shear strain rate over the Amery Ice Shelf (East Antarctica). . . . . color

3.86 Time series of ice surface velocity derived from ERS SAR interferometry for a profile along Monacobreen (Svalbard). 272

3.87	(a) Line-of-sight velocity of PIG (West Antarctica), November 11, 1995. (b) Increase in along-track velocity between February 15, 1995 and November 11, 1995. Increase in line-of-sight velocity measured between (c) November 11, 1995 and November 20, 1999, (d) November 11, 1995 and February 24, 1996, and (e) November 20, 1999 and March 4, 2000	color
3.88	(a) Image of the ice-velocity difference of the Drygalski Glacier, Antarctic Peninsula, superimposed on an ERS SAR amplitude image of November 1, 1995. Changes in the magnitude of the velocity vector are given along (b) the longitudinal profile A (east–west) and (c) the transverse profile B (north–south), shown in (a)	color
3.89	Surge of Boydell and Sjögren Glaciers, shown in a comparison of images	275
3.90	The central panel shows a cloud-free MODIS image from November 1, 2003 of the Larsen B Ice Shelf region	color
3.91	(a) Landsat 7 image of the Hektoria Glacier from February 20, 2003. (b) ICESat elevation profiles projected upstream	color
3.92	An InSAR image of the tidal motion of Thwaites Glacier (Antarctica) from 1996	color
3.93	Examples of the satellite-aliasing problem	278
3.94	Normalized tidal displacements of PIG, West Antarctica, recorded with ERS differential interferometry, and color-coded from magenta to yellow and blue	color
3.95	A vector velocity map of Petermann Gletscher (NW Greenland), derived from ERS InSAR ascending and descending track data, and superimposed on a SAR amplitude image	color
3.96	Satellite-derived products used in the computation of the mass balance of the David Glacier, Victoria Land (Antarctica)	color
3.97	Distribution maps of (a) the hydrostatic height anomaly, and (b) the thickness of marine ice computed for the Amery Ice Shelf, Antarctica	color
3.98	A map of the Earth's gravity anomalies based on 111 days of data collected during the commissioning phase of the GRACE data	color
3.99	Color-coded maps of Antarctic mean-monthly surface temperatures for 1992 derived from NOAA AVHRR data	color
3.100	Color-coded monthly mean-anomaly maps of Antarctic surface temperatures in July from 1979 to 1998	color
3.101	Views from the MISR of the Amery Ice Shelf–Lambert Glacier system in East Antarctica on October 25, 2002, illustrating ice surface textures and cloud-top heights	color
3.102	Comparison of the annual pattern of the air temperature measured by the AWS at Dome C in East Antarctica	297
3.103	Hemispheric spectral albedo for very new snow, new snow, the onset of melt, and wet snow in the spectral range of 300–2500 nm	300
3.104	Schematic representation of the BRDF terminology	302
3.105	Monthly averaged surface albedo for August 1989 over the Greenland Ice Sheet and derived from cloud-free NOAA AVHRR visible and near-infrared radiances	color
3.106	A three-dimensional map of the surface albedo for the ablation zone of the Jakobshavn Isbrae (Glacier) region of West Greenland on July 29, 1997 from APP products	color

xx   Figures

3.107   (a) The spatial variability of snow surface grain size on the Antarctic Ice Sheet derived from ERS-2 ATSR-2 radiance data, and (b) its temporal variability at Dome C in East Antarctica .... color
3.108   (a) A cloud-free AVHRR visible band image of a large Antarctic megadune field. (b) Locations of megadune fields identified in satellite imagery .... color
3.109   Streamlines of surface inversion winds on the Antarctic Ice Sheet inferred from NSCAT data, and superimposed onto the Radarsat 5-km-resolution Antarctic mosaic from the AMM 317
3.110   Maps of East Antarctica showing (a) the amplitude of the azimuthal variation of the mean backscatter of the snow/firn at C-band using ERS-1 wind scatterometer data, and (b) the orientations of the directional anisotropy of the backscatter coefficient of the snow cover, as defined by the antenna look direction for minimum backscatter coefficient .... color
3.111   An 800-km-long strip of ERS C-band SAR images extending across NE Greenland from the ice sheet dry-snow zone (image bottom) through the percolation zone to the rocky coast (top). .... color

Table

3.1 Summary table of ice-sheet-related applications of major satellite sensor classes   138

Abbreviations

AABW	Antarctic Bottom Water
AAR	Accumulation Area Ratio
AATSR	Advanced Along-Track Scanning Radiometer
ADD	Antarctic Digital Database
ADEOS	Japanese ADvanced Earth Observing Satellite
ALI	Advanced Land Imager
ALOS	Advanced Land Observing Satellite
AMI	Active-Microwave Instrument (ESA)
AMISOR	Australian Amery Ice Shelf Ocean Research project
AMM	CSA/NASA Antarctic Mapping Mission
AMSR	Advanced Microwave Scanning Radiometer
APP	AVHRR Polar Pathfinder program
ASAR	Envisat Advanced SAR
ASCAT	ESA's Advanced Scatterometer
ASF	NASA Alaska Satellite Facility
ASTER	EOS Terra Advanced Spaceborne Thermal Emission and Reflection radiometer
ATSR	ERS-1 Along-Track Scanning Radiometer
ATSR/2	ERS-2 Along-Track Radiometer/2
AUV	Autonomous Underwater Vehicle
AVHRR	Advanced Very High Resolution Radiometer
AVNIR(-2)	Advanced Visible and Near Infrared Radiometer (-2)
AWS	Automatic Weather Station
BAPS	Berg Analysis and Prediction Subsystem
BPDF	Bidirectional Polarization Distribution Function
BRDF	Bidirectional Reflectance Distribution Function
CDW	Circumpolar Deep Water
CFAR	Constant False Alarm Rate method

CHAMP	CHAllenging Mini-satellite Payload
CIS	Canadian Ice Service
CLIC	CLImate and Cryosphere Project
CMIS	Conical Microwave Imager/Sounder
CNES	Centre National d'Études Spatiales
COSMO	Constellation Of Small satellites for Mediterranean basin Observation (Italy)
CSA	Canadian Space Agency
CTD	Conductivity Temperature Density
DAAC	Distributed Active Archive Center (NASA)
DEM	Digital Elevation Model
DESCW	Display Earth remote-sensing Swath Coverage for Windows
DInSAR	Differential SAR interferometry
DISORT	DIScrete-Ordinate Radiative Transfer model
DISP	Declassified Intelligence Satellite Photography data
DLR	Deutsche Forschungsanstalt für Luft und Raumfahrt
DMI	Danish Meteorological Institute
DMSP	Defense Meteorological Satellite Program satellites
DORIS	Doppler Orbitography and Radiopositioning Integrated by Satellite
ECMWF	European Centre for Medium-Range Weather Forecasts
EISMINT	European Ice Sheet Modelling INiTiative
ELA	Equilibrium Line Altitude
EMISAR	ElectroMagnetics Institute Synthetic Aperture Radar (Denmark)
ENSO	El Niño–Southern Oscillation
EO-1	U.S. Earth Observing-1 satellite
EOS	Earth Observing System
EROS	Earth Resources Observation System
ERS	Environmental Research Satellite
ERTS	Earth Resources Technology Satellite
ESA	European Space Agency
EScat	ERS-1 and ERS-2 scatterometer
ESMR	Nimbus-5 Electrically Scanning Microwave Radiometer
ETM+	Enhanced TM+ (Landsat)
FFT	Fast Fourier Transform
FOV	Field-Of-View
FRISP	Filchner-Ronne Ice Shelf Programme
GC-Net	Greenland Climate Network
GCM	General Climate (Circulation) Model
GCP	Ground Control Point
GGM01	GRACE Gravity Model 01
GIS	Geographic Information System
GLAS	ICESat Geoscience Laser Altimeter System
GLI	GLobal Imager
GLIMS	Global Land Ice Measurement from Space project

GMM	Global Monitoring Mode (Envisat ASAR)
GNSS	Global Navigation Satellite System
GOCE	ESA Gravity field and steady-state Ocean Circulation Explorer
GPS	Global Positioning System
GR	Gradient Ratio
GRACE	U.S.-German Gravity Recovery and Climate Experiment
GRACE-FO	GRACE Follow-On
Greenmap	VECTRA Greenland Mapping Project
GSFC	NASA Goddard Space Flight Center
H	Horizontal polarization
HRG	SPOT High Resolution Geometric
HRVR	SPOT High Resolution Visible Radiometer
HSSW	High Salinity Shelf Water
HYDROS	NASA's HYDROsphere State mission
IASC	International Arctic Science Committee
IBE	Inverse Barometer Effect
ICEMON	European ICE MONitoring in the Polar Regions initiative
InSAR	SAR interferometry
IPCC	Intergovernmental Panel on Climate Change
IPR	Ice Penetrating Radar
IRS	Indian Remote Sensing satellite
ISAPP	NASA Ice Sheet Altimeter Pathfinder Program
ISMASS	Ice Sheet Mass Balance and Sea Level program
ISST	Ice Sheet Surface Temperature
IST	Ice Surface Temperature
ISW	Ice Shelf Water
ITASE	International Trans-Antarctic Scientific Expedition project
JAXA	Japanese Aerospace eXploration Agency
JERS	Japanese Earth Resources Satellite
LGM	Last Glacial Maximum
LIDAR	LIght Detection And Ranging
LISS	Linear Imaging Self-scanning Sensor
LOS	Line Of Sight
MAMM	Radarsat-1 Modified Antarctic Mapping Mission
MiniMAMM	Mini Modified Antarctic Mapping Mission (Radarsat-1)
MCF	Minimum Cost Flow
MERIS	MEdium Resolution Imaging Spectrometer
MetOp	Meteorological Operational satellite series
MIMOSA	Mapping of Ice and MOnitoring of Sub Arctic areas mission
MIRAS	Microwave Imaging Radiometer by Aperture Synthesis (SMOS)
MISR	Terra Multi-angle Imaging Spectro Radiometer
MOD02QKM	Terra MODIS level-1b 250-m data
MODIS	MODerate resolution Imaging Spectroradiometer
MRIN!	More Research Is Necessary
MSS	Multi-Spectral Scanner

NAO	North Atlantic Oscillation
NASA	National Aeronautics and Space Administration
NASDA	National Space Development Agency of Japan
NAT	North American Trough
NCAR	National Center for Atmospheric Research
NCEP	National Centers for Environmental Prediction
NIC	NOAA/U.S. Navy/Coastguard National Ice Center
NOAA	National Oceanic and Atmospheric Administration
NPOESS	U.S. National Polar-orbiting Operational Environmental Satellite System
NPP	NPOESS Preparatory Project
NSCAT	NASA Scatterometer
NSIDC	U.S. National Snow and Ice Data Center
OLS	Optical Linescan System
OPS	JERS-1 OPtical System
PALSAR	ADEOS-II Phased Array type L-band Synthetic Aperture Radar
Pan	Panchromatic
PARASOL	Polarization and Anisotropy of Reflectances for Atmospheric Sciences coupled with Observations from a LIDAR
PARCA	NASA's Program for Arctic Regional Climate Assessment
PIG	Pine Island Glacier
PLF	Pulse Limited Footprint
PMR	Power-to-Mean-Ratio method
PMW	Passive MicroWave
Pol-InSAR	Polarimetric SAR Interferometry
POLDER	POLarization and Directionality of the Earth's Reflectances
PolSAR	SAR polarimetry
PR	Polarization Ratio
PRARE	Precise RAnge and Range-rate Equipment
PRISM	Panchromatic Remote-sensing Instrument for Stereo Mapping
RA	Envisat Radar Altimeter
RAMP	Radarsat-1 Antarctic Mapping Project
RBV	Landsat 3 Return Beam Vidicon
RES	Radio Echo Sounding
RIGGS	Ross Ice Shelf Geophysical and Glaciological Survey
SAM	Southern Annular Mode
SAR	Synthetic Aperture Radar
SASS	Seasat A Scatterometer
SCAR	Scientific Committee on Antarctic Research
SCP	Scatterometer Climate Record Pathfinder group
SCR	Signal-to-Clutter Ratio
SIR-C	Shuttle Imaging Radar C-band mission
SIRAL	ESA's Synthetic Aperture Interferometric Radar ALtimeter
SIRF	Scatterometer Image Reconstruction with Filtering algorithm
SLA	Snow Line Altitude

SLC	Single-Look Complex
SMMR	Nimbus-7 Scanning Multi-channel Microwave Radiometer
SMOS	Soil Moisture and Ocean Salinity mission
SNR	Signal-to-Noise Ratio
SOI	Southern Oscillation Index
SPOT	Système Pour l'Observation de la Terre
SRTM	Space Shuttle Radar Topography Mission
SSM/I	Special Sensor Microwave/Imager (DMSP)
THIR	Nimbus-7 Temperature Humidity Infrared Radiometer
TIR	Thermal InfraRed
TM	Landsat Thematic Mapper
TOA	Top-Of-Atmosphere
USGS	U.S. Geological Survey
V	Vertical polarization
VIIRS	Visible Infrared Imaging Radiometer Suite
WAIS	West Antarctic Ice Sheet
WCRP	World Climate Research Programme
WiFS	Wide-Field Sensor
WMO	World Meteorological Organization
XPGR	Cross-Polarization Gradient Ratio

About the author

ROB MASSOM

Rob Massom has participated in nine Antarctic and three Arctic sea ice research campaigns, including the Arctic Marginal Ice Zone Experiment (MIZEX, 1983–84) and the First Winter Weddell Sea Project cruise of 1986. After his bachelor's degree in Physical Geography from London University, Rob earned a Ph.D. in Glaciology/Remote Sensing in 1989 from the Scott Polar Research Institute (University of Cambridge). After that, he spent three years at NASA Goddard Space Flight Center as a U.S. National Research Council postdoc (working with Joey Comiso),

before joining the Antarctic Cooperative Research Centre (CRC) in Hobart (Tasmania, Australia) in 1992 as a sea ice research scientist. He is currently a senior research scientist with the Antarctic Climate and Ecosystems CRC in Hobart. Rob's research interests include remote sensing, sea ice and polar oceans and their ecological significance, precipitation over the Antarctic Ice Sheet, and the impact of modes of large-scale anomalous atmospheric circulation on sea ice and ecology. Rob lives in Hobart with his wife Yuko and daughter Adelle. Even after 25 years' involvement in polar research, Rob is still drawn by the magical lure of polar regions, driven by the fact that the more he learns, the more he realizes he doesn't know.

Publisher credits

The following publishers have given their kind permission to reprint or adapt materials that appear in the following figures. In addition, we have attempted to reach the senior authors to obtain their permission as well; we trust that those we have not been able to reach will not object to their work being shown and discussed herein as advances in the field. Complete citations of these sources are given in the reference sections at the end of the respective chapters.

Publisher	*Figure No.*
Alaska Satellite Facility	2.1, 2.8, 3.10, 3.11, 3.36
American Academy for the Advancement of Science	2.4, 2.18, 3.17, 3.26, 3.59a, 3.78, 3.82, 3.83, 3.89, 3.94
American Geophysical Union	3.4, 3.21, 3.27, 3.30, 3.31, 3.34, 3.43, 3.44, 3.45, 3.49, 3.50, 3.58, 3.66, 3.70, 3.71, 3.79, 3.86, 3.95, 3.97, 3.108
American Meteorological Society	3.22, 3.99, 3.100, 3.105
Antarctic Meteorological Research Center, University of Wisconsin	1.3
Applied Physics Laboratory at Johns Hopkins University	3.62
Blackwell Publishing	3.61
Cambridge Journals	3.40, 3.90, 3.91
Canadian Aeronautics and Space Institute/Canadian Remote Sensing Society	2.19, 2.20, 2.21
Canadian Space Agency and Radarsat International	2.22, 3.36, 3.63
Commonwealth of Australia	3.39, 3.110

xxxii Publisher credits

Publisher	Figure No.
Danish Centre for Remote Sensing	2.6
Elsevier Publishing	1.1, 1.2, 3.5, 3.59b, 3.102
European Space Agency	2.5, 2.10, 2.11, 2.12, 2.23, 3.41, 3.52, 3.55a, 3.73, 3.110
Institut Français de Recherche pour l'Exploitation de la Mer	3.47
Institute of Electrical and Electronics Engineers	2.3, 2.13, 2.15, 3.3, 3.12, 3.18, 3.19, 3.20, 3.57, 3.69, 3.109, 3.111
International Association of Hydrological Sciences	3.104
International Glaciological Society	2.14, 3.2, 3.6, 3.7, 3.8, 3.13, 3.15, 3.16, 3.24, 3.25, 3.35, 3.37, 3.46, 3.60, 3.67, 3.72, 3.74, 3.75, 3.76, 3.80, 3.81, 3.84, 3.85, 3.87, 3.88, 3.92, 3.93, 3.96
International Society for Photogrammetry and Remote Sensing	2.7
National Aeronautics and Space Administration	1.2, 3.9, 3.14, 3.32, 3.33, 3.38, 3.51, 3.55b, 3.63, 3.64, 3.65, 3.98, 3.101
National Oceanographic and Atmospheric Administration	3.42, 3.48
National Snow and Ice Data Center	3.56, 3.68, 3.106
Nature Magazine	2.16, 2.17, 3.77
Norwegian Polar Institute	3.29
NPA Group	2.2
Springer Science and Business Media	3.53, 3.103
Taylor & Francis, Ltd.	2.9, 3.54
University of Bergen Geophysical Institute:	3.28
U.S. Geological Survey	3.1

1

Polar ice sheets: Introduction

In Volume 2 of this book, we review the immensely important and unique role that satellites are playing in addressing key ice-sheet-related issues, with illustrations of the exciting results that are emerging. These paint a picture of complex regional variability in properties, processes, and phenomena, although large-scale patterns are beginning to materialize. Not only the strengths but also the weaknesses of the various sensors and techniques are highlighted. This review is by no means exhaustive, and the interested reader is urged to follow up on the references provided. We concentrate here on the two major polar ice sheets. Smaller ice caps and temperate glaciers are particularly sensitive to small climate perturbations (Warrick et al., 1996), but are outside the scope of this book (please see Haeberli, 2004; Houghton et al., 2001; and the World Glacier Monitoring Service ⟨*http://www.geo.unizh.ch/wgms/*⟩ for information on monitoring these icemasses). Except for the question of resolution, however, many of the satellite techniques outlined in this chapter can also be applied to small ice caps and glaciers. Please refer to Chapter 5 of Lubin and Massom (Volume 1 of this book) for a separate discussion of the basic principles of each remote-sensing class as applied to the measurement of snow and ice, as well as additional information of sensor characteristics (see the Appendix).

1.1 THE GLOBAL IMPORTANCE OF POLAR ICE SHEETS

The vast polar ice sheets of Antarctica and Greenland hold a major and enduring fascination, and remain the world's last great wildernesses. They are notoriously inhospitable and remote, and are shrouded in darkness for a significant part of the year. They play a central role in the global climate system, interacting in a complex fashion with the atmosphere and ocean, acting as major hemispheric heat sinks as a result of the radiatively induced Equator-to-pole temperature difference, and dominating the high-latitude radiation balance by virtue of their high albedo (Allison et al., 2001; Fitzharris et al., 1996; Weller, 1967). Ice sheets have profound direct and indirect impacts on patterns of oceanic and atmospheric temperature and circulation (King and Turner, 1997; Schwerdtfeger, 1984), and also global biogeochemical cycles. For one thing, intense oceanic and atmospheric circulations are driven by the high Equator-to-pole temperature gradient, an effect which is most pronounced in the Southern Hemisphere. Due to their high elevation, the Greenland and Antarctic Ice Sheets also represent major topographic barriers to synoptic-scale patterns of atmospheric circulation (King and Turner, 1997). Moreover, they create their own regional climate conditions—e.g., *katabatic wind*[1] regimes (Parish and Bromwich, 1987, 1991). They form deserts which are among the driest on Earth—precipitation rates of <7 cm per annum occur in parts of Antarctica, for example (Vaughan et al., 1999).

Strong and direct interactions between ice sheet and ocean occur primarily under floating ice shelves (Grosfeld et al., 1997; Hellmer and Jacobs, 1992; Jacobs et al., 1992; Holland and Jenkins, 1999; Holland et al., 2003; Jenkins et al., 2003). The major components and processes of the *ice shelf–ocean system*, which represents a key component of the global climate system and one that is highly vulnerable to change (Huybrechts, 2001), are illustrated schematically in Figure 1.1. The *grounding line* or *zone* is a critical parameter in that it demarcates the boundary between continental ice that is grounded and floating—i.e., it is the region where the ice sheet loses contact with the sub-glacial bedrock and begins to float as an ice shelf/glacier tongue. In winter, salt rejected by sea ice growth forms dense, high-salinity water (known as High-Salinity Shelf Water or HSSW) which sinks and flows under the ice shelf. Melt occurs when this water comes into contact with deep continental ice, due to the strong temperature contrast between the ice-shelf base and seawater, with the resultant freshened (meltwater) plume rising under the ice-shelf base. The mixing of glacial meltwater with ambient water masses leads to the formation of cold, fresh and oxygen-rich Ice Shelf Water (ISW) (Fahrbach et al., 1994). This either refreezes (accretes) to the ice-shelf base as "marine ice", formed by the ascending water attaining its *in situ* freezing point, or mixes with relatively warm Circumpolar Deep Water (CDW) at the continental shelf break and modifies the water mass properties of local shelf waters to critically contribute to the formation of Deep and Bottom Water (Grosfeld et al., 1997; Foldvik et al., 2004; Foldvik and Gammelsrød, 1988; Rintoul, 1998; Williams and Bindoff, 2003). As discussed in

---

[1] *Katabatic winds* are gravity-driven winds induced by intense radiative cooling of air masses on the interior ice-sheet plateau, with their strength being derived from downslope acceleration of the air as it drains seawards under gravity. They are characterized by their persistence in both strength and direction, with the former being largely controlled by orographic conditions. They are typically channeled by the ice-sheet topography to emerge over the coastal zone via outlet glacier valleys and ice shelves (Parish and Bromwich, 1987, 1991; Yu et al., 2005).

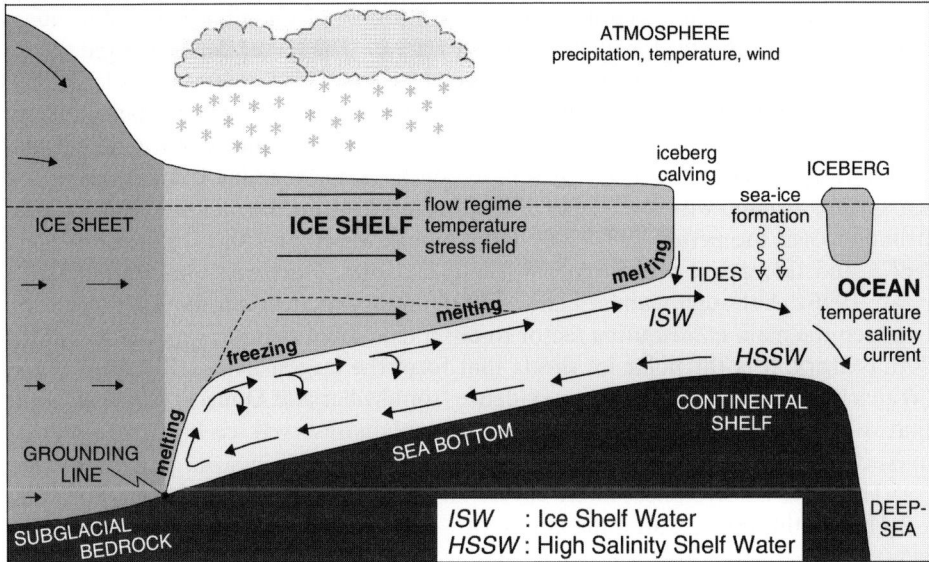

Figure 1.1. A simplified two-dimensional schematic cross-section of a coupled Antarctic ice shelf–ocean system, depicting the grounding line, ocean circulation in the sub-ice shelf cavity, and the complex relationship between ice flow, accumulation, melting, sea ice, water-mass formation/modification, and iceberg calving. This relationship determines the rate of advance or retreat of the ice sheet or grounding zone. Note that moderate freeze rates occur towards the ice front, while the melt rate is high towards the grounding zone. See Allison (2003) and Grosfeld and Sandhäger (2004) for a more in-depth discussion of these processes and interactions.

From Grosfeld and Sandhäger (2004). Copyright Elsevier 2004, reproduced with permission.

Chapter 5 of Lubin and Massom (Volume 1 of this book), Antarctic Bottom Water (AABW) ventilates much of the World's deep ocean and is a major driving force behind *global ocean thermohaline circulation*. Moreover, ice shelf and iceberg melting is estimated to account for 5% of the current total oceanic heat loss south of latitude 60°S, providing a major heat sink for global-ocean circulation and a major component of the *ocean freshwater budget* (Beckmann and Goosse, 2003; Houghton et al., 2001). The complex processes and feedbacks involved in ice shelf–ocean interactions are evaluated by Allison (2003) and Grosfeld and Sandhäger (2004), while Williams et al. (1998a) provide an overview of ice-shelf-modeling concepts. The sensitivity of the ice shelf–ocean interaction system to changes in oceanic boundary conditions is further evaluated by Grosfeld et al. (2001). Model results suggest a high vulnerability of ice-shelf regions to ocean warming in particular (Grosfeld and Sandhäger, 2004; Williams et al., 1998b, 2002).

Ice sheets and glaciers are sensitive indicators and modulators of *climate variability and change*. Being integrators of climate, they are viewed as reflecting a long-term response to climate change (Payne and Bamber, 2004). Of immediate practical concern is uncertainty in the contribution of changes in ice-sheet mass balance to the observed rise in *global sea level*. The latter has occurred at an estimated rate of 1.5–2 mm per annum over the past 100 years (Douglas, 1997, 2001; Houghton et al., 2001) and somewhat faster over the past decade (Nerem and Mitchum, 2001a, b). This rate is less than during past epochs of rapid deglaciation, but significant nonetheless. It is based on the satellite radar altimeter record of sea-surface height, with the Topex-Poseidon radar altimeter making a major contribution over the period 1993–2001 (Nerem and Mitchum, 2001a, b; Nerem et al., 1997), and tide gauge measurements (Douglas, 2001; Hunter et al., 2003; Warrick et al., 1996). Although Cabanes et al. (2001) argued that ocean thermal expansion has been the main contributing factor to observed global eustatic sea level rise in the 20th century, it is the polar ice sheets that form the largest potential source of rise (Warrick et al., 1996). Indeed, new evidence from Miller and Douglas (2004) suggests that volume changes alone are insufficient to explain observed sea-level rise, and that mass change is a key contributing factor.

The *Antarctic Ice Sheet* alone covers an area of $\sim 12.4 \times 10^6$ km^2, and averages ~ 2.4 km in thickness, with a maximum of ~ 4.7 km (in the Wilkes sub-glacial basin between Casey and Vostok) and a volume of $\sim 25.7 \times 10^6$ km^3 (Houghton et al., 2001). It stores $\sim 90\%$ of the world's ice, equivalent to $\sim 70\%$ of its freshwater or an approximately 65-m rise in global sea level were it to melt. The *Greenland Ice Sheet*, on the other hand, is $\sim 1.7 \times 10^6$ km^2 in area, has an average thickness of 1.6 km, and a volume of $\sim 3 \times 10^6$ km^3 (for the grounded ice)—i.e., the equivalent to a sea level rise of ~ 7.2 m (Thomas et al., 2001a, b). Note that the ice volumes given above are best estimates based on available data, as laid out in the Intergovernmental Panel on Climate Change (IPCC) Third Assessment Report (Houghton et al., 2001), and modifications are likely as data improve in terms of both quality and quantity (availability). Current IPCC estimates put the contribution of Antarctica to 20th century global sea-level rise at -0.2 to 0.0 mm per annum (Church et al., 2001). Of particular concern, however, is the likely response of the great ice sheets to a predicted increase of 1 to 4°C in mean global temperature by the end of this century (ACIA, 2004; Gregory et al., 2004; Houghton et al., 2001; Schiermeier, 2004). This follows on from the observed temperature increase over the last 150 years (Jones et al., 1999). Moreover, recent results suggest a possible irreversibility of a Greenland deglaciation with major climatic implications (Gregory et al., 2004; Ridley et al., 2005; Toniazzo et al., 2004). For further in-depth discussion on global sea-level rise, contributions, and issues relating to uncertainties affecting the accuracy of current sea-level change observations and future predictions, please see Chao et al. (2002), Church (2001), Church et al. (2001), Cabanes et al. (2001), Douglas and Peltier (2002), Douglas et al. (2001), Warrick et al. (1996), and White et al. (2005).

In the past, a lack of accurate and consistent data on key variables over the vast distances involved has drastically affected our ability to unambiguously measure and predict ice-sheet response to climate variability and/or change, and the role of ice sheets in such change. This has resulted in considerable uncertainty in estimates of ice-sheet mass balance, and poor knowledge of whether the great ice sheets of Antarctica and Greenland have been losing or gaining net ice mass (Alley, 2002).

The tide is turning, however, thanks in large part to recent advances in satellite remote-sensing technology and the emergence of important new techniques for processing the data. These developments have led to major improvements in our ability to observe, monitor, and model ice-sheet properties and processes, and extraordinary new findings are coming to light as a result. It now appears that mass loss is occurring in Greenland, while considerable regional variations are apparent in Antarctica (Rignot and Thomas, 2002; Thomas, 2004). A major challenge in unraveling information contained in satellite signals is that current changes are often (though, as we shall see, not always) small within interannual to decadal timeframes (Zwally et al., 2000), involve complex interactions and processes, and occur over vast and widely separated areas. Until recently, the requisite observational precision exceeded that achievable, contributing to a high level of uncertainty in our knowledge of the current state of the great ice sheets and undermining predictions of future change driven by climate variability/change. As a result, glaciological estimates of ice-sheet net mass balance have exhibited a considerable range (Bentley and Giovinetto, 1991; Houghton et al., 2001; Jacobs et al., 1992; Ohmura and Reeh, 1991). This is equivalent to a eustatic sea-level rise of between $\sim$1 mm per annum and 0.4 mm per annum for Antarctica and Greenland, respectively (ESA, 2003; Houghton et al., 2001)—i.e., uncertainty in the current mass balance of Antarctica is of the order of approximately half of the current estimated sea-level rise. Major uncertainties and unknowns in predicting future ice-sheet behavior, as outlined by Allison et al. (2001), Fitzharris et al. (1996), and Goodison et al. (1999), translate into the largest uncertainty in the source of global sea-level rise (Houghton et al., 2001). As the annual mass exchange between the ice sheets and ocean is estimated to be equivalent to $\sim$8 mm per annum of global sea level equivalent (Zwally et al., 2002), even a small fractional change in the volume of ice stored in the ice sheets would have a significant impact on sea level. Satellite-based techniques are enabling a substantial reduction of these uncertainties. As we shall see, startling results are emerging, many of which are forcing a radical re-evaluation of previously held beliefs and engendering renewed concern about the rapidity of apparent change and its wider implications.

A recent IPCC comparison of different model projections of the rise in global average sea level from 1990 to 2100, and driven by the same greenhouse gas (IS92a) scenario, yielded estimates in the range 0.11 to 0.77 m (Houghton et al., 2001). This range reflects the systematic uncertainty in modeling and observation, a factor which has again limited our ability to both critically test models and develop suitable adaptive strategies. The predictive capability of modern glacier and ice-sheet models has been affected both by uncertainty in model boundary conditions and deficiencies in the physics of model parameterizations (Payne and Bamber, 2004). Over 100 million people of the world's ever-expanding population currently inhabit coastal zones at altitudes within 1 m of mean sea level (Leatherman, 2001). Moreover, increased coastal erosion related to sea-level rise is becoming apparent across the globe (Day, 2004). Clearly, an urgent consideration is an improved definition of the current and future contributions by ice sheets and other sources (see Chao, 1994; Cubasch et al., 1992; Houghton et al., 2001; Peltier, 2001) to global sea-level rise, and a reduction of current uncertainties. This involves an improved understanding of the dynamics of both major polar ice sheets. In addition, the increased input of freshwater (meltwater) into the oceans may induce a number of other important changes. These include changes in the strength and pattern of global ocean thermohaline circulation (Clark et al., 2002; Fichefet et al., 2003; Houghton et al., 2001; Peterson et al., 2002; Rind et al., 2001; Seidov et al., 2001), impacting climate patterns, biogeochemical cycles, and high-latitude ecology. High levels of uncertainty, together with poor knowledge of the complex feedbacks involved, have in the past again tended to undermine prediction of these impacts. Once again, satellite remote sensing is playing a critical role in reducing uncertainties.

Given their relatively small ice volume, *mountain glaciers* respond rapidly to climate change on short timescales of <100 years. See Dyurgerov and Meier (2004), Houghton et al. (2001), Meier et al. (2003), and Oerlemans et al. (1998) and references therein for an assessment of recent change in glacier regimes and their contribution to sea level rise. Indeed, many have exhibited substantial retreat over the past century (Arend et al., 2002; Dowdeswell et al., 1997; Dyurgerov and Meier, 2000, 2004; Fitzharris et al., 1997; Haeberli and Hoelzle, 1995; Hoelzle et al., 2003; McDowell, 2002). Dyurgerov and Carter (2004) have assessed the contribution of Arctic mountain and sub-polar glaciers to an observed recent inflow of freshwater to the Arctic Ocean. Information on current knowledge of the dynamics and mass balance of Arctic glaciers and ice sheets in relation to sea level and climate change is available from the International Arctic Science Committee (IASC) Working Group on Arctic Glaciology at ⟨http://www.phys.uu.nl/~wwwimau/research/ice_ climate/ iasc_wag/massbalance.html⟩.

Compared with glaciers, the *dynamic response time* t_R of the vast polar ice sheets to climate change is much longer. It has traditionally been approximated by the relationship:

$$t_R \approx \sim \frac{H}{a_0} \tag{1.1}$$

where H is the maximum ice thickness and a_0 the ablation rate (Paterson, 1994). This yields a present-day value of $\sim$3,000 years for Greenland, for example. Due to this apparently slow response time, predicting the future evolution of an ice sheet requires thorough knowledge of its dynamics under past, present, and future forcing regimes. The key research tool in this respect is the ice sheet dynamical model, which generally consists of components describing the ice flow, the mass balance at the ice–ocean and ice–atmosphere interfaces, and the solid-Earth response (Huybrechts et al., 2004). A complicating factor is that any current or future change in ice-sheet mass depends on the past dynamic and climate history over a range of timescales—from decadal through millennial or longer (Reeh, 1999). Ice discharge–change response times are determined by thermal and physical processes at the ice–bedrock interface, the ratio of ice thickness to annual mass turnover, processes affecting ice viscosity, and isostasy (Houghton et al., 2001; Paterson, 1994). Indeed, the large polar ice sheets appear to be still adjusting to their past history, and the transition from glacial to inter-glacial conditions between 20,000 and 10,000 years before present (BP) in particular (Houghton et al., 2001). The present and future contribution of polar ice sheets to global sea-level change therefore has a component relating to past climate changes (ESA, 2003; Houghton et al., 2001). It should also be remembered that abrupt climate change has occurred in the past (Alley, 2000, 2001; Alley et al., 1999, 2003; Clark et al., 2002; Cuffey and Clow, 1997; Lowell, 2000; Mayewski et al., 2001; Morgan et al., 2002; Sabadini, 2002; Severinghaus and Brook, 1999; Severinghaus et al., 1998; Taylor et al., 2001). A notable example based on its impact on mankind and biota was the Last Glacial Maximum (LGM) at the beginning of the Holocene, ~20,000 BP. Please see NOAA (2004), and the references therein, for a paleo-perspective on abrupt climate change.

Given the short timespan of the satellite data record—i.e., approximately 30 years at most—and the factors outlined above, observed ice-volume changes are best evaluated within the context of ice-sheet evolution on century to longer timescales, as determined from ice-sheet dynamic model computations and the glacial geologic record, combined with information on short- and long-term accumulation rates. The latter are best derived from airborne/*in situ* high-resolution radar sounding and/or ice core measurements. *Ice-core analysis* techniques have reached a level of maturity whereby a broad range of environmental variables can be tracked at resolutions as short as seasonal over glacial–interglacial timescales (see the journal *Annals of Glaciology*, **35** [2002] for up-to-date information on polar ice-core research). In particular, ice cores provide records of accumulation rate, temperature and atmospheric circulation that may be directly compared with remotely sensed data from the last three decades. Airborne radar sounding measurements, on the other hand, offer much wider geographical coverage. Regarding models, performance is constrained by trade-offs between resolution, scales, and computing cost, although considerable benefits have accrued from extraordinary advances in available computing power over the past two decades.

Having stated above that current conditions should be evaluated within a long-term context, and that ice sheet response times are themselves long, it is becoming increasingly apparent that substantial and unexpected changes are, however, occurring in certain regions on much shorter timescales than had been previously anticipated (Oppenheimer and Alley, 2004). An alarming recent example is the extraordinary collapse of the Larsen B Ice Shelf in 2002, which was unprecedented in that it occurred over the space of a few weeks. This and other new discoveries have resulted largely from analysis of satellite data, and increasingly from the emergence of Synthetic Aperture Radar (SAR) interferometry (InSAR) as a key glaciological research tool (see Chapter 2), and the refinement of existing techniques—e.g., radar altimetry (recently supplemented by satellite laser altimetry). In Antarctica, for example, complementary InSAR, radar altimeter, and laser altimeter measurements have shown that rapid and extensive thinning is taking place in part of the West Antarctic Ice Sheet (WAIS: Thomas et al., 2004), while thickening is apparent in other regions (see Section 3.6). Moreover, the last 20 years has witnessed the wholesale collapse of Antarctic Peninsula ice shelves (in addition to Larsen B) that had remained intact for millennia (see Section 3.4). These startlingly dramatic breakups have coincided with an extraordinary warming episode in the region, with a temperature increase of 2.5°C over the past 50 years (King and Comiso, 2003; King et al., 2003; Skvarca et al., 1999; Vaughan et al., 2003). This compares with a rise in global mean temperature of 0.6 ±0.2°C over the 20th century (Houghton et al., 2001). Recent research—e.g., by Thompson and Solomon (2002)—has suggested links between the warming trend and a trend towards the positive index state of the Southern Annular Mode (the dominant pattern of atmospheric variability in the Southern Hemisphere—see Chapter 5 of Volume 1 of this book). The wider continental-scale picture is highly complex, however, with certain regions experiencing cooling over the same timeframe. Antarctic climate change over the past half-century is discussed by Turner et al. (2005). The key present day factor is that anthropogenically induced greenhouse warming may be increasing the likelihood of abrupt regional and global climate-change events (Houghton et al., 2001). A major challenge is to separate recent change from the long-term background signal and natural variability in available data (Rignot and Thomas, 2002).

Another major concern is that substantial changes have indeed been observed in ice-stream velocity and behavior in recent years, again based on satellite data analysis. Reasons for these changes are the subject of intense debate, with a concensus yet to be reached. It is apparent, however, that ice-stream stability is again determined by a complex interplay of mechanisms related to the thermal and force balance and sub-glacial sediment and hydrological conditions—i.e., water flow (Payne and Bamber, 2004). Of great concern is the proposition that, if the ice streams were to enlarge and/or speed up, then an acceleration in outward (horizontal) ice flux could change the ice-sheet configuration and even bring about ice-sheet collapse—with dramatic consequences for global sea level (Mercer, 1968, 1978; Thomas 1979a, b). The *West Antarctic Ice Sheet (WAIS)* has been under particular scrutiny in this respect (Alley and Bindschadler, 2001; Bindschadler, 1998a; Bindschadler et al., 1998; Hughes, 1973; Thomas et al., 2004; WAIS, 2003—see also Anderson et al., 2001 for an historical perspective). Comprising ~13% of the continent's ice (equivalent to a sea-level rise of ~6 m), this sector differs from the more stable East Antarctic Ice Sheet in that it is largely "marine"—i.e., grounded below sea level—a configuration that may make it particularly susceptible to dynamic instabilities and rapid disintegration under a climate-change scenario (Hughes, 1977; Mercer, 1978; Weertman, 1974). Uncertainty has engendered considerable ongoing debate—e.g., Alley and Bindschadler (2001), Bentley (1997, 1998), Bindschadler (1997), MacAyeal (1992), Oppenheimer (1998), and Vaughan and Spouge (2002). This is nicely summarized in stimulating reviews by Bindschadler and Bentley (2002) and Oppenheimer and Alley (2004), the latter evaluating recent findings (largely from remote-sensing studies) in the light of long-term climate policy strategies and implications. Moreover, modeling studies, for example, by Warner and Budd (1998) and Huybrechts and de Wolde (1999), have predicted that even a moderate change in ocean temperature will have a profound effect on ice-shelf basal melt. Indeed, an increase in basal melting due to contact with a warmer ocean provides possibly the most feasible mechanism by which climate warming could increase mass loss from the Antarctic Ice Sheet, and substantial changes in the geometry of ice shelves may have profound consequences for the dynamics of grounded ice sheets. This comes on top of the recent and far-reaching discovery that *ice-shelf basal melting* may account for up to one-third of the loss of grounded ice from the Antarctic Ice Sheet (Jacobs et al., 1996; SCAR, 2002), with similar findings in Greenland (Rignot and Thomas, 2002). Clearly, predictions of future sea level rise, and the oceanic impact of enhanced freshwater input, require substantial improvements in our understanding of the interactions between ice shelves and the underlying ocean.

As noted previously, mass changes to ice shelves have no direct effect on sea level, as they are already floating. However, it has been argued that ice shelves act as buttresses which constrain the grounded ice, and that their removal could dramatically and abruptly accelerate the rate of ice discharge from inland regions of the ice sheet (upstream of the grounding line or zone) to the ocean (Hughes, 1977; Mercer, 1978; Thomas, 1979a; Warner and Budd, 1998; Weertman, 1974). The buttressing effect of ice shelves, and the role of their removal in the collapse of the grounded ice sheet, has been disputed in certain quarters—e.g., by Hindmarsh (1993), Huybrechts and de Wolde (1999), and Alley and Whillans (1991). Evidence is, however, emerging from new satellite data analysis of a rapid acceleration in grounded ice discharge following the recent collapse of the Larsen and Prince Gustav Channel Ice Shelves on the Antarctic Peninsula (De Angelis and Skvarca, 2003; Rignot et al., 2004; Rott et al., 2002; Scambos et al., 2004a)—see Section 3.8.6—with major implications for the rate of ice-sheet loss into the ocean and associated sea-level rise. Using new results from SAR interferometric analyses as well as other satellite data, these and other studies are again forcing a re-evaluation of the potential contribution of ice masses on the Antarctic Peninsula to global sea-level rise (Rignot, 2004; Thomas et al., 2004), a factor that appears to have been significantly underestimated in previous studies.

Evidence is also coming to light of similar change in the Arctic. Recent rapid acceleration of one of the fastest glaciers in Greenland and indeed the world—namely, Jakobshavn Isbrae—has been observed after thinning and breakup of its floating tongue (Joughin et al., 2004a; Thomas et al., 2003). Changes in elevation, dynamic behavior, and seasonal melt extent have also recently been observed in Greenland (Abdalati, 2001; Abdalati et al., 2001; Krabill et al., 2004; Paterson and Reeh, 2001; Thomas, 2001), although their cause remains largely unexplained. Of additional concern is that current General Climate Models (GCMs) indicate an amplification of warming at high latitudes, based largely upon the ice albedo–temperature feedback (Shine et al., 1984), with the likelihood of regional warming over Greenland being three times the global average (Houghton et al., 2001). Magnitudes vary, however, from model to model (Holland and Bitz, 2003). Clearly, there is a critical need to accurately detect and monitor short-term changes in key ice sheet parameters and determine longer term trends in ice volume (Zwally et al., 2002a). Lengthening time series of key satellite data, and technological advances, are greatly improving our ability to make more accurate and meaningful estimates of changes in ice sheet mass balance. Abdalati et al. (2004) provide an up-to-date strategy for the collection and synthesis of Antarctic Ice Sheet mass balance data and understanding the processes involved, taking into account the new technologies available. This initiative is carried out within the Scientific Committee for Antarctic Research (SCAR) Ice Sheet Mass Balance and Sea Level (ISMASS) program.

1.2 ICE-SHEET MASS BALANCE—BACKGROUND INFORMATION

As *mass balance* is the unifying theme throughout this chapter, and serves as an indicator of ice sheet/glacier state of "health", we begin by giving brief background information on this important concept. It was noted above that uncertainty in the response of the large ice sheets, particularly the Antarctic Ice Sheet, to climate change represents the largest single unknown in the determination of future sea level change. In order to understand past sea level change and to predict future change, it is essential to measure and explain the current state of balance of glaciers and ice sheets, and especially to resolve the large uncertainties in the mass budgets of the Antarctic and Greenland Ice Sheets (Abdalati et al., 2004).

Figure 1.2 is a schematic representation of a polar ice sheet, and is included to introduce the main parameters involved. Polar ice-sheet mass balance and equilibrium state are complex functions of internal ice dynamics and thermodynamics processes and external climate forcing, with ice mass responding dynamically to the latter—e.g., climate variations (Van der Veen, 2001). These in effect cause ice

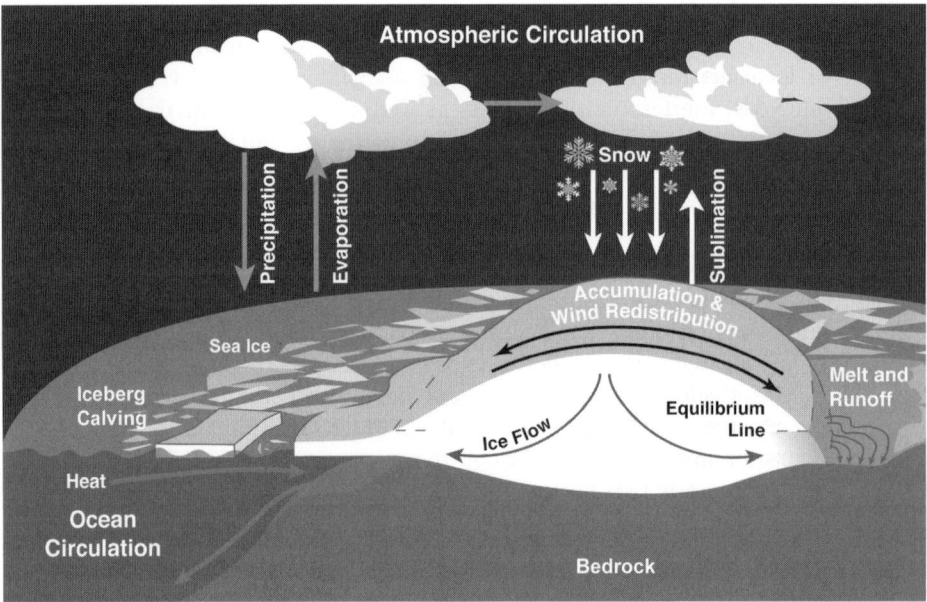

Figure 1.2. A schematic diagram of an ice sheet. Snow accumulation on polar ice sheets is approximately balanced by gravitational ice flow towards the ice-sheet margins, where ice below the equilibrium line seasonally melts or calves into the ocean as icebergs. The equilibrium line marks the location where snow accumulation is exactly balanced by ablation over 1 year. Changes in ice mass result from an imbalance between ice mass gain/inputs (snowfall, condensation, and occasional rainfall) and loss/outputs (iceberg discharge, melt runoff, sublimation/evaporation, and eolian removal of snow).

Adapted from Zwally et al. (2002). Courtesy of Jay Zwally, NASA Goddard Space Flight Center. Graphic by Deborah McLean. Copyright Elsevier 2002, reprinted with permission.

masses to approach a new equilibrium which is compatible with new environmental conditions (Van der Veen and Payne, 2004). Put simply, mass balance describes the state of the ice mass in terms of its gain (accumulation) versus loss (ablation), with a glacier or ice sheet said to be in a state of negative mass balance if loss exceeds gain. More precisely, mass balance traditionally has two different meanings (Hagen and Reeh, 2004). The *local mass balance* refers to the local change in the mass or ice thickness in a vertical column through the glacier/ice sheet. At a given location, the *specific mass balance* describes the sum of accumulation and ablation, and may be either positive or negative depending on the predominance of accumulation or ablation, respectively. Note that the sign relays no information on the local change in ice thickness or mass, as the specific mass balance may also be counteracted by mass input/loss due to horizontal ice flux (ice export). Regarding the contribution of ice sheets to global sea-level rise, only the mass balance of the grounded part of the ice sheet requires consideration. Once again, mass changes in ice shelves, by melt or freezing (Williams et al., 2002), have little or no direct effect on sea level as the shelves are already floating. Meltwater from ice shelves (and melting icebergs) does, however, alter the ocean freshwater balance, with important climatic and ecological consequences.

Following Hagen and Reeh (2004), Reeh (1999), and Reeh and Gundestrup (1985), the local mass balance at a given geographical locality on the ice sheet at time t is given by:

$$\frac{\partial H}{\partial t} = b_S + b_B - F\left(H\left[\frac{\partial u_S}{\partial x} + \frac{\partial v_S}{\partial y}\right] + u_S\frac{\partial H}{\partial x}\right) \quad (1.2)$$

where H is ice thickness, b_S and b_B are specific mass balances at the ice-sheet surface and bottom respectively, u_S and v_S are horizontal components of the surface velocity (with u_S being in the direction of flow), $F = \bar{u}/u_S$ where $\bar{u}$ is the depth-averaged velocity, and the flow is assumed to be in the x-direction. Equation (1.2) is an expression of mass conservation as a version of the equation of continuity (Paterson, 1994; Reeh, 1999). All quantities are expressed in terms of ice equivalent. For floating glaciers and ice shelves, b_B can be a dominant term, both in terms of melt and freeze-on/accretion (Foldvik and Kvinge, 1977; Jenkins et al., 1997; Oerter et al., 1992; Rignot and Jacobs, 2002; Robin, 1979). The calculation of local mass balance from (1.2) requires information on ice thickness and its gradient in the direction of ice flow, horizontal surface velocity, the ratio of the depth-averaged horizontal velocity to the horizontal surface velocity, the total surface strain rate—i.e., the sum of the transverse and longitudinal strain rates—and specific mass balance (neglecting changing density profiles and bedrock/till vertical motion). In effect, strain rates describe how the shape of an ice segment changes as a result of deformation (please see Pattyn and Naruse, 2003; Van der Veen, 1999a; and the review by Van der Veen and Payne, 2004 for the governing equations).

A key parameter that can be measured from space is the change in ice sheet surface elevation over time. At a fixed position on the glacier/ice sheet, this gives both the local *net mass balance* and a minimum value of the *specific mass balance* (Hagen and Reeh, 2004). It is determined by the kinematic boundary condition, which relates the rate of change $\partial H/\partial t$ in surface elevation to the vertical ice-particle velocity w_S, the horizontal ice velocity vector $\vec{u}_S$, the surface gradient $\mathrm{grad}\, H$, and the specific mass balance b_S by:

$$\frac{\partial H}{\partial t} = b_S + w_S - \vec{u}_S \cdot \mathrm{grad}\, H \tag{1.3}$$

Here, the term $w_S - \vec{u}_S \cdot \mathrm{grad}\, H$ is the *emergence velocity*, representing the vertical flow of ice relative to the surface (Paterson, 1994). It follows that the *net balance* can be estimated with knowledge of the emergence velocity and under the assumption that the ice density remains invariant with depth through the period of interest (Hagen and Reeh, 2004). As $h = H + B$, where B is the bedrock elevation, the local mass balance is:

$$\frac{\partial H}{\partial t} = b_S - \frac{\partial B}{\partial t} + w_S - \vec{u}_S \cdot \mathrm{grad}\, h \tag{1.4}$$

Use of this equation to estimate local mass balance requires knowledge of both the horizontal and vertical components of the ice velocity, as well as knowledge or neglect of $\partial B/\partial t$.

On the wider scale, the *total mass balance* is an aggregate term used to describe the total mass change of a glacier or ice sheet, or a region thereof (e.g., a drainage basin). It is determined by integrating either the specific or local mass balance over the wider area and subtracting the loss through vertical boundary surfaces—e.g., iceberg-calving fronts (Hagen and Reeh, 2004). The total mass balance is generally given by (Paterson, 1994; Reeh, 1999):

$$\frac{\Delta V}{\Delta t} = Q_a - Q_m - Q_c \pm Q_B \tag{1.5}$$

where V is the ice volume, Δt is one year, Q_a the net annual surface accumulation (precipitation and horizontal drift of snow by wind-blown—eolian—processes minus evaporation/sublimation and erosion by eolian redistribution), Q_m the annual loss by surface meltwater runoff, Q_c the annual loss by iceberg calving, and Q_B the annual balance at the ice base—i.e., freeze-on (positive) or melt (negative) of ice—with all volumes again expressed in terms of ice equivalent. An ice sheet is said to be in a state of balance if the loss terms exactly match the input terms.

Another approach to determining both ice sheet mass balance and its sensitivity to climate variability and/or change involves combining *in situ* measurements of climate variables and high-resolution GCMs with degree-day models or surface energy-balance models (Paterson, 1994; Reeh, 1999). As their name suggests, degree-day models are based upon parameterization of the number of days at a given location that the surface air temperature exceeds the zero-degree threshold—i.e., the melting point. *Energy-balance models* are based on energy and mass budget equations, and derive each of the relevant energy fluxes between the atmosphere and ice surface. Mass budget is defined as the (annual) difference between mass gained through accumulation and mass lost by ablation. The total energy flux from the atmosphere towards the ice surface is defined as:

$$Q_{tot} = (1-\alpha)F_{sw} + F_{lw}^{\Downarrow} + F_{lw}^{\Uparrow} + F_{sens} + F_{lat} \qquad (1.6)$$

where Q_{tot} is the energy flux available for melt, F_{sw} the incoming shortwave radiative flux, α the surface albedo (which determines the proportion of incoming shortwave radiation that is absorbed), $F_{lw}^{\Downarrow}$ the incoming longwave radiative flux, $F_{lw}^{\Uparrow}$ the outgoing longwave radiative flux—i.e., emitted from the surface in accordance with the Stefan–Boltzmann law (see Chapter 5 of Volume 1 of this book), F_{sens} the sensible-heat flux, and F_{lat} the latent-heat flux (Cawkwell and Bamber, 2002). By convention, energy fluxes directed away from the surface are negative, while those directed towards the surface are positive. Energy balance models are forced by observations of meteorological variables such as near-surface wind speed, temperature and humidity (e.g., from satellite-interrogated automatic weather stations), and cloud amount (and type/optical properties), radiative fluxes and precipitation, or by modeled approximations where measurements do not exist (which is typically the case). The surface energy balance determines the extent and magnitude of melt (ablation) at an ice-sheet surface, and energy balance models will produce estimates of the ablation term and mass balance if assumptions are made on the state of the sub-surface/surface, or if they are linked to a sub-surface module (Greuell and Genthon, 2004). An example of a recent mass balance estimate produced in this manner, and of the Greenland Ice Sheet, is provided by Box et al. (2004). Please see the review by Greuell and Genthon (2004) and the references therein for detailed information on each of these components.

Uncertainties in these variables and wind velocity, temperature, surface roughness lengths and their spatial variability, precipitation, sublimation, and albedo lead to considerable uncertainties in mass balance estimates derived by means of energy balance modeling, as reviewed by Greuell and Genthon (2004). As with sea ice (Chapter 5 of Volume 1 of this book), albedo is a major factor controlling shortwave radiative fluxes during daylight hours. Given that it is generally high—i.e., ~0.87—for dry, fine-grained snow in Antarctica (Grenfell et al., 1994), a small change can have a large impact on the amount of energy absorbed and thus the surface temperature (Stroeve, 2001). Accurate albedo estimates are therefore essential. Although the albedo–melt–temperature feedback mechanism is the major driving force behind predicted change in numerical simulations, it remains poorly characterized. Polar clouds and aerosols play an important role; these effects are discussed further in Chapter 4 of Volume 1 of this book. Currently, little is known about how the atmospheric contribution to the mass balance of ice sheets—i.e., energy fluxes and precipitation minus evaporation $(P - E)$—will change with the effects of global warming. These and other important issues are raised by Allison et al. (2001) and Walsh et al. (2001). Improved parameterization is required of storm tracks and moisture pathways over the ice sheet, drainage (katabatic) winds, meteorological conditions at the ice sheet surface, and boundary-layer stability. One research focus has been the estimation of the spatio-temporal variability in precipitation/accumulation over ice sheets from meteorological data (Connolley and King, 1996; Cullather et al., 1998). As with sea ice, a key factor in evaluating the dynamic response of the ice sheets to climate change/variability is understanding the impact of changes in major modes of large-scale climate variability, and teleconnections with lower latitudes, as they affect patterns of precipitation, accumulation, and melt. These modes, which include the Arctic Oscillation (AO), North Atlantic Oscillation (NAO), Southern Annular Mode (SAM), Antarctic Dipole, Antarctic Circumpolar Wave (ACW), and El Niño–Southern Oscillation (ENSO), are described in more detail in Chapter 5 of Volume 1 of this book. Once again, high-resolution analysis of ice core records provides an important means of determining the longer term impact of these cyclical modes and their variability, and of placing modern observations in an historical context.

It is apparent from the discussion above that *ice dynamics* is a key consideration in ice-sheet modeling in that it determines the horizontal redistribution of ice mass (Paterson, 1994; Thomas et al., 1985). In-depth background information on modeling land ice dynamics is given by Van der Veen and Payne (2004). Following Robin et al. (1983), a full description of ice-sheet dynamics requires knowledge of surface geometry, ice thickness H, velocity field, accumulation or ablation (surface and basal), fluctuations in the ice-sheet dimensions—e.g., in H—and the ice margin location, temperature field (ice flow is thermo-mechanically coupled) and the boundary between accumulation and ablation (snow-line position). Glaciers, and ice streams or ice sheets, flow due to gravity and in the direction of the greater surface slope. In the so-called *shallow-ice approximation*, the *driving stress* (*basal shear stress*) τ—i.e., the force per unit planimetric area—controls the ice-sheet flow. This is modeled from the product of ice surface slope φ and H as follows:

$$\tau = \rho_i g H \sin \varphi \qquad (1.7)$$

where ρ_i is ice density and g is gravitational acceleration (Robin et al., 1983). The ice responds to gravity by moving over its substrate and by deforming internally (Paterson, 1994). The rate of internal deformation increases with the cube of ice stress and with temperature (Paterson, 1994), and further depends on the ice deformation history (Alley, 1992). The rate of ice deformation, characterized by the strain rate tensor (Nye, 1959), is described by gradients in ice velocity (see Van der Veen and Payne, 2004 for a review of the governing equations). Basal velocity occurs by deformation of sub-glacial material (fabric) and/or basal sliding (Alley and Whillans, 1991; Kamb, 1991). According to Abdalati et al. (2004) and Payne and Bamber (2004), an area in major need of development is that of quantifying basal slip, either by sliding over a rigid substrate or by deformation of an intervening layer of sediment, and understanding the mechanisms that control it. Given the complexity involved, a major challenge is to develop a realistic and widely applicable flow law to relate strain rate to ice temperature, stress, and properties (Alley, 1992). The relationship between the horizontal and vertical components of ice velocity is given by the combination of the surface boundary condition and continuity equations (Glen, 1958; Nye, 1953; Van der Veen and Payne, 2004).

Acceleration can usually be ignored, and therefore equations of ice motion are equilibrium equations that balance internal stresses with gravity. A special case is where *surge behavior* occurs. Surging ice flow is characterized by long periods (often decades to centuries) of slow flow punctuated by short periods (1 to a few years) of rapid flow, which has the effect of transporting significant volumes of ice to lower altitudes (Joughin et al., 1996; Murray et al., 2002; Strozzi et al., 2002). This is an example of an important nonlinear cryospheric process, another being the deformation of ice grains (Li et al., 1996).

Ice motion is restrained by forces opposed to the gravitational driving force. For grounded ice, these forces emanate primarily from shear between ice and bedrock—the basal shear stress (see above, and the review by Van der Veen and Payne, 2004 for the governing equations). As a result, maps of driving stress can potentially yield proxy information on conditions at the ice base–bedrock interface (Thomas et al., 1985). Sub-glacial environments are characterized by complex interactions between mechanical, hydrological, and thermodynamic processes. The magnitude of basal resistance to ice motion, which is a major influence on ice sheet velocity and mass balance, is determined by the coupling of these processes. In spite of low surface slopes and associated low driving stresses, rapid flow occurs within ice streams, due to lubrication of the bed by weak sub-glacial till and pressurized water (Kamb, 2001; Tulaczyk et al., 2000a, b). For floating ice shelves, basal stress is zero, and restraints to ice motion take the form of shear between the shelf and its margins and also at grounded ice rises where present (Robin et al., 1983).

Realistic modeling of ice sheets requires a fully coupled solution of momentum and force conservation equations (Bindschadler et al., 1998). In-depth discussion of the theory is beyond the scope of this book, and the interested reader is encouraged to consult Greuell and Genthon (2004), Huybrechts (2004), Huybrechts et al. (2004), Paterson (1994), Van der Veen and Payne (2004), Van der Veen and Whillans (1989a, b), Van de Wal (2004) and Whillans et al. (1989), and the references therein, for more detailed information. While the flow law for ice deformation is fairly well-understood, considerable uncertainty exists due to poor knowledge of values of crystal orientation fabrics and temperature profiles (Alley, 1992). The vertical ice temperature profile is an example of an important parameter that cannot be measured from space, yet is critical to both the initiation and cessation of streaming flow (Scambos et al., 2004a, b). Basal sliding processes also remain poorly understood. As stated above, understanding the driving forces at work requires accurate information on the ice sheet geometry—i.e., both the surface and bedrock topography. While surface boundary conditions can to some extent be derived from satellite data, GCMs, climate parameterizations, and *in situ* measurements, basal boundary conditions—e.g., geothermal heat flux, basal hydrology and sediment strength—are less well-characterized. This is an important deficiency, given that changes in ice-stream physical properties, such as patterns of sub-glacial water flow, are presumed to represent important feedback mechanisms controlling rapid changes in ice-sheet characteristics and behavior.

1.2.1 Comparative characteristics of the Greenland and Antarctic Ice Sheets

Rather than forming single entities, the Greenland and Antarctic Ice Sheets are composed of series of individual drainage basins. These are in contact with different climate regimes, and even different parts of the same basin may have a different state of mass budget, given the long residence time of ice and snow falling inland—i.e., thousands of years. Drainage basins are large in Antarctica— i.e., typically of the order of 200,000 km^2, and about half as large in Greenland (GLAS Science Team, 1997). A high priority is the need to measure long-term changes in key ice-sheet parameters (e.g., surface elevation, ice velocity, ablation, accumulation rates) over all major drainage basins, and in sufficient detail to detect any substantial redistributions of ice volume. This will aid determination of whether the changes are caused by long-term dynamical effects on the ice flow or by shorter term climatic effects on the surface mass balance (GLAS Science Team, 1997). It represents a major challenge, but one that is well-suited to satellite remote sensing.

Due to their different climatic regimes, balance-process characteristics differ for the Greenland versus the Antarctic Ice Sheet. In Greenland, ice flow rates vary from a few meters per annum near the ice-sheet summit to several kilometers per annum near the termini of large outlet glaciers such as Jakobshavn Isbrae (Joughin et al., 1998). Seaward discharge occurs via the numerous fast-flowing outlet glaciers, which drain the ice sheet through coastal mountains. Ice here is mainly "lost" by iceberg calving and summer-time runoff of surface meltwater into the sea (Hanna et al., 2005), apart from in the north where substantial basal melting occurs near the grounding lines of deep outlet glaciers (Rignot and Thomas, 2002; Thomas et al., 2000, 2001a). Discharge from the Antarctic Ice Sheet occurs mainly via fast-flowing outlet glaciers and ice streams, some of which flow at speeds of $\sim$0.5 to 3 km per annum—i.e., up to 100 times faster than the surrounding ice. Ice streams form major conduits between the ice sheet interior and margins (Alley and Whillans, 1991), discharging an estimated 90% of snow accumulating inland (McIntyre, 1985) and mainly feeding into floating ice shelves. The latter comprise $\sim$40% of the Antarctic coastline. Ice loss in Antarctica is predominantly by (i) basal melting of ice shelves (Hellmer et al., 1998; Jacobs et al., 1996; Jenkins et al., 1997; Rignot and Jacobs, 2002; Williams et al., 1998a, b, 2002) and (ii) iceberg calving at the ice fronts (Jacobs et al., 1992). With the notable exception of a few regions—e.g., the Antarctic Peninsula—temperatures typically remain sufficiently low throughout the year in Antarctica that there is virtually no surface melting.

1.2.2 Approaches to the measurement of ice-sheet mass balance

Based upon the concepts briefly outlined above, two basic primary approaches have been used for measuring or estimating current ice-sheet mass balance using time series of satellite data (Abdalati et al., 2004; Bingham and Rees, 1997; Frezzotti,

1997; Hagen and Reeh, 2004; Reeh, 1999; Rémy and Legrésy, 1998; Rémy et al., 2001; Rignot and Thomas, 2002). These are:

(1) the *integrated* approach, involving the direct measurement of changes in ice sheet mass by monitoring surface elevation change—i.e., inferring $\partial H/\partial t$—on the basis of equating elevation change to a change in ice volume; and
(2) the *component* or *flux* method, involving measurement or estimation of each term on the right-hand side of (1.2)—i.e., ice mass in minus mass out for a given region.

The integrated approach is suitably driven by continuous measurement of ice sheet elevation by satellite altimeters. These measurements can be used to estimate changes in ice sheet mass if changes in near-surface *firn*[2] density (densification/compaction) and bedrock elevation (due to post-glacial rebound) are small or can be accurately determined (Houghton et al., 2001). Bedrock vertical motion is typically estimated from coupled models of lithosphere flexure and ice sheet dynamics—e.g., Huybrechts and Le Meur (1999), Le Meur and Huybrechts (1998), and Zweck (1998). The prediction of changes in mass balance requires an improved understanding of surface elevation coupled with information on changes in meteorological and climate forcings on seasonal to interannual timescales (Abdalati et al., 2004). While the integrated approach involves measurement of mass changes without separate determination of the input and output fluxes, the latter are individually measured or estimated with the component method. The latter is particularly effective when applied to individual drainage basins (Abdalati et al., 2004).

The two approaches are largely independent and thus complementary. Only the component approach, however, can be used to estimate not only present but also past and future *mass budgets*, in that it addresses the causes of observed changes in ice sheet areal extent and thickness H:

$$\frac{\partial H}{\partial t} = Q_a - Q_L - \nabla \cdot (H \langle \bar{u} \rangle) \tag{1.8}$$

where Q_a is the net accumulation rate, Q_L is net mass loss rate, and $\langle \bar{u} \rangle$ is the depth-averaged velocity (ice flux $F = H\langle \bar{u} \rangle$). As such, this approach alone can contribute to an understanding of the physical causes of observed ice sheet changes (Abdalati et al., 2004). The mass input can be calculated by the integration of specific mass balance across the ice sheet, while calculation of the mass flux in turn has two components, requiring measurement of (i) the column-mean ice-flow velocity across the grounding line, and (ii) the ice thickness along the grounding line. Ideally, both measurements should be carried out around the entire perimeter of the ice sheet. The outcome of this component approach is an ice-mass imbalance (GLAS Science Team, 1997). While this approach suffers less from uncertainties in near-surface processes than the integrated approach, it is more susceptible to errors associated with uncertainty

---

[2] Firn is partially consolidated, granular snow that is not yet glacial ice yet has passed through one summer melt season.

in ice thickness and velocity and accumulation rates. Indeed, errors have in the past tended to be large using the flux method, due to a lack of large-scale data and the difference in the timescales and non-uniformity of input and output terms (Ohmura et al., 1999; Warrick et al., 1996). For example, even a small error in locating the grounding line can translate into a large error in the calculation of grounding-line flux, as the gradient of ice thickness is typically large close to this boundary (Bamber and Kwok, 2004). The tide is again turning, however, due largely to revolutionary advances in remote-sensing technologies and analysis techniques, including SAR interferometry, enhanced radar altimetry, and laser altimetry. These are outlined in Section 1.3.

A third and emergent method of estimating mass balance involves measuring mass change, using new satellite-derived measurements of *changes in the Earth's gravity field*. Launched in March 2002, the joint US–German Gravity Recovery and Climate Experiment (GRACE) is the first in a series of satellite missions devoted to measurement of the Earth's time-variant gravity field. The others are GOCE (the European Space Agency Gravity field and steady-state Ocean Circulation Explorer, launch 2006) and the GRACE-FO (Follow-On, launch 2007). In addition, the German (DLR or Deutsche Forschungsanstalt für Luft und Raumfahrt) CHAMP (CHAllenging Mini-satellite Payload) mission has been providing information on the Earth's gravity field since 2000 (Reigber et al., 2003). With GRACE, two identical satellites circle the Earth in the same orbital plane but separated by 220 km. With this configuration, the GRACE mission measures the Earth's gravitational field to a precision 100–1,000 times greater than previously possible, and at a spatial resolution of 100 km. Data from GRACE have been used to create both new global maps of the mean gravity field (with gravity anomaly and geoid map products) and high-resolution maps of the monthly average gravity field (GRACE provides global coverage every 30 days) (Tapley et al., 2004a, b). The latter are useful for detecting and tracking changes in time-variable gravity effects. Of major interest from a polar ice sheet and global sea level perspective is that these data can be used to estimate temporal variations in the distribution of surface mass. The challenge in data analysis is to separate the various contributions to the signal, including ice, ocean, solid Earth, and atmosphere (Dietrich et al., 2004). When combined with modern satellite altimeter-derived estimates of ice-sheet elevation change (see Section 3.6), GRACE provides an important constraint on contemporary total mass imbalance estimates (Bentley and Wahr, 1998; Wu et al., 2002). Combined with measurements of ice sheet elevation by radar/laser altimetry, these relatively high spatial resolution measurements of time-variable gravity provide a means of separating changes in ocean volume resulting from thermal expansion versus changes in mass, and their relative contributions to observed and projected sea-level change (Wahr et al., 2000; Wu et al., 2002). Further details of this application are given in Section 3.12.

The GOCE mission (⟨<*http://www.esa.int/export/esaLP/ESAYEK1VMOC_goce_0.html*>⟩) will achieve its scientific goals by combining satellite gradiometry with Global Positioning System (GPS) inter-satellite tracking at low altitude. It will result in an improved model, at a spatial resolution of ~100 km, of the geoid and static gravity field to an accuracy of ~2 cm and 1 mGal (10^{-5} m s^{-2}), respectively. The beneficial outcomes will be:

- an accurate and unified global height reference for ice sheet and ocean elevations;
- an improved marine geoid (surface of equal gravitational potential) from which to derive absolute ocean currents from satellite radar altimetry, leading to an improved understanding of global ocean circulation and the transfer of heat; and
- improved constraints for lithospheric modeling.

Note that measurement and monitoring of ice sheet surface elevation requires an accurate geoid as a reference surface.

Another approach to the measurement/estimation of ice mass balance, extensively used on mountain glaciers but uncommonly on ice sheets, involves monitoring the location of the *Equilibrium Line Altitude* (ELA)—i.e., the average elevation of the line at which ablation exactly balances accumulation—at the end of the ablation season. At this time, the equilibrium line location approximates that of the snow line (Paterson, 1994), and forms a measure of the relative mass balance of a glacier for a given year (Bamber and Kwok, 2004). Following Benn and Lehmkuhl (2000), this concept requires some qualification when used in concert with glacier mass balance. For one thing, the definition of the ELA does not imply that the glacier or ice mass as a whole is in equilibrium. The value of the ELA-associated zero annual mass balance for the entire glacier is termed the steady-state ELA. Ice geometry and mass are taken to be in equilibrium with climate, and the glacier is neither shrinking nor growing, when the steady-state ELA coincides with the annual ELA (Benn and Lehmkuhl, 2000). Although this rarely happens in nature, the steady-state ELA is a valuable concept which provides an important measure of climate means associated with a given glacier geometry. Due to the sensitivity of the ELA to perturbations in air temperature and snowfall, it entails an important indicator of glacier response to climate change and/or variability, and knowledge of the ELA also allows reconstruction of past climate conditions. Please see Benn and Lehmkuhl (2000) for an in-depth review of the relationship between ELA and glacier mass balance.

Associated with the ELA, which has traditionally been measured with stakes but more recently with satellite data combined with *in situ* measurements, is the *Snow Line Altitude* (SLA) (Khalsa et al., 2004). This separates the accumulation and ablation zones, and is roughly equivalent to the ELA at the end of the annual melt season, although it may be higher where superimposed ice (refrozen meltwater) occurs (Khalsa et al., 2004). Another parameter related to glacier mass balance, and potentially measurable by satellite, is the *Accumulation Area Ratio* (AAR). This is defined as the ratio of a glacier's accumulation zone to its total area (Khalsa et al., 2004). In principle, the apportionment of area between the ablation and accumulation zones can be derived from knowledge of the ELA, enabling determination of the AAR. A theoretical basis for relating remotely sensed glacier parameters to annual mass balance is given by Khalsa et al. (2004).

Based upon measurements from a "benchmark" glacier that is representative of glaciers in a given region, the *annual mass balance* of a glacier b_a can be expressed as:

$$b_a = a_1(AAR[ELA] + a_2) \qquad (1.9)$$

where b_a is the specific mass balance $b_i(z)$ summed over all elevations for the glacier and over an entire year, and a_1 and a_2 are constants determined by fitting *in situ* measurements acquired over several years (Khalsa et al., 2004).

This has been a brief introduction to ice sheets and their mass balance. For more in-depth information on ice sheets as a pivotal element of the global cryosphere and global climate, and their potential response to climate change, see Allison et al. (2001), Fitzharris et al. (1996), Goodison et al. (1999), and Houghton et al. (2001). A detailed assessment of the performance of current ice sheet models is given by the IPCC report (Houghton et al., 2001), with Wild et al. (2003) modeling the impacts of polar ice sheets on global sea level under typical enhanced greenhouse-gas scenarios. For background information on ice-sheet glaciology, including descriptions of the transformation of snow to firn then to ice, see Alley and Bindschadler (2001), Drewry (1986), Hambrey (1994), Hughes (1998), Paterson (1994), Menzies (1995). Rémy et al. (2001) give excellent information on the surface and sub-surface physical characteristics of ice sheets and their relationship with weather and climate. The field of glacier mechanics is covered by Hooke (2005), while a full discussion of glacier/ice sheet thermodynamics is given by Paterson (1994) and Van der Veen (1999a). Bamber and Payne (2004), and the papers therein, provide a review of observations and modeling of both ice sheet and glacier mass balance. Thomas (2004) and Bentley (2004) give up-to-date reviews of observational aspects of mass balance estimates of the Greenland and Antarctic Ice Sheets. Hagen and Reeh (2004) provide information on *in situ* measurement techniques that complement satellite remote sensing, while Greuell and Genthon (2004) cover modeling. Models are required to provide a better understanding of the complex relationship between mass balance and climate, given the temporal and spatial variability in the effects of the latter. Please see Massom (1991) for a detailed review of earlier satellites, payloads, polar applications, and data availability (and Kramer, 2002 for detailed information on satellite missions and payloads in general). Reviews of remote sensing of the cryosphere are also provided by Bamber and Kwok (2004), Hall and Martinec (1985), and Massom (1995), with Bindschadler (1998b) and König et al. (2001) focusing on glacier and ice sheet applications. Armstrong et al. (1973) provided an illustrated glossary of snow and ice terminology.

An effort is made here not to treat ice sheets in isolation. As discussed in Chapter 5 of Volume 1 of this book, intimate and subtle associations occur between ice sheets and the surrounding sea ice and ocean, underlining the need to treat the cryosphere as a whole rather than a collection of individual unrelated elements. For example, ice sheet precipitation and accumulation rates over ice sheets are affected by the concentration and extent of the sea ice cover, via the impact of the latter on atmospheric moisture availability (Rind et al., 1995). This impact on water isotope fractionation further affects the interpretation of ice cores collected on ice sheets (Noone and Simmonds, 2004). Another example is the impact of ice shelf melt on sea ice (Hellmer, 2004). Moreover, ice sheet coastal configuration has a first-order effect on fast ice and polynya formation and distribution, as do grounded icebergs (both large and small) (Arrigo et al., 2002; Massom, 2003; Massom et al., 1998, 2001). The latter also have a major impact on the dynamics and thickness distribution of pack ice in the Antarctic coastal zone, and out to a distance of ~300 km in certain regions—and often thousands of kilometers from their point of calving (Massom, 2003; Massom et al., 2001)—see Volume 1 of this book. The impact of ice sheets on the wider climate system is felt via their modification of ocean water masses and their input of freshwater into the ocean via ice-sheet/-shelf basal melt and surface meltwater runoff (the latter in Greenland in particular). Iceberg melt is also a key factor affecting the ocean freshwater budget, and again one that is poorly quantified.

1.3 THE RECENT REVOLUTION IN SATELLITE REMOTE SENSING

The successful reduction of the uncertainties in estimates of ice sheet mass balance and concomitant sea level rise is heavily dependent upon the accurate merging of satellite data with modeled and observed atmospheric and oceanic data, and the further refinement of ice-sheet models constrained by improved observations (Houghton et al., 2001). While precise and important, *in situ* measurements are spatially and temporally limited and are logistically expensive. Satellite remote sensing, on the other hand, is ideally suited to ice-sheet research in that it can collect data over vast regions in a regular, repetitive, systematic, and cost-effective fashion, and over relatively long periods in a consistent manner. In the case of microwave sensors, this is irrespective of weather conditions and polar darkness. Over the past decade, satellite remote sensing has matured to the stage where it is revolutionizing our ability to examine ice sheet processes both on the basin-wide to continental scale and in extraordinary detail (Bindschadler, 1998b). This is a rapidly evolving and exciting field, and one which is again producing surprising and often ground-breaking results which are forcing re-evaluations of previously held estimates and concepts. Our previously blurred vision of ice-sheet characteristics, behavior, and variability is gradually coming more into focus to reveal startling facets and discoveries on a number of levels.

Not only are improved satellites and sensors and new technologies being launched, some are specifically designed to measure ice sheets—e.g., ICESat and CryoSat (note that while the launch of CryoSat failed in October 2005, it is retained throughout the book in the hope that a follow-on mission will occur). Improved processing techniques are also resulting in enhanced and indeed innovative products. In addition to their sensing capabilities, satellites continue to play a major role in tracking and transmitting data from important surface instrument packages deployed in remote regions—e.g., by the Argos system onboard current NOAA polar-orbiting satellites (⟨http://www.cls.fr/html/argos/welcome_en.html⟩) and Global Positioning System (GPS). Please see Hofmann-Wellnhof et al. (1995), Keller et al. (1997), Leick (2003), Parkinson and Spilker (1996), and Seeber (2003) for general background information on GPS. Precise GPS positioning is also of importance in enabling the navigation of aircraft in real time, and also provides important

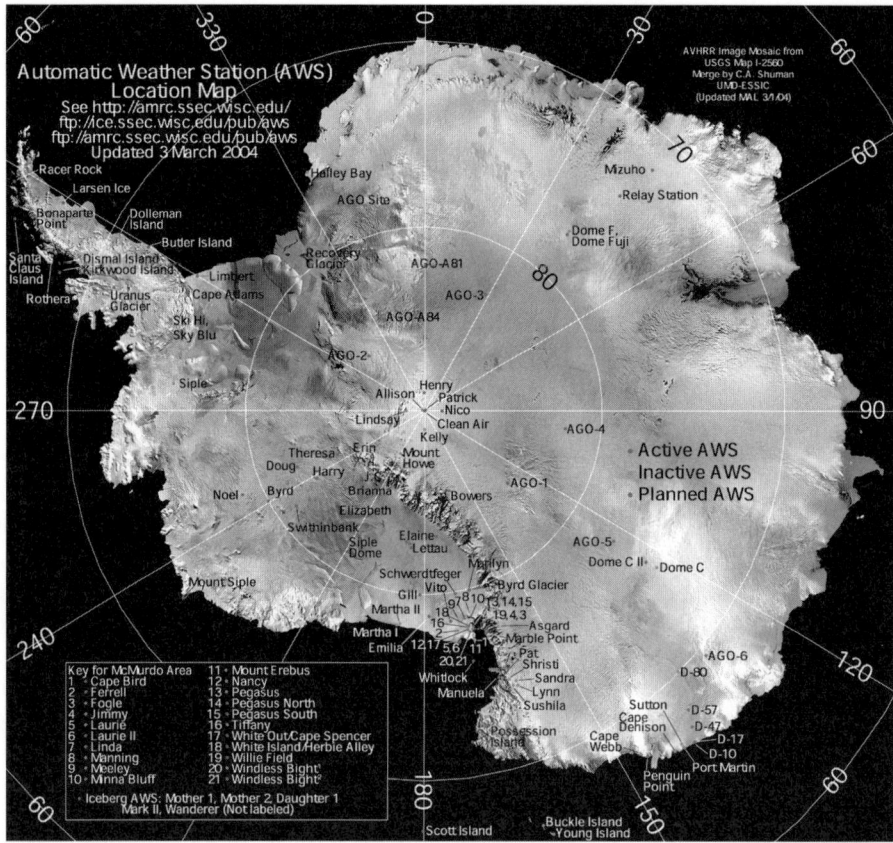

Figure 1.3. The location of Automatic Weather Stations (AWSs) in Antarctica in March 2004, superimposed on a NOAA (National Oceanic and Atmospheric Administration) Advanced Very High Resolution Radiometer (AVHRR) image mosaic of the ice sheet (USGS Map I-2560: Ferrigno et al., 2000). Basic AWSs measure wind speed and direction and air temperature at a nominal height of 3 m above the surface, and air pressure at the electronics enclosure height. Others also measure relative humidity and are equipped with sounding sensors to measure accumulation at the AWS.

From ⟨http://uwamrc.ssec.wisc.edu/aws.html⟩. Courtesy of the Automatic Weather Stations Project and Antarctic Meteorological Research Center, Space Science and Engineering Center, University of Wisconsin, Madison, WI.

ground-control points for validating satellite data. Of particular importance are networks of remote *Automatic Weather Stations (AWSs)*, which yield unique information on surface air temperature, pressure, wind velocity, and even accumulation rate (see ⟨http://cires.colorado.edu/steffen/gc-net/gc-net.html⟩ and ⟨http://uwamrc.ssec.wisc.edu/greenland.html⟩ for information on Greenland and ⟨http://uwamrc.ssec.wisc.edu/aws.html⟩ for the Antarctic Ice Sheet). The location map of current AWSs in Antarctica is shown in Figure 1.3. Important meteorological data are also collected at manned stations.

Satellite remote sensing now forms a cornerstone of much of ice sheet research. It plays a central role in major international research programs devoted to mass balance measurement. These include NASA's Program for Arctic Regional Climate Assessment (PARCA) (Abdalati, 2001; Thomas, 2004; Thomas et al., 2001a; ⟨http://cires.colorado.edu/parca.html⟩) and the NASA/US National Science Foundation West Antarctic Ice Sheet (WAIS) Initiative (WAIS, 2003; ⟨http://igloo.gsfc.nasa.gov/wais⟩). Crucially, it provides both a unique and fresh perspective on existing problems and the ability to unearth new issues and phenomena. Important though remote sensing is, models in fact represent the only practical means of determining the overall ice sheet response to climate change forcing (both past and future), as ice sheets are highly nonlinear systems (DeConto and Pollard, 2002; Paterson, 1994; Warner and Budd, 1998). Remote sensing has a key role to play in this respect by providing data with which to constrain, run, and validate models. For background information on current ice-sheet-modeling techniques, see Hulbe and Payne (2001), Houghton et al. (2001), Van der Veen (1999b), Van der Veen and Payne (2004), and comparisons within the European Ice Sheet Modelling INiTiative (EISMINT)—e.g., Huybrechts (1998) and Payne et al. (2000). Recommendations for ice-sheet model improvements are given by Abdalati et al. (2004).

Satellite remote sensing will never entirely replace *in situ* measurements, however; rather, it complements and extends them in the spatio-temporal domain. *Surface* and *aircraft measurements* remain essential, not only to validate and calibrate the satellite data/products but also to provide information that is currently unattainable from space—e.g., internal structure and vertical flow characteristics, and the absolute thickness of grounded ice. Direct measurement of the latter can at present only be carried out by surface or airborne *radio echo-sounding*, or *ice-penetrating radar* (Blankenship et al., 2001; Braaten and Gogineni, 2002; Braaten et al., 2003; Drewry et al., 1980; Gogineni et al., 2001, 2003; Robin, 1983), as current satellite-borne radars operate at frequencies which are too high to penetrate the entire ice sheet to the bedrock (Doake et al., 2003). At radio frequencies of 50 to 500 MHz, an ice sheet has a sufficiently low dielectric loss to allow observation of the ice–bedrock interface (Bogorodskii et al., 1985). With knowledge of the density of the ice/firn column, and the speed of the radiowave, it is possible to estimate ice thickness—to an estimated accuracy of 10 m for airborne measurements in Greenland (Gogineni et al., 1998). Ice-penetrating radar also provides a means of tracking internal layers (Kanagaratnam et al., 2001), which can also be related to past ablation and accumulation conditions (Nereson et al., 2000; Siegert, 1999) as well as changes in ice flow direction and past ice deformation (Siegert et al., 2004). Moves are afoot, however, to develop a new spaceborne ice-thickness sounder (see Section 3.13). Laboratory ice deformation studies, involving ice crystal orientation fabrics, remain the key to improving the parameterization of the physical basis for the ice flow laws used in numerical models (Jacka, 1994; Li et al., 1996). Moreover, *ice mechanics* research is required to better understand the mechanism of the rapid flow of ice streams and the physical variables that control it and couple it to global change—e.g., Engelhardt et al. (1990), Kamb (2001), MacAyeal et al. (1995), and

Smith (1997). Determination of *basal conditions* is critical in this respect (Anandakrishnan et al., 2001; Christofferson and Tulaczyk, 2003; MacAyeal, 1992; Pattyn, 1996), and is theoretically possible with multifrequency ice-penetrating radar (in concert with laboratory experiments on sub-glacial sediments). Again, such systems are currently under development, with a long-term view to deployment in space (⟨*http://www.ku-prism.org*⟩). Siegert and Ridley (1998) also used satellite radar in combination with airborne radio echo-sounding to determine basal ice-sheet conditions at Dome C (central East Antarctica).

In Chapter 2, we examine the extraordinary new technique of Synthetic Aperture Radar (SAR) interferometry, and evaluate its immense impact on ice sheet research. In Chapter 3, we then present a review of the application of remote sensing to the measurement of key ice-sheet processes and parameters, from the micro- to large-scale. Up-to-date examples are presented where possible to illustrate the extraordinary role that satellites now play in modern ice-sheet and related climate research.

1.4 REFERENCES

Abdalati, W. (ed.) (2001). PARCA: Mass balance of the Greenland Ice Sheet. *Journal of Geophysical Research*, **106**(D24), 33689–34058.

Abdalati, W., W. Krabill, E. Frederick, S. Manizade, C. Martin, J. Sonntag, R. Swift, R. Thomas, W. Wright, and J. Yungel (2001). Outlet glacier and margin elevation changes: Near-coastal thinning of the Greenland Ice Sheet. *Journal of Geophysical Research*, **106**(D24), 33729–33742.

Abdalati, W., I. Allison, F. Carsey, G. Casassa, M. Fily, M. Frezzotti, H. A. Fricker, C. Genthon, I. Goodwin, Z. Guo et al. (2004). Recommendations for the collection and synthesis of Antarctic ice sheet mass balance data. *Global and Planetary Change*, **42**(1–4), 1–15.

ACIA (Arctic Climate Assessment) (2004). *Impacts of a Warming Arctic*. Cambridge University Press, Cambridge, UK, 139 pp.

Alley, R. B. (1992). Flow-law hypotheses for ice-sheet modeling. *Journal of Glaciology*, **38**, 245–256.

Alley, R. B. (2000). Ice-core evidence of abrupt climate changes. *Proceedings of the National Academy of Sciences USA*, **97**(4), 1331–1334.

Alley, R. B. (ed.) (2001). *Abrupt Climate Change*. National Research Council, National Academy Press, Washington, DC, 238 pp.

Alley, R. B. (2002). On thickening ice? *Science*, **295**, 451–452.

Alley, R. B. and R. A. Bindschadler (eds.) (2001). *The West Antarctic Ice Sheet: Behavior and Environment* (AGU Antarctic Research Series No. 77). American Geophysical Union, Washington, DC, 296 pp.

Alley, R. B. and I. M. Whillans (1991). Changes in the West Antarctic Ice Sheet. *Science*, **254**, 959–963.

Alley, R. B., J. Lynch-Stieglitz, and J. P. Severinghaus (1999). Global climate change. *Proceedings of the US Academy of Science*, **96**(18), 9987–9988.

Alley, R. B., J. Marotzke, W. D. Nordhaus, J. T. Overpeck, D. M. Peteet, R. A. Pielke Jr., R. T. Pierrehumbert, P. B. Rhines, T. F. Stocker, L. D. Talley et al. (2003). Abrupt climate change. *Science*, **299**, 2005–2010.

Allison, I. (2003). *The AMISOR Project: Ice Shelf Dynamics and Ice–Ocean Interaction of the Amery Ice Shelf* (FRISP Report 14). Geophysical Institute, University of Bergen, Norway, pp. 1–9.

Allison, I., R. G. Barry, and B. E. Goodman (2001). *Climate and Cryosphere (CLIC) Project: Science and Co-ordination Plan, Version 1* (WCRP-114, WMO/TD 1053). World Climate Research Programme, World Meteorological Organization, Geneva, Switzerland.

Anandakrishnan, S., R. B. Alley, R. W. Jacobel, and H. Conway (2001). The flow regime of Ice Stream C and hypotheses concerning its recent stagnation. In: R. B. Alley and R. A. Bindschadler (eds.), *The West Antarctic Ice Sheet: Behavior and Environment* (AGU Antarctic Research Series No. 77). American Geophysical Union, Washington, DC, 283–294.

Anderson, J. B., J. S. Wellner, A. L. Lowe, A. S. Mosola, and S. S. Shipp (2001). Footprint of the expanded West Antarctic Ice Sheet: Ice stream history and behavior. *GSA Today*, **11**(10), 4–9.

Arend, A. A., K. A. Echelmeyer, W. D. Harrison, C. S. Lingle, and V. B. Valentine (2002). Rapid wastage of Alaska glaciers and their contribution to rising sea level. *Science*, **297**, 382–386.

Armstrong, T., B. Roberts, and C. Swithinbank (1973). *Illustrated Glossary of Snow and Ice*. Scott Polar Research Institute, University of Cambridge, Cambridge, UK, 69 pp.

Arrigo, K. R., G. L. van Dijken, D. G. Ainley, and M. A. Fahnestock (2002). Ecological impact of a large Antarctic iceberg. *Geophysical Research Letters*, **29**(7), doi:10.1029/2001GL014160.

Bamber, J. L. and R. Kwok (2004). Remote-sensing techniques. In: J. L. Bamber and A. J. Payne (eds.), *Mass Balance of the Cryosphere*. Cambridge University Press, Cambridge, UK, pp. 59–113.

Bamber, J. L. and A. J. Payne (eds.) (2004). *Mass Balance of the Cryosphere*. Cambridge University Press, Cambridge, UK, 644 pp.

Beckmann, A. and H. Goosse (2003). A parameterization of ice shelf–ocean interaction for climate models. *Ocean Modelling*, **5**, 157–170.

Benn, D. I. and F. Lehmkuhl (2000). Mass balance and equilibrium-line altitudes of glaciers in high-mountain environments. *Quaternary International*, **65–66**, 15–29.

Bentley, C. R. (1997). Rapid sea-level rise soon from West Antarctic Ice Sheet collapse? *Science*, **275**, 1077.

Bentley, C. R. (1998). Ice on the fast track. *Nature*, **394**, 21–22.

Bentley, C. R. (2004). Mass balance of the Antarctic Ice Sheet: Observational aspects. In: J. L. Bamber and A. J. Payne (eds.), *Mass Balance of the Cryosphere*. Cambridge University Press, Cambridge, UK, pp. 459–489.

Bentley, C. R. and M. B. Giovinetto (1991). Mass balance and sea-level change. *Proceedings of International Conference on the Role of the Polar Regions in Global Change, June 11–15, 1990, Fairbanks, Alaska*. University of Alaska, pp. 481–488.

Bentley, C. R. and J. M. Wahr (1998). Satellite gravity and the mass balance of the Antarctic Ice Sheet. *Journal of Glaciology*, **44**(147), 207–213.

Bindschadler, R. A. (1997). Actively surging West Antarctic ice streams and their response characteristics. *Annals of Glaciology*, **24**, 409–414.

Bindschadler, R. (1998a). Future of the West Antarctic Ice Sheet. *Science*, **282**(5388), 428–429.

Bindschadler, R. A. (1998b). Monitoring ice-sheet behavior from space. *Reviews of Geophysics*, **36**, 79–104.

Bindschadler, R. A. and C. R. Bentley (2002). On thin ice?: Impact of global warming. *Scientific American*, **287**(6), 98–106.

Bindschadler, R. A., R. B. Alley, J. Anderson, S. Shipp, H. Borns, J. Fastook, S. Jacobs, C. F. Raymond, and C. A. Shuman (1998). What is happening to the West Antarctic Ice Sheet. *EOS, Transactions of the American Geophysical Union*, **79**(22), 257, 264–265.

Bingham, A. W. and W. G. Rees (1997). Satellite data synergies for monitoring Arctic ice. *Proceedings of 3rd ERS Scientific Symposium, March 17–21, Florence, Italy* (SP-414, Vol. 2). ESA, Frascati, Italy, pp. 867–870.

Blankenship, D. D., D. L. Morse, C. A. Finn, R. E. Bell, M. E. Peters, S. D. Kempf, S. M. Hodge, M. Studinger, J. C. Behrendt, and J. M. Brozena (2001). Geologic controls on the initiation of rapid basal motion for West Antarctic ice streams: A geophysical perspective including new airborne radar sounding and laser altimetry results. In: R. Alley and R. Bindschadler (eds.), *The West Antarctic Ice Sheet: Behavior and Environment* (AGU Antarctic Research Series No. 77). American Geophysical Union, Washington, DC, pp. 105–121.

Bogorodskii, V. V., C. R. Bentley, and P. E. Gudmandsen (1985). *Radioglaciology: Glaciology and Quaternary Geology*. D. Reidel, Dordrecht, The Netherlands, 254 pp.

Box, J. E., D. H. Bromwich, and L.-S. Bai (2004). Greenland Ice Sheet surface mass balance for 1991–2000: Application of Polar MM5 mesoscale model and in-situ data. *Journal of Geophysical Research*, **109**(D16), doi:D16105, 10.1029/2003JD004451.

Braaten, D. and S. P. Gogineni (2002). Radar measurements of ice sheet thickness of outlet glaciers in Greenland. *Proceedings of IGARSS '02, Toronto, 24–28 June 2002*. Institute of Electrical and Electronic Engineers, Piscataway, NJ, Vol. 4, pp. 2188–2189.

Braaten, D., P. Kanagaratnam, T. Akins, and S. Gogineni (2003). *Measurement of Thickness of the Greenland Ice Sheet and High-resolution Mapping of Internal Layers* (Technical Report RSL 20780-2). Radar Systems and Remote Sensing Laboratory, University of Kansas, Lawrence, KS.

Cabanes, C., A. Cazenave, and C. Le Provost (2001). Sea level change from Topex-Poseidon altimetry for 1993–1999 and possible warming of the Southern Oceans. *Geophysical Research Letters*, **28**(1), 9–12.

Cawkwell, F. G. L. and J. L. Bamber (2002). The impact of cloud cover on the net radiation budget of the Greenland Ice Sheet. *Annals of Glaciology*, **34**, 141–149.

Chao, B. F. (1994). Man-made lakes and global sea level. *Nature*, **370**, 258.

Chao, B. F., T. Farr, J. LaBrecque, R. Bindschadler, B. Douglas, E. Rignot, C. K. Shum, and J. Wahr (2002). Understanding sea level changes. *Proceedings of IGARSS '02, Toronto, 24–28 June 2002*. Institute of Electrical and Electronic Engineers, Piscataway, NJ, Vol. 1, pp. 125–127.

Christofferson, P. and S. Tulaczyk (2003). Response of subglacial sediments to basal freeze-on: 1. Theory and comparison to observations from beneath the West Antarctic Ice Sheet. *Journal of Geophysical Research*, **108**(B4), 2222, DOI:10.1029/2002JB001935.

Church, J. A. (2001). How fast are sea levels rising? *Science*, **294**, 802–803.

Church, J. A., J. M. Gregory, P. Huybrechts, M. Kuhn, K. Lambeck, M. T. Nhuan, D. Qin, and P. L. Woodworth (2001). Changes in sea level. In: J. T. Houghton, Y. Ding, D. J. Griggs, M. Noguer, P. Van der Linden, X. Dai, K. Maskell, and C. I. Johnson (eds.), *Climate Change 2001: The Scientific Basis* (contribution of Working Group 1 to the Third Assessment Report of the Intergovernmental Panel on Climate Change). Cambridge University Press, Cambridge, pp. 639–694.

Clark, P. U., N. G. Pisias, T. F. Stocker, and A. J. Weaver (2002). The role of thermohaline circulation in abrupt climate change. *Nature*, **415**, 863–869.

Connolley, W. M. C. and J. C. King (1996). A modelling and observational study of East Antactic surface mass balance. *Journal of Geophysical Research*, **101**(D1), 1335–1343.

Cubasch, U., K. Hasselmann, H. Höck, E. Maier-Reimer, U. Mikolajewicz, B. D. Santer, and R. Sausen (1992). Time-dependent greenhouse warming computations with a coupled ocean–atmosphere model. *Climate Dynamics*, **8**, 55–69.

Cuffey, K. M. and G. D. Clow (1997). Temperature, accumulation, and ice sheet elevation in central Greenland through the last deglacial transition. *Journal of Geophysical Research*, **102**(C12), 26383–26396.

Cullather, R. I., D. H. Bromwich, and M. L. Van Woert (1998). Spatial and temporal variability of Antarctic precipitation from Antarctic methods. *Journal of Climate*, **11**(3), 334–367.

Day, C. (2004). Sea-level rise exacerbates coastal erosion. *Physics Today*, **57**(2), 24–26.

De Angelis, H. and P. Skvarca (2003). Glacier surge after ice shelf collapse. *Science*, **299**, 1560–1562.

DeConto, R. and D. Pollard (2002). *The Antarctic Climate Evolution (ACE) Paleoclimate and Ice Sheet Modeling Workshop, Northampton, Massachusetts, May 30–June 2* (draft summary report). Scientific Committee on Antarctic Research, Cambridge, UK, 20 pp. (Available at: ⟨http://www.ace.scar.org/Workshops/UMass/nhamp.html⟩).

Dietrich, R., H. Miller, and M. Wiehl (2004). Mass distribution and mass changes in Antarctica: The role of recent and upcoming satellite missions. *Terra Nostra* (Schriften der Alfred-Wegener-Stiftung 2004/4, Abstract Volume, Abstract S12/O21). GeoUnion Alfred-Wegener-Stiftung, Berlin, pp. 284.

Doake, C. S. M., H. F. J. Corr, A. Jenkins, K. W. Nicolls, and C. Stewart (2003). Interpretation of polarisation behaviour of radar waves transmitted through Antarctic ice shelves. *Proceedings of ESA Workshop, POLInSAR: Applications of SAR Polarimetry and Polarimetric Interferometry, ESRIN, Frascati, Italy, January 14–16* (ESA SP-529). Available online at ⟨http://www.earth.esa.int/polinsar⟩.

Douglas, B. (1997). Global sea level rise: A redetermination. *Surveys of Geophysics*, **18**(2–3), 279–292.

Douglas, B. (2001). Sea level changes in the era of the recording tide gauges. In: B. Douglas, M. Kearney, and S. Leatherman (eds.), *Sea Level Rise*. Academic Press, San Diego, pp. 37–64,

Douglas, B. and W. Peltier (2002). The puzzle of global sea level rise. *Physics Today*, **55**, 35–41.

Douglas, B., M. Kearney, and S. Leatherman (eds.) (2001). *Sea Level Rise*. Academic Press, San Diego.

Dowdeswell, J. A., J. O. Hagen, H. Bjornsson, A. F. Glazovsky, W. D. Harrison, P. Holmlund, J. Jania, R. M. Koerner, B. Lefauconnier, S. S. L. Ommanney et al. (1997). The mass balance of circum-Arctic glaciers and recent climate change. *Quaternary Research*, **48**, 1–14.

Drewry, D. (1986). *Glacial Geologic Processes*. Edward Arnold, London, 288 pp.

Drewry, D. J., D. T. Meldrum, and E. Jankowski (1980). Radio echo and magnetic sounding of the Antarctic Ice Sheet, 1978–1979. *Polar Record*, **20**, 43–57.

Dyurgerov, M. B. and C. L. Carter (2004). Observational evidence of increases in freshwater inflow to the Arctic Ocean. *Arctic, Antarctic, and Alpine Research*, **36**(1), 117–122.

Dyurgerov, M. B. and M. F. Meier (2000). Twentieth century climate change: Evidence from small glaciers. *Proceedings of the National Academy of Sciences USA*, **97**(4), 1406–1411.

Dyurgerov, M. B. and M. F. Meier (2004). Glaciers and the study of climate and sea-level change. In: J. L. Bamber and A. J. Payne (eds.), *Mass Balance of the Cryosphere*. Cambridge University Press, Cambridge, UK, pp. 579–622.

Engelhardt, H. F., Humphrey, N., Kamb, B., and Fahnestock, M. (1990). Physical conditions at the base of a fast-moving Antarctic ice stream. *Science*, **248**, 57–59.

ESA (2003). *CryoSat Science Report* (ESA SP-1272). ESA, Noordwijk, The Netherlands.

Fahrbach, E., R. G. Peterson, G. Rohardt, P. Schlosser, and R. Bayer (1994). Suppression of bottom water formation in the southeastern Weddell Sea. *Deep-Sea Research*, **41**, 389–411.

Ferrigno, J. G., J. L. Mullins, J. A. Stapleton, P S. Chavez Jr., M. G. Velasco, R. S. Williams Jr., G. F. Delinski Jr., and D. A. Lear (2000). *Satellite Image Map of Antarctica* (2nd edn., prepared by the U.S. Geological Survey with support from the National Science Foundation, Miscellaneous Investigations Series, Map I-2560—scale 1:5,000,000). U.S. Geological Survey, Reston, VA.

Fichefet, T., C. Poncin, H. Goosse, P. Huybrechts, I. Janssens, and H. Le Treut (2003). Implications of changes in freshwater flux from the Greenland Ice Sheet for the climate of the 21st century. *Geophysical Research Letters*, **30**(17), 1911, doi:10.1029/2003GL017826.

Fitzharris, B. B., I. Allison, R. J. Braithwaite, J. Brown, P. Foehn, W. Haeberli, K. Higuchi, V. M. Kotlakov, T. D. Prowse, C. A. Rinaldi et al. (1996). The cryosphere: Changes and their impacts. *Climate Change 1995: Impacts, Adaptations, and Mitigation of Climate Change—Scientific–Technical Analyses* (contribution of Working Group 11 to the Second Assessment Report of the Intergovernmental Panel on Climate Change). Cambridge University Press, Cambridge, UK, pp. 241–265.

Fitzharris, B. B., T. J. Chinn, and G. N. Lamont (1997). Glacier balance fluctuations and atmospheric circulation patterns over the Southern Alps, New Zealand. *International Journal of Climatology*, **17**, 745–763.

Foldvik, A. and T. Gammelsrød (1988). Notes on Southern-Ocean hydrography, sea-ice and bottom water formation. *Palaeogeography, Palaeoclimatology, Palaeoecology*, **67**, 3–17.

Foldvik, A. and T. Kvinge (1977). Thermohaline convection in the vicinity of an ice shelf. In: M. J. Dunbar (ed.), *Polar Oceans*. Arctic Institute of North America, University of Calgary, Calgary, Canada, pp. 247–255.

Foldvik, A., T. Gammelsrød, S. Østerhus, E. Fahrbach, G. Rohardt, M. Schröder, K. W. Nicholls, L. Padman, and R. A. Woodgate (2004). Ice shelf water overflow and bottom water formation in the southern Weddell Sea. *Journal of Geophysical Research*, **109**, C0201, doi:10.1029/2003JC002008.

Frezzotti, M. (1997). Ice front fluctuation, iceberg calving flux and mass balance of Victoria Land glaciers (Antarctica). *Antarctic Science*, **9**(1), 61–73.

GLAS Science Team (1997). *Geoscience Laser Altimeter System Science Requirements, Version 2.01*, 71 pp. Available online at ⟨http://www.csr.utexas.edu/glas/pdf/sci_reqs_v15.pdf⟩.

Glen, J. W. (1958). *The Flow Law of Ice: A Discussion of the Assumptions Made in Glacier Theory, Their Experimental Foundations and Consequences* (IAHS Publication No. 147). International Association of Hydrological Sciences, Christchurch, New Zealand, pp. 171–183.

Gogineni, S., T. Chuah, C. Allen, K. Jezek, and R. Moore (1998). An improved coherent radar depth sounder. *Journal of Glaciology*, **44**(148), 659–669.

Gogineni, S., D. Tammana, D. Braaten, C. Leuschen, T. Akins, J. Legarsky, P. Kanagaratnam, J. Stiles, C. Allen, and K. Jezek (2001). Coherent radar ice thickness measurements over the Greenland Ice Sheet. *Journal of Geophysical Research*, **106**(D24), 33761–33772.

Gogineni, S., G. Prescott, D. Braaten, C. Allen, K. Jezek, and the PRISM Research Team (2003). Polar radar for ice sheet measurements. *Proceedings of IGARSS '03, Toulouse, France, July*.

Goodison, B. E., R. D. Brown, and R. G. Crane (eds.) (1999). Cryospheric systems. In: M. D. King (ed.), *EOS Science Plan: The State of Science in the EOS Program* (NASA NP-1998-12-069-GSFC). NASA Goddard Space Flight Center, Greenbelt, MD, pp. 261–307.

Gregory, J. M., P. Huybrechts, and S. C. B. Raper (2004). Threatened loss of the Greenland Ice Sheet, *Nature*, **428**, 616.

Grenfell, T. C., S. G. Warren, and P. C. Mulen (1994). Reflection of solar radiation by the Antarctic snow surface at ultraviolet, visible and near-infrared wavelengths. *Journal of Geophysical Research*, **99**(D9), 18669–18684.

Greuell, W. and C. Genthon (2004). Modelling land ice surface mass balance. In: J. L. Bamber and A. J. Payne (ed.), *Mass Balance of the Cryosphere*. Cambridge University Press, Cambridge, UK, pp. 117–168.

Grosfeld, K., and H. Sandhäger (2004). The evolution of a coupled ice shelf–ocean system under different climate states. *Global and Planetary Change*, **42**, 107–132.

Grosfeld, K., R. Gerdes, and J. Determann (1997). Thermohaline circulation and interaction between ice shelf cavities and the adjacent ocean. *Journal of Geophysical Research*, **102**(C7), 15595–15610.

Grosfeld, K., H. Sandhäger, and M. A. Lange (2001). Sensitivity of a coupled ice-shelf/ocean system to changed oceanographic boundary conditions. In: H. Oerter (ed.), *Filchner-Ronne-Ice-Shelf-Programme* (Report 14). Available online at ⟨http://earth.uni-muenster.de/projekte/FRISP15.pdf⟩.

Haeberli, W. (2004). Glaciers and ice caps: Historical background and strategies of world-wide monitoring. In: J. L. Bamber and A. J. Payne (eds.), *Mass Balance of the Cryosphere*. Cambridge University Press, Cambridge, UK, pp. 559–578.

Haeberli, W. and M. Hoelzle (1995). Application of inventory data for estimating characteristics of and regional climate effects on mountain glaciers: A pilot study with the European Alps. *Annals of Glaciology*, **21**, 206–212.

Hagen, J. O. and N. Reeh (2004). *In situ* measurement techniques: Land ice. In: J. L. Bamber and A. J. Payne (eds.), *Mass Balance of the Cryosphere*. Cambridge University Press, Cambridge, UK, pp. 11–42.

Hall, D. K. and J. Martinec (1985). *Remote Sensing of Snow and Ice*. Chapman & Hall, New York, 189 pp.

Hambrey, M. J. (1994). *Glacial Environments*. University College London Press, 296 pp.

Hanna, E., P. Huybrechts, I. Janssens, J. Cappelen, K. Steffen, and A. Stephens (2005). Runoff and mass balance of the Greenland Ice Sheet: 1958–2003. *Journal of Geophysical Research*, **110**, D13108, doi:10.1029/2004JDOO5641.

Hellmer, H. H. (2004). Impact of Antarctic ice shelf melting on sea ice and deep ocean properties. *Geophysical Research Letters*, **31**(10), L10307, doi:10.1029/2004GL019506.

Hellmer, H. H. and S. S. Jacobs (1992). Ocean Interactions with the Base of Amery Ice Shelf, Antarctica. *Journal of Geophysical Research-Oceans*, **97**(C12): 20305–20317.

Hellmer, H. H., S. S. Jacobs, and A. Jenkins (1998). Oceanic erosion of a floating Antarctic glacier in the Amundsen Sea. In: S. S. Jacobs and R. F. Weiss (eds.), *Ocean, Ice, and Atmosphere: Interactions at the Antarctic Continental Margin* (AGU Antarctic Research Series No. 75). American Geophysical Union, Washington, DC, pp. 83–99.

Hindmarsh, R. C. A. (1993). Qualitative dynamics of marine ice sheets. In: W. R. Peltier (ed.), *Ice in the Climate System* (NATO ASI Series, Series I, 12). Springer-Verlag, Berlin, pp. 67–99.

Hoelzle, M., W. Haeberli, M. Dischl, and W. Peschke (2003). Secular glacier mass balances derived from cumulative glacier length changes. *Global and Planetary Change*, **36**(4), 77–89.

Hofmann-Wellenhof, H. Lichtenegger, and J. Collins (1995). *GPS: Theory and Practice* (3rd edn.). Springer-Verlag, New York, 384 pp.

Holland, D. M. and A. Jenkins (1999). Modelling thermodynamic ice–ocean interactions at the base of an ice shelf. *Journal of Physical Oceanography*, **29**, 1787–1800.

Holland, D. M., S. S. Jacobs, and A. Jenkins (2003). Modelling the ocean circulation beneath the Ross Ice Shelf. *Antarctic Science*, **15**(1), 13–23.

Holland, M. M. and C. M. Bitz (2003). Polar amplification of climate change in coupled models. *Climate Dynamics*, **21**, 221–232, doi:10.1007/s00382-003-0332-6.

Hooke, R. LeB. (2005). *Principles of Glacier Mechanics* (2nd edn.). Cambridge University Press, Cambridge, UK, 448 pp.

Houghton, J. T., Y. Ding, D. J. Griggs, M. Noguer, P. Van der Linden, X. Dai, K. Maskell, and C. I. Johnson (eds.) (2001). *Climate Change 2001: The Scientific Basis* (contribution of Working Group 1 to the Third Assessment Report of the Intergovernmental Panel on Climate Change). Cambridge University Press, Cambridge, UK, pp. 639–694.

Hughes, T. J. (1973). Is the West Antarctic Ice Sheet disintegrating? *Journal of Geophysical Research*, **78**(33), 7884–7910.

Hughes, T. J. (1977). West Antarctic ice streams. *Reviews of Geophysics*, **15**, 1–46.

Hughes, T. J. (1998). *Ice Sheets*. Oxford University Press, New York, 343 pp.

Hulbe, C. L. and A. J. Payne (2001). The contribution of numerical modelling to our understanding of the West Antarctic Ice Sheet. In: R. Alley and R. Bindschadler (eds.), *The West Antarctic Ice Sheet: Behavior and Environment* (AGU Antarctic Research Series No. 77). American Geophysical Union, Washington, DC, pp. 201–219.

Hunter, J., R. Coleman, and D. Pugh (2003). The sea level at Port Arthur, Tasmania, from 1841 to the present. *Geophysical Research Letters*, **30**(7), 54–1 to 54–4, doi:10.1029/2002GL016813.

Huybrechts, P. (1998). *Report of the 3rd EISMINT Workshop on Model Intercomparison*. European Science Foundation, Strasbourg, France, 120 pp.

Huybrechts, P. (2001). Changes of the polar ice sheets. In: J. Lozan, H. Grassl, and P. Hupfer (eds.), *Climate of the 21st Century: Changes and Risks*. GEO Wissenschaftliche Auswertungen, Hamburg, pp. 221–226.

Huybrechts, P. (2004). Antarctica: Modelling. In: J. L. Bamber and A. J. Payne (eds.), *Mass Balance of the Cryosphere: Observations and Modelling of Contemporary and Future Changes*. Cambridge University Press, Cambridge, UK, pp. 491–523.

Huybrechts, R. and J. de Wolde (1999). The dynamic response of the Greenland and Antarctic Ice Sheets to multiple-century climatic warming. *Journal of Climate*, **12**(8), 2169–2188.

Huybrechts, P. and E. Le Meur (1999). Predicted present-day evolution patterns of ice thickness and bedrock elevation over Greenland and Antarctica. *Polar Research*, **18**(2), 299–306.

Huybrechts, P., J. Gregory, I. Janssens, and M. Wild (2004). Modelling Antarctic and Greenland volume changes during the 20th and 21st centuries forced by GCM time slice integrations. *Global and Planetary Change*, **42**, 83–105.

Jacka, T. H. (1994). Investigations of discrepancies between laboratory studies of the flow of ice: Density, sample shape and size, and grain size. *Annals of Glaciology*, **19**, 146–154.

Jacobs, S. S., H. H. Helmer, C. S. M. Doake, A. Jenkins, and R. M. Frolich (1992). Melting of ice shelves and the mass balance of Antarctica. *Journal of Glaciology*, **38**(130), 375–387.

Jacobs, S. S., H. H. Hellmer, and A. Jenkins (1996). Antarctic Ice Sheet melting in the southeast Pacific. *Geophysical Research Letters*, **23**(9), 957–960.

Jenkins, A., D. G. Vaughan, S. S. Jacobs, H. H. Hellmer, and J. R. Keys (1997). Glaciological and oceanographic evidence of high melt rates beneath Pine Island Glacier, West Antarctica. *Journal of Glaciology*, **43**(143), 114–121.

Jenkins, A., H. Corr, K. Nicholls, C. Stewart, and C. Doake (2003). Interactions between ice and ocean near an ice-shelf grounding line. *EOS Transactions of the American Geophysical Union*, Fall Meeting Supplement, Abstract C21A-07.

Jones, P. D., M. New, D. E. Parker, S. Martin, and I. G. Rigor (1999). Surface air temperature and its changes over the past 150 years. *Reviews of Geophysics*, **37**(2), 173–199.

Joughin, I., S. Tulaczyk, M. Fahnestock, and R. Kwok (1996). A mini-surge on the Ryder Glacier, Greenland observed via satellite radar interferometry. *Science*, **274**(5285), 228–230.

Joughin, I. R., R. Kwok, and M. A. Fahnestock (1998). Interferometric estimation of three-dimensional ice-flow using ascending and descending passes. *IEEE Transactions on Geoscience and Remote Sensing*, **36**(1), 25–37.

Joughin, I., W. Abdalati, and M. Fahnestock (2004). Large fluctuations in speed on Greenland's Jakobshavn Isbrae glacier. *Nature*, **432**, 608–610.

Kamb, B. (1991). Rheological non-linearity and flow instability in the deforming bed mechanism of ice stream motion. *Journal of Geophysical Research*, **96**, 16585–16595.

Kamb, B. (2001). Basal zone of the West Antarctic ice streams and its role in lubrication of their rapid motion. In: R. Alley and R. Bindschadler (eds.), *The West Antarctic Ice Sheet: Behavior and Environment* (AGU Antarctic Research Series No. 77). American Geophysical Union, Washington, DC, pp. 157–199.

Kanagaratnam, P., S. P. Gogineni, N. Gundestrup, and L. Larsen (2001). High-resolution radar mapping of internal layers at the North Greenland Ice Core Project. *Journal of Geophysical Research*, **106**(D24), 33799–33812.

Keller, K., R. Forsberg, and C. S. Nielsen (1997). Kinematic GPS for ice sheet monitoring and SAR interferometry in Greenland. *Proceedings of International Symposium on Kinematic Systems in Geodesy, Geomatics and Navigation, Banff, Canada, June 3–6* (KIS97). University of Calgary, pp. 525–528.

Khalsa, S. J. S., M. B. Dyurgerov, T. Khromova, B. H. Raup, and R. G. Barry (2004). Space-based mapping of glacier changes using ASTER and GIS tools. *IEEE Transactions on Geoscience and Remote Sensing*, **42**(10), 2177–2183.

King, J. C. and J. C. Comiso (2003). The spatial coherence of interannual temperature variations in the Antarctic Peninsula. *Geophysical Research Letters*, **30**, 1040, doi:10.129/2002GL015580.

King, J. C. and J. Turner (1997). *Antarctic Meteorology and Climatology*. Cambridge University Press, Cambridge, UK, 409 pp.

King, J. C., J. Turner, G. J. Marshall, W. M. Connolley, and T. A. Lachlan-Cope (2003). Antarctic Peninsula climate variability and its causes as revealed by analysis of instrument records. In: E. Domack, A. Leventer, A. Burnett, R. Bindschadler, P. Conley, and M. Kirby (eds.), *Antarctic Peninsula Climate Variability: Historical and Palaeoenvironmental Perspective* (AGU Antarctic Research Series No. 79). American Geophysical Union, Washington DC, pp. 17–30.

König, M., J.-G. Winther, and E. Isaksson (2001). Measuring snow and glacier ice properties from satellite. *Reviews of Geophysics*, **39**(1), 1–28.

Krabill, W., E. Hanna, P. Huybrechts, W. Abdalati, J. Cappelen, B. Csatho, E. Frederick, S. Manizade, C. Martin, J. Sonntag et al. (2004). Greenland Ice Sheet: Increased coastal thinning. *Geophysical Research Letters*, **31**, L24402, doi:10.1029/2004GL021533.

Kramer, H. J. (2002). *Observation of the Earth and Its Environment: Survey of Missions and Sensors* (4th edn.). Springer-Verlag, Berlin, 1540 pp.

Leatherman, S. (2001). Societal and economic costs of sea level rise. In: B. Douglas, M. Kearney, and S. Leatherman (eds.), *Sea Level Rise*. Academic Press, San Diego, pp. 181–223.

Leick, A. (2003). *GPS Satellite Surveying* (3rd edn.). John Wiley & Sons, New York, 464 pp.

Le Meur, E. and P. Huybrechts (1998). Present-day uplift patterns over Greenland from a coupled ice-sheet/visco-elastic bedrock model. *Geophysical Research Letters*, **25**(21), 3951–3954.

Li., J., T. H. Jacka, and W. F. Budd (1996). Deformation rates in combined compression and shear for ice which is initially isotropic and after the development of strong anisotropy. *Annals of Glaciology*, **23**, 247–252.

Lowell, T. V. (2000). As climate changes, so do glaciers. *Proceedings of the National Academy of Sciences*, **97**(4), 1351–1354.

MacAyeal, D. R. (1992). Irregular oscillations of the West Antarctic ice sheet. *Nature*, **359**, 29–32.

MacAyeal, D. R., R. A. Bindschadler, and T. A. Scambos (1995). Basal friction of Ice Stream E, West Antarctica. *Journal of Glaciology*, **41**(138), 247–262.

Massom, R. (1991). *Satellite Remote Sensing of Polar Regions: Applications, Limitations and Data Availability*. Belhaven Press, London/Lewis Publishers (CRC Press), Boca Raton, FL, 307 pp.

Massom, R. (1995). Satellite remote sensing of polar snow and ice: Present status and future directions. *Polar Record*, **31**(177), 99–114.

Massom, R. A. (2003). Recent iceberg calving events in the Ninnis Glacier region, East Antarctica. *Antarctic Science*, **15**(2), 303–313.

Massom, R. A., P. T. Harris, K. J. Michael, and M. J. Potter (1998). The distribution and formative processes of latent heat polynyas in East Antarctica. *Annals of Glaciology*, **27**, 420–426.

Massom, R. A., K. L. Hill, V. I. Lytle, A. P. Worby, M. J. Paget, and I. Allison (2001). Effects of regional fast-ice and iceberg distributions on the behaviour of the Mertz Glacier polynya, East Antarctica. *Annals of Glaciology*, **33**, 391–398.

Mayewski, P. A., J. Souney, K. J. Kreutz, V. I. Morgan, T. van Ommen, and I. Goodwin (2001). Ice core proxy for Antarctic atmospheric circulation over the last several hundred years. *EOS, Transactions of the American Geophysical Union*, **82**(20), S234, Spring Meeting Suppl (Abstract OS51B-11).

McDowell, M. (2002). Melting ice triggers Himalayan flood warning. *Nature*, **416**, 776.

McIntyre, N. F. (1985). The dynamics of ice-sheet outlets. *Journal of Glaciology*, **31**, 99–107.

Meier, M. F., M. B. Dyurgerov, and G. J. McCabe (2003). The health of glaciers: Recent changes in glacier regime. *Climate Change*, **59**(1–2), 123–135.

Menzies, J. (ed.) (1995). *Modern Glacial Environments: Processes, Dynamics and Sediments*. Butterworth-Heinemann, Oxford, 621 pp.

Mercer, J. H. (1968). Antarctic ice and Sangamon Sea level. *International Association of Scientific Hydrology Symposia*, **79**, 217–225.

Mercer, J. H. (1978). West Antarctic ice sheet and CO_2 greenhouse effect: A threat of disaster. *Nature*, **271**, 321–325.

Miller, L. and B. C. Douglas (2004). Mass and volume contributions to 20th century global sea level rise. *Nature*, **248**, 407–409.

Morgan, V., M. Delmotte, T. van Ommen, J. Jouzel, J. Chappellaz, S. Woon, V. Masson-Delmotte, and D. Raynaud (2002). Relative timing of deglacial climate events in Antarctica and Greenland. *Science*, **297**, 1862–1864.

Murray, T., T. Strozzi, A. Luckman, H. Pritchard, and H. Jiskoot (2002). Ice dynamics during a surge of Sortebrae, East Greenland. *Annals of Glaciology*, **34**, 323–329.

Nerem, R. S. and G. T. Mitchum (2001a). Observations of sea level change from satellite altimetry. In: B. C. Douglas, M. S. Kearney, and S. P. Leatherman (eds.), *Sea Level Rise: History and Consequences*. Academic Press, New York, pp. 121–163.

Nerem, R. S. and G. T. Mitchum (2001b). Sea level change. In: L. Fu and A. Cazenave (eds.), *Satellite Altimetry and Earth Sciences: A Handbook of Techniques and Applications*. Academic Press, New York, pp. 329–349.

Nerem, R. S., B. J. Haines, J. Hendricks, J. F. Minster, G. T. Mitchum, and W. B. White (1997). Improved determination of global mean sea level variations using TOPEX/Poseidon altimeter data. *Geophysical Research Letters*, **24**(11), 1331–1334.

Nereson, N. A., C. F. Raymond, R. W. Jacobel, and E. D. Waddington (2000). The accumulation pattern across Siple Dome, West Antarctica, inferred from radar-detected internal layers. *Journal of Glaciology*, **46**(152), 75–87.

NOAA (2004). *A Paleo Perspective on Abrupt Climate Change* (digital media). National Oceanic and Atmospheric Administration Climatic Data Center. Available online at ⟨http://www.ncdc.noaa.gov/paleo/abrupt/⟩.

Noone, D. and I. Simmonds (2004). Sea ice control of water isotope transport to Antarctica and implications for ice core interpretation. *Journal of Geophysical Research*, **109**, D07105, DOI:10.1029/2003JD004228.

Nye, J. F. (1953). The flow law of ice from measurements in glacier tunnels, laboratory experiments and the Jungfraufirn borehole experiments. *Proceedings of the Royal Society of London, Series A* **219**, 477–489.

Nye, J. F. (1959). A method of determining the strain-rate tensor at the surface of a glacier. *Journal of Glaciology*, **3**(25), 409–419.

Oerlemans, J., B. Anderson, A. Hubbard, P. Huybrechts, T. Johannesson, W. H. Knap, M. Schmeits, A. P. Stroeven, R. S. W. Van de Wal, J. Wallinga, and Z. Zuo (1998). Modelling the response of glaciers to climate warming. *Climate Dynamics*, **14**, 267–274.

Oerter, H., J. Kipfstuhl, J. Determann, H. Miller, D. Wagenbach, A. Mimikin, and W. Graf, (1992). Evidence for basal marine ice in the Filchner-Ronne Ice Shelf. *Nature*, **358**, 399–401.

Ohmura, A. and N. Reeh (1991). New precipitation and accumulation maps for Greenland. *Journal of Glaciology*, **37**, 140–148.

Ohmura, A., P. Calanca, M. Wild, and M. Anklin (1999). Precipitation, accumulation, and mass balance of the Greenland Ice Sheet. *Zeitschrift für Gletscherkunde und Glazialgeologie*, **35**(1), 1–20.

Oppenheimer, M. (1998). Global warming and the stability of the West Antarctic Ice Sheet. *Nature*, **393**, 325–332.

Oppenheimer, M. and R. B. Alley (2004). The West Antarctic Ice Sheet and long term climate policy. *Climatic Change*, **64**(1–2), 1–10.

Parish, T. and D. Bromwich (1987). Surface windfield over the Antarctic Ice Sheet. *Nature*, **328**(6125), 51–54.

Parish, T. R. and D. H. Bromwich (1991). Continental-scale simulation of the Antarctic katabatic wind regime. *Journal of Climate*, **4**, 135–146.

Parkinson, B. W. and J. J. Spilker Jr. (1996). *Global Positioning System: Theory and Applications*. American Institute of Aeronautics and Astronautics, Boston, 643 pp.

Paterson, W. S. B. (1994). *The Physics of Glaciers* (3rd edn.). Butterworth-Heinemann, Oxford, UK, 480 pp.

Paterson, W. S. B. and N. Reeh (2001). Thinning of the ice sheet in north-west Greenland over the past forty years. *Nature*, **414**, 60–62.

Pattyn, F. (1996). Numerical modeling of a fast flowing outlet glacier: Experiments with different basal conditions. *Annals of Glaciology*, **23**, 237–246.

Pattyn, F. and R. Naruse (2003). The nature of complex ice flow in Shirase Glacier catchment, East Antarctica. *Journal of Glaciology*, **49**(166), 429–436.

Payne, A. J. and J. L. Bamber (2004). Conclusions, summary and outlook. In: J. L. Bamber and A. J. Payne (eds.), *Mass Balance of the Cryosphere*. Cambridge University Press, Cambridge, UK, pp. 623–639.

Payne, A. J., P. Huybrechts, A. Abe-Ouchi, R. Calov, J. L. Fastook, R. Greve, S. J. Marshall, I. Marsiat, C. Ritz, L. Tarasov et al. (2000). Results from the EISMINT model intercomparison: The effects of thermomechanical coupling. *Journal of Glaciology*, **46**(153), 227–238.

Peltier, W. (2001). Global glacial isostatic adjustment and modern instrumental records of relative sea level history. In: B. Douglas, M. Kearney, and S. Leatherman (eds.), *Sea Level Rise*. Academic Press, San Diego, pp. 65–95.

Peterson, B. J., R. M. Holmes, J. W. McClelland, C. J. Vorosmarty, I. A. Shiklomanov, A. I. Shiklomanov, R. B. Lammers, and S. Rahmstorf (2002). Increasing river discharge to the Arctic Ocean. *Science*, **298**, 2171–2173.

Reeh, N. (1999). Mass balance of the Greenland Ice Sheet: Can modern observation methods reduce the uncertainty? *Geografiska Annaler: Special Issue. Methods of Mass Balance Measurements and Modelling*, **81A**(4), 735–742.

Reeh, N. and N. Gundestrup (1985). Mass balance of the Greenland Ice Sheet at Dye 3. *Journal of Glaciology*, **31**(108), 198–200.

Reigber, C., H. Lühr, and P. Schwintzer (eds.) (2003). *First CHAMP Mission Results for Gravity, Magnetic and Atmospheric Studies*. Springer-Verlag, Berlin, 563 pp.

Rémy, F. and B. Legrésy (1998). Antarctic non-stationary signals derived from Seasat-ERS-1 altimetry comparison. *Annals of Glaciology*, **27**, 81–85.

Rémy, F., B. Legrésy, and L. Testut (2001). Ice sheet and satellite altimetry. *Surveys in Geophysics*, **22**(1), 1–29.

Ridley, J., P. Huybrechts, J. Gregory, and J. Lowe (2005). Elimination of the Greenland Ice Sheet in a high-CO_2 climate. *Journal of Climate*, **18**(17), 3409–3427.

Rignot, E. (2004). Contribution to sea level rise from Antarctic Peninsula glaciers. *European Geosciences Union, Nice, April 25–30, Nice, France* (abstract).

Rignot, E. and S. S. Jacobs (2002). Rapid bottom melting widespread near Antarctic Ice Sheet grounding lines. *Science*, **296**, 2020–2023.

Rignot, E. and R. H. Thomas (2002). Mass balance of polar ice sheets. *Science*, **297**, 1502–1506.

Rignot, E., G. Casassa, P. Gogineni, W. Krabill, A. Rivera, and R. Thomas (2004). Accelerated ice discharge from the Antarctic Peninsula following the collapse of Larsen B Ice Shelf. *Geophysical Research Letters*, **31**, L18401, doi:10.1029/2004GL020697.

Rind, D., R. Healy, C. Parkinson, and D. Martinson (1995). The role of sea-ice in $2 \times CO_2$ climate model sensitivity: 1. The total influence of sea-ice thickness and extent. *Journal of Climate*, **8**(3), 449–463.

Rind, D., P. deMenocal, G. Russell, S. Sheth, D. Collins, G. Schmidt, and J. Teller (2001). Effects of glacial meltwater in the GISS coupled atmosphere–ocean model: Part I. North Atlantic Deep Water response. *Journal of Geophysical Research*, **106**, 27335–27354.

Rintoul, S. R. (1998). On the origin and influence of Adélie Land Bottom Water. In: S. Jacobs and R. Weiss (eds.), *Ocean, Ice and the Atmosphere: Interactions at the Antarctic Continental Margin* (AGU Antarctic Research Series No. 75). American Geophysical Union, Washington, DC, pp. 151–171.

Robin, G. de Q. (1979). Formation, flow, and disintegration of ice shelves. *Journal of Glaciology*, **24**(90), 259–271.

Robin, G. de Q. (ed.) (1983). Radio-echo studies of internal layering of polar ice sheets. *The Climatic Record in Polar Ice Sheets*. Cambridge University Press, Cambridge, UK, pp. 89–93.

Robin, G. de Q., D. J. Drewry, and V. A. Squire (1983). Satellite observations of polar ice fields. *Philosophical Transactions of the Royal Society of London*, **A309**, 447–461.

Rott, H., W. Rack, P. Skvarca, and H. D. Angelis (2002). Northern Larsen Ice Shelf Antarctica: Further retreat after collapse. *Annals of Glaciology*, **34**, 277–282.

Sabadini, R. (2002). Ice sheet collapse and sea level change. *Science*, **295**, 2376–2377.

Scambos, T. A., J. A. Bohlander, C. A. Shuman, and P. Skvarca (2004a). Glacier acceleration and thinning after ice shelf collapse in the Larsen B embayment, Antarctica. *Geophysical Research Letters*, **31**, L18402, doi:10.1029/2004GL020670.

Scambos, T., J. Bohlander, B. Raup, and T. Haran (2004b). Glaciological characteristics of Institute Ice Stream using remote sensing. *Antarctic Science*, **16**(2), 205–213.

SCAR (2002). *Antarctic Ice Sheet Mass Balance and Sea Level (ISMASS)* (brief report of a workshop, SCAR Bulletin 146). Scientific Committee for Antarctic Research, Cambridge, UK.

Schiermeier, Q. (2004). A rising tide. *Nature*, **428**, 114–115.

Schwerdtfeger, W. (1984). *Weather and Climate of the Antarctic*. Elsevier Science, Amsterdam, 262 pp.

Seeber, G. (2003). *Satellite Geodesy: Foundations, Methods and Applications* (2nd edn.). Walter de Gruyter, Berlin, 589 pp.

Seidov, D., B. J. Haupt, E. J. Barron, and M. Maslin (2001). Ocean bi-polar seesaw and climate: Southern versus northern meltwater impacts. In: D. Seidov, B. J. Haupt, and M. Maslin (eds.), *The Oceans and Rapid Climate Change: Past, Present, and Future*. American Geophysical Union, Washington, DC, pp. 147–167.

Severinghaus, J. S. and E. J. Brook (1999). Abrupt climate change at the end of the Last Glacial Period inferred from trapped air in polar ice. *Science*, **286**(5441), 930–934.

Severinghaus, J. S., T. Sowers, E. J. Brook, R. B. Alley, and M. L. Bender (1998). Timing of abrupt climate change at the end of the Younger Dryas Interval from thermally fractionated gases in polar ice. *Nature*, **391**, 141–146.

Siegert, M. J. (1999). On the origin, nature and uses of Antarctic ice-sheet radio-echo layering. *Progress in Physical Geography*, **23**, 159–179.

Siegert, M. J. and J. K. Ridley (1998). Determining basal ice sheet conditions at Dome C, central East Antarctica, using satellite radar altimetry and airborne radio-echo sounding information. *Journal of Glaciology*, **44**, 1–8.

Siegert, M. J., B. Welch, D. Morse, A. Vieli, D. D. Blankenship, I. Joughin, E. C. King, G. Leysinger Vieli, A. J. Payne, and R. Jacobel (2004). Ice flow direction change in interior West Antarctica. *Science*, **305**, 1948–1951.

Skvarca, P., W. Rack, H. Rott, and T. Ibarzábal y Donángelo (1999). Climatic trend and the retreat and disintegration of ice shelves on the Antarctic Peninsula: An overview. *Polar Research*, **18**, 151–157.

Smith, A. M. (1997). Variations in basal conditions on Rutford Ice Stream, West Antarctica. *Journal of Glaciology*, **43**, 245–255.

Stroeve, J. (2001). Assessment of Greenland albedo variability from the Advanced Very High Resolution Radiometer Polar Pathfinder data set. *Journal of Geophysical Research*, **106**(D24), 33989–34006.

Strozzi, T., A. Luckman, T. Murray, U. Wegmüller, and C. L. Werner (2002). Glacier motion estimation using SAR offset-tracking procedures. *IEEE Transactions on Geoscience and Remote Sensing*, **40**(11), 2384–2391.

Tapley, B. D., S. Bettadpur, M. Watkins, and C. Reigber (2004a). The Gravity Recovery and Climate Experiment: Mission overview and early results. *Geophysical Research Letters*, **31**(9), L09607, doi:10.1029/2004GL019920.

Tapley, B. D., S. Bettadpur, J. C. Ries, P. F. Thompson, and M. Watkins (2004b). GRACE measurements of mass variability in the Earth system. *Science*, **305**(5683), 503–505.

Taylor, K. C., R. Alley, M. Bender, E. Brook, P. Mayewski, E. Meyerson, J. Severinghaus, and J. White (2001). Rapid climate change at Siple Dome, Antarctica. *EOS, Transactions of the American Geophysical Union*, **82**(47), Fall Meeting Supplement (Abstract U42A-0012).

Thomas, R. H. (1979a). The dynamics of marine ice sheets. *Journal of Glaciology*, **24**, 167–177.

Thomas, R. H. (1979b). Effect of climatic warming on the West Antarctic Ice Sheet. *Nature*, **277**, 355–358.

Thomas, R. H. (2001). Remote sensing reveals shrinking Greenland Ice Sheet. *EOS, Transactions of the American Geophysical Union*, **82**(34), 369–373.

Thomas, R. H. (2004). Greenland: Recent mass-balance observations. In: J. L. Bamber and A. J. Payne (eds.), *Mass Balance of the Cryosphere*. Cambridge University Press, Cambridge, UK, pp. 393–436.

Thomas, R. H., R. A. Bindschadler, R. L. Cameron, F. D. Carsey, B. Holt, T. J. Hughes, C. W. M. Swithinbank, I. M. Whillans, and H. J. Zwally (1985). *Satellite Remote Sensing for Ice Sheet Research* (NASA Technical Memorandum 86233). NASA, Washington, DC, 32 pp.

Thomas, R. H., W. Abdalati, T. L. Akins, B. M. Csatho, E. B. Frederick, S. P. Gogineni, W. B. Krabill, S. S. Manizade, and E. J. Rignot (2000). Substantial thinning of a major east Greenland outlet glacier. *Geophysical Research Letters*, **27**(9), 1291–1294.

Thomas, R. H. and PARCA Investigators (2001a). PARCA 2001, Program for Arctic Regional Climate Assessment (PARCA): Goals, key findings, and future directions. *Journal of Geophysical Research*, **106**(D24), 33691–33706.

Thomas, R., B. Csatho, C. Davis, C. Kim, W. Krabill, S. Manizade, J. McConnell, and J. Sonntag (2001b). Mass balance of higher-elevation parts of the Greenland Ice Sheet. *Journal of Geophysical Research*, **106**(D24), 33707–33716.

Thomas, R. H., W. Abdalati, W. B. Krabill, S. Manizade, and K. Steffen (2003). Investigation of surface melting and dynamic thinning of Jakobshavn Isbrae, Greenland. *Journal of Glaciology*, **49**(165), 231–239.

Thomas, R., E. Rignot, G. Casassa, P. Kanagaratnam, C. Acua, T. Akins, H. Brecher, E. Frederick, P. Gogineni, W. Krabill et al. (2004). Accelerated sea-level rise from West Antarctica. *Science*, **306**(5694), 255–258.

Thompson, D. W. J. and S. Solomon (2002). Interpretation of recent Southern Hemisphere climate change. *Science*, **296**(5569), 895–899.

Toniazzo, T., J. Gregory, and P. Huybrechts (2004). Climatic impact of a Greenland deglaciation and its possible irreversibility. *Journal of Climate*, **17**(1), 21–33, doi:10.1175/1520-0442(2004)017.

Tulaczyk, S., B. Kamb, and H. F. Engelhardt (2000a). Basal mechanics of Ice Stream B, West Antarctica: 1. Till mechanics. *Journal of Geophysical Research*, **105**, 463–481.

Tulaczyk, S., B. Kamb, and H. Engelhardt (2000b). Basal mechanics of Ice Stream B, West Antarctica: 2. Undrained plastic bed model. *Journal of Geophysical Research*, **105**, 483–494.

Turner, J., S. R. Colwell, G. J. Marshall, T. A. Lachlan-Cope, A. M. Carleton, P. D. Jones, V. Lagun, P. A. Reid, and S. Iagovkina (2005). Antarctic climate change during the last 50 years. *International Journal of Climatology*, **25**(3), 279–294.

Van de Wal, R. S. W. (2004). Greenland: Modelling. In: J. L. Bamber and A. J. Payne (eds.), *Mass Balance of the Cryosphere*. Cambridge University Press, Cambridge, UK, pp. 437–458.

Van der Veen, C. J. (1999a). *Fundamentals of Glacier Dynamics*. A. A. Balkema, Rotterdam, 462 pp.

Van der Veen, C. J. (1999b). Evaluating the performance of cryospheric models. *Polar Geography*, **23**, 83–96.

Van der Veen, C. J. (2001). Greenland Ice Sheet response to external forcing. *Journal of Geophysical Research*, **106**(D24), 34047–34058.

Van der Veen, C. J. and A. Payne (2004). Modelling land ice dynamics. In: J. L. Bamber and A. J. Payne (eds.), *Mass Balance of the Cryosphere*. Cambridge University Press, Cambridge, UK, pp. 167–226.

Van der Veen, C. J. and I. M. Whillans (1989a). Force budget: I. Theory and numerical methods. *Journal of Glaciology*, **35**, 53–60.

Van der Veen, C. J. and I. M. Whillans (1989b). Force budget: II. Application to two-dimensional flow along the Byrd Station Strain Network, Antarctica. *Journal of Glaciology*, **35**, 61–67.

Vaughan, D. G. and J. R. Spouge (2002). Risk estimation of collapse of the West Antarctic Ice Sheet. *Climatic Change*, **52**, 65–91.

Vaughan, D. G., J. L. Bamber, M. Giovinetto, J. Russell, and A. P. R. Cooper (1999). Reassessment of surface mass balance in Antarctica. *Journal of Climate*, **12**, 933–946.

Vaughan, D. G., G. J. Marshall, W. M. Connolley, C. Parkinson, R. Mulvaney, D. A. Hodgson, J. C. King, C. J. Pudsey, and J. Turner (2003). Recent rapid regional climate warming on the Antarctic Peninsula. *Climate Change*, **60**(3), 243–274.

Wahr, J., D. Wingham, and C. Bentley (2000). A method of combining ICESat and GRACE satellite data to constrain Antarctic mass balance. *Journal of Geophysical Research*, **105**(B7), 16279–16294.

WAIS (2003). *The West Antarctic Ice Sheet Initiative: A Multi-disciplinary Study of Rapid Climate Change and Future Sea Level*. Available online at ⟨http://igloo.gsfc.nasa.gov/wais/⟩.

Walsh, J. E., J. Curry, M. Fahnestock, M. C. Kennicutt II, A. D. McGuire, W. B. Rossow, M. Steele, C. J. Vorosmarty, R. Wharton, C. Elfring et al. (2001). *Enhancing NASA's Contributions to Polar Science*. National Academy Press, Washington, DC, 138 pp.

Warner, R. C. and W. E. Budd (1998). Modelling the long-term response of the Antarctic ice sheet to global warming. *Annals of Glaciology*, **27**, 161–168.

Warrick, R., C. Le Provost, M. Meier, J. Oerlemans, and P. Woodworth (1996). Changes in sea level. In: J. T. Houghton, L. G. Meira Filho, B. A. Callander, N. Harris,

A. Kattenberg, and K. Maskell (eds.), *Climate Change, 1995: The Science of Climate Change*. Cambridge University Press, Cambridge, UK, pp. 359–405.

Weertman, J. (1974). Stability of the junction of an ice sheet and an ice shelf. *Journal of Glaciology*, **13**(67), 3–11.

Weller, G. E. (1967). *Radiation Fluxes over an Antarctic Ice Surface: Mawson, 1961–1962* (ANARE Scientific Report 96). Australian Antarctic Division, Melbourne, 106 pp.

Whillans, I. M., Y. H. Chen, C. J. Van der Veen, and T. J. Hughes (1989). Force budget: III. Application to three-dimensional flow of Byrd Glacier, Antarctica. *Journal of Glaciology*, **35**, 68–80.

White, N. J., J. A. Church, and J. M. Gregory (2005). Coastal and global averaged sea-level rise for 1950 to 2000. *Geophysical Research Letters*, **32**(1), L01601, doi:10.1029/2004GL021391.

Wild, M., P. Calanca, S. C. Scherrer, and A. Ohmura (2003). Effects of polar ice sheets on global sea level in high-resolution greenhouse scenarios. *Journal of Geophysical Research*, **108**(D5), 4165, doi:10.1029/2002JD002451.

Williams, G. D. and N. L. Bindoff (2003). Wintertime oceanography of the Adélie Depression. *Deep-Sea Research Part II*, **50**(8–9), 1373–1392.

Williams, M. J. M., A. Jenkins, and J. Determann (1998a). Physical controls on ocean circulation beneath ice shelves revealed by numerical models. In: S. S. Jacobs and R. F. Weiss (eds.), *Ocean, Ice, and Atmosphere: Interactions at the Antarctic Continental Margin* (AGU Antarctic Research Series No. 75). American Geophysical Union, Washington, DC., pp. 285–299.

Williams, M. J. M., R. C. Warner, and W. F. Budd (1998b). The effects of ocean warming on melting and ocean circulation under the Amery Ice Shelf, East Antarctica. *Annals of Glaciology*, **27**, 75–80.

Williams, M. J. M., R. C. Warner, and W. F. Budd (2002). Sensitivity of the Amery Ice Shelf, Antarctica, to changes in the climate of the Southern Ocean. *Journal of Climate*, **15**, 2740–2757.

Wu, X., M. Watkins, R. Kwok, E. Ivins, and J. Wahr (2002). Measuring present-day secular change in Greenland ice mass with future GRACE gravity data. *Journal of Geophysical Research*, **107**(B11), 2291, doi:10.1029/2001JB000543.

Yu, Y., X. Cai, J. C. King, and I. A. Renfrew (2005). Numerical simulations of katabatic jumps in Coats Land, Antarctica. *Boundary-layer Meteorology*, **114**(2), 413–437.

Zwally, H. J., B. Schutz, W. Abdalati, J. Abshire, C. Bentley, A. Brenner, J. Bufton, J. Dezio, D. Hancock, D. Harding et al. (2002). ICESat's laser measurements of polar ice, atmosphere, ocean, and land. *Journal of Geodynamics*, **34**, 405–445.

Zweck, C. (1998). Glacial isostasy and the crustal structure of Antarctica. *Annals of Glaciology*, **27**, 321–326.

2

Synthetic aperture radar interferometry and related techniques

2.1 INTRODUCTION AND BACKGROUND

No recent development in the field of polar remote sensing has engendered more excitement and interest than Synthetic Aperture Radar (SAR) *interferometry*, or *InSAR*. Over the past decade, this technique, which uses measured differences in the phase of the return radar signal acquired from two or more slightly different positions in space to detect slight changes on the Earth's surface, has revolutionized our ability to measure two key ice sheet and glacier parameters—namely, surface elevation and flow (motion). It has the extraordinary capacity to measure on the large scale but to an unprecedented level of detail and precision—of millimeters to centimeters for surface displacement and meters for surface elevation (Joughin et al., 2000; Madsen and Zebker, 1998). While the precision of the latter may not be as great as that achievable using other satellite techniques—e.g., laser altimetry (see Chapter 3)—a major strength of SAR is its unique ability to simultaneously measure both ice motion and elevation if suitable data are available (Mohr et al., 1998). As a result, satellite InSAR is now firmly established as a powerful glaciological research tool, and some of the emergent results are nothing short of astounding. Regarding the measurement of ice motion, InSAR has another unique capacity—the ability to acquire high-density data not only in outlet glaciers and ice streams but also in featureless areas such as ice-sheet plateaux (e.g., Goldstein et al., 1993; Joughin et al., 1995; Kwok and Fahnestock, 1996; Rignot et al., 1997). Conventional feature-tracking techniques using SAR and optical data largely fail in the latter regions (see Chapter 3), as they rely on the presence of identifiable surface-tracking features (e.g., crevasses) and on optimal solar illumination conditions (for optical systems). The ability of InSAR to acquire data over large regions, irrespective of cloud cover and solar illumination conditions, and at a high horizontal resolution is a major advantage.

The availability of this extraordinary new dataset has led to major advances in a number of important areas of ice sheet research, including the following:

- InSAR has revolutionized our ability to measure and monitor outlet glacier discharge—e.g., Rignot et al. (1997) and Joughin et al. (1999a)—and changes in ice discharge rates.
- Being sensitive to centimeter-scale movement, InSAR is an ideal tool for determining the location of the grounding line of outlet glaciers and for monitoring changes in its location (Goldstein et al., 1993; Gray et al., 2002; Rignot, 1996). This is achieved by detecting the limit of tidal flexure in floating glacier tongues and ice shelves—a task that was much more difficult to achieve prior to InSAR.
- These applications have led to improved estimates of regional mass budgets— e.g., Rignot and Thomas (2002)—using the component approach (see Chapter 1).
- The high temporal resolution of InSAR compared with conventional feature-tracking techniques has enabled improved detailed studies of changes in ice flow characteristics, and even the detection of glacier surge behavior—e.g., Fischer et al. (2003); Joughin et al. (1996a); and Strozzi et al. (2000).
- Other applications of InSAR include analysis of ice-shelf dynamics (MacAyeal et al., 1998), and the monitoring of ice-shelf rift propagation leading to large iceberg calving events (Fricker et al., 2002).

Results from these and a plethora of other InSAR studies represent an immensely important source of new information on the current state of ice sheet and glacier mass balance (Rignot and Thomas, 2002). Indeed, they are helping to reduce current uncertainty in projections of the likely response of ice sheets to predicted global warming and their contribution to global sea-level rise. Moreover, they are unearthing extraordinary changes in ice-sheet dynamics and thermodynamics that are occurring at a much more rapid rate than had previously been anticipated. An improved understanding of ice dynamics is important in the context of ice-sheet stability (Bindschadler et al., 1998). The overall importance of such measurements and their immense benefit to the numerical modeling community are discussed in Chapter 1.

In this chapter, we first examine the basic principles of InSAR as applied to ice-sheet research, then evaluate the inherent limitations and error sources impacting the technique. We end with an assessment of the related techniques of SAR speckle tracking and maximum coherence tracking, and look at possible future developments. Due to its highly dynamic behavior, pack ice is generally not suited to study by conventional repeat pass InSAR, due to the breakdown of interferometric coherence[1] between image acquisitions. The technique has, however, been success-

[1] *Coherence* is a measure of the correlation of the phase information contained in corresponding signals.

fully used to map more stable fast ice (Hirose and Vachon, 1998) and detect fast ice deformation (Dammert et al., 1998; Li et al., 1996; Morris et al., 1999; Sharov et al., 2002). We concentrate here on InSAR as applied to polar glaciers and ice sheets, and using satellite data only. It should be noted that the techniques outlined here are also generally applicable to non-polar ice masses. Aircraft and surface measurements still have a key role to play, however, not least in InSAR calibration and validation and in spatially limited but high-resolution regional studies—e.g., by Danish researchers in Greenland (Mohr and Madsen, 1996a; ⟨http://www.emi.dtu.dk/research/DCRS/Science/Landice/glacier.html⟩). Although *in situ* measurements—e.g., by GPS (Global Positioning System)—are generally more accurate than corresponding InSAR measurements, the former are logistically challenging, expensive, and extremely sparse in space and time compared with satellite SAR coverage. Satellite SARs provide regular repetitive coverage, unaffected by clouds and darkness, over a number of years to enable monitoring of change and/or variability. In terms of InSAR, this results in measurements over a dense regular net that are compatible with modern ice-sheet and glacier model simulations. Case studies using satellite InSAR and further discussion of the unique benefits being gained are presented in Chapter 3.

We cannot possibly do full justice to this complex and rapidly evolving technique here. The interested reader is encouraged to follow up on a number of excellent review articles on InSAR. These include the following references: Abdelfattah and Nicolas (2002); Allen (1995); Bamler (1997); Bamler and Hartl (1998); Cloude (2004); Dixon (1994); Gens and van Genderen (1996a); Hellwich (1999a, b); Klees and Hanssen (2000); Klees and Massonnet (1999); Madsen and Zebker (1998); Massonnet (1997); Massonnet and Feigl (1998); Rocca et al. (1997); Rodríguez and Martin (1992); Rosen et al. (2000); Smith (2002); van Genderen and Gens (1996); Vettore et al. (2002); and Zebker (2000). Signal processing of SAR is covered by Soumekh (1999). Although much of the discussion in the above articles centers on fields other than glaciology—e.g., crustal dynamics and seismic events—the basic principles are the same or similar.

2.2 UNDERLYING PRINCIPLES AND TERMINOLOGY OF INSAR

In this section, we provide a brief outline of the terminology and main principles of InSAR. We saw in Chapter 5 of Volume 1 of this book how conventional SARs produce all-season, day–night maps of the radar reflectivity of illuminated scenes on the Earth's surface—i.e., they use information contained in the amplitude of the backscattered energy. The intensity of the returned echoes is strongly dependent on both the physical (surface and near-surface morphology and roughness) and electrical (e.g., dielectric constant and absorption) characteristics of the target medium (again see Volume 1). The amplitude[2] of a backscattered wave is derived

---

[2] *Amplitude* is defined here as the strength or "height" of an electromagnetic wave (in units of voltage).

from the ratio of the power transmitted by a radar to the power of the received signal reflected by the surface within the area viewed. SAR is, however, a *coherent* imaging system, retaining information not only on magnitude but also *phase* (or angle of a complex number) in the radar echo during data acquisition. Electromagnetic waves transmitted by a radar are considered to be "in phase" if their origin is perfectly superimposable. One cycle of phase has occurred when the sine wave begins to repeat itself—i.e., the phase angle is $>360°$. The concept of phase is outlined schematically in Figure 2.1. By the technique of SAR interferometry, two or more complex-valued SAR scenes of the same surface area are combined to derive additional information compared with a single image. The InSAR technique exploits this additional information, using the phase measurements to infer differential *range*[3] and range change in two or more SAR images of the same surface area. By this means, it can measure the radial velocity of moving scatterers, survey height information from the illuminated scene, and detect and track subtle changes in both motion and scene content (Allen, 1995). A major advantage of InSAR is its unique ability to provide densely spaced measurements over large areas, and in a near-instantaneous fashion that is ideally suited to the monitoring of change. For a more thorough treatment of the underlying principles of radar remote sensing, and SAR in particular, the reader is referred to Curlander and McDonough (1991); Elachi (1987, 1988); Fitch (1988); Henderson and Lewis (1998); Hein (2004); Levanon (1988); Olmsted (1993); Rees (2001); and Ulaby et al. (1982).

The key to InSAR is that differences in phase measured between two coherent images represent fraction-of-a-wavelength differences in the round-trip travel time of radar pulses between the satellite and surface scatterers within each *resolution cell*. These in turn represent extremely precise measures of the change in *path length*[4] between the surface pixels and spacecraft, enabling the precise inference of change in the sensor range direction if the interferometric geometry is accurately known and assuming that atmospheric propagation delay effects and other errors (noise sources) are negligible. The basis of satellite SAR interferometry is illustrated in a simple schematic in Figure 2.2. For ice sheets and glaciers, the measured phase difference then results from a combination of (a) a parallax effect due to surface topography, and (b) a shift in the location of the scatterers due to ice motion. In this chapter, we present techniques that have recently been developed to separate these important quantities—i.e., to produce both Digital Elevation Models (DEMs) and maps of ice surface velocity.

For InSAR to work, two (or more) SAR images of the same area must be acquired such that the same points on the ground are viewed from slightly different angles. This can be achieved by a number of means, and interferometric

---

[3] *Range* is the line-of-sight distance between the radar and each illuminated scatterer (target).
[4] *Path length* is the distance between the satellite sensor and the surface.

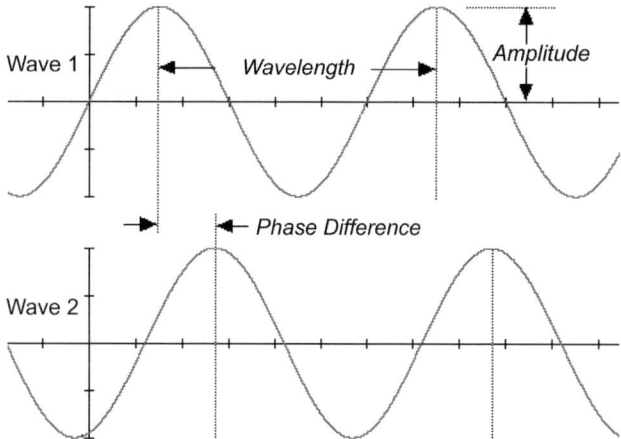

Figure 2.1. A schematic representation showing the relationship between the *wavelength*, *amplitude*, and *phase* characteristics of electromagnetic radiation. A synthetic aperture radar illuminates the Earth with a beam of coherent microwave radiation, and measures the echo backscattered from the surface. The wave can be described by these three properties. Wavelength is the distance separating two peaks on the wave, and is fixed for SAR. Amplitude is the displacement of the wave at its peak. For SAR, this corresponds to the brightness of the resultant image. The phase describes the shift of one wave relative to another, and is usually measured in angular units of degrees or radians (where $360° = 2\pi$ radians). Where two waves oscillate together "in phase", they are not shifted with respect to one another—i.e., they have a phase shift of $0°$. Where the two waves are oriented in an opposite fashion—i.e., the crest of one corresponds to the trough of the other—then they are said to be "out of phase". With SAR, the phase of the echoing signal is compared with a reference wave, and the resultant phase of a SAR image comprises the *phase difference* between the two. In effect, the phase is directly related to the distance traveled by the radiation—i.e., between the sensor and surface. Due to the sub-meter wavelength of current Earth-observing SARs, and the distance between adjacent pixels of several meters, the phase of a single SAR image does not yield any useful information. Precise knowledge of phase properties in radar signal data is, however, a key element in interferometric (as well as in polarimetric) SAR. The subtraction of the phase information of one SAR image from another from approximately the same area results in the construction of an interferogram (if the data are suitable).
After ⟨http://www.asf.alaska.edu/apd/software/insar/phase.html⟩.

systems can be either *single-pass* or *repeat-pass* (*multipass*). With single-pass systems, two SAR images are simultaneously acquired of the same region by two antennae fixed on an aircraft or spacecraft and separated by a known and constant *baseline*[5] (Gabriel and Goldstein, 1988). In this case, only one of the SAR antennae transmits radar pulses, while both antennae receive the radar echoes (Bamler et al., 2003). Such a system has only flown once in space so far—onboard the joint NASA-DLR (Deutsche Forschungsanstalt für Luft und Raumfahrt) Space Shuttle Radar

---

[5] *Baseline* is defined as the physical separation between the two antennae positions.

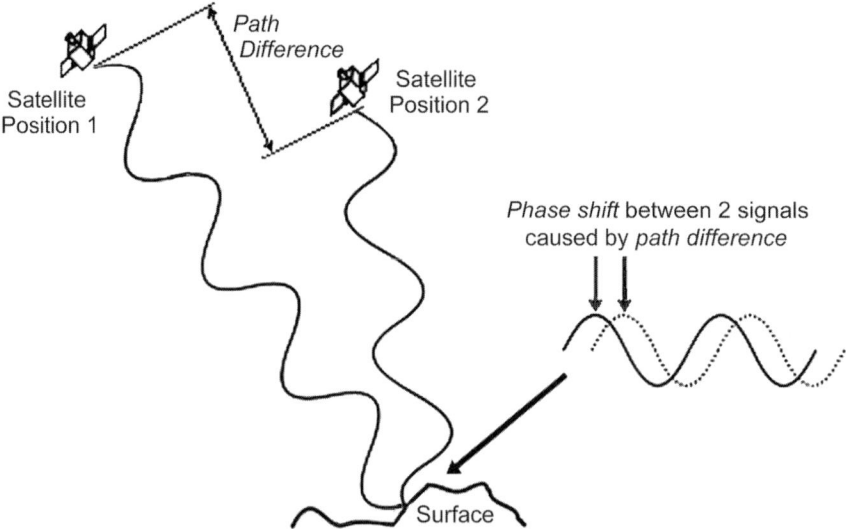

Figure 2.2. A schematic showing the basic principle of repeat-pass inteferometric phase measurement from space. In practice, the position of the satellite between two image acquisitions is never identical. The corresponding difference in the path (distance between satellite and ground) means that a difference in phase exists between the two signals—i.e., a phase shift. The phase is directly related to the distance covered by the radar signal—i.e., between the sensor and the surface. The physical path difference can be expressed as an integer number of wavelengths plus the fraction of one wavelength. It can also be expressed as a difference in phase angle between the two signals. SAR interferometry makes use of this phase information by subtracting the phase value in one image from that of the other, for the same point on the ground. In effect, this is generating the interference between the two phase signals and is the basis of interferometry. Again, a full interferometric cycle is 2π.
After Capes and Haynes (2004). Reproduced with permission.

Topography Mission (SRTM) (Jordan et al., 1996; Rabus et al., 2003). In this instance, the receiving antennae were fixed at either end of a 60-m-long boom. Operating from February 11 to 23, 2000, SRTM acquired data in the latitudinal band between 56°S and 60°N only (see ⟨http://www.jpl.nasa.gov/srtm/⟩ for further information). Aircraft single-pass systems have been extensively used in Greenland (e.g., Bindschadler et al., 1999; Dall et al., 2000, 2001; Gens and van Genderen, 1996a; Madsen et al., 1995, 1996; Mohr and Madsen, 1996a) and alpine regions (e.g., Cumming et al., 1996a, b). To measure surface velocity with single-pass systems, the antennae must be separated in the *along-track* flight direction (i.e., to achieve a *temporal baseline*[6]). In other words, the two antennae must follow each other a short distance apart on the same platform (to create a time lag of the order of a few milli-seconds) or on the same orbit (Bamler et al., 2003). Alternatively, the antennae must be displaced in the *across-track* direction to measure surface elevation

---

[6] *Temporal baseline* is the time interval between successive satellite observations.

by achieving a *spatial baseline* (Mohr, 1997). The importance of these baseline distinctions, and of the length of baselines, will become apparent later. So far, no dedicated along-track interferometric system has flown in space, although the Shuttle Imaging Radar C-band mission (SIR-C) acquired a few data takes in experimental along-track mode in 1994 (Bamler et al., 2003; Rabus et al., 2003). Planned future spaceborne systems making use of these technologies are outlined in Section 2.10.

Most large-scale ice sheet research has so far been carried out using interferometer data collected by conventional single-antenna SAR systems on polar-orbiting satellites used in repeat-pass mode. With repeat-pass InSAR, the two or more SAR images required for interferometric processing are acquired from the same surface area but from marginally different positions in space (i.e., on different satellite overpasses) separated by an interferometric baseline and a temporal baseline that typically ranges from a day to weeks. To date, this has been achieved either by using data acquired either by a single satellite SAR, or by identical SARs on different satellites flying in specially designed orbits—e.g., the Tandem Mission of ERS-1 and -2, which occurred over a 9-month period from October 1995 to June 1996, was the first satellite polar-orbiting mission specifically designed for SAR interferometry (Coulson, 1993; Solaas, 1994). Taking this a step further, *differential InSAR* involves the acquisition of three SAR images of the same area acquired on three different passes to generate two *interferograms* (see Section 2.2.3 for an explanation of interferogram). By this technique, surface displacements (i.e., ice-surface motion) can be measured by subtracting a topography-only interferogram from a second one containing both a topographic and deformation/motion signal. Again, these concepts are expanded upon in Section 2.2. Detailed comparison of the relative strengths and weaknesses of these different approaches, and between aircraft versus satellite InSAR, is beyond the scope of this chapter, and the interested reader should consult Mohr (1997).

The first application of radar interferometry to the study of the Earth's surface occurred in the 1970s (Madsen and Zebker, 1998). Early efforts focused on aircraft application (Graham, 1974), with the first spaceborne proof of concept occurring with Seasat L-band SAR (HH-pol) data acquired in 1978 (Goldstein et al., 1988). Further developments in the 1980s largely focused on terrestrial topographic-mapping applications (e.g., Zebker and Goldstein, 1986). Gabriel et al. (1989) then introduced the technique of differential interferometry (see above), showing that surface displacements of a few centimeters or less can be measured in this fashion. It was not until the launch of ERS-1 in 1991, however, that a relatively large amount of satellite SAR data suitable for interferometry became available (although the satellite was not specifically designed for InSAR application). Using these data, Goldstein et al. (1993) provided the first demonstration of repeat-pass interferometry for ice-sheet velocity mapping. Indeed, this represented the first direct measurement of the large-scale ice velocity field from space without ground control points, and sparked a large and ever-growing number of other glaciological studies. Further impetus was provided by the adjustment by ESA of the orbits of ERS-1 and -2 to give their identical SARs a 1-day separation between coverage. This "Tandem

Mission", . Importantly, it not only provided a wealth of unique data but also spawned detailed investigations of InSAR performance under a wide range of conditions.

The success of satellite repeat-pass and differential InSAR is dependent on a number of stringent conditions and inherent constraints. Importantly, the SAR images must be highly correlated in terms of their phase characteristics, and both temporal and spatial coherence must also remain high. This is a major factor limiting the quality and amount of SAR images suitable for InSAR processing, given that atmospheric/ionospheric propagation conditions and physical characteristics on an ice sheet/glacier surface can change significantly between image acquisitions. These and other key limiting factors will be further evaluated in subsequent sections.

2.2.1 The basis of ice sheet surface elevation and motion measurement by satellite repeat-pass interferometry

In order to lay the foundations for subsequent discussion of the application of satellite repeat-pass InSAR to the measurement of ice sheet surface elevation and motion, we begin with a further short elaboration of the basic principles of InSAR. Figure 2.3 is a schematic representation of the geometry of satellite repeat-pass InSAR, whereby two SARs fly on (ideally) parallel tracks to view the surface from slightly different directions—i.e., the images are acquired from different orbits. Once again, this configuration is referred to as *across-track interferometry*.

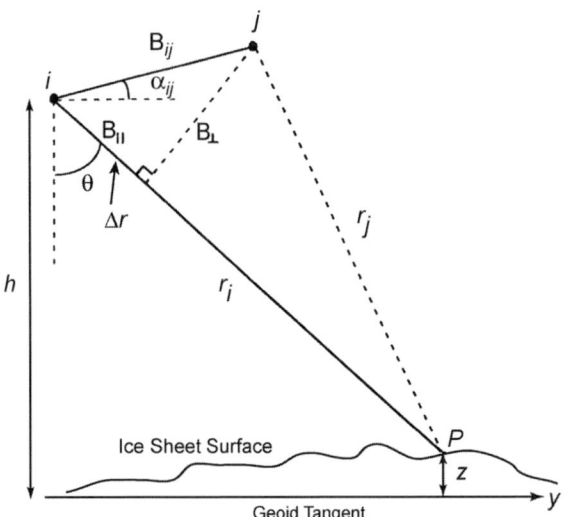

Figure 2.3. Geometric configuration for satellite repeat-pass interferometry, showing the spatial baseline B_{ij} created between the ith and jth repeat SAR observations of the same surface point P. At a virtually identical look angle θ, the differential range to point P is given by the interferometric phase.

After Kwok and Fahnestock (1996) and McLeod et al. (1998). Copyright IEEE 1996, reproduced with permission.

Separation of the flight-paths between the acquisition location/times (epochs) *i* and *j* is termed the *interferometric baseline* B_{ij}, while its component perpendicular to the sensor look direction is the *perpendicular* or *effective baseline* $B_\perp$. More specifically, B_{ij} is defined as the (vector) separation or (scalar) distance between image-acquisition points (i.e., antennae) on two orbital trajectories. With accurate information on the two slant ranges r_i and r_j and the sensor/satellite orbital location, the elevation of points on the Earth's surface can in theory be mapped from the images back into space by geometric triangulation (Bamler and Hartl, 1998).

In principle, InSAR is similar to conventional stereoscopy (see Chapter 3) as a means of generating DEMs from space (please see Toutin and Gray, 2000 and USGS, 1993 for general information on constructing DEMs from satellite data). By utilizing stereo methods similar to those developed for optical sensors, InSAR exploits the observed phase difference information of every pixel to measure the stereo parallaxes $\Delta r = r_i - r_j$. As noted earlier, this is feasible by virtue of the fact that SAR is a coherent imaging system that records not only the amplitude but also the phase of the received electromagnetic wave and preserves it throughout all processing steps (Bamler et al., 2003). The major strengths of the interferometric technique is that the slant-range difference can be determined very accurately from the interferogram—i.e., to fractional-wavelength precision—and also that the technique is highly effective in relatively featureless terrain such as that which occurs in inland ice-sheet regions. Conventional stereoscopy, on the other hand, is limited to the spatial resolution of the imagery and requires sufficient image contrast and the identification of homologous surface in the imagery points to be effective (Rosen et al., 2000). Enhanced-resolution techniques have been devised using satellite optical data—e.g., photoclinometric processing (Scambos and Fahnestock, 1998; Scambos and Haran, 2002)—although these are limited by cloud cover (see Chapter 3). Moreover, interferometry can in principle provide an absolute three-dimensional location of each image pixel, by simultaneously measuring the *slant range*,[7] *elevation angle*, and *azimuth*[8] angle (or at least by making flow-related assumptions). As a result, the images can be accurately transformed to a cartographic *ground-range* format.

As a starting point, let us consider the interaction of a radar beam with an ice-sheet surface, and the signal return of elemental scatterers comprising each resolution cell, or pixel, in the SAR image. In a single SAR image acquired at epoch *i* and from altitude *h* above a datum, the phase ϕ_i of a distributed target *P*—i.e., each sample at a distance (or slant range) r_i from the antenna—is given by (Zebker et al., 1994a):

$$\phi_i = \frac{4\pi}{\lambda} r_i + \phi_{scatt,i} + \phi_{noise,i} + \phi_{atm,i} \qquad (2.1)$$

where λ is the radar wavelength ($\sim$3.1 cm for X-band, $\sim$5.6 cm for C-band, and $\sim$23 cm for L-band); $\phi_{scatt,i}$ is the wavelength-dependent contribution from scatterers

---

[7] *Slant range* represents the distance between the sensor (radar antenna) and the target on the surface.

[8] *Azimuth* is the image scale or linear distance in the along-track direction.

within the resolution cell—i.e., the phase shift experienced by the waves when scattered; $\phi_{noise,i}$ is the phase contribution from system-related noise (e.g., quantization, thermal, etc.: Massonnet, 1995); and $\phi_{atm,i}$ is phase noise related to varying propagation delays through the atmosphere and ionosphere (Bamler and Hartl, 1998; Gray et al., 2000; Hanssen, 2001; Joughin et al., 1996b; Zebker and Villasenor, 1992). Noise contributions and error sources are covered in the next section. For interferometry, a resolution cell is represented as a complex *phasor* of the coherent backscatter from the scattering elements on the ground and the propagation phase delay (Rosen et al., 2000). In effect, the backscatter-phase delay entails the net phase of the coherent sum of contributions from all elemental scatterers within the resolution cell. Each is characterized by individual backscatter phases and differential path delays relative to a reference surface normal to the radar look or range direction (Rosen et al., 2000). Interferometry is based on the underlying assumption that the constituent phasors have amplitude and phase that are statistically independent from each other.

It can be seen from (2.1) that the phase information contained in one SAR image alone is not useful in the geometrical sense, as both r_i and $\phi_{scatt,i}$ are unknown. The basis and power of SAR interferometry is its capacity to accurately retrieve the *path length* between sensor and surface. It achieves this from phase difference information measured on a pixel-by-pixel basis between two interferometrically compatible (i.e., coherent) *Single-Look Complex* (SLC) SAR images of the same area, but acquired from marginally different orbital locations or/and at different times. These data contain signal information in a complex number. As stated previously, the separation between the latter is termed the *interferometric baseline* B_{ij}. Considering the InSAR geometric configuration shown in Figure 2.3, and assuming that the surface has remained motionless over the period between image acquisitions (a reasonable assumption for single-pass but not necessarily for repeat-pass systems), the phase difference $\Delta\phi_{ij}$ (or *interferometric phase*) for a given pixel is proportional to the *range parallax* $\Delta r = (r_j - r_i)$. This is given by (Zebker et al., 1994a):

$$\Delta\phi_{ij} = \phi_j - \phi_i = \frac{4\pi}{\lambda}(\Delta r) + (\phi_{scatt,j} - \phi_{scatt,i}) + (\phi_{noise,j} - \phi_{noise,i}) + \phi_{atm,ij} \quad (2.2)$$

where subscripts i and j indicate image acquisition epochs, and $\phi_{noise,ij}$ is *phase noise* from a variety of sources combined, which will be evaluated in Section 2.3. As such, the range difference between the two image acquisitions can in theory be computed to an accuracy of a fraction of a wavelength—depending on the system Signal-to-Noise Ratio (SNR). This can only be achieved if the phase contributions on the right-hand side of (2.2) remain constant or can be minimized by the choice of suitable image pairs (Sharov and Gutjahr, 2002). As noted above, this is a fundamental constraint on satellite repeat-pass interferometry and its universal application to glacier and ice sheet studies. The assumption of invariant scattering phase (i.e., $\phi_{scatt,i} = \phi_{scatt,j}$), for example, imposes an effective limit on the time interval between image-acquisition epochs i and j. This is defined as the *temporal baseline*, or $\Delta T = (T_j - T_i)$ (2.3). If ΔT is too long, this will cause *temporal decorrelation*, or a loss of phase coherence in the second image relative to the first (Zebker and

Villasenor, 1992). This relates to physical changes in the surface or the troposphere/ionosphere over the period separating image acquisitions. Coherence is defined as the mutual correlation coefficient between two images (Just and Bamler, 1994; Zebker and Villasenor, 1992), and ranges from 0.0 (total decorrelation—i.e., no phase information) to 1.0 (complete correlation—i.e., no phase noise). Temporal decorrelation, which results from changes in the distribution and/or characteristics of surface scatterers, is a key factor over snow and ice surfaces and again effectively limits the amount of suitable SAR images available for InSAR processing. This and other inherent constraints, error sources, and limitations will be addressed in more detail in Section 2.3.

Recalling the previous section and in the ideal case, the optimal configuration for the measurement of surface elevation alone is across-track interferometry using a single spacecraft or aircraft with two SAR antennae configured across-track, separated by a known (small) distance (baseline), and operating simultaneously (Bamler et al., 2003). As the interferometric baseline B_{ij}, and thus the look angle between the primary and secondary antennae, on such single-pass systems is typically small, then the simultaneous imaging of any surface object guarantees that the atmospheric conditions are virtually identical for both channels. Moreover, the scattering properties will also be almost identical. As a result, and neglecting system-related noise, the range parallax is then proportional to the phase difference of the corresponding pixels in the interferometric SAR pair. It follows that (2.2) then simplifies to:

$$\Delta \phi_{ij} = \frac{4\pi}{\lambda} (\Delta r) \qquad (2.4)$$

In this case, $\Delta \phi_{ij}$ depends on terrain height alone. Following Bamler (1997), Bamler et al. (2003), and Zebker et al. (1994a), the *phase-to-height sensitivity* of a repeat-pass across-track SAR interferometer is given by:

$$\frac{\partial \Delta \phi_{ij}}{\partial z} = \frac{4\pi}{\lambda} \frac{B_\perp}{r_i \sin \theta} \qquad (2.5)$$

where $\partial \Delta \phi_{ij}$ is the *interferometric phase uncertainty*, z is the surface height (elevation), and θ is the look angle of the interferometer. It is clear from (2.5) that any phase noise present translates to a random height error in the InSAR-derived DEM. It is also apparent that the sensitivity of the interferometer to surface height increases with decreasing radar wavelength but increasing baseline. Larger baselines, however, create correspondingly larger differences in look angle, which in turn engender spatial decorrelation of the image (Gatelli et al., 1994). In practical terms, there are therefore upper limits for interferometric baselines, ranging from a few hundred meters to a few kilometers. These and other issues and trade-offs relating to InSAR coherence and decorrelation are covered in more detail in Section 2.3.

In effect, *interferometric fringes*[9] can be interpreted as contour lines in topography-only interferograms. The fixed height difference on the surface producing a phase shift of 2π is termed the *altitude of ambiguity* or *ambiguity height* (Massonnet and Feigl, 1998; Mohr, 1997). In other words, the ambiguity height is the height difference equivalent to a target moving half a wavelength—i.e., one fringe in the interferogram, away from the radar—from one image acquisition to the next (Mohr et al., 1998). The ambiguity height $z2\pi$ is a function of the perpendicular baseline $B_\perp$, radar wavelength and incidence angle, and satellite altitude. It can be formulated from (2.5) as:

$$z2\pi = \frac{\lambda}{2} \frac{r \sin \theta}{B_\perp} \qquad (2.6)$$

The altitude of ambiguity is in effect a measure for the sensitivity of an interferometric image pair to topography (Pattyn and Derauw, 2002). An *ambiguity velocity* can be defined in a similar fashion.

From the relationship in (2.2) and (2.4), the range parallax Δr for each pixel within a coherent SAR image pair theoretically constitutes an accurate measure of the off-nadir viewing angle θ, assuming that no motion of the surface has occurred between the ith and jth image acquisitions. This in turn depends on the terrain height z above a datum geoid. Following McLeod et al. (1998), if range r_i and range difference Δr are accurately known, then θ can be determined from:

$$\sin(\alpha_{ij} - \theta) = \frac{(r_i + \Delta r)^2 - r_i^2 - B_{ij}^2}{2 r_i B_{ij}^2} \qquad (2.7)$$

where α_{ij} is the tilt of the baseline B_{ij} with respect to the horizontal. Given the known antenna altitude h above the datum geoid, the elevation above the datum z at each sample is then:

$$z = h - r \cos \theta \qquad (2.8)$$

The above relationships hold for single-pass across-track interferometers in general, and for satellite repeat-pass interferometers when viewing surfaces that remain stationary over the temporal baseline. An additional term is required, however, when applying satellite repeat-pass InSAR techniques to ice sheets and glaciers: this term is ice motion. In this case, the interferometric phase difference at each sample $\Delta \phi_{ij}$ comprises contributions both from the topography relative to B_{ij} and target displacement over time—i.e., ice motion. Methods to separate these two contributions, in order to derive useful information on both ice-sheet elevation and motion from SAR data from polar-orbiting satellites, are outlined in the next section, and are elaborated upon in Section 2.5.

2.2.2 Measurement of ice-surface displacement/motion

A unique feature of satellite repeat-pass InSAR is its sensitivity to very small movements of the surface. Local centimeter-scale relative motions in the slant

---

[9] Isophase lines/contours in an interferogram—see Section 2.2.3.

range generate large local phase shifts in the interferogram (Joughin et al., 1996c). If extreme and/or random, such motions can lead to complete *phase decorrelation* (this effect will be discussed further in Section 2.3). However, if the scattering centers across a number of neighboring pixels move in a coherent fashion, and the movement is constant, then this opens up the exciting possibility of determining the ice motion field to an unprecedented precision. Wu and Thiel (1996) estimate that, under optimal conditions, an accuracy for surface movement estimation of 2.3 mm in slant range can be attained with InSAR. For an interferogram constructed from two SAR images acquired from the same look direction, which has been most commonly the case, the phase difference $\Delta\phi_{M,ij}$ is proportional to the ratio of the relative motion in the direction of the radar *Line-Of-Sight* (LOS) only (either towards or away from the sensor), ΔM, to the transmitted radar wavelength λ. Under these circumstances, the following expression holds for the measured phase difference $\Delta\phi_{ij}$ (on a pixel-by-pixel basis):

$$\Delta\phi_{ij} = \Delta\phi_{T,ij} + \Delta\phi_{M,ij} = \frac{4\pi B_\perp n_\rho}{\lambda r} + \frac{2\Delta M}{\lambda} + \phi_{noise,ij} \qquad (2.9)$$

where $\Delta\phi_{T,ij}$ is the topographic contribution to the measured phase difference; $B_\perp$ is the *effective baseline* where $B_\perp = B_{ij}\cos(\theta - \alpha_{ij})$; n_ρ is the relative displacement of point scatterers normal to the slant range direction; r is the sensor-to-target distance (Rocca et al., 1997); and $\phi_{noise,ij}$ is the phase noise.

For the sake of simplicity, the unwrapped interferometric phase ϕ is generally presented as a linear combination of independent contributions from several phase terms (Bamler and Hartl, 1998), including imaging geometry (the flat-Earth phase) ϕ_o, topography ϕ_{topo}, ice motion ϕ_{motion}, tropospheric plus ionospheric propagation delay ϕ_{atm}, and system noise ϕ_{noise} as follows:

$$\phi = \phi_o + \phi_{topo} + \phi_{motion} + \phi_{atm} + \phi_{noise} \qquad (2.10)$$

An underlying principle of InSAR is that the terms ϕ_{atm} and ϕ_{noise} are minimized by the selection of suitable interferometric image pairs, such that (2.10) can be rewritten after the flat-Earth correction as a function of the topographic and motion phase only.

The primary issue in applying repeat-pass InSAR to studies of ice sheet/glacier surface motion is that motion and topography cannot be unambiguously determined from (2.9). As a result, an interferogram formed from a single satellite image pair comprises fringes due to both differential surface motion over time ΔT and surface topography (Joughin et al., 1995). To retrieve the ice motion component, the topographic component must be neglected or removed, and vice versa. This can be achieved by one of three techniques, again assuming that the ice motion is constant (Mohr, 1997; Mohr and Madsen, 1996a):

(1) Ice motion information can only be extracted directly from a single interferogram alone if the look angles of the multiple data acquisitions are identical, which is equivalent to having a *zero baseline*. This scenario, whereby there is no sensitivity to topography, is seldom if ever satisfied by current or past

spaceborne SARs. An exception is presented by Goldstein et al. (1993), who were able to base their ERS InSAR analysis of the Rutford Ice Stream in Antarctica on passes so close ($\sim$4 m) that the sensitivity to topography was practically zero, under the assumption that the scene was relatively featureless (see Section 2.3.5 for information on the effect of topography on optimal baseline). As such, they were able to eliminate the parallax effect and directly measure LOS displacements and convert these to horizontal velocities of $\sim$390 m per annum (Figure 2.4, see color section), which were in close agreement with surface measurements. When the spatial baseline is significant, as is generally the case with current satellite systems (but not with single-pass interferometers), then additional information is required to resolve the ambiguities between phase differences introduced both by scene topography and surface displacement (Madsen and Zebker, 1998). This is a crucial aspect of InSAR applied to the measurement of moving ice masses.

(2) The topographic term can be supplied and removed by incorporating an existing, independently derived DEM (Joughin et al., 1995, 1996a; Massonnet et al., 1993, 1996; Rignot et al., 1995). For this to be accomplished, the DEM must first be transformed to azimuth slant-range coordinates and scaled proportionally to the baseline of the single calculated interferogram in order to represent a sufficiently realistic estimate of the phase component $\Delta\phi_{T,ij}$. The topography resulting in this geometry must be converted from length to phase units for this purpose. Accurate registration of the DEM to the viewing geometry of the interferogram results in a synthesized interferogram. The difference of the original and synthesized interferograms then becomes the differential interferogram, with the DEM correcting for topographic effects and the phase signal relating solely to ice surface displacement/motion.

(3) Alternatively, the third technique is required for regions where sufficiently accurate DEMs are unavailable. This involves at least three coherent SAR images acquired on separate passes—i.e., with equal temporal baselines. Two separate interferograms are then derived: one "mixed" topographic/phase, the other "topography-only" (Gabriel et al., 1989; Joughin et al., 1996c; Kwok and Fahnestock, 1996; Mohr et al., 1998; Rignot, 1996). The subtraction of the latter from the former effectively cancels out phase due to common topography, such that the remaining phase map relates solely to relative surface displacement/motion (in the direction of the sensor LOS only). An underlying assumption is that the ice flow within the SAR scenes remains invariable over the time interval between the image acquisitions. Fortunately, this tends to be the case for most areas of the ice sheet. In other words, the same set of motion-induced fringes will typically be present in both interferograms, provided that the temporal baselines are equal and not too great (Rosen et al., 2000).

For Option 2 above, the quality of the DEM has a substantial bearing on the accuracy of the InSAR product, as does the baseline and radar wavelength (Gens and van Genderen, 1996a, b). The requirement of an accurate DEM for interferogram geo-coding is a critical factor, given that vertical height errors translate into significant horizontal-positioning errors—e.g., by a factor of 2.5 times greater for

ERS-1/-2 (Mohr et al., 1998). An advantage of this technique is that it requires only one interferogram—i.e., one coherent image pair—thereby minimizing temporal decorrelation (which is a key issue over ice sheets). The major disadvantage is that accurate large-scale DEMs are not available for all regions (although great advances are being made—see Chapter 3), and those that are available tend to be limited by a relatively poor horizontal resolution (typically of the order of 1–2 km). This introduces the technical challenge of having to accurately co-register the interferogram and DEM (a particularly difficult task in featureless ice-sheet inland regions), and the possibility of local topography-induced phase errors (Mohr et al., 2003). Referring back to (2.5), an approximate estimate of the required accuracy of the DEM can be obtained through the relation between interferometric *phase uncertainty* $\partial \Delta \phi_{ij}$ and unknown elevation ∂z. This is given by:

$$\partial \Delta \phi_{ij} = \frac{4\pi B_\perp}{\lambda r \sin \theta} \partial z \qquad (2.11)$$

It follows that a phase error $\partial \Delta \phi_{ij}$ translates into an error in the retrieved horizontal motion, $\partial y = \lambda/(4\pi \sin \theta) \partial z$, over the time interval (temporal baseline) ΔT (Mohr, 1997). As such, a reference-height error causes a velocity uncertainty which is inversely proportional to the temporal baseline. In their study of the Greenland Ice Sheet, Joughin et al. (1995) estimated the error to be π radians at most with a perpendicular baseline ($B_\perp$) of ~50 m. Where $\Delta T = 3$ days, this corresponds to an equivalent horizontal velocity of ~4 m per annum, but ~12 m per annum when $\Delta T = 1$ day—i.e., during the ERS Tandem Mission (Mohr, 1997).

Option 3 (above) was developed in an effort to be independent of the need for an external DEM and the limitations described above (Gabriel et al., 1989; Zebker et al., 1994b). As ΔT is accurately known, the displacement terms can be merged into one velocity component, assuming invariant ice motion. By using two highly correlated interferograms of the same scene, motion and topography can be separated. The major disadvantage of this technique, which is expanded upon in Section 2.5, is the requirement of multiple coherent image pairs, thereby decreasing the availability of suitable data while increasing the probability of temporal decorrelation and the margin for error. Kwok and Fahnestock (1996) used this technique to determine the relative topography and motion of an ice stream that was recently discovered in NE Greenland in satellite imagery, and Thiel et al. (1995) retrieved information on topography and relative tidal-induced motion of an Antarctic ice shelf. The derivation of absolute motion and elevation requires additional information (Joughin et al., 1996c; Rignot et al., 1996; and Mohr and Madsen, 1996b). Furthermore, Rignot (1996) was able to estimate not only the absolute velocity but also the tidal motion and melt rate of the Petermann Gletscher, Greenland, using three interferograms combined with a model of tidal motion. These and other case studies are discussed further in Chapter 3.

Options 2 and 3 are both forms of *differential SAR interferometry*, or *DInSAR*, with the multiple Option (3) often being referred to as *double-differencing*. The latter will be examined more closely in Section 2.5. It should be noted that while a single pair of repeat-pass InSAR images can yield ice surface displacement in the look direction of the SAR only, the data are not useful if this direction deviates from the direction of ice flow by more than about 60° (Bamber et al., 2000). This issue, and methods to derive more useful two- and full three-component ice velocity fields from InSAR data, is again discussed in Section 2.5.

2.2.3 What is an interferogram?

An *interferogram* is formed from a pair of interferometric images as the product of the complex values of the second SAR image with the *complex conjugate* of the first, or reference, image—i.e., the corresponding phases have to be differenced and the corresponding amplitudes averaged at each point in the image. This effectively cancels out the common backscatter phase in each resolution element while leaving a phase difference term (i.e., the interferometric phase) which is proportional to the differential path delay. This is a geometric quantity directly related to the elevation angle of the resolution element. Ignoring any slight difference in backscatter phase in the two coherent images treats each resolution cell as a *point scatterer*. In general, interferometry makes the assumption of point scatterers to consider geometry only. As such, the interferogram represents a map of phase difference at each co-registered pixel location. It also contains the amplitude information of the SAR image pair, as corresponding amplitudes are averaged for each point in the scene. The steps involved in retrieving useful geophysical quantities from SAR image pairs are outlined in Section 2.4.

In an interferogram, each complex element of the image matrix is represented by a pixel within which the phase is represented in color and the magnitude is mapped to a gray-scale shade. For phase information, an interferogram comprises a map of isophase lines referred to as *fringes*, and usually displayed in a color wheel fashion, where one complete fringe cycle represents a 2π phase shift (i.e., one complete interferometric phase cycle from 0 to 2π). As the microwave signal must travel both from and back to the satellite, one interferometric phase cycle in the interferogram is equivalent to a LOS displacement of half the radar wavelength—e.g., 2.8 cm for ERS and Radarsat (C-band SARs)—regardless of baseline. In other words, from the precise knowledge of the acquisition geometry and radar system parameters, ground-points characterized by a constant phase difference within a given fringe represent a constant path-length difference. As such, the fringe pattern effectively represents a contour map of the LOS surface displacement, with a contour interval of half the satellite SAR wavelength (Goldstein et al., 1993). Unfortunately, a number of additional challenges complicate direct analysis of the observed interferometric phase $\Delta\phi_{ij}$. The first occurs because $\Delta\phi_{ij}$ can only be measured *modulo* 2π, whereas it is the absolute phase, which is proportional to range, that is generally required in the analysis. The technique used to solve this problem, and to restore the correct multiple of 2π to each point in the interferometric phase image, is known as *phase unwrapping* (Goldstein et al., 1988; Robinson, 1993). Phase-unwrapping techniques are outlined in Section 2.4.4. A further problem relates to the technical difficulty in accurately determining the baseline length (and orientation). Another challenging factor is that InSAR only provides a measure of the relative rather than abolute ice-surface elevation and/or motion and, in the case of the latter, in the LOS direction of the SAR only. These issues, and means of solving them, are expanded upon in subsequent sections.

2.3 INHERENT CONSTRAINTS AND SOURCES OF AMBIGUITY AND ERROR

Before we examine the processing steps involved in retrieving quantitative ice-sheet/ glacier-surface elevation and velocity information from InSAR data, we present a brief summary of the inherent constraints and major factors contributing to uncertainty in the measurement of phase and thus derived elevation and motion. This is not meant to be a deterrent! Rather, the aim is to prepare the reader for the complexities involved in what is an immensely powerful technique, and one that is revolutionizing our ability to quantify ice-sheet and glacier elevation, shape, and dynamics. We follow this with a description of the stages involved in producing and calibrating an interferogram. Both aspects are well-documented in the literature, and will only be outlined here.

Initially, a number of strict constraints are imposed by satellite repeat-pass InSAR (Massonnet and Rabaute, 1993). Images must belong to the same orbital cycle, and also be collected by SARs operating at identical wavelengths (Massonnet and Feigl, 1998). Moreover, the baseline attitude knowledge requirement is 1 arcsecond (Madsen and Zebker, 1998). Accurate knowledge of sensor location and orbit location is critical. Other constraints, including those relating to decorrelation and baseline length (spatial and temporal), will become apparent in the following discussion.

From (2.2), it is apparent that the success of the InSAR technique depends on the phase of the target scatterers remaining largely unchanged—i.e., coherent—from one image to another, and also invariance in system- and environment-related noise. In reality, this is often not the case, and a range of error sources act to corrupt the measurement of $\Delta\phi_{ij}$, and thus estimates of surface height and/or displacement (McLeod et al., 1998; Zebker and Villasenor, 1992). In the most extreme cases, they can lead to a breakup of fringes in the interferogram. First and foremost, the quality of an interferogram depends on the degree of correlation between the two complex images used to generate it. *Decorrelation* is a measure of the degree of phase coherence in complex radar returns from a target pixel of combined image 2 (and 3) relative to image 1, which in the case of satellite repeat-pass InSAR are acquired at different times and from different locations. Phase decorrelation comprises four uncorrelated components (Rosen et al., 2000; Zebker and Villasenor, 1992), namely:

- *thermal decorrelation* $\gamma_{thermal}$, caused by radar system (i.e., receiver) noise;
- *temporal decorrelation* $\gamma_{temporal}$, related, for example, to (a) the random change of the physical properties and position of the ground scatterers and (b) changes in

atmospheric and ionospheric propagation effects, over the interval between image acquisitions;
- *spatial* or *baseline decorrelation* $\gamma_{spatial}$, resulting from the difference in viewing geometries between the two image acquisitions; and
- *processing decorrelation* $\gamma_{process}$, due to imprecise interferometric processing and/or misalignment of the images forming the interferogram.

If the satellite tracks can be considered as parallel, then the absolute value of coherence (correlation coefficient) γ can be given by:

$$\gamma = \gamma_{thermal} \gamma_{temporal} \gamma_{spatial} \gamma_{process} \qquad (2.12)$$

While some of these contributions can be readily minimized by the careful choice of suitable SAR image pairs, others are highly variable in space and time and present a considerable challenge. Their overall effect is to diminish the quality and accuracy of the resultant interferogram by decreasing the interferometric phase coherence—i.e., increasing the decorrelation (Zebker and Villasenor, 1992). This leads at times to complete loss of coherence (i.e., total decorrelation), making the image pair unusable for InSAR analysis. To reiterate, such constraints limit the number of suitable SAR images available. We first examine each factor, then show how the overall level of coherence between the SAR image pair can be calculated. Please refer to the following for more in-depth analyses of error sources, and their impact on InSAR measurement sensitivity and accuracy: Gens (1999a); Gens and van Genderen (1996b); Joughin et al. (1998a); Massonnet and Feigl (1998); Mohr and Madsen (1999); Rodríguez and Martin (1992); Rosen et al. (2000); Zebker et al. (1994b); and Zebker and Villasenor (1992).

2.3.1 Thermal decorrelation

Thermal decorrelation $\gamma_{thermal}$ refers to radar receiver and quantization noise as they affect the Signal-to-Noise Ratio (SNR) of the instrument. As its name suggests, SNR is the ratio of the level of the information-bearing signal to the level of the noise power (the maximum SNR of a sensor is termed its *dynamic range*). As noise corrupts the phase measurement, the SNR will contribute to determining the measurement uncertainty (Allen, 1995; Rodríguez and Martin, 1992; Zebker et al., 1994b). Following Zebker and Villasenor (1992) and Engdahl (1996), $\gamma_{thermal}$ relates to the SNR of the instrument in the following manner:

$$\gamma_{thermal} = \frac{1}{1 + SNR^{-1}} \qquad (2.13)$$

Low SNR is a limitation for satellite repeat-pass InSAR. Additional system-related errors include uncertainties in system-clock timing and data-sampling clock jitter, for example (Massonnet and Vadon, 1995). System instabilities such as clock drift can be an issue when processing long strips of ERS data interferometrically, for example (Eineder, 2003; Massonnet and Vadon, 1995). Thermal noise from the radar receiver can be minimized by increasing the power of the transmitted signal, and also if the SAR system phase reference remains stable and the SAR processor is phase-preserving (Prati et al., 1990). The thermal decorrelation term is generally small compared with the other decorrelation sources laid out below. In general, thermal noise-related ambiguities are treated as additive noise as part of the overall noise floor of the system (Rosen et al., 2000).

2.3.2 Temporal decorrelation

Temporal decorrelation $\gamma_{temporal}$ is a major limiting factor, and occurs when the sub-resolution properties of imaged scatterers and/or their distribution change in an incoherent manner over the unavoidable time interval between SAR image acquisitions (for satellite repeat-pass InSAR). This is a major issue for InSAR studies of glaciers and ice sheets (Hoen and Zebker, 2000a). It generally limits usable repeat-pass image acquisition intervals to 1 to 3 days for conventional InSAR studies of the outer margins of ice sheets in particular (Cumming and Zhang, 1999; Wu and Thiel, 1996), although longer temporal baselines can be used for interior regions where less changeable conditions occur. Alternative techniques have, however, been developed to utilize image pairs with longer temporal baselines (see Sections 2.6 through 2.8). Phase decorrelation is particularly severe during the ice-sheet/glacier melt season around the margins of Greenland and parts of Antarctica (including the northern Antarctic Peninsula). For melting snow and ice with even a small freewater content, the dominant backscatter mechanism is surface scattering, compared with volume scattering under cold, dry conditions (Rott and Siegel, 1996; Ulaby et al., 1982). Although this can potentially be turned to an advantage by enabling InSAR mapping of wet snow covers (Strozzi et al., 1999), it is a serious limitation for all other applications. The impact of variable penetration depth and volume scattering on the accuracy of InSAR is discussed in Section 2.3.6.

Outside the season and regions of melt, a multitude of temporally and spatially varying processes can potentially contribute to a loss of interferometric phase coherence, depending on the temporal baseline, radar wavelength, and other factors. For instance, strong winds can contribute to phase decorrelation by redistributing snow across the surface, again introducing an apparent range displacement (Guneriussen et al., 2001). Eolian erosion and redistribution of unconsolidated snow is a major process affecting ice-sheet surface properties (Goodwin, 1990), and one which can intensify or abate depending on the prevailing meteorological conditions (Frezzotti et al., 2002; Fujii and Kusunoki, 1982; Massom et al., 2004; Young et al., 1996). This leads to the formation of sastrugi and dunes as well as glazed (icy) surfaces. *Sastrugi* are ridges of wind-sculpted snow, typically a meter or so high and meters to tens of meters long, which are formed when wind erodes and drifts the snow. *Surface glazing*, whereby a thin, icy layer forms on the snow surface, occurs in summer by radiational effects, and in winter by the formation of regelation ice films due to kinetic heating of wind-driven saltating drift snow (Goodwin, 1990). Katabatic wind activity (Jezek and Rignot, 1994; Parish and Bromwich, 1991) may also affect phase coherence, particularly in ice sheet margins where atmospheric conditions are more variable.

Precipitation events can also significantly alter the surface physical and thus wavelength-dependent backscatter characteristics over the time interval between InSAR image acquisitions. Snowfall is often also accompanied by an increase in surface wind speed (Massom et al., 2004). Even a light snowfall or small changes in snow/firn properties can introduce path-length distortions and cause a significant height error in InSAR DEMs derived over glaciers and ice sheets (Guneriussen et al., 2001; Mohr et al., 1998, 2003). This can apparently even occur when and where coherence remains high. It follows that, even under clear-sky conditions, periodic surface changes related to *hoar frost formation* and destruction may also have an impact. Dense carpets of low-density hoar frost crystals form under clear calm conditions (Shuman and Alley, 1993; Steffen et al., 1999), and can cover large or localized areas of an ice sheet surface. This carpet, which is typically centimeters thick, forms in a matter of hours, and can be destroyed equally rapidly by an increase in wind speed. Another factor may be *snow metamorphism* driven by the vertical temperature gradient in the upper snow column (Colbeck, 1987), and particularly the formation of horizons of relatively large (centimeter-scale) depth hoar crystals. As such, changes can occur on short timescales, ranging from only a few hours to synoptic and beyond, with significant spatial variability across a given SAR scene. Such effects are poorly understood at present in terms of their potential impact on phase decorrelation, and require further validation with coincident *in situ* measurements.

The negative impact of such changes in limiting the number of suitable images available is highlighted by Goldstein et al. (1993), who encountered only 2 cases providing useful results from 12 cases attempted. Problems associated with temporal decorrelation can be avoided to some extent by choosing SAR images (a) with as short a temporal baseline as possible, and (b) acquired during periods of low, and preferably zero, accumulation and/or ablation (Guneriussen et al., 2001; Reeh et al., 1999). Such information may be derived from other sources—e.g., surface observations, meteorological analyses, atmospheric model moisture-flux calculations (Chen et al., 1997; Cullather et al., 1998; Turner et al., 1999; Vaughan et al., 1999) and other satellite datasets. The latter include radar scatterometer, SAR amplitude (Fahnestock et al., 1993), and passive-microwave polarization data (Massom et al., 2004). Another solution is to acquire and process more than one dataset from the region of interest (if possible), in order to maximize the probability of attaining coherent data.

The processes (decorrelation effects) outlined above are wavelength-dependent. In general L-band SARs are less affected by small-scale changes in ice sheet surface and near-surface properties than shorter wavelength C-band SARs, but important trade-offs occur. These, and the other relative attributes of different wavelength SARs, are compared more fully in Section 2.4.

2.3.3 Decorrelation due to excessive target motion/rotation

Temporal decorrelation also includes decorrelation due to excessive surface motion and/or the rotation of individual scattering centers within each pixel (Zebker and Villasenor, 1992). For successful reconstruction of ice motion, the ensemble of scatterers over several neighboring pixels must remain coherent and take part in the same movement. Ice flowing "too slowly" can result in phase differences which are largely indistinguishable from noise. Moreover, if the ice, comprising individual sub-scatterers, moves "too quickly" over the time interval between acquisitions, then the scattering-phase contributions $\phi_{scatt,ij}$ in the two images no longer cancel, resulting in image decorrelation. There is in fact a limit on the highest velocity detectable with an interferometric system, over which it is impossible to unwrap the phase (Kwok and Fahnestock, 1996). This is defined by the *spatial gradient of the velocity* as:

$$\delta_\nu = \frac{\lambda}{2\beta_r \Delta T \sin\theta} \quad (2.14)$$

where β_r is the surface resolution of the multi-look[10] data. In the example given by Kwok and Fahnestock (1996), this limit is ~23 cm day^{-1} (82.5 m per annum) for ERS data over a distance of 1 km where $\Delta T = 3$ days. Bamber and Kwok (2004) further noted that a significant phase ambiguity exists if the lateral displacement between consecutive pixels translates to a change in phase of $>2\pi$. This can again undermine the determination of ice velocity by InSAR.

Where the glacier/ice-sheet motion is too fast, then differential InSAR interferograms can be characterized by very dense (>1 per pixel) and convoluted phase fringes (Costantini, 1999). Costantini (1999) has developed a method of phase unwrapping in an attempt to retrieve meaningful data from such extreme cases. In their study of the Jakobshavn Isbrae (Glacier) in Greenland—one of the fastest flowing glaciers in the world (~20 m day^{-1}: Iken et al., 1993)—using tandem pairs of ERS-1/-2 images, Goldstein and Werner (1997) were able to unwrap the interferograms for the upper reaches using a scene-dependent filter. They did, however, encounter low-phase coherence in spite of the short temporal baseline of 1 day.

Scatterer rotation effects are again generally most pronounced in outlet glacier regions, and along glacier/ice-stream margins in particular (Joughin, 2002; Joughin et al., 1996a). Phase noise can be extreme towards glacier/ice-shelf fronts, and particularly in tidewater glaciers (Sharov et al., 2002). The latter are characterized by maximum ice velocity and longitudinal strain rates (i.e., velocity gradients) at their seaward margins (Forster et al., 1999). As a result, phase coherence tends to break down, resulting in a dearth of interferometric measurements of ice velocity along the seaward margins of glaciers and ice sheets. The special case of the effect of vertical

---

[10] *Looks* are groups of signal samples in a SAR processor that splits the full synthetic aperture into several sub-apertures (sub-intervals in pulse compression processing), each representing an independent look of the identical scene. The image formed by incoherently summing of these looks is called a *multi-look* image.

motion of floating ice shelves and glacier tongues due to tidal motion, mentioned earlier, is further evaluated in Section 2.5. To reiterate, this information can, however, be used to advantage by enabling uniquely accurate detection and monitoring of grounding-line location (Rignot, 1998a, b).

2.3.4 Atmospheric and ionospheric propagation delay

Satellite repeat-pass interferometry depends upon the constancy of wave propagation conditions within the atmosphere—see (6.2). Tropospheric water vapor and ionospheric effects can introduce varying propagation phase and group delays to the SAR signal. These can result in path-length changes of up to a few wavelengths (Goldstein, 1995; Gray et al., 2000; Hanssen, 1998, 2001; Jezek and Rignot, 1994; Joughin et al., 1996b; Mattar et al., 1999; Mohr and Madsen, 1999; Tarayre and Massonnet, 1994, 1996). This effect is particularly complex in that the refractive indexes of the ionosphere and troposphere exhibit a high degree of spatial and temporal heterogeneity. Ionospheric refraction, for example, can result in both pixel misregistration and artifacts in the phase difference. Ionospheric disturbances appear in InSAR data as kilometer-scale modulations in the pixel shift in the azimuth required for optimal registration of the interferometric pair (Mattar and Gray, 2002). These modulations also occasionally affect the InSAR phase, and may appear as streaks (coined *azimuth streaks*) in the coherence map (Gray et al., 1999, 2001; Jezek and Rignot, 1994) (Figure 2.5). The effect is apparently more pronounced for L-band systems—e.g., the Japanese Earth Resources Satellite-1 (JERS-1) and Japanese Advanced Land Observing Satellite (ALOS) Phased Array type L-band Synthetic Aperture Radar (PALSAR)—but can also be significant even at the C-band—e.g., ERS-1/-2, Envisat and Radarsat-1 and -2. Gray et al. (1999) present evidence that the effect is caused by small-scale ionospheric disturbances associated with polar auroral regions, and propose that the streaks are related to a two-dimensional "shadow" of the along-track gradient of the integrated electron density.

For DEM generation, the ionospheric phase contribution can be approximated, and removed from the interferogram phase, by the scaled and filtered integral of this azimuth shift map. For displacement measurement applications, appropriate filtering of the distinct patterns of the azimuth streaks is the only resort. As ionospheric delays are dispersive, frequency-diverse measurements can potentially help to mitigate their impact. Tropospheric delays, on the other hand, are non-dispersive and tend to mimic topographic and/or surface displacement effects in interferograms (Goldstein, 1995). Schemes for distinguishing tropospheric effects have been proposed (e.g., Massonnet and Feigl, 1995), as have methods for reducing atmospheric noise by averaging interferograms (e.g., Zebker et al., 1997) and stacking/averaging phase gradients (Sandwell and Price, 1998). Unfortunately, however, no systematic correction algorithm currently exists (Rosen et al., 2000), and this remains a research issue.

Sec. 2.3] **Inherent constraints and sources of ambiguity and error** 61

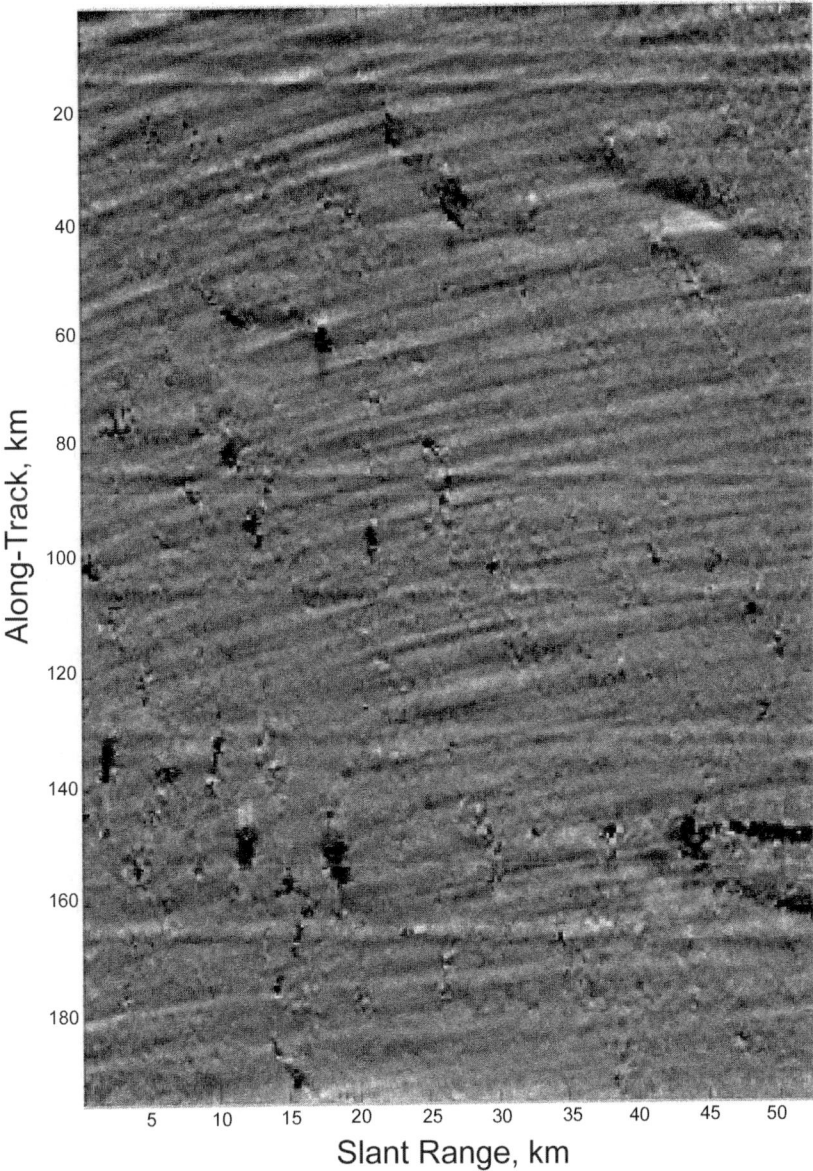

Figure 2.5. An example of "azimuth streaks" in an azimuth shift image for a Radarsat-1 Antarctic Mapping Mission (AMM) image pair. The data have been high-pass filtered to minimize the azimuth component of ice motion and accentuate the streaking. This image is presented in slant range, across a ground range swath of $\sim$110 km. In this example, the peak-to-peak modulation in azimuth pixel shift is typically 1 m. The phase errors associated with these streaks, of a few tenths of a radian, can lead to errors in InSAR velocity estimates of a few meters per annum for a tandem ERS-1/-2 InSAR pair (Joughin et al., 1998a).

From Gray et al. (1999). Copyright ESA 1999, reproduced with permission.

2.3.5 Baseline (spatial) decorrelation

Baseline length and orientation represent another limiting factor for coherence. With InSAR, a precise knowledge of the position and length of baseline B_{ij}, and angle α_{ij}, is an important prerequisite for successful InSAR processing. Baseline length is also commonly taken as a proxy measure of the coherence expected for interferometric data—i.e., their suitability for a given application. *Baseline*, or *spatial/geometric, decorrelation* γ_{spatial} refers to a loss of coherence due to the effect of viewing a given pixel from marginally different angles (i.e., at the two image acquisition points). This changes the relative range to each distributed scatterer and produces slightly different interference results (Gatelli et al., 1994; Li and Goldstein, 1990; Rodríguez and Martin, 1992).

Unfortunately, accurate determination of baseline length is technically difficult (Small et al., 1993; Mohr, 1997). In theory, satellite sensor trajectories should be known to an accuracy of millimeters to centimeters to enable the detection of similar range changes with InSAR. In reality, however, errors in the determination of orbital parameters of spaceborne SAR satellites such as ERS-1 and -2 are of the order of tens of centimeters using precision orbit ephemeris data (Bamler, 1997; Rosen et al., 2000), and baseline length can only be estimated from satellite ephemeris data to an accuracy of a few meters (Solaas, 1994). This can introduce significant residual errors into InSAR-derived elevations and ice motion fields (Joughin et al., 1996c). Indeed, baseline errors often represent the largest error source in interferometric estimates of these key ice sheet parameters, a factor that can be minimized by the use of short baselines (for ice motion InSAR) and tie-points (Joughin et al., 1996c).

Recent improvements in satellite-orbital-tracking technology have led to improvements in baseline determination (Gens and van Genderen, 1996a, b). Precision satellite-state vector data are available for ERS satellites with an absolute accuracy of some decimeters (Massman et al., 1998). The follow-on ESA Envisat satellite is equipped with a sophisticated onboard microwave tracking device, the Doppler Orbitography and Radio-positioning Integrated by Satellite (DORIS) system, which provides orbital-tracking data to better than 10-cm accuracy (Guijarro et al., 2000; Williams, 2001). In spite of these advances, satellite orbital-state vector parameters are still not known with sufficient accuracy to directly determine B and α to the required accuracy (Madsen and Zebker, 1998). As part of the interferometric analysis process, it is therefore generally necessary to produce refined estimates of the baseline geometry (Zebker et al., 1994a). Fortunately, baseline uncertainties can be corrected by incorporating independently derived tie-points with known position, elevation, and/or velocity (Gudmundsson et al., 2002; Joughin et al., 1996b; Massonnet and Feigl, 1998; Mohr et al., 1998; Zebker et al., 1994b). These can take the form of a DEM, bedrock outcrops or *in situ* GPS measurements. Without this step, baseline errors result in a near-linear phase ramp in the interferogram (Bamber et al., 2000). According to Mohr (1997), at least four accurate and well-distributed tiepoints are needed to estimate linear baseline errors in the repeat-pass ERS-1/-2 system. Problems arise, however, where no tie-points are available and corrections must be extrapolated. This can act to significantly compromise the accuracy of absolute velocity estimates, for example (Joughin et al., 1996c).

Recalling (2.5), it can be seen that the sensitivity of the interferometric phase is proportional to the perpendicular baseline $B_\perp$, and that large baselines are desirable for height retrieval. Complex trade-offs come into play, however. For example, the two complex SAR images forming the interferogram tend to decorrelate with increasing baseline length (Li and Goldstein, 1990; Rodriguez and Martin, 1992; Zebker and Villasenor, 1992; Zebker et al., 1994a, b). This coherence loss can be partially compensated for by *spectral shift filtering* of the individual SLC images (Gatelli et al., 1994). An interferogram becomes completely decorrelated, and therefore of no use for topographic or surface-motion reconstruction, once the spectral shift exceeds the SAR system bandwidth W (the bandwidth is a measure of the frequency-limiting stages in the system).

For spaceborne SARs, the coherence (correlation) between radar echoes varies from unity (perfect correlation) at zero baseline to zero (complete decorrelation to make the interferogram useless) at some effective *critical baseline*, $B_{\perp crit}$, which approximates:

$$B_{\perp crit} = \frac{\lambda W R \tan(\theta - \tau_y)}{c} \qquad (2.15)$$

where τ_y is the component of the local terrain slope in the range direction; $\theta - \tau_y$ is the local incidence angle; R is the distance from the center of the pixel target and the sensor; and c is the speed of light (Bamler, 1997). As such, an increase in the spatial decorrelation from 0 to 1 corresponds to an increase in $B_\perp$ from 0 to $B_{\perp crit}$ (Mohr, 1997). In effect, the critical baseline is attained when the baseline becomes longer than half of the reflected beamwidth. For the parameters of the ERS-1/-2 C-band SARs, for example, $B_{\perp crit} = \sim 1,030$ m (Bamler, 1997; Sandwell and Price, 1998). This is another fundamental constraint for InSAR, and $B_\perp$ should be kept below its critical length in order to obtain optimum sensitivity and facilitate phase unwrapping. Results from Hoen and Zebker (2000a) further indicate that the critical baseline is greatly reduced over the Greenland Ice Sheet due to strong volume scatter effects. In general, a baseline of 150 m is typically agreed to be optimal for topographic recovery, while a zero or near-zero baseline is best for ice motion estimation (change detection) but is usually unavailable/unattainable by conventional satellite repeat-pass InSAR. Note from (2.9) that the optimal baseline length is in practice terrain-dependent, as moderate to steep slopes can generate an aliased phase rate or a phase that is difficult to process in the phase-unwrapping stage (Massonnet and Feigl, 1998; Toutin and Gray, 2000). As a result, a smaller baseline is generally appropriate in mountainous regions.

2.3.6 Volume decorrelation, and ambiguities relating to radar penetration depth and volume scattering

Another potentially significant yet poorly understood error source, contributing to interferometric phase decorrelation, is the impact of wavelength-dependent *penetration depth* within the snow–firn–ice mass (Dall et al., 2001; Hoen and Zebker, 2000a, b; Madsen et al., 1999; Meyer, 2004; Reeh et al., 1999). From an InSAR perspective, this is defined as the phase center of the interferometric depth/scattering volume, and depends on the physical properties of the snow/ice. Working in Greenland, Rignot et al. (2001) measured penetration depths that varied significantly with surface type (i.e., facies—see Chapter 3). At C-band ($\lambda = 5.6$ cm), penetration of the radar signal was found to be minimal (1–2 m) on exposed ice, although corresponding L-band measurements ($\lambda = 23.5$ cm) were generally 5–10 m greater. Rignot et al. (2001) argue that such depths are reasonable for the measurement of ice-sheet volume changes, and show that where the phase center of the radar data is several meters below the surface, then InSAR-derived topographies generally follow the ice surface profile quite accurately (Bindschadler et al., 1999). Joughin et al. (1996b) further conclude that C-band ERS InSAR-derived topography agrees with true topography to within a penetration depth offset of at worst a few meters over seasonally melting ice facies in Greenland.

Penetration depths are, however, significantly greater in perennially dry firn both at higher elevations in Greenland and over most of Antarctica. Dall et al. (2001) report penetration depths of up to 13 m at C-band on the Geikie Ice Cap summit in the percolation zone of the Greenland Ice Sheet (Figure 2.6), while Rignot et al. (2001) measured values of up to 10 m at Greenland Summit. Through analysis of

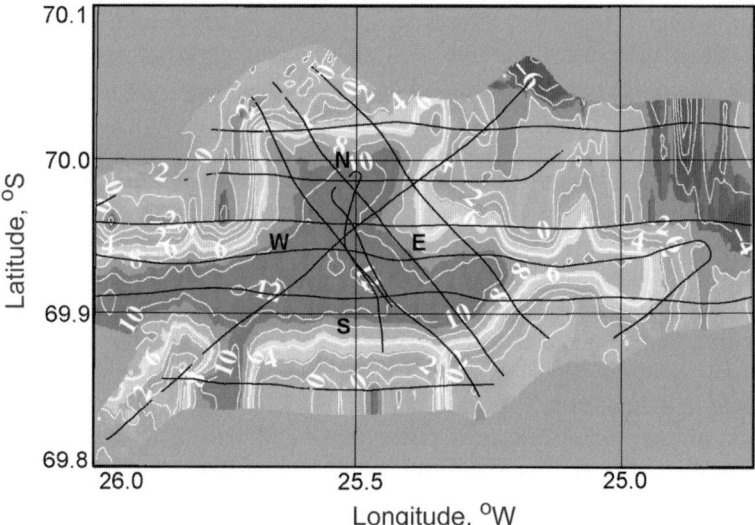

Figure 2.6. Map of the C-band penetration depth of the Geikie Ice Cap (Greenland), measured as the difference between the EMISAR-derived InSAR-derived DEM (measured by the Danish EMISAR [Electromagnetics Institute Synthetic Aperture Radar] aircraft InSAR system: Dall et al., 2000) and coincident airborne laser altimeter elevation profiles (aircraft flight tracks marked as black lines).

Image courtesy of Jørgen Dall, Danish Centre for Remote Sensing.

images of ERS interferometric correlation (coherence images—see p. 69), Hoen and Zebker (2000a) computed penetration depths of 12 to 35 m, including unexpectedly large depths for the percolation zone (see Chapter 3). For Antarctica, Rott et al. (1993) report a depth of ~20 m for permanently dry firn which is strongly layered (at C-band). This problem is exacerbated at L-band, with penetration depths of up to 60–120 m over cold, smooth, exposed ice—i.e., a relatively small proportion of ice sheets—as measured in Greenland by Rignot et al. (2001). These authors suggest that such extreme levels of penetration can significantly bias geophysical product retrieval from radar data.

InSAR mapping of perennially dry snow zones in Greenland and Antarctica may also involve significant baseline-dependent decorrelation due to volume scattering from a variable range of depths within the firn/ice (Rott and Siegel, 1996; Winebrenner et al., 1997). Hoen and Zebker (2000a, b) further show that volume scattering is the main scattering mechanism affecting both scattering center locations and InSAR coherence for the dry snow and percolation zones of the Greenland Ice Sheet. Winebrenner et al. (1997) examine this issue by modeling the depth distribution of scattering in perennially dry firn. They conclude that, if backscattering is indeed dominated by scattering from ice grains but modulated by random, centimeter-scale density layering, then resultant phase decorrelation should be reduced for horizontal polarization (HH-pol) compared with vertical polarization (VV-pol).

The net effect of a time-dependent penetration depth is an underestimate of surface elevation from InSAR analysis. For simplicity, penetration depth is typically considered constant in time in InSAR processing, and as part of the topography component of the interferometric phase (Meyer, 2004). The impact of ϕ_{pd} on the latter increases with the interferometric baseline B. The dependence of ϕ_{pd} on penetration depth d and B_{ij} is shown in Figure 2.7. Following Meyer (2004), the resultant *topographic height error* Δz_{pd} is given by:

$$\Delta z_{pd} = \frac{\lambda r \sin(\theta)}{4\pi B_\perp} \phi_{pd} \qquad (2.16)$$

Further issues relating to the relationship between baseline and penetration depth are discussed by Hoen and Zebker (2000a).

Although phase decorrelation is a limiting factor, Hoen and Zebker (2000b) and Rott and Siegel (1996) point out that its investigation can in fact also provide valuable insights into the interaction mechanisms of microwaves with snow and ice. This can potentially form a means of target discrimination—e.g., blue ice fields versus firn. Still further insight can potentially be gained by using the combination of polarimetric SAR and SAR interferometry, known as Pol-InSAR (Dall et al., 2003, 2004). This emerging technique is covered in more detail in Section 2.9. Moreover, it is expected that combining SAR with other data—e.g., accurate ice sheet surface elevation profiles from the ICESat Geoscience Laser Altimeter System (GLAS, see Chapter 3)—will provide additional important information. It is apparent that the relationship between radar wavelength, imaging conditions, and snow/ice physical and electrical properties is complex (see Chapter 5 of Volume 1 of

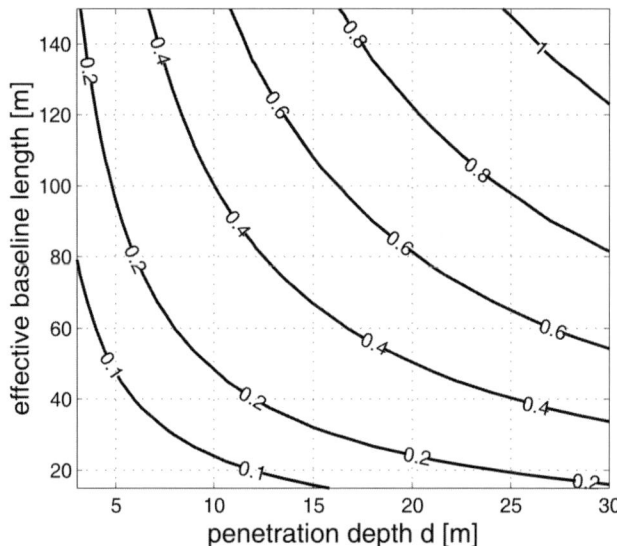

Figure 2.7. Effect of penetration depth d and effective baseline length on the interferometric phase (in radians).

From Meyer (2004). Copyright ISPRS 2004, reproduced with permission.

this book and Chapter 3 of this volume), and has important implications for the use of InSAR (and other radar) data. This is another case where "More Research Is Necessary"—MRIN!

2.3.7 Image speckle

Speckle in SAR imagery is a scattering phenomenon resulting from random coherent interference caused by the presence of multiple scatterers within a pixel—i.e., it occurs because the sensor resolution is insufficient to resolve individual scatterers. While strictly speaking not noise in the physical sense, speckle will, like noise, degrade the phase measurement. In fact, it is a major limiting factor in InSAR, in that it affects the performance of phase-unwrapping algorithms (Joughin et al., 1998a). As we shall see in Section 2.4, phase unwrapping is a key stage in the production of an interferogram. Speckle is less of an issue over ice-sheet inland regions after the phase has been unwrapped, as it can be reduced by filtering which retains a sufficient spatial resolution—i.e., 100–200 m—so as to not drastically affect the level of spatial detail present in the interferogram (Joughin, 1995). In this sense, speckle is a more significant problem for mountain glaciers and some outer margins of ice sheets, where the features of interest are on a smaller scale (Joughin et al., 1998a). As will be seen in Section 2.6, speckle can in fact be exploited to advantage, as the basis for an alternative important means of measuring ice motion where only SAR image pairs with a relatively long temporal baseline are available, as is the case for Radarsat-1 (24 days in general).

2.3.8 Geometric distortions and steep terrain effects

Being range sensors, SAR systems suffer from information loss or low coherence in high-relief (mountainous) regions and areas of steep slope, due to geometric effects of *foreshortening, layover*, and *shadowing* (Curlander and McDonough, 1991). These effects, which are illustrated schematically in Figure 2.8, depend upon the complex relationship between SAR look direction, incidence angle, and platform altitude, as well as terrain (object) configuration, slope angle, and orientation relative to the radar beam orientation. *Foreshortening* is the spatial distortion whereby terrain slopes facing the SAR illumination are mapped as having a compressed scale relative to their actual area, and targets in high terrain "tilt" towards the radar in a two-dimensional image. This geometric effect is more pronounced for steeper slopes and for radars employing steeper incidence angles. Slopes steeper than the radar look angle will tend to *layover*, an extreme form of foreshortening whereby the top of the slope is imaged over the lower slope, causing loss of information in that area. Foreshortening compresses the backscattered signal energy, with the affected area appearing brighter on the image. *Image layover* tends to undermine successful phase unwrapping in mountainous regions (depending on the viewing geometry) (Rufino et al., 1996). Steep slopes facing away from the radar—e.g., slope bc in Figure 2.8c—will either be in shadow (causing a loss of data) or have a lower SNR (causing substantial errors in interferometric phase determination). In the example shown in Figure 2.8c, backscatter information is lost not only from area bc but also from cd. *Shadowing* effects are most prominent with large incidence angle illumination, but tends to be less of a problem for satellite SARs with steep incidence angles than slope *foreshortening* and *layover*. The complex impact of these factors can sometimes be reduced if data from different look directions are available—e.g., ascending and descending orbit data combined (Kropatsch and Strobl, 1990). In addition to the well-known geometric distortions outlined above, the surface-slope angle has a direct impact on the quality of phase unwrapping (Gens and van Genderen, 1996b). In steep terrain, the 2π ambiguity of the phase requires additional information in order to be solved.

2.3.9 Estimation of phase decorrelation

Fortunately, the overall degree of decorrelation due to the causes outlined above can be estimated between two complex SAR images (Touzi et al., 1999). This is termed the interferometric *coherence* γ—see (2.12). Indeed, coherence must be measured to assess the suitability of given datasets for InSAR processing—i.e., it provides an important quality check prior to the production of an interferogram. Assuming constancy in interferometric phase over a small region, the value of the total interferometric phase γ between two co-registered complex SAR images S_1 and S_2 used to generate the interferogram is defined as the ratio of the averaged interferogram magnitude to the geometric mean of the averaged powers in the two

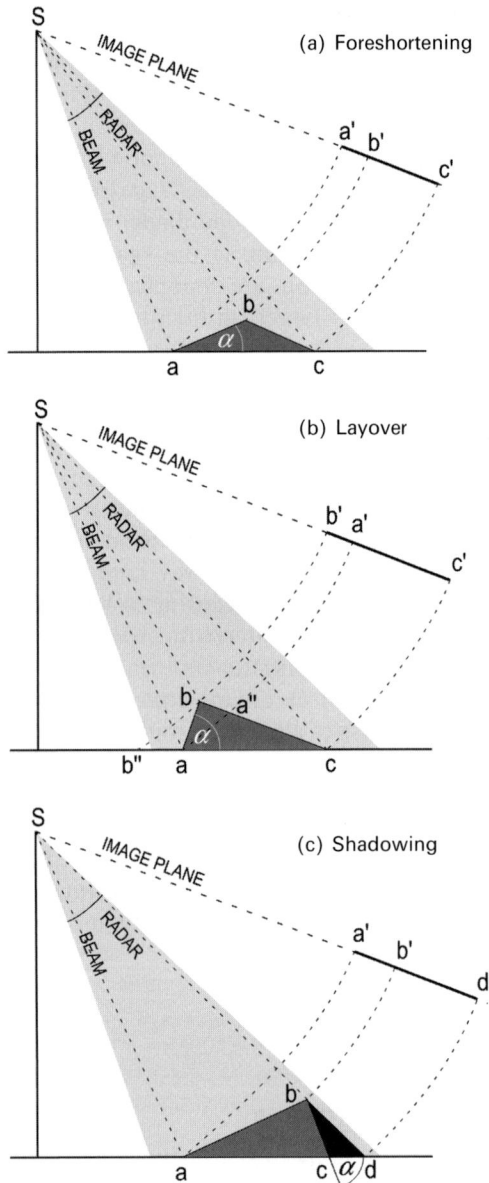

Figure 2.8. A simplified schematic illustration of forms of distortion in radar images of the Earth's surface induced by projecting the terrain feature (hill/mountain) abc along equal slant range arcs into a'b'c' in the image plane, where S is the sensor (satellite). These are (a) foreshortening, (b) layover, and (c) shadowing. In (a), the terrain feature abc is illuminated by the radar beam (lightly shaded) and the echoes from points a, b, and c are received at travel times corresponding to a', b', and c' in the slant-range image plane.

From Olmsted (1993). Copyright Alaska Satellite Facility 1993, reproduced with permission.

images (Hoen and Zebker, 2000a). It is given by:

$$\gamma = \frac{|\langle S_1 S_2^* \rangle|}{\sqrt{\langle S_1 S_1^* \rangle \langle S_2 S_2^* \rangle}} \quad (2.17)$$

where * denotes complex conjugation and $\langle \cdot \rangle$ denotes statistical expectation, determined by spatial (ensemble) averaging. In practice, the latter is determined by averaging adjacent pixels—i.e., by taking looks—such that:

$$\bar{\gamma} = \frac{\left|\sum_{i=1}^{N} S_i(i) S_2^*(i)\right|}{\sqrt{\sum_{i=1}^{N} S_i(i) S_1^*(i) \sum_{i=1}^{N} S_2(i) S_2^*(i)}} \quad (2.18)$$

where $N = n_{ra} \times n_{az}$ is the total number of pixels averaged in range and azimuth, respectively. In practice, the window size used must be neither too large nor too small. In the former case, loss of detail would result, whereas the use of too small a window would cause the computed signal to be too unstable. Issues relating to *correlation bias* are discussed by Hoen and Zebker (2000a).

Recalling earlier discussion, interferometric coherence ranges from 0.0 (no correlation), where there is no useful information in the interferogram, to 1.0 (perfect correlation), where there is no noise in the interferogram. Both extremes are rarely seen, and most interferometric image combinations lie somewhere in-between. As a rule of thumb, values of 0.7–1.0 indicate excellent coherence, 0.5–0.7 good coherence, and 0.3–0.5 noisy but usable coherence. Once again, the importance of the absolute value of interferometric coherence γ is that it provides a fundamental measure of the noise level, and thus quality, of the resultant interferogram (Just and Bamler, 1994; McLeod et al., 1998; Touzi et al., 1999). The magnitude of γ is intimately related to the local standard deviation of differential phase. A high magnitude indicates that (i) the images have a good SNR, (ii) any surface motion is spatially "organized", and (iii) the phase centers of scatterers are stable. As seen later in Section 2.4, standard practice in the InSAR processing chain is to produce a *coherence image*. The quality of the coherence product itself depends on the resampling method(s) used (Joughin et al., 1994; Lin et al., 1992; Massonnet and Feigl, 1998; Small et al., 1993). The relationship between SNR and γ (2.12) is as follows (Rocca et al., 1997):

$$SNR = \frac{|\gamma|}{1 - |\gamma|} \quad (2.19)$$

Given the factors outlined above, it is essential, though not always possible, to avoid coherence loss during the interferogram generation process. Issues compromising this situation are covered in Section 2.4. It should be noted that maps of interferometric correlation, which are essentially byproducts of InSAR data processing, are themselves emerging as a useful and highly promising means of detecting subtle changes in the surface and near-surface characteristics of ice sheets—e.g., Hoen and

Zebker (2000b). This and other applications of InSAR are covered more fully in Chapter 3.

2.3.10 Processing errors and other sources of ambiguity

Interferometric *processing errors* also contribute to phase decorrelation. These can occur at a number of stages in the processing sequence, starting with inaccurate image co-registration. This will result in degraded coherence and interferometric phase measurement at the pixel level (Allen, 1995). Ideally, the two paired interferometric images must be co-registered to sub-pixel-level accuracy in order to achieve the level of accuracy in phase difference measurement required to compute Δr. Moreover, and as stated in Section 2.2.3, $\Delta\phi_{ij}$ can only be measured *modulo* 2π. To calculate the relative elevation/motion for each point in the interferogram, it is therefore necessary to accurately add the correct integer number of phase cycles to each phase measurement. As stated earlier, the technique of solving this 2π ambiguity is called *phase unwrapping*. Unfortunately, this problem can be unambiguous without *a priori* assumptions. Phase unwrapping is discussed further in Section 2.4. Please see Massonnet and Feigl (1998) for a more in-depth assessment of constraints and error sources inherent to SAR interferometry.

2.3.11 Summary of baseline-related trade-offs

Not only does the baseline length B_{ij} play a key role in interferometric geometry and the successful generation of an interferogram. It also serves as a standard measure of both the quality of the interferogram and the suitability of SAR image pairs for a given application, prior to processing. Equally important for satellite repeat-pass InSAR is the temporal baseline ΔT, in that scatterers must remain coherent for InSAR to work at all. What constitutes an optimal spatial or temporal interferometric baseline combination ultimately depends on the application (Mohr and Madsen, 1996b). With this in mind, certain rules-of-thumb apply regarding ice-sheet and glacier research. All else being equal, these can be summarized as follows:

- For *topographic mapping*, errors can be minimized by choosing an image configuration with a large interferometric baseline (Mohr and Madsen, 1996b). Good height accuracy is also required for InSAR-derived DEMs used for differential InSAR mapping of ice motion (see Sections 2.2.2 and 2.5), in order to minimize any slope-related errors (Joughin et al., 1998a). If the baseline is too small, then the sensitivity to topography will be low, and phase noise may become dominant. As such, the baseline must be sufficiently long to yield the desired topographic sensitivity without causing too much baseline decorrelation (Li and Goldstein, 1990; Madsen and Zebker, 1998; Rodríguez and Martin, 1992). For ERS data, for example, the requisite baseline length should be >50 m. Baseline decorrelation increases, however, with increasing baseline (Kwok and Fahnestock, 1996). If the baseline is too large, then phase aliasing

may occur, leading to a decrease in coherence. As such, the baseline length should ideally be <300 m for ERS data.
- For *ice motion mapping*, B_{ij} should be as small as possible to obtain accurate velocity estimates (Joughin et al., 1996b). This provides a large altitude of ambiguity (defined as the elevation change in an interferogram corresponding to a 2π phase shift), thereby minimizing residual topographic errors.
- Increasing the temporal baseline ΔT will decrease the uncertainty in InSAR ice motion retrievals, but increase the likelihood of temporal decorrelation (for both ice motion and height retrievals).
- Longer temporal baselines are in general better suited to regions of slow-moving ice, such as inland ice sheets, while short temporal baselines are necessary with fast outlet glaciers. According to Joughin et al. (1998a), the 1-day temporal baseline of the ERS-1/-2 Tandem Mission, for example, is well-suited to measuring ice velocities in the range of 100–1,500 m per annum. By the same token, Mohr et al. (1997) found little correlation in 35-day repeat interval ERS data over glaciers in NE Greenland. Once again, temporal decorrelation is one of the major constraints on repeat-pass interferometry on a case-by-case basis, yet is difficult to assess and model. With a single-satellite SAR, increased flexibility is obviously afforded by a short, exact repeat period of the orbit, enabling the construction of a range of different temporal baselines as required. The trade-off for this is a reduction in the extent of areal coverage (Joughin et al., 1998a).
- In the case of differential InSAR using SAR data only (see Section 2.5), matching pairs of mixed short- and long-temporal-baselength images are ideally required to retrieve both motion and height.
- For differential InSAR using an independently derived DEM, the "criticality" of the perpendicular baseline $B_\perp$ depends upon the resolution of the DEM used in the differential process—the coarser the DEM, the smaller $B_\perp$ needs to be.

Clearly, no single set of parameters is optimal for all ice-sheet locations (Bindschadler, 1998), and important trade-offs come into play. In reality, the lack of suitable coherent image pairs may preclude the use of datasets that are otherwise suitable for InSAR processing, as discussed in the next section. As noted by Joughin et al. (1998a), the majority of the baselines from the ERS-1/-2 Tandem Mission, for example, are too long and therefore sub-optimal for InSAR processing. This situation will hopefully change in the not-too-distant future, with the launch of dedicated InSAR missions (see Section 2.10).

2.4 SATELLITE-SAR INTERFEROMETRIC DATA REQUIREMENTS AND PROCESSING STEPS

Interferometric pre-processing and processing of SAR data is a complex procedure, with the quality of the final product depending both on the quality of the datasets

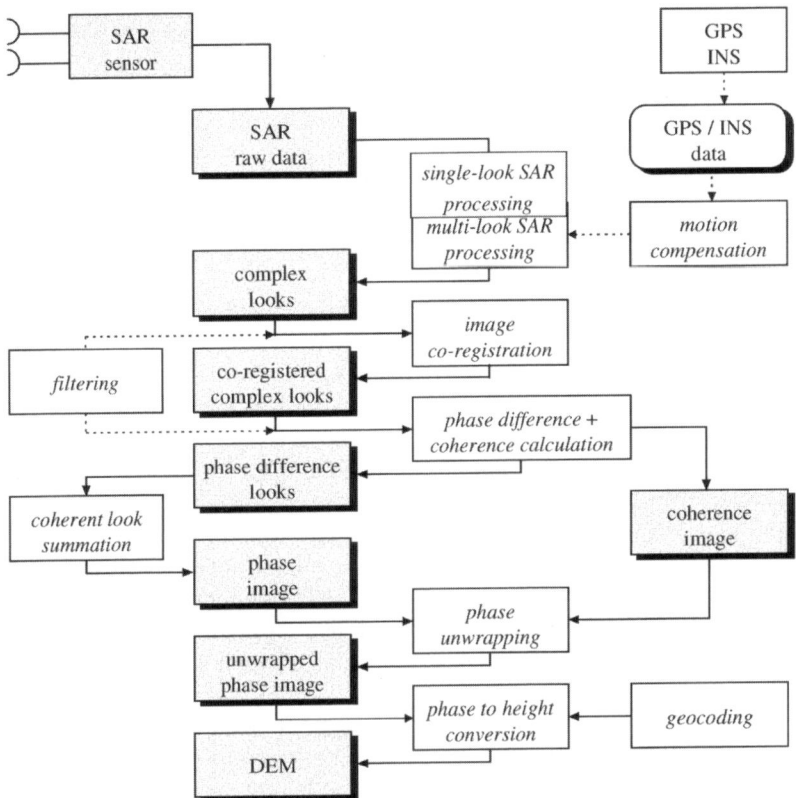

Figure 2.9. The typical chain of processing steps involved in the retrieval of a Digital Elevation Model (DEM) from a coherent interferometric SAR image pair. Note that this sequence includes processing from raw data onwards. In fact, the user can generally access ready-processed Single-Look Complex (SLC) data. The degree of algorithm automation/ operator intervention varies depending on the software package chosen and the quality of the interferometric image pair.
After Gens and van Genderen (1996a). Copyright Taylor & Francis 1996, reproduced with permission.

used and the performance of each individual processing step. This section provides brief and simplified background information on data requirements and interferometric processing. Please see the following references for a more detailed discussion of issues relating to large-scale mapping with satellite interferometric data and how to construct an interferogram: Franceschetti and Fornaro (1999); Gens (1998); Gens and van Genderen (1996a, b); Madsen et al. (1993, 1999); Massonnet and Feigl (1998); Mohr and Madsen (2000); and van Genderen and Gens (1996). Interferometric processing of stripmap Radarsat-1 data is discussed by Geudtner et al. (1997, 1998).

The following sequence of interferometric processing steps, portrayed schematically in Figure 2.9, is generally followed to retrieve ice-sheet geophysical quantities (in this case a DEM) from a coherent repeat-pass SAR image pair (Dixon, 1994):

i. Selection of a suitable pair of SAR images, and conversion of raw data to single-look complex images if necessary. Information is required on precise orbit and calibration parameters, the time reference and intervals of each image, and the chosen temporal and spatial resolutions.
ii. Common *band filtering*.
iii. High-accuracy *image co-registration*, to an accuracy of up to 1/8 of a pixel.
iv. *"Master" SAR image filtering*, in azimuth (along columns) and in range (along rows). Filter lengths vary, although a typical length is 32 samples.
v. *"Slave" SAR image filtering* and resampling. The filtering is similar to that applied to the master image. The slave image is resampled to be "in register" with the master image, using a suitable interpolation method.
vi. *Interferogram generation*. This is generally a simple pixel-by-pixel complex multiplication of the master image by the conjugate of the co-registered slave image.
vii. Generation of an associated *coherence image*. This involves a simple complex cross-correlation between the master and co-registered slave images, and is typically carried out in a small local area surrounding each pixel in the interferogram—e.g., 5×5 to 5×25 samples.
viii. Range and azimuth *low-pass filtering* and resampling. This is a coherent spatial filter operation on the interferogram. Filtering is applied to improve phase coherence, but at the expense of horizontal spatial resolution.
ix. *Removal of flat-Earth fringes* (required because the fringes in an interferogram result not only from the surface topography but also the Earth's curvature).
x. Interferogram *phase unwrapping*, from 360° to an absolute phase.
xi. *Phase interpolation*, to fill in gaps in the unwrapped phase due to noise and decorrelation of the interferometric phase.
xii. *Conversion of phase* to ice sheet elevation. This involves a simple pixel-by-pixel scaling to convert the unwrapped phase image to a height image (DEM) in units of meters.
xiii. *Terrain distortion correction*. This entails a pixel-repositioning step in the range direction.
xiv. *Geo-coding* (of all map products), to transform the image to map projection space.

Further details of the main steps are given below. Note that both commercial and freeware software packages are available for interferometric processing (Gens, 1998, 1999b; Kampes et al., 2003; Rosen et al., 2004; Vincent and Rundle, 1998). While significant user intervention is currently required, work is underway to develop a more automated approach (Guritz et al., 1998). Excellent freeware information is available from the Alaska Satellite Facility at ⟨http://www.asf.edu⟩.

The typical sequence required to derive ice sheet/glacier motion from suitable SAR imagery is illustrated in Figure 2.10 (see color section), in this case using ERS-1 and-2 image pairs of Pine Island Glacier (West Antarctica) separated by 1 day and acquired during the ERS Tandem Mission. The resultant map of ice motion derived in this fashion is shown in Figure 2.11 (see color section). More detailed information on the derivation of ice sheet/glacier topography and motion by InSAR and associated techniques is given in Sections 2.5 to 2.8 inclusive. Note that other examples are given in Chapter 3.

2.4.1 Selection of suitable SAR images

Datasets from several spaceborne SAR systems, including those onboard ERS-1 and -2, Radarsat-1, and Envisat, and to a lesser extent JERS-1, are suitable for interferometric studies of glaciers and ice sheets, with more datasets in the near future (e.g., from ADEOS-II and Radarsat-2). Images from ERS-1 and -2 can be readily combined for InSAR analysis, as they are acquired by identical-wavelength SARs and belong to the same orbital cycle. It was previously not thought possible to combine ERS-2 and Envisat data interferometrically, due to the slightly different frequencies of their SARs (see the Appendix), but this was recently achieved (ESA, 2003). This opens up a potentially exciting, albeit more complex, avenue of radar cross-interferometry (Gatelli et al., 1994).

Data selection itself is a non-trivial matter. It involves a number of criteria, including the availability of suitable image pairs, their date, and orbits. Again, these must be either raw or *SLC* format. These data contain signal information in a complex number—i.e., the intensity (amplitude a) and phase ϕ are stored in *real* or Re (cosine) and *imaginary* or Im (sine) parts—with the complex signal being represented as a vector whose length refers to the amplitude and whose orientation represents the phase. This information can be extracted from the complex values by:

$$\phi = \arctan \frac{\text{Im}}{\text{Re}} \qquad (2.20)$$

$$a = \sqrt{\text{Im}^2 + \text{Re}^2} \qquad (2.21)$$

As stated in the previous section, the interferometric baseline and time interval between image acquisitions are major considerations when selecting image pairs suitable for a given InSAR application. In the case of ERS-1 and -2, for example, the optimal baseline may be taken to be $\sim$150–300 m for height recovery, 30–50 m for surface change detection, and 0–5 m for surface feature movement studies— e.g., glaciers (Engdahl, 1996; Sandwell and Price, 1998). Once again, Joughin et al. (1996b) recommend that interferograms with the shortest available baselines should be used to estimate ice velocities. Although satellite orbits converge towards the poles, few image pairs with very short spatial baselines are available. To reiterate, temporal separation is also of central importance in that scatterers must remain coherent for InSAR to work at all.

Fortunately, a number of online Internet software packages have been developed to help the user select suitable image pairs. An example is ESA's Display Earth remote-sensing Swath Coverage for Windows (DESCW) data selection tool (D'Elia and Biasutti, 1999; for further information and software download, please see ⟨http://earthnet.esrin.esa.it/descw⟩). It is hoped that similar software devoted to other datasets will become widely available as the volume of interferometric data increases. Generally, it is not possible to gauge the actual interferometric phase coherence (or lack thereof) between an image pair until processing commences, although Monti Guarnieri and Prati (1997, 1999) provide a means of estimating coherence for data-browsing purposes.

Regarding orbital parameters, the actual position and velocity of a satellite at a particular time along its flightpath, or orbit, is given by means of *orbit-state vectors*. The latter are usually provided in the header file attached to the complex radar image file. Where available, *precision orbit* information should be acquired to minimize the uncertainty in baseline determination (see Section 2.3.7.1) and maximize the precision of image co-registration and interferogram geo-coding (see below). Precision orbit data are available from ESA for use with ERS-1 and -2 (Gruber et al., 1996; Zhu et al., 1996), and online from the Delft Institute for Earth-oriented Space Research in the Netherlands at ⟨http://www.deos.tudelft.nl/ers/precorbs/⟩. Moreover, interferometric baseline listings for ERS data of interest are given at ⟨http://earth1.esrin.esa.it/eeo4.588;internalandsk=06597095⟩. Further information on orbit listings is given online at ⟨http://envisat.esa.int/rootcollection/eeo4.10075/0010d.html⟩.

Another excellent source of information is the NASA Alaska Satellite (formerly SAR) Facility (ASF) website at ⟨http://www.asf.alaska.edu/apd/documents/index.html⟩. The ASF is one of NASA's Distributed Active Archive Centers (DAACs) with large data holdings including ERS-1, ERS-2, JERS-1, and Radarsat-1, as well as associated information on interferometric baselines—e.g., for Radarsat-1 at ⟨http://www.asf.alaska.edu/baselines/⟩. A sizable collection of ERS Tandem Mission data has been acquired both at ASF and McMurdo Station, Antarctica (Gurwitz et al., 1999). Information on interferometry using JERS-1 SAR data is available from the National Space Development Agency of Japan (NASDA, 1995; ⟨http://www.eorc.nasda.go.jp/EORC/DataLibrary/CDROM.html⟩). Baseline information on JERS-1 image pairs is available on ⟨http://www.eorc.nasda.go.jp/JERS-1/Result/⟩. These and other similar websites also provide indispensable background information and practical help to aid interferometric processing. Without doubt, the Internet is revolutionizing our ability to not only choose and access suitable data, but also to access key ancillary information and even software. Information on the latter, and a tutorial on the use of InSAR, is given in Gens and Logan (2003).

The orbital configuration of current satellites prevents satellite SARs from fully covering the Earth's polar regions. The ERS satellites leave a latitudinal gap of about 8° at the poles, while the nominally right-looking SAR onboard Radarsat-1 offers coverage in the latitude band 78–90°N but leaves a significant gap over the Antarctic continent. In early September 1997, however, Radarsat-1 was maneuvered to a south-looking orientation to collect the first complete SAR coverage of Antarctica, and at a high resolution (25 m) with radiometrically and geometrically corrected SAR data (Gray et al., 1998; Jezek, 1999; Jezek et al., 1998). Data were collected from September 19 to October 14, 1997, during this first joint Canadian Space Agency (CSA)/NASA Antarctic Mapping Mission (AMM) (Jezek, 1999;

Jezek et al., 1998), under the auspices of the Radasat Antarctic Mapping Project (RAMP). With the nominal 24-day repeat cycle of Radarsat-1, AMM provided 6 days of repeat-pass interferometric data. These remain the only satel-lite InSAR data collected south of 80°S (Joughin, 2002). They are available from the NASA Alaska Satellite Facility (⟨http://www.asf.alaska.edu/4_4_1.html⟩). Forster et al. (1998) presented a preliminary analysis of an InSAR pair for the Recovery Glacier, East Antarctica, and carried out an estimate of strain rates (see Chapter 3).

Polar InSAR studies have so far tended to be largely regional, partly due to the lack of suitable larger scale datasets. This has been particularly true in Antarctica. The follow-up Radarsat-1 Modified Antarctic Mapping Mission (MAMM) was conducted from September 3 to November 14, 2000 to address this issue (Jezek, 2002). Unfortunately, this mission was only able to collect data from a north-looking orientation due to concerns about the health of the spacecraft, which limited coverage to north of 80°S (Smith et al., 2002). Nevertheless, three full cycles of Fine-1 beam data (see ⟨www.rsi.ca⟩ for details) were collected along both ascending and descending orbits, making this the most comprehensive InSAR dataset collected for ice motion to date (Joughin, 2002). A third mapping mission, MiniMAMM, was carried out in September–December 2004 to replicate MAMM over selected areas—i.e., the David and Pine Island Glaciers, the Filchner Ice Streams and parts of the Antarctic Peninsula (Harbin, 2005). Planned products include coastline estimates. Repeat-pass data were acquired along the same orbits as MAMM from three Radarsat-1 repeat cycles from September–December 2004, over various image beam modes (Fine 1, Standard 1, 2, and 6, and Extended Low) and for both ascending and descending orbits. Planned products include interferometric coherence maps and ice-velocity maps. Analysis of these exciting datasets is ongoing at the NASA Alaska Satellite Facility and Byrd Polar Research Center (⟨http://www-bprc.mps.ohio-state.edu/radarsat⟩). Considerable effort is also being put into producing an ERS InSAR mosaic of the Greenland Ice Sheet (Shepherd and Mohr, 2001)—see Section 2.4.3.

A major current issue is that no satellite SAR system has so far been launched that is dedicated to InSAR. As such, existing and past systems are not optimized for interferometry, and all have inherent strengths and weaknesses in this regard (Massonnet and Feigl, 1998). The following is a short and by no means exhaustive evaluation of issues affecting the suitability of the different, existing satellite datasets to InSAR analysis for glacier/ice-sheet research (detailed descriptions of sensor characteristics are given in the Appendix). Note again that plans for future, improved systems are evaluated in Section 2.10.

- The orbital history of ERS-1 includes seven distinct phases using five different cycles of 3, 35, and 168 days.[11] Data from any of the 35-day phases can be combined to form an interferogram, whereas those from the two different 3-day phases (in 1991 and 1994) cannot. The same applies to the two different 168-day cycles (Scharroo and Visser, 1998). Shortcomings in the use of ERS-1 SAR data for DEM generation and change detection are orbit control and temporal

---

[11] For ERS-1, the 3-day (ice) phases were from December 28, 1991 to April 2, 1992 and December 23, 1993 to April 10, 1994. 35-day orbits lasted from April 2, 1992 to December 23, 1993 and March 21, 1995 until the mission end on March 10, 2000. The 168-day (geodetic) phases were from April 10, 1994 to March 21, 1995.

decorrelation for data from the 35-day and greater orbits. ERS-1 orbits are in general insufficiently well-known to estimate interferometric baselines to the accuracy required to estimate ice motion and generate DEMs (Zebker et al., 1994b). As a result, the baseline must be determined using independent tie-points within the interferogram (Joughin et al., 1996c; Zebker et al., 1994b).
- While the SAR onboard ERS-2 is identical to that on ERS-1, and the former was placed in a 35-day orbit, its orbit can be more accurately determined by means of the onboard Precise RAnge and Range-rate Equipment (PRARE) (Adam et al., 1999; Bedrich, 1997; Zandbergen et al., 1997). Temporal decorrelation remains a major limiting factor for these data when used alone. The fixed viewing geometry of the ERS-1/-2 SAR system ($\sim 23°$), for example, is sub-optimal for InSAR ice motion retrieval (Mohr et al., 1997). Latterly, ERS-2 has experienced yaw-pointing uncertainties of $\pm 10°$, making the data unsuitable for InSAR analysis (Smith, 2002).
- During the *Tandem Mission* (October 1995–June 1996), ERS-1 and -2 used the same 35-day orbital cycle, but shifted by 1 day. This minimized temporal decorrelation, and produced a dataset eminently suitable for, and widely applied to, ice-sheet and glacier studies. For ERS, ascending-orbit spatial baseline conditions are superior to those of descending-orbit data (Reeh et al., 2003).
- Both ERS satellites exhibited high radiometric stability, to 0.5 dB (Sanchez and Laur, 1997).
- Radar wavelengths strongly determine the degree to which surface change affects InSAR correlation. The JERS-1 L-band SAR is less susceptible to surface change and temporal decorrelation than the shorter wavelength C- and X-band SARs—i.e., snow and ice coherence is better preserved (Frölind and Ulander, 2002; Rignot et al., 2001). In theory, this enables use of a longer temporal baseline for L-band data (Joughin et al., 1998a). Since phase is proportional to baseline length and inversely proportional to wavelength (2.5), lower frequency SARs require a larger baseline, and are therefore better suited to height than motion retrieval (Werner et al., 1992). Moreover, the JERS-1 SAR has a relatively large incidence angle of $\sim 35°$ (compared with $\sim 23°$ for ERS). Pitted against these advantages is the likelihood of larger penetration depth effects (see Section 2.3.4). Unfortunately, JERS-1 images also tend to be of a poorer quality than ERS images because of a lower SNR ratio, and significant temporal decorrelation results from the 44-day repeat cycle (Rossi et al., 1996). Moreover, the orbit was less well-maintained (Madsen and Zebker, 1998). As such, relatively few JERS-1 images suitable for ice-sheet/glacier InSAR analysis are available, resulting in few studies in polar regions. Note that while JERS-1 was equipped with an onboard tape recorder (with limited capacity), the lack of similar recorders onboard ERS-1 and -2 limited data acquisition by them to regions within $\sim 3,400$ km of dedicated receiving stations (Massonnet and Feigl, 1998).
- With a repeat cycle of 14 days and due to the improved snow and ice coherence at L-band compared with higher frequency SARs, ESA's TerraSAR-L will be more suited to InSAR measurement of ice sheets and glaciers than JERS-1 (Zink, 2004). The launch of TerraSAR-L is planned for 2007.
- The interferometric capabilities of the new X-band SAR to be launched on TerraSAR-X in 2006 are assessed by Eineder et al. (2004). Although this mission is not designed specifically for InSAR, it offers a number of innovative

developments that may be of interest for InSAR applications. These features include: (a) a high resolution of 3 m in both spotlight and stripmap modes; (b) a high-range bandwidth to allow large baselines; (c) a burst-synchronized ScanSAR mode; and (d) an along-track interferometric mode based on an innovative split-antenna mode. See the Appendix to Volume 1 and Chapter 5 in Volume 1 of this book for details of the TerraSAR missions.

- Modern SARs—e.g., Radarsat-1 and -2, Envisat ASAR, ALOS PALSAR, and TerraSAR-X and -L—are characterized by flexible (programmable) scanning capabilities, compared with the fixed viewing angles of ERS-1/-2 and JERS-1. Details are again given in the Appendix to Volume 1 and Chapter 5 of Volume 1 of this book. Radarsat-1, for example, gives three options for slant-range resolution which vary with scanning mode (Parashar et al., 1993). Of particular importance is the fact that the use of fine-resolution mode relaxes the criterion for the critical baseline outlined in Section 2.3.5, making it possible to use baselines of >1 km (Toutin and Gray, 2000). This increases the sensitivity to topography and can result in more accurate DEM products than are achievable with the lower resolution modes of Radarsat and ERS, for example (Vachon et al., 1995a, b). Joughin (2002) further suggests that fine-beam Radarsat data are preferable, as standard beam data have approximately one-third the range resolution and have errors in the range component that are larger by a factor of 3. A limiting factor overall is that the Radarsat orbit is too imprecisely known (i.e., neither laser tracking nor GPS data are available for accurate orbit determination). Moreover, Radarsat-1 has not been subject to tight orbit maintenance for provision of suitable interferometric baselines (Bamler et al., 1998; Geudtner et al., 1998), although the orbit constraint was improved from ±5 km to ±2 km in 2001. Another limiting factor with Radarsat-1 is the lack of zero Doppler steering, which is required to keep the radar beam perpendicular to the satellite flight-path (Smith, 2002). Although some ice-sheet studies have been carried out using Radarsat-1 standard beam data collected during the Antarctic Mapping Mission (e.g., Gray et al., 1999), temporal decorrelation is a limiting factor in many regions, given the long repeat cycle of 24 days. All is not lost, however, as an alternative technique known as *speckle tracking* has been devised to retrieve useful ice motion estimates from such data (see Section 2.6). Planned for launch in late 2005, Radarsat-2 features improved resolution, the ability to routinely switch between left- and right-looking, and offers some interferometric compatibility with Radarsat-1 (Caves et al., 2002). Issues relating to InSAR processing specific to wide-swath ScanSAR data are discussed in the next section. These issues also apply to Envisat-1 and ALOS PALSAR (see below).

- While modern and future SARs have the ability to operate over a wide range of user-specified imaging modes, thereby increasing their overall flexibility and reducing the imaging revisit interval, this is not necessarily advantageous for InSAR application. In fact, this may act to jeopardize the routine acquisition of InSAR data and the consistency of the spaceborne interferometric archive, given that a stringent requirement of SAR interferometry is that images should be acquired with an identical viewing geometry (Bamler et al., 2003). This in effect means that usable InSAR images can generally only be acquired from a revisit time forced to the natural satellite orbit repeat cycle—e.g., 24 days from Radarsat-1 and -2, 35 days for Envisat, 46 days for PALSAR, 11 days for TerraSAR-X, and 14 days for TerraSAR-L. The COSMO-Skymed program consists of several satellites in order to reduce revisit time to less than 1 day (COSMO is Constellation Of Small satellites for Mediterranean basin Observation). Such a multi-satellite system can be reconfigured to form an across-track or an along-track interferometer (Bamler et al., 2003)—see Section 2.10. Note that the next generation of SAR satellites have tight orbit control and high-precision orbit determination—i.e., to 5 cm (Zink, 2004). The "stripmap" single-polarization and ScanSAR modes of TerraSAR-L are specifically designed for InSAR applications requiring maximum bandwidth (in this case 80 MHz) to enable multi-looking for phase noise reduction and precise—i.e., <5 μs—burst synchronization allowing repeat-pass ScanSAR interferometry. The TerraSAR-L ScanSAR mode offers data at an azimuth resolution of 20 m over a 200-km swath.

- Given the caveats outlined above, Envisat-1 has the potential to build upon and even improve the SAR interferometric polar dataset established by ERS-1 and -2. With similar characteristics to its precursors to ensure data continuity, Envisat-1 is equipped with an advanced C-band SAR (ASAR), and flies in a 35-day repeat cycle (ESA, 2001; ⟨http://envisat.esa.it/⟩). It is also equipped with three different atmospheric sensors, which should provide important simultaneous information for the correction of atmospheric effects (Gens and van Genderen, 1996a). Improved orbit tracking is also achieved by inclusion of an onboard DORIS system (Guijarro et al., 2000), which allows the spacecraft position to be fixed to an accuracy of a few centimeters—a significantly higher level of accuracy than that achievable with previous spacecraft. The more accurate the orbit determination, the more accurate (and possibly automated) the InSAR data processing. Envisat, like Radarsat-1 and -2 and ALOS PALSAR (below), has an electronically steerable radar antenna. This enables the scanning of wider areas of 400 km in ScanSAR mode, again enabling faster revisit times (of 4 days). This may to some extent counteract temporal decorrelation, which is a major factor given the long nominal repeat cycle. The ASAR can also record images at different wave polarizations. While this may help to discriminate different ice surface types, it also increases the complexity of InSAR data processing and interpretation. Please see McLeod et al. (1998) for an evaluation of the interferometric capabilities of Envisat.

- The PALSAR onboard the Japanese ALOS satellite, the L-band follow-on to JERS-1 (still to be launched), operates with a variable incidence angle and a low-resolution ScanSAR mode. Precision orbit determination is carried out to facilitate interferometry, but the application is somewhat limited by the long revisit time of 46 days (which can again, however, potentially be shortened by steering the antenna). The duty cycle of this sensor is dominated by the operational requirement to detect earthquake activity. Please see Fiedler et al. (2002)

for a further assessment of the interferometric performance and capabilities of ALOS PALSAR and Envisat ASAR.

Regarding optimal radar frequency, polarization, and look angle, this is a complex issue. In general, a shorter wavelength (e.g., C- versus L-band) is generally desirable for glaciological interferometric studies as the signal level is higher (Rosen et al., 2000). For a given ice surface velocity and SNR, C-band systems are approximately four times more sensitive than L-band (Zink, 2004). Due to substantially greater penetration depth at L- than C-band noted in Section 2.3.4, shorter wavelength systems are also better suited to InSAR studies of ice elevation/volume. On the other hand, deeper penetration can be preferable for velocity mapping, in that temporal decorrelation is diminished by the greater temporal stability of scatterer behavior at depth—i.e., coherence is better preserved (Rignot et al., 1996; Michel and Rignot, 1999; Zink, 2004). Rignot et al. (2001) further suggest that longwavelength SARs are preferable for use over warm (melting) ice masses (in spring–summer). Wavelength-dependent penetration depth issues are discussed further in Section 2.3.7.4. Note that the upcoming TerraSAR-L has an improved sensor design (higher resolution and better SNR) and shorter repeat cycle (14 days) to compensate for the loss of sensitivity for L-band SARs noted above (Zink, 2004).

Once suitable image pairs have been chosen, the subsequent processing steps involve a multitude of tasks and considerations. Prior to examining these tasks in more detail, we briefly evaluate a number of special requirements and issues that are specific to the processing of ScanSAR data—an important consideration given that ScanSAR is becoming a standard feature of spaceborne SARs.

2.4.1.1 ScanSAR interferometry: tradeoffs between flexibility and complexity

ScanSAR systems, such as those on Radarsat-1 and -2, Envisat, and ALOS, image several sub-swaths in a *burst-mode* fashion quasi-simultaneously by steering the antenna beam (Holzner and Bamler, 2002; Moore et al., 1981; Tomiyasu, 1981). Compared with fixed incidence-angle systems, such as those on ERS-1 and -2 and JERS-1, ScanSAR systems achieve wider coverage (over 400- to 500-km versus ~100-km swaths) and potentially shorter revisit times, although this is achieved at the expense of a decrease in azimuth resolution. By this technique, each resolution element on the ground is "seen" by the radar whenever a burst of its echoes is received. The sequence of bursts (i.e., "radar on") and pauses ("radar off") is referred to as the azimuth-scanning pattern. Compared with conventional "stripmap" systems, repeat-pass interferometric observations are more complex for ScanSAR systems due to the beam-switching scenario (Bamler and Holzner, 2004). For an interferometric image pair to be coherent, the azimuth-scanning patterns of both acquisitions must be accurately synchronized with respect to the aspect angle—i.e., the same target point must be imaged from the same along-track positions for both image acquisitions (Bamler et al., 1999; Cumming and Wong, 1998; Vachon et al., 1998). As a result, interferometric processing of burst-mode (ScanSAR) data is also more complex, including co-registration (Monti Guarnieri and Prati, 1996; Monti Guarnieri et al., 1995). Another inherent constraint is that the critical baseline length is shorter for ScanSAR interferometry (e.g., 400 m in wide-swath mode) than for "standard" (e.g., ERS) interferometry due to the reduced range bandwidth of the former (Monti Guarnieri et al., 1998).

In spite of these constraints, Bamler et al. (1998) demonstrate the feasibility of ScanSAR interferometry both by combining Radarsat-1 ScanSAR Narrow datasets with a Standard beam dataset and by combining two ScanSAR Narrow datasets, and conclude that the two-beam ScanSAR mode of Radarsat is preferred for interferometry. See Raney et al. (1991) and Chapter 5 of Volume 1 of this book for a fuller description of Radarsat-1 scanning modes. Further information on phase preserving, burst-mode processing, and the required synchronization of the azimuth-scanning pattern is given by Bamler et al. (1999) and Holzner and Bamler (2002). Details of ScanSAR interferometry and Radarsat-2 and -3 are discussed by Bamler and Holzner (2004), while signal and spectral properties specific to ScanSAR burst-mode data are covered by Bamler and Eineder (1996) and Monti Guarnieri and Prati (1996).

2.4.2 Geometric co-registration

A key initial step in InSAR processing is the co-registration of the input images (both intensity and phase) to a high precision prior to interferogram generation (Gens and van Genderen, 1996b; Just and Bamler, 1994; Massonnet, 1993; Mohr and Madsen, 2001). This is a non-trivial task, given the differing look angles and skewed SAR trajectories involved (Allen, 1995; Migliaccio and Bruno, 2002). Co-registration involves the geometric alignment of one image (the *slave*) to the other reference image (the *master*). Generally, this process entails two steps—namely, (i) changing each pixel location in the slave image with respect to the master image; and (ii) recalculating the phase and amplitude information of the phasor by interpolation of each pixel in the slave image (Gens, 1998; Massonnet and Feigl, 1998; Samson, 1996).

Accurate co-registration requires a precise knowledge of the offset between the two scenes, such that the *slave* can be "superposed" onto the *master* while maintaining the phase information content of the pixels. Image offsets occur as a result of uncertainty in sensor location, differences in the look angle between repeat passes, and diverging or converging orbits (Kwok and Fahnestock, 1996). Although techniques vary in detail, co-registration is generally carried out in two stages. The first involves coarse matching, whereby the shifts in both range and azimuth are estimated to an accuracy of a few pixels—either by using satellite orbit ephemeris information (Small et al., 1993) or manually by using tie-points, if available (Kwoh et al., 1994). With an absolute accuracy of a few decameters for ERS satellites (Massman et al., 1998), precision satellite state-vector data enable more accurate image co-registration (Gens and van Genderen, 1996b). Coarse matching is followed by fine co-registration, whereby the image matching (alignment) is refined to a fraction of a pixel, typically by comparing roughly corresponding areas and solving for a set of local transformation parameters (either in the frequency or spatial domain) (Massonnet and Feigl, 1998; Samson, 1996). A large variety of techniques have been put forward for this purpose. These include searching for the minimum of the average fluctuation of the phase difference image (Lin et al., 1992), the maximum correlation (Geudtner, 1996; Small et al., 1993), and the maximum SNR (Gabriel and Goldstein, 1988). An alternative technique for the determination of the relative mis-registration between two interferometric SAR images is given by Scheiber and Moreira (2000). This method is based on the spectral properties of the complex SAR signal, and differs from conventional co-registration methods in that it requires neither interpolation/cross-correlation nor coherence/fringe optimization procedures. Instead, the phase information of different spectral looks is evaluated to give misregistration information on a pixel-by-pixel basis. This technique is at least as accurate as the conventional algorithms while its implementation is simple. In general, a trade-off exists between the accuracy of the technique applied and processing time. As a rule of thumb, the quality requirement for image co-registration in order to avoid phase errors is of the order of one-eighth of a pixel.

Working in an area of the Greenland Ice Sheet with no ground-control points, Kwok and Fahnestock (1996) adopted the method of Gabriel and Goldstein (1988), whereby subpixel-offset corrections are based on visibility of the fringes in an interferogram. While the accuracy of the baseline is generally sufficient to align the image pair, the baseline must subsequently be adjusted using tie-points, however. Improvements in co-registration are possible if identifiable ground-control points are present within the scene. Madsen et al. (1999), for example, used automatic tie-pointing to achieve an enhanced level of geometric fidelity for ERS data over Greenland. This technique is again facilitated by the availability of precision satellite-state vector data. Tie-points can take the form of data from GPS surveys (Keller et al., 1997), DEMs, and aircraft laser altimeters, or alternatively bedrock outcrops—i.e., known points of zero motion (Mohr et al., 1998). It is possible to locate the tie-points within each SLC image by attaining geometric fidelity and incorporating precision orbit data (Mohr and Madsen, 1999).

After its parameters have been determined, the transformation is applied to the slave image. Pixel values in the latter must now be interpolated in order to attain the new value according to the transformation applied (Usai, 2001). It should be noted that the interpolation method used can have a significant impact on the quality of InSAR products (Bamler and Hanssen, 1997; Hanssen and Bamler, 1999). Co-registration is generally preceded by image filtering in both the range and azimuth directions, the aim being to improve the SNR (Bamler and Just, 1993; Geudtner, 1996). Samson (1996) and Hanssen and Bamler (1999) provide detailed comparisons of the different co-registration techniques, and also an assessment of parameters affecting image co-registration.

2.4.3 Pre-processing and interferogram generation

After co-registration, the complex interferogram is formed by multiplying each complex pixel of the first (master) image by the complex conjugate of the same pixel in the second (slave) image, as discussed in Section 2.2.1. The resultant interferogram contains both amplitude and phase information. Fringes depicting phase difference are at this stage a measure of the combined surface elevation, displacement in the range displacement, and error terms (over the repeat observation interval).

In the interferogram, a phase shift of one wavelength (e.g., 5.6 cm for the ERS C-band SARs) corresponds to half-a-wavelength range displacement (owing to the two-way journey of the radar signal). Assuming no or minimal surface relief (or zero-baseline), each fringe can be converted to a relative horizontal surface displacement (velocity) of $2.8/(\Delta T \sin\theta)\pi$ cm day^{-1} in the range direction, where ΔT is the time separation between image acquisitions (the temporal baseline) in days and θ is the radar look angle (Goldstein et al., 1993; Kwok and Fahnestock, 1996). In reality, such cases are an exception to the rule, however, and topography must be separated from displacement. The technique of interferogram *double-differencing*, introduced on p. 52, to carry this out will be explained in Section 2.5.1.2.

After multiplication, the resultant interferometric phase products—i.e., the phase and coherence images—are typically multi-looked. By this process, the values of the phase/coherence of neighboring pixels within a window of fixed dimensions are averaged. Multi-looking is an important step in facilitating the subsequent phase unwrapping, in that it reduces image speckle (which degrades the phase measurement—see Section 2.3.7) (Joughin and Winebrenner, 1994; Wegmüller et al., 2002). Multi-looking results not only in an improved SNR but also increased computational efficiency by creating more manageable datasets. This is an important step, given the size of SAR datasets acquired at a very high spatial resolution—e.g., five-look ERS interferograms. The major disadvantage of multi-looking is that it results in degraded spatial resolution. Another limitation relates to possible under-sampling in cases where relatively high-phase gradients are present. As such, multi-looking is more appropriately applied to relatively smooth phase surfaces—e.g., ice sheet interior regions (Wegmüller et al., 2002).

Filtering is also typically carried out to reduce phase noise while maintaining spatial sampling, again to optimize coherence (Cumming and Zhang, 1999; Kwok and Fahnestock, 1996; Schwäbisch and Geudtner, 1995; Wegmüller et al., 2002). When successful, filtering also results in a large reduction in the number of residues (see below), which can significantly reduce the complexity of phase unwrapping and increase its efficiency (Wegmüller et al., 2002). Limitations include unwanted effects in the case of high-phase gradients—i.e., loss of fringes—and phase discontinuities (e.g., layover) potentially leading to errors in phase unwrapping (Wegmüller et al., 2002).

Another important product is a corresponding map of interferometric coherence, defined as the amplitude of complex coherence γ (see Sections 2.2.1 and 2.3.8). This provides a fundamental measure of the "exploitability" of the interferogram for a given application—i.e., it is a measure of the quality of the phase correlation or coherence between the two constituent InSAR images. Further information on interferogram generation is given in the references in Section 2.2 and Geudtner (1996) and Vachon et al. (1995a). Specific techniques and issues relating to InSAR processing of ScanSAR data are discussed by Bamler et al. (1998).

Compared with mapping a small region containing ground-control points, the production of a large mosaic map composed of multiple scenes poses significant challenges for InSAR processing. Datasets comprising large sections of an orbital strip must be processed as a homogeneous unit to minimize discontinuities in phase, radiometry, or sampling grid (Eineder et al., 1999). Madsen et al. (1999) and ESA (2001) discuss issues relating to the processing of long datastrips comprising ERS SAR scene mosaics, and interferometric adjustment. Figure 2.12 (see color section) is an example of continental-scale mapping by the VECTRA Greenland Mapping Project (Greenmap), a collaborative project involving groups from Denmark, the UK, the USA, Germany, and ESA (⟨*http://www.emi.dtu.dk/research/projects/ greenlandmapping/Research.html*⟩) using ERS SAR data. The latter are limited to scenes of only ~100 × 100 km in size. Such projects are playing a key role not only in improving our understanding of the complexities of ice sheet InSAR but also in developing enhanced processing techniques (e.g., Mohr and Madsen, 1999, 2000). The hope is that the increased availability of ScanSAR InSAR data will greatly facilitate basin-wide and even ice sheet mapping for interferometry. Note that the example in Figure 2.12 also shows the relationship between interferometric correlation, which exhibits a wide spatial variability across Greenland, and resultant maps of interferogram phase.

2.4.4 Phase unwrapping

As previously discussed, a key issue in SAR interferometry is that the phase information in the interferogram, which relates directly to surface topography (Graham, 1974), is only known to *modulo* 2π ($360°$). If the phase variation exceeds $360°$, it wraps around to $0°$. As an example, if one rotation of phase ($360°$) represents a 40-m change in height across the image, then it is impossible to differentiate between a height of, for example, 45 m and 85 m in the interferogram since both elevations will have the same phase value. Consequently, it is necessary to add the correct integer number of phase cycles to each interferometric phase measurement in order to determine the actual elevation at each point. Accurate phase unwrapping is also required for ice sheet velocity interferograms. The process of solving this 2π ambiguity and recovering the continuous phase information from the discrete wrapped phase is known as *phase unwrapping*.

Although phase unwrapping is the core technique enabling the derivation of useful geophysical quantities from InSAR data (Gabriel et al., 1989; Zebker and Goldstein, 1986), it remains one of the most problematical areas in interferometric processing, and a standardized procedure has yet to be developed. Indeed, it is a major research topic. In an idealized case where the phase field is well-behaved and smooth, all phase differences between adjacent interferogram samples lie between $+\pi$ and $-\pi$, and no discontinuities (local errors in the measured phase) exist in the phase data. Under these circumstances, the unwrapped phase can simply be obtained by a path-independent integration of the phase gradients (phase differences of adjacent wrapped phases) over the entire dataset, starting from a reference location and with the assumption that all phase differences are in the interval $(-\pi, +\pi)$ (Werner et al., 2002). In reality, such an approach typically fails as the phase can often step outside this interval—i.e., local phase gradients of $>\pm\pi$ can be present (Huntley, 1989). This occurs for a number of reasons. Primarily, phase noise is generally present (please refer back to Section 2.3), and can conspire with phase aliasing and phase discontinuities to complicate or even destroy the ability to unwrap the phase correctly (Carballo and Fieguth, 2002). Phase aliasing results from an insufficient sampling rate, defined by the *Nyquist rate*[12] (Gens, 2003; Spagnolini, 1993). The phase is locally undersampled when the phase gradient is greater than half a fringe (phase cycle) per sample (Werner et al., 2002). Note that undersampling already occurs in the presence of phase noise—i.e., at lower gradients. Phase discontinuities, or local errors in the measured phase, are typically present in areas where discontinuities in the interferometric phase occur due to SAR image layover (see Section 2.3.8) or discontinuous surface deformation. The latter commonly occurs, for example, at glacier–bedrock interfaces (Werner et al., 2002; Young and Hyland, 2002). As a result of these factors, the relationship between discretely measured wrapped phase values and the geometric information is highly nonlinear, and phase unwrapping is thus a nonlinear inverse problem.

The process of phase unwrapping involves searching for the correct integer number of phase cycles that need to be added to each phase measurement in order to obtain the correct slant range distance (between the antenna and the surface scattering center). Once again, although a large variety of algorithms have been put forward to solve the two-dimensional phase-unwrapping problem, and with differing levels of complexity, no consensus has so far been reached as to the optimal approach. An in-depth evaluation and comparison of these algorithms is beyond the scope of this book. Rather, in this section, we present basic background and comparative information on the major algorithm classes available and under development. For more detailed information on phase unwrapping, please consult Gens (2003), who provides a welcome overview of the major developments and status quo of this complex and potentially confusing field. The reader is also referred to Ghiglia and Pritt (1998) and Robinson (1993) for excellent introductions to phase unwrapping and the basic terms used. Following these and other texts—e.g., Baldi (2001)—phase-unwrapping algorithms are classified into two broad categories, namely:

- *path-following methods*, which use a local approach to solving the phase-unwrapping problem; and
- *minimum-norm methods*, which are based on a path-independent or "global" approach.

---

[12] The reciprocal of the *Nyquist interval*—i.e., the minimum theoretical sampling rate required to fully describe a given signal—enabling its faithful reconstruction from the samples.

An important factor common to a large proportion of phase-unwrapping algorithms, and leading to their improved performance, has been the incorporation of weighting factors (quality maps). These are arrays of values that define the quality of the phase data on the pixel level (Gens, 2003). In order to apply the phase-unwrapping techniques to a given dataset in an optimal fashion, a detailed knowledge of the performance of each method is required (Chen and Zebker, 2000). Please refer to Gens (2003), Ghiglia and Pritt (1998), Hartl and Wu (1993), Rosen et al. (2000), and Wegmüller et al. (2002) for an in-depth evaluation of the constituent algorithms outlined below, and Fornaro et al. (1997) for a detailed analysis of the links between local and global phase-unwrapping algorithms.

2.4.4.1 Path-following techniques

Path-following methods integrate wrapped-phase differences over a path that covers the entire phase dataset. For phase discontinuity, the sum of the phase differences in a clockwise direction about a set of four adjacent points indicates a residue of negative or positive polarity. These residues (local errors in the measured phase caused by signal noise or actual discontinuities) are connected by so-called *branch cuts*, which are implicitly or explicitly generated by the algorithm to prevent any integration (unwrapping) path from crossing the branch cuts or lines (Gens, 2003; Goldstein et al., 1988). In other words, an attempt is made to locate phase discontinuities of $> \pm \pi$, and to connect and mark them with branch cuts in order that the phase is not integrated across the discontinuities (Goldstein et al., 1988). As such, residues and branch cuts are a key component of path-following methods. The classical path-following algorithm is *Goldstein's branch cut algorithm* (Goldstein et al., 1988). This method defines branch cuts between all detected residues preventing any integration path from crossing these cuts. All residues need to be balanced, either by a residue of opposite polarity or by a branch cut connected to the image border. The approach minimizes the sum of the branch cut lengths. This algorithm has the advantage of being computationally fast and requiring little computer memory. It entails an efficient and robust unwrapping solution that works well for images characterized by high-phase correlation (Werner et al., 2002). Its major weakness relates to the lack of weighting factors for guiding the placement of branch cuts, resulting in poor performance in regions with a high density of branch cuts and of low coherence. Improvements to this method have been suggested by Ghiglia and Pritt (1998), Goldstein and Werner (1998), and Huntley (1989).

An alternative approach is the *minimum-discontinuity algorithm* of Flynn (1997). This aims to find a solution that minimizes the phase discontinuities by adding multiples of 2π to the phase values enclosed by loops formed by paths detected by a tree-growing approach (i.e., with branches forming tree-like networks) that traces paths of discontinuity. This is achieved at the expense of high computational requirements. Uniquely, this algorithm operates with or without weighting factors. The minimum discontinuity algorithm is an adaptation of the *fringe-line detection technique* of Lin et al. (1992, 1994), which attempts to locate fringes by using edge detection methods. In practical terms, this algorithm can only be applied where the SNR is high and the fringes are well-separated. Unfortunately, the least squares curve-fitting technique employed tends to fail when closely spaced fringes are present.

A different approach to defining branch cuts is provided by the *Minimum Cost Flow (MCF) networks* algorithm developed by Costantini (1998). This method has emerged from the demand for the optimized and automated unwrapping of disconnected areas of high coherence (Werner et al., 2002). It exploits the fact that the phase differences of neighboring pixels can be estimated with a potential error that is an integer multiple of 2π to formulate the phase-unwrapping problem as a global minimization problem with integer variables (Ahuja et al., 1993; Gens, 2003). In other words, it aims to locate a phase-unwrapping path through the interferogram that minimizes the number of global phase-unwrapping errors. A growing interest in phase-unwrapping methods using this relatively accurate and computationally efficient approach has stimulated a large number of studies on potential weighting factors in order to define the cost function used for this specific approach—e.g., Wilkinson (1997); Eineder et al. (1998); Carballo and Fieguth (1999, 2000, 2002); and Refice et al. (1999). Please see Gens (2003) for further details on this and other sources from which weighting factors can be derived. Other path-following methods include the *minimum spanning tree algorithm* (Chen and Zebker, 2000, 2002; Ching et al., 1992). This adapts the Goldstein et al. (1988) algorithm and approximates a *single minimum Steiner tree* (see Chen and Zebker, 2000 for in-depth information). Note that path-following algorithms are sometimes sub-divided in the literature into three sub-classes—namely, (i) *residue compensation methods*; (ii) *path-dependent methods*; and (iii) *quality-guided path methods*. Once again, a full explanation is beyond the scope of this chapter, and the reader is referred to Ghiglia and Pritt (1998), Chen et al. (2002), Herráez et al. (2002), and Baldi (2001).

2.4.4.2 Minimum-norm techniques

Minimum-norm methods differ from path-following methods in that they approach the solution of the phase-unwrapping problem in a mathematically formal manner, and are based on a global approach. While robust, algorithms in this class are computationally expensive (Baldi, 2001; Hung and Yamada, 1998). An example of a global algorithm is the *least-squares method*, which iteratively estimates and improves a best fit of an unwrapped phase model to a local wrapped phase image sample (Pritt, 1996; Zebker and Lu, 1997). Using this technique in both its weighted and unweighted forms, phase-unwrapping problems are solved in an efficient manner by discretized partial differential equations (Gens, 2003). Least-squares methods differ from path-following methods in that they obtain solutions by integrating through the residues to minimize the gradient differences. Various techniques have been developed to solve the unweighted form, including *Discrete Cosine Transforms* or DCTs (Ghiglia and Romero, 1994), *Fast Fourier Transforms* or FFTs (Pritt and Shipman, 1994), and the *unweighted multi-grid technique* (Pritt, 1996). The solution for the weighted form of least-squares phase unwrapping is more complex, however, as it requires iterative methods (Ghiglia and Romero, 1994; Pritt, 1996). While the accuracy depends on the choice of weighting mask, computational efficiency decreases for the weighted compared with the unweighted forms. Please see Gens (2003) and Ghiglia and Pritt (1998) for details.

An alternative global phase-unwrapping algorithm is given by *minimum L^p-norm phase unwrapping* (Ghiglia and Romero, 1996; Chen and Zebker, 2000). Here, L is the difference between a pair of corresponding unwrapped and wrapped phase gradients, and p is the power to which L must be raised in order to establish an error matrix (Chen and Zebker, 2000; Gens, 2003). This method is a generalization of the weighted least squares approach, and requires the solution of a nonlinear partial differential equation which is implemented in an iterative scheme. It is computationally very intensive. Other global phase-unwrapping techniques include an algorithm based on Green's formulation (Fornaro et al., 1996a). This algorithm is computationally efficient given its use of FFT techniques within the iteration process (Gens, 2003), and is mathematically equivalent to the least squares solution (Fornaro and Sansosti, 1999; Fornaro et al., 1996b).

The number and variety of algorithms outlined above, and present in the literature, is testament to the difficulty and complexity of phase unwrapping. It remains a difficult and challenging problem (Chen and Zebker, 2002; Gens, 2003; Massonnet et al., 1996), and the focus of intensive ongoing research. Although significant progress has been made, a number of issues remain to be resolved. For example, research on the definition of optimal weighting factors is still in its infancy. Another factor relates to the shear size of datasets. While computing capabilities have improved dramatically, so has the volume of SAR data to be processed. The ever-expanding size of SAR datasets for InSAR processing necessitates the development of algorithms that are both robust and ever-more computationally efficient. Recent research has indeed focused on developing such algorithms—e.g., Chen and Zebker (2001, 2002). A significant amount of human interaction is still required, however, depending on the characteristics and size of the dataset being processed (Eineder et al., 1999). An ongoing aim is to minimize this intervention with a view to eventually creating a truly operational interferometric processing system (Gens, 2003). Given these and other factors laid out above and in the literature, it has become apparent that the optimal approach at present may be to combine phase-unwrapping algorithms in order to take advantage of their relative attributes (Chen and Zebker, 2000; Ghiglia and Pritt, 1998; Mohr, 1997).

It should be noted that the success of a given phase-unwrapping procedure also depends on the quality of the interferogram. While phase unwrapping may be successful in areas of high coherence, the extraction of accurate topography can be very difficult where decorrelation or layover contaminates the phase. Eineder and Suchandt (2003) propose a means of improving InSAR phase unwrapping and DEM reconstruction by recovering radar shadow in mountainous regions. Phase-unwrapping algorithms can also often leave gaps, within which they cannot determine the actual phase. Issues related to the unwrapping of surface displacement (change) signals are discussed by Gudmundsson et al. (2002) and examined in

Section 2.5.2. Once again, a trade-off generally exists between computational requirements and accuracy of solution.

In summary, errors in phase unwrapping lead directly to errors in InSAR estimates of both ice elevation and velocity. Such errors most commonly occur where strong phase gradients occur, such as at shear margins of ice streams or at bull's-eye patterns (Joughin et al., 1998a). These gradients, and thus phase-unwrapping errors, can be reduced, and interferometric coherence (correlation) increased, by shortening the temporal baseline. Moreover, phase gradients related to vertical displacement can be minimized by increasing the radar incidence angle. Coherence can also be increased by decreasing the interferometric baseline and/or by increasing the range resolution (Zebker and Villasenor, 1992).

2.4.5 Phase-to-height conversion

Depending on the application and processing steps taken, the interferogram now contains information either related to surface elevation and/or motion. Several methods are available for converting phases into heights (Bürgmann et al., 2000). These include the *normal baseline method*, the *baseline rotation method*, and the *integrated incidence angle method*. Most techniques require at least three ground-control points within the area of the interferogram, and knowledge of their absolute phase and position in the image, in order to derive absolute measurements. This stage also involves adjustment of the range coordinate in the elevation computations, again on a pixel-by-pixel basis, to compensate for surface relief displacement effects. The importance of tiepoints, and the steps required to convert phase to ice motion for the case of motion-only interferograms is discussed in Section 2.5.

2.4.6 Interferogram geo-coding

At this stage, the geometrically calibrated interferogram is in *slant range* geometry—i.e., it consists of arrays of pixels fixed into a geometry corresponding to the acquisition parameters of the satellite. For the data to be used quantitatively, they must be *geo-coded* (geo-referenced)—i.e., transformed into object space (Geudtner and Schwäbisch, 1996). This process involves assigning pixels within a given image with ground coordinates. The act of geo-coding and subsequent transforming into a map projection puts the image into *ground range*, and creates a raster-based DEM. This both enables and facilitates the co-registration of interferometric results with data from other sources—e.g., GPS and airborne laser altimetry—and also the correction of radar artifacts such as image foreshortening and layover in regions of steep terrain (as discussed in Section 2.3.9; Madsen and Zebker, 1998). A variety of techniques have been put forward for interferogram geo-coding, which are evaluated by Hellwich (1999a). For detailed descriptions of geo-coding methods in general, please see Lillesand and Kiefer (2000), Richards and Jia (2005), and Williams (2001).

2.5 THE MEASUREMENT OF ICE-SHEET ELEVATION AND SURFACE VELOCITY USING DIFFERENTIAL INSAR TECHNIQUES

In this section, we draw upon the work of Kwok and Fahnestock (1996) and Joughin et al. (1996b, c) to examine the retrieval of ice-sheet elevation and surface motion by InSAR *double-differencing*. As stated previously, the double-differencing method has the advantage of providing both topographic and motion fields while not requiring any information apart from the SAR data themselves (unless absolute measurements are required). In addition to the limitations and constraints laid out in Section 2.3, two conditions must be satisfied for the technique to be successful. These are that:

i. the displacement gradient across a pixel must fall within a threshold value— i.e., changes must not be too large; and
ii. the root-mean-square (r.m.s.) position of surface scatterers within each pixel must remain constant to within a fraction (e.g., 10–20%) of the radar wavelength—i.e., radar-scattering characteristics within each pixel should remain similar in time.

Using a series of four repeat-pass ERS-1 SLC slant-range images spaced every 3 days and working in NE Greenland, Kwok and Fahnestock (1996) demonstrate that the topography is separable from the surface displacement field if a sequence of suitable correlated radar images is available (see also Joughin et al., 1999b). By the double-differencing method, two 3-day interferograms are first differenced to remove the phase contribution of the displacement field and obtain a topography-only interferogram. This topography-only interferogram can then be subtracted from the mixed interferograms to separate the displacement fringe patterns from those due to relief. A constraint for this approach is that at least one of the interferograms must have a large enough baseline to enable extraction of elevation (i.e., three-dimensional position) data with sufficient accuracy for geo-coding (Mohr, 1997).

Following Kwok and Fahnestock (1996), we begin with a brief review of notation used, and discuss ice-sheet motion in terms of displacement, position, and velocity. An ice particle on the surface at time t_0, and initially at position $\vec{x}$, has at some time later moved to a position $\vec{x}(t_i)$. Displacement $\vec{\delta}$ refers to the finite difference in the particle position at times t_i and t_j, with subscripts i and j referring to the satellite image acquisition epochs in Figure 2.3:

$$\vec{\delta} = \vec{x}(t_j) - \vec{x}(t_i) \tag{2.22}$$

Assuming that the displacement field is slowly varying spatially, it follows that the average particle velocity $\vec{\nu}$ over the time interval $\Delta T = t_j - t_i$ is:

$$\vec{\nu} = \vec{\delta}/\Delta T \tag{2.23}$$

In reality, the flow within outlet glacier and ice sheet margins can and does change with time (Reeh et al., 2003). In Greenland, for example, higher velocities are observed in summer than in winter. Moreover, glacier surges periodically occur in both hemispheres (Fatland and Lingle, 1998; Frolich and Doake, 1998; Joughin et al., 1996a; Reeh et al., 1994). However, in the absence of any additional information, the underlying assumption of invariant ice flow over the relatively short interval between image acquisitions is reasonable for ice-sheet inland regions at least, and particularly where short temporal baselines can be achieved. A major advantage of InSAR is that it does not rely on the presence and identification of surface features, as is the case with the conventional feature-tracking techniques discussed in Chapter 3. As such, it is well-suited to use in relatively featureless terrain—e.g., ice-sheet inland regions—if suitable imagery is available. Please see Fatland and Lingle (1998), Joughin et al. (1996c), Mohr and Madsen (1999), and Mohr et al. (2003) for additional information on the impact of non-steady-state flow on velocity retrieval accuracy.

We again refer to the geometry and coordinate system for repeat-pass satellite interferometry presented in Figure 2.3. For a given pair of repeat observations of the same scene, from the ith and jth epochs, with SAR look angle θ (e.g., 23° for ERS-1 and -2), separated by the baseline B_{ij}, and assuming that phase decorrelation effects are minimal, the phase difference at each sample (pixel) is:

$$\Delta\phi_{ij} = \frac{4\pi}{\lambda} B_{ij} \sin(\theta - \alpha_{ij}) \qquad (2.24)$$

where λ is the radar wavelength and α_{ij} the baseline tilt relative to the horizontal.

Following Kwok and Fahnestock (1996), the phase convention used in complex notation is $e^{-j\phi\Delta}$, and the negative sign is omitted in the equations. Moreover, the difference in pathlength in the slant range direction ($|\vec{r}_i| - |\vec{r}_j|$) is approximated by $\vec{B} \cdot \vec{r}$, which contains phase contributions from the orientation of the Earth's surface relative to the baseline. It follows that for a displacement of the scatterers of $\vec{v}\Delta T \cdot \vec{r}$ in the direction of the radar LOS, the observed phase will include a displacement contribution of $\frac{4\pi}{\lambda}\vec{v}\Delta T \cdot \vec{r}$, or:

$$\Delta\phi_{ij} = \frac{4\pi}{\lambda} B_{ij} \sin(\theta - \alpha_{ij}) + \frac{4\pi}{\lambda}\Delta\rho_{ij} + \Phi_{ij} = \phi_{top} + \phi_{displ} \qquad (2.25)$$

where Φ_{ij} is phase noise (from sources discussed in Section 2.3 combined); $\Delta\rho_{ij}$ is an abbreviation of $\vec{v}\Delta T \cdot \vec{r}$; and ϕ_{top} and ϕ_{displ} are displacement due to surface topography and motion, respectively. While this additional term is independent of the spatial baseline, it is dependent on the time separation between image acquisitions, or *temporal baseline*, ΔT. For practical purposes, ε_{ij} is typically taken to be negligible.

From (2.25) and for repeat-pass satellite interferometry, it can be seen that $\Delta\phi_{ij}$ contains contributions from both topography (ϕ_{top}) and any displacement of the target scatterers (ϕ_{displ}) toward or away from the radar LOS direction (slant range) over ΔT (Joughin et al., 1996c). Assuming that error terms are negligible, the resultant interferogram therefore represents a measure of both the surface topography and displacement. As a result and recalling earlier discussion, the retrieval of the true ice motion field requires (i) the removal of the contribution of topography to the measured phase of each pixel, and (ii) additional information to derive the two or three components of the ice-flow velocity vector. Converting relative to absolute motion requires additional information.

2.5.1 Ice-sheet topography from DInSAR

Separation of motion from topography using SAR data alone—i.e., in the absence of an independent DEM—is achieved by the double-differencing technique described above, and pioneered by Gabriel et al. (1989). Again following Kwok and Fahnestock (1996), the procedure involves (i) scaling the baselines (to normalize each interferogram such that the phase measurements are on identical scales), and (ii) differencing the two interferograms to detect small changes in the fringe patterns. Scaling is essential because the interferograms have different baselines, and is typically carried out prior to unwrapping (Wegmüller and Strozzi, 1998).

The phase from two interferograms from three passes (images 1, 2, and 3) is:

$$\Delta\phi_{12} = \frac{4\pi}{\lambda} B_{12} \sin(\theta - \alpha_{12}) + \frac{4\pi}{\lambda}\Delta\rho \tag{2.26}$$

and

$$\Delta\phi_{23} = \frac{4\pi}{\lambda} B_{23} \sin(\theta - \alpha_{23}) + \frac{4\pi}{\lambda}\Delta\rho \tag{2.27}$$

where subscripts 1 through 3 refer to images 1 through 3. If $\Delta\rho$ remains constant over the interval between observations, then differencing the two interferograms yields a single interferogram with the properties:

$$\Delta\phi'_{13} = \Delta\phi_{12} - \Delta\phi_{23} \tag{2.28}$$

$$= \frac{4\pi}{\lambda}(B_{12}\sin(\theta - \alpha_{12}) - B_{23}\sin(\theta - \alpha_{23})) \tag{2.29}$$

In this fashion, the contribution of the displacement field to the phase of the resultant interferogram, $\Delta\phi'_{13}$, has been removed.

The next step is to resolve the individual baselines (B_{12}, B_{23}) into horizontal and vertical components, B^y and B^z, respectively, giving:

$$\Delta\phi'_{13} = \frac{4\pi}{\lambda}((B^y_{12} - B^y_{23})\sin\theta - (B^z_{12} - B^z_{23})\cos\theta) \tag{2.30}$$

$$= \frac{4\pi}{\lambda}(B'^y_{13}\sin\theta - B'^z_{13}\cos\theta) \tag{2.31}$$

$$= \frac{4\pi}{\lambda}B'_{13}\sin(\theta - \alpha'_{13}) \tag{2.32}$$

In effect, this is equivalent to a topography-only interferogram with the baseline formed between image acquisition point 1 and point 3—i.e., the interferometric sensitivity to the variation in ice sheet topography now depends on B'_{13}.

As the interferometric phase is measured only to *modulo* 2π, differencing the interferograms necessitates that the phase be *unwrapped* to remove this ambiguity, as discussed in Section 2.4.4. This is carried out prior to the differencing (Kwok and

Fahnestock, 1996). Since the phase constant is only known approximately after unwrapping (from the approximate baseline), only relative changes in surface relief can be derived without the presence of known reference targets within the interferogram. Absolute surface elevation can only be retrieved if tie-points of known elevation (but not necessarily known velocity) are present, and can be identified, in the interferogram (Mohr, 1997; Sun et al., 2000; Zebker et al., 1994a), or by fitting the data to an existing DEM (Joughin et al., 1996c). Moreover, baseline errors can be minimized by incorporating independent ground-control points within the interferogram (Joughin et al., 1998a; Zebker et al., 1994). As discussed earlier, tie-points can take the form of GPS measurements on the ice (Keller et al., 1997), or points on bedrock outcrops, and InSAR retrieval accuracy varies with the quality of the tie-points used. Even minor errors in baseline can result in long-wavelength errors in the interferometric elevation field.

An example is given in Figure 2.13a (see color section). The resultant DEM is characterized by high-frequency spatial features superimposed on the smooth terrain. These features were not visible in the original SAR intensity images, underlining the value of interferometry as a means of mapping subtle variability in surface roughness characteristics in great detail and over large and often relatively featureless areas. As illustrated in Figure 2.13a–b, the illuminated topographic image is comparable with an enhanced AVHRR image produced using the technique described in Scambos et al. (1999) but unlimited by cloud and/or darkness. Kwok and Fahnestock (1996) also note the existence of long-wavelength residuals in the derived topography, which they suggest are inherent to interferograms even where the baseline is accurately known.

An example of what can be achieved is given in Figure 2.14. Taken from Joughin et al. (1996b), this shows the stunning result of fitting a high-resolution swath of InSAR-derived elevation data to a DEM derived from radar altimetry. For this particular case, comparison of InSAR-derived elevations with airborne laser altimetry shows the overall accuracy (to 1σ) of the former to be ~ 2 m. Information on the relative strengths and weaknesses of InSAR-derived ice sheet elevation retrievals versus those from other satellite techniques is given in Chapter 3.

2.5.2 Retrieval of ice-sheet surface motion

In this section, we outline the further steps necessary to retrieve maps of the surface motion of ice sheets from InSAR data using the double-differencing technique. Following on from the previous section, the topography-only interferogram, $\Delta\phi'_{13}$, is subtracted from the mixed interferogram comprising the first and third images (i.e., with a 6-day separation in Kwok and Fahnestock, 1996), after phase unwrapping:

$$\Delta\phi_{13} - k\Delta\phi'_{13} = \frac{4\pi}{\lambda}\Delta\rho \qquad (2.33)$$

where k is the factor scaling the B'_{13} fringe pattern into the B_{13} interferogram (see Kwok and Fahnestock, 1996 for a further explanation). Scaling is essential to

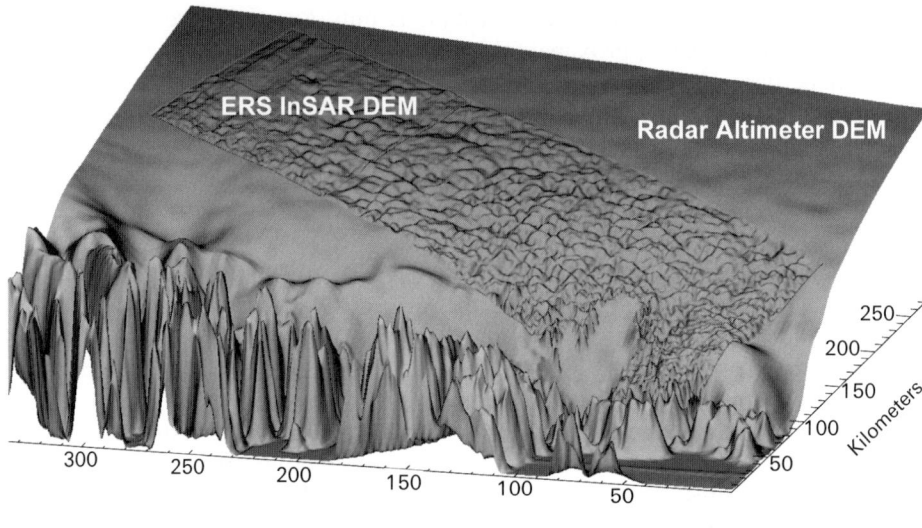

Figure 2.14. A 3-D shaded relief representation of the surface elevation of part of West Greenland, showing a swath derived from ERS InSAR data superimposed on a smoother ice sheet surface derived from pulse-limited radar altimetry. While the vertical precision of radar altimeters is greater than that achievable by InSAR (at least over lower sloping ice-sheet inland regions), the horizontal resolution of InSAR is significantly finer. This enables derivation of much greater detail in surface elevation, as shown in this image. A comparison of the strengths and weaknesses of different satellite techniques of measuring ice-sheet elevation is given in Chapter 3.

From Joughin et al. (1996b). Copyright International Glaciological Society 1996, reproduced with permission.

compensate for the fact that the sensitivity to topographic variation in the two interferometric combinations generally differs, as the baselines are rarely the same. As a result, the phase differences in the topographic pair must be scaled to match the frequency of variability in the topography-change pair.

As with the computation of topography, the phase differences, which are typically displayed in the interferogram as fringes with a periodicity of 2π, must be unwrapped to generate a continuous-phase field (Goldstein et al., 1988). The ice-surface motion field at this stage is relative only—i.e., it represents a direct measure of the velocity-gradient field, and in the direction of the sensor LOS only. Once again, fixed (stationary) points within the fringe pattern, such as rock outcrops or nunataks, are beneficial as a means of adjusting the field to an absolute datum and improving retrieval accuracy (Kwok and Fahnestock, 1996). In the absence of such points, then coincident velocity measurements at points on the surface—e.g., by GPS—can be used to control the velocity field. Joughin et al. (1996c) showed that baseline estimates derived from GPS tie-points on a moving ice sheet are less accurate than those from bedrock, making the velocity error larger. At least one velocity control point is required to convert the unwrapped phase data to a two-dimensional velocity field (Bindschadler, 1998; Goldstein et al., 1993). Additional control points are desirable, however, to both counteract unwrapping ambiguities due to noise in the phase field and to account for uncertainty in the baseline. Rosen et al. (2000) state that four such points are generally sufficient to derive a velocity map covering tens of thousands of square kilometers in this fashion. Fortunately, tie-points from GPS measurements (if present), for example, need not necessarily be identified in the images, due to the use of navigation data with absolute positions synchronized to the SAR data (Mohr, 1997). In an interferogram, a location can be calculated by combining knowledge of the absolute tie-point positions, the processing reference line, and the processing Doppler information.

Note that the ice velocity field can only be calculated by InSAR in regions where the phase is continually "unwrappable" from a given control point. This is a major disadvantage of InSAR in that large regions of the ice sheet can be characterized by low-phase coherence in currently available SAR images. This is due, for example, to areas of large differential ice velocity, shear margins, and low SNRs, and potentially impedes the measurement of ice sheet motion over large areas (Forster et al., 2003). As we shall see in Section 2.6, variations on this technique, and alternative techniques, have been developed in an effort to overcome this deficiency.

Unfortunately, absolute datum points are often unavailable, particularly in the vast and remote interior regions of the Greenland and Antarctic Ice Sheets. Under these circumstances, maps of balance velocities derived from satellite altimetric DEMS (e.g., Ekholm, 1996) can provide an alternative, though less accurate, source of velocity tiepoints in regions of slow flow (Bamber et al., 2000; Joughin et al., 1998b, 2001). Balance velocities are defined as the depth-averaged ice velocities required to maintain the steady-state shape of an ice sheet (Budd and Warner, 1996; Joughin et al., 1997), and are estimated from accumulation, ice thickness, and surface slope information (see Chapter 3 for more in-depth information). Joughin et al. (1998b) were able to achieve an ice velocity accuracy of $\sim$3 m per annum using this method, and showed that balance velocity error impacts can be minimized by selecting the control points from slow-moving regions.

Extensive work has been carried out in Greenland using satellite and aircraft InSAR data combined with GPS and laser altimetry to study ice sheet elevation and dynamics, particularly by groups from Denmark (Mohr and Madsen, 1996a) and NASA (Krabill et al., 2000). Such studies have provided very accurate tie-points with which to "calibrate" and register InSAR output—e.g., GPS measurements along the 2000-m contour to provide absolute measurements—as part of the Program for Arctic Regional Climate Assessment (PARCA) project (Thomas, 2004) and in the summit region (Hvidberg et al., 1997). Joughin et al. (1996a, b) noted some difficulty, however, in obtaining accurate tie-points from DEMs in steep coastal margins where the DEM is derived from satellite radar altimeter data. Data from the GLAS instrument onboard ICESat yield improved measurements (compared with radar altimetry) over such regions (Zwally et al., 2002)—see Chapter 3. These data are also likely to have a large impact in Antarctica, where relatively few aircraft and *in situ* measurements are available. Similar effects in West Antarctica are documented by Thomas et al. (2004).

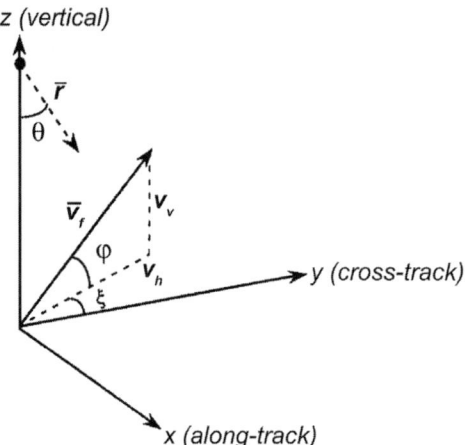

Figure 2.15. Schematic of the geometry for the measurement of ice-sheet surface velocity from InSAR, where $\bar{r}$ is the radar line-of-sight, φ is the ice-surface slope in the direction of ice flow, $\bar{V}_f$ is the ice velocity in the direction of flow, and V_h and V_v are the horizontal and vertical components of the ice surface velocity, respectively.

After Kwok and Fahnestock (1996) and Forster et al. (2003). Copyright IEEE 1996, reproduced with permission.

A general consideration, noted earlier, is that baseline estimates are relatively insensitive to the ground-control height error for short baselines—i.e., <50 m. This enables accurate velocity estimates to be made even though the ground control may be poor (Goldstein et al., 1993; Joughin et al., 1996c). It should be noted that certain studies do not require knowledge of absolute values—e.g., strain rates can be computed from relative motion fields (Forster et al., 2003; Hyland and Young, 2002; Young and Hyland, 2002)—see Chapter 3. A further complication for floating ice masses, such as ice shelves and glacier tongues, is that InSAR-derived velocities contain a significant vertical component due to tidal flexure. The removal of this component is covered in the next section (2.5.3).

Kwok and Fahnestock (1996) and Forster et al. (2003) adopted the following approach to estimating the ice-surface velocity in the direction of ice flow from InSAR data. On an ice sheet, where surface relief is present, the range (line-of-site) displacement or velocity measured by InSAR is a function of the radar look angle θ, the surface slope ψ, and the angle (in the x–y-plane) formed between the look direction projected onto the x–y-plane and the ice flow directions ξ. This is depicted schematically in Figure 2.15 (please see Joughin et al., 1995 and Forster et al., 2003 for further details). From this geometry, the ice-sheet surface velocity in the radar LOS (slant range) direction V_r only is given by:

$$V_r = \frac{\phi \lambda}{4\pi \Delta T} + V_0 \sin\theta \cos\xi \qquad (2.34)$$

where ϕ is the interferometric phase due to ice displacement; λ is the radar wavelength; ΔT is the time interval between image acquisitions (temporal baseline); and

V_0 is the spatially constant component of the surface velocity. Recalling earlier discussion, the first term is based on the satellite-derived phase difference due to ice-surface displacement between image acquisitions, and represents the spatial variability in velocity. Note once more that this is only a relative rather than absolute velocity, as it includes no spatially constant velocity. The second term is analogous to a baseline velocity, with the sine and cosine functions projecting the constant ice surface velocity to the radar LOS (Forster et al., 2003).

The ice surface velocity in the direction of actual flow V_f can then be written (Forster et al., 2003; Kwok and Fahnestock, 1996):

$$V_f = \frac{V_r}{\cos\xi \cos\varphi \sin\theta - \cos\theta \sin\varphi} \quad (2.35)$$

As the measurement of velocity requires observations at two or more points (i.e., from two or more look directions), the radar look angle θ and the angle between the radar look and actual ice flow direction ξ change with one measurement of V_f. Changes are negligible between samples for high-resolution data, however, and the angles θ and ξ are assumed to remain constant over spatial (measurement) intervals of <100 m (Forster et al., 2003).

In the above relationships, the surface slope is taken to be in the direction of the flow, with the implicit assumption that the flow vector is tangential to the surface topography. Moreover, $V_v = V_f \sin\varphi$ and $V_h = V_f \sin\varphi$ are the vertical and horizontal components, respectively, of the velocity vector in the flow direction (Figure 2.15), and both are present in the InSAR-derived ice motion. The fact that a repeat-pass interferometer is sensitive to surface displacement directed along the LOS from the radar to the ground only (Joughin et al., 1998a) has the effect of emphasizing vertical relative to horizontal displacement (Joughin et al., 1995). As a result, it is not possible to unambiguously separate the mixed vertical and horizontal displacement signals in an interferogram without additional information or assumptions (Joughin et al., 1998a). After Kwok and Fahnestock (1996), the relative contribution of these two terms (V_v and V_h) to V_f is given by their ratio:

$$R = \cot\varphi \tan\theta \cos\xi \quad (2.36)$$

Kwok and Fahnestock (1996) note that the fractional contribution of the velocity components can be computed if the three angles (θ, φ, ξ) are known. In fact, θ is a system parameter, and φ can be computed from the differential InSAR-derived DEM $(\Delta\phi'_{13})$. The main error source is ambiguous knowledge of the flow direction. This is typically assumed to be parallel to ice stream margins, or in the direction of visible flowlines (Joughin et al., 1996c, 1998a; Mohr et al., 1998). Kwok and Fahnestock (1996) give the error as:

$$\frac{\partial R}{\partial \xi} = -\cot\varphi \tan\theta \sin\xi \quad (2.37)$$

As such, the error increases with ξ, because interferometry is less sensitive to horizontal motion as ξ approaches $\pi/2$. Even with excellent independent knowledge of the ice-flow direction, the accuracy of the resultant velocity estimate may be poor when this direction is close to that of the sensor (satellite) along-track direction—i.e., where there is little or no sensitivity to displacement (Rosen et al., 2000). Given these factors, the ability to obtain the full three-component flow vector from single satellite track data is limited, and requires another approach using additional SAR imagery or information.

It is apparent from the discussion above that a key limiting factor for repeat-pass InSAR acquiring LOS-only observations from along a single-orbit track is that it yields one velocity component only (Goldstein et al., 1993; Joughin et al., 1996c; Rignot, 1996). However, ice sheet/glacier mass balance assessments by flux divergence calculations require at least the two-dimensional horizontal surface velocity pattern to be known (Mohr and Madsen, 1996b; Reeh and Gundestrup, 1985). Moreover, knowledge of the full three-dimensional flow pattern is required for ice-dynamics studies and the validation of ice-flow models. As a result, additional steps are necessary to determine the vertical and horizontal components of the velocity vector. If the interferogram is constructed with repeat-pass data from one look direction only, the two-dimensional flow field is typically constructed by assuming that the ice is in steady state and that the horizontal flow component occurs in the direction of maximum-averaged downhill surface slope—i.e., over a horizontal scale of 10–20 ice thicknesses or several kilometers (Joughin et al., 1996c, 1998a; Madsen et al., 1999; Paterson, 1994). According to Joughin et al. (1998a), this results in an averaged flow direction that fails to resolve perturbations on scales of <1–10 km. With its relatively high degree of accuracy, the InSAR-derived DEM can be employed to define the direction of maximum slope if no other information is available. While this assumption holds for large regions of the ice-sheet interior (Paterson, 1994), where the velocity perpendicular to the slope is typically <1 m per annum (Bindschadler, 1998), it can break down around the margins. Here, the general ice flow direction can deviate quite significantly from the regional slope aspect (Bamber et al., 1999; Mohr and Madsen, 1996b). Under these circumstances, flowlines present in the SAR amplitude imagery or contemporary high-resolution (cloud-free) visible imagery can provide another proxy indication of regional ice-flow direction, albeit over limited areas due to the featureless nature of much of the ice sheets (Joughin et al., 1998a; Rosen et al., 2000).

The direct InSAR measurement of the full three-component flow velocity vector ideally requires SAR data from at least three different look directions. This has only been achieved once so far by current satellite systems—i.e., during a limited (6-day) period south-looking phase of the Radarsat-1 Antarctic Mapping Mission (Gray et al., 1998). Spaceborne SAR systems such as ERS have typically acquired interferometric data from a north-looking configuration. It is not possible, however, to obtain both south- and north-looking coverage at latitudes higher than $\sim 80°$ with these spacecraft. This situation will change in the near future with the launch of satellites such as Radarsat-2 (planned for 2006) with the ability to routinely switch look direction. Current satellite SAR systems can at best generally only offer two different look directions—i.e., from images acquired on ascending and descending orbits. If these data are available and well-correlated, then the 2-D flow pattern can be derived from the InSAR measurements from two different look directions

(Madsen et al., 1999; Mohr et al., 1998; Vachon et al., 1996). Such information is particularly valuable in regions where the direction of ice flow deviates from the regional slope aspect. Reeh et al. (2003) recommend the use of ascending orbit data only (at least for the ERS-1/-2 Tandem Mission) to construct the DEM for surface slope and velocity calculation, as the spatial baseline conditions are superior to those of descending orbit data. Please note that two-dimensional flow patterns can be directly derived from an interferometric pair by another technique called speckle tracking, although at a reduced resolution (see Section 2.6 for an explanation).

Given the lack of suitable polar InSAR data from three or more look directions, a technique has been developed by Joughin et al. (1998a) and Mohr et al. (1998) to derive all three components of the velocity vector using satellite data acquired from two look directions. This method exploits SAR data from both ascending and descending orbit tracks combined with surface slope information from an InSAR-derived DEM. Deriving the third velocity component—i.e., vertical velocity—from InSAR data acquired from only two look directions imposes a key assumption, noted above, that ice flow is parallel to the surface. By the surface-parallel flow assumption, the vertical velocity component is expressed in terms of the horizontal velocity vector.

An example map of InSAR-derived horizontal ice displacement derived in this fashion, and from NE Greenland, is given in Figure 2.16 (see color section). The corresponding map of ice flow vectors superimposed on InSAR-derived elevations is shown in Figure 2.17 (see color section). In this case, the smooth, large-scale variation apparent in the InSAR-derived flow field closely agrees with *in situ* measurements (carried out at GPS stakes numbered in the image). It does, however, often deviate from the local surface gradient. Mohr et al. (1998) make the point that this unusual behavior could not be predicted by glacier/ice-sheet dynamics models, illustrating the value of satellite-derived measurements of both flow direction and magnitude.

Inaccuracy in three-component velocity estimates results from a combination of errors in image co-registration (see Section 2.3.5), slope, the DEM used, baseline accuracy (Section 2.3.10), phase unwrapping (Section 2.4.4), and phase noise (Section 6.3) (Joughin et al., 1998a). Accurate estimates of the vertical and horizontal components of velocity necessitate accurate surface slope information, with slope errors contributing to inaccuracy in measurements of fine-scale details of the velocity field. Slopes derived from InSAR DEMs are also affected by other sources of phase errors with length-scales equivalent to topographic variation (Joughin et al., 1996b). Image speckle is a significant potential source of slope error, but can be largely eliminated by the filtering of interferometric DEMs, the latter enabled by the relatively smooth topography of ice sheets (Joughin et al., 1998a). In general, slope errors can be reduced by averaging multiple DEMs (if possible). As the sensitivity of velocity estimates to errors in the interferometric DEM is proportional to baseline length, the shorter the baseline the better. Joughin et al. (1998a) further elaborated upon the impact of long-wavelength errors in the InSAR DEM, which they suggest introduce errors of a few meters per annum in velocity estimates. Long-wavelength (20–100 km) phase errors of up to 10 radians have also been noted in the along-track direction in tandem ERS interferograms from Greenland by Joughin et al. (1996b). These are possibly related to clock drift or nonlinear variation to the baseline (Joughin et al., 1998a).

Strictly speaking, ice flow is generally inclined slightly downward in accumulation zones but upward in ablation zones (Joughin et al., 1998a; Paterson, 1994). This deviation from surface-parallel flow is termed the *submergence/emergence velocity*, which in steady state is equal to the local mass balance (Joughin et al., 1998a). While estimates of the horizontal components of motion are relatively unaffected by deviations from surface-parallel flow, such deviations may introduce significant errors into InSAR estimates of vertical motion. Reeh et al. (1999) further argue that while the assumption of surface-parallel flow holds for much of the time, it neglects two potentially important factors. These are: (i) the contribution of the local mass balance to the vertical velocity component, and (ii) a contribution arising if the ice sheet/glacier is not in steady state. Both factors can be significant. On Storstrømmen Glacier (NE Greenland), for example, Reeh et al. (1999) estimated non-steady-state vertical velocities to be up to 5 m per annum in excess of the vertical surface-parallel flow component over large areas of the ablation zone. Based on these results, Reeh et al. (1999, 2003) and Mohr and Reeh (2002) propose replacing the surface-parallel flow assumption with a more realistic relationship between the surface-velocity components. They specifically suggest application of the principle of mass conservation (see Chapter 3) to derive an equation relating the vertical to the horizontal surface-velocity vectors. Obviously, this adds a level of complexity to the InSAR analysis, but is worthwhile if additional information on quantities such as ice thickness and surface mass balance is available (Mohr et al., 1998, 2003; Reeh et al., 1999). Ice thickness cannot be directly measured from space, and requires measurement by airborne or surface radio echo-sounding (as discussed in Chapter 3). An exciting aspect of this new approach combining InSAR and the principle of mass conservation is that it imposes direct (satellite) measurements of the ice surface displacement field as a boundary condition in ice-dynamics models (Mohr, 1997).

Other constraints apply to the technique outlined above. An important, over-riding issue is that data from each viewing direction must again be co-registered with a high degree of accuracy. Mohr (1997) further notes that for Sun-synchronous satellites such as ERS-1 and -2, however, power constraints severely limit data collection from ascending orbits on the Earth's shadow side. Moreover, while the LOS vectors from different viewing angles should ideally be perpendicular to each other for optimal sensitivity, this is typically not achievable. An intrinsic limitation on interferometric measurements is that no information can be derived on the displacements parallel to the flight direction. Mohr and Madsen (1996a) show how aircraft InSAR data can be used to augment satellite-derived InSAR motion data when the latter are perpendicular to the satellite SAR LOS. Another factor to be considered is that the relatively small incidence angle of most spaceborne SARs—e.g., 23° for ERS—results in the amplification of vertical motions by at least a factor of 2 relative to horizontal motions (Joughin et al., 1996c).

2.5.3 Tidal effects on InSAR retrievals over floating ice

Problems arise when using InSAR techniques to determine the two-dimensional ice flow velocity pattern of ice shelves and glacier tongues which are floating on the ocean. These complications relate to significant vertical velocities due to tidal flexure effects downstream from the *grounding line* (the boundary or zone separating floating from grounded ice), which cause "instantaneous" LOS velocities which are substantially greater than horizontal flow velocities (Goldstein et al., 1993). In effect, this corrupts the decomposition of the displacement into two horizontal components (Mohr and Madsen, 1996b; Rignot, 1996). Furthermore, the decomposition of the interferometric phase into displacement and topography is corrupted if the uplift is not exactly the same in the two interferograms. Correcting for tidal effects is of key importance in this respect, and generally involves the removal of range-displacement variations due to vertical tidal-related motion between image acquisitions using data from regional tidal models (e.g., Padman et al., 2000, 2002; Rignot et al., 2000) or coincident GPS measurements (Keller et al., 1997). Errors can be introduced at this stage due to mis-mapped grounding lines and poor bathymetric information (Gray et al., 2001), and also the "inverse barometer effect" (Rignot et al., 2000). This effect is explained in detail in Chapter 3.

The other side of the coin is that the location of the InSAR-measured change between tidal flexure and "normal" two-dimensional flow can be used advantageously to accurately detect, map, and monitor the grounding line or hinge-line location (Goldstein et al., 1993; Gray et al., 2002; Rignot, 1996, 1998a, b). An example is shown in Figure 2.18 (see color section). This is an important and poorly understood parameter in ice sheet dynamics and mass balance, and one that may be sensitive to global change—i.e., grounding-line migration appears to be occurring in places in response to a warming of the ocean (see Chapter 3). The advent of InSAR opens up a unique opportunity to systematically monitor changes in grounding-line location for the first time from space, and to a high precision (Rignot, 2002). Moreover, InSAR enables more accurate estimates of ice fluxes and mass balance when combined with knowledge of ice thickness and accumulation rates. These applications are again covered in more detail in Chapter 3. For a more in-depth analysis of overall error sources in the retrieval of ice sheet motion by InSAR, including those associated with inaccurate estimates of the interferometric baseline, please see Joughin et al. (1996c), Mohr and Madsen (1999), Mohr et al. (1998), and Sharov et al. (2002).

2.6 ICE-SHEET SURFACE MOTION RETRIEVAL BY SPECKLE TRACKING

Although the three Radarsat-1 Antarctic Mapping Missions to date (see pp. 75–76) provided a wealth of data over extensive areas, significant challenges are involved in traditional interferometric processing of Radarsat data to estimate ice velocity. These relate largely to the problem of temporal decorrelation over the nominal 24-day orbit repeat period. Reasonable and usable (though relatively low) phase coherence can be maintained in low-accumulation (<15 cm per annum) regions such as much of central Antarctica (Gray et al., 1999; Joughin, 2002). In regions of higher accumulation (e.g., >30 cm per annum), however, temporal decorrelation typically occurs where ΔT is greater than 1–3 days. This is a major limitation, given the large repeat interval of current and future satellites—e.g., 24 days for Radarsat-1, 35 days for ERS-2 and Envisat, and 46 days for ALOS—and severely complicates the retrieval of radial ice motion from differential phase (Gray et al., 1999).

Another factor affecting repeat-pass interferometric correlation is the nature of ice motion, as discussed in Section 2.2. A 24-day repeat cycle provides a strong sensitivity to displacement, with one interferometric fringe being roughly equivalent to 1 m per annum of horizontal ice motion perpendicular to the satellite track. Joughin et al. (2002) show that, while the 24-day repeat cycle of Radarsat is virtually ideal for measuring slow ice motion (i.e., <100 m per annum) by InSAR, its application is severely limited in faster flowing regions where coherence breaks down—e.g., around the ice sheet perimeter. In Antarctica, for example, the speed of many outlet glaciers exceeds 300 m per annum, and the gradients in speed also tend to be high (Gray et al., 2001). Data from the 1-day repeat ERS-1/2 Tandem Mission are much better matched to fast motion, but are not always available. Moreover, although ice-sheet margins are subject to seasonal melt, interior regions are expected to remain coherent for significantly longer periods.

Fortunately, an alternative technique has been devised to determine ice displacement (motion) from coherent pairs of SAR amplitude images in regions where either the ice motion and/or temporal decorrelation rules out the use of conventional InSAR techniques (Gray et al., 1998). Termed *speckle tracking*, this method can still yield dense nets of large-scale ice motion estimates under such circumstances. It is based upon an image correlation approach that depends on the coherent pattern in the image pair being correlated (Derauw, 1999; Gray et al., 1998; Joughin, 2002; Joughin et al., 1999b; Michel and Rignot, 1999; Thiel et al., 1996). Although less accurate than conventional InSAR, speckle tracking can yield complementary and additional two-dimensional ice motion information in areas where InSAR analysis breaks down (Gray et al., 1998, 1999). With adequate coherence, it is also suitable for areas of high velocity (e.g., ice shelves and outlet glaciers) and for image pairs with long repeat cycles—e.g., of 24 days (Gray et al., 2001). Gray et al. (1998), for example, used this technique to measure velocities in the range of ~250–800 m per annum for the Slessor Glacier flowing into the Filchner Ice Shelf in West Antarctica. They also noted that the vector direction corresponds closely with the glacier flowlines apparent in the amplitude imagery, adding confidence to the retrievals.

Speckle tracking requires a coherent pair of SAR images, an accurate though coarsely sampled DEM, and, for floating ice shelves and glacier tongues, an estimate of the height-difference range due to tidal flexure (Gray et al., 2001). In effect, speckle tracking is a refinement of the image-registration stage used in InSAR, but involves the correlation of speckle across a coherent pair of SAR amplitude images rather

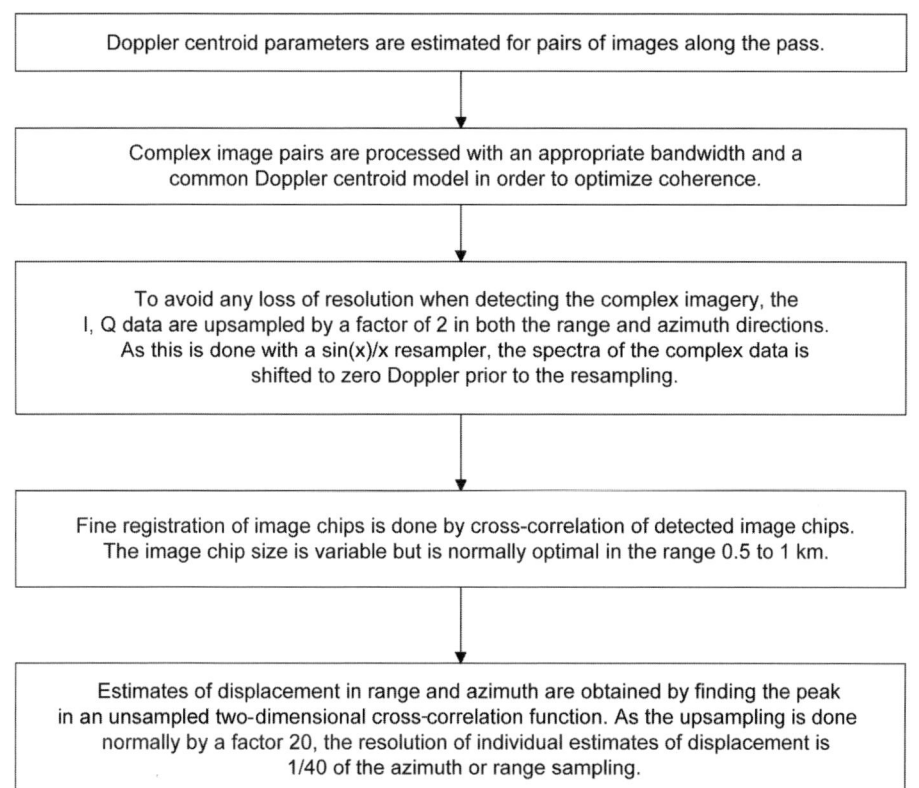

Figure 2.19. The data-processing sequence to retrieve range and azimuth displacement data by speckle tracking from a coherent pair of SAR images.
From Gray et al. (2001). Copyright Canadian Remote Sensing Society (CRSS) 2001, reproduced with permission.

than counting the number of fringes in phase-derived interferograms. Compared with the incoherent cross-correlation (feature-tracking) approach applied to optical images, which relies on identification of image features such as crevasses (as outlined in Chapter 3), the accuracy with which image registration can be achieved is an order of magnitude better with speckle tracking (Gray et al., 1998). This can be to better than 1/10 of a pixel at best in areas of high coherence. The SAR data-processing steps required to estimate azimuth and range displacements are summarized in Figure 2.19. After initial processing, ice flow speed and direction are determined to sub-pixel accuracy by the cross-correlation of small "chips" across the SAR imagery (Gray et al., 1999, 2001; Joughin et al., 2002; Pattyn and Derauw, 2002). Knowledge of the spatial baseline is again necessary. The DEM (Bamber and Bindschadler, 1997; Liu et al., 1999) is used to calculate the look and incidence angles and the terrain slope (Gray et al., 1998, 1999). Calibration of the ice displacement retrievals is still required, and is typically carried out using one or two reference points of known velocity within the image, and preferably "zero-

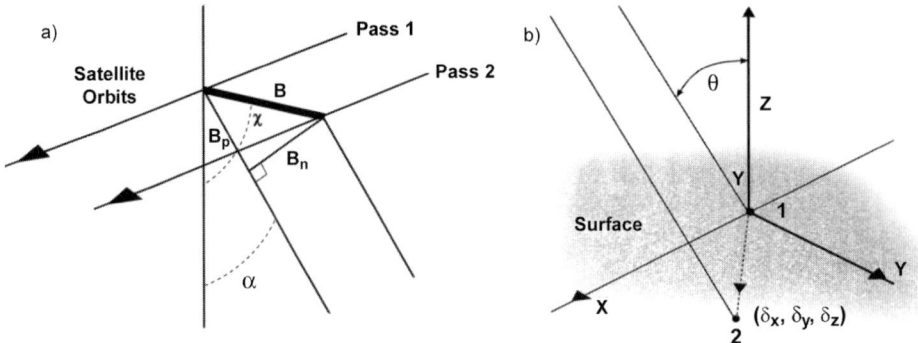

Figure 2.20. A schematic of (a) the satellite geometry on two passes separated by baseline B, and (b) the ground geometry (where the surface is gray) for the calculation of ice sheet surface displacement in three dimensions using the SAR speckle-filtering technique.

From Gray et al. (2001), after Joughin et al. (1996c). Copyright CRSS 2001, reproduced with permission.

velocity" points—e.g., exposed rock outcrops or mountains. This is necessary because of the limited accuracy of the Radarsat-1 orbit-state data. Such points are required to derive the absolute velocity field, but are again unnecessary for the measurement of velocity gradients.

As outlined by Gray et al. (2001), the three components of ice sheet surface velocity can be estimated by SAR speckle tracking from the corrected range and azimuth displacements (δ_r and δ_a, respectively) using the formalism given by Joughin et al. (1996c). The satellite and ground geometry involved is presented in Figure 2.20. With this configuration, δ_a and δ_r are related to displacements in a local Cartesian coordinate system (where δ_y is locally horizontal and parallel to the range plane, δ_x parallel to the azimuth direction, and δ_z the local vertical) by:

$$\delta_x = \delta_a \qquad (2.38)$$

$$\delta_r = B_p + \delta_y \sin\theta - \delta_z \cos\theta \qquad (2.39)$$

where $B_p = B\cos(\chi - \alpha)$ is the parallel baseline; χ is the baseline angle; α is the SAR look angle at the satellite; and θ is the local incidence angle. In Figure 2.20b, the x-plane is parallel to the SAR along-track direction, and the y–z-plane is in the plane of incidence. It follows that the three-dimensional ice displacements over time ΔT are described within this framework by $(\delta_x, \delta_y, \delta_z)$.

Following Joughin et al. (1996c) and assuming surface-parallel ice flow (see Section 2.5.2 for the approximations implicit in this assumption), the vertical displacement δ_z can be related to the horizontal displacements by:

$$\delta_z = \delta_x S_a + \delta_y S_r \qquad (2.40)$$

where S_a and S_r are the terrain slopes in the azimuth and range directions, respectively (relative to the local horizontal plane). It is possible from the above

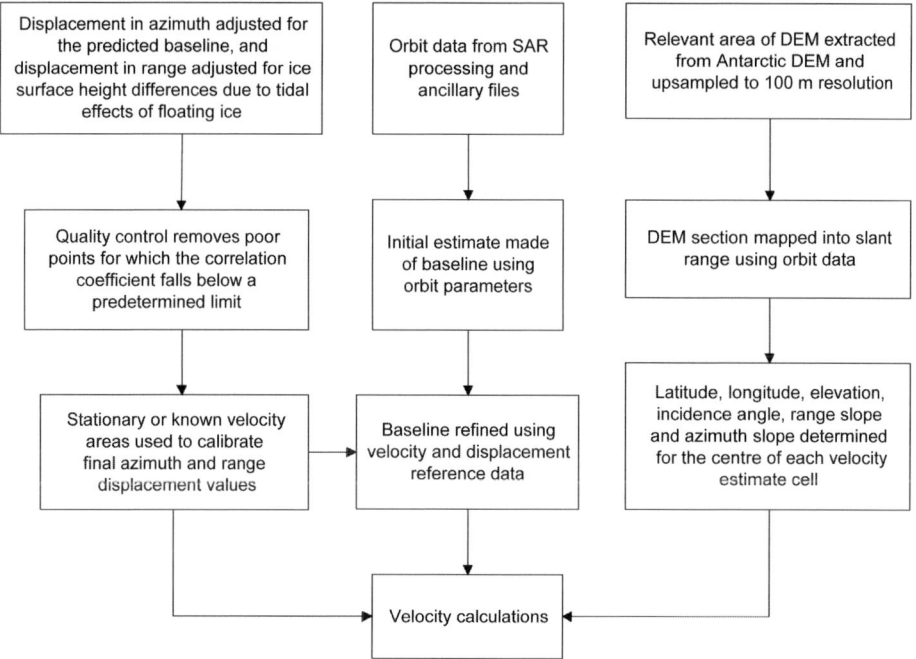

Figure 2.21. Schematic of the procedure for calculating ice surface velocity by speckle tracking, using displacement, DEM, satellite orbit data, and (for floating glaciers and ice shelves) tidal models.
From Gray et al. (2001). Copyright CRSS 2001, reproduced with permission.

equations to derive the local horizontal velocity in the range plane δ_y as:

$$\delta_y = \frac{(\delta_r - B_p + \delta_a S_a \cos\theta)}{(\sin\theta - S_r \cos\theta)} \quad (2.41)$$

It follows that the total (i.e., three-dimensional) ice displacement δ is:

$$\delta = \sqrt{\delta_x^2 + \delta_y^2 + \delta_z^2} \quad (2.42)$$

and, assuming invariant flow between image acquisitions, the ice velocity V over time interval ΔT in meters per annum is:

$$V = \frac{365}{\Delta T}\delta \quad (2.43)$$

The direction of motion can be estimated from δ_x and δ_y, and the angle between true north and local azimuth (Gray et al., 2001). The processing steps necessary required for combining the ice displacement data, elevation data, orbit information, and, for floating ice masses, the tidal model, are presented schematically in Figure 2.21. In this case, where the technique is applied to a floating ice shelf—i.e., the Filchner Ice

Shelf—a correction to the range displacement due to tidal motion is made using an Antarctic tidal model (Padman et al., 2000; Rignot et al., 2000) combined with an ice shelf flexure model. Grounding line position—i.e., the limit of tidal flexure—can be determined by examining the phase fringes from InSAR data—e.g., Goldstein et al. (1993); Rignot (1996). The tidal model is typically used to predict heights at the time of satellite overpasses (data acquisistion times). See Gray et al. (2001) for further explanation.

Like conventional interferometry, speckle tracking is applicable to vast featureless ice sheet regions where conventional feature-tracking techniques using time series of optical images fail, by virtue of the fact that matches are based on image speckle rather than visible surface features—e.g., crevasses. Indeed, it is generally more accurate than any incoherent image feature-tracking method, and yields a dense net of measurements that lend themselves well to modern glaciological studies (Gray et al., 2001). The relative advantages of speckle tracking applied to ice sheets, and compared with conventional InSAR, can be summarized as follows (Gray et al., 1998, 1999, 2001):

- It measures ice surface displacement in both the range and azimuth directions—i.e., from one suitable SAR image pair only. Recall that InSAR estimation is limited to displacement in the sensor LOS only for a single coherent image pair, or requires coherent data from both ascending and descending orbits to measure the two-dimensional motion field.
- The technique is uniquely useful when either the temporal separation of the image pair or the ice motion are large—i.e., in regions of low interferometric coherence.
- The technique foregoes the need for phase analysis. As a result, errors due to incorrect phase unwrapping and/or phase aliasing are avoided.
- As phase unwrapping is not necessary, velocities can be derived for regions separated by incoherent data.
- The retrieval of ice speed from range and azimuth displacement data is relatively straightforward, if the orbits and topography are accurately known (although uncertainties in these parameters generally require that ice velocity ground-control points are present to calibrate the derived speed).
- Local errors do not propagate through the image, as they can with phase analysis techniques.
- Accuracy requirements for topographic data and satellite orbital parameters are not as stringent as they are for conventional InSAR techniques for measuring ice motion.
- In certain cases, the use of phase information can be used to improve the estimation of velocity gradients or range velocities.

The main disadvantage of speckle tracking is that the relative velocity in the range direction is less accurate than that achievable with unwrapped differential phase. Moreover, some of the same limitations affecting InSAR apply to this technique. These include uncertainties in satellite orbital parameters and resulting baselines, errors in the DEM used, ambiguities introduced by tidal motion of floating ice (please see the previous section), and errors due to ionospheric and atmospheric propagation delays (including *azimuth streaking*) (Gray et al., 1999). In addition, errors tend to occur in the fine registration of the image chips employed, although averaging can reduce this contribution, which tends to be random (Gray et al., 1999). Errors in applying the speckle-tracking method vary with region and with the availability of reliable reference values (Gray et al., 1999). Following Gray et al. (2001), the uncertainty in ice velocity estimates ΔV resulting from both bias and random errors is related to uncertainty in the displacement estimates $\Delta \delta_X$, $\Delta \delta_Y$, and $\Delta \delta_Z$ by:

$$\Delta V = \left[\frac{c^2 \delta_X}{V}\right] \Delta \delta_X + \left[\frac{c^2 \delta_Y}{V}\right] \Delta \delta_Y + \left[\frac{c^2 \delta_Z}{V}\right] \Delta \delta_Z \qquad (2.44)$$

where $c = 365/\Delta T$ (2.45). Under ideal circumstances, errors are of the order of ~3 cm per day in azimuth and 10 cm per day in range (Gray et al., 1998). Error sources are evaluated in detail by Gray et al. (2001) and Bamler (1997).

Results from a study by Gray et al. (1999, 2001) in applying speckle tracking to Radarsat-1 interferometric data collected during the Radarsat-1 AMM mission over two important Antarctic regions—namely, tributaries of the West Antarctic Ice Streams and the Filchner Ice Shelf—are shown in Figure 2.22 (see color section). Traditional InSAR analysis relying on phase unwrapping could not be used in the latter due to patchy coherence and the possibility of fringe aliasing due to occasional very high fringe rates (due to the nominal 24-day acquisition separation for Radarsat-1 SAR images).

While speckle-tracking estimates are of a lower resolution (comparable with the window size used for matching) and poorer accuracy, they are extremely valuable as a means of filling in gaps in areas where no suitable data are available for conventional interferometric processing (Joughin, 2002). In his mass balance study of East Antarctic glaciers, Rignot (2002), for example, found that most ERS (descending) passes observed the glaciers in a direction perpendicular to ice flow, which is least favorable for InSAR mapping. He completed the InSAR measurements by applying a speckle-tracking technique in the along-track direction of ERS. Rignot (2002) estimates the nominal precision of this technique to be 1/30 of a pixel (i.e., 4 m along-track), leading to a velocity precision of 50 m per annum with 1-day repeat-pass data, which is acceptable for fast-moving outlet glaciers. The precision improves to a few meters per year with Radarsat-1 data collected on a 24-day repeat-pass cycle with a 5-m pixel size (Joughin, 2002; Joughin et al., 1999a, b).

Joughin et al. (1999b) also used a combination of speckle tracking and phase interferometry to monitor the catchment area for the ice streams feeding the West Antarctic Ice Sheet. By this means, they identified more complex and previously unknown regional ice flow patterns. Taking this a step further, Joughin (2002) developed algorithms to merge interferometric and speckle-tracking data from multiple swaths to form a single, seamless, large-scale velocity mosaic. At each point in the mosaic, available data are combined to produce estimates of the velocity and the associated error. This technique was applied to Radarsat-1 data collected over Lambert Glacier, Antarctica, during AMM and MAMM, as shown in

Figure 3.80 (in Chapter 3). Joughin (2002) suggests that these algorithms are equally applicable to other sources of InSAR data, and that the only thing currently preventing their wider application, even to the entire ice sheet, is a lack of suitable data. Further diagrams and discussion of errors are given in Chapter 3.

Again, there is an additional error contribution for floating ice masses due to the unknown tidal height difference between the two data acquisitions in the absence of tidal information (Gray et al., 1999). Joughin (2002) evaluates this error source in his study of the Amery Ice Shelf, where the range of tidal displacement is about ±1 m (Padman et al., 2002), which is significantly less than for Filchner-Ronne Ice Shelf or the Siple Coast section of the eastern Ross Ice Shelf. Joughin (2002) argues that while this could lead to worst-case errors of 30–50 m per annum, examination of overlapping velocity estimates indicates that the 24-day tidal differences in these data lead to considerably smaller errors of a few meters per annum. He concludes that, in this case, neglect of the tidal effect has a minimal impact on ice motion retrieval accuracy, with averaging of multiple swaths further reducing this error. However, tidal models are reaching such levels of sophistication and accuracy that InSAR- and speckle-tracking-derived ice shelf velocities can readily incorporate tidal corrections (e.g., Padman and Kottmeier, 2000; Rignot et al., 2000).

2.7 ICE-SHEET SURFACE MOTION RETRIEVAL BY COHERENCE TRACKING

Coherence tracking, which is also referred to as the *coherence optimization* or *fringe visibility* algorithm (Derauw, 1999; Pattyn and Derauw, 2002; Strozzi et al., 2002), is comparable with speckle tracking (Gray et al., 1998, 2001). It differs, however, in that it uses phase information in the complex-valued single-look SAR images rather than speckle information in the real-valued amplitude images. Each procedure searches for the best match or correlation between two samples of data taken from two co-registered images. By coherence tracking, small data patches (kernels) are selected throughout the images, series of small interferograms with changing offsets are constructed, and the coherence is estimated (see Strozzi et al., 2002 for more in-depth information). In theory, the accuracy of the unwrapped range velocity is very high, depending on the radar wavelength. Following Pattyn and Derauw (2002), the error σ_u' is given by:

$$\sigma_u' = \frac{\lambda}{4\pi \sin \alpha} \sigma_\phi \qquad (2.46)$$

where α is the SAR incidence angle (e.g., 23° for ERS); λ is the radar wavelength; and the error estimate in phase σ_ϕ is given by:

$$\sigma_\phi = \frac{1}{\sqrt{2N}} \frac{\sqrt{1-\gamma^2}}{\gamma} \qquad (2.47)$$

where γ is the coherence and N the interferometric multi-looking factor—i.e., the number of full-resolution samples employed for coherence estimation (Just and

Bamler, 1994; Rodriguez and Martin, 1992). In an example given by Pattyn and Derauw (2002), using ERS data and where $N = 50$, (2.47) yields an error in unwrapped range velocity of ~ 0.425 m per annum for an average coherence $\gamma = 0.7$. Additional errors again occur, however, due to atmospheric effects. These could be of the order of 5 m per annum (Joughin et al., 2000). Another error source for floating ice masses is tidal displacement.

Young and Hyland (2002) applied this technique to a study of the Amery Ice Shelf in East Antarctica using Radarsat-1 AMM image pairs acquired with 24-day separation. They used fixed features within the image pairs—e.g., rock outcrops—as reference points of zero velocity in order to derive a dense set of ice shelf horizontal velocity measurements. They further incorporated information from tide models of Fricker et al. (2002) and Padman et al. (2002) to remove the tidal component in the derived motion (Gray et al., 2001). In this case, the precision of the match using maximum coherence was estimated to be ~ 0.1 pixel, or a precision in velocity of about 8 m per annum along-track and 26 m per annum across-track, which represents the random component of the error budget for each velocity estimate.

Significant improvements are again expected with the launch of Radarsat-2 (see Chapter 5 of Volume 1 of this book and the Appendix to this volume for specifications and viewing geometry). In theory, the routinely interchangeable left- and right-imaging modes of Radarsat-2 will enable complete views of both the Antarctic and Arctic on a regular basis. Radarsat-2 will also have an ultra-fine-resolution mode (~ 3 m) that will reduce the stringent baseline requirements for interferometry and will make the speckle-tracking technique suitable for smaller and slower glaciers (Short and Gray, 2004). This new high-resolution mode could potentially reduce the magnitude of errors in the speckle-tracking method to tens of centimeters. A trade-off is reduced coverage compared with the reduced resolution ScanSAR mode—i.e., a swath of only 20 km versus ~ 500 km.

2.8 RELATIVE STRENGTHS AND WEAKNESSES OF DIFFERENTIAL INSAR, COHERENCE TRACKING, AND INTENSITY TRACKING

Strozzi et al. (2002) carried out an error analysis on the three ice motion retrieval techniques outlined above (see also Mohr and Madsen, 1999). This led to the conclusion that they are complementary, and that the formulation of a useful application strategy depends on the availability of suitable SAR data, the degree of image coherence, the size, orientation, and flow rate of the glacier, the accuracy and spatial resolution required, and the computation time involved. According to Strozzi et al. (2002), the strategy for mapping glacier flow can be summarized as follows:

- Differential InSAR is the most accurate method available to measure ice displacement in the slant-range direction, and offers dense spatial coverage (at a typical horizontal resolution of 20 m). The most important limiting factors are the feasibility of phase unwrapping (which is only reliable in areas of high-phase coherence and low fringe rates) and coherence itself. Isolated areas of high

coherence separated by low coherence often cannot be utilized due to the lack of a reference point.
- Maximum-coherence tracking is also suitable for regions of high-phase coherence. A major difference is that it provides absolute motion estimates, and can therefore be applied to disconnected regions of high coherence—i.e., separated by areas of low coherence. Moreover, this approach yields estimates of motion in two horizontal directions (InSAR does not measure displacement in the azimuth direction). The disadvantages are a poorer accuracy and resolution than DInSAR, and a greater computational expense.
- Intensity (speckle) tracking can be applied to image pairs lacking in coherence—e.g., in regions of fast/incoherent flow or separated by acquisition intervals of longer than a few days (i.e., longer than the period over which phase coherence is typically maintained). This is an important consideration, given that current and future satellite SAR missions all have nominal revisit intervals in excess of 24 days. As with coherence tracking, this approach enables measurement of both the azimuth and slant range components of surface displacement. Disadvantages include the relatively poor spatial resolution (dictated by the large image patch size) and the low accuracy (compared with InSAR).

The overall conclusion of Strozzi et al. (2002) was that a combined approach is the most efficient and effective means of measuring ice motion. The special case of using InSAR data to detect and measure ice-flow acceleration/deceleration and glacier surge events is discussed in Section 3.8.5.

Strozzi et al. (2002) also proposed a means of retrieving two-dimensional displacement maps when SAR data are only available from one orbital configuration—i.e., where suitable data from both ascending and descending passes are unavailable. This entails combining DInSAR in the slant-range direction with an offset-tracking procedure in the azimuth direction. According to the authors, the accuracy achievable is greater than 5 times that using speckle-/coherence-tracking techniques (which are often referred to collectively as *offset tracking*), due to the use of the high-resolution SAR data. It is, however, less accurate than DInSAR applied to dual-azimuth data. These trade-offs are, however, offset by the availability of a larger suitable dataset.

It should be noted that alternative approaches to the retrieval of ice motion by the differential processing of repeat-pass interferograms are currently being developed and tested. Two new techniques have been devised by Sharov et al. (2002), neither of which involve the need for interferometric phase unwrapping. This step was taken to help reduce areal error propagation, thereby improving retrieval accuracy. The first algorithm, called *GINSAR*, is an unsupervised technique based on (i) the calculation of interferometric phase gradients (Sandwell and Price, 1998), (ii) the generation of ice-surface slope maps, and (iii) the analysis of differences between multi-temporal slope maps. The second algorithm, which adopts a so-called *transferential approach* and derives ice motion from single interferograms, is applicable to tidewater glaciers. Due to their dynamism, the latter are difficult to measure by conventional InSAR techniques. The transferential approach is based on the interferometric measurement of glacier forcing of adjacent fast ice (where present). Another advantage of the new techniques is that they remain usable even under significant phase noise. In initial testing, these new algorithms performed reasonably well when applied to ERS-1/-2 InSAR data acquired over large tidewater glaciers in the European Arctic. Discrepancies in output are apparent, however, underlining the need for further validation and development.

2.9 POLARIMETRIC INTERFEROMETRIC SAR (POL-INSAR)

An emerging field in polar remote sensing is the symbiosis of *SAR polarimetry* (PolSAR) and interferometry (InSAR) as *Pol-InSAR* (Cloude and Pottier, 1997; Papathanassiou and Cloude, 2001). The discussion has so far centered on spaceborne SARs operating at fixed polarizations. Current and near-future satellite SAR systems—e.g., Envisat-1 ASAR, Radarsat-2 (⟨*http://www.radarsat2.info/*⟩), ALOS PALSAR (⟨*http://alos.nasda.go.jp/index-e.html*⟩), TerraSAR-X (Moreira, 2003; ⟨*http://www.infoterra-global.com/terrasar.htm*⟩), and TerraSAR-L (Zink, 2003)—have the enhanced capability of acquiring data at different polarizations. All of these new systems can operate in user-selectable single- and dual-polarization modes (see Volume 1 of this book and the Appendix to this volume for more details). In addition, the Radarsat-2, ALOS, and TerraSAR-X/-L systems can operate as polarimeters and measure the amplitude and phase of the backscattered signals in the four available transmit-and-receive linear polarization combinations (i.e., HH, HV, VV, VH). Collectively, these measurements enable determination of target backscatter response in any possible transmit-and-receive polarization combination, and support Pol-InSAR (Cloude, 2004; Cloude and Papathanassiou, 1998; Papathanassiou and Cloude, 2001). This provides the potential for much-improved classification of sea ice, for example.

The Pol-InSAR technique, which is a model-based parameter estimation method (Dall et al., 2004), is based upon the acquisition of two polarimetric SAR images which satisfy the repeat-pass interferometric conditions of time lapse and baseline (spatial separation). The complete polarimetric/interferometric information is then stored in three 3×3 complex matrices, the coherence matrix of each image, and the analogous matrix formed from the product of the scattering vectors of images 1 and 2. Phase-preserving transformations, optimization procedures, and target decomposition methods are then employed to create interferograms between all possible elliptical polarization states, maximize the coherence, and derive information on scattering mechanisms from which target information can be inferred. Please see Cloude (2004), Cloude and Papathanassiou (1998), and Papathanassiou and Cloude (2001) for more in-depth information.

A Pol-InSAR dataset consists of fully polarimetric datasets coherently acquired from both ends of one or more interferometric baselines (Cloude and Papathanassiou, 1998; Dall and Skriver, 2003). While InSAR responds primarily to the distribution and location of scatterers within a resolution cell, the Pol-SAR response is primarily a function of their orientation and shape, enabling discrimination between different scattering mechanisms—e.g., volume versus surface scattering. As a result, polarimetric interferometry opens up the possibility of enhanced detection of surface phenomena and understanding of scattering processes, as a result of the effect of the latter on interferometric combinations from different polarizations (Hellwich, 1999a, b; Papathanassiou and Cloude, 2001; Ulbricht et al., 1998). The potential and applicability of Pol-InSAR has yet to be fully realized for ice sheet research, but has been demonstrated by current research—e.g., in Greenland (Dall et al., 2003, 2004; Dall and Skriver, 2003). Danish scientists collected Pol-InSAR data over Greenland in 1997 and 1998 using the airborne EMISAR system (Dall et al., 2001; Linz Christensen et al., 1998; ⟨http://www.emi.dtu.dk/research/DCRS/Emisar/emisar.html⟩). Recalling Section 2.3.6, penetration depths up to 13 m were measured at C-band, and a major problem with radar observations of ice sheets is the facies- and wavelength-dependent penetration of the signal into the firn/ice. While this problem is less of an issue for radar altimeters than for InSAR systems, recent work by Dall et al. (2003) and Dall and Skriver (2003) suggests that spaceborne Pol-InSAR methods can be used to estimate the penetration depth and correct for the corresponding bias in both single-channel InSAR-derived and radar-altimeter-derived ice sheet elevation data. Dall et al. (2004) introduce simple scattering models for this purpose. Moreover, Pol-InSAR might provide information about the physical characteristics of the ice and its near-surface structure, thereby enabling a classification of ice sheet facies (Rignot, 1995). Another possibility is that Pol-InSAR data could be combined with ICESat GLAS data to derive maps of radar penetration depth, given that the GLAS is a laser (optical) system that measures ice sheet topography with minimal penetration into the snow—i.e., it provides a true approximation of the actual surface topography (see Chapter 3).

Dall and Skriver (2003) and Cloude (2004) discuss issues relating to the extraction of geophysical information from Pol-InSAR data. Processing consists of two stages—namely, pre- and post-processing. *Pre-processing* includes *baseline estimation, focusing, calibration, motion compensation*, and *phase unwrapping*. *Post-processing* applies polarimetric operations such as *coherence optimization* and extracts geophysical parameters by inverting an electromagnetic backscattering model. This model must provide an adequate description of the backscattering properties of the scene in question while being sufficiently simple to be inverted. The model used by Hoen and Zebker (2000b), for example, fits these criteria.

2.10 DEVELOPMENTS AND OUTSTANDING ISSUES

The discussion above has hopefully given an indication of what satellite SAR interferometry and related technologies can and cannot achieve. In spite of the considerable complexities involved in processing, careful application of InSAR is producing highly encouraging and indeed exciting results, which compare very favourably in terms of their precision with independent data—e.g., from GPS and airborne laser altimeters. Compared with conventional *in situ* and aircraft techniques, spaceborne InSAR and associated maximum-coherence and speckle-tracking techniques entail a very cost-effective tool for comprehensive regional surveys and monitoring of remote glaciers and ice sheets with good accuracy. Indeed, repeat-pass InSAR has a key role to play in greatly improving estimates of both current ice mass balance and dynamics, by providing suitable high-density and large-scale datasets that are highly compatible with modern ice-sheet models. The data themselves are also enabling a revolutionary re-evaluation of ice-stream and drainage basin mass balance, with many surprising results. As stated by Mohr (1997), InSAR provides a completely new basis for modeling the spatially and temporally inhomogeneous ice sheet environment. Examples and other issues relating to the extraordinary contribution of InSAR to glacier and ice-sheet research are described in Chapter 3.

Uniquely, these techniques can yield "near-instantaneous" measurements of glacier and ice sheet velocities, compared with velocities measured by conventional satellite feature-tracking techniques which yield averages over long observation intervals (often by a year or more). In certain areas, subtleties in surface elevation and fine-scale velocity structure are being seen for the first time (see Chapter 3). Given its all-weather, day–night capability, SAR interferometry also has a major role to play in the detection and monitoring of unpredicted dynamic manifestations of ice sheets (Bindschadler, 1998), including glacier and ice-stream surges (see Section 3.8.5). Such changes in ice dynamics are particularly important in that they could be indicative of ice-sheet instability. Of major importance is the ability to accurately map and monitor grounding-line locations from space for the first time. Series of such measurements may unearth significant variability in ice dynamics behavior through time. Combined with InSAR- and altimeter-derived DEMs, these new techniques are providing glaciologists with extraordinary datasets that are directly applicable to state-of-the-art ice sheet models, both as input and for validation purposes.

While InSAR is an immensely exciting new development in polar research, a good deal of work lies ahead before the technique becomes fully operational and its full potential is realized. Current research is addressing weak links, including phase unwrapping (Gens, 1998), difficulty of automation in processing, limitations due to environmental factors (tropospheric/ionospheric and temporal decorrelation), the development of consistent image-mosaicking methods, and baseline estimation by data fusion (Klees, 1999). Error analysis remains a key current research area of satellite interferometry, as the latter becomes more extensively applied to mass-balance estimates which require a high degree of precision. Other work is also aiming to develop techniques to effectively assimilate InSAR data with data from other sources—e.g., GPS and altimeters (Williams, 2001)—and with ice models. Moreover, it is focusing on developing a strategy for obtaining the ice-flow vector in situations where ascending and descending data are not both available. The synergistic merging of InSAR data with data from the ICESat laser altimeter (GLAS) shows immense promise in the generation of ice-sheet DEMs to an unprecedented accuracy, given the high precision of GLAS in the outer ice-sheet margins and its minimal penetration depth (irrespective of ice-sheet facies). The synergy of various sensors onboard satellite SAR missions is also being investigated as a means of eliminating tropospheric effects. An example is the MEdium Resolution Imaging Spectrometer (MERIS) flying alongside the Advanced SAR (ASAR) onboard Envisat. The ESA VECTRA Project has brought together glaciological and interferometric experts to address the challenges (Shepherd and Mohr, 2001; ⟨http://vectra.mssl.ucl.ac.uk⟩). There is a clear need for long-term measurement continuity.

Temporal decorrelation related to short-term variations in ice sheet/glacier surface characteristics and tropospheric water vapor remains the major error source in repeat-pass interferometry using current conventional satellite–SAR systems with relatively long orbital periods—e.g., ERS, Envisat, and Radarsat. Once again, these satellites were not designed for interferometry. Overall, this powerful technique is constrained by the fact that no satellite dedicated to InSAR has been launched to date, resulting in a somewhat limited dataset. While anticipated new developments in phase unwrapping may reduce decorrelation impacts, what are really required are multiple datasets and the launch of dedicated InSAR missions with revisit intervals and other characteristics optimized for interferometric analysis. A major breakthrough for the measurement of ice motion would also be the launch of a satellite–SAR system capable of routinely acquiring data of the same surface from three look directions.

According to Bamler et al. (2003), it should be possible in future to largely bypass the limitations associated with decorrelation and to achieve a new level of accuracy by acquiring simultaneous observations using larger, or even multiple, baselines. Such a system can comprise a dedicated constellation of satellites flying in close formation. For example, one conventional transmitting SAR satellite could be augmented by a set of low-cost and receive-only micro-satellites equipped with small antennae to produce radar images that can be coherently combined. This configuration is known as the *interferometric cartwheel* concept (Amiot et al., 2002; Krieger et al., 2002; Massonnet, 2001; Massonnet et al., 2000), and operates by means of the primary satellite illuminating the ground while the receive-only systems point at the same area. In fact, a cartwheel constellation of three receive-only micro-satellites is under consideration for flight in tandem with TerraSAR-L (Zink et al., 2004). Possible candidate orbit configurations are shown in Figure 2.23. Another solution would be to fly two SAR satellites in tandem formation (along the lines of the ERS Tandem Mission) and separated by a short timelag—i.e., just large enough to avoid interference between the transmitters—or to fly the two satellites in a close and cooperative constellation. In fact, the Canadian Space Agency (with interest from ESA) are investigating the feasibility of launching Radarsat-3 some 1–2 years after Radarsat-2 (due for launch in 2006), and placing the two satellites in orbits suitable for global InSAR measurement—i.e., in a tandem mission (Caves et al., 2002; Evans and Girard, 2002; Girard et al., 2002).

A recent initiative has been put forward through the NASA Jet Propulsion Laboratory to launch a dedicated InSAR satellite in the 2009/2010 timeframe (Smith, 2004). The main thrust of this proposed project is to provide a dedicated spaceborne radar interferometry mission that can acquire measurements from multiple directions in order to determine actual surface-displacement vectors—not

Sec. 2.10] Developments and outstanding issues 115

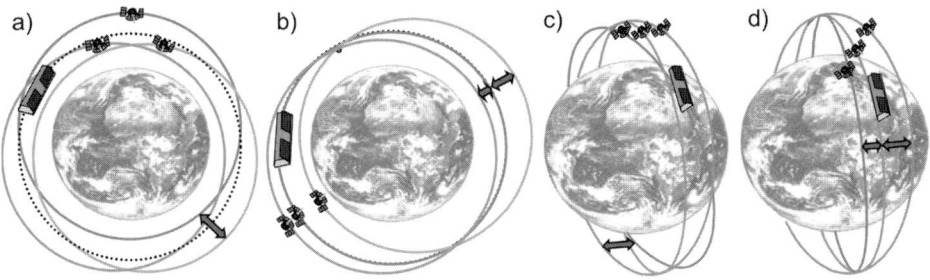

Figure 2.23. Examples of interferometric satellite constellations investigated by CNES and DLR for a possible TerraSAR-L inteferometric cartwheel configuration. These are: (a) *interferometric cartwheel* (proposed by CNES); (b) *two-scale cartwheel* (CNES); (c) *cross-track pendulum* (DLR); and (d) *trinodal pendulum* (DLR). Of these configurations, the trinodal Pendulum and the two-scale cartwheel enable the simultaneous acquisition of multiple interferometric baselines at a constant baseline ratio. In this figure, TerraSAR-L is the large satellite, while the smaller spacecraft are the passive micro-satellites.
From Zink et al. (2004). Copyright ESA 2004, reproduced with permission.

only of ice sheets/glaciers but also of volcanic and tectonically active areas (Massonnet et al., 1995). The payload will be an L-band (1.25 GHz) HH-polarization SAR designed for repeat-pass interferometry. An L-band system is generally preferred to shorter wavelength InSAR systems for Earth science applications due to its lower susceptibility to the effects of temporal decorrelation (compared with higher frequency systems). The derivation of three-dimensional vector information will be achieved by acquiring data in both right- and left-looking modes and on descending and ascending orbits/passes, and from an exact repeat orbit with a repeat frequency of ≤9 days. This system will operate in three modes—namely, (i) stripmap mode (three possible beams, 35-m resolution, 120-km swath width; (ii) high-resolution mode in one of seven beams, trading swath coverage (~40 km) for improved resolution (10 m); and (iii) an experimental ScanSAR mode (100-m resolution over a 350-km swath width). The SAR incidence angle ranges from 20° (18° look angle) to 43° (37° look angle). Please see Smith (2004) for more detailed information on instrument and viewing parameters.

Alternatively, Moccia and Vetrella (1992) have proposed the concept of a tethered system of satellites which are separated by a known short baseline. Smaller, lighter, and cheaper SAR missions with shorter revisit cycles are also being considered. Séguin and Girard (2002), for example, suggested the launch of a small SAR satellite into the same orbit as the ALOS PALSAR to create a tandem mission and enable the acquisition of more frequent repeat-pass InSAR data over polar regions. Madsen and Zebker (1998) further proposed the launch of a single satellite equipped with two displaced antennae (similar to current airborne systems). The concept of tight formation orbits for an along-track SAR interferometer is covered by Gill and Runge (2003). Phase difference in the along-track geometry has been used to monitor ocean currents (Goldstein and Zebker, 1987; Shemer et al., 1993). A closely spaced along-track interferometer configuration, or the use of the *split-antenna mode* of TerraSAR-X, may also permit the interferometric measurement of sea-ice motion.

There is no doubt as to the power and unique capabilities of satellite SAR inteferometry as a means of advancing our knowledge of ice-sheet mass balance and dynamics, and it is anticipated that this extraordinary technique will become increasingly important over the next decade. It is hoped that the major space agencies will, in the not-too-distant future, combine resources and expertise to launch compatible SAR systems in orbital configurations optimized for interferometry. In particular, future missions should be designed with the capability of maintaining the interferometric baseline in a specified range to supply larger and more consistent datasets and enable efficient application of the data. The benefit would be immense, not only to glaciology but also to a wide variety of other scientific and operational endeavors. These include terrestrial elevation mapping (Toutin and Gray, 2000) and hazard monitoring—e.g., of earthquakes, tectonic and volcanic activity, and ground subsidence (Bürgmann et al., 2000; Massonnet and Feigl, 1998; Massonnet et al., 1993, 1995; Rosen et al., 2000; Zebker et al., 1994a). Emergent applications of InSAR in hydrology and geomorphology are reviewed by Smith (2002). Some of the upcoming, sophisticated high-resolution SARs will allow for innovative orbit configurations that can potentially be optimized for InSAR application. Future missions will benefit from important hardware development such as improved antenna design, RF electronics, digital electronics, and data processors. Aircraft InSAR and *in situ* data collection campaigns will continue to play a leading role in gathering high-precision data with which to calibrate and validate the satellite measurements. In a parallel development, interferometric processing is also being incorporated into ice-sheet elevation measurement using advanced spaceborne radar altimetry, with the ESA's innovative *Synthetic Aperture Interferometric Radar ALtimeter (SIRAL)* onboard CryoSat (see Chapter 5 in Volume 1 of this book and the next chapter for details). Moreover, it is expected that the emerging field of Pol-InSAR, as outlined in Chapter 3, will also gain momentum as suitable data become available.

2.11 REFERENCES

Abdelfattah, R. and J. M. Nicolas (2002). Topographic SAR interferometry formulation for high-precision DEM generation. *IEEE Transactions on Geoscience and Remote Sensing*, **40**(11), 2415–2426.

Adam, N., M. Eineder, H. Breit, and S. Suchandt (1999). Quality of the D-PAF ERS orbits before and after the inclusion of PRARE data. *2nd International Workshop on ERS SAR Interferometry, "FRINGE99", Liège, Belgium, November 10–12.*

Ahuja, R. K., T. L. Magnanti, and J. B. Orlin (1993). *Network Flows: Theory, Algorithms and Applications.* Prentice Hall, Englewood Cliffs, NJ.

Allen, C. T. (1995). Interferometric Synthetic Aperture Radar. *IEEE Geoscience and Remote Sensing Society Newsletter*, September 3–13.

Amiot, T., F. Douchin, E. Thouvenot, J. C. Souyris, and B. Cugny (2002). The interferometric cartwheel: A multi-purpose formation of passive radar microsatellites. *Proceedings of International Geoscience and Remote Sensing Symposium IGARSS '02, 24–28 June 2002*. Institute of Electrical and Electronic Engineers, Piscataway, NJ, Vol. 1, pp. 435–437.

Baldi, A. (2001). Two-dimensional phase unwrapping by quad-tree decomposition. *Applied Optics*, **40**, 1187–1194.

Bamber, J. L. and R. A. Bindschadler (1997). An improved elevation dataset for climate and ice-sheet modelling: Validation with satellite imagery. *Annals of Glaciology*, **25**, 438–444.

Bamber, J. L. and R. Kwok (2004). Remote-sensing techniques. In: J. L. Bamber and A. J. Payne (eds.), *Mass Balance of the Cryosphere*. Cambridge University Press, Cambridge, UK, pp. 59–113.

Bamber, J. L., I. Joughin, and R. J. Hardy (1999). A comparison of balance velocities, and InSAR-derived velocities for the Greenland Ice Sheet. *Proceedings of ESA Fringe '99 Meeting, November 10–12, Liège, Belgium* (CD-ROM, ESA SP-478). ESA, Noordwijk, The Netherlands.

Bamber, J. L., R. J. Hardy, and I. Joughin (2000). An analysis of balance velocities over the Greenland ice sheet and comparison with synthetic aperture radar interferometry. *Journal of Glaciology*, **46**(152), 67–74.

Bamler, R. (1997). Digital terrain models from radar interferometry. In: D. Fritsch and D. Hobbie (eds.), *Photogrammetric Week '97*. Wichman, Heidelberg, Germany, pp. 93–105. Available online at ⟨http://www.ifp.uni-stuttgart.de/publications/phowo97/bamler.pdf⟩.

Bamler, R. and M. Eineder (1996). ScanSAR processing using standard high precision SAR algorithms. *IEEE Transactions on Geoscience and Remote Sensing*, **34**(1), 212–218.

Bamler, R. and R. Hanssen (1997). Decorrelation induced by interpolation errors in InSAR. *Proceedings of International Geoscience and Remote Sensing Symposium IGARSS '97, Singapore, 3–8 August 1997*. Institute of Electrical and Electronic Engineers, Piscataway, NJ, Vol. 4, pp. 1710–1712.

Bamler, R. and P. Hartl (1998). Synthetic aperture radar interferometry. *Inverse Problems*, **14**, R1–R54.

Bamler, R. and J. Holzner (2004). ScanSAR interferometry for Radarsat-2 and Radarsat-3. *Canadian Journal of Remote Sensing*, **30**(3), 437–447.

Bamler, R. and D. Just (1993). Phase statistics and decorrelation in SAR interferograms. *Proceedings of International Geoscience and Remote Sensing Symposium IGARSS '93, Tokyo, 18–21 August 1993*. Institute of Electrical and Electronic Engineers, Piscataway, NJ, pp. 980–984.

Bamler, R., D. Geudtner, B. Schättler, U. Steinbrecher, J. Holzner, J. Mittermayer, H. Breit, A. Moreira, P. W. Vachon, K. E. Mattar et al. (1998). Radarsat SAR interferometry using standard, fine, and ScanSAR modes. *Proceedings of the Radarsat ADRO Final Symposium, Montreal*.

Bamler, R., D. Geudtner, B. Schättler, P. W. Vachon, U. Steinbrecher, J. Holzner, J. Mittermayer, H. Breit, and A. Moreira (1999). Radarsat ScanSAR interferometry. *Proceedings of International Geoscience and Remote Sensing Symposium IGARSS '99, 28 June–2 July 1999*. Institute of Electrical and Electronic Engineers, Piscataway, NJ, Vol. 3, pp. 1517–1521.

Bamler, R., M. Eineder, B. Kampes, H. Runge, and N. Adam (2003). SRTM and beyond: Current situation and new developments in spaceborne SAR and InSAR. *Proceedings of the ISPRS Workshop on High Resolution Mapping from Space, 6–8 October 2003,*

Universität Hannover, Germany. International Society for Photogrammetry and Remote Sensing, Istanbul, Turkey.

Bedrich, S. (1997). *PRARE System Performance* (Earthnet online, ESA). Available online at ⟨http://earth1.esrin.esa.it/florence/papers/program-details/data/bedrich/index.html⟩.

Bindschadler, R. (1998). Monitoring ice sheet behavior from space. *Reviews of Geophysics*, **36**(1), 79–104.

Bindschadler, R. A., R. B. Alley, J. Anderson, S. Shipp, H. Borns, J. Fastook, S. Jacobs, C. F. Raymond, and C. A. Shuman (1998). What is happening to the West Antarctic Ice Sheet. *EOS, Transactions of the American Geophysical Union*, **79**(22), 257, 264–265.

Bindschadler, R., M. Fahnestock, and A. Sigmund (1999). Comparison of Greenland Ice Sheet topography measured by TOPSAR and airborne laser altimetry. *IEEE Transactions on Geoscience and Remote Sensing*, **37**(5), 2530–2535.

Budd, W. F. and R. C. Warner (1996). A computer scheme for rapid calculations of balance-flux distributions. *Annals of Glaciology*, **23**, 21–27.

Bürgmann, R., P. A. Rosen, and E. J. Fielding (2000). Synthetic aperture radar interferometry to measure Earth's surface topography and its deformation. *Annual Review Earth Planetary Sciences*, **28**, 169–209.

Capes, R. and M. Haynes (2004). *InSAR Basics: A Brief Guide to SAR Interferometry (InSAR)* (electronic media). Available online at ⟨http://www.npagroup.co.uk/insar/whatisinsar/insar_simple.htm⟩.

Carballo, G. F. and P. W. Fieguth (1999). Probabilistic cost functions for network flow phase unwrapping. *Proceedings of International Geoscience and Remote Sensing Symposium IGARSS '99, Hamburg, Germany, 28 June–2 July 1999*.

Carballo, G. F. and P. W. Fieguth (2000). Probabilistic cost functions for network flow phase unwrapping. *IEEE Transactions on Geoscience and Remote Sensing*, **38**(5), 2192–2201.

Carballo, G. F. and P. W. Fieguth (2002). Hierarchical network flow phase unwrapping. *IEEE Transactions on Geoscience and Remote Sensing*, **40**(8), 1695–1708.

Caves, R., A. P. Luscombe, P. K. Lee, and K. James (2002). Topographic performance evaluation of the Radarsat-2/-3 tandem mission. *Proceedings of International Geoscience and Remote Sensing Symposium IGARSS '02, 24–28 June 2002, Toronto, Canada*.

Chen, C. W. and H. A. Zebker (2000). Network approaches to two-dimensional phase unwrapping: Intractability and two new algorithms. *Journal of the Optical Society of America A*, **17**(3), 401–414.

Chen, C. W. and H. A. Zebker (2001). Two-dimensional phase unwrapping with use of statistical models for cost functions in nonlinear optimization. *Journal of the Optical Society of America A*, **18**, 338–351.

Chen, C. W. and H. A. Zebker (2002). Phase unwrapping for large SAR interferograms: Statistical segmentation and generalised network models. *IEEE Transactions on Geoscience and Remote Sensing*, **40**(8), 1709–1719.

Chen, Q., D. H. Bromwich, and L. Bai (1997). Precipitation over Greenland retrieved by a dynamic method and its relation to cyclonic activity. *Journal of Climate*, **10**(5), 839–870.

Ching, N. H., D. Rosenfeld, and M. Braun (1992). Two-dimensional phase unwrapping using a minimum spanning tree algorithm. *IEEE Transactions on Image Processing*, **1**(3), 355–365.

Cloude, S. R. (2004). Radar polarimetry and interferometry: A tutorial introduction. *IEEE Geoscience and Remote Sensing Society Newsletter*, **131**, 25–29.

Cloude, S. R. and K. P. Papathanassiou (1998). Polarimetric SAR interferometry. *IEEE Transactions on Geoscience Remote Sensing*, **36**(5), 1551–1565.

Cloude, S. R. and E. Pottier (1997). Polarimetric SAR interferometry. *IEEE Transactions on Geoscience and Remote Sensing*, **36**(5), 68–78.

Colbeck, S. C. (1982). An overview of seasonal snow metamorphism. *Reviews of Geophysics and Space Physics*, **20**(1), 45–61.

Costantini, M. (1998). A novel phase unwrapping method based on network programming. *IEEE Transactions on Geoscience and Remote Sensing*, **36**(3), 813–821.

Costantini, M. (1999). A method for phase unwrapping in the presence of dense fringe patterns. *Proceedings of ESA Fringe '99 Meeting, November 10–12, Liège, Belgium* (CD-ROM, ESA Special Publication SP-478). ESA, Noordwijk, The Netherlands.

Coulson, S. N. (1993). SAR interferometry with ERS-1. *ESA Earth Observation Quarterly*, **40**, 20–23.

Cullather, R. I., D. H. Bromwich, and M. L. Van Woert (1998). Spatial and temporal variability of Antarctic precipitation from atmospheric methods. *Journal of Climate*, **11**, 334–368.

Cumming, I. and F. Wong. (1998). The effect of ScanSAR parameters on interferogram quality. *Proceedings of International Geoscience and Remote Sensing Symposium IGARSS '98, Seattle, WA, 6–10 July 1998*. Institute of Electrical and Electronic Engineers, Piscataway, NJ, Vol. 5, pp. 1635–1637.

Cumming, I. and J. Zhang (1999). Measuring the 3-D flow of the Lowell Glacier with InSAR. *Proceedings of ESA Fringe '99 Meeting, November 10–12, Liège, Belgium* (CD-ROM, ESA Special Publication SP-478). ESA, Noordwijk, The Netherlands.

Cumming, I. G., J.-L. Valero, P. W. Vachon, M. Brugman, K. Mattar, D. Geudtner, M. S. Seymour, and A. L. Gray (1996a). Results of ERS interferometry measurements of alpine glaciers. *Proceedings of 18th Annual Symposium of the Canadian Remote Sensing Society, Vancouver, B.C., March 25–29*. Canadian Remote Sensing Society, Canadian Aeronautics and Space Institute, pp. 324–329.

Cumming, I., J.-L. Valero, M. Seymour, P. Vachon, K. Mattar, D. Geudtner, and L Gray (1996b). Glacier flow measurements with ERS tandem mission data. *Proceedings of "Fringe 96" Workshop on ERS SAR Interferometry, September 30–October 2, Zurich, Switzerland* (ESA SP-406). Available online at ⟨http://www.geo.unizh.ch/rsl/fringe96/⟩.

Curlander, J. C. and R. N. McDonough (1991). *Synthetic Aperture Radar: Systems and Signal Processing*. John Wiley & Sons, New York, 647 pp.

Dall, J. and H. Skriver (2003). *Pol-InSAR*. Available online at ⟨http://www.emi.dtu.dk/research/DCRS/Technology/polinsar.html⟩.

Dall, J., S. N. Madsen, K. Keller, and R. Forsberg (2000). Using airborne SAR interferometry to measure the elevation of a Greenland Ice Cap. *Proceedings of IEEE 2000 International Geoscience and Remote Sensing Symposium, Honolulu, July*. Institute of Electrical and Electronics Engineers, Piscataway, NJ, pp. 1125–1127.

Dall, J., S. N. Madsen, K. Keller, and R. Forsberg (2001). Topography and penetration of the Greenland Ice Sheet measured with airborne SAR interferometry. *Geophysical Research Letters*, **28**(9), 1703–1706.

Dall, J., K. P. Papathanassiou, and H. Skriver (2003). Polarimetric SAR interferometry applied to land ice: First results. *Proceedings of International Geoscience and Remote Sensing Symposium IGARSS '03, Toulouse, France, July*. Institute of Electrical and Electronics Engineers, Piscataway, NJ, pp. 1432–1434.

Dall, J., K. P. Papathanassiou, and H. Skriver (2004). Polarimetric SAR interferometry applied to land ice: Modeling. *Proceedings of EUSAR 2004 Conference, Ulm, Germany, 25–27 May 2004*. VDE Verlag, Frankfurt, Germany, pp. 247–250.

Dammert, P. B. G., M. Leppäranta, and J. Askne (1998). SAR interferometry over Baltic Sea ice. *International Journal of Remote Sensing*, **19**(16), 3019–3037.

D'Elia, S. and R. Biasutti (1999). *DESCW: PC Software Supporting Remote Sensing Data* (ESA Bulletin 97). Available online at ⟨http://esapub.esrin.esa.it/bulletin/bullet97/delia.pdf⟩.

Derauw, D. (1999). DInSAR and coherence tracking applied to glaciology: The example of Shirase Glacier. *Proceedings of ESA Fringe '99 Meeting, November 10–12, Liège, Belgium* (CD-ROM, ESA Special Publication SP-478). ESA, Noordwijk, The Netherlands.

Dixon, T. (ed.) (1994). SAR interferometry and surface change detection. *Workshop in Boulder, February 3 and 4* (a report). Available online at ⟨ttp://southport.jpl.nasa.gov/scienceapps/dixon/index.html⟩.

Eineder, M. (2003). Oscillator clock drift compensation in bistatic interferometric SAR. *Proceedings of International Geoscience and Remote Sensing Symposium IGARSS '03*. Institute of Electrical and Electronics Engineers, Washington, D.C.

Eineder, M. and S. Suchandt (2003). Recovering radar shadow to improve interferometric phase unwrapping and DEM reconstruction. *IEEE Transactions on Geoscience and Remote Sensing*, **41**(12), 2959–2962.

Eineder, M., M. Hubig, and B. Milcke (1998). Unwrapping large interferograms using minimum cost flow algorithm. *Proceedings of International Geoscience and Remote Sensing Symposium IGARSS '98, Seattle, WA, 6–10 July 1998*. Institute of Electrical and Electronic Engineers, Piscataway, NJ, Vol. 1, pp. 83–87.

Eineder, M., B. Schättler, M. Hubig, W. Knöpfle, N. Adam, and H. Breit (1999). Operational processing large areas of interferometric SAR data. *Proceedings of ESA Fringe '99 Workshop, Liège, Belgium, 10–12 November 1999*. ESA, Noordwijk, The Netherlands.

Eineder, M., H. Runge, E. Boerner, R. Bamler, N. Adam, B. Schättler, H. Breit, and S. Suchandt (2004). SAR interferometry with TerraSAR-X. *Proceedings of Fringe '03 Workshop, 1–5 December 2003, Frascati, Italy* (ESA SP-550). ESA, Noordwijk, The Netherlands.

Ekholm, S. (1996). A full coverage, high-resolution, topographic model of Greenland, computed from a variety of digital elevation data. *Journal of Geophysical Research*, **B10**(21), 961–972.

Elachi, C. (1987). *Introduction to the Physics and Techniques of Remote Sensing*. John Wiley & Sons, New York, 413 pp.

Elachi, C. (1988). *Spaceborne Radar Remote Sensing: Applications and Techniques*. Institute of Electrical and Electronics Engineers Press, New York.

Engdahl, M. (1996). Phase unwrapping in SAR interferometry using instantaneous frequency estimation. MSc thesis, Faculty of Information Technology, Department of Technical Physics, Helsinki University of Technology.

ESA (2001). ERS SAR interferometry-derived information on Antarctica and Greenland: The VECTRA Cooperation Project. *ESA Earth Observation Quarterly*, **68**, Supplement, 4 pp.

ESA (2003). *All that Glitters: The First ERS/Envisat Interferogram*. Available online at ⟨http://www.esa.int/export/esaCP/SEM97H5V9ED_index_0.html⟩.

Evans, N. and R. Girard (2002). Radarsat-2/-3: An innovative opportunity for a global topographic mapping mission. *53rd International Astronautical Congress, 12–15 October 2002, Houston, TX* (Paper IAC-02-B.05.03). American Institute of Aeronautics and Astronautics, Reston, VA.

Fahnestock, M., R. Bindschadler, R. Kwok, and K. Jezek (1993). Greenland ice-sheet surface properties and ice dynamics from ERS-1 SAR imagery. *Science*, **262**(5139), 1530–1534.

Fatland, D. R. and C. S. Lingle (1998). Analysis of the 1993–95 Bering Glacier (Alaska) surge using differential SAR interferometry. *Journal of Glaciology*, **44**(148), 532–546.

Fiedler, H., G. Krieger, F. Jochim, M. Kirschner, and A. Moreira (2002). *Analysis of Satellite Configurations from Spaceborne SAR Interferometry*. Available online at ⟨http://www.weblab.dlr.de/rbrt/pdf/ISFF_0223.pdf⟩.

Fischer, A., H. Rott, and H. Björnsson (2003). Observation of recent surges of Vatnajökull, Iceland, by means of ERS SAR interferometry. *Annals of Glaciology*, **37**, 69–76.

Fitch, J. P. (1988). *Synthetic Aperture Radar*. Springer-Verlag, New York.

Forster, R. R., K. Jezek, L. Gray, and K. Mattar (1998). Analysis of glacier flow dynamics from preliminary Radarsat InSAR data of the Antarctic Mapping Mission. *International Geoscience and Remote Sensing Symposium, Seattle, WA, July.*

Flynn, T. J. (1997). Two-dimensional phase unwrapping with minimum weighted discontinuity. *Journal of the Optical Society of America A*, **14**, 2692–2701.

Fornaro, G. and E. Sansosti (1999). A two-dimensional region growing least squares phase unwrapping algorithm for interferometric SAR processing. *IEEE Transactions on Geoscience and Remote Sensing*, **37**(5), 2215–2226.

Fornaro, G., G. Franceschetti, and R. Lanari (1996a). Interferometric SAR phase unwrapping using Green's formulation. *IEEE Transactions on Geoscience and Remote Sensing*, **34**, 720–727.

Fornaro, G., G. Franceschetti, R. Lanari, and E. Sansosti (1996b). Robust phase unwrapping techniques: A comparison. *Journal of the Optical Society of America A*, **13**, 2355–2366.

Fornaro, G., G. Franceschetti, R. Lanari, E. Sansosti, and M. Tesauro (1997). Global and local phase unwrapping technique: A comparison. *Journal of the Optical Society of America A*, **14**, 2702–2708.

Forster, R. R., K. C. Jezek, H. G. Sohn, A. L. Gray, and K. E. Matter (1998). Analysis of glacier flow dynamics from preliminary Radarsat InSAR data of the Antarctic Mapping Mission. *Proceedings of International Geoscience and Remote Sensing Symposium IGARSS '98, Seattle, WA, 6–10 July 1998*, Institute of Electrical and Electronic Engineers, Piscataway, NJ, pp. 2225-2227.

Forster, R. R., E. Rignot, B. L. Isacks, and K. C. Jezek (1999). Interferometric radar observations of glaciers Europa and Penguin, Hielo Patagónico Sur, Chile. *Journal of Glaciology*, **45**(150), 325–337.

Forster, R. R., K. C. Jezek, L. Koenig, and E. Deeb (2003). Measurement of glacier geophysical properties from InSAR wrapped phase. *IEEE Transactions on Geoscience and Remote Sensing*, **41**(11), 2595–2604.

Franceschetti, G. and G. Fornaro (1999). Synthetic aperture radar interferometry. In: G. Franceschetti and R. Lanari (eds.), *Synthetic Aperture Radar Processing*. CRC Press, Boca Raton, FL, pp. 167–223,

Frezzotti, M., S. Gandolfo, F. La Marca, and S. Urbini (2002). Snow dunes and glazed surfaces in Antarctica: New field and remote sensing data. *Annals of Glaciology*, **34**, 81–88.

Fricker, H. A., N. W. Young, I. Allison, and R. Coleman (2002). Iceberg calving of the Amery Ice Shelf, East Antarctica. *Annals of Glaciology*, **34**, 241–246.

Frolich, R. M. and C. S. M. Doake (1998). Synthetic aperture radar interferometry over Rutford Ice Stream and Carlson Inlet, Antarctica. *Journal of Glaciology*, **44**(146), 77–92.

Frölind, P.-O. and L. M. H. Ulander (2002). Digital elevation map generation using VHF-band SAR data in forested areas. *IEEE Transactions on Geoscience and Remote Sensing*, **40**(8), 1769–1776.
Fujii, Y. and K. Kusunoki (1982). The role of sublimation and condensation in the formation of ice sheet surface at Mizuho Station, Antarctica. *Journal of Geophysical Research*, **87**(C6), 4293–4300.
Gabriel, A. K. and R. M. Goldstein (1988). Crossed orbit interferometry: Theory and experimental results from SIR-B. *International Journal of Remote Sensing*, **9**(5), 857–872.
Gabriel, A. K., R. M. Goldstein, and H. A. Zebker (1989). Mapping small elevation changes over large areas: Differential radar interferometry. *Journal of Geophysical Research*, **94**(B7), 9183–9191.
Gatelli, F., A. Monti Guarnieri, F. Parizzi, P. Pasquali, C. Prati, and F. Rocca (1994). The wavenumber shift in SAR interferometry. *IEEE Transactions on Geoscience and Remote Sensing*, **32**(4), 855–865.
Gens, R. (1998). Quality assessment of SAR interferometric data. PhD thesis, University of Hannover, Germany, 141 pp.
Gens, R. (1999a). Quality assessment of interferometrically derived digital elevation models. *International Journal of Applied Earth Observation and Geoinformation*, **1**(2), 102–108.
Gens, R. (1999b). SAR interferometry: Software, data format and data quality. *Photogrammetric Engineering and Remote Sensing*, **65**(12), 1375–1378.
Gens, R. (2003). Two-dimensional phase unwrapping for radar inteferometry: Developments and new challenges. *International Journal of Remote Sensing*, **24**(4), 703–710.
Gens, R. and T. Logan (2003). *Alaska Satellite Facility Software Tools Manual*. Available online at ⟨http://www.asf.alaska.edu/apd/documents/pdf/asf_software_tools.pdf⟩.
Gens, R. and J. L. van Genderen (1996a). SAR interferometry: Issues, techniques, applications. *International Journal of Remote Sensing*, **17**(10), 1803–1835.
Gens, R. and J. L. van Genderen (1996b). Analysis of the geometric parameters of SAR interferometry for spaceborne systems. *International Archives of Photogrammetry and Remote Sensing* (Vol. XXXI, Part B2). ISPRS, Vienna, pp. 107–110.
Geudtner, D. (1996). *The Interferometric Processing of ERS-1 SAR Data* (ESA Technical Report ESA-TT-1341, translation of DLR-FB 95-28). ESA, Noordwijk, The Netherlands.
Geudtner, D. and M. Schwäbisch (1996). An algorithm for precise reconstruction of InSAR imaging geometry: Application to "flat Earth" phase removal, phase-to-height conversion, and geocoding of InSAR-derived DEMs. *Proceedings of EUSAR '96, Germany*. VDE Verlag, Frankfurt, Germany, pp. 249–252.
Geudtner, D., P. W. Vachon, K. E. Mattar, and A. L. Gray (1997). Radarsat repeat-pass SAR interferometry: Results over an Arctic test site. *Proceedings of Geomatics in the Era of Radarsat (GER '97), May 27–30, Ottawa* (CD-ROM).
Geudtner, D., P. W. Vachon, K. E. Mattar, and A. L. Gray (1998). Radarsat repeat-pass SAR interferometry. *Proceedings of International Geoscience and Remote Sensing Symposium IGARSS '98, Seattle, WA, 6–10 July 1998*. Institute of Electrical and Electronic Engineers, Piscataway, NJ, Vol. 3, pp. 1635–1637.
Ghiglia, D. and M. Pritt (1998). *Two-dimensional Phase Unwrapping: Theory, Algorithms, and Software*. John Wiley & Sons, New York, 494 pp.
Ghiglia, D. C. and L. A. Romero (1994). Robust two-dimensional weighted and unweighted phase unwrapping that uses fast transform and iterative methods. *Journal of the Optical Society of America A*, **11**, 107–117.

Ghiglia, D. C. and L. A. Romero (1996). Minimum L^p-norm two-dimensional phase unwrapping. *Journal of the Optical Society of America A*, **13**, 1999–2013.

Girard, R., P. F. Lee, and K. James (2002). The RADARSAT-2 and -3 topographic mission: An overview. *Proceedings of International Geoscience and Remote Sensing Symposium IGARSS '02, Toronto, Canada*. Institute of Electrical and Electronic Engineers, Piscataway, NJ, Vol. 3, pp. 1477–1479.

Gill, E. and H. Runge (2003). Tight formation flying for an along-track SAR interferometer. *54th International Astronautical Congress, 29 September–3 October 2003, Bremen, Germany*. International Astronautical Federation, Paris, IAC-03-B.4.01, 11 pp.

Goldstein, R. (1995). Atmospheric limitations to repeat-track radar interferometry. *Geophysical Research Letters*, **22**(18), 2517–2520.

Goldstein, R. and C. Werner (1997). Radar ice motion interferometry. *Proceedings of 3rd ERS Symposium on Space, Florence, Italy, March 17–21*. ESA, Noordwijk, The Netherlands, Vol. 2, pp. 969–972.

Goldstein, R. M. and C. L. Werner (1998). Radar interferogram filtering for geophysical applications. *Geophysical Research Letters*, **25**, 4035–4038.

Goldstein, R. M. and H. A. Zebker (1987). Interferometric radar measurement of ocean surface currents. *Nature*, **328**, 707–709.

Goldstein, R. M., H. A. Zebker, and C. L. Werner (1988). Satellite radar interferometry: Two-dimensional phase unwrapping. *Radio Science*, **23**(4), 713–720.

Goldstein, R. M., H. Engelhardt, B. Kamb, and R. M. Frolich (1993). Satellite radar interferometry for monitoring ice sheet motion: Application to an Antarctic ice stream. *Science*, **262**(5139), 1525–1530.

Goodwin, I. D. (1990). Snow accumulation and surface topography in the katabatic zone of eastern Wilkes Land, Antarctica. *Antarctic Science*, **2**(3), 235–242.

Graham, L. C. (1974). Synthetic interferometer radar for topographic mapping. *Proceedings of the IEEE*, **62**(6), 763–768.

Gray, A. L., K. E. Mattar, P. W. Vachon, R. Bindschadler, K. C. Jezek, R. Forster, and J. P. Crawford (1998). InSAR results from the Radarsat Antarctic Mapping Mission data: Estimation of glacier motion using a simple regression procedure. *Proceedings of International Geoscience and Remote Sensing Symposium IGARSS '98*. Institute of Electrical and Electronic Engineers, Piscataway, NJ, Vol. 3, pp. 1638–1640.

Gray, A. L., K. E. Mattar, and N. Short (1999). Speckle tracking for 2-dimensional ice motion studies in polar regions. *Proceedings of 2nd International Workshop on ERS SAR Interferometry, "Fringe '99", Liège, Belgium, November 10–12* (CD-ROM, ESA Special Publication SP-478). ESA, Noordwijk, The Netherlands, pp. 1–8.

Gray, L., K. E. Mattar, and G. Sofko (2000). Influence of ionospheric electron density fluctuations on satellite radar interferometry. *Geophysical Research Letters*, **27**(10), 1451–1454.

Gray, A. L., N. Short, K. E. Mattar, and K. C. Jezek (2001). Velocities and flux of the Filchner Ice Shelf and its tributaries determined from speckle tracking interferometry. *Canadian Journal of Remote Sensing*, **27**(3), 193–206.

Gray, L., N. Short, R. Bindschadler, I. Joughin, L. Padman, P. Vornberger, and A. Khananian (2002). Radarsat interferometry for Antarctic grounding-zone mapping. *Annals of Glaciology*, **34**, 269–276.

Gruber, T., F.-H. Massmann, and C. Reigber (1996). *ERS D-PAF Altimeter and Orbit Global Products Manual* (Technical Report ERS-D-GPM-31200, GFZ/D-PAF). Zentrum der Deutschland für Luft- und Raumfahrt, Oberpfaffenhofen, Germany.

Gudmundsson, S., J. M. Carstensen, and F. Sigmundsson (2002). Unwrapping ground displacement signals in satellite radar interferograms with aid of GPS and MRF regularization. *IEEE Transactions on Geoscience and Remote Sensing*, **40**(8), 1743–1754.

Guijarro, J., A. Auriol, M. Costes, C. Jayles, and P. Vincent (2000). *MWR and DORIS—Supporting Envisat's Radar Altimery Mission* (ESA Bulletin 104, pp. 41–46). Available online at ⟨http://esapub.esrin.esa.it/bulletin/bullet104/guijarro104.pdf⟩.

Guneriussen, T., K. A. Hogda, H. Johnsen, and I. Lauknes (2001). InSAR for estimation of changes in snow water equivalent of dry snow. *IEEE Transactions on Geoscience and Remote Sensing*, **39**(10), 2101–2108.

Guritz, R., O. Lawlor, T. Logan, R. Fatland, J. Groves, S. Li, and V. Kaupp (1998). Automated digital elevation model (DEM) production using ERS SAR Tandem pairs. *Proceedings of International Geoscience and Remote Sensing Symposium IGARSS '98, Seattle, WA*. Institute of Electrical and Electronic Engineers, Piscataway, NJ, poster.

Hanssen, R. (1998). *Atmospheric Heterogeneities in ERS Tandem SAR Interferometry* (DEOS Report 98.1). Delft University Press, The Netherlands.

Hanssen, R. (2001). *Radar Interferometry: Data Interpretation and Error Analysis*. Kluwer Academic, Dordrecht, The Netherlands.

Hanssen, R. and R. Bamler (1999). Evaluation of interpolation kernels for SAR interferometry. *IEEE Transactions on Geoscience and Remote Sensing*, **37**(1), 318–321.

Harbin, M. (2005). Additional Antarctic acquisition opportunities. *Alaska Satellite Facility News and Notes*, **2**(1), 2 [Alaska Satellite Facility, Fairbanks, AK].

Hartl, P. and X. Wu (1993). SAR interferometry: Experiences with various phase unwrapping methods. *Proceedings of 2nd ERS-1 Symposium, Hamburg, Germany*. ESA, Noordwijk, The Netherlands, pp. 727–732.

Hein, A. (2004). *Processing of SAR Data: Fundamentals, Signal Processing, Interferometry* (Signals and Communication Technology series). Springer-Verlag, Berlin.

Hellwich, O. (1999a). Basic principles and current issues of SAR interferometry. *Proceedings of ISPRS Joint Workshop Sensors and Mapping from Space, September 27–30* (No. 18, CD-ROM). Institut für Photogrammetrie und Ingenieurvermessungen, Universität Hannover.

Hellwich, O. (1999b). SAR interferometry: Principles, processing, and perspectives. In: C. Heipke and H. Mayer (eds.), *Festschrift für Prof. Dr.-Ing. Heinrich Ebner zum 60 Geburtstag*. Technische Universität, Munich, Germany, pp. 109–120.

Henderson, F. M. and A. J. Lewis (eds.) (1998). *Principles and Applications of Imaging Radar, Manual of Remote Sensing* (3rd edn., Vol. 2). John Wiley & Sons, New York, 896 pp.

Hirose, T. and P. W. Vachon (1998). Demonstration of ERS Tandem Mission SAR interferometry for mapping land fast ice evolution. *Canadian Journal of Remote Sensing*, **24**(1), 89–92.

Herráez, M. A., D. R. Burton, M. J. Lalor, and M. A. Gdeisat (2002). Fast two-dimensional phase-unwrapping algorithm based on sorting by reliability following a noncontinuous path. *Applied Optics*, **41**(35), 7437–7444.

Hoen, E. W. and H. A. Zebker (2000a). Penetration depths inferred from interferometric volume decorrelation observed over the Greenland Ice Sheet. *IEEE Transactions on Geoscience and Remote Sensing*, **38**(6), 2571–2583.

Hoen, E. W. and H. A. Zebker (2000b). Topography-driven variations in backscatter strength and depth observed over the Greenland Ice Sheet with InSAR. *Proceedings of International Geoscience and Remote Sensing Symposium IGARSS '00*. Institute of Electrical and Electronic Engineers, Piscataway, NJ, Vol. 2, pp. 470–472.

Holdsworth, G. (1969). Flexure of a floating ice tongue. *Journal of Glaciology*, **8**(54), 385–397.

Holzner, J. and R. Bamler (2002). Burst-mode and ScanSAR interferometry. *IEEE Transactions on Geoscience and Remote Sensing*, **40**(9), 1917–1934.

Hung, K. M. and T. Yamada. (1998). Phase unwrapping by regions using least-squares approach. *Optical Engineering*, 37, 2965–2970.

Huntley, J. M. (1989). Noise-immune phase unwrapping algorithm. *Applied Optics*, **23**, 3268–3270.

Hvidberg, C. S., K. Keller, N. S. Gundestrup, C. C. Tscherning, and R. Forsberg (1997). Mass balance and surface movement of the Greenland Ice Sheet at Summit, Central Greenland. *Geophysical Research Letters*, **24**(18), 2307–2310.

Hyland, G. and N. W. Young (2002). Velocity and strain rates derived from InSAR analysis over the Amery Ice Shelf, East Antarctica. *Annals of Glaciology*, **34**, 228–234.

Iken, A., K. Echelmeyer, W. Harrison, and M. Funk (1993). Mechanisms of fast flow in Jakobshavns Isbrae, Greenland: Part 1. Measurements of temperature and water level in deep boreholes. *Journal of Glaciology*, **39**(131), 15–25.

Jezek, K. C. (1999). Glaciological properties of the Antarctic ice sheet from Radarsat-1 synthetic aperture radar imagery. *Annals of Glaciology*, **29**, 286–290.

Jezek, K. (2002). Radarsat-1 Antarctic Mapping Project: Change-detection and surface velocity campaign. *Annals of Glaciology*, **34**, 263–268.

Jezek, K. and E. Rignot (1994). Katabatic wind processes on the Greenland Ice Sheet. *EOS, Transactions of the American Geophysical Union*, **75**(Suppl. 212), 44.

Jezek, K., K. H. Sohn, and K. Noltimier (1998). The Radarsat Antarctic Mapping Project. *Proceedings of International Geoscience and Remote Sensing Symposium IGARSS '98, Seattle, WA, 6–10 July 1998.* Institute of Electrical and Electronic Engineers, Piscataway, NJ, Vol. 5, pp. 2462–2464.

Jordan, R. L., Caro, E. R., Kim, Y., Kobrik, M., Shen, Y., and Stuhr, F. V. (1996). Shuttle radar topography mapper (SRTM). In: G. Franceschetti, C. J. Oliver, F. S. Rubertone, and S. Tajbakhsh (eds.), *Microwave Sensing and Synthetic Aperture Radar*. International Society for Optical Engineering, Bellingham, WA, pp. 412–422.

Joughin, I. (1995). Estimation of ice sheet topography and motion using interferometric synthetic aperture radar. PhD thesis, University of Washington, Seattle, WA.

Joughin, I. (2002). Ice-sheet velocity mapping: A combined interferometric and speckle-tracking approach. *Annals of Glaciology*, **34**, 195–201.

Joughin, I. R. and D. P. Winebrenner (1994). Effective number of looks for a multi-look interferometric phase distribution, *Proceedings of International Geoscience and Remote Sensing Symposium IGARSS '94, Pasadena, CA.* Institute of Electrical and Electronic Engineers, Piscataway, NJ, pp. 2276–2278.

Joughin I., D. P. Winebrenner, and D. B. Percival (1994). Probability density functions for multilook polarimetric signatures. *IEEE Transactions on Geoscience and Remote Sensing*, **32**, 562–574.

Joughin, I. R., D. P. Winebrenner, and M. A. Fahnestock (1995). Observations of ice-sheet motion in Greenland using satellite radar interferometry. *Geophysical Research Letters*, **22**, 571–574.

Joughin, I., S. Tulaczyk, M. Fahnestock, and R. Kwok (1996a). A mini-surge on the Ryder Glacier, Greenland observed via satellite radar interferometry. *Science*, **274**, 1525–1530.

Joughin, I., D. Winebrenner, M. Fahnestock, R. Kwok, and W. Krabill (1996b). Measurement of ice-sheet topography using satellite-radar interferometry. *Journal of Glaciology*, **42**(140), 10–22.

Joughin, I., R. Kwok, and M. Fahnestock (1996c). Estimation of ice-sheet motion using satellite radar interferometry: Method and error analysis with application to Humboldt Glacier, Greenland. *Journal of Glaciology*, **42**(142), 564–575.

Joughin, I., M. Fahnestock, S. Ekholm, and R. Kwok (1997). Balance velocities for the Greenland Ice Sheet. *Geophysical Research Letters*, **24**(23), 3045–3048.

Joughin, I. R., R. Kwok, and M. A. Fahnestock (1998a). Interferometric estimation of three-dimensional ice-flow using ascending and descending passes. *IEEE Transactions on Geoscience and Remote Sensing*, **36**(1), 25–37.

Joughin, I., M. Fahnestock, R. Thomas, and R. Kwok (1998b). Ice flow in northeast Greenland using balance velocities as control. *Proceedings of International Geoscience and Remote Sensing Symposium IGARSS '98, Seattle, WA*. Institute of Electrical and Electronic Engineers, Piscataway, NJ, Vol. 4, pp. 2246–2248.

Joughin, I., M. Fahnestock, R. Kwok, P. Gogineni, and C. Allen (1999a). Ice flow of Humboldt, Petermann and Ryder Gletscher, northern Greenland. *Journal of Glaciology*, **45**(150), 231–241.

Joughin, I., L. Gray, R. Bindschadler, S. Price, D. Morse, C. Hulbe, K. Mattar, and C. Werner (1999b). Tributaries of West Antarctic ice streams revealed by Radarsat interferometry. *Science*, **286**(5438), 283–286.

Joughin, I., M. Fahnestock, and J. L. Bamber (2000). Ice flow in the northeast Greenland ice stream. *Annals of Glaciology*, **31**, 141–146.

Joughin, I., M. Fahnestock, D. MacAyeal, J. L. Bamber, and P. Gogineni (2001). Observation and analysis of ice flow in the largest Greenland ice stream. *Journal of Geophysical Research*, **106**(D24), 34021–34034.

Joughin, I., S. Tulaczyk, R. Bindschadler, and S. Price (2002). Changes in West Antarctic ice stream velocities: Observation and analysis. *Journal of Geophysical Research*, **107**(B11), 2289, doi:10.1029/2001/JB001029.

Just, D. and R. Bamler (1994). Phase statistics of interferograms with applications to synthetic aperture radar. *Applied Optics*, **33**(20), 4361–4368.

Kampes, B. M., R. F. Hanssen, and Z. Perski (2003). Radar interferometry with public domain tools. *Proceedings of ESA Fringe '03 Workshop, December 1–5*. ESA/ESRIN, Noordwijk, The Netherlands. Available online at ⟨http://earth.esa.int/workshops/fringe03/⟩.

Keller, K., R. Forsberg, and C. S. Nielsen (1997). Kinematic GPS for ice sheet monitoring and SAR interferometry in Greenland. In: M.E. Cannon and G. Lachapelle (eds.), *Proceedings of International Symposium on Kinematic Systems in Geodesy; Geomatics and Navigation, Banff, Canada, 3–6 June 1997*. Department of Geomatics Engineering, University of Calgary, Calgary, Canada, pp. 525–528.

Klees, R. (1999). SAR interferometry technology. *XXII IUGG General Assembly, Birmingham, UK, July 19–30* (IAG Special Study Group 2.160, final report, period 1995–1999). International Union of Geodesy and Geophysics, University of Colorado, Boulder, CO.

Klees, R. and R. Hanssen (2000). Basics of synthetic aperture radar interferometry and applications. *Nordic Geodetic Commission (NKG) Autumn School, Nordic Geodesy towards the 21st Century, August 28–September 2, Fevik, Norway*. Finnish Geodetic Institute, Masala, Finland, pp. 169–211.

Klees, R. and D. Massonnet (1999). Deformation measurements using SAR interferometry: Potential and limitations. *Geologie en Mijnbouw*, **77**, 161–176.

Krabill, W. B., R. H. Thomas, C. F. Martin, R. N. Swift, and E. B. Frederick (1995). Accuracy of airborne laser altimetry over the Greenland Ice Sheet. *International Journal of Remote Sensing*, **16**(7), 1211–1222.

Krabill, W., Abdalati, W., Frederick, E., Manizade, S., Martin, C., Sonntag, J., Swift, R., Thomas, R., Wright, W., and J. Yungel (2000). Greenland Ice Sheet: High-elevation balance and peripheral thinning. *Science*, **289**, 428–430.

Krieger, G., M. Wendler, H. Fiedler, J. Mittermayer, and A. Moreira (2002). Performance analysis for bistatic interferometric SAR configurations. *Proceedings of International Geoscience and Remote Sensing Symposium IGARSS '02, Toronto, Canada*. Institute of Electrical and Electronic Engineers, Piscataway, NJ, Vol. 1, pp. 650–652.

Kropatsch, W. G. and D. Strobl (1990). The generation of SAR layover and shadow maps from digital elevation models. *IEEE Transactions on Geoscience and Remote Sensing*, **28**, 98–107.

Kwoh, L. K., E. C. Chang, W. C. A. Heng, and H. Lim (1994). DEM generation from 35-day repeat pass ERS-1 interferometry. *Proceedings of International Geoscience and Remote Sensing Symposium IGARSS '94, Pasadena, CA*. Institute of Electrical and Electronic Engineers, Piscataway, NJ, pp. 2288–2290.

Kwok, R. and M. A. Fahnestock (1996). Ice sheet motion and topography from radar interferometry. *IEEE Transactions on Geoscience and Remote Sensing*, **34**(1), 189–200.

Levanon, N. (1988). *Radar Principles*. John Wiley & Sons, New York.

Li, F. and R. M. Goldstein (1990). Studies of multi-baseline interferometric synthetic aperture radars. *IEEE Transactions on Geoscience and Remote Sensing*, **28**(1), 88–97.

Li, S., L. Shapiro, L. McNutt, and A. Jeffers (1996). Application of satellite radar interferometry to the detection of sea ice deformation. *Journal of the Remote Sensing Society of Japan*, **16**(2), 67–77.

Lillesand, T. M. and R. W. Kiefer (2000). *Remote Sensing and Image Interpretation* (4th edn.). John Wiley & Sons, New York.

Lin, Q., J. F. Vesecky, and H. A. Zebker (1992). New approaches in interferometric SAR data processing. *IEEE Transactions on Geoscience and Remote Sensing*, **30**(3), 560–567.

Lin, Q., J. F. Vesecky, and H. A. Zebker (1994). Phase unwrapping through fringe-line detection in synthetic aperture radar interferometry. *Applied Optics*, **33**, 201–208.

Linz Christensen, E., N. Skou, J. Dall, K. W. Woelders, J. H. Jørgensen, J. Granholm, and S. N. Madsen (1998). EMISAR: An absolutely calibrated polarimetric L- and C-band SAR. *IEEE Transactions on Geoscience and Remote Sensing*, **36**(6), 1852–1865.

Liu, H., K. C. Jezek, and B. Li (1999). Development of an Antarctic digital elevation model by integrating cartographic and remotely sensed data: A geographic information system based approach. *Journal of Geophysical Research*, **104**(B10), 23199–23214.

MacAyeal, D. R., E. Rignot, and C. L. Hulbe (1998). Ice-shelf dynamics near the front of the Filchner-Ronne Ice Shelf, Antarctica, revealed by SAR interferometry: Model/interferogram comparison. *Journal of Glaciology*, **44**(147), 419–428.

Madsen, S. and H. A. Zebker (1998). Imaging radar interferometry. In: F. Henderson and A. J. Lewis (eds.), *Principles and Applications of Imaging Radar: Manual of Remote Sensing* (3rd edn., Vol. 2, R. A. Ryerson, editor-in-chief). John Wiley & Sons, New York, pp. 359–380.

Madsen, S. N., H. A. Zebker, and J. Martin (1993). Topographic mapping using radar interferometry: Processing techniques. *IEEE Transactions on Geoscience and Remote Sensing*, **31**(1), 246–256.

Madsen, S. N., J. M. Martin, and H. A. Zebker (1995). Analysis and evaluation of the NASA/JPL Across-track Interferometric SAR System. *IEEE Transactions on Geoscience and Remote Sensing*, **33**(2), 383–391.

Madsen, S. N., N. Skou, J. Granholm, K. Woelders, and E. L. Christensen (1996). A system for airborne SAR interferometry. *International Journal of Electronics and Communications*, **50**(2), 106–111, 1996.

Madsen, S. N., J. J. Mohr, and N. Reeh (1999). Mapping Greenland by ERS-1/2 InSAR for ice mass balance and dynamics studies. *Proceedings of ESA Fringe '99 Meeting, November 10–12, Liège, Belgium* (CD-ROM, ESA Special Publication SP-478). ESA, Noordwijk, The Netherlands. Available online at ⟨http://www.esa.int/fringe99⟩.

Massman, F. H., J. C. Neymeyer, K. Raimondo, K. Enninghorst, and H. Li (1998). Quality of the D-PAF orbits before and after the inclusion of PRARE data. *Proceedings of 3rd ERS Symposium* (Vol. 3, SP-414). ESA, Noordwijk, The Netherlands, pp. 1655–1660.

Massom, R. A., M. Pook, J. C. Comiso, N. Adams, J. Turner, T. Lachlan-Cope, and T. Gibson (2004). Precipitation over the interior East Antarctic Ice Sheet related to mid-latitude blocking-high activity. *Journal of Climate*, **17**(10), 1914–1928.

Massonnet, D. (1995) Limitations to SAR interferometry due to instrument, climate or target geometry instabilities. *EARSeL Advances in Remote Sensing, Topography from Space*, **4**(2), 19–25.

Massonnet, D. (1997). Satellite radar interferometry. *Scientific American*, February, 46–53.

Massonnet, D. (2001). The interferometric cartwheel, a constellation of low cost receiving satellites to produce radar images that can be coherently combined. *International Journal of Remote Sensing*, **22**(12), 2413–2430.

Massonnet, D. and K. L. Feigl (1995). Discrimination of geophysical phenomena in satellite radar interferograms. *Geophysical Research Letters*, **22**, 1537–1540.

Massonnet, D. and K. L. Feigl (1998). Radar interferometry and its application to changes in the Earth's surface. *Reviews of Geophysics*, **36**(4), 441–500.

Massonnet, D. and T. Rabaute (1993). Radar interferometry: Limits and potential. *IEEE Transactions on Geoscience and Remote Sensing*, **31**(2), 455–464.

Massonnet, D. and H. Vadon (1995). ERS-1 internal clock drift measured by interferometry. *IEEE Transactions on Geoscience and Remote Sensing*, **33**(2), 401–408.

Massonnet, D., M. Rossi, C. Carmona, F. Adragna, G. Peitzer, K. Feigl and T. Rabaute (1993). The displacement field of the Landers Earthquake mapped by radar interferometry. *Nature*, **364**(8), 138–142.

Massonnet, D., P. Briole, and A. Arnaud (1995). Deflation of Mount Etna monitored by spaceborne radar interferometry. *Nature*, **375**, 567–570.

Massonnet, D., Vadon, H., and M. Rossi (1996). Reduction of the need for phase unwrapping in radar interferometry. *IEEE Transactions on Geoscience and Remote Sensing*, **32**, 489–497.

Massonnet, D., Thouvenot, E., Ramongassie, S., Phalippou, L. (2000). A wheel of passive radar microsats for upgrading existing SAR projects. *Proceedings of International Geoscience and Remote Sensing Symposium IGARSS '00, Honolulu, HI, 24–28 July 2000*. Institute of Electrical and Electronic Engineers, Piscataway, NJ, Vol. 3, pp. 1000–1003.

Mattar, K. E. and A. L. Gray (2002). Reducing ionospheric electron density errors in satellite radar interferometry applications. *Canadian Journal of Remote Sensing*, **28**(4), 593–600.

Mattar, K. E., A. L. Gray, D. Geudtner, and P. W. Vachon (1999). Interferometry for DEM and terrain displacement: Effects of inhomogeneous propagation. *Canadian Journal of Remote Sensing*, **25**(1), 60–69.

McLeod, I. H., I. G. Cumming, and M. S. Seymour (1998). Envisat ASAR data reduction: Impact on SAR interferometry. *IEEE Transactions on Geoscience and Remote Sensing*, **36**(2), 589–602.

Meyer, F. (2004). Topography and displacement of polar glaciers from multi-temporal SAR interferograms: Potentials, error analysis and validation. *Proceedings of the XXth ISPRS Congress, July 12–23, Istanbul* (Commission 3). International Society for Photogrammetry and Remote Sensing, Istanbul, Turkey.

Michel, R. and E. Rignot (1999). Flow of Glaciar Moreno, Argentina, from repeat-pass Shuttle imaging radar images: Comparison of the phase correlation method with radar interferometry. *Journal of Glaciology*, **45**(149), 93–100.

Migliaccio, M. and F. Bruno (2002). Coherence loss minimization in SAR interferometric registration. *International Geoscience and Remote Sensing Symposium, Toronto, Canada, June 24–28*.

Moccia, A. and S. Vetrella (1992). A tethered interferometric SAR for a topographic mission. *IEEE Transactions on Geoscience and Remote Sensing*, **30**(1), 103–109.

Mohr, J. J. (1997). Repeat track SAR interferometry: An investigation of its utility for studies of glacier dynamics. PhD thesis, Technical University of Denmark, Copenhagen.

Mohr, J. J. and S. N. Madsen (1996a). Application of interferometry to studies of glacier dynamics. *Proceedings of International Geoscience and Remote Sensing Symposium IGARSS '96, Lincoln, NE, 27–31 May 1996*. Institute of Electrical and Electronic Engineers, Piscataway, NJ, pp. 276–278.

Mohr, J. J. and S. N. Madsen (1996b). Multi-pass interferometry for studies of glacier dynamics. *Proceedings of "Fringe 96" Workshop on ERS SAR Interferometry, Zurich, Switzerland* (ESA SP-406). ESA, Noordwijk, The Netherlands, pp. 345–352.

Mohr, J. J. and S. N. Madsen (1999). Error analysis for interferometric SAR measurements of ice sheet flow. *Proceedings of International Geoscience and Remote Sensing Symposium IGARSS '99, Hamburg, Germany, June 28–July 2*. Institute of Electrical and Electronics Engineers, Piscataway, NJ, pp. 98–100.

Mohr, J. J. and S. N. Madsen (2000). Processing interferometric ERS-1/2 Tandem data cast to cast in Greenland. *ESA ERS–Envisat Symposium, Gothenburg, Sweden, October 16–20*.

Mohr, J. J. and S. N. Madsen (2001). Geometric calibration of ERS satellite SAR images. *IEEE Transactions on Geoscience and Remote Sensing*, **39**(4), 842–850.

Mohr, J. J. and N. Reeh (2002). Glacier surface velocity measurements from radar interferometry and the principle of mass conservation. *Proceedings of International Geoscience and Remote Sensing Symposium IGARSS '02, 22–28 June 2002, Toronto, Canada*. Institute of Electrical and Electronic Engineers, Piscataway, NJ, Vol. 2, pp. 1054–1056.

Mohr, J. J., S. N. Madsen and N. Reeh (1997). ERS Tandem study of glacier dynamics in NE-Greenland. *Proceedings of the 3rd ERS Symposium, ESA SP-414, Florence, Italy, 17–21 March 1997*. ESA, Noordwijk, The Netherlands, Vol. II, pp. 989–994.

Mohr, J. J., N. Reeh, and S. N. Madsen (1998). Three dimensional glacial flow and surface elevation measured with radar interferometry. *Nature*, **391**(6664), 273–276.

Mohr, J. J., N. Reeh, and S. N. Madsen (2003). Accuracy of three-dimensional glacier surface velocities derived from radar interferometry and ice-sounding radar measurements. *Journal of Glaciology*, **49**(165), 210–222.

Monti Guarnieri, A. and C. Prati (1996). ScanSAR focusing and interferometry. *IEEE Transactions on Geoscience and Remote Sensing*, **34**, 1029–1038.

Monti Guarnieri, A. and C. Prati (1997). SAR interferometry: A "quick and dirty" coherence estimator for data. *IEEE Transactions on Geoscience and Remote Sensing*, **35**(3), 660–669.

Monti Guarnieri, A. and C. Prati (1999). An interferometric quick-look processor. *IEEE Transactions on Geoscience and Remote Sensing*, **37**(2), 861–866.

Monti Guarnieri, A., C. Prati, and F. Rocca (1995). Interferometry with ScanSAR, *Proceedings of International Geoscience and Remote Sensing Symposium IGARSS '95, Florence, Italy, 10–14 July 1995*. Institute of Electrical and Electronic Engineers, Piscataway, NJ, pp. 550–552.

Monti Guarnieri, A., C. Prati, F. Rocca, and Y.-L. Desnos (1998). Wide baseline interferometry with very low resolution SAR systems. *Proceedings of European Conference on Synthetic Aperture Radar, EUSAR '98, Berlin*. VDE Verlag, Frankfurt, Germany, pp. 361–364.

Moore, R. K., J. P. Claassen, and Y. H. Lin (1981). Scanning spaceborne synthetic aperture radar with integrated radiometer. *IEEE Transactions on Aerospace and Electronic Systems*, **17**(3), 410–420.

Moreira, A. (2003). TerraSAR-X: Upgrade to a fully polarimetric imaging mode. *Proceedings of POLinSAR 2003, Workshop on Applications of SAR Polarimetry and Polarimetric Interferometry, Frascati, Italy, 14–16 January*. ESA/ESRIN, Noordwijk, The Netherlands. Available online at ⟨http://earth.esa.int/polinsar/se_spacebornesar.html⟩.

Morris, K., S. Li, and M. Jeffries (1999). Meso- and micro-scale sea-ice motion in the East Siberian Sea as determined from ERS-1 SAR data. *Journal of Glaciology*, **45**(150), 370–383.

NASDA (1995). *JERS-1 Science Program Application Data Sets: JERS-1/SAR Interferometry* (CD-ROM, Vol. 1). National Space Development Agency of Japan, Tokyo.

Olmsted, C. (1993). *Alaska SAR Facility Scientific SAR User's Guide* (ASF-SD-003). Alaska SAR Facility, Juneau, 53 pp.

Padman, L. and C. Kottmeier (2000). High-frequency ice motion and convergence in the Weddell Sea. *Journal of Geophysical Research*, **105**(19), 3379–3400.

Padman, L., D. MacAyeal, and E. Rignot (2000). *Analysis of Sub-ice-shelf Tides in the Weddell Sea Using SAR interferometry* (Filchner-Ronne Ice Shelf Programme (FRISP) Report 13). Alfred-Wegener-Institut für Polar- und Meeresforschung, Bremerhaven, Germany, pp. 60–65.

Padman, L., H. A. Fricker, R. Coleman, S. Howard, and L. Erofeeva (2002). A new tide model for the Antarctic ice shelves and seas. *Annals of Glaciology*, **34**, 247–254.

Papathanassiou, K. P. and S. R. Cloude (2001). Single baseline polarimetric SAR interferometry. *IEEE Transactions on Geoscience and Remote Sensing*, **39**(11), 2352–2363.

Parashar, S., E. Langham, J. McNally, and S. Ahmed (1993). Radarsat mission requirements and concepts. *Canadian Journal of Remote Sensing*, **18**(4), 280–288.

Parish, T. R. and D. H. Bromwich (1991). Continental-scale simulation of the Antarctic katabatic wind regime. *Journal of Climate*, **4**(2), 135–146.

Paterson, W. S. B. (1994). *The Physics of Glaciers* (3rd edn.). Pergamon/Elsevier Science, Oxford, UK, 480 pp.

Pattyn, F. and D. Derauw (2002). Ice-dynamic conditions of Shirase Glacier, Antarctica, inferred from ERS SAR interferometry. *Journal of Glaciology*, **48**(163), 559–565.

Prati, C., F. Rocca, A. Guarnieri, and E. Damonti (1990). Seismic migration for SAR focusing: Interferometric applications. *IEEE Transactions on Geoscience and Remote Sensing*, **28**(4), 627–640.

Pritt, M. D. (1996). Phase unwrapping by means of multigrid techniques for interferometric SAR. *IEEE Transactions on Geoscience and Remote Sensing*, **34**, 728–738.

Pritt, M. D. and J. S. Shipman (1994). Least-squares two-dimensional phase unwrapping using FFTs. *IEEE Transactions on Geoscience and Remote Sensing*, **32**, 706–708.

Rabus, B., M. Eineder, A. Roth, and R. Bamler (2003). The Shuttle Radar Topography Mission: A new class of digital elevation models acquired by spaceborne radar. *ISPRS Journal of Photogrammetry and Remote Sensing*, **57**, 241–262.

Reeh, N. and N. Gundestrup (1985). Mass balance of the Greenland Ice Sheet at Dye 3. *Journal of Glaciology*, **31**(108), 198–200.

Reeh, N., Bøggild, C. E., and Oerter, H. (1994). *Surge of Storstrømmen, a large outlet glacier from the inland ice of North-East Greenland*, Grønlands Geologiske Undersøgelse Rapport, 162, 201–209, Copenhagen.

Reeh, N., S. N. Madsen, and J. J. Mohr (1999). Combining SAR interferometry and the equation of continuity to estimate the three-dimensional glacier surface velocity. *Journal of Glaciology*, **45**(151), 533–538.

Reeh, N., J. J. Mohr, S. N. Madsen, H. Oerter, and N. Gundestrup (2003). Three-dimensional surface velocities of the Storstrømmen Glacier derived from radar interferometry and ice-sounding radar measurements. *Journal of Glaciology*, **49**(165), 201–209.

Rees, W. G. (2001). *Physical Principles of Remote Sensing* (2nd edn). Cambridge University Press, Cambridge, UK, 372 pp.

Refice, A., G. Satalino, S. Stramaglia, M. T. Chiaradi, and N. Veneziani (1999). Weights determination for minimum cost flow InSAR phase unwrapping. *Proceedings of International Geoscience and Remote Sensing Symposium IGARSS '99, Hamburg, Germany, 28 June–2 July 1999*. Institute of Electrical and Electronic Engineers, Piscataway, NJ, Vol. 2, pp. 1342–1344.

Richards, J. A. and X. Jia (2005). *Remote Sensing Digital Image Analysis* (4th edn.). Springer-Verlag, Berlin.

Rignot, E. (1995). Backscatter model for the unusual radar properties of the Greenland Ice Sheet. *Journal of Geophysical Research*, **100**(E5), 9389–9400.

Rignot, E. (1996). Tidal motion, ice velocity and melt rate of Petermann Gletscher, Greenland, measured from radar interferometry. *Journal of Glaciology*, **42**(142), 476–485.

Rignot, E. (1998a). Radar interferometry detection of hinge-line migration on Rutford Ice Stream and Carlson Inlet, Antarctica. *Annals of Glaciology*, **27**, 25–32.

Rignot, E. (1998b). Hinge-line migration of Petermann Gletscher, north Greenland, detected using satellite radar interferometry. *Journal of Glaciology*, **44**, 469–476.

Rignot, E. (2002). Mass balance of East Antarctic glaciers and ice shelves from satellite data. *Annals of Glaciology*, **34**, 217–227.

Rignot, E. and R. H. Thomas (2002). Mass balance of polar ice sheets. *Science*, **297**, 1502–1506.

Rignot, E. J., K. C. Jezek, and H. G. Sohn (1995). Ice flow dynamics of the Greenland ice sheet from SAR interferometry. *Geophysical Research Letters*, **22**, 575–578.

Rignot, E., R. Forster, and B. Isacks (1996). Interferometric radar observations of Glaciar San Rafael, Chile. *Journal of Glaciology*, **42**, 279–291.

Rignot, E. J., S. P. Gogineni, W. B. Krabill, and S. Ekholm (1997). North and northeast Greenland ice discharge from satellite radar interferometry. *Science*, **276**, 934–937.

Rignot, E., L. Padman, D. R. MacAyeal, and M. Schmeltz (2000). Observation of ocean tides beneath the Filchner-Ronne Ice Shelf using SAR interferometry. *Journal of Geophysical Research*, **105**(C8), 19615–19630.

Rignot, E., K. Echelmeyer, and W. B. Krabill (2001). Penetration depth of interferometric synthetic-aperture radar signals in snow and ice. *Geophysical Research Letters*, **28**(18), 3501–3504.

Robinson, D. (1993). Phase unwrapping methods. In: W. R. Robinson and G. T. Reid (eds.), *Interferogram Analysis: Digital Fringe Pattern Measurement Techniques*. Institute of Physics, Bristol, pp. 194–229.

Rocca, F., C. Prati, and A. Ferretti (1997). An overview of SAR interferometry. In: T. D. Guyenne (ed.), *Proceedings of 3rd ERS Symposium, Space at the Service of Our Environment, Florence, Italy, 17–21 March 1997*. ESA, Noordwijk, The Netherlands. Available online at ⟨http://earth.esa.int/symposia/program-details/speeches/rocca-et-al/⟩.

Rodríguez, E. and J. M. Martin, Theory and design of interferometric synthetic aperture radars, *IEEE Proceedings F*, **139**(2), 147–159.

Rosen, P. A., S. Hensley, I. R. Joughin, F. K. Li, S. N. Madsen, E. Rodríguez, and R. M. Goldstein (2000). Synthetic aperture radar interferometry. *Proceedings of the IEEE*, **88**(3), 333–382.

Rosen, P. A., S. Hensley. G. Peltzer, and M. Simons (2004). Updated repeat orbit interferometry package released. *EOS, Transactions of the American Geophysical Union*, **3**(47).

Rossi, M., B. Rogron, and D. Massonnet (1996). JERS-1 SAR image quality and interferometric potential. *IEEE Transactions on Geoscience and Remote Sensing*, **34**(3), 824–827.

Rott, H. and A. Siegel (1996). Glaciological studies in the Alps and in Antarctica using ERS interferometric SAR. *Proceedings of "Fringe '96" Workshop on ERS SAR Interferometry, Zurich, Switzerland, September 30–October 2* (Vol. 2, ESA SP-406). ESA, Noordwijk, The Netherlands, pp. 149–159.

Rott, H., K. Sturm, and H. Miller (1993). Active and passive microwave: Signatures of Antarctic firn by means of field measurements and satellite data. *Annals of Glaciology*, **17**, 337–343.

Rufino, G., A. Moccia, and S Asposito (1996). DEM generation by means of ERS tandem data. *Proceedings of "Fringe '96" Workshop on ERS SAR Interferometry, Zurich, Switzerland, September 30–October 2*. ESA, Noordwijk, The Netherlands.

Samson, J. (1996). Coregistration in SAR interferometry. Master's thesis, Faculty of Geodetic Engineering, Delft University of Technology.

Sanchez, J. I. and H. Laur (1997). *The ERS SAR Products, Systems and Performances* (Earthnet Online, ESA). Available online at ⟨http://earth1.esrin.esa.it/florence/papers/program-details/data/sanchez/index.html⟩.

Sandwell, D. and E. Price (1998). Phase gradient approach to stacking interferograms. *Journal of Geophysical Research*, **103**(B12), 30183–30204.

Scambos, T. A. and M. A. Fahnestock (1998). Improving digital elevation models over ice sheets using AVHRR-based photoclinometry. *Journal of Glaciology*, **44**(146), 97–103.

Scambos, T. A. and T. Haran (2002). An image-enhanced DEM of the Greenland Ice Sheet. *Annals of Glaciology*, **34**, 291–298.

Scambos, T. A., G. Kavaran, and M. A. Fahnestock (1999). Improving AVHRR resolution through data cumulation for mapping polar ice sheets. *Remote Sensing of Environment*, **69**(1), 56–66.

Scharroo, R. and P. N. Visser (1998). Precise orbit determination and gravity field improvement for the ERS satellites. *Journal of Geophysical Research*, **103**(C4), 8113–8127.

Scheiber, R. and A. Moreira (2000). Coregistration of interferometric SAR images using spectral diversity. *IEEE Transactions on Geoscience and Remote Sensing*, **38**(5), 2179–2191.

Schwäbisch, M. and D. Geudtner (1995). Improvement of phase and coherence map quality using azimuth pre-filtering: Examples from ERS-1 and X-SAR. *Proceedings of Inter-*

national Geoscience and Remote Sensing Symposium IGARSS '95, Florence, Italy, 10–14 July 1995. Institute of Electrical and Electronic Engineers, Piscataway, NJ, pp. 205–207.

Séguin, G. and R. Girard (2002). Interferometric missions using small satellite SAR satellites: Pecora 15. Land Satellite Information IV, ISPRS Commission I. *Proceedings of Future Intelligent Earth Observation Satellites (FIEOS) 2002 Conference, 10–15 November 2002, Denver, CO*. Available online at ⟨http://www.isprs.org/commission1/proceedings/paper/00053.pdf⟩.

Sharov, A. I. and K. Gutjahr (2002). Some methodological enhancements to InSAR surveying of polar ice caps. *Observing Our Environment from Space: Proceedings of 21st EARSel Symposium*. European Association of Remote Sensing Laboratories, Hannover, Germany, pp. 65–72.

Sharov, A. I., K. Gutjahr, F. Meyer, and M. Schardt (2002). Methodical alternatives to the glacier motion measurement from differential SAR interferometry. *International Archives of Photogrammetry and Remote Sensing*, **XXXIV**(3A), 324–329.

Shemer, L., M. Marom, and D. Markman (1993). Estimates of currents in the nearshore ocean region using interferometric synthetic aperture radar. *Journal of Geophysical Research*, **98**(C4), 7001–7010.

Shepherd, A. and J. J. Mohr (2001). ERS SAR interferometry-derived information on Antarctica and Greenland: The VECTRA Cooperation Project. *Earth Observation Quarterly*, **68**, 1–4.

Short, N. H. and A. L. Gray (2004). Potential for Radarsat-2 interferometry: Glacier monitoring using speckle tracking. *Canadian Journal of Remote Sensing*, **30**(3), 504–509.

Shuman, C. A. and R. B. Alley (1993). Spatial and temporal characterization of hoar formation in central Greenland using SSM/I brightness temperatures. *Geophysical Research Letters*, **20**(23), 2643–2646.

Small, D., C. Werner, and D. Nüesch (1993). Baseline modelling for ERS-1 SAR interferometry. *Proceedings of International Geoscience and Remote Sensing Symposium IGARSS '93, Tokyo, Japan, 18–21 August 1993*. Institute of Electrical and Electronic Engineers, Piscataway, NJ, pp. 1204–1206.

Smith, B. D., B. E. Engelhardt, and D. H. Mutz (2002). The Radarsat-MAMM Automated Mission Planner. *Artificial Intelligence Magazine*, **23**(2), 25–36.

Smith, J. A. (ed.) (2004). *InSAR: Interferometric Synthetic Aperture Radar Concept Study Report*. NASA Jet Propulsion Laboratory, Pasadena, CA.

Smith, L. C. (2002). Emerging applications of interferometric synthetic aperture radar (InSAR) in geomorphology and hydrology. *Annals of the Association of American Geographers*, **92**(3), 385–398.

Solaas, G. A. (1994). *ERS-1 Interferometric Baseline Algorithm Verification* (Technical Report ES-TN-DPE-OM-GS02, version 1.0. ESA/ESRIN, Noordwijk, The Netherlands. Available online at ⟨http://gds.esrin.esa.it:80/CEFB565F/CORBITS⟩.

Soumekh, M. (1999). *Synthetic Aperture Radar Signal Processing with MATLAB Algorithms*. John Wiley & Sons, New York, 648 pp.

Spagnolini, U. (1993). 2-D phase unwrapping and phase aliasing. *Geophysics*, **58**(9), 1324–1334.

Steffen, K., W. Abdalati, and I. Sherjal (1999). Hoar development on the Greenland ice sheet. *Journal of Glaciology*, **45**(148), 63–68.

Strozzi, T., A. Luckman, and T. Murray (2000). The evolution of a glacier surge observed with the ERS satellites. *Proceedings of ERS–Envisat Symposium, Gothenburg, Sweden, October 16–20*. ESA, Noordwijk, The Netherlands.

Strozzi, T., U. Wegmüller, and C. Mätzler (1999). Mapping wet snow covers with SAR interferometry. *International Journal of Remote Sensing*, **20**(12), 2395–2403.

Strozzi, T., A. Luckman, T. Murray, U. Wegmüller, and C. L. Werner (2002). Glacier motion estimation using SAR offset-tracking procedures. *IEEE Transactions on Geoscience and Remote Sensing*, **40**(11), 2384–2391.

Sun, G., K. J. Ranson, J. Bufton, and M. Roth (2000). Requirement of ground tie points for InSAR DEM generation. *Photogrammetric Engineering and Remote Sensing*, **66**, 81–85.

Tarayre, H. and D. Massonnet (1994). Effects of a refractive atmosphere on interferometric processing. *Proceedings of International Geoscience and Remote Sensing Symposium IGARSS '94, Pasadena, CA*. Institute of Electrical and Electronic Engineers, Piscataway, NJ, pp. 717–719.

Tarayre, H. and D. Massonnet (1996). Atmospheric propagation heterogeneities revealed by ERS-1 interferometry. *Geophysical Research Letters*, **23**(9), 989–992.

Thiel, K. H., P. Hartl, and X. Wu (1995). Monitoring the ice movements with ERS SAR interferometry in the Antarctic region. *Proceedings of 2nd ERS Applications Workshop, London, December* (ESA SP-383). ESA, Noordwijk, The Netherlands, pp. 219–223.

Thiel, K. H., A. Wehr, and X. Wu (1996). Ice sheet and ice shelf movement detection in test area: Antarctic. *Proceedings of Workshop on Glaciological Applications of Satellite Radar Interferometry, March 28–29*. NASA Jet Propulsion Laboratory, Pasadena, CA.

Thomas, R. H. (2004). Greenland: Recent mass-balance observations. In: J. L. Bamber and A. J. Payne (eds.), *Mass Balance of the Cryosphere*. Cambridge University Press, Cambridge, UK, pp. 393–436.

Thomas, R., E. Rignot, G. Casassa, P. Kanagaratnam, C. Acua, T. Akins, H. Brecher, E. Frederick, P. Gogineni, W. Krabill et al. (2004). Accelerated sea-level rise from West Antarctica. *Science*, **306**(5694), 255–258.

Tomiyasu, K. (1981). Conceptual performance of a satellite borne, wide swath synthetic aperture radar. *IEEE Transactions on Geoscience and Remote Sensing*, **GE-19**(2), 108–116.

Toutin, T. and L. Gray (2000). State-of-the-art of elevation extraction from satellite SAR data. *International Society for Photogrammetry and Remote Sensing Journal of Photogrammetry and Remote Sensing*, **55**, 13–33.

Touzi, R., A. Lopes, J. Bruniquel, and P. W. Vachon (1999). Coherence estimation for SAR imagery. *IEEE Transactions on Geoscience and Remote Sensing*, **37**(1), 135–149.

Turner, J., W. M. Connolley, S. Leonard, G. J. Marshall, and D. G. Vaughan (1999). Spatial and temporal variability of net snow accumulation over the Antarctic from ECMWF re-analysis project data. *International Journal of Climatology*, **19**(7), 697–724.

Ulaby, F. T., Moore, R. K., and A. K. Fung (1982). *Radar Remote Sensing and Surface Scattering and Emission Theory: Vol. 2. Microwave Remote Sensing—Active and Passive*. Addison-Wesley, Reading, MA, 1069 pp.

Ulbricht, A., K. P. Papathanassiou, and S. R. Cloude (1998). Polarimetric analysis of SAR-interferograms in L- and P-band. *Proceedings of International Geoscience and Remote Sensing Symposium IGARSS '98, Seattle, WA*. Institute of Electrical and Electronic Engineers, Piscataway, NJ, Vol. 3, pp. 1647–1649.

Usai, S. (2001). A new approach for long term monitoring of deformations by differential SAR interferometry. PhD thesis, Delft University of Technology.

USGS (1993). *Digital Elevation Models: Data User's Guide*. U.S. Geological Survey, Reston, VA.

Vachon, P. W., Gray, A. L., and Geudtner, D. (1995a). ERS-1 SAR repeat-pass interferometry: Temporal coherence and implications for Radarsat. *Proceedings of 17th*

Canadian Symposium on Remote Sensing, Saskatoon, Saskatchewan, Canada, Canadian Remote Sensing Society, Ottawa, pp. 803–898.

Vachon, P. W., Geudtner, D., Gray, A. L., and Touzi, R. (1995b). ERS-1 synthetic aperture radar repeat-pass interferometry studies: Implications for Radarsat. *Canadian Journal of Remote Sensing*, **21**, 441–454.

Vachon, P. W., D. Geudtner, K. Mattar, A. L. Gray, M. Brugman, and I. G. Cumming (1996). Differential SAR interferometry measurements of Athabasca and Saskatchewan Glacier flow rate. *Canadian Journal of Remote Sensing*, **22**(3), 287–296.

Vachon, P. W., A. L. Gray, and K. E. Mattar (1998). Radarsat and ERS repeat-pass interferometry at CCRS. *Proceedings of 2nd Workshop on SAR Interferometry, Tsukuba, Japan, 18–20 November 1997*, pp. 173–182.

van Genderen, J. L. and R. Gens (1996). SAR interferometry: Principles and research trends. *Proceedings of Geoinformatics '96 Symposium, 16–18 October 1996, Wuhan, China*. Wuhan Technical University of Surveying and Mapping, Wuhan, China, Vol. 1, pp. 99–106.

Vaughan, D. G., J. L. Bamber, M. B. Giovinetto, J. Russell, and A. P. R. Cooper (1999). Reassessment of net surface mass balance in Antarctica. *Journal of Climate*, **12**(4), 933–946.

Vettore, A., S. Ponte, and N. Crocetto (2002). Space-based surface change detection with differential (SAR) interferometry: Potentiality, dimulations and preliminary investigations. *Symposium on Geospatial Theory, Processing and Applications: International Society for Photogrammetry and Remote Sensing Commission IV, Ottawa, July 9–12*. ISPRS, Istanbul, Turkey.

Vincent, P. and J. Rundle (1998). Synthetic aperture radar interferometry capability now available to universities. *EOS, Transactions of the American Geophysical Union*, **79**, 34.

Wegmüller, U. and T. Strozzi (1998). Characterization of different interferometry approaches. *Proceedings of EUSAR '98, May 25–27, Friedrichshafen, Germany*. VDE Verlag, Frankfurt, Germany, pp. 237–240.

Wegmüller, U., C. L. Werner, T. Strozzi, and A. Wiesmann (2002). *Phase Unwrapping with Gamma ISP* (technical report). Gamma Remote Sensing AG, Bern, Switzerland. Available online at ⟨http://www.gamma-rs.ch/docs/unwrapping.pdf⟩.

Werner, C., S. Hensley, R. M. Goldstein, P. A. Rosen, and H. A. Zebker (1992). Techniques and applications of SAR interferometry for ERS-1: Topographic mapping, change detection, and slope measurement. *1st ERS-1 Symposium: Space at the Service of Our Environment, Cannes, France, November 4–6* (ESA SP-359). ESA, Noordwijk, The Netherlands, pp. 205–210.

Werner, C. L., U. Wegmüller, and T. Strozzi (2002). Processing strategies for phase unwrapping for INSAR applications. *Proceedings of 4th European Conference on Synthetic Aperture Radar EUSAR 2002, Cologne, Germany, 4–6 June 2002*. VDE Verlag, Frankfurt, Germany, Vol. 1, pp. 353–356.

Wilkinson, A. J. (1997). Techniques for 3-D surface reconstruction using synthetic aperture radar interferometry. Doctoral thesis, University College London, 395 pp.

Williams, J. (2001). *GIS Processing of Geocoded Satellite Data*. Springer-Praxis, Chichester, UK, 327 pp.

Winebrenner, D. P., I. R. Joughin, and M. A. Fahnestock (1997). Interferometric SAR for observation of glacier motion and firn penetration. *Proceedings of 3rd ERS Scientific Symposium, March 17–21, Florence, Italy* (ESA SP-414). ESA, Frascati, Italy.

Wu, X. and Thiel K.-H. (1996). The use of tandem data in the Antarctic area. *Proceedings of "Fringe '96" Workshop on ERS SAR Interferometry, Zurich, Switzerland* (ESA SP-406). ESA, Noordwijk, The Netherlands.

Young, N. W. and G. Hyland (2002). Velocity and strain rates derived from InSAR analysis over the Amery Ice Shelf, East Antarctica. *Annals of Glaciology*, **34**, 228–234.

Young, N. W., D. Hall, and G. Hyland (1996). Directional anisotropy of C-band backscatter and orientation of surface microrelief in East Antarctica. *Proceedings of 1st Australian ERS Symposium, University of Tasmania, Hobart, February 6* (COSSA Publication 037). CSIRO Office of Space Science and Applications, Canberra, pp. 117–126.

Zandbergen, R., J. M. Dow, F. Martínez Fadrique, M. Romay Merino, and R. Píriz (1997). *Progress in ERS Orbit and Tracking Data Analysis* (Earthnet Online, ESA). Available online at ⟨http://earth1.esrin.esa.it/florence/papers/program-details/data/zandbergen/index.html⟩.

Zebker, H. A. (2000). Studying the Earth with interferometric radar. *Computing in Science and Engineering*, **2**(3), 52–60.

Zebker, H. A. and R. M. Goldstein (1986). Topographic mapping from interferometric synthetic aperture radar observations. *Journal of Geophysical Research*, **91**(B5), 4993–4999.

Zebker, H. A., Rosen, P. A., and Hensley, S. (1997). Atmospheric effects in interferometric synthetic aperture radar surface deformation. *Journal of Geophysical Research*, **102**, 7547–7563.

Zebker, H. A., P. A. Rosen, R. M. Goldstein, A. Gabriel, and C. L. Werner (1994a). On the derivation of coseismic displacement fields using differential radar interferometry: The Landers Earthquake. *Journal of Geophysical Research*, **99**(B10), 19617–19634.

Zebker, H. A., C. L. Werner, P. A. Rosen, and S. Hensley (1994b). Accuracy of topographic maps derived from ERS-1 interferometric radar. *IEEE Transactions on Geoscience and Remote Sensing*, **32**(4), 823–836.

Zebker, H. A. and J. Villasenor (1992). Decorrelation in interferometric radar echoes. *IEEE Transactions on Geoscience and Remote Sensing*, **30**(5), 950–959.

Zebker, H. A. and Y. Lu (1997). Phase unwrapping algorithms for radar interferometry: Residue-cut, least squares, and synthesis algorithms. *Journal of the Optical Society of America A*, **15**(3), 586–598.

Zhu, S., C. Reigber, and F.-H. Massmann (1996). *ERS Standards Used at D-PAF: The German PAF for ERS* (Technical Report ERS-D-STD-31101 Rev. B, GFZ/D-PAF). Zentrum der Deutschland für Luft- und Raumfahrt, Oberpfaffenhofen, Germany.

Zink, M. (2004). The TerraSAR-L interferometric mission objectives. *Proceedings of Fringe '03 Workshop, December 1–5* (ESA SP-550). ESA/ESRIN, Frascati, Italy.

Zink, M., G. Krieger, and T. Amiot (2004). Interferometric performance of a cartwheel constellation for TerraSAR-L. *Proceedings of Fringe 2003 Workshop, December 1–5* (ESA SP-550). ESA/ESRIN, Frascati, Italy.

Zwally, H. J., B. Schutz, W. Abdalati, J. Abshire, C. Bentley, A. Brenner, J. Bufton, J. Dezio, D. Hancock, D. Harding et al. (2002). ICESat's laser measurements of polar ice, atmosphere, ocean and land. *Journal of Geodynamics*, **34**, 405–445.

3

Satellite remote sensing of ice sheet parameters and processes

3.1 INTRODUCTION

In this chapter, we outline and evaluate the important role of satellite remote sensing in deriving important glaciological parameters. For convenience, the chapter is split into sections which deal with the measurement of individual parameters or processes; in reality, these are intimately interrelated. A summary of satellite sensor classes used to measure these parameters is provided in Table 3.1, with the parameters listed in the order in which they are subsequently discussed in the text. This highlights the complementarity of the different data sources. Indeed, it will become apparent that the synergistic application of data from multiple sources is once again invariably more powerful than the use of individual datasets alone. Dynamics information can be obtained on surface topography (slope and elevation), ice motion (velocity, flow patterns, strain rate), fracture (crevasse and rift) patterns, ice thickness (for floating ice), and iceberg calving, size, and drift. Surface property measurements include: surface and near-surface temperature, albedo, snow-grain size, roughness, patterns of surface melt and freeze, snow accumulation rate, and inference of surface wind patterns. It should be noted that satellite imagery is also extensively used to plan field operations, assess route safety—i.e., avoid crevasse fields—and characterize the surface glaciology at sampling sites. The International Trans-Antarctic Scientific Expedition (ITASE) project, for example, has used Radarsat-1 SAR (Synthetic Aperture Radar) imagery (ground resolution 25 m) obtained in October 1997 during the first Antarctic Mapping Mission (AMM: see ⟨http://www-bprc.mps.ohio-state.edu/GDG/itase_image.htm⟩).

3.2 IMPROVED DETECTION AND MAPPING OF SURFACE FEATURES

3.2.1 High-resolution visible–thermal infrared methods

A fundamental application of satellite imagery is the mapping and monitoring of ice sheet surface morphological detail. Based upon three decades of experience, trained

Table 3.1. Summary table of ice-sheet-related applications of major satellite sensor classes.

Ice sheet parameter	PMW	SAR	Low-res. visible–TIR#	Hi-res. visible–TIR#	Radar Scatt	Radar Alt	Laser Alt#
Ice sheet morphology/extent	–	P	P	P	–	S	S
Ice sheet facies	S	P	S	S	P	S	–
Melt/Refreeze extent and onset	P	P	S	S	P	S	–
Changes in ice sheet margins	–	P	P	P	–	S	S
Iceberg detection, tracking and statistics	S	P	P	P	P	–	–
Iceberg thickness	–	–	–	–	–	P	P
Topography and elevation change	–	P	P	S	–	P	P
Accumulation (rate)	R	R	–	–	R	–	–
Ice flow (motion)	–	P	S	P	–	–	–
Balance velocity and flux	–	P	P	–	–	P	P
Strain rate	–	P	–	P	–	–	–
Surge behavior	–	P	–	P	–	–	–
Tidal motion	–	P	–	–	–	P	P
Grounding zone detection and monitoring	–	P	–	S	–	S	S
Ice discharge flux	–	P	–	P	–	P	P
Ice shelf basal melt and refreeze	–	P	–	–	–	P	P
Surface/Near-surface temperature	P	–	P	S	–	–	–
Surface albedo	–	–	P	S	–	–	S
Snowgrain size	S/R	–	P	S	S	–	–
Snow layering/volume characteristics	R	R	–	–	R	R	–
Snow impurity content	–	–	P	–	–	–	–
Surface roughness	–	P	P	P	P	P	P
Proxy wind patterns	–	–	–	–	P	–	–

TIR = thermal infrared; PMW = passive microwave; SAR = synthetic aperture radar; Scatt = radar scatterometer; Alt = radar altimeter; # = cloud-affected (note that visible to near-infrared sensors are also darkness-affected); P = primary data source; S = secondary data source; and R = research and development.

glaciologists can routinely extract much unique information from carefully processed satellite imagery. This information has proved to be of key importance in both extending limited surface observations and in planning and executing field measurement campaigns. Indeed, many subtle large-scale features which cannot be identified at ground level are readily apparent from space, and numerous discoveries have resulted. Surface features mirror "current" ice-flow characteristics, and are also a manifestation of basal/bedrock characteristics and conditions (Budd, 1970; McIntyre, 1986). They also contain imprints of past conditions that, when combined with other information—e.g., from ice-penetrating radar—can provide unique insight into changes in ice sheet flow behavior over long periods of time (centuries through millennia). In addition, features such as *ice rumples* may be useful as climate-change indicators (Scambos et al., 2004a). Rumples are surface manifestations of an ice shelf floating up and over bed obstructions and, as such, are potentially sensitive indicators of ice-shelf thickness changes related, for example, to changes in sub-shelf ocean temperatures (Scambos et al., 2004a). As we shall see in Section 3.8, surface features such as crevasse patterns also provide an important means of determining ice-flow velocity in satellite image time series. Cloud cover is a major limiting factor, as is polar darkness for shortwave channels. Cloud not only obscures the surface but also degrades the quality of data, in that the discrimination of cloud from snow/ice is difficult.

The unique ability of satellites to accurately detect and map ice-sheet surface morphological features, and systematically monitor ice-sheet margin change—i.e., areal extent (see Section 3.4)—dates back to the launch of Landsat-1 (formerly ERTS-1) in 1972. Simultaneously imaging in four bands (visible to near-IR) at a high spatial resolution of ~79 m across a 185-km swath width, initial cloud-free Multi-Spectral Scanner (MSS) data revealed numerous subtle ice sheet surface features for the first time, and had immediate value as a cost-effective means of mapping morphological detail and coastal configuration. Not only were these early images breathtaking in their beauty, but they also captured the imagination of a generation of glaciologists with their scientific content (Figure 3.1). The availability of multi-spectral datasets also heralded the dawn of digital image enhancement techniques. These include band ratioing for extracting surface property information by reducing the effects of variable terrain slope on the reflectivity signature—e.g., Orheim and Lucchitta (1987, 1988)—and principal components analysis to enhance surface morphology—e.g., Bindschadler and Scambos (1991). For a fuller assessment of the immense contribution of earlier datasets, please see Massom (1991).

These early data form the start of a unique time series which now extends over 30 years, enabling change detection over a variety of scales (Mika, 1997). This time series comprises data from increasingly sophisticated sensors, with a trend towards higher spatial and radiometric resolution greatly benefiting ice-sheet research. Important sensors include the Landsat Thematic Mapper (TM) and Enhanced TM+ (ETM+), Système Pour l'Observation de la Terre (SPOT) High Resolution Visible Radiometer (HRVR), High Resolution Geometric (HRG) and High Resolution Stereoscopic (HRS), and EOS Terra Advanced Spaceborne Thermal Emission and Reflection radiometer (ASTER). The Landsat TM and later sensors added Thermal Infrared (TIR) bands, giving them the additional ability of acquiring images during periods of polar darkness. Modern datasets can be accurately geo-referenced, even in the absence of Ground-Control Points (GCPs), by means of geographical coordinates derived from the orbit ephemeris data. The estimated geo-location error (1σ) of the ETM+, for example, is 50 m (Frezzotti et al., 2002a). Current SPOT sensors—e.g., the HRS on SPOT-5—also have an

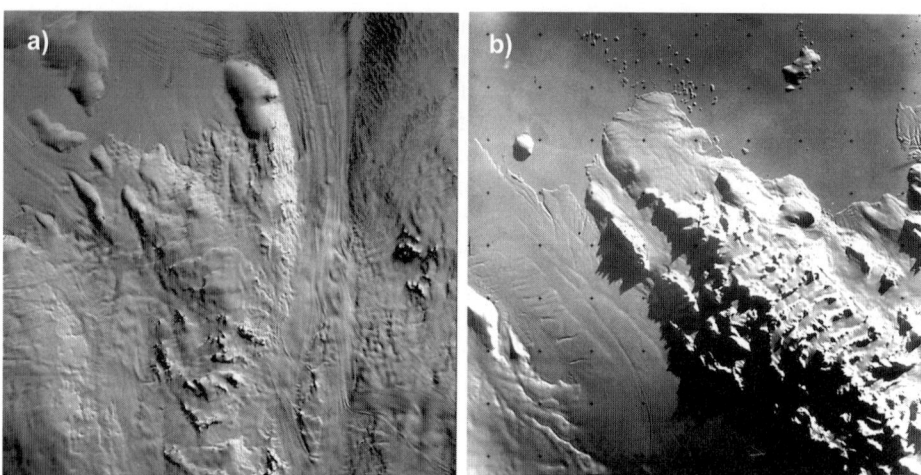

Figure 3.1. (a) Landsat 2 Multi-Spectral Scanner (MSS) image of the Jutulstraumen Ice Stream, Queen Maud Land, East Antarctica, acquired on October 30, 1975. The ice stream joins the Fimbul Ice Shelf in the upper part of the image. Flowlines, blue ice, and nunataks (rock outcrops surrounded by the ice sheet) are visible. NASA Landsat image 2281-07474. (b) Landsat 3 Return Beam Vidicon (RBV) image 30927-20382-C of the Rennick Glacier and the Oates Coast, northern Victoria Land, Antarctica, acquired on September 17, 1980. The MSS acquired data simultaneously in four spectral bands (0.5–0.6, 0.6–0.7, 0.7–0.8, and 0.8–1.1 µm). RBV collected data simultaneously in three spectral bands (0.475–0.575, 0.58–0.68, and 0.69–0.83 µm). The spatial resolution and swath width of both sensors is 80 m and 185 km, respectively (Massom, 1991).

Imagery courtesy of the U.S. Geological Survey Image Processing Facility, Flagstaff, AZ, obtained from ⟨http://woodshole.er.usgs.gov/project-pages/glacier_studies/gl_slide/⟩. (a) Figure 70 (p. B95) in Swithinbank (1988). (b) Figure 40 (p. B50) in Swithinbank (1988).

enhanced viewing capability (compared with the fixed angle viewing of Landsat sensors), with stereo-viewing for topographic mapping purposes (Baudoin et al., 2003; Reinartz et al., 2004). This capability, enabling the creation of a three-dimensional model of a surface based on the parallax between two images, has also been incorporated into the Terra ASTER (Toutin and Cheng, 2001), as well as high-resolution sensors onboard commercial Ikonos-II, Earth Resources Observation System-A (EROS-A) OrbView, and QuickBird satellites (Toutin, 2004). SPOT sensors, however, lack the capability of ASTER and the ETM+ to operate during periods of polar darkness—i.e., they do not possess a TIR channel. The follow-on program to SPOT, Pleïades, will be launched in 2008 (see Volume 1 of this book for details).

Similar sensors have been launched within other programs (see Chapter 5 of Volume 1 of this book). For example, the U.S. Earth Observing-1 (EO-1) satellite (launch November 2000; ⟨http://eo1.usgs.gov/⟩) carries the Advanced Land Imager (ALI), which is in effect an economical test-bed copy of the Landsat TM. It operates in nine multi-spectral bands (0.433–2.35 µm, resolution 30 m) and one panchromatic band (0.48–0.69 µm, resolution 10 m). Due to its 12-bit radiometric resolution, ALI is better suited than Landsat sensors to detecting surface targets of low brightness contrast (Raup et al., 2005). Although the swath width is only 37 km, future ALI sensors are planned to be equipped with a 185-km swath width equivalent to that of

Landsat sensors. The Indian Remote Sensing satellite IRS-1D (1997–present) is the latest in a series of satellites starting with the launch of IRS-1A in 1988. It carries three sensors—namely, the Linear Imaging Self-scanning Sensor III (LISS III), Panchromatic (Pan), and Wide Field Sensor (WiFS). While IRS Pan data are acquired at a high spatial resolution of 5.8 m (at 0.5–0.75 µm, over a nominal ~70 × 70-km swath), compared with 10 m for SPOT 1–4 Pan data and latterly 5 m for SPOT-5 HRG data (swathwidth 55 km), their radiometric resolution is lower—i.e., 64 versus 256 gray levels. The LISS-III offers imagery at a resolution of 23.5 m and 70 m over four multispectral bands (from 0.52 to 1.70 µm) for scenes of ~130 × 130 km. The WiFS is a two-channel instrument (0.62–0.68 and 0.77–0.86 µm) with a 190-m resolution over a 692-km swath. In future, the joint Chinese–Brazilian CBERS-2 satellite will operate two sensors offering multispectral data at resolutions of 20 and 80 m. To reiterate, cloud cover is a major limiting factor for all sensors operating at visible to TIR wavelengths.

Glaciological applications of Landsat-7 ETM+ imagery have been evaluated by Bindschadler et al. (2001a) and Gasch et al. (2000). Of glaciological interest with ASTER are its high-spatial-resolution (15 m) visible and near-IR channels, and its stereo and pointing capabilities, but limited to a 60-km swath (Khalsa et al., 2004). As such, this and similar sensors are best suited to detailed studies of limited areas that require high resolution—e.g., valley glaciers. Having said this, a long-term Antarctic coastal-mapping project is underway using mosaics of Landsat and ancillary data collected during two time periods—namely, the early- to mid-1970s and late-1980s to early-1990s (Ferrigno et al., 1998; Swithinbank et al., 1997; Williams et al., 1995, 1997; Williams and Ferrigno, 1998). These data provide an important means of determining ice-sheet coastal changes (Ferrigno and Gould, 1987; Swithinbank et al., 2003). Satellite-image atlases have also been produced by the U.S. Geological Survey (USGS), for both Antarctica (Swithinbank, 1988) and Greenland (Williams and Ferrigno, 1995). Ferrigno et al. (1994) used Landsat imagery to produce a high-resolution image map of regions of the West Antarctic Ice Sheet (WAIS). Please see Section 3.4 for further discussion of satellite detection of change in ice sheet margins. High-resolution Landsat and SPOT imagery have also been of immense value in planning and executing field programs, enabling glaciologists to home in on important features which are barely visible from ground level. These and other images are especially useful when combined with Global Positioning System (GPS) measurements—another technological development that has revolutionized field glaciology. Unfortunately, data coverage is "spotty", being severely limited by cloud cover, failures of data transmission links, and expense (both the SPOT and Landsat programs are run as commercial enterprises). Such data have typically been collected not routinely but by individual research programs. Nevertheless, both offer a highly cost-effective means of mapping remote regions, and are particularly well-suited to localized rather than ice-sheet-wide studies. Moreover, analysis of high-resolution visible satellite data has cast considerable light on the identification of flow patterns and surface morphological features associated with basal topography, for example. This information combined has greatly enhanced our knowledge of ice dynamics and the interpretation of *in situ* data (Doake et al., 2001). For ice-sheet research, SPOT data have been limited to more localized studies than Landsat imagery (Bindschadler, 1993; Merry and Whillans, 1993; Vaughan et al., 1988).

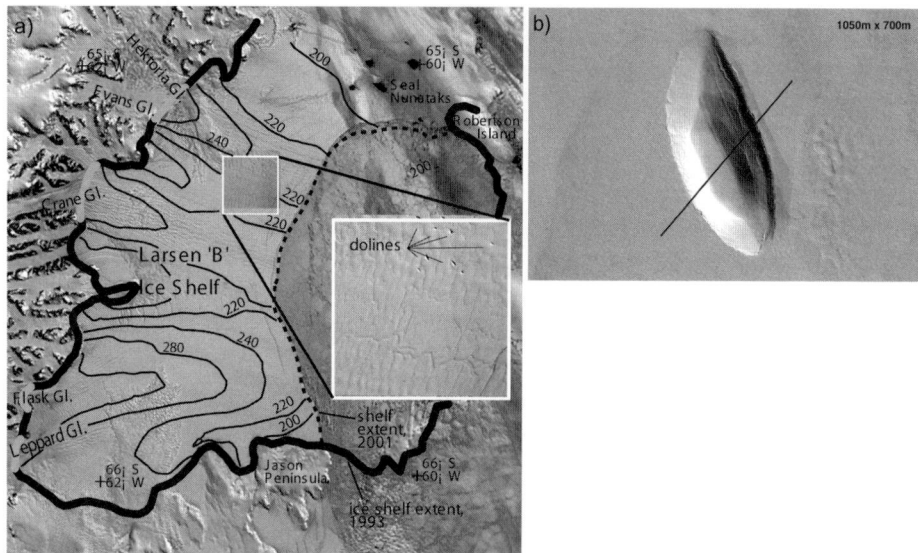

Figure 3.2. (a) A MODerate-resolution Imaging Spectroradiometer (MODIS) channel-1 image of the Larsen B Ice Shelf, Antarctic Peninsula from November 19, 2000. The thick line delineates the ice shelf boundary in 1993, while thin lines are contours of estimated ice thickness. An Ikonos image from October 5, 2000 is superimposed at the correct location. (b) Enlarged Ikonos panchromatic scene to show one of the dolines.

From (a) Scambos et al. (2000) and (b) Bindschadler et al. (2002). Copyright International Glaciological Society 2002, reproduced with permission.

Glaciologists now also have access to optical data from a range of commercial satellites with a remarkable meter- to submeter-scale resolution (see Chapter 5 in Volume 1 of this book). Although greatly limited in their areal coverage, over swaths which are typically ~10 km wide, these images offer an extraordinary level of detail for mapping purposes, as shown in the study by Bindschadler et al. (2002) of *ice dolines* on the Larsen Ice Shelf, Antarctica. These poorly understood features are elongated depressions a few hundred meters across and up to 20 m deep. Bindschadler et al. (2002) used 1-m resolution Ikonos imagery to quantify doline dimensions, and to derive elevation profiles by *photoclinometry* (a technique described in Section 3.6.3). The amount of information obtainable is apparent in Figure 3.2. Landsat and ERS SAR imagery were also used to show that dolines are associated with meltponds or lakes, and can form in a single melt season and persist for years. Once again, satellite-based analysis of such features provides unique insight into the state of the surface.

A dominant characteristic of ice sheet inland regions (in particular) is a nearly flat topography and bright surface (Bindschadler, 2003). As a result, the range of radiance within the Field-Of-View (FOV) of spaceborne optical sensors tends to be small. Not only spatial but also *radiometric resolution S* is a key to determining how much detail of a relatively featureless ice sheet surface can be resolved from space.

Following Bindschadler and Vornberger (2000), S is given by:

$$S = (R_{max} - R_{min})/(DN_{max} - DN_{min}) \tag{3.1}$$

where R_{max} and R_{min} denote the radiometric range over which a given sensor band operates, and DN_{max} and DN_{min} are the *quantization number* corresponding to these radiances. The lower the value of S, the greater the radiometric sensitivity. Much of the subtle topographic detail remains unresolved by optical imagers which are confined to an 8-bit quantization of the satellite-received radiance (i.e., a radiometric range of 0–255) and relatively low Signal-to-Noise Ratios (SNRs)—e.g., the Landsat-7 ETM+, Landsat TM, and SPOT sensors. Bindschadler and Vornberger (2000) carried out a comparative assessment of the relative merits and limitations of Landsat TM, SPOT, National Oceanic and Atmospheric Administration (NOAA) Advanced Very High Resolution Radiometer (AVHRR), and Russian Resurs-01 visible-band sensors as means of detecting ice-sheet topographic detail. They concluded that the TM tends to saturate under conditions of either a highly reflective surface or the Sun at a high elevation above the horizon, due to the application of 8 bits over a relatively narrow radiometric range. In this case, saturation can be quantified by assuming the Lambertian reflectance condition R (Bindschadler and Vornberger, 2000), such that:

$$R = \alpha \cos \theta_I E_{Sun} \tag{3.2}$$

where α is the surface reflectivity; θ_I is the illumination angle—i.e., the angle between the surface normal and the vector to the Sun; and E_{Sun} is the exo-atmosphere solar irradiance within the corresponding spectral band (Neckel and Labs, 1984). Surface slopes are approximately horizontal over ice sheet interiors, and are approximately equal to the solar zenith angle. In this case, image saturation occurs when $R > R_{max}$ or $\alpha \cos \theta_I > R_{max}/E_{Sun}$. SPOT HRV sensors are less prone to saturation than the TM, but are more limited in their areal coverage and operate at a lower radiometric resolution. Against this is balanced their higher spatial resolution (of 10–20 m).

Recent advances in sensor technology and increased data communication bandwidths have heralded a significant increase in the amount and quality of information obtainable. Panchromatic data from the Ikonos sensor (0.4–1.0 µm), such as those shown above and also in Figure 3.3, are supplied with an 11-bit linear quantization. Bindschadler (2003) showed that with a 12-bit quantization, in this case of the NASA EO-1 ALI sensor, it is possible to detect very subtle variations in surface reflectance associated with fields of sastrugi,[2] for example. Importantly, this is possible even though individual sastrugi are smaller than the sensor pixel size—i.e., 10 m for the panchromatic band (0.48–0.69 µm) (⟨*http://eo1.gsfc.nasa.gov/*⟩). Once again, a trade-off is that data collection is restricted to a narrow swath of ∼10 km. As such, high-resolution sensors are unsuitable for regional studies, but are best suited to detailed localized studies or the validation/interpretation of coarser resolution satellite data. The repeat interval is also long unless the sensor has a tilt

---

[1] Sastrugi are low, linear snow dunes formed by the erosion and deposition of snow by the wind, and form in the direction of the prevailing wind.

Figure 3.3. An Ikonos image of part of the Bindschadler Ice Stream (formerly Ice Stream D), West Antarctica, on February 9, 2001, showing fields of sastrugi. These elongated wind-sculpted snow dune features are typically a few tens of centimeters high. This image is a 1 × 1-km portion of the original 11 × 11-km scene, with a pixel size of 1 m. It shows the extraordinary amount of detail that can be obtained from space, but at the expense of spatial (and temporal coverage). Indeed, such images resemble large-scale aerial photographs.
From Bindschadler (2003). Copyright IEEE 2003, reproduced with permission.

capability. Note that the ALI bands were selected to match and enable cross-comparison with Landsat-7 ETM+ bands (Ungar et al., 2003).

While the Landsat time series dates back to the early 1970s, the declassification of imagery from the US spy satellite program in 1995 has enabled examination of change dating back to 1962 (McDonald, 1995a, b). This program included fairly extensive high-resolution photographic coverage of both major ice sheets, but over a number of short periods and again limited by cloud. These Declassified Intelligence Satellite Photography (DISP) data are available from the USGS EROS Data Center at ⟨http://edcsns17.cr.usgs.gov/EarthExplorer/⟩. When cloud-free, these images reveal substantial surface detail (Bindschadler and Vornberger, 1998; Paulsen and Wilson, 1998). Although non-digital data, they can be digitized for direct comparison with later satellite data. By this approach, Bindschadler and Vornberger (1998) compared West Antarctic data acquired by the Corona satellite in 1963 (ground resolution 150 m) with AVHRR data from 1992 and SPOT imagery collected between 1989 and 1995. Skilled glaciological detective work showed that substantial changes had occurred in the region's velocity field, with Whillans Ice Stream (formerly Ice Stream B) widening by 4 km and decreasing in speed by 50%. Moreover, the ice ridge separating the Whillans and Kamb Ice Streams (formerly B and C) had eroded by 14 km. Bindschadler and Seider (1998) provided a valuable summary of all DISP photography of Antarctica, and from three missions during 1962 and 1963. This study included listings and maps of data collected south of 60°S.

In a similar study, Jezek (1998) compared DISP images from the Corona satellite acquired in 1963 with SPOT and Radarsat-1 imagery from the 1990s to determine how different Antarctic flow regimes on the Ross Ice Shelf and Crary Ice Rise (at ∼83°S, 174°W) have changed over the 35-year period. This study supports the concept that the ice rise, and currently its effect on local ice flow (Bindschadler, 1993), is shifting. The record of large-scale flow patterns visible in satellite imagery shows that West Antarctic ice streams draining into the Ross Ice Shelf exhibit a history of major flow variability over century to millennial timescales, believed to be related to changes in the internal dynamics of the ice sheet. Other satellite-based studies, when combined with field measurements, have given tremendous insight into changes in ice-sheet flow regimes. These are examined in Section 3.8.

3.2.2 Moderate-resolution visible/thermal infrared

Although somewhat limited by moderate spatial resolution (∼1.1 km at nadir), the AVHRR onboard NOAA polar-orbiting satellites (Kidwell, 2000) offers wide-swath coverage and a cost-effective alternative to the high-resolution data discussed above. Moreover, the latter are limited by their orbital configuration to coverage equatorward of ∼82.5°N/S for Landsat and 86°N/S for SPOT, and by relatively small scenes (up to 185 km wide). The comparable Operational Linescan System (OLS) series onboard Defense Meteorological Satellite Program (DMSP) satellites offers a better spatial resolution of ∼0.6 km, but is limited by relatively poor radiometric resolution which largely precludes the detection of subtle ice sheet surface features (Scambos et al., 1999). Bindschadler and Vornberger (2000) also evaluated Russian Resurs-01 data, which are quantized to 8 bit. With a spatial resolution of 160 m and a swath width of 600 km, this sensor is intermediate between the AVHRR and SPOT/Landsat series sensors. Its poorer radiometric resolution and higher noise, however, restricts its applicability in ice-sheet research.

While persistent cloud cover is a major limiting factor with visible to TIR sensors such as the NOAA AVHRR, the acquisition of many overlapping scenes

146 Satellite remote sensing of ice sheet parameters and processes   [Ch. 3

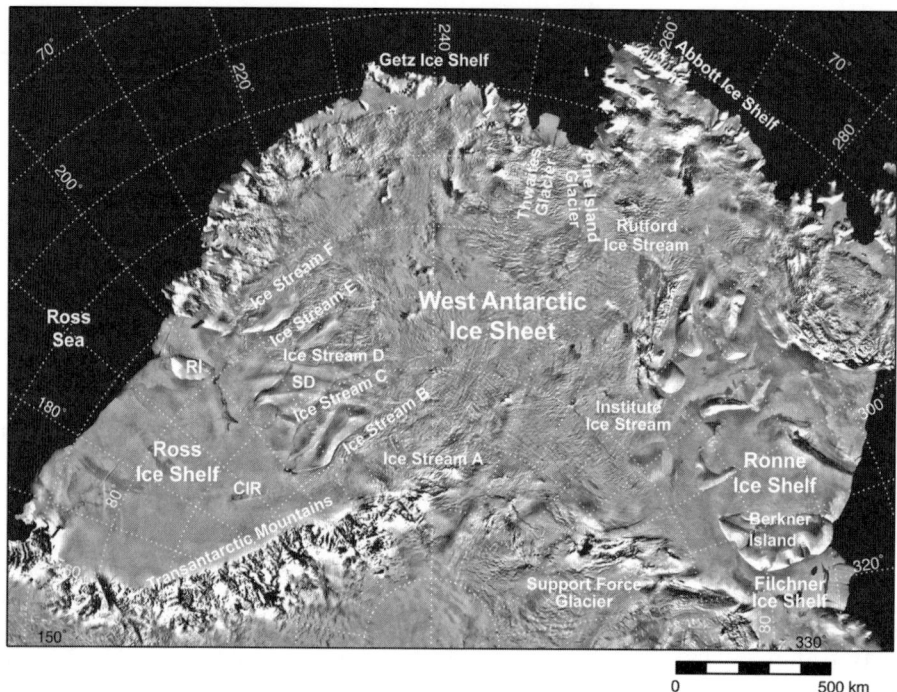

Figure 3.4. The West Antarctic sector of the AVHRR cloud-free image mosaic of Antarctica produced by Ferrigno et al. (2000), with major features marked. Note the strong contrast between the ice streams, ridges, and domes on the one hand and the flatter, undulating interior regions on the other. Note also how flat the ice shelves are. Note that Ice Streams B, C, and D have recently been renamed the Whillans, Kamb, and Bindschadler Ice Streams, respectively, after eminent glaciologists. RI = Roosevelt Island; CIR = Crary Ice Rise; and SD = Siple Dome.

After Fahnestock and Bamber (2001). Copyright AGU 2001, reproduced with permission.

at high latitudes increases the chance of obtaining cloud-free coverage. Moreover, the AVHRR operates at a relatively high radiometric resolution (i.e., 10-bit quantization) combined with a low noise level of ~ 0.4 uncalibrated sensor units on a scale of 0–1,024 (ITT, 1982; Schowengerdt, 1987). This makes it an important tool for mapping subtle yet important ice sheet surface features (Bindschadler and Vornberger, 2000; Fujii et al., 1987; Scambos et al., 1999). These qualities were exploited by Ferrigno et al. (2000), and earlier by Merson (1989), to construct an AVHRR mosaic map of the entire continent of Antarctica using cloud-free data (available from ⟨http://terraweb.wr.usgs.gov/TRS/projects/ Antarctica/AVHRR.html⟩)—see Figure 1.3. This map, as discussed by Fahnestock and Bamber (2001), gives an effective "snapshot" of the complex ice sheet morphology (see Figure 3.4) with an r.m.s. (root mean square) error of 2.5 km in geometric accuracy. Other studies have taken advantage of the ease of AVHRR data availability to map large ice dynamics-related features related to ice streams—

e.g., Bindschadler and Vornberger (1990). As discussed later, satellite analysis has also enabled improved understanding of remnant surface features related to past flow behavior.

Unfortunately, many important ice sheet morphological features cannot be readily resolved at the ~1.1-km resolution of the AVHRR. These include undulations, fine-scale ridges, and flow stripes, which provide important clues on the flow history, mass balance, and stability of ice sheets (Bindschadler and Vornberger, 2000; Bindschadler et al., 2001b; Jacobel et al., 1996, 2000; Scambos et al., 1999). *Flow stripes*, which are also known as *flow bands* or *streak lines*, are elongated ridges or troughs formed as grounded ice flows over bedrock features which can persist as surface features to serve as a proxy indicator of flow stability (Gudmundsson et al., 1998; Scambos et al., 2004a). They serve as indicators of ice flow direction from a localized source (Casassa and Whillans, 1994). With a long-axis scale of tens to hundreds of kilometers, a width of tens to hundreds of meters, and a surface elevation of only a few meters, these subtle features are difficult if not impossible to detect, yet alone map, on the ground. In response to this challenge, and the expense and difficulty of obtaining large areal coverage with SPOT and Landsat data, glaciologists have developed an innovative technique to improve the spatial and radiometric resolution of AVHRR data while taking advantage of its excellent areal coverage over a wide swath. Termed *data cumulation* (Kvaran et al., 1996; Scambos et al., 1999), this image-processing technique combines several scenes of the same area, acquired under similar illumination conditions (i.e., within a maximum period of a few weeks) and imaging geometry, into a single scene with enhanced spatial and radiometric resolution. The process depends first and foremost on accurate cloud masking, and image co-registration to a sub-pixel level (Scambos et al., 1992, 1999). The images are resampled to a smaller pixel size and combined using the registration vector, and then the average pixel value at each of the new, smaller pixels is determined. For AVHRR imagery, data cumulation can improve the spatial resolution to 0.7 km, and achieves a fourfold improvement in radiometric resolution, using principal components and their cumulation. This enables enhanced detection of subtle ice-sheet topographic features not apparent in the original images (Figure 3.5). Scambos et al. (1999) recommend using visible rather than near-IR band data, as snow reflectance at the latter is sensitive to changes in snow-grain size (Dozier et al., 1981; Wiscombe and Warren, 1980). For Greenland, data are best used from the early boreal spring, when surface melting is minimal and snow surface conditions remain relatively uniform (Shuman et al., 1997).

Fahnestock et al. (2000a) used data cumulation to produce a new and enhanced composite image of the entire Ross Ice Shelf (area $\sim$525,000 km^2) (Figure 3.6). This enabled much-improved identification of a range of regional-scale surface relief features through their distinctive morphology and pattern. These included crevasse or rift sets associated with zones of former grounded ice, and ice stream margin "scars" connecting upflow to existing or former shear margins (Figure 3.7). Fahnestock et al. (2000a) could also detect and map flow stripes in the enhanced imagery. Analysis of the patterns of these flow-related features—i.e., topographic features preserved on the ice shelf—allows inference of the complex history of

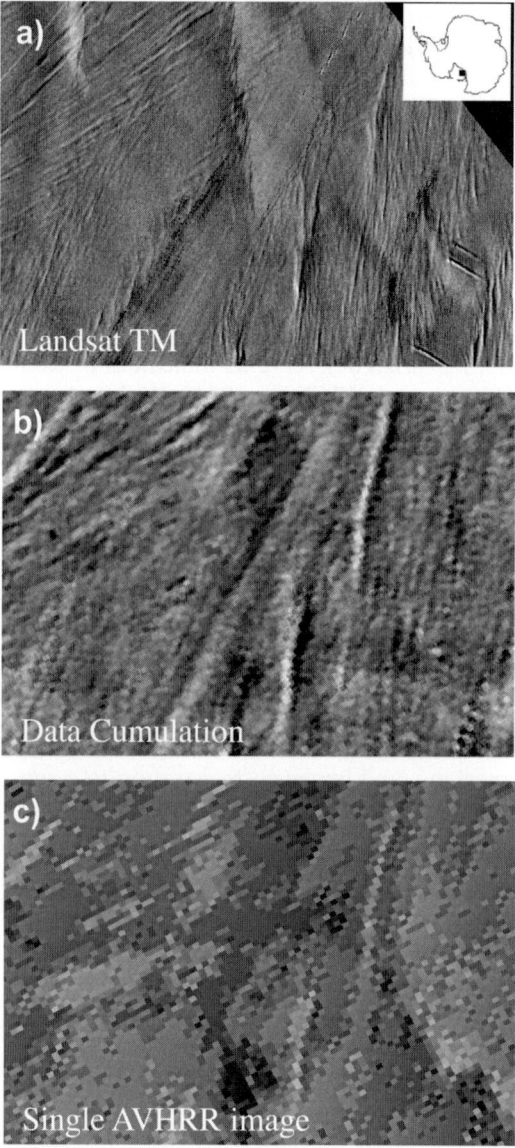

Figure 3.5. Comparison of (a) a Landsat TM scene (Path 40, Row 119, acquired on 20 January 1989), (b) an AVHRR image processed by the data cumulation method (a composite of 14 images acquired in November and early December 1992), and (c) a single AVHRR scene (one of the components used to construct the data-cumulated scene), acquired on November 14, 1992. The scenes cover the region of the western Ross Ice Shelf, Antarctica, $\sim$57 × 82 km in area (shown in the inset). Surface features associated with flowline traces, ice-bottom crevasses and shear flow are more clearly visible in the data-cumulated compared with the normal AVHRR image.

From Scambos et al. (1999). Copyright Elsevier 1999, reproduced with permission.

Sec. 3.2] Improved detection and mapping of surface features 149

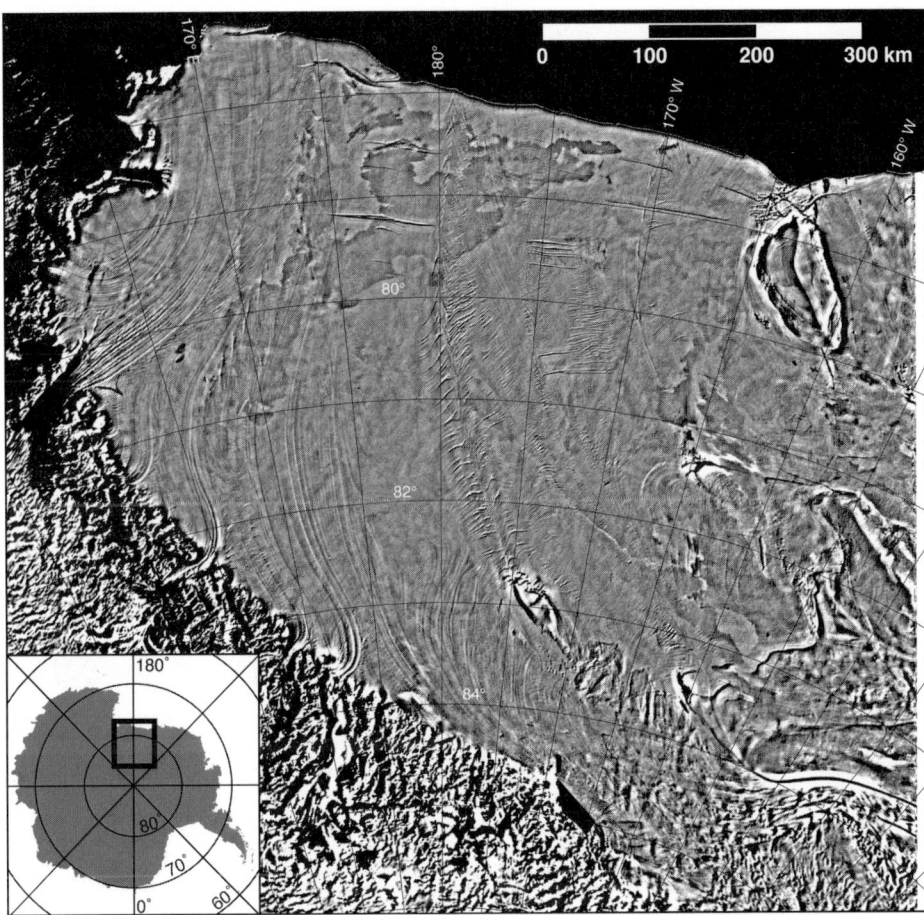

Figure 3.6. Data-cumulated (enhanced resolution) NOAA AVHRR image composite of the Ross Ice Shelf, derived from 15 images from November–December 1992 selected from the NSIDC polar 1-km level-1B AVHRR dataset (⟨http://www-nsidc.colorado.edu/NASA/GUIDE/AVHRR⟩), showing a wealth of information on features related to the ice-flow history from outlet glaciers and ice streams. The Trans-Antarctic Mountains are to the left and bottom left.
From Fahnestock et al. (2000a). Copyright IGS 2000, reproduced with permission.

variations in the ice stream flow and discharge into the Ross Ice Shelf over the last millennium. The rationale is that the ice shelf is sufficiently large enough to contain such a record. Other studies that have applied data cumulation techniques include those of Hulbe and Fahnestock (2004), Kwok and Fahnestock (1996), Scambos and Nereson (1997), and Scambos et al. (1999).

A key feature of the newer satellite datasets is their much-improved radiometric resolution—i.e., sensitivity to subtle surface reflectance variability. The launch of MODerate resolution Imaging Spectroradiometer (MODIS) sensors onboard the NASA EOS satellites Terra (EOS AM, in December 1999) and Aqua (EOS PM,

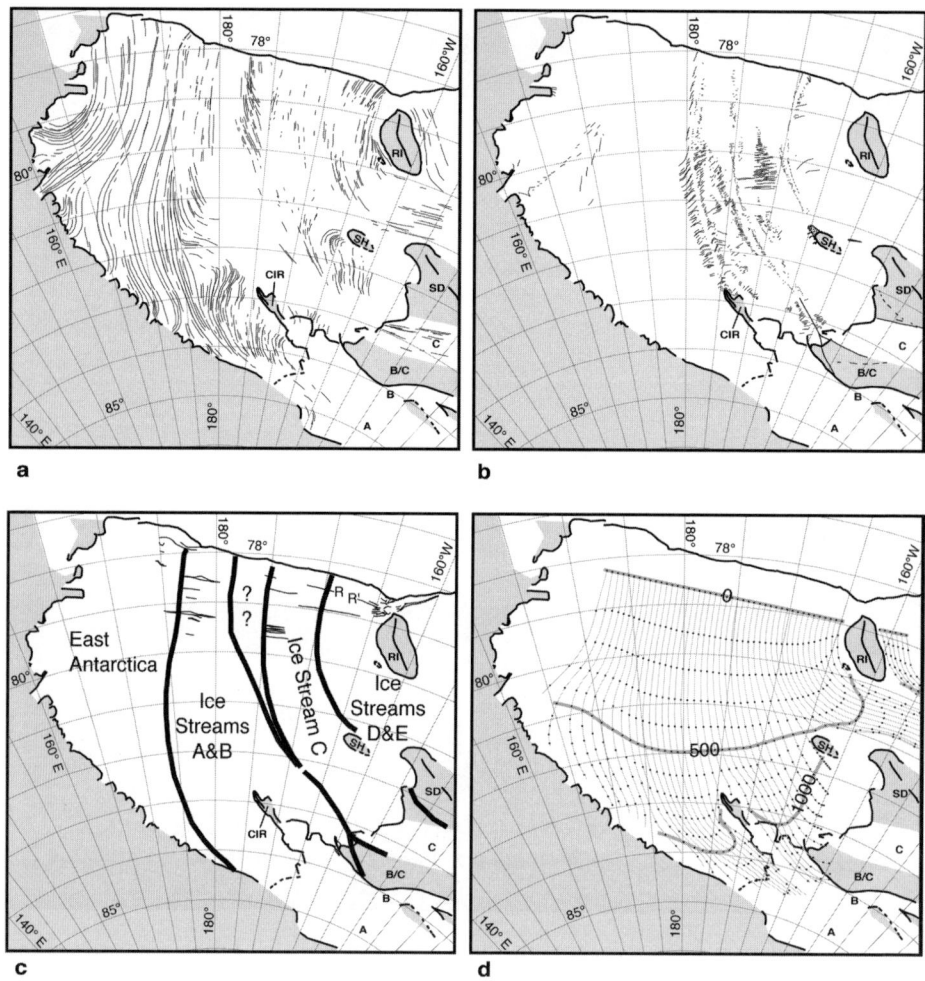

Figure 3.7. Dynamic features derived from Figure 3.6: (a) flowstripes, (b) rifts and crevasse series, (c) provenance of the ice determined from tracing flow boundary tracks, and (d) flowlines and 100-year particle motion dots calculated from the Ross Ice Shelf Geophysical and Glaciological Survey (RIGGS) velocity field (Thomas et al., 1984).

From Fahnestock et al. (2000a). Copyright International Glaciological Society, reproduced with permission.

in May 2002) opened up a new era of large-scale satellite coverage of the polar regions at visible to TIR wavelengths. This instrument provides higher radiometric resolution (12 bit) in 36 spectral bands ranging in wavelength from 0.405 to 14.385 μm (Barnes et al., 1998), compared with 5–6 broad bands for the AVHRR. The MODIS sensors are also equipped with sophisticated onboard calibration. Two bands are imaged at a nominal resolution of 250 m at nadir, 5 at 500 m, and the remaining 29 bands at 1 km. The 2,330-km swath provides almost complete global coverage on a daily basis at high latitudes. As such, this sensor is much-improved compared with the AVHRR, but yields substantially larger datasets and requires a more sophisticated X-band satellite data reception facility for direct data download. However, data products are again readily available, and are making an immense contribution to ice-sheet research, both in terms of mapping surface features and monitoring coastal margin changes. For example, subtle features such as relict flow stripes are visible as minor topographic undulations (Scambos et al., 2004a). An example, after Hulbe and Fahnestock (2004), is given in Figure 3.8. Moreover, MODIS data are being used to complete the poleward extent of Landsat maps of Antarctica (Ted Scambos, NSIDC, pers commun, 2003). These data also have a valuable role to play as a means of interpreting coincident microwave data, particularly from the new generation of SARs with multi-polarization and polarimetric capabilities—e.g., Envisat Advanced SAR (ASAR, launched on March 1, 2002; Zink et al., 2004), ALOS PALSAR (Advanced Land Observing Satellite Phased Array L-band Synthetic Aperture Radar to be launched in late-2005), and Radarsat-2 SAR (launch planned in 2006, see below). As can again be seen in Figure 3.9, which is a MODIS image of the entire Amery Ice Shelf (East Antarctica), the quality and information content of cloud-free MODIS imagery is superb. Moreover, it illustrates another important use—namely, the planning of field operations—in this case for the Australian Amery Ice Shelf Ocean Research (AMISOR) project. The prospect of enhancing the resolution of MODIS imagery by data cumulation is indeed enticing (see Figure 3.8). Standard snow and ice data products from MODIS (Hall et al., 2002) are routinely available from the NSIDC at ⟨http://nsidc.org/daac/modis/index.html⟩. Raup et al. (2005) further demonstrate the usefulness of photoclinometrically processed EO-1 ALI data in studyng subtle topographic variations associated with streak lines on the Amery Ice Shelf.

The new *Medium Resolution Imaging Spectrometer* (MERIS) sensor onboard ESA's Envisat also has the potential to become a first-class ice sheet research tool that complements MODIS. This sensor operates over 15 spectral bands in the visible and near-IR (spectral coverage 0.412–1.05 µm), with 300-m ground resolution over a 1,150-km swathwidth (⟨<http://envisat.esa.int/instruments/meris/⟩). The bandwidth is programmable between 0.0025 and 0.03 µm. As with MODIS, it offers superb quality data with substantially more detail on ice sheet characteristics than was previously achievable using poorer resolution sensors, but lacks a night-time viewing capability—i.e., TIR channels.

3.2.3 Synthetic aperture radar

Used extensively in ice sheet research since the launch of ERS-1 in 1991, spaceborne Synthetic Aperture Radar (SAR) is a powerful research tool by virtue of its ability to penetrate the shroud of darkness and cloud cover and acquire high-resolution measurements (Rott and Nagler, 1993; Rott et al., 1995). It offers information that complements that from visible and TIR sensors due to its sensitivity to surface roughness and its ability to penetrate the ice volume and yield information on the internal structure, properties, and state of the upper ice layers. Although unraveling this information can be a complex task, a number of broad relationships are apparent.

As outlined in Chapter 5 of Volume 1 of this book, the strength of the radar backscatter signal reflected by a dry snowpack is wavelength-, polarization-, and

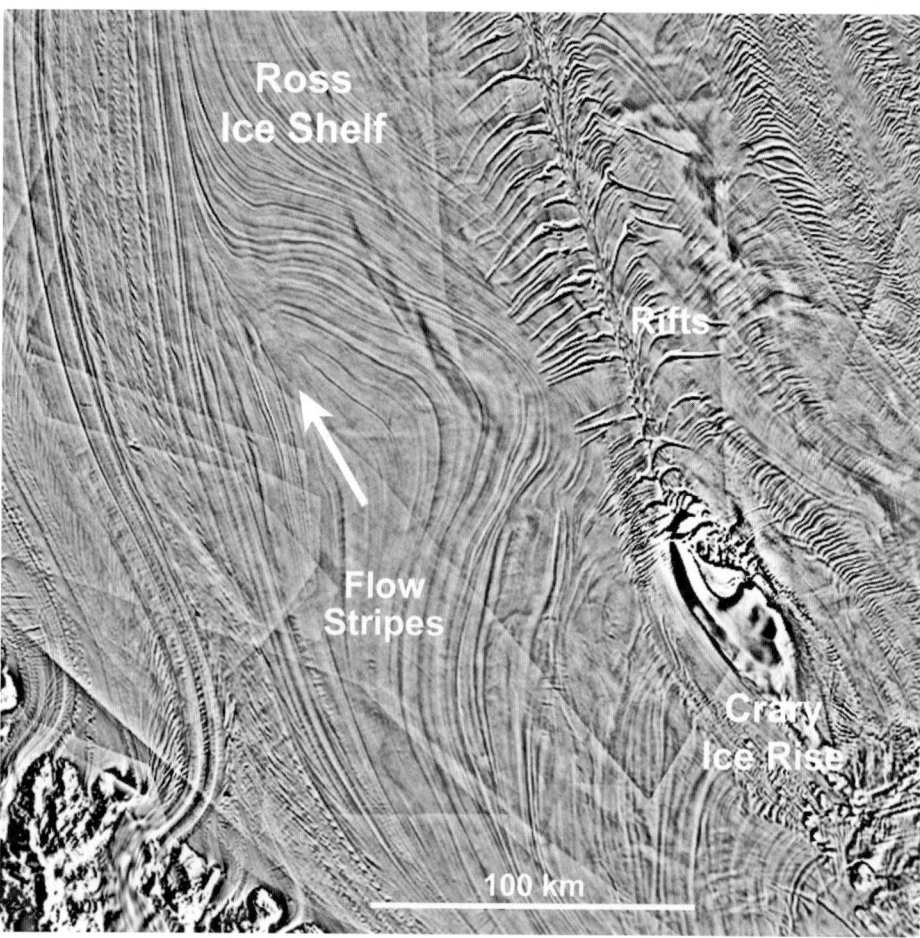

Figure 3.8. An enhanced resolution MODIS multi-image composite of the southeastern section of the floating Ross Ice Shelf, created from about 10 images by the data cumulation technique (see pp. 147–148). Images are from band 1 of the EOS Terra MODIS, and were acquired in December 2001. Straight lines are edges of cloud masks. Note the extraordinary amount of detail apparent in the ice flow field (in the form of flow stripes and rifts) and the complexity of the flow patterns created by the confluence of ice streams feeding into the region and ice deformation, in this case the Whillans Ice Stream (B) entering from the bottom right of the image and the Kamb Ice Stream (C) entering from the middle right. The white arrow shows the approximate flow direction. The Transantarctic Mountains are to the bottom left of the image. Note also the major impact of Crary Ice Rise (83°S, 174°W) on the ice flow and deformation, with long rifts propagating downstream. This feature is an ice-capped island that causes the ice to rise ∼50 m compared with the surrounding ice surface. A satellite data-based study by Jezek (1998), and outlined in Section 3.2.1, suggests that the orientation of Crary Ice Rise may be changing, with implications for the large effect that the feature has on local ice flow.

After Hulbe and Fahnestock (2004). Copyright International Glaciological Society 2004, reproduced with permission.

Sec. 3.2] Improved detection and mapping of surface features 153

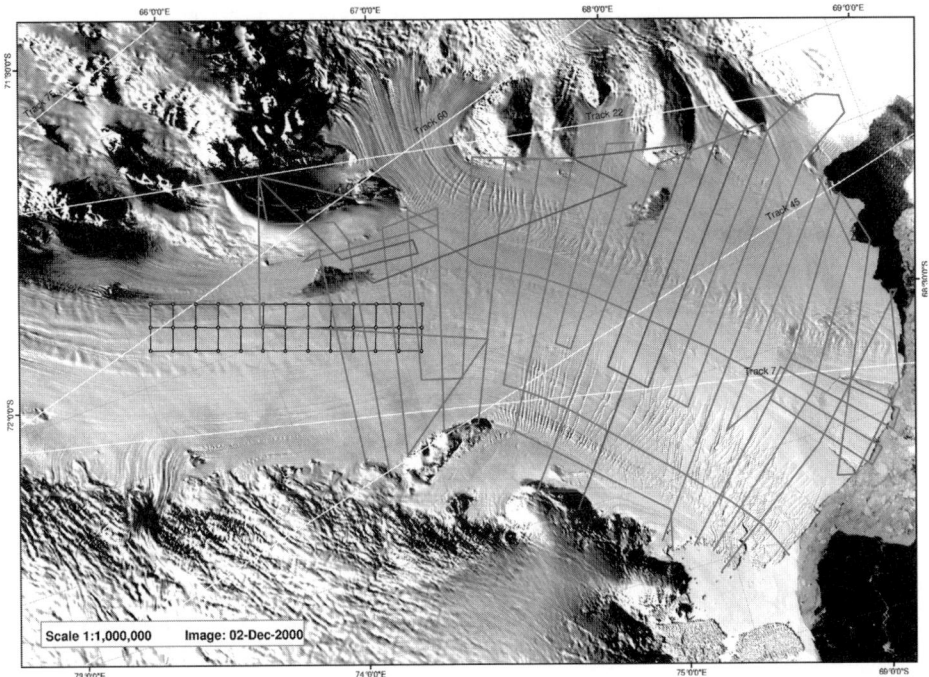

Figure 3.9. A mapped cloud-free MODIS band-2 image of the Amery Ice Shelf, at a full spatial resolution of 250 m and acquired on December 2, 2000. Apart from showing immense detail within the ice shelf and feeding glaciers, including complex crevasse and rumple patterns together with major rift formation at the ice shelf terminus (see Section 3.4.2), such images are of immense value in the planning and execution of field experiments. Each segment represents a proposed radar echo-sounding aircraft flight, carried out as part of a joint Italian–Australian research project in 2003/4. The white lines represent ICESat paths (for extraction of GLAS laser altimeter heights), while the black grid below Jetty Peninsula shows the location of an earlier ground-based GPS survey (October–November 1995). The image was prepared as part of a joint Italian and Australian study of the mass budget of a large sector of the East Antarctic Ice Sheet.

Image courtesy of Neal Young and Glenn Hyland (Antarctic Climate and Ecosystems CRC and Australian Antarctic Division). MODIS image © NASA 2000.

incidence-angle-dependent, and is primarily affected by its *dielectric* and *geometric* properties. These include *surface roughness* characteristics, *morphology*, *grain size* and *shape*, and the presence of *internal reflectors*—e.g., ice lenses and pipes and crevasses (Bardel et al., 2002; Fahnestock et al., 1993; Jezek et al., 1993, 1994). Again recalling Chapter 5 of Volume 1, the degree of *scattering* by a target medium is proportional to its *(complex) dielectric constant* (ε')—i.e., the medium's electrical property influencing radar returns from it. The dielectric constant of snow depends on a number of factors, including crystal shape and snow density (Hallikainen and Winebrenner, 1992). For a perfectly homogeneous and dry snowpack, *reflectivity*[2] increases with ε' (Zahnen et al., 2002). Values of $\varepsilon' = 1.72 \pm 0.05$ have been measured by Kärkäs et al. (2002) on the surface, and $\varepsilon' = 1.81 \pm 0.02$ in the upper meter, of the Antarctic Ice Sheet. From measurements collected in Dronning Maud Land (East Antarctica), Kärkäs et al. (2002) constructed an empirically derived relationship between ε' and snow density ρ_s (in $g\,cm^{-3}$) of $\varepsilon' = 0.9 + 2.2\rho_s$ (where the mean wetness in per cent by volume = $0.9 \pm 0.1\%$), which is similar to that provided by Ulaby et al. (1996)—i.e, $\varepsilon' = 1.0 + 2.1\rho_s$ (3.3). This relationship is given by Mätzler (1996) to be $\varepsilon' = 1.0 + 1.60\rho_s + 1.86\rho_s^3$ (3.4).

Although the bulk permittivity is important, more important are snow/firn stratification and the distribution, size, and number density of scatterers (Zahnen et al., 2002). While fresh snowfall is homogeneous and largely transparent to radar waves, the uppermost meters of ice sheets and polar glaciers are typically characterized by multiple layering comprising snow, firn, and ice of variable density and grain size, shape, and orientation. The degree of complexity and discontinuity within the surface layers as a function of time depends on prevalent meteorological conditions, the local accumulation rate, temperature, and the presence/absence of melt–refreeze cycling. Grain size is affected by firn temperature and accumulation rate (Alley, 1988), with regions of low accumulation having larger grain sizes due to intense temperature-gradient metamorphism over the radar *penetration depth*[3] (Mätzler, 1987; Ulaby et al., 1982). Significant grain-size growth also occurs adjacent to icy layers in the firn, by depth-hoar formation.

Given the relationship between snow density and its dielectric constant (Glen and Paren, 1975), the formation over time of snow layers of different density (Colbeck, 1987) causes reflections internal to the snowpack (Wiesmann and Mätzler, 1999). Differences in the granularity of the snow, characterized by grain size and shape, also cause changes in radar scattering (Rott, 1989; Rott et al., 1993; Shi, 1993). Strong changes in this respect are caused by *melt–refreeze cycling*, resulting in enhanced backscatter and a very bright signal at higher frequencies in particular—e.g., X- and C-band. The dielectric constant is strongly affected by the liquid water content of the snow (Jezek et al., 1993; Mätzler, 1987, 1996). For ice sheets, liquid water appears when the temperature approaches 0°C. *Wet snow* is highly *absorptive*, leading to a low backscatter and a dark signature in the radar amplitude image. Returning to geometric effects on radar backscatter, the amount of backscatter is as a general rule proportional to the surface roughness on scales equivalent/relative to the wavelength of the radiation. *Surface roughness* is itself determined by a number of factors, including accumulation rate and wind speed

---

[2] The fraction of the incident radiant energy reflected by a surface that is exposed to uniform radiation from a source that fills its FOV. In effect, it is a measure of the efficiency of a radar target in intercepting and returning a radar signal. This depends upon the size, shape, aspect, and the dielectric properties of the target medium/object and includes the effects of not only reflection but also scattering and diffraction.

[3] The depth below the surface of a material by which the incident radiation has been attenuated to $1/e$ of its original field strength.

and direction (Furukawa and Young, 1997; Long and Drinkwater, 2000). For more in-depth information on the dielectric and other microwave properties of snow and freshwater ice, see Glen and Paren (1975); Lytle and Jezek (1994); Matsuoka et al. (1998); Mätzler (1996, 1998, 2001); Mätzler and Wegmüller (1987); Mätzler and Wiesmann (1999); Moore and Fujita, 1993; Rott et al. (1993); Warren (1984); and Wiesmann and Mätzler (1999).

As such, the interpretation of SAR data collected over ice sheets is complicated by radar penetration into the snow–ice mass and strong contributions to the observed backscatter both from volume scattering and surface roughness (Fahnestock et al., 1993; Mätzler and Schanda, 1984). These data, however, reveal features which are not always apparent in visible and TIR images, including crevassed shear margins on ice streams and adjacent to ice rises on ice shelves (Doake et al., 2001; Sievers et al., 1993). These appear bright in SAR imagery. Internal fracturing is also thought to contribute to enhanced backscattering. The radar penetration depth at C-band ($\lambda = \sim 6$ cm—i.e., for ERS, Envisat, and Radarsat) for dry polar firn is of the order of tens of meters (Bingham and Drinkwater, 2000; Partington, 1998), 6–7 m at Ku-band (estimated from satellite radar altimeter data by Legrésy and Rémy, 1998), but substantially greater at L-band ($\lambda = \sim 23$ cm)—e.g., for JERS-1 and ALOS PALSAR (Rignot et al., 2000a). The wavelength of Ku-band is ~ 2 cm. The longer wavelength SAR is more effective at detecting shear margins, particularly in regions where the firn contains more icy near-surface layers (Doake et al., 2001). The important role of ice-stream margins in ice-sheet dynamics is underlined by Echelmeyer et al. (1994). Vaughan (1993a) and Vaughan et al. (1994) further noted a strong relationship between the SAR-derived normalized backscatter coefficient σ^0 and strain rate (related to the velocity gradient) across shear margins. Major discoveries in SAR imagery include a large fast-flow feature (ice stream) in NE Greenland (in ERS-1 data) by Fahnestock et al. (1993). Problems can occur in mountainous regions due to *radar shadowing, layover, or foreshortening* (see Chapter 2).

Another factor affecting SAR applicability is coverage. For example, ERS (Environmental Research Satellite) and JERS (Japanese Earth Resources Satellite) SAR coverage is limited to a narrow swath of 100 km maximum and a fixed incidence angle. Regional-scale mapping requires the mosaicking of available scenes, examples being the ERS-1 map of the Filchner-Ronne Ice Shelf of Jonas and Vaughan (1996) and a map of Greenland by Fahnestock et al. (1993). The advent of wide-swath SARs (with ScanSAR capability), beginning with Radarsat-1 in 1995, has greatly enhanced coverage over an approximately 500-km-wide area. Taking advantage of this, the first joint NASA/CSA (Canadian Space Agency) Radarsat-1 *Antarctic Mapping Mission* (AMM) completed the first ever high-resolution "snapshot" map of the entire Antarctic Ice Sheet in October 1997 after a 30-day period (Jezek, 1999; Jezek et al., 1998a, b), under the auspices of the Radarsat Antarctic Mapping Project (RAMP). This involved temporarily rotating the normally right- (i.e., north-) looking SAR antenna to a left-looking configuration, thereby filling in the existing data gap around the South Pole. The resultant map, which comprises a 30-m-resolution image mosaic, is shown in Figure 3.10, and represents a unique

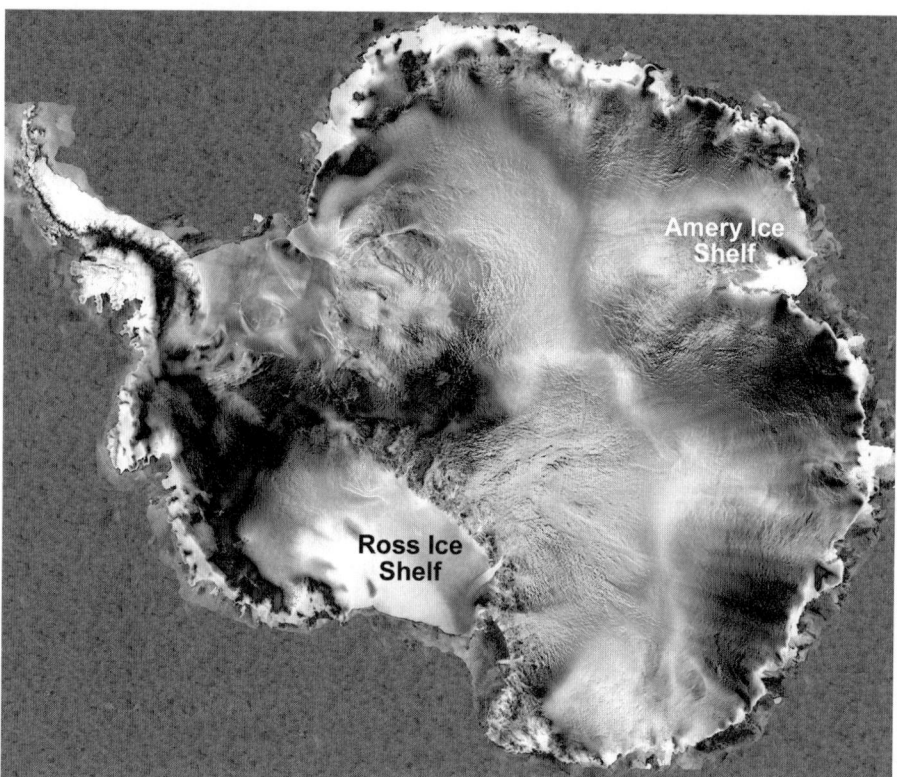

Figure 3.10. High-resolution Radarsat-1 image mosaic of Antarctica (resolution 125 m) collected during AMM between 19 September and 14 October 1997. This comprises over 6,000 image frames. The significant variations in radar backscatter, or brightness, relate to variable near-surface properties—e.g., snow-grain size, density, moisture content, stratigraphy, and surface roughness. In general, bright areas indicate a relatively rough surface, caused by crevassing, etc., uneven sub-glacial topography, and the presence of sub-surface ice melt–refreeze features and enlarged snow-grain size. Darker areas are indicative of relatively smooth surfaces. These physical properties are in turn related to larger scale properties and ice sheet regimes—e.g., surface melt patterns, accumulation rate, ice flow, ice-divide location, and topography (Stearns and Hamilton, 2004).

Acquired from the Byrd Polar Research Center (⟨*http://www-bprc.mps.ohio-state.edu/*⟩). Image courtesy NASA, CSA, Radarsat International, processed by the NASA Alaska Satellite Facility. Image copyright CSA/Radarsat International 1997, reproduced with permission.

baseline map of calibrated radar backscatter data. These image data were combined with a state-of-the-art Digital Elevation Model (DEM) to remove geometric distortion and facilitate radiometric normalization. The resultant terrain-corrected and seamless SAR mosaic reveals an extraordinary wealth of glaciological detail, including extensive networks of ice streams (including previously unknown streams in East Antarctica), and the positions of ice divides and the ice margins (Figure 3.11). Significant regional-scale backscatter variability is apparent, and is the

Sec. 3.2]   Improved detection and mapping of surface features   157

Figure 3.11. Two Radarsat-1 images derived from the Radarsat-1 AMM dataset shown in Figure 3.10, and homing in on the Lambert Glacier and Amery Ice Shelf region of East Antarctica to show the extraordinary and unprecedented level of detail obtainable from these data. The white box in image (a) delineates the northern (i.e., right-hand) part of the region shown in image (b). The spatial resolution of image (b) is 200 m. Satellite SAR images—e.g., from Radarsat—have revealed more complex ice dynamics than previously known. Ice velocities in the major outlet glaciers/ice streams can be more than ten times greater than the flow of the adjacent slow-moving ice, and the resulting shear margins at the ice stream boundaries are often heavily crevassed and clearly detectable from space by SAR due to their strong backscatter. Surface flow features are well-preserved, and flowstripes can be used to delineate the contributions of different glaciers to the ice flow regime. In this case, the complexity of flow activity around the confluence of major glaciers as they avoid rock outcrops and flow down from the ice-sheet plateau to feed the Amery Ice Shelf is fully apparent. Major research initiatives such as the Australian Amery Ice Shelf Ocean Research (AMISOR) project (Allison, 2001, 2003) are making extensive use of these and complementary data, including those from MODIS (Figure 3.9).

Image courtesy NASA, CSA, Radarsat International, processed by the NASA Alaska Satellite Facility. Imagery copyright CSA/Radarsat International 2005.

subject of ongoing research—e.g., Kärkäs et al. (2002) and Zahnen et al. (2002). Current research is combining the attributes of different satellite sensors to studying ice sheet glacio-morphology. For example, Scambos and Jezek (1999) compared Radarsat-1 AMM and enhanced resolution AVHRR images to gain a broad overview of a large part of the WAIS drainage system. Information on the AMM is available from ⟨www-bprc.mps.ohio-state.edu/radarsat⟩, and data are available from ⟨http://nsidc.org/daac/ramp/index.html⟩.

The follow-on *Modified Antarctic Mapping Mission* (MAMM) planned to repeat this exercise in late 2000, but the antenna did not switch due to concerns about the "health" of the satellite. Nonetheless, the campaign acquired highest resolution Radarsat-1 data over the period September 3 to November 4, 2000 of Antarctica's fast glaciers over the entire viewable region, which extends north of 82.1°S latitude (Jezek, 2002; Jezek et al., 2003). Results from this mission are evaluated in this chapter. A third mapping mission, MiniMAMM, was carried out in September–December 2004 to replicate MAMM over selected areas—i.e., the David and Pine Island Glaciers, the Filchner Ice Streams and parts of the Antarctic Peninsula (Harbin, 2005). Planned products include calibrated mosaics and coastline estimates.

The follow-on Radarsat-2 satellite, which is a commercial enterprise, has the enhanced capability to routinely switch between right- and left-looking (⟨http://www.radarsat2.info/⟩: Van der Sanden and Ross, 2001), and includes an ultra-high-resolution (3 m) mode (over a 10–20-km swath). This offers the mouthwatering prospect of further future comparison to detect both small- and large-scale change in surface and near-surface properties and ice sheet margins, and for direct merging with other high-resolution data. The new era of polarimetric SAR, ushered in by the launch of the Envisat Advanced SAR (ASAR) in 2002 with its alternating polarization mode and developing to full polarimetric capability with the ALOS PALSAR (in 2006), Radarsat-2 (in 2006), and TerraSAR-X and -L (Moreira, 2003) in 2006 and 2007, respectively, also opens up the exciting prospect of enhanced detection of subtle ice sheet features. The application of polarimetric SAR to ice-sheet research is in its infancy, however, and much validation work remains before these new signatures of different snow and ice surfaces are fully understood. Please see Chapter 5 of Volume 1 of this book for additional background information. Due to their poor spatial resolution, radar scatterometers and Passive Microwave (PMW) radiometers are limited in their ability to resolve detailed ice sheet morphological features. They are, however, ideally suited to the mapping of large-scale zones. New satellite polarimetric microwave radiometers—e.g., WindSat (Gaiser et al.)—show potential as a means of gaining improved characterization of the surface.

3.2.4 The mapping of ice-sheet facies

Ice sheets and glaciers are characterized by series of distinctive altitudinal snow zones or *facies*, which form as a result of climate differences with increasing elevation and distance from the ocean. First delineated by Benson (1962), these facies represent zones of different snow and ice diagenesis determined by the amount of snow accumulation and melt and their seasonal behavior. They are best-developed in Greenland and broadly comprise (from the highest elevation outwards to the ice sheet margin):

- a high altitude, interior *dry-snow zone*, where little if any surface melt occurs;
- a *percolation zone* characterized by episodic melting in the summer only, with the meltwater percolating downwards through the firn to refreeze and form ice lenses and pipes within the snow–firn mass;
- a *wet*, or soaked, *snow zone* within which melting saturates the depth of annual accumulation; and
- an *ablation*, or ice, *zone* around the ice sheet margin—i.e., at the lowest elevations—with a substantial proportion of the year's accumulation melting.

Boundaries between these facies are determined with respect to the timing and expression of maximum seasonal melt (Long and Drinkwater, 1999). Ablation and accumulation zones are separated by the *equilibrium line*, where the net balance is zero. The *firn line* delineates the upper limit of the ablation zone. With the exception of a number of regions, including the Antarctic Peninsula, the Antarctic Ice Sheet is largely dominated by dry-snow facies.

As these zones are highly sensitive to environmental factors such as air temperature, they are viewed as proxy indicators of climate change/variability. Due to the gentle surface slope of the ice sheet interior, even a small change in climate will result in a large lateral shift in the borders separating facies. Importantly, such changes can be readily detected in even low-resolution satellite data (Long and Drinkwater, 1994; Wismann and Boehnke, 1996, 1997; Wismann et al., 1996). Williams et al. (1991) investigated optical satellite remote sensing (Landsat) as a means of mapping ice sheet facies, but concluded that boundaries—e.g., between wet-snow and percolation facies—are often indistinguishable or ambiguous at visible to TIR wavelengths. Moreover, fresh snowfall can obscure differences (see also Hall et al., 1987; Parrot et al., 1993; Rott and Markl, 1989; and Winther, 1993a). Persistent cloud cover is also a limiting factor. Recent field programs—e.g., NASA's PARCA initiative in Greenland (Abdalati, 2001; Thomas et al., 2001a)—have made it possible to closely link the ice sheet radar response to physical models of snow and ice facies, enabling their mapping and monitoring in time series of radar data. This capability has spawned a number of large-scale studies using both satellite radar scatterometer and SAR data to identify facies and monitor their seasonal and interannual variability in both the Arctic (e.g., Bindschadler and Vornberger, 1992; Brown et al., 1999; Drinkwater et al., 2001; Engeset et al., 2002; Fahnestock et al., 1993; Jezek et al., 1991, 1993, 1994; Long and Drinkwater, 1994, 1999; Rees et al., 1995; Vornberger and Bindschadler, 1992; Wismann and Boehnke, 1996) and Antarctic (e.g., Bardel et al., 2002; Braun et al., 2000; Rau et al., 2000a, b). König et al. (2002) used SAR data to detect superimposed ice layers in Svalbard. This ice type contains small air bubbles, which act as scattering centers at higher frequencies—e.g., C-band—to create a medium-bright SAR signature. Superimposed ice is of importance in mass-balance calculations as it represents meltwater that refreezes onto a sub-freezing glacier surface. Arigony et al. (2004) show the considerable benefits of combining SAR and optical data (Landsat and ASTER) to monitor glaciological parameters on the Antarctic Peninsula. Other studies have concentrated on the detection and monitoring of the location of the equilibrium line—e.g., Adam et al. (1997) and Brown et al. (1999). Bingham and Rees (1997) demonstrated the effectiveness of a multi-sensor approach for this purpose. Mass balance studies using the equilibrium line approach are limited to some extent by the difficulty in distinguishing the equilibrium line from the snowline in satellite imagery—e.g., using both Landsat and SAR data (Bamber and Kwok, 2004).

Long and Drinkwater (1999) showed that large-scale information on ice-sheet facies and their seasonal and interannual variability can be routinely extracted from the operational radar scatterometer data-stream (see Chapter 5 in Volume 1 of this book for an evaluation of the relative strengths and weaknesses of scatterometers compared with SARs). The example shown in Figure 3.12 depicts the sensitivity of Ku- and C-band values of the average backscatter coefficient in each pixel normalized at the 40° incidence angle (σ_{40}°, termed dataset A—again see Chapter 5 of Volume 1) to spatio-temporal changes in surface and near-surface properties of the Greenland Ice Sheet. In the winter, A is largely determined by inhomogeneities and layering related to variable patterns of accumulation and melt–refreeze (Rott et al., 1993). During summer, on the other hand, the signature primarily results from the appearance of liquid water in those facies experiencing melt. As such, σ_{40}° data over ice sheets indicate large seasonal to interannual variability in areas experiencing ablation and diagenetic changes. The transition from summer melt to autumnal freeze-up is shown in Figure 3.12, and the scatterometer-derived delineation of facies is shown schematically in Figure 3.13. See Long and Drinkwater (1999) for a more in-depth examination of the backscatter response of different facies in Greenland and their seasonal evolution. Decadal-scale variability is illustrated in Figure 3.14 (see color section), which compares Ku-band A data from September 1996 (ADEOS NSCAT) with those from 1978 (Seasat). Melt detection is covered in the next section.

Working in the Antarctic Peninsula, Rau et al. (2000a, b) and Rau and Braun (2002) mapped and monitored (with increasing elevation) bare ice, wet-snow, frozen percolation, and dry-snow zones based on their distinctive SAR backscatter signatures. Results are shown in Figures 3.15 (see color section) and 3.16. The dry-snow zone is restricted to the highest elevations which remain untouched by the 0°C air temperature isotherm in the austral summer, and comprises a moderately layered snowpack characterized by small snow crystals and a lack of significant internal ice layers. This results in a consistently low radar backscatter (typically in the range of -14 to -20 dB) due to the large penetration depth and dominance of volume scattering (Partington, 1998; Rau et al., 2000a, b). Further downslope is the frozen percolation zone. Within this zone, the refreezing of downward-percolating meltwater results in permanent ice lenses and pipes, leading to considerable complexity in microwave signature (Jezek et al., 1994; Rignot et al., 1993). In winter, these inhomogeneities produce one of the highest (brightest) of all natural backscatter signals (Bindschadler and Vornberger, 1992; Scambos et al., 2003)—e.g., of >-8 dB (Drinkwater et al., 2001 give a mean value of ~ 0 dB in Greenland for satellite radar scatterometer data). This high backscatter is accompanied by large Polarization Ratios (PRs: Rignot, 1995). Scambos et al. (2003) evaluated the overall importance of firn impermeability for Antarctic ice shelves, in the relationship of radar backscatter to *surface melt season length*. The percolation zone is separated from the dry-snow zone by a *transition zone* with backscatter values of -14 to -8 dB. An increase in the liquid-water content during melt events has a major and

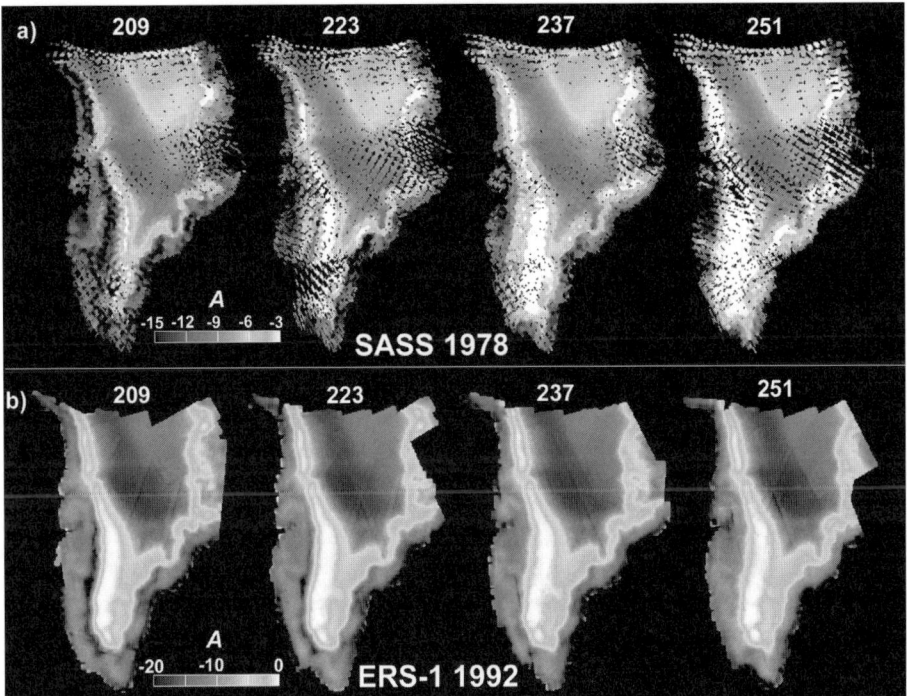

Figure 3.12. Daily image time series of satellite radar scatterometer A data of Greenland for four periods (days 209, 223, 237, and 251) in the boreal summer to autumn freeze-up period from (a) 1978 Seasat-A radar scatterometer (SASS) Ku-band data; and (b) 1992 ERS-1 C-band data. The data were enhanced using the Scatterometer Image Reconstruction with Filtering (SIRF) algorithm (Long, 2003; see Chapter 5 in Volume 1 of this book), and the imaging period is 2 weeks. The gray scale indicates values of σ^{o}_{VV} (40°) at Ku- and C-band, respectively.

From Long and Drinkwater (1999). Copyright IEEE 1999, reproduced with permission.

immediate effect on the radar backscatter (see Section 3.3.2), with high absorption of the microwave beam and a penetration depth reduced to the uppermost few centimeters (see also Chapter 5 in Volume 1 of this book). The wet-snow zone is characterized by specular reflection if the surface is smooth, resulting in a dark appearance in the SAR image. At the lowest elevation, the ablation (bare ice) zone has a higher backscatter signal due to enhanced surface scatter. In Greenland, the snow cover in the ablation zone melts entirely in summer to produce substantial runoff (Hanna et al., 2005). Such surface meltwater runoff is less prevalent in Antarctica, but has been detected in satellite radar altimeter data on the Amery Ice Shelf by Phillips (1998), for example. It will become apparent (in Section 3.3.4) that determination of the absolute rate of melt by remote sensing remains a major challenge.

It is anticipated that the availability of polarimetric SAR data from Envisat and near-future satellites will improve the detection and interpretation of ice-sheet facies

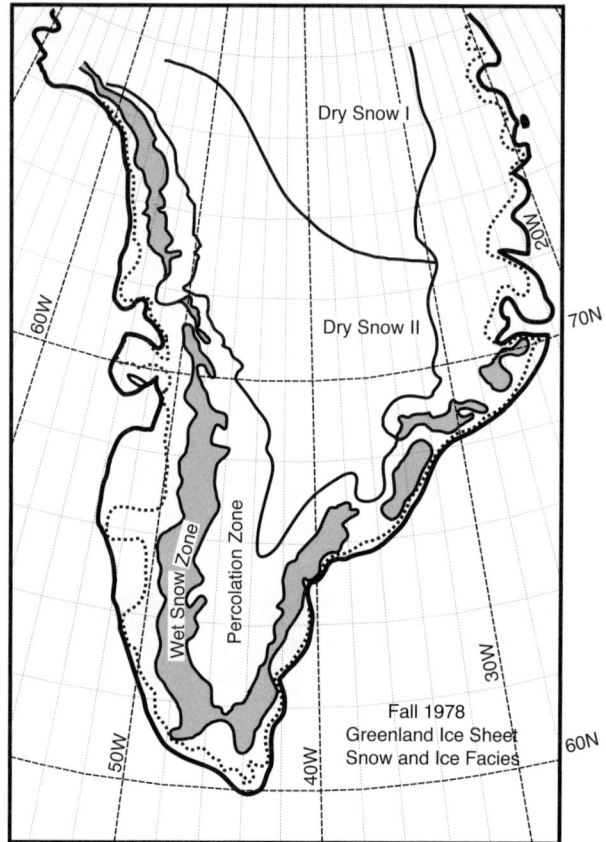

Figure 3.13. Ice-sheet facies in Greenland observed in Seasat SASS data for the boreal autumn of 1978.

From Long and Drinkwater (1994). Copyright International Glaciological Society 1994, reproduced with permission.

and related boundaries. For example, HV (Horizontal–Vertical) and VH combinations should provide information on polarization properties based largely on volume scattering, while VV and HH data should largely be indicative of surface-scattering contributions (Rott and Davis, 1993; Shi et al., 1992, 1994). This is an emerging field, and considerable work again lies ahead to evaluate and interpret the data from new satellite polarimetric systems. It should also be noted that useful information on snowpack/firn structure can be obtained from radar altimeter data—e.g., from the dual-frequency Envisat instrument (Legrésy et al., 2005). Work by Brisset and Rémy (1996), Féménias et al. (1993), Legrésy and Rémy (1997), and Rémy et al. (1993, 1995, 1996a) has shown that satellite-borne radar altimetry can quantify backscattering mechanisms in polar firn at vertical incidence and provide temporal histograms of the return energy from the surface and potentially the underlying volume, the latter given the substantial penetration depth at typical altimeter

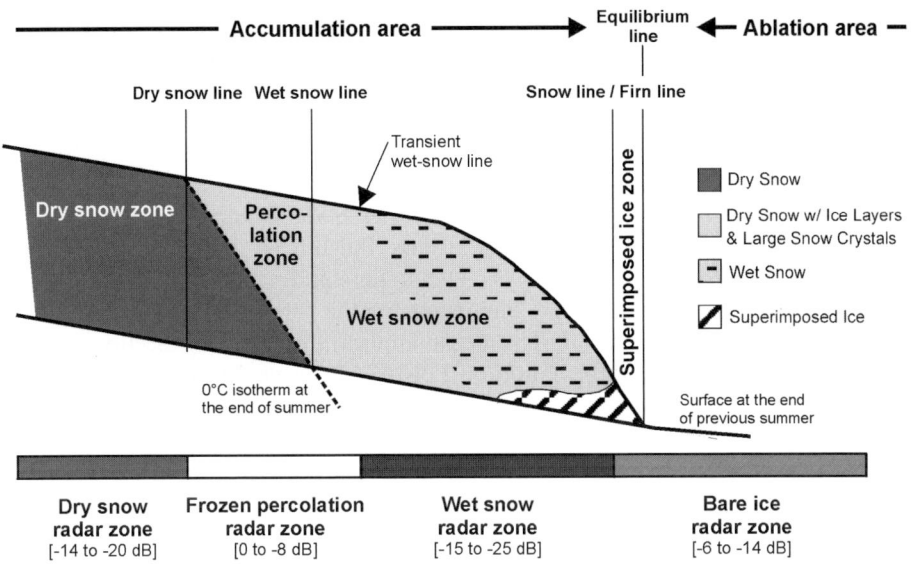

Figure 3.16. Ice-sheet facies and corresponding radar zones, with typical backscatter values (ERS C-band, VV-pol) for the Antarctic Peninsula region during the ablation season.

From Rau and Braun (2002), after Paterson (1994) and Benson (1962). Copyright International Glaciological Society 1994, reproduced with permission.

frequencies (which can also be determined—see Section 3.2.3). Such estimates are possible due to the apparent separability of the volume and ground (surface) echo over dry ice sheets. This topic is covered further in Section 3.16.4.1.

In addition to measuring amplitude, the new generation of *polarimetric SARs* measure the phase difference in the return of the HH and VV polarization signals. This difference contains information on the structural characteristics of the scatterers. The phase information in the products from these SARs can be used to derive an estimate of the correlation coefficient for the HH and VV returns, which can be considered as a measure of how alike the HH and VV scatterers are. While polarimetric SARs are expected to enhance the detection and monitoring of ice sheet facies and other characteristics, the exact benefits are currently unknown due to a lack of data, and further research is essential. The advent of satellites which routinely acquire these data has also led to the development of *polarimetric SAR interferometry* (*Pol-InSAR*), which is the merging of SAR interferometry and polarimetric SAR (Cloude and Papathanassiou, 1998; Cloude et al., 2001; Papathanassiou and Cloude, 2001). This exciting new technique, which is a model-based parameter estimation technique (Dall et al., 2004), combines the analysis of the phase differences with varying the radar signal polarization to produce differential range and range change measurements (interferometry) from two or more SAR images. This technology shows great potential when applied to studies based upon the radar penetration characteristics of snow and ice masses, as an advantage of this symbiosis is gaining the ability to more accurately separate surface- and volume-scattering effects and contributions. As a result, Pol-InSAR can potentially provide information about the physical characteristics of the ice and its near-surface structure (e.g., density of ice inclusions, grain size, and water content), thereby enabling improved mapping and monitoring of ice-sheet facies (Dall et al., 2003, 2004; Rignot, 1995). Again, this field is in the research and developmental stage at present, due to the current scarcity of suitable data—i.e., limited to a few airborne campaigns. Next-generation spaceborne SAR missions such as ALOS, TerraSAR-X and -L will support repeat-pass Pol-InSAR imaging, with repeat periods varying from 11 to 46 days. The understanding and modeling of the temporal decorrelation will be very important for the further development of both Pol-InSAR and conventional interferometric applications. See also Section 3.16.4.3 and Chapter 2 for further information on Pol-InSAR applied to ice sheet research.

3.3 DETECTION, MAPPING AND MONITORING OF ICE-SHEET SURFACE MELT AND REFREEZING

Understanding ice sheet climate and melt characteristics is of prime importance to the estimation of ice-sheet mass balance, the interpretation of observed elevation change, and the impact of change. Summer season melting releases large quantities of freshwater into the ocean around Greenland in particular (Thomas, 2004; Thomas et al., 2001a), and interannual variations are likely to have an impact on global sea level and ocean thermohaline circulation. Meltwater can in addition enhance the flow of outlet glaciers by lubricating the base–bedrock interface (Zwally et al., 2002b). Surface melt also affects ice-sheet isotopic, particulate, and chemical records contained in ice-core measurements used in the reconstruction of past climate. Ablation responds rapidly to regional climate change. Due to the positive feedback associated with wet snow (Steffen, 1995), whereby the latter absorbs up to three times more incident solar radiation than dry snow to cause further melt (the ice–albedo–temperature feedback mechanism), ice-sheet melt characteristics also play a central role in both energy and mass exchanges at the ice sheet surface (the close association between albedo and melt is examined more fully in Section 3.14).

3.3.1 Visible and thermal infrared techniques

The onset of melt can be detected as a change in reflectance at visible to near-IR wavelengths (e.g., Hall et al., 1987; Parrot et al., 1993; Pattyn and Declair, 1993; Williams et al., 1991), enabling sensors such as the Landsat TM and ETM+, AVHRR, Envisat AATSR (Advanced Along-Track Scanning Radiometer), and Aqua/Terra MODIS to map melt and ablation characteristics at a high to medium resolution (Figure 3.17). The collection of uninterrupted time series is, however, severely limited by cloud and darkness (although most melt occurs during the summer, and is indeed greatly affected by cloud cover). Moreover, the melt signature at optical wavelengths is less distinct than it is at microwavelengths. As such, the uninterrupted detection and mapping of surface melt at large spatial

Sec. 3.3] Detection, mapping and monitoring of ice-sheet surface melt 165

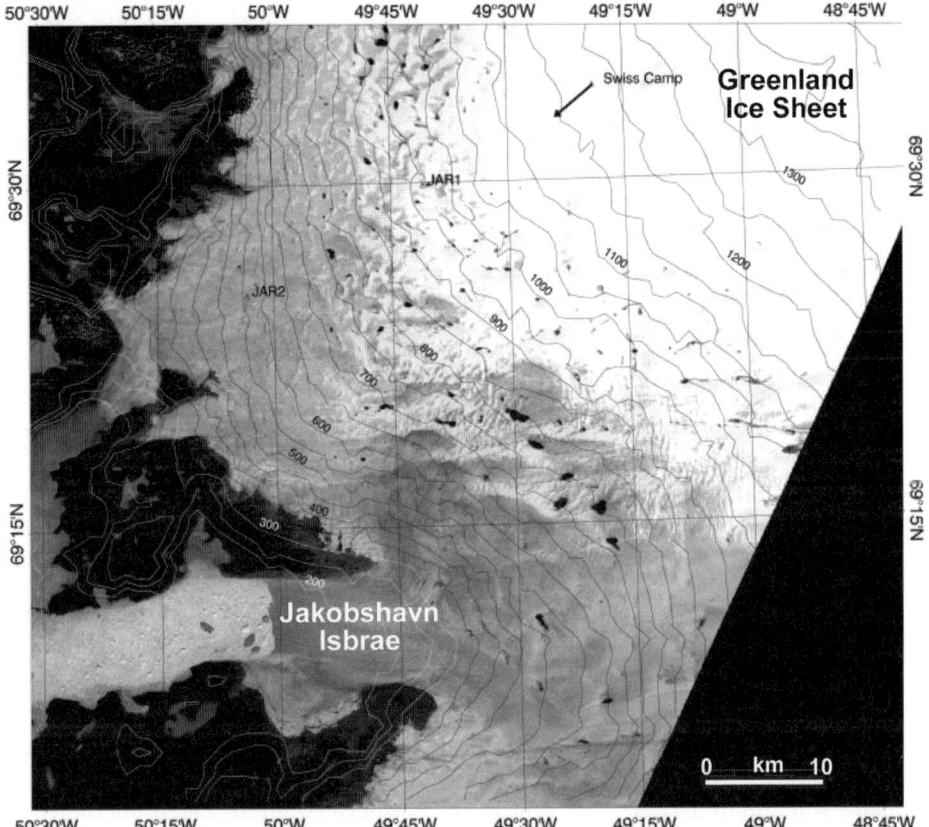

Figure 3.17. A cloud-free Landsat Thematic Mapper image (channel 3) of an area of the west-central Greenland Ice Sheet, acquired on June 22, 1990—i.e., approximately one-third of the way through the annual melt season. Elevation contours are superimposed at intervals of 50 m. The grayer regions at lower elevations are bare ice. The firn–ice boundary typically migrates to the approximate elevation of the equilibrium line near Swiss Camp (marked) by the end of the melt season—i.e., late August–early September. The dark patches are surface melt lakes, with some showing inflow channels (dark lines). Melt-lake formation extends to the region above the equilibrium line later in the summer melt season. Jakobshavn Isbrae (Glacier), one of the fastest flowing glaciers in the world, is at the lower left.

From Zwally et al. (2002b). Copyright IEEE 2002, reproduced by permission.

scales is most effectively accomplished by satellite microwave remote sensing, both active and passive. These methods are based upon the sensitivity of radar backscatter response and PMW emission to the appearance of even small amounts of liquid water in the previously dry-snow/firn mass.

Satellite-based techniques complement and extend traditional methods that use surface energy balance models (as reviewed by Greuell and Genthon, 2004) and positive-degree temperature (degree-day) models to infer ice sheet melt rates—

e.g., Braithwaite (1995); Drewry and Morris (1992); Ritz et al. (1996). For degree-day models, input information in the form of surface temperature fields can be obtained from satellite TIR data from sensors such as the AVHRR, MODIS, ERS-2, ATSR-2, and AATSR (see Section 3.13). Regarding the determination of ablation rates using the surface energy balance approach, the latent- and sensible-heat terms in (1.6) (in Chapter 1) cannot be measured directly from space. Rather, this technique requires significant independent information on near-surface air temperature, wind speed, and humidity.

3.3.2 Active-microwave techniques

As stated above, radar backscatter is highly sensitive to the appearance of liquid water. This transforms the dielectric properties of the snow/firn medium by regulating the reflection/transmissivity at the surface and limiting the volume-scattering contribution, by absorption and extinction within the upper layers (Long and Drinkwater, 1994). While much of the continental Antarctic Ice Sheet remains free from seasonal melt, the northern sector of the Antarctic Peninsula undergoes summer melt. This region has been under close scrutiny for signs of responses in the extent of austral-summer melt area to the extraordinary regional warming trend, of $\sim 0.5°C$ per decade observed since 1945 (Comiso, 2000; King, 1994a; King et al., 2003; Morris and Vaughan, 2003; Skvarca et al., 1998; Skvarca and De Angelis, 2003; Smith et al., 1996; Vaughan and Doake, 1996; Vaughan et al., 2001a, 2003). Rau and Braun (2002) used calibrated Radarsat-1 ScanSAR mosaics to identify and map the dry-snow zone of the northern Antarctic Peninsula (north of 70°S), imposing a threshold backscatter value of $-14\,dB$. Monitoring the dry snowline in SAR time series enables the detection of (a) extreme high-temperature events via their impact on the migration of the line, and (b) periods of relative stability in meteorological parameters and their impact on snow cover properties (Rau et al., 2000a, b). Rau and Braun (2002) in fact revealed an upward shift in the dry snowline on the eastern flank of the peninsula between 1992 and 1998, which they attributed to the increasing number of high-temperature events observed during the 1990s (Fox and Cooper, 1998; Scambos et al., 2000). Shi and Dozier (1995, 2000a, b) further demonstrated the potential of enhanced melt mapping using polarimetric SAR data.

With their excellent coverage and rapid revisit interval, *radar scatterometers* are particularly well-suited to the mapping of seasonal snow-melt zones (Drinkwater and Lin, 2000; Drinkwater and Long, 1998; Long and Drinkwater, 1999). The value of $A(\sigma_{40})$ for snow/firn decreases with increasing wetness (Wismann, 2000; Wismann et al., 1996) (Figure 3.18), with reductions exceeding $15\,dB$ in Seasat A Scatterometer (SASS) data (Long and Drinkwater, 1994) and melting surfaces appearing dark in processed images. Subsequent refreezing creates a drier granular and icy medium, which appears brighter due to greater backscatter. Thomas et al. (1985) used SASS data to demonstrate the capability of scatterometers as a means of

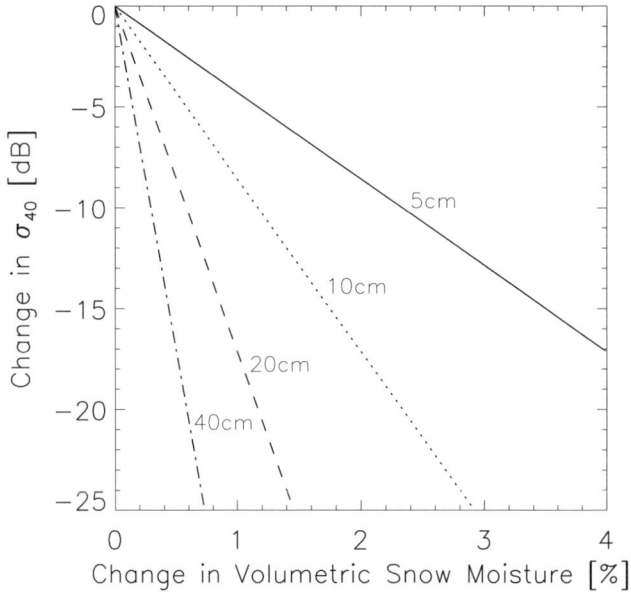

Figure 3.18. Plot of the change in σ_{40} as a function of *volumetric snow moisture content* for different thicknesses of a wet-snow layer. The curves are based on the model in Winebrenner et al. (1994), with the assumption that the snow is uniformly distributed through the layer. As discussed in Chapter 5 of Volume 1 of this book, with a progression in snow wetness, a wet upper layer can attenuate the microwaves to such an extent that little or no information is obtained from the underlying snow layers. Due to the ambiguity between snow wetness and the thickness of the wet-snow layer, it is not possible to retrieve absolute snow moisture content from scatterometer data (Wismann, 2000). It is, however, possible to monitor relative temporal and spatial changes in snow-melt intensity.

From Wismann (2000). Copyright AAAS 2000, reproduced with permission.

mapping ice-sheet-wide melting. Although short, the Seasat dataset represents an important yardstick against which to gauge change. Subsequent scatterometers have been characterized by improved resolution, enhanced performance, and data continuity. Moreover, the poor nominal resolution of 25–50 km has been significantly improved through the development of enhancement algorithms (Early and Long, 2001; Long, 2003; Long et al., 2001). NASA Scatterometer (NSCAT), SASS, and ERS-1 and ERS-2 scatterometer (EScat) data have been processed through the NASA-sponsored *Scatterometer Climate record Pathfinder* (SCP) program to construct time series of images of the radar backscatter of the Greenland Ice Sheet. Links to these data are available at ⟨http://nsidc.org/data/scatterometry⟩. Monitoring of interannual variability in melt onset and extent is possible because the surface melt response is similar at Ku- and C-band. The dataset continues with the SeaWinds sensor onboard QuikScat and ADEOS-II (Japanese ADvanced Earth Observing

Satellite). Importantly, these instruments provide ice-sheet-wide coverage on a daily basis, and at a spatial resolution of up to 4.5 km (after enhancement). Please see the SCP website (⟨http://www.scp.byu.edu/⟩) for further information. The near-term future of this important time series is assured with the launch in 2005 of the ESCAT follow-on, ESA's Advanced Scatterometer (ASCAT) onboard the first of three MetOp satellites—each with a planned 5-year lifetime (⟨http://earth.esa.int/METOP.html⟩: Figa-Saldaña et al., 2002).

The continuous availability of scatterometer data since the launch of ERS-1 in 1991 provides a consistent means of examining long-term variability in the extent of seasonal ice sheet melt—e.g., Ashcraft and Long (2000); Bingham and Drinkwater (2000); Drinkwater and Long (1999); Long and Drinkwater (1994, 1999); and Wismann (2000). Drinkwater and Long (1998) noted a general trend toward increased melt and a resultant 20% shrinkage in the areal extent of the dry-snow zone in Greenland since 1978. The large interannual variability in melt extent in Greenland, and the apparent increase in areal melt extent since the early 1990s, is illustrated in Figure 3.19 (see color section). This shows temporally integrated representations of the decrease in calibrated σ_{40} associated with seasonal snow melt, for the years 1992, 1995, 1996, and 1999. These measurements are based on ERS-1 and -2 C-band scatterometer data (frequency 5.3 GHz, VV-pol), at a spatial resolution of 50 km both along-track and across-track, and form a subset of a time series analysis from 1992 to 1999 by Wismann (2000). The maximum integrated normalized radar cross-section for each grid cell during the respective boreal-summer period is used as a measure of the intensity of summer snow melt. While relative minima of snowmelt were encountered in 1992 and 1996, the years 1995 and 1999 are examples of years in which the spatial extent of snow melt was large. The melt affected anomalously high altitudes (of >2,000 m) in southern Greenland, with intense melt (characterized by a large decrease in σ_{40}) occurring on the southwestern slope of the ice sheet. During the years of extensive melt, the seasonal melt zone invaded the dry-snow zone border along the west coast (see also Long and Drinkwater, 1999). The intensification of melt is depicted schematically in Figure 3.20 (see color section). Note also the intensification of melt in northern and northeastern Greenland in Figure 3.19. Similar scatterometer-based melt studies have been carried out in Antarctica. Young and Hyland (1998), for example, created a time series of the areal extent of Antarctic Ice Sheet seasonal melt from analysis of ERS wind scatterometer data. Over the period 1991–1998, maximum melt extent and duration occurred in the two summer seasons of 1991–1992 and 1997–1998.

While the above studies have been largely concerned with seasonal to interannual variability, a study by Nghiem et al. (2001) further demonstrated the ability of the QuikScat SeaWinds scatterometer to detect snow-melt regions on the Greenland Ice Sheet over shorter timescales. They used twice-daily vertical polarization data, acquired at ~06:20 and 18:20 GMT, to study diurnal variations of melt regions over a full summer melt to autumn freeze-up period (in 1999). The data from the early morning (t_a) in an ascending orbit pass and late afternoon (t_p) in a descending pass were co-located for each day, then

Sec. 3.3]   Detection, mapping and monitoring of ice-sheet surface melt   169

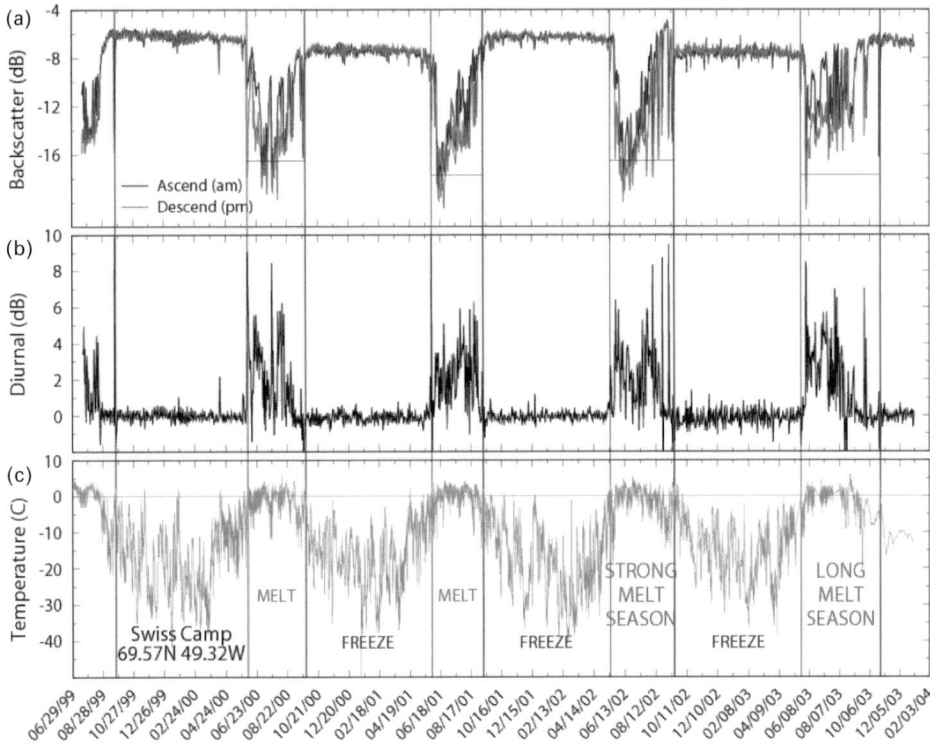

Figure 3.21. A record of melt duration at the ETH/CU Camp station on the Greenland Ice Sheet from 1999 to 2003: (a) time series of daily QuikScat backscatter data for both ascending and descending orbits, (b) the QuikScat diurnal signature, and (c) bottom panel for air temperature measured at an AWS site. Each year, first melt dates are marked with the first vertical line, and last melt dates with the second vertical line. Horizontal bars in (a) mark a backscatter level at 10 dB lower than that before the first melt.
From Steffen et al. (2004). Copyright AGU 2004, reproduced with permission.

differenced by:

$$\Delta\sigma_{VV} = \sigma_{VV}(t_p) - \sigma_{VV}(t_a) \tag{3.5}$$

where σ_{VV} is vertical-polarization backscatter and all quantities are in dB. The criterion for the detection of melt is then based on $|\Delta\sigma_{VV}| > 1.8\,\text{dB}$ for melt and less than $>1.0\,\text{dB}$ for refreezing (Nghiem et al., 2001; Steffen et al., 2004). Results for the period 1999 to 2003 are shown in Figure 3.21.

3.3.3 Passive-microwave techniques

By virtue of their ability to measure the dramatic rise in *microwave emissivity* associated with the onset of melt, Passive Microwave (PMW) sensors are also used to monitor the spatial extent and duration of ice sheet seasonal melt. Dating back to the launch of the Nimbus-7 Scanning Multi-channel Microwave Radiometer (SMMR) in 1978 and offering complete daily coverage (with the Special Sensor Microwave/Imager or SSM/I onboard DMSP satellites since 1987) uninterrupted by clouds and darkness, multi-frequency and multi-polarization PMW data are well-suited to routine time series studies of seasonal and interannual change in melt extent (albeit at a poor spatial resolution of 25 km). These data are available from the U.S. National Snow and Ice Data Center (NSIDC: ⟨http://nsidc.org⟩). Note that complete polar coverage with SMMR took place once every 3 days.

A number of studies have exploited this information to map the onset, extent, and duration of surface melt, both in Greenland—e.g., Abdalati and Steffen (1995); Mote et al. (1993); and Mote and Anderson (1995)—and Antarctica—e.g., Ridley (1993) and Zwally and Fiegles (1994). Techniques include determining melt by applying threshold values to the observed microwave emission of single-channel data (Mote and Anderson, 1995; Ridley, 1993). Zwally and Fiegles (1994) determined that any pixel exhibiting brightness–temperature (T_B) increases of >30 K above the annual mean value was affected by melt. Torinesi et al. (2003) used an algorithm based on 19 GHz H-polarization T_B data to detect and quantify trends and variability in the partial melting of Antarctic Ice Sheet margins over the period 1980–1999. This study further identified interannual signatures in the melt indices associated with the Antarctic Oscillation and possibly the Southern Oscillation. Figure 3.22 (see color section) shows their map of mean annual melt duration over 18 years within the period 1980–1999.

Steffen et al. (1993a) used the normalized ratio of the difference between 19-GHz H-polarization (19H) and 37-GHz H-polarization (37H) T_Bs to detect the onset and length of the melt season. This technique was further developed into the *Cross-Polarization Gradient Ratio* (*XPGR*), defined by Abdalati and Steffen (1995, 1997a) as:

$$XPGR = \frac{19H - 37V}{19H + 37V} \quad (3.6)$$

Ratioing in this manner reduces the depolarization effects of water and temperature dependency. Compared with single-channel approaches, the multi-channel technique produces a higher inertia signal that dampens some of the day-to-day variability and is well-suited to longer term monitoring (Thomas et al., 2001a). For XPGR, near-surface melt (or the appearance of ∼1% free water in the upper meter of firn) is assumed to occur in a pixel when this ratio exceeds a threshold value established by theory and field observations.

Joshi et al. (1998) further utilized the *Polarization Ratio* (*PR*) at 37 GHz:

$$PR = \frac{37V - 37H}{37V + 37H} \quad (3.7)$$

They first used SAR data to define signature pixels which are representative of ice sheet facies, then produced a scatterplot of XPGR versus PR for signature pixels (this technique shows maximum separability of the facies). This results in a clustering

Sec. 3.3] Detection, mapping and monitoring of ice-sheet surface melt 171

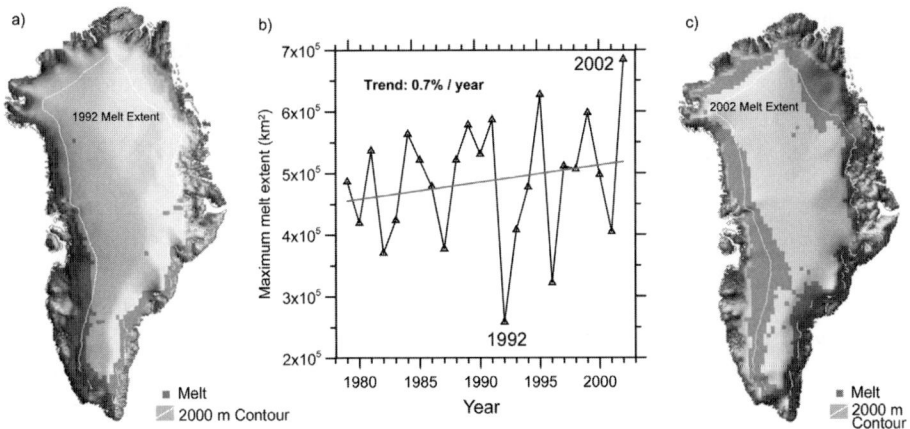

Figure 3.23. A map of the maximum areal extent of surface melt (darker gray) on the Greenland Ice Sheet for (a) 1992 and (c) 2002, derived from analysis of Nimbus-7 SMMR and DMSP SSM/I data. (b) Time series of maximum annual melt extents for the Greenland Ice Sheet derived from the same data. While 1992 represents the lowest extent over the 24-year period, seasonal melt in 2002 began earlier than usual and progressed higher up on the ice sheet than at any time over the period analyzed. In NE Greenland, the melt extended beyond an elevation of 2,000 m (marked as a white line)—i.e., into a region that normally remains too cold for melting. The summit of the ice sheet is at ~3,340 m above mean sea level.

Imagery courtesy Konrad Steffen and Russell Huff, University of Colorado at Boulder. Further information on satellite-derived Greenland maximum melt extent from 1978 to 2002 is given at ⟨http://cires.colorado.edu/steffen/melt/index.html⟩.

of pixels representing each facies, from which the latter are delineated using a Gaussian maximum-likelihood classification scheme. The results, from Greenland for the period 1987–1996, are comparable with those of Long and Drinkwater (1994). From these data, Joshi et al. (1998) observed a slight warming trend from 1987 to 1991 (i.e., an increase in melt), followed by a sudden cooling—attributed to the major volcanic eruption of Mount Pinatubo in 1991 (Abdalati and Steffen, 1997b)—then a decrease in area of the dry-snow facies, suggesting an overall warming trend.

In their analysis, Abdalati and Steffen (2001) showed a positive melt trend for Greenland of 0.97% per annum over the period 1979–1999, coinciding with a mean boreal-summer temperature increase of 0.25°C and dominated by the behavior of the western flank of the ice sheet. Extreme melt years were 1991, 1995 (Abdalati and Steffen, 2001), and again in 2002 (Steffen and Huff, 2003), with the latter representing a new record for the 24-year period (surpassing the previous record areal extent by 9%). As shown in Figure 3.23, maximum surface melt area increased on average by 16% from 1979 to 2002, with the extent of 685,000 km^2 in 2002 being 2.6 times greater than that in 1992 and significantly greater than the average melt extent of 455,000 km^2 from 1979–2003 (Steffen and Huff, 2003; Steffen et al., 2004). Note that trend analyses based on relatively short time series are highly dependent on the values of the end members. A statistical analysis of the melt time series by Steffen et al. (2004) confirmed an increased melt extent earlier in the melt season. The record exhibits a high temporal variability (Abdalati and Steffen, 2001), however, with a significant cooling in 1992 again associated with the eruption of Mount Pinatubo. These findings have been confirmed by contemporary hourly meteorological observations from Automatic Weather Stations on the ice sheet (AWSs: ⟨http://cires.colorado.edu/steffen/index.html⟩). Mote (1998a) combined satellite PMW data with a synoptic climatology of geo-potential heights at 700 hPa to measure the variable effect of mid-tropospheric circulation patterns on surface melt extent in Greenland from 1979 to 1989. His findings suggested that the strength and location of the major regional circulation features—namely, the *Baffin Bay Low* and the *North American Trough* (*NAT*)—are strongly related to the downstream airflow over southern Greenland and therefore melt extent. Increased melting is associated with a westward-displaced NAT, while an eastward-displaced NAT produces reduced melt. Further work showed that melt extent is also related to the *North Atlantic Oscillation* (*NAO*: Mote, 1998a, b). Moreover, Joshi et al. (2001) inferred the annual melt extent, duration, and season length in Greenland for 1979–1997 using an edge-detection algorithm.

In the Southern Hemisphere, Scambos et al. (2000) attributed the extraordinary recent collapse of the Larsen B Ice Shelf (see Section 3.4.1) to an increase in surface melt, producing surface ponding during longer melt seasons observed in the 1990s. This prompted Fahnestock et al. (2002) to examine melt season duration on Antarctic Peninsula ice shelves using a combined SMMR–SSM/I T_B dataset from 1978–2000. They used two techniques to detect surface melt—namely, XPGR and bimodal histograms (Figure 3.24)—and found an increase of ~1 day per annum in the summer melt season over the period 1978–2000 (Figure 3.25, see color section). These results provided independent support for the cause-and-effect hypothesis of Scambos et al. (2000, 2003). They showed that unusually long melt seasons over northern regions of the Larsen Ice Shelf—e.g., in 1992/3 and several later years—preceded the large-scale disintegrations observed in high-resolution satellite image sequences from the 1990s (Vaughan, 1993b; Rott et al., 1996, 1998a; Doake et al., 1998; Lucchitta and Rosanova, 1998)—see Section 3.4.1. They also provided additional evidence for the regional warming discussed in King (1994a), King and Comiso (2003), Smith et al. (1996), Vaughan and Doake (1996), Skvarca et al. (1998), and Comiso (2000).

Steffen et al. (2004) showed that more realistic representations of surface melt extent and duration in Greenland can be derived by combining active- and PMW satellite data. An analysis of daily QuikScat diurnal-difference radar scatterometer data from 1999 to 2004 confirmed the extreme melt situation in 2002 detected in PMW data (see Figure 3.21). In addition, QuikScat data showed a significant increase in melt-season length over several areas of the Greenland Ice Sheet in 2003. A comparison of SSM/I and QuikScat melt data by Steffen et al. (2004) revealed that the latter is more sensitive to surface melt and detects melt earlier. In fact, QuikScat can detect a weaker stage of melt while PMW sensors can detect a later or stronger stage of melt. A further finding was that the PMW-derived melt extent is approximately confined to the QuikScat-derived melt areas experiencing a melt duration of 2 weeks or longer.

Sec. 3.3] Detection, mapping and monitoring of ice-sheet surface melt 173

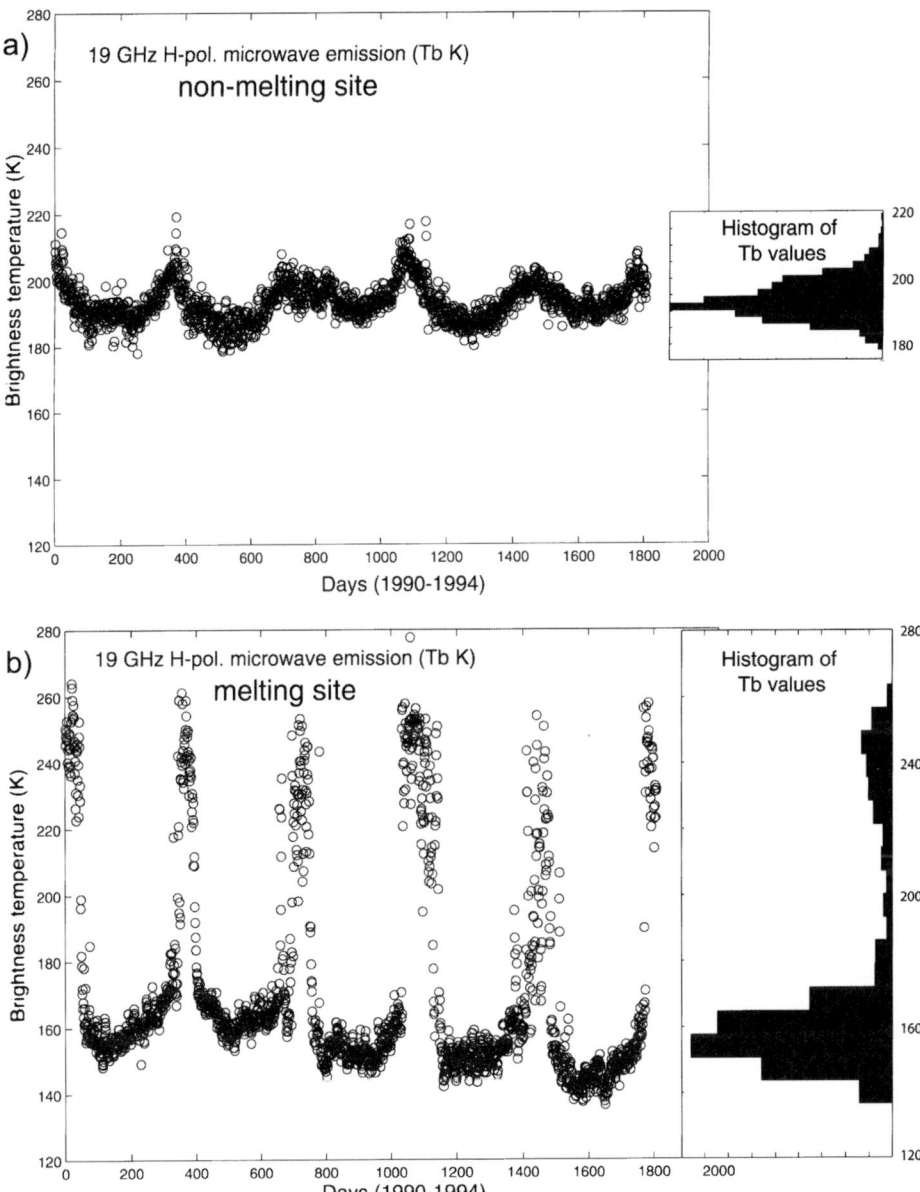

Figure 3.24. Microwave emission at 19-GHz H-polarization measured by the DMSP SSM/I, expressed as brightness–temperature, for two pixels on the Antarctic Peninsula, 1 January 1990 to 31 December 1994—one within the non-melting zone (a) and one within the melting zone (b). The frequency of occurrence for each time series is shown by the histograms. The abrupt jumps in emission in the austral summer in (b) are caused by melt effects, leading to a bimodal T_b distribution.

From Fahnestock et al. (2002). Copyright International Glaciological Society 2002, reproduced with permission.

3.3.4 Estimation of surface melt rates

The techniques described above are relatively straightforward means of detecting and mapping *melt onset* and *season length*. However, the accurate determination of melt rate, either directly or through measurement of components of the surface energy balance, is more problematical as it requires significant additional information (Walsh et al., 2001). In other words, the area of melt does not simply equate to mass loss, and additional information is required on the magnitude of melt and rates of ablation. This is an important deficiency in our understanding of ice sheets and their possible response to climate warming (Morris, 1999). Liston and Winther (2004) addressed this issue by combining satellite-derived maps of Antarctic blue-ice and snow areas, coincident meteorological station data, and a high-resolution meteorological distribution model to estimate daily simulated melt values on a 1-km grid covering Antarctica. Their findings suggest that the simulated melt has little current effect on the mass balance of Antarctica in that most of the meltwater refreezes close to where it was produced. The melt contribution is, however, important for the surface energy balance in terms of modifying surface and near-surface snow and ice properties—e.g., grain size and density—and thus albedo.

Mote (2003) used annual melt duration information derived from SSM/I data in a surface mass-balance model to estimate mass balance, surface-elevation changes, and meltwater runoff rates for Greenland. Precipitation-rate estimates were based on modeled precipitation from European Centre for Medium-Range Weather Forecasts (ECMWF) analysis data—e.g., Genthon and Krinner (1998). For the energy balance approach, measurement uncertainties of individual components make the combined errors large. This in turn limits the accuracy of estimates of both melt-related changes in surface elevation and meltwater runoff into the ocean. One approach is to combine the binary melt estimates derived from the techniques outlined above with positive degree-day probability distributions in order to estimate ablation rates for given ice sheet regions—e.g., Zwally and Fiegles (1994) in Antarctica and Mote (2000) in Greenland. *In situ* meteorological data from networks of satellite-interrogated AWSs are an important additional source of information in this respect (Shuman and Stearns, 2001; Steffen et al., 1996). For example, the Greenland Climate Network (GC-Net) was established in 1994 to monitor coupled glaciological and climatological parameters at various ice sheet locations (Box and Steffen, 2001; Steffen and Box, 2001). Moreover, AWS data have been used in Antarctic Ice Sheet energy balance calculations (Van den Broeke et al., 2004).

These analyses represent an important first step towards quantitatively estimating ablation rates from PMW satellite data. Moreover, they have been enhanced by the availability of improved data from new-generation Advanced Microwave Scanning Radiometers (AMSR-E and AMSR) launched onboard NASA Aqua and ADEOS-II in May and December of 2002, respectively, offering a spatial sampling interval of 12.5 km compared with 25 km for the DMSP SSM/I and Nimbus-7 SMMR (for lower frequency data). The potential also exists for combining observations from PMW, scatterometer, SAR, and visible/near-IR imagery for a more complete description of the ice sheet melt characteristics—e.g., the degree of melt (Thomas et al., 2001a).

According to Abdalati and Steffen (2001), the relationship between coastal temperatures and ice-sheet melt extent implies an increase in surface runoff contribution to sea level rise of 0.31 mm per annum for a 1°C temperature rise. While absolute measurements of loss by surface melt are also complicated by the refreezing of downward-percolating meltwater within the ice sheet volume (Pfeffer et al., 1991), this is not always the case. Using GPS measurements, Zwally et al. (2002b) discovered an acceleration in ice flow within the equilibrium zone of west-central Greenland (at $\sim$1,200-m elevation) from a mid-winter average of 31.3 cm d^{-1} to 40 cm d^{-1} at the height of the summer melt period in 1996–1999. They attributed this acceleration to rapid downward percolation of meltwater through the entire ice mass (Figure 3.26), to increase basal water pressures and accelerate basal-sliding velocities. The apparent coupling between surface melting (as detected in satellite and *in situ* measurements) and ice sheet flow entails a mechanism for the rapid and large-scale response of the ice sheet to climate warming. An acceleration of ice flow may lead to ice thinning upstream, with the lowering of the surface elevation establishing a feedback to more melting. Similar increases have been noted in outlet glaciers in summer—e.g., Joughin et al. (1996a). Change-in-elevation measurements from the ICESat (launched January 2003) altimeter mission will enable the examination of these processes over a wider area and longer time period, especially when combined with satellite-derived time series of surface albedo (see Section 3.15), temperature (Section 3.14), melt-season length and areal extent (Section 3.3), and ice motion (Section 3.8). Unfortunately, the CryoSat mission failed on launch in October 2005.

3.4 DETECTION AND MAPPING OF CHANGES IN ICE SHEET MARGINS

Glacier tongues, iceberg tongues (comprising lines of grounded icebergs), *ice shelves*, and *ice fronts* are the most dynamic and changeable features of the ice sheet margin, with substantial changes occurring on annual to decadal timescales and spatial scales of kilometers to tens of kilometers (Lange and Kohnen, 1985; Williams and Ferrigno, 1998). The natural variability of the boundary conditions of these features is, however, poorly known (Rémy and Legrésy, 2004; Rémy and Minster, 1997). Given their remoteness and vast span, the only feasible means of mapping and monitoring coastal margins is satellite remote sensing (Williams et al., 1995). Being in direct contact with the ocean as well as atmosphere, ice shelves and floating glacier tongues, which account for 11% of the area and 2.5% of the ice volume of Antarctica (Drewry, 1983), are particularly sensitive to ocean warming (Huybrechts and de Wolde, 1999; Jacobs et al., 1992; Van der Veen, 1991; Warner and Budd, 1998; Williams et al., 1998a, b). Ice shelves also influence regional ocean circulation (Wong et al., 1998), and the formation of Antarctic Deep and Bottom Water in places (Foldvik and Gammelsrød, 1988; Foldvik et al., 2004; Grosfeld et al., 1997). As a result, their breakup could cause significant changes to a large range of local,

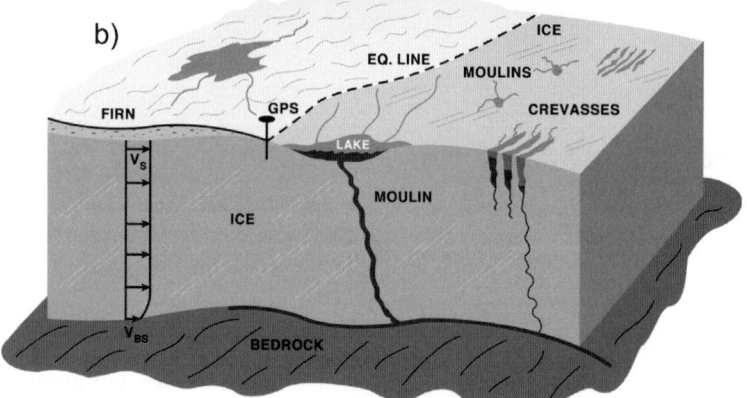

Figure 3.26. (a) Photograph of a meltwater stream flowing into a large moulin in the ablation zone of the Greenland Ice Sheet in summer. Moulins are openings of up to 10 m in diameter through which surface meltwater drains to the bedrock. (b) Schematic of glaciological features in the ablation and equilibrium zones of the West Greenland Ice Sheet. These include surface lakes, crevasses, meltwater in-flow channels, and moulins. Meltwater is shown draining downwards through the ice to its base (the bedrock). Basal ice flow at the pressure melting point is partly from basal sliding and partly from shear deformation. The magnitude of sliding is difficult to establish without borehole measurements.

(a) Courtesy of Roger J. Braithwaite (University of Manchester, England); and (b) from Zwally et al. (2002b). (b) Copyright AAAS 2002, reproduced with permission.

regional, and global environmental parameters. The calving of icebergs is also a key factor, both in terms of ice-sheet mass balance and their impact on ocean properties and circulation (iceberg detection and tracking is covered in Section 3.5). While ice shelves do not make a direct contribution to sea-level rise (as they are floating), we saw in Chapter 1 how they are thought to play an important indirect role via their buttressing effect in front of encroaching ice streams and outlet glaciers (Hughes, 1992; Mercer, 1978).

In Antarctica, a long-term project is underway to detect and map circumpolar coastal change using a range of satellite data, but based initially on Landsat data (Ferrigno et al., 2002; Lucchitta et al. 1991; Williams and Ferrigno, 1998; Williams et al., 1997). The planned outcome is a series of twenty-four $1:10^6$-scale glaciological and coastal change maps (see Williams and Ferrigno, 1998). Working in Greenland, Sohn et al. (1998) used a combination of techniques, including satellite data from the past 30 years, to show that Jakobshavn Isbrae has systematically retreated over the past 150 years. Sohn and Jezek (1999) have developed a technique to automatically map ice sheet margins using SPOT and ERS-1 SAR data, which they applied to a study of margin advance/retreat in the vicinity of Jakobshavn Isbrae (West Greenland). Satellite image-processing techniques used for ice-margin detection and mapping include dynamic thresholding, edge enhancement, edge detection, edge following, and region growing. Ice-sheet margin change in Greenland has also been detected using SAR data—e.g., by Fahnestock and Bindschadler (1993). A recent dramatic event in the Arctic has been the major breakup of the Ward Hunt Ice Shelf (the largest northern ice shelf) skirting the northern coast of Ellesmere Island. This event was detected in Radarsat-1 SAR data, which revealed a clear fracture in April 2000 and widespread decay by September 2002 (Mueller et al., 2003; Vincent et al., 2004).

Zhou and Jezek (2002) constructed two ortho-rectified mosaics of Greenland from DISP data acquired in 1962 and 1963. They estimated the relative geo-location accuracy to be $\sim$200 m and the absolute accuracy to be $\sim$450 m (compared with ERS SAR data from 1992), which they showed is sufficient for quantitative comparisons of ice sheet extent (marginal change) over a 30-year interval. Detailed information on the processing steps involved is also given in Zhou et al. (2002). These particular data can be accessed at the U.S. NSIDC at ⟨http://nsidc.org/data/docs/daac/nsidc0118_greenland_disp.gd.html⟩.

3.4.1 The collapse of the Larsen Ice Shelf

Satellite remote sensing has been responsible for the discovery of a number of remarkable recent changes in ice fronts and iceberg calvings. Perhaps the most dramatic and unprecedented event detected and monitored in this way has been the recent collapse of the Larsen Ice Shelf, located along the eastern flank of the Antarctic Peninsula between $\sim$74° and 65°S, in a series of major breakups over the last decade. The latest episode, revealed by analysis of 250/500-m-resolution MODIS imagery, entailed the "disintegration" of about 3,320 km^2 of the Larsen B Ice Shelf over a 35-day period beginning on January 31, 2002. Not only did this event entail a

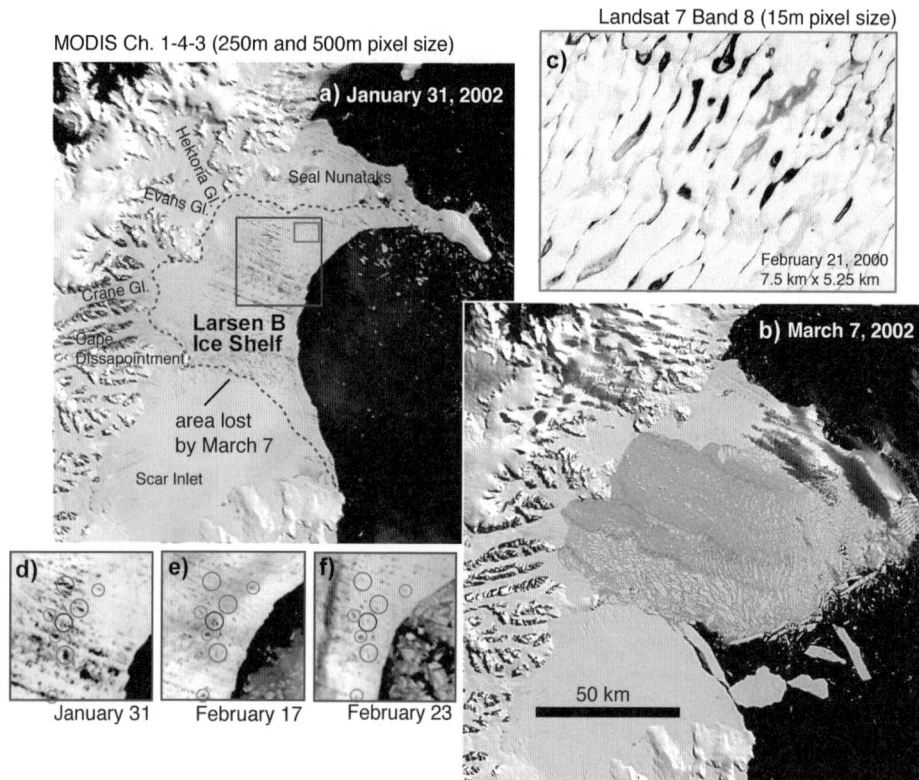

Figure 3.27. Cloud-free satellite images of the breakup and surface melt ponding on the Larsen B Ice Shelf. Images (a) and (b) are derived from channel-1, -2, and -3 data from the Terra MODIS, with the dashed line indicating the extent of disintegration on March 7, 2002. (c) A Landsat 7 ETM+ band-8 image, showing a high-resolution view of surface meltponds/lakes from a previous summer for the small boxed region in (a). The small inset images in (d) to (f)—of MODIS channel-1 images corresponding to the larger box in (a)—track the disappearance of ponds over the weeks prior to the breakup. Circled in (d) are ponds that survived the 23-day period after January 31, 2002, while those circled in (e) and (f) denote ponds that disappeared by February 17 and 23, 2002, respectively.

From Scambos et al. (2003). Copyright AGU 2003, reproduced with permission.

major change to the east Antarctic Peninsula coast, but it also released thousands of icebergs into the Weddell Sea (Figure 3.27) (Scambos et al., 2003). Prior to this, the main part of the ice shelf in Larsen Inlet broke away between March 1986 and November 1989 (Skvarca, 1993, 1994). Another major collapse, of the Larsen A Ice Shelf, occurred in early 1995, when 4,200 km^2 disintegrated over a few days in January (Rott et al., 1996) followed by further breakup in the subsequent weeks (Rack et al., 2000; Skvarca et al., 1999a, b). The northern Larsen Ice Shelf then lost 1,185 km^2 in November 1998 and a further 200 km^2 by February 1999. These and other breakups in the peninsula region, involving the loss of ~13,500 km^2 in total

Sec. 3.4] Detection and mapping of changes in ice sheet margins 179

from seven ice shelves since 1974, have been further documented by Vaughan (1993b); Rott et al. (1998a, 2002); Lucchitta and Rosanova (1998); Scambos et al. (2000, 2003); Skvarca and De Angelis (2003); Skvarca et al. (1999a, b); and Vaughan and Doake (1996). In all cases, satellite data have revealed both the magnitude of these events, and also have given clues as to their causal mechanisms. The sequence of breakup events since 1993 and determined by analysis of ERS SAR data by Rack and Rott (2003, 2004) is illustrated in Figure 3.28. Scambos et al. (2003) and Skvarca

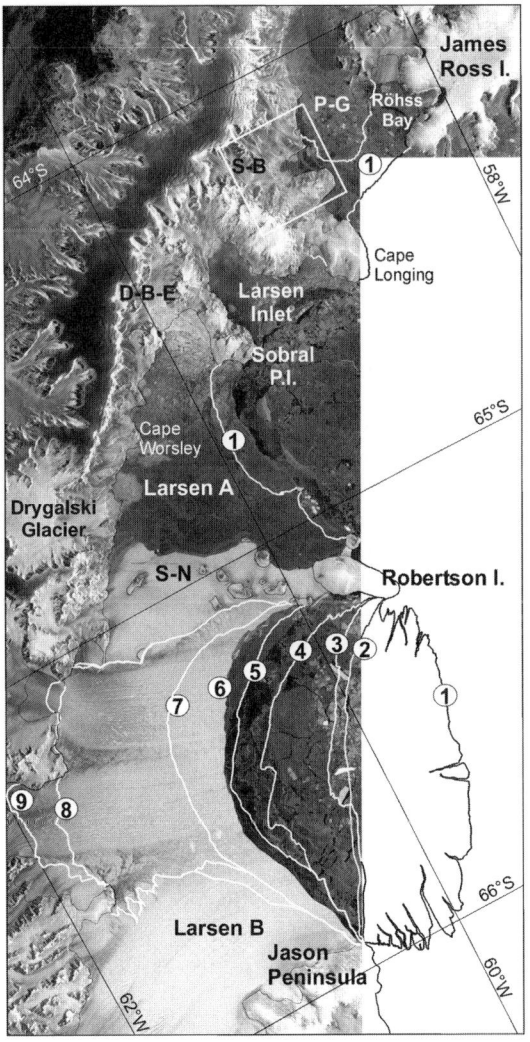

Figure 3.28. An ERS SAR image composite of the northern Larsen Ice Shelf, from October 2000, with ice edge positions marked over the period August 1993 to March 2002.
From Rack and Rott (2003). ERS image copyright ESA 2003. Copyright University of Bergen Geophysical Institute 2003, reproduced with permission.

and De Angelis (2003) provide excellent summaries and analyses of recent shelf breakup events in the peninsula region.

The ice-shelf demise has coincided with the major warming trend of 0.5°C per decade in the Antarctic Peninsula region over the past 50 years and noted earlier. This has shifted the shelves, which are already close to their climatic limit, into zones of ablation, resulting in an increase in the annual summer melt season by 2–3 weeks over the last 20 years (Rott et al., 1998a; Skvarca et al., 1999a, b). Skvarca et al. (1999a) used a variety of satellite data to monitor the Larsen Ice Shelf over a 30-year period, including DISP photographs (from 1963), Landsat images, and ERS-1/2 and Radarsat-1 SAR images, and revealed major changes in both the surface characteristics and dynamic behavior. Figure 3.29 provides a further example of combining different datasets to monitor the recession of an ice front, in this case of the Larsen A Ice Shelf between 1975 and 1997. Based upon satellite and model analysis of the rapid collapse of the Larsen A Ice Shelf in 1995, Doake et al. (1998) and Rott et al. (1996) suggested that ice shelves may disintegrate rapidly once they retreat beyond a certain critical limit, as a result of perturbed mass balance.

The unprecedented nature of the Larsen B breakup prompted a satellite data and modeling analysis by Scambos et al. (2000). They hypothesized a failure mechanism connected to enhanced localized melting, whereby meltwater filling crevasses would wedge them open by the forcing effect of water pressure. This would weaken the ice shelf (Hughes, 1983; Van der Veen, 1998; Weertman, 1973) and make it more susceptible to disintegration or even wholesale failure. Sequential MODIS images showed extensive meltponding over the Larsen B Ice Shelf in late January 2002 (see Figure 3.27), consistent with an unusually warm summer and extended melt season (see below). With a lower albedo than the surrounding ice, meltponds are analogous to their sea ice counterparts (see Chapter 5 of Volume 1 of this book) in that they contribute to locally enhanced melt, in this case supplying more water to crevasses. The meltponds disappeared in a series of images from February, indicating meltwater drainage through the crevasses (Figure 3.27). Scambos et al. (2000, 2003) highlighted a strong correlation between long melt seasons and ice shelf breakup, as also discussed in Fahnestock et al. (2002) (see Section 3.4.3.1). Doake et al. (1998) further hypothesized that downward heat transport by meltwater and expansion of the latter on refreezing at depth may contribute to the observed catastrophic failure. In addition, MacAyeal et al. (2003) developed a fragment-capsize model to help explain the most dramatic aspects of the January 1995 and March 2002 breakup events. Shepherd et al. (2003) also estimated that a thinning of the Larsen Ice Shelf occurred between 1992 and 2001, due to enhanced basal melting from a warmer ocean and measured by SAR interferometry (see Section 3.11), and that this contributed to the observed collapse.

A summary of the recent warming trend in the region of the Larsen B Ice Shelf, the presence/absence of surface meltponds and breakup events of ice shelves in the Larsen A and B, Larsen Inlet, and Prince Gustav Channel regions is presented in Figure 3.30 (see color section). This indicates a strong relationship between the number of melt days, monitored by satellite PMW emission changes (Fahnestock et al., 2002), and the extent of ice-shelf breakup. According to Scambos et al. (2003),

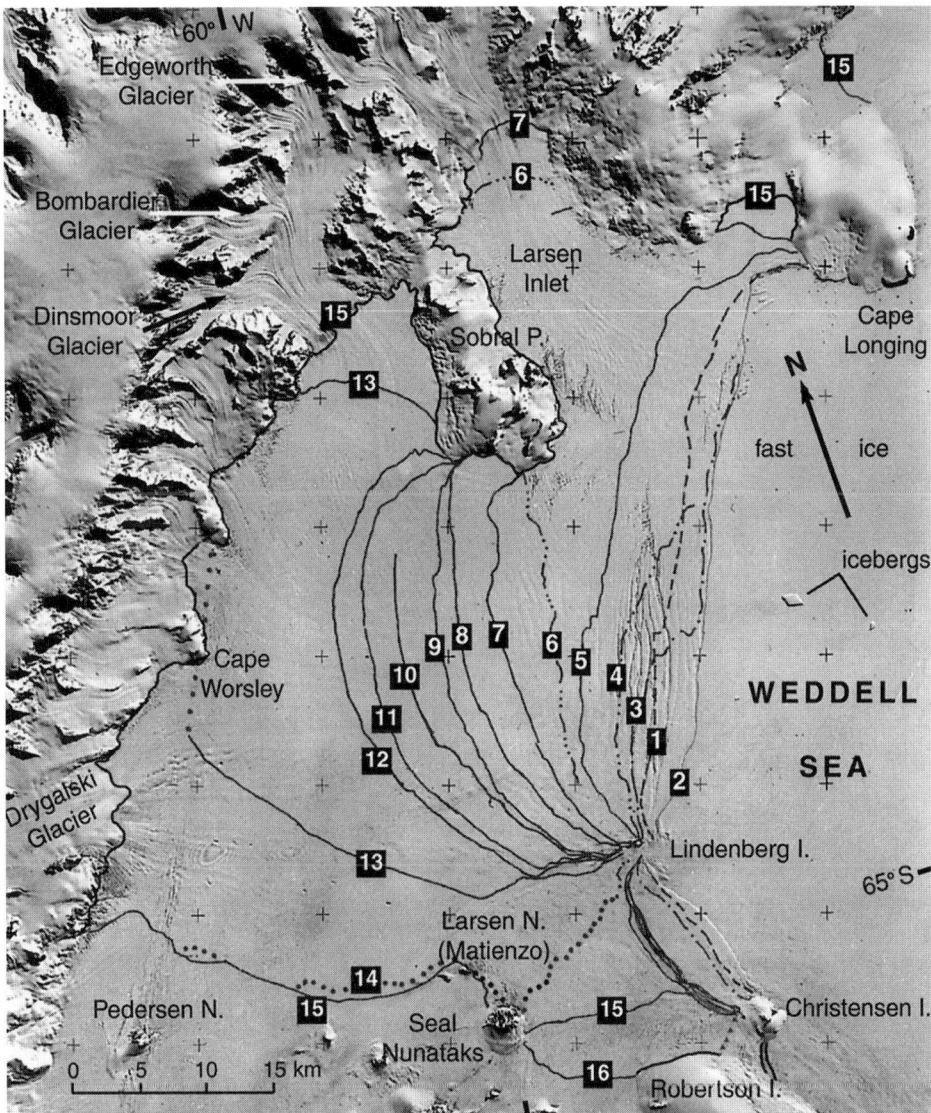

Figure 3.29. Ice front positions for the Larsen A Ice Shelf and Larsen Inlet, eastern Antarctic Peninsula, from (1) August 1963 to (16) July 1997, superimposed on a cloud-free Kosmos KATE-200 photograph from October 3, 1975. Other dates are (2) October 1975, (3) November 1978, (4) February 1979, (5) March 1986, (6) November 1989, (7) December 1992, (8) January 1993, (9) February 1993, (10) October 1994, (11–13) January 1995, (14) March 1995, and (15) March 1997. Locations prior to 1992 were derived from Argon (position 1), Kosmos KATE-200 (2), Landsat MSS (3 and 4), and Landsat TM (5 and 6) images. Those after 1992 were determined from ERS-1 SAR imagery, apart from position 10, which was surveyed by GPS.

From Skvarca et al. (1999b). Reprinted with permission from *Polar Research*. Copyright Norwegian Polar Institute 1999, reproduced with permission.

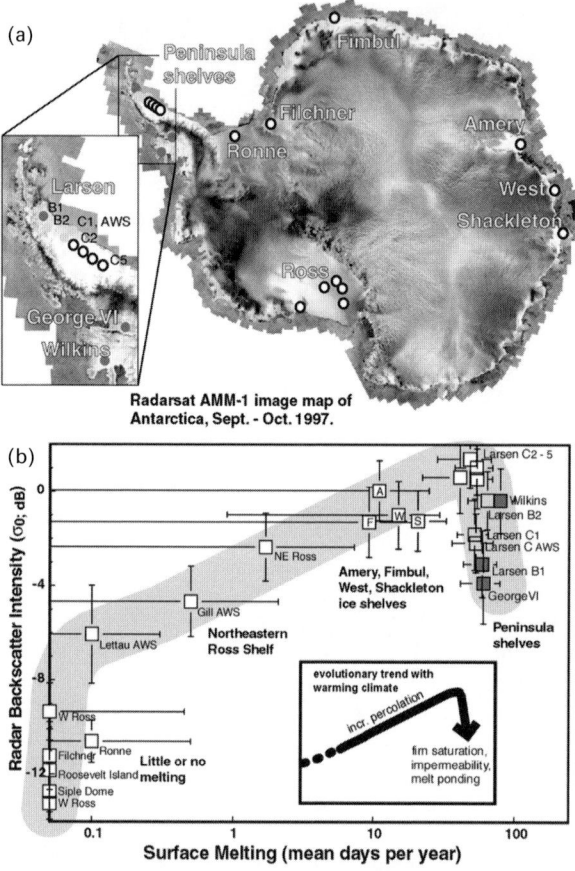

Figure 3.31. (a) The Radarsat-1 Antarctic Mapping Mission (AMM) image mosaic of Antarctica with ice shelf locations shown in (b) marked as circles, and showing approximate radar backscatter intensity. (b) Plot of backscatter intensity—the mean value from 100-km^2 regions at each circle in (a)—versus mean melt days per annum determined from satellite passive microwave data from 1978–1999 (Fahnestock et al., 2002), for ice sheet and shelf areas. The inset in (b) shows the evolutionary trend associated with a warming climate. This figure clearly indicates the importance of firn impermeability as a condition contributing to melt-related ice shelf breakup.

From Scambos et al. (2003). Copyright AGU 2003, reproduced with permission.

a long melt season equates to the generation of surface meltponds required in the "fracture deepening from meltwater infiltration" model of ice-shelf breakup outlined above. Another requirement of this model is for the shelf firn to be initially impermeable to support ponding. The relationship between annual surface melt-season length and satellite radar backscatter of ice-shelf firn was exploited by Scambos et al. (2000, 2003) to detect and monitor the susceptibility of ice shelves to the breakup process (Figure 3.31). Using these parameters, they suggested that the remaining Larsen B and C Ice Shelves have firn conditions indicative of impending breakup. It appears from the assembled evidence that, once the melt-season length threshold is reached, then ice-shelf collapse occurs within approximately two decades (Scambos et al., 2003). Similar conditions are observed for the currently retreating George VI and Wilkins Ice Shelves—the latter retreated nearly 1,100 km^2 in early March 2002 (⟨*http://nsidc.org/iceshelves/larsenb2002/*⟩). At present, only ice shelves on the Antarctic Peninsula are at or approaching a state of firn saturation, even though significant melt duration occurs around much of the continental perimeter. As shown in Figure 3.31, the Fimbul, Amery, Shackleton and West Ice Shelves are closest to the limit designated by Scambos et al. (2003) for breakup to occur. The process can be expected to become more widespread, however, if Antarctic summer temperatures continue to increase. Clearly, ice shelves are sensitive climatic indicators, and the nature of their most recent demise needs to be examined in the light of the paleo-record of past fluctuations (Gilbert and Domack, 2003; Pudsey and Evans, 2001). Indeed, it is of major concern that shelf breakups around the Antarctic Peninsula over the past two decades have greatly diminished features that appeared from the geological record to have remained stable (in place) over the previous several millennia (Domack et al., 2002; Gilbert and Domack, 2003; Scambos et al., 2003). Indeed, sediment–core analysis in front of the Larsen B Ice Shelf suggests that this shelf had not disintegrated prior to the present event since the Last Glacial Maximum, or ~12,000 BP (Domack et al., 2002; Gilbert and Domack, 2003; Scambos et al., 2003). The U.S. NSIDC at the University of Colorado has an excellent website devoted to the satellite monitoring of these and other Antarctic ice shelves using MODIS imagery (thermal IR during polar darkness): ⟨*http://nsidc.org/sotc/iceshelves.html*⟩.

3.4.2 Examples of recent change in the configuration of other glacier systems

Another major recent focus has been the Pine Island Glacier–Thwaites Glacier system in West Antarctica (Vaughan et al., 2001b), given the rapid changes that appear to be occurring there. These fast-moving ice streams are of major importance, as they drain a significant part of the WAIS, which is thought to be susceptible to collapse (see Chapter 1). As we shall see in later sections, this system has undergone a steady loss of elevation combined with grounding line retreat in recent decades. Bindschadler and Rignot (2001) have examined recent rift evolution of the Pine Island Glacier (PIG, 75°S, 102°W) using data from ERS and Radarsat-1 SAR, MODIS, Landsat 7 ETM+, and ASTER (Figure 3.32). Rifts form large-scale detachment boundaries (Joughin and MacAyeal, 2002), and their formation and propagation is a precursor to iceberg calving (Fricker et al., 2005; Joughin and MacAyeal, 2005; Larour et al., 2004). ASTER data also have a spatial resolution of 15 m at best (Kaeaeb, 2001; Yamaguchi et al., 1998, 2001), and as both ETM+ and ASTER are nadir-viewing, their imagery can be readily co-registered and directly compared. The major rift apparent in Figure 3.32 is thought to be the dynamic manifestation of major recent changes in the system. It "completed" in early November of 2001, as seen in the Terra Multi-angle Imaging Spectro

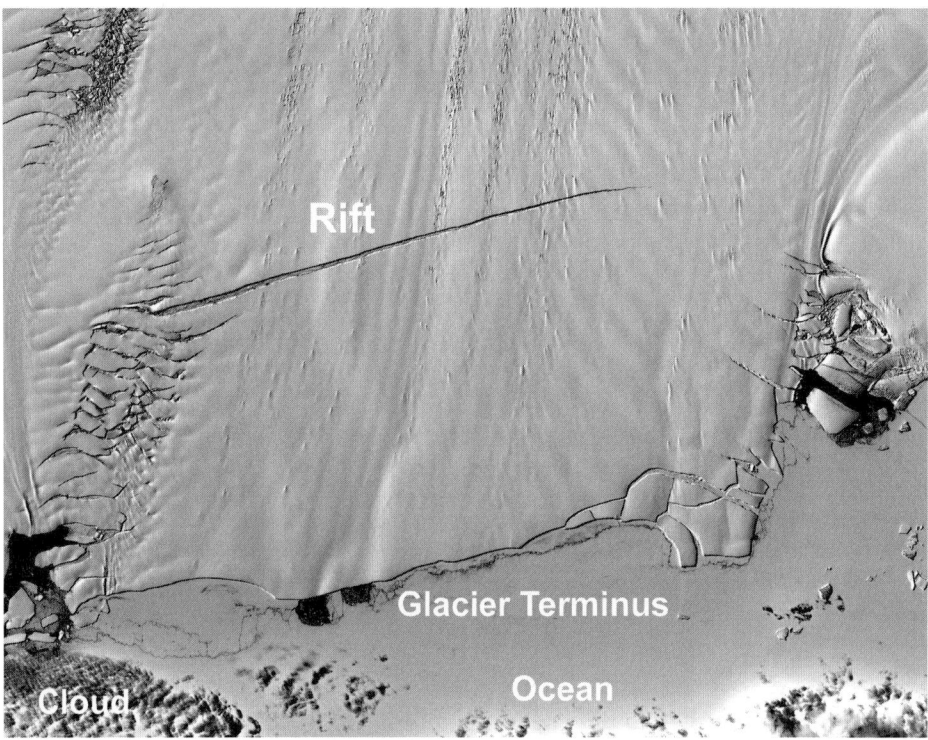

Figure 3.32. An Advanced Spaceborne Thermal Emission and Reflection Radiometer (ASTER) image of the Pine Island Glacier, West Antarctica, acquired on December 12, 2000. The image dimension is 48 × 38 km.

Image courtesy NASA/GSFC/MITI/ERSDAC/JAROS, and U.S./Japan ASTER Science Team (⟨http://www.asterweb.jpl.nasa.gov⟩).

Radiometer (MISR) image sequence in Figure 3.33. Vaughan et al. (2001b) catalogued earlier terminus variability between 1966 and 1998 by comparing aerial photography and high-resolution satellite imagery (Figure 3.34). For similar analyses of terminus variability through time, see Ferrigno et al. (1993), Kellogg and Kellogg (1987); and Bindschadler (2001), who examined a 28-year Landsat record (1973–2001) of the PIG's floating tongue.

Other studies have also combined satellite with ancillary data to monitor ice front changes—e.g., Ferrigno et al. (1993); Frezzotti and Mabin (1994); Frezzotti et al. (1998a); Kargel et al., 2005; Skvarca (1994); Vaughan and Doake (1996); and Wendler et al. (1996). In his study of ice-front fluctuations of numerous outlet glaciers in Victoria Land (Antarctica), Frezzotti (1997) found an apparent cyclical behavior of the order of 40–50 years. Frezzotti and Polizzi (2002) used a combination of maps, aerial photographs, and satellite images, carefully co-registered, to determine changes in the ice front location of 18 glaciers along the Wilkes Land Coast, East Antarctica (~126–145°E) from 1947 to 1997. Satellite data used included DISP (from 1963), Landsat MSS (1973–1974), Soyuz Kosmos KATE-200 (1985),

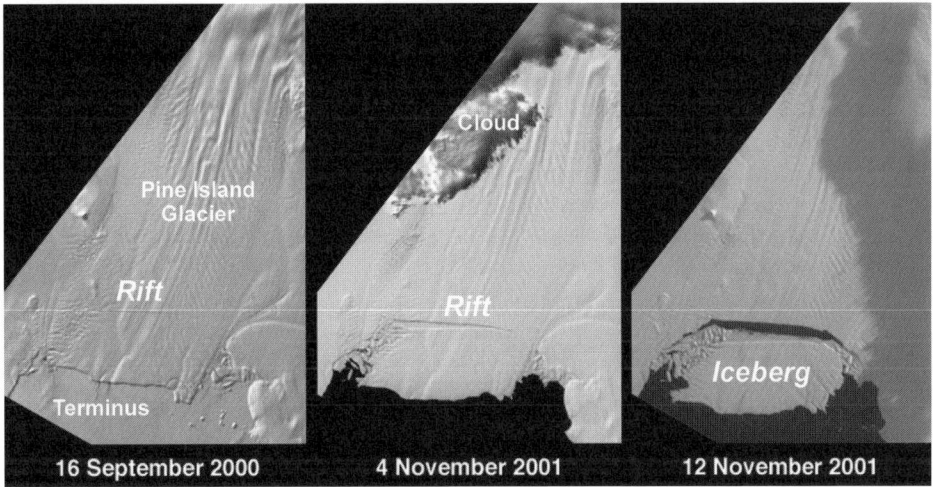

Figure 3.33. Images of the Pine Island Glacier acquired by the Terra Multi-angle Imaging Spectro Radiometer (MISR), showing the birth of a large tabular iceberg (42 × 17 km) some time between November 4 and 12, 2001. This event was preceded by the formation of the large rift (crack) shown in Figure 3.32 and originally discovered in Landsat-7 ETM+ imagery (Bindschadler and Rignot, 2001). Analysis of multi-sensor satellite datasets revealed the crack to be propagating through the shelf ice at a rate averaging ~15 m per day, accompanied by a slight rotation of ~1°.

Images courtesy of NASA/GSFC/LaRC/JPL, MISR Team.

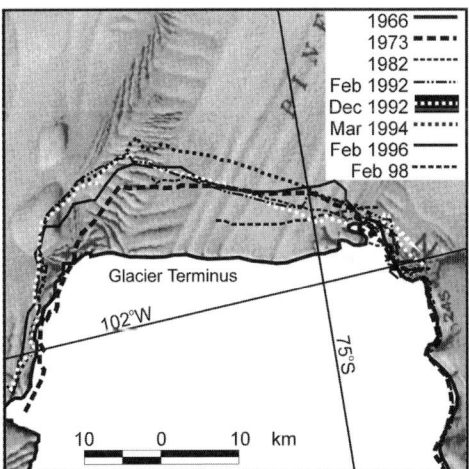

Figure 3.34. Map of ice front positions for the Pine Island Glacier (West Antarctica), overlain on a sketchmap drawn from aerial photography collected in 1966 (USGS, 1993). Other image sources are: January 24, 1973—Landsat 1 (path 246, row 114); January 15, 1983—Landsat; February 9, 1992, December 4, 1992, March 15, 1994, and February 1996—ERS-1 SAR; February 13, 1998, airborne survey.

From Vaughan et al. (2001b). Copyright AGU 2001, reproduced with permission.

Landsat TM (1989–1990), ERS-1/-2 SAR (1995–1996), and Radarsat-1 SAR (1997). From this time series, Frezzotti and Polizzi (2002) estimated that the total area of the floating glaciers decreased from 3,035 km^2 in 1963 to 2,785 km^2 in 1989, with the majority exhibiting a cyclical behavior of 5 to 50 years without a marked trend. Further around the coast, a fascinating study by Keys et al. (1998) combined modern satellite data with the historical record (ship observations) to monitor the Ross Ice Shelf frontal position over the period 1841–1997 (Figure 3.35). Such studies are enlightening in that they place recent major iceberg-calving events into an historical context (see the next section).

Comparison of data from the three Radarsat-1 Antarctic Mapping Missions has enabled the creation of a unique "snapshot" assessment of circumpolar changes in ice margin advance/retreat over a 7-year period. One detail, of the Shirase Glacier in East Antarctica (at $\sim$70°S, 40°E), is shown in Figure 3.36 (see color section). This fast-flowing outlet glacier attains speeds at its terminus (calving front) of up to 2.7 km per annum (Pattyn, 1996). Such glaciers and ice streams play a key role in the dynamic behavior of the ice sheet and its response to changing climatic conditions, and are also a major source of small icebergs. A similar study was carried out on the Amery Ice Shelf by Fricker et al. (2002a), again using a variety of data. The resultant map of changes in ice-front location is shown in Figure 3.37. This study also notes that a large iceberg-calving event is imminent—the first since late 1963 to early 1964—as a result of longitudinal-to-flow rift formation at the ice front (Fricker et al., 2005). This and other studies—e.g., Rignot and MacAyeal (1998)—have also shown that InSAR (SAR interferometry) analysis provides a unique means of detecting rift formation and propagation. Such work is also combining satellite data with *in situ* observations, using GPS techniques in particular, to enhance our currently poor knowledge of rift formation and evolution.

3.5 ICEBERGS

3.5.1 Iceberg detection and size statistics

Together with an estimate of the basal melt rate of floating ice masses, the "average" iceberg production (calving) rate represents an important "direct" though challenging measure of the mass flux from the ice sheet at its perimeter—given the vast scale involved (Bigg, 1999; Jacobs et al., 1992; Reeh, 1994). Joughin and MacAyeal (2002) estimated that the three large icebergs calved from the Ross and Filchner-Ronne Ice Shelves in 2000, and first detected in satellite imagery, contained approximately twice the annual accumulation of the entire Antarctic Ice Sheet. One of these bergs had an area of $\sim$10,600 km^2—approximately the size of Jamaica (Young, 2001). Iceberg calving rates are not constant (Frezzotti, 1997; Higgins, 1991), however, and calving flux remains a major uncertainty in current estimates of ice-sheet mass balance. Earlier mass-budget estimates of Antarctica have suggested that calving accounts for up to $\sim$80% of the total mass output, but with an associated error that is larger than the net mass budget estimate itself

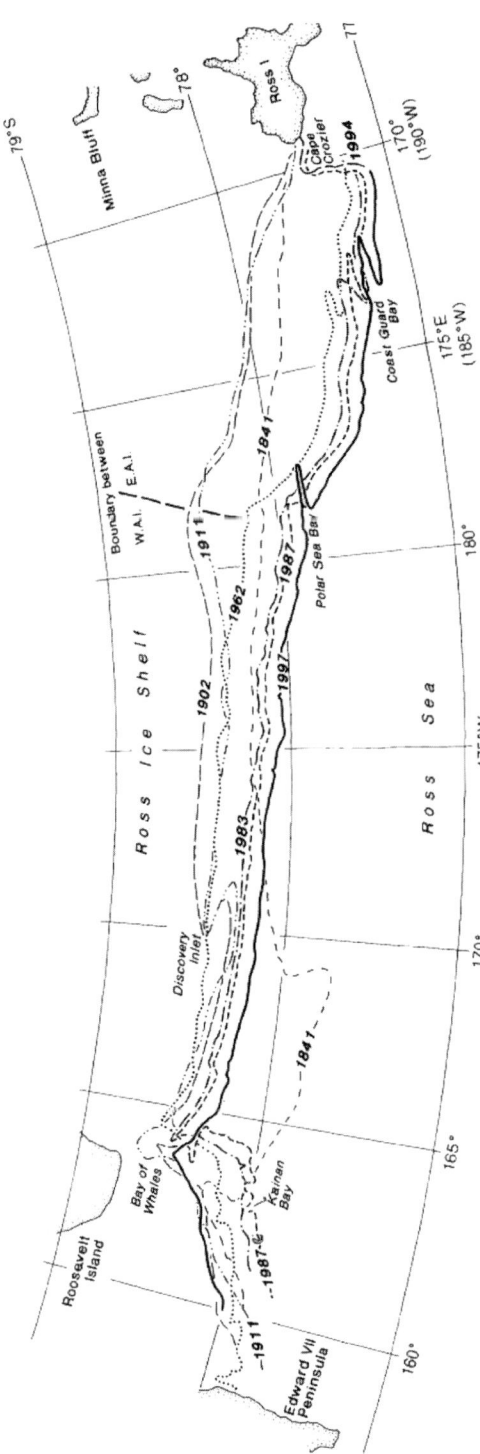

Figure 3.35. The location of the ice front of the Ross Ice Shelf, 1841–1997, derived from satellite observations and historical *in situ* measurements.

From Keys et al. (1998), after Jacobs et al (1986). Copyright International Glaciological Society 1998, reproduced with permission.

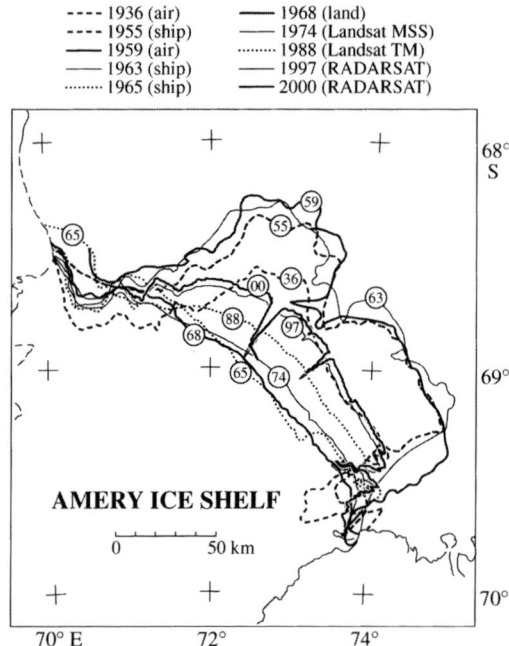

Figure 3.37. Frontal positions of the Amery Ice Shelf for ten epochs from 1936 to 2000, obtained using different remote-sensing (and ship/surface) datasets (outlined in the legend). From Fricker et al. (2002a). Copyright International Glaciological Society 2002, reproduced with permission.

(Jacobs et al., 1992, 1996; Orheim, 1985). In addition, icebergs represent a significant hazard to polar shipping and operations, particularly in the Arctic (Bertoia et al., 2004; Committee on Earth Observation Satellites, 2000). They also play a key role in determining the distribution of both pack ice and fast ice (Massom, 2003) and are sites of polynya formation (Chapter 5 in Volume 1 of this book), are important agents of ocean sedimentation (Lisitzin, 2002), and have an important effect on benthic (ocean floor) faunal communities at ocean depths of $<\sim 400\,m$ (Teixidó et al., 2004).

As they melt and drift with the ocean currents, icebergs provide a significant source of freshwater input to the surface layer of the ocean, enough to affect the stability of stratification in the upper ocean (Gladstone et al., 2001). Any increase in iceberg discharge, and/or ice shelf melt, is likely to have a major impact on the ocean by delivering sufficient freshwater to potentially disrupt the overturning circulation of the ocean (Drinkwater et al., 2004a). Indeed, icebergs may contribute to decadal-scale salinity anomalies in polar and sub-polar ocean gyres (Walsh et al., 2001). As noted earlier, an important current unknown is the distribution and variability of freshwater input to the oceans from icebergs, and its impact on the stability of upper-ocean circulation and ocean circulation, is currently a major unknown. This occurs along preferred corridors related to the paths of the main currents. As such, improved information is required on iceberg drift rates and trajectories as well as melt, fracture, and dissolution rates. Icebergs may also be a primary source of iron input (Löscher et al., 1997), thereby potentially impacting ocean primary productivity and the rate of carbon sequestration. Once again, these impacts are poorly parameterized and understood under current conditions, let alone in a global change scenario.

Icebergs form either by the calving of the seaward margins of floating glacier tongues or ice shelves (as shown in Figure 3.27), or by the fragmentation of existing icebergs. Iceberg production rates vary considerably in both time and space, depending on complex interactions involving outlet glacier velocity, degree of crevassing, temperature, sea ice conditions and extent, tidal forcing, and ocean waves and swell and their variability (Loset and Carstens, 1993; Vinje, 1989). In the Arctic, it is thought that maximum production occurs in summer, when temperatures and wave action are at their maximum and sea ice extent is at its minimum (Vinje, 1989).

Accurate determination of the calving rates/frequency and size distribution of these icebergs, and their variability, is a key to more accurately constraining the ice sheet mass loss term using the mass budget approach (Jacobs, 1992). This is a major challenge, given that Orheim (1984, 1988) estimated that there are of the order of 200,000 icebergs in the Southern Ocean and south of the Antarctic Convergence at any one time, ranging in size from tens of meters to hundreds of kilometers across—e.g., Lazzara et al. (1999). While the latter are star attractions in the media, the myriads of small icebergs are equally important (Massom, 2003), yet frequently fall through the observational net and go undetected and unmonitored.

Following Hagen and Reeh (2004), the *iceberg-calving flux* M_C is defined as:

$$M_C = 2WH \sum b_i / T \qquad (3.8)$$

where $2W$ is the calving front width; H is the ice thickness averaged across this front; b_i is an average value of regressions of the ice front location due to the calving of icebergs; and T is a time interval that is large compared with the mean interval separating major calving events. In practice, it is simpler to estimate the *ice flux* M_f at a floating glacier or ice shelf front, from knowledge of $2W$ and H together with ice velocity u_f by:

$$M_f = 2WHu_f \qquad (3.9)$$

where u_f is determined from observations separated by periods of a few days to weeks. The two fluxes are related by:

$$M_C = M_f - 2WH\, \partial L/\partial t \qquad (3.10)$$

where L is the location of the calving front (Hagen and Reeh, 2004; Meier, 1994). It is apparent that the term $\partial L/\partial t$ must be understood as the time derivative of the ice front position history smoothed over the period between observations, and which may exceed the mean interval between calving events (Hagen and Reeh, 2004). This factor has been evaluated in studies of the calving of Greenland glaciers by Weidick (1994) and Weidick et al. (1996).

Shipborne iceberg observations are historically useful, and provide supplementary information on iceberg freeboard (elevation above sea level). Antarctic iceberg observations from such a program, and dating back to 1978, are available from ⟨http://www.antcrc.utas.edu.au/~jacka/IceData/html/icedata.html⟩. They are, however, spatially and temporally biased, and cannot provide the "big picture" estimates required by modern mass balance research. Satellites provide a unique means of monitoring the advance of ice shelves and glacier tongues (e.g., Ferrigno and Gould, 1987; Jacobs et al., 1986), their retreat by calving (e.g., Lazzara et al., 1999; Sohn et al., 1998), and the size distribution and behavior of resultant icebergs, although this is itself a major challenge requiring intensive effort. Routine surveys of ice sheet/shelf margins are required to monitor not only the calving of icebergs but also to monitor their drift behavior and eventual melt. Due to their availability and excellent coverage, medium-resolution visible to TIR sensors—e.g., AVHRR (1.1-km resolution), OLS (0.56–2.7-km resolution), and latterly MODIS (0.25–1.0-km resolution) and Global Land Imager (GLI, 0.5–1.0-km resolution)—are widely used to operationally detect large, iceberg-calving events and monitor large iceberg drift (Brunt, 2003). Major discoveries have been made in this fashion, including the sequence of large-scale calving events from the Ross Ice Shelf in 2000 (Young, 2001) (Figure 3.38). This approach is limited, however, by cloud cover and the difficulty in distinguishing bergs from sea ice, given their similar albedos and thermal signatures. The characteristic signature of leeside (i.e., downwind) areas of open water, due to the differential drift response of icebergs and sea ice to the same wind forcing, can also serve as a means of detection, as can shadowing (which is particularly pronounced at low-Sun angles). Enhanced detection can also take place in summer when much of the sea ice melts back (at least in Antarctica). An example of the detection of icebergs using 250-m resolution MODIS imagery under ideal conditions—i.e., near-minimum sea ice extent—is shown in Figure 3.39.

Satellite-borne SARs are well-suited to the detection and size determination of small to medium icebergs by spatial and texture analysis, as these generally form bright radar targets that can generally be distinguished from the background sea ice or ocean signal (Gill, 2001; Haykin et al., 1994; Viehoff and Li, 1995; Willis et al., 1996). For information on the microwave signature of icebergs and its variability, please see Young and Hyland (1997) and Young et al. (1998). The distinctive radar-backscatter signature from icebergs results from surface- and volume-scattering characteristics (Haykin et al., 1994). Strong volume scattering is due to the low-absorption characteristics of non-saline glacial ice. This enables frequency-dependent penetration of the radar energy to a considerable depth (up to meters to decameters) into the iceberg volume, where it is scattered by trapped air pockets, crevasses, ice pipes and lenses, and other dielectric discontinuities. Surface scattering is dependent on the surface-structural characteristics and orientation of the iceberg relative to the sensor viewing geometry. Such is the strength of the return signal from icebergs that the latter are only indistinguishable from the background speckle in SAR amplitude imagery in instances with high clutter and where the bergs are small relative to the image resolution (Van der Sanden and Ross, 2001). Surface melt also reduces the radar penetration and leads to a backscatter reduction and darkening in the SAR image. This decreases the contrast between berg and background, as does rough (windy) open-ocean conditions. Analysis of various modes of Radarsat-1 data

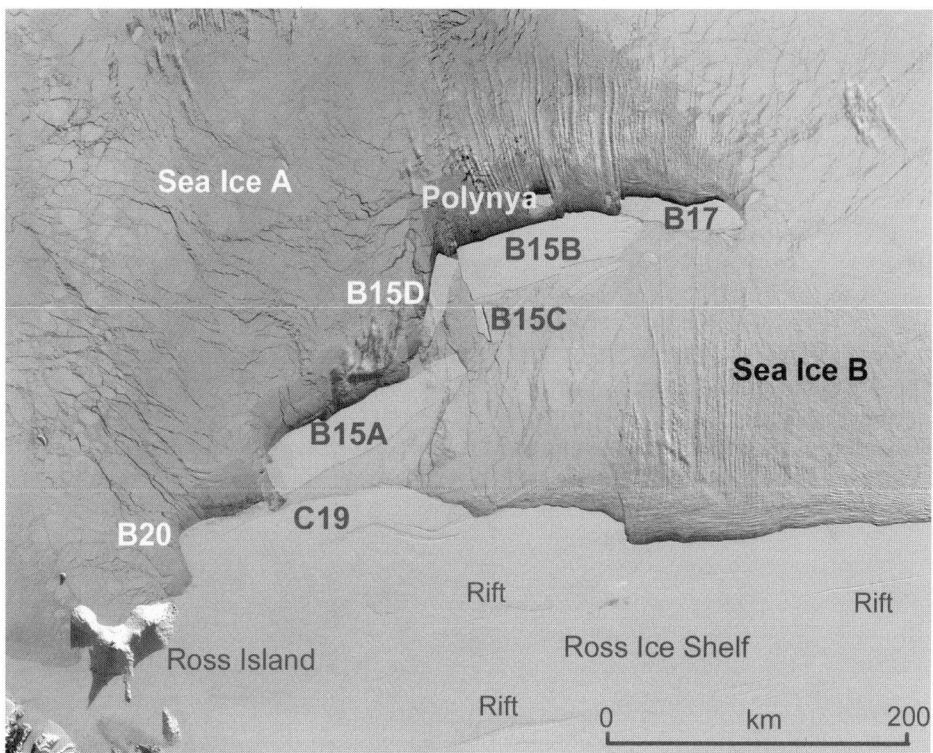

Figure 3.38. A Moderate Resolution Imaging Spectroradiometer (MODIS) image (250-m resolution) of the Ross Sea and Ross Ice Shelf acquired on September 17, 2000 (20:50 UTC), showing sections of the massive iceberg B15 that calved from the Ross Ice Shelf in March 2000. B15 is the largest berg ever observed, with dimensions of ~295 × 37 km when it calved (Young, 2001). The outer section of ice shelf labeled C19 (15 × 43 km) calved a few days later and subsequently drifted westwards to ground with B15A (35 × 161 km) adjacent to Ross Island, where they had a major impact on the distribution and movement of sea ice, and thus on shipping access to McMurdo Sound and wildlife, such as the important emperor penguin colony at Cape Crozier (see Chapter 5 of Volume 1 of this book). Note the wider impact of these massive barriers on regional sea ice conditions in the current image and soon after their calving. Significantly thicker ice (characterized by a higher albedo) occurs to the right (marked Sea Ice B) compared with the left of the image (Sea Ice A), as a result of the convergence of ice drift against the bergs due to forcing by the prevailing wind and upper-ocean current fields. The region to the left is dominated by a more divergent and thinner ice cover (characterized by a lower albedo), on the leeside of the berg barrier, with polynya conditions occurring adjacent to bergs B15A, B15B, and B17. The remainder of the icebergs drifted slowly northwestwards across the Ross Sea towards Cape Adare before heading westwards around the coast to the region adjacent to the Mertz Glacier (see Figure 3.39). The impact of bergs on sea ice and ecology (including ocean primary production) is discussed more fully in Chapter 5 of Volume 1 of this book.

Image courtesy of Jacques Descloitres, MODIS Land Rapid Response Team, NASA Goddard Space Flight Center, Greenbelt, MD. Obtained from NASA's Visible Earth website ⟨http://visibleearth.nasa.gov⟩.

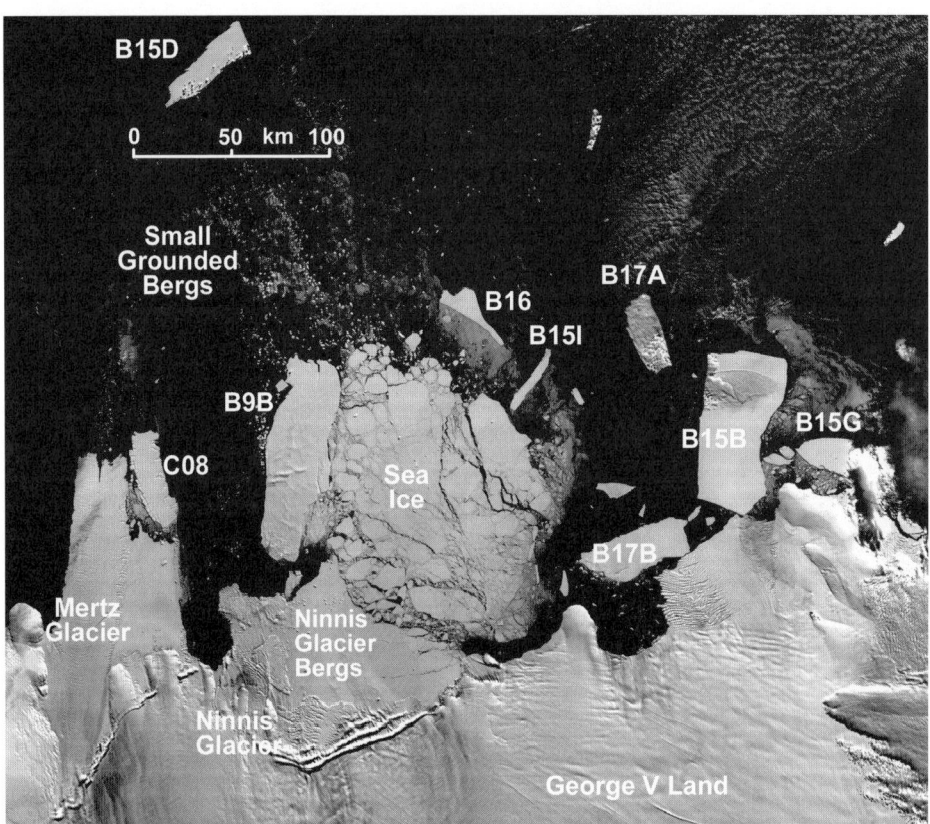

Figure 3.39. A MODIS image (250-m resolution) of large icebergs along the coast of George V Land (East Antarctica, ~145–155°E) and acquired by NASA's EOS Terra satellite on March 9, 2002. The Mertz Glacier Tongue can be seen to the left (west), as well as grounded iceberg B9B and others that have calved from the Ninnis Glacier in recent years. Fragments of the icebergs B15, B16, and B17 that calved from the Ross Ice Shelf (see Figure 3.38) have drifted into the region immediately to the east, apart from B15D which has already drifted past the Mertz Glacier on a trajectory that is likely to take it around the continent and eventually through the Weddell Sea. Analysis of image time series determines whether icebergs are moving or grounded. Icebergs show up clearly here, as they are not surrounded by sea ice (this is the time of minimum seasonal sea-ice extent).

Data provided by NASA/GSFC/DAAC. After Young (2002). Copyright Commonwealth of Australia, reproduced by permission.

by Power et al. (2001) indicates that icebergs with sizes on the order of the resolution of the particular SAR mode can generally be detected at SAR incidence angles of greater than ~35°. Larger icebergs will be detected more consistently, and even during rough-sea states.

The recent emergence of wide-swath (400–500 km) ScanSAR technology in space (e.g., Radarsat-1, Envisat, and ALOS) has greatly enhanced coverage—an

Sec. 3.5]　　　　　　　　　　　　　　　　　　　　　　　　　　　Icebergs　193

important factor from an operational perspective (Bertoia et al., 2004; Gill, 2001; Power et al., 2001), but at the expense of degraded spatial resolution. An example of a Radarsat-1 ScanSAR Wide imagery of Antarctic icebergs, and showing their movement and grounding characteristics, is shown in Figure 3.40. A comparison of these SAR images with the equivalent visible image in Figure 3.39 shows how relatively clearly the bergs stand out in the former. This is particularly true with small bergs (marked SB), which are virtually impossible to detect in satellite visible to TIR data when surrounded by sea ice. Although speckle noise can limit the effectiveness of SAR in unambiguously detecting small bergs (Willis et al., 1996), it can be minimized in the SAR processor by obtaining several looks, albeit at the expense of spatial resolution. Ocean-surface roughening under high wind-speed conditions can also impact detection, by providing a bright radar background target equivalent to consolidated sea ice. Williams et al. (1999) developed a segmentation algorithm to automate the detection of icebergs and analyze their size distribution in ERS SAR images. This technique was also applied to a study of iceberg distributions along the coast of East Antarctica. A complementary manual technique has been used by Gladstone and Bigg (2002) in the Weddell Sea. Such studies greatly enhance standardized ship-based observations (e.g., Orheim, 1990, 1993). Of considerable operational potential in the detection (and tracking) of larger icebergs is the Global Monitoring Mode (GMM) of Envisat (Figure 3.41), with the routine collection of reduced-resolution (1 km) data over a wide (400 km) swath (see Chapter 5 of Volume 1 of this book). These data are available to ESA-accredited researchers in near-real time via the Internet.

In all of the above, the success rate for iceberg detection drops off markedly with a decrease in berg size (Willis et al., 1996). This is both scientifically and operationally important (after all, the *RMS Titanic* was sunk by a small berg). Gladstone and Bigg (2002) set an effective lower size limit for iceberg detection in ERS SAR imagery of $2–5 \times 10^4 \, \text{km}^2$, equivalent to a length scale of $\sim$150–200 m. As for the optimal spaceborne SAR configuration for iceberg detection, Willis et al. (1996) recommended a high-frequency system operating over large incidence angles and at horizontal polarization. The Radarsat-1 C-band, HH-polarization SAR, which operates over a wide swath (to 500 km) at incidence angles varying from 20–60°, has been extensively used by the Canadian Ice Service or CIS (⟨*http://www.cis.ec.gc.ca/home.html*⟩) and Danish Meteorological Institute (DMI) for operational iceberg detection in the Arctic (Bertoia et al., 2004). The CIS uses a Berg Analysis and Prediction Sub-system (BAPS) to detect icebergs in SAR data (Ou, 2004). The algorithm detects possible iceberg targets in a SAR image, and generates an iceberg message and graphics (in the form of a set of shape files). Manual editing tools are then used to edit and validate iceberg detections. Analysis of Radarsat-1 data has shown that while the ScanSAR Wide (30-m resolution) and Narrow (50-m resolution) modes provide a reasonable compromise between detection and swath coverage for icebergs, the Fine mode (8-m resolution) is optimal for iceberg and ship discrimination. Some difficulties have occurred in accessing high-resolution SAR coverage of the Canadian Eastern Seaboard during the winter and spring using Radarsat-1 alone, due to the higher priority requirements of the CIS to capture

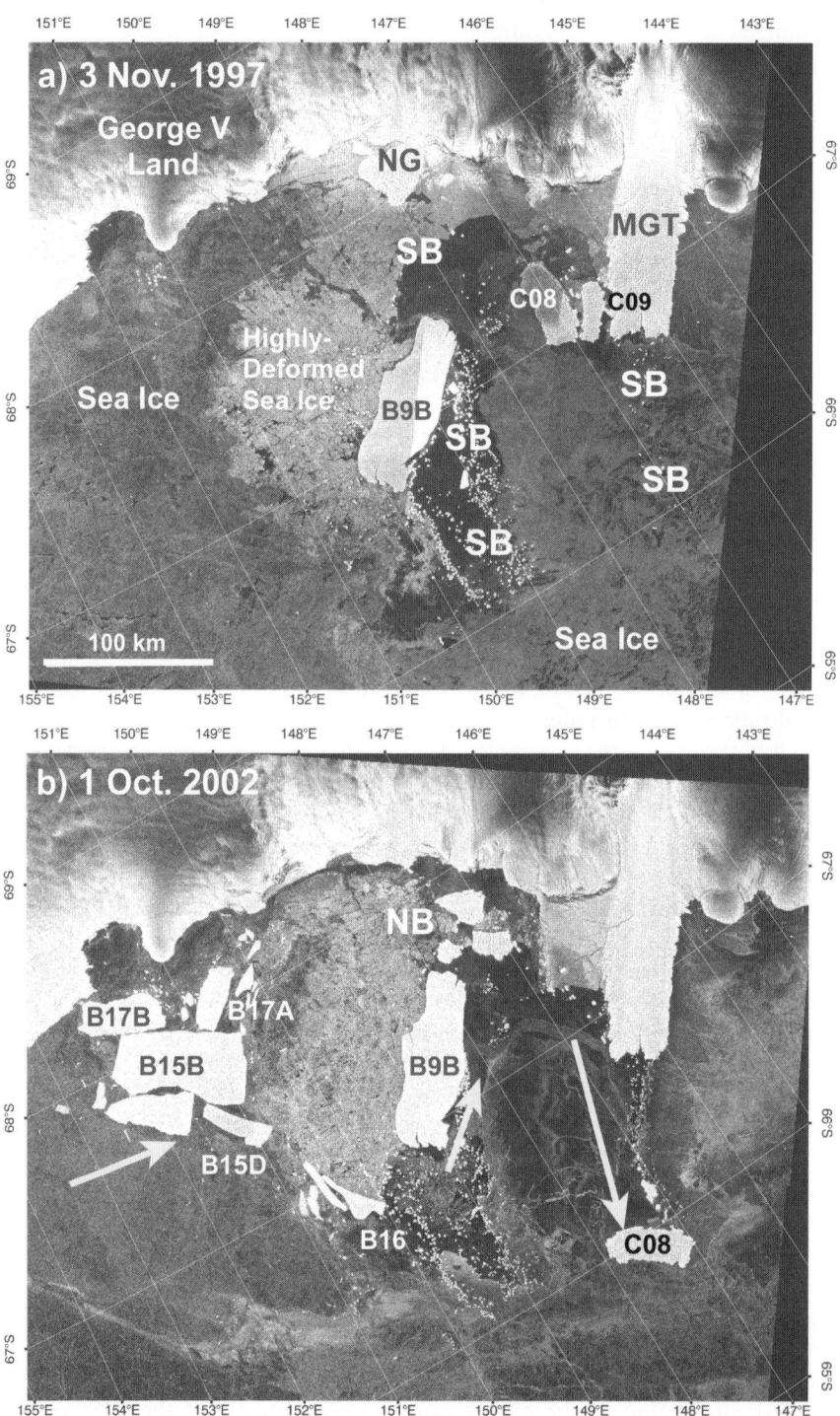

◀ **Figure 3.40.** Radarsat-1 ScanSAR imagery (100-m resolution) of the Mertz Glacier region of East Antarctica from (a) November 3, 1997, and (b) October 1, 2002. The later image shows the most recent calving of the Ninnis Glacier, which occurred in January 2000 (Massom, 2003), and the appearance of large icebergs to the east—i.e., from a calving of the Ross Ice Shelf shown in Figure 3.38 and carried into the region by the East Wind Drift coastal current. Of the other large bergs, C08 and C09 are remnants of the previous calving of the Ninnis Glacier between 1980 and 1982, and resided in the location shown in (a) from 1989–1991 until 1998–2002 (Frezzotti et al., 1998a). Iceberg B9 calved off the Ross Ice Shelf in 1987, before part of it (B9B) grounded in this region in mid-1992 (Keys, 1994). Groups of small bergs (<5 km in dimension) are marked SB. As seen in Chapter 5 of Volume 1 of this book, such bergs have an immense impact on regional fast- and pack-ice conditions (Massom et al., 2001). Due to the prevailing ocean currents, bathymetry, and fast-ice conditions, this region acts as a major trap for icebergs (Frezzotti et al., 1998a; Massom, 2003). As a result, icebergs tend to reside in the region for periods of ∼5 to 15 years before escaping to drift westwards (Massom, 2003). The arrows denote the direction of iceberg drift during sporadic ungrounding phases. Note that B9B sporadically ungrounded to drift ∼50 km to the SW over the 5-year period between the two images. Icebergs C09 and C08, on the other hand, ungrounded to drift to the NNE (C09 in two fragments in early 1998 and late 2000, and C08 in early 2002) (Massom, 2003). All temporarily grounded on a shoal at the tip of the line of small grounded bergs emanating from the terminus of the Mertz Glacier Tongue—see (b)—where they resided for some months to temporarily enlarge the important Mertz Glacier Polynya (Massom, 2003), a feature that is described in Chapter 5 of Volume 1 of this book. NG = Ninnis Glacier, MGT = Mertz Glacier Tongue, and SB = groups of small grounded icebergs.

After Massom (2003). Radarsat imagery copyright CAS/Radarsat International. Copyright Cambridge Journals 2003, reproduced with permission.

lower resolution (i.e., wide swath) imagery for its sea-ice-monitoring program. The recent availability of complementary Envisat ASAR data has, however, eased this problem by providing access to more high-resolution orbits.

Information on iceberg (and ice edge) locations around Greenland is derived operationally from SAR data by the DMI using two complementary techniques (Bertoia et al., 2004). These are the *Constant False Alarm Rate method* (CFAR) and the *Power-to-Mean-Ratio method* (PMR). The main problem in detecting numerous small icebergs in SAR imagery is of distinguishing point targets from the background sea ice/ocean clutter and minimizing the occurrence of "false alarms". With the CFAR technique (Gill, 2001), the water and sea ice background signal is described by a statistical distribution, and data are assigned to this background if they comply with the distribution. If not, they are identified as an iceberg "target". The detection threshold, which is controlled by the statistical probability that the detection is false, is adjusted to maintain a constant "false alarm rate" (Gill, 2001). The PMR technique, on the other hand, is a statistical parameter that aids the identification of targets with brightness characteristics that differ from the background sea ice/ocean. Values of PMR are determined by moving windows across the image, one of 20×20 pixels in size and another of 4×4 pixels for inter-window spacing (Gill, 2001). The advantage of using two techniques simultaneously is to address the known errors in automatic iceberg detection

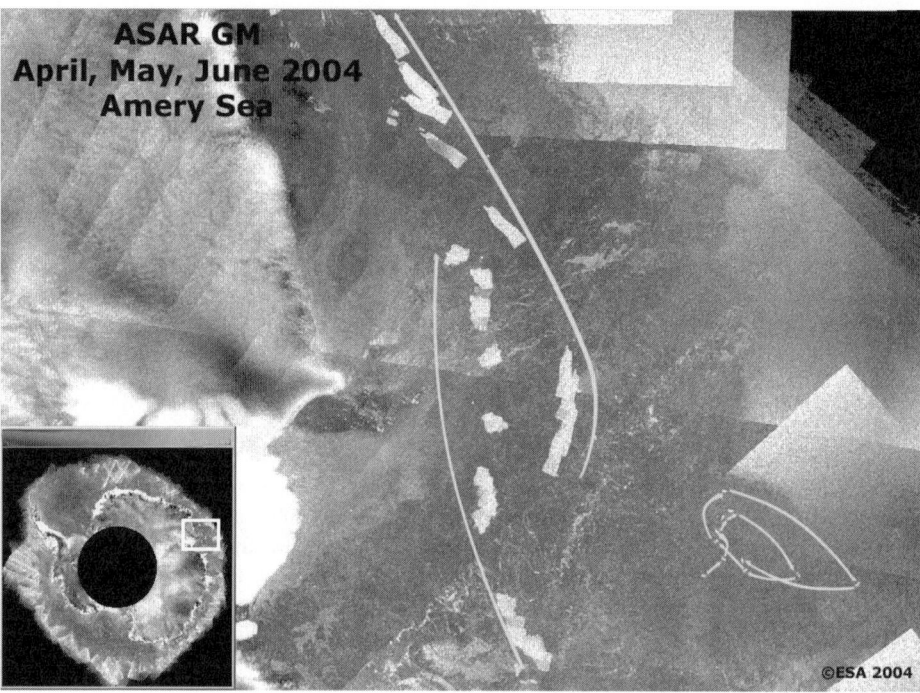

Figure 3.41. A sequence of SAR amplitude images from the new 1-km-resolution Global Monitoring Mode (GMM) of Envisat for the period April through June 2004, showing the displacement over time in the image mosaic of two large tabular icebergs in the region of the Amery Ice Shelf, East Antarctica. The longer more rectangular iceberg is B15D. This is a fragment of the immense B15 iceberg, which originated from a major calving of the Ross Ice Shelf in March 2000 (see Figure 3.38). Over a 2-month timeframe, B15D (50 × 12 km in size) drifted westwards by about 400 km at an average speed of ∼5 m per minute. From its calving in 2000 to mid-2004, it had drifted ∼25% of the way around the Antarctic coast, carried westwards in the near-shore East Wind Drift. The other iceberg (round shape) is probably D14. Such information provides considerable insight into the behavior of ocean-current regimes in regions where measurements of the latter are extremely sparse.

Image courtesy of ESA (ESA, 2004), reproduced with permission. Envisat imagery copyright 2004.

algorithms. In other words, targets identified in both products are assumed by the DMI analyst to represent real icebergs. While yielding unreliable or no information on iceberg size, these methods have proved to be extremely useful in an operational sense for estimating densities and populations of small icebergs in regions where the latter are numerous (Bertoia et al., 2004). In eastern Baffin Bay alone, 20 outlet glaciers produce an estimated 10,000 icebergs per year—each potentially a hazard to shipping operations. An example of a DMI iceberg detection product, from West Greenland waters, is shown in Figure 3.42. Such analyses also rely upon supplementary information from aircraft overflights to help interpret the satellite imagery.

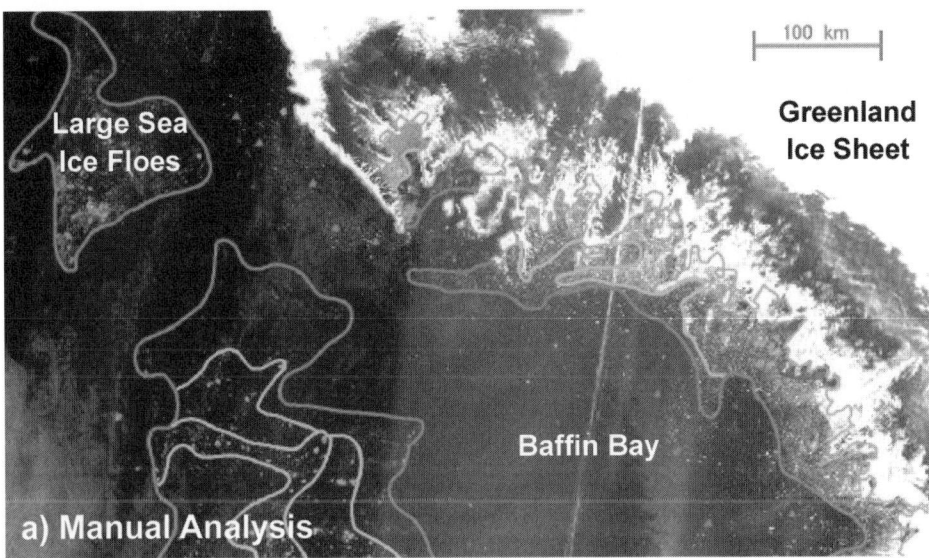

Figure 3.42. An iceberg detection product derived operationally by the Danish Meteorological Institute. It was produced by applying the CFAR detection algorithm to Radarsat-1 imagery, in combination with the Power-to-Mean-Ratio filter to enhance target detection. This example is from the Eastern Baffin Bay region of the Arctic (at ~75°N, 60°W), on July 27, 2001. Each dot represents a possible iceberg.

From Bertoia et al. (2004). Copyright NOAA 2004, reproduced with permission.

Although SAR is the optimal sensor for regional iceberg detection, data availability is limited outside the operational realm, and particularly in Antarctica. A useful alternative, albeit at a poorer spatial resolution, is the radar scatterometer (Stephen and Long, 2000; Young and Hyland, 1997). While their spatial resolution is nominally poor—i.e., 25–50 km—an extensive series of enhanced resolution radar backscatter data has been produced by combining data from multiple passes in a resolution enhancement algorithm (Early and Long, 2001). An example is shown in Figure 3.43 (see color section). The major advantages of spaceborne scatterometers are their excellent wide-swath coverage, low data rate and all-weather, day–night imaging capability, leading to near-daily repeat coverage at high latitudes compared with the 35-day repeat cycle of the ERS SARs, for example. The major diasadvantage is the ability to detect large icebergs only.

One factor complicating estimates of ice mass loss by iceberg calving is that these events occur irregularly and up to decades apart (Frezzotti et al., 1998a; Massom, 2003), a notable recent exception being the wholesale calving of a large section of the Larsen B Ice Shelf in 2002. This makes it difficult to separate natural variability of a stable ice shelf from possible longer term trends based upon the short satellite time series available. One way of addressing this issue is to develop an improved understanding of iceberg-calving processes. As shown in Section 3.5, satellite remote sensing also gives insight into the poorly understood and complex mechanisms by which icebergs calve. Joughin and MacAyeal (2005) used InSAR data to study the rift/detachment process involved in the calving of large icebergs from the Ross Ice Shelf. Fricker et al. (2002a) used Radarsat-1 InSAR data to infer information about the dynamic development of the ice-front rifts shown in Figure 3.44 (see color section). They concluded that the Amery Ice Shelf is a stable system that is currently undergoing changes that are part of its natural advance–calve–advance cycle which they suggest has a repeat period of 65–70 years. Fricker et al. (2005) further used ERS and Radarsat-1 SAR, Landsat ETM+, and Terra MISR images to analyze changes in the rift characteristics over the period 1996 to 2004 (Figure 3.44). The combination of these data enabled resolution of a seasonal signal in rift propagation rates—i.e., more pronounced in "summer"—superimposed on a steady rate (Figure 3.45, see color section). ICESat Geoscience Laser Altimeter System (GLAS) data were used to estimate the thickness of the ice mélange[41] contained within the rifts, based on the hydrostatic relationship.

In their analysis of ice-shelf front behavior using radar altimeter data, Zwally et al. (2002c) further suggested that large-embayment Antarctic ice shelves tend to break up in response to the cumulative effects of tide-induced flexing, while the calving of thinner, marginal ice shelves occurs in response to the cumulative effects of swell-induced flexing. Another study, by Massom (2003), used ERS and JERS-1 SAR and Radarsat-1 ScanSAR data to monitor the calving of the Ninnis Glacier (~68.15°S, 147.5°E) in 2000, and found that this event was far from instantaneous. Rather, it took 10 years to complete, due in part to complex bathymetric and pack/fast ice conditions around the glacier tongue (with fast ice

---

[4] A conglomeration of sea ice, wind-blown snow, fragments from the ice shelf and marine ice (Fricker et al., 2005). A term coined by MacAyeal et al. (1998).

tending to "lock" bergs in place over extended periods). Frezzotti et al. (1998a) studied the same region using a variety of satellite data dating back to Argon U.S. reconnaissance satellite photography from 1963, and were able to piece together the origin of icebergs and their birth from past calving events (Figure 3.46).

It is anticipated that the advent of polarimetric spaceborne SARs, such as those onboard Envisat-1, ALOS, and Radarsat-2, will improve iceberg detection. Flett (2003) reported enhanced iceberg–ship discrimination in the Canadian Arctic using cross-polarized data. Work underway has shown that target features such as the maximum/mean/variance of intensity, along with target shape characteristics— such as compactness, roughness, and elongation—provide a reasonable discrimination of icebergs versus vessels in Envisat data (Lane et al., 2003). Further research by Howell et al. (2004) has investigated two methods for target discrimination, namely a *multi-polarized area ratio* and *HV Signal-to-Clutter Ratio* (SCR). It appears that HV data are particularly advantageous for iceberg detection, particularly at steeper incidence angles ($<35°$) and in high-sea states in sea-ice-free ocean (Van der Sanden and Ross, 2001). Analysis of Envisat data has confirmed that Alternating Polarization (AP) mode targets offer more information than single polarization with respect to radar-scattering mechanisms (Howell et al., 2004). The availability of such data from Radarsat-2 in all currently available Radarsat-1 beam modes, for example, also gives flexibility of coverage, with the choice of mode depending on the required resolution and region of interest. It should be noted that while certain new modes, such as the ultra-fine-resolution (3 m) mode, will further enhance iceberg detection, their coverage is limited to narrow swaths—e.g., 20 km wide in this case.

3.5.2 Monitoring iceberg drift

The drift behavior of bergs is another important yet poorly understood parameter. Due to their immense draft of tens to hundreds of meters, *iceberg drift* is generally driven more by the climatological pattern of upper-ocean currents rather than short-term variability in winds (Crepon et al., 1988). Iceberg drift is also affected by the concentration and thickness of surrounding sea ice (Lichey and Hellmer, 2001). Developing an improved understanding of the patterns of iceberg drift (and grounding), and their variability, is important from a number of perspectives. Primarily, iceberg drift and eventual decay represents an important term in the ocean freshwater budget, redistributing large quantities of freshwater which in turn affect local water masses and convective overturning of the ocean. Iceberg melt also releases entrained material and nutrients into the water column, which in the case of the latter may enhance ocean primary production. Knowledge of circumpolar iceberg melt rates and freshwater fluxes is an important factor (Budd et al., 1980; Hamley and Budd, 1986), but is currently largely lacking, not in least part due to the magnitude and complexity of the problem. Improving this knowledge is a challenging task, given that many thousands of icebergs are calved or are present in high-latitude and even mid-latitude waters each year, and these cover a wide range of sizes. From an operational perspective, iceberg drift is a major potential hazard to shipping and hydrocarbon operations. Finally, the drift and grounding behavior of icebergs, both large and small, has a first-order effect on the distribution of pack and

200 Satellite remote sensing of ice sheet parameters and processes [Ch. 3

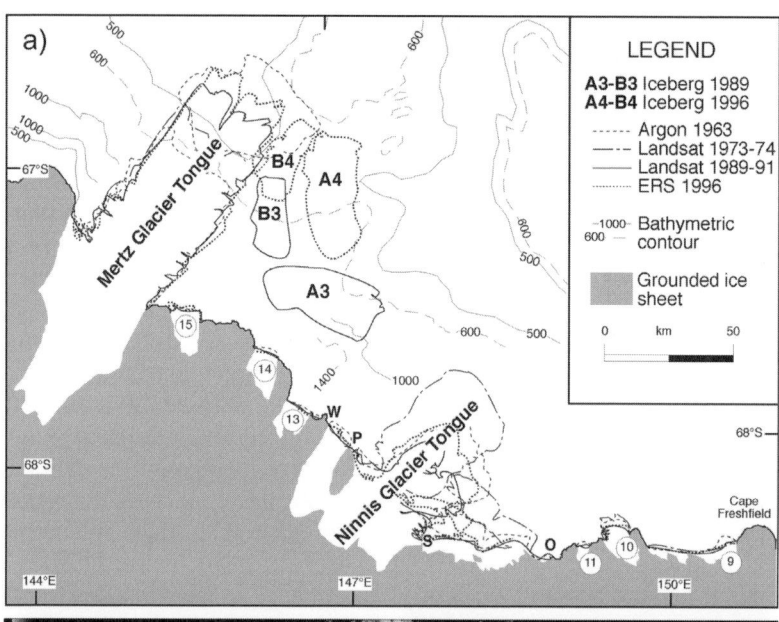

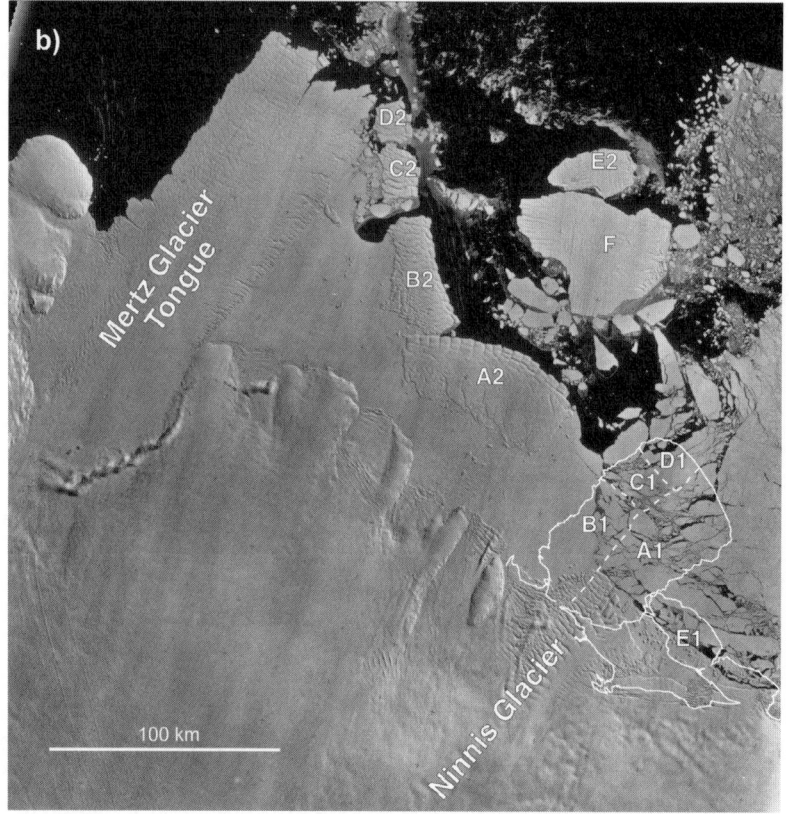

◀ **Figure 3.46.** (a) Schematic map of the western George V Land coastline (East Antarctica) showing variations in the ice front and the movement of icebergs, derived from satellite images collected from 1963 to 1996 and listed in the legend. (b) A Soyuz Kosmos KATE-200 image of the same region from February 14, 1984. The solid line to the bottom right depicts the Ninnis Glacier front in 1974 (determined from Landsat imagery). The dashed line is the iceberg limit, and A1–E1 are the original positions of icebergs A2–E2 in 1974 and prior to the calving of the Ninnis Glacier in 1984. Iceberg F calved from the Cook Ice Shelf in 1984. See also Figure 3.40 (note that A2 and B2 are equivalent to C08 and C09 in the latter figure).
From Frezzotti et al. (1998a). Copyright International Glaciological Society 1998, reproduced with permission.

fast ice (see Massom et al., 2001 and Chapter 5 of Volume 1 of this book). Given the importance of iceberg drift, and the recent increase in satellite-derived information combined with improved meteorological data, recent research has spawned numerical models to simulate the tracks of icebergs under a range of conditions—e.g., Gladstone et al. (2001) and Lichey and Hellmer (2001). While initial results are promising, there is a need for additional data with which to drive and validate the models and improve their performance.

Early tracking efforts involved placing satellite-tracked beacons on the bergs—e.g., Tchernia and Jeannin (1984). More recently, a range of different satellite imaging methods have been used to improve our baseline knowledge of the birth, journey through life, and eventual demise of icebergs (Gladstone and Bigg, 2002; Vinje, 1980). As is apparent throughout this book, each technique is characterized by trade-offs related to sensor characteristics, and its efficacy is determined by whether the application is operational or research-oriented. Phillips and Laxon (1995) investigated the use of high-frequency SSM/I PMW data, at 85 GHz and a spatial resolution of 12.5 km, to detect and track large icebergs. Radar scatterometers are, however, more extensively used for iceberg tracking over large areas. They are particularly well-suited to operational monitoring of large-iceberg drift, given their attributes (as outlined above). The measured backscatter varies approximately linearly with incidence angle and as a bi-sinusoidal function of sensor look direction, defined by an amplitude and orientation or phase angle (Young and Hyland, 1997; Young et al., 1998). The look-direction anisotropy over the sea ice "background" is typically small (Young, 1998, 1999). Satellite-derived information on large-iceberg drift provides a proxy measure of ocean currents averaged over the draft of the iceberg. This constitutes useful ocean circulation information in regions where such information is sparse or non-existent—e.g., within the sea-ice zone (Tchernia and Jeannin, 1984; Young, 1998, 1999). An example is given in Figure 3.47 (see color section), showing the tracks (drift from east to west) and drift speed of several large icebergs relative to the bathymetry of the coastal sector of East Antarctica between 20°E and 120°E. These tracks, which were derived from ERS scatterometer data, tend to follow the continental shelf break (at ∼1,000 m). Drift speed is a function of the time interval between images; in this case, the speed is given by distance made good in 20-day intervals. Typical drift speeds of 0–10 km per day occurred between 130 and 45°E, increasing to ∼20 km per day farther to the west. The strong relationship between iceberg drift, ocean bathymetry, and currents is apparent in the retroflection of one of the bergs in Figure 3.47. The icebergs in this example have a draft of 200–350 m. Even after image-resolution enhancement, scatterometry is restricted to the tracking of large tabular icebergs, with the many small bergs present remaining unresolved. Moreover, manual analysis is necessary to determine whether bergs are moving or grounded (on the many banks in the near-coastal zone).

In the Southern Ocean, the *NOAA/U.S. Navy/Coastguard National Ice Center* (NIC) maintains an operational iceberg database dating back to 1976 (⟨http://www.natice.noaa.gov/southbergs.htm⟩ and ⟨http://www.natice.noaa.gov/products/iceberg/⟩). The NIC exploits a range of satellite sensors to operationally detect and track icebergs greater than 10 nautical miles in size along the long axis, using mainly AVHRR, DMSP OLS and radar scatterometer data, supplemented infrequently by Radarsat SAR data. An example of an OLS image product is

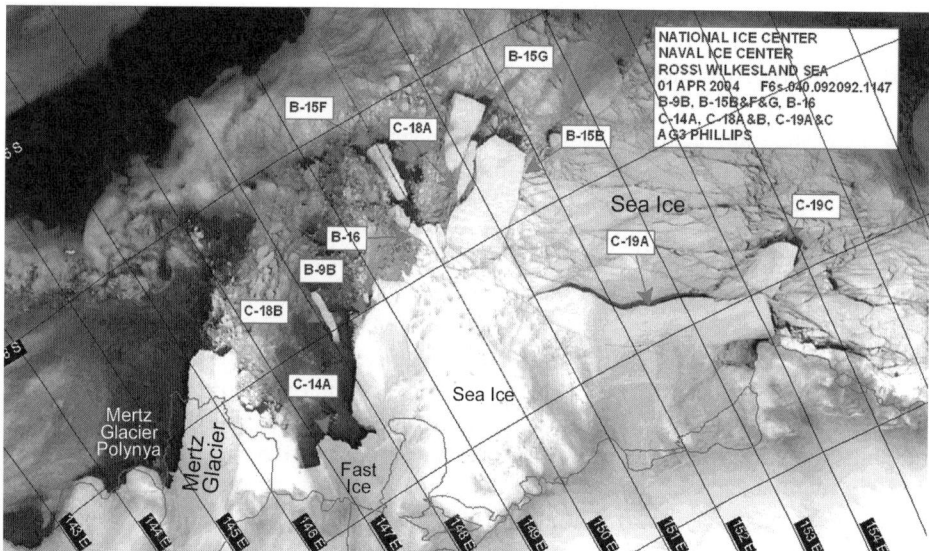

Figure 3.48. An example of a DMSP OLS-based operational iceberg image product from the U.S. Navy/NOAA/Coastguard National Ice Center (NIC). This image is from the George V Land region of East Antarctica adjacent to the Mertz Glacier and was acquired on April 1, 2004 (11 : 47 GMT). It depicts named icebergs from various calvings of the Ross Ice Shelf that have drifted westwards around the coast within the climatological East Wind Drift. Clouds are a major limitation to the use of imagery collected at visible to thermal infrared wavelengths, and, since early 2000, the NIC have incorporated radar scatterometer data into their iceberg analyses. Note also the mis-match between the coastline mask and the actual coastline in the image; this is due to a combination of inaccuracies in the former and in satellite navigation parameters. This is a common factor affecting AVHRR and OLS data at high latitudes, where few if any control points of known latitude and longitude are visible in the imagery. About 55% of all iceberg locations incorporated into the NIC database are now based upon QuikSCAT SeaWinds scatterometer data (Long et al., 2002).

Image obtained from the NIC website at ⟨http://www.natice.noaa.gov/products/iceberg/index.htm⟩. Reproduced with permission.

shown in Figure 3.48. While temporal coverage offered by wide-swath, medium-resolution visible to TIR sensors is eminently well-suited to iceberg tracking, and the data are readily available in real/near-real time, cloud cover is a severe limiting factor. Another factor, again alluded to in the last section, is that icebergs are often difficult to distinguish from the surrounding sea ice at these wavelengths. The nomenclature applied by the NIC to the naming of icebergs divides the Antarctic region into four quadrants, and designates each quadrant with a letter A to D. Here, A is from the 0° to 90°W (the Atlantic Ocean sector), B is from 90°W to 180° (the Pacific Ocean sector), C is from 180° to 90°E (the Australian sector), and D is from 90°E to 0° (the Indian Ocean sector). When a named berg splits into two or more fragments, each is given the designation of the parent berg plus an additional letter. In this case, for example, C19A and C19C are the first and third fragments from berg C19.

The NIC reports an Antarctic iceberg position only once every 15–20 days, and confined to certain regions only and south of latitude 60°S. The SCP group at Brigham Young University (Utah), on the other hand, systematically track the circumpolar drift of Antarctic icebergs once every 1–5 days using enhanced-resolution radar scatterometer data. Examples are given in Figure 3.49. The SCP database covers the periods 1978 (using Seasat SASS data) and 1992 to the present (using data from ERS-1/-2 Active Microwave (AMI), ADEOS NSCAT, and QuikSCAT SeaWinds) (Ballantyne, 2002; Ballantyne and Long, 2003). Further operational monitoring is planned with ASCAT on the MetOp series from 2005 onwards. In all cases, resolution-enhancement algorithms are applied to optimize iceberg detection, given the poor nominal resolution of the scatterometer data. The resolution is still poor even after enhancement—i.e., ~2.225–25 km—depending on the sensor, meaning that only the larger bergs can be tracked. The initial location for each iceberg is determined either from (i) a position reported by the NIC webpage (⟨http://www.natice.noaa.gov/icebergs.htm⟩) or (ii) by the detection of a moving iceberg in a time series of images. The iceberg is then tracked from its starting point in the image time series. Data gaps occur due to missing data and occasional loss of contrast between the iceberg and surrounding area during summer months. This comprehensive database is available online at ⟨http://www.scp.byu.edu/data/ iceberg/database.html⟩. As with SAR, surface melting can reduce the contrast between icebergs and pack ice during summer months to complicate iceberg detection and identification/tracking. Having said this, these data remain an important source of iceberg drift information. Another major archive of satellite-derived iceberg information is the *Antarctic Meteorological Research Center* (AMRC) at the University of Wisconsin (⟨http://amrc.ssec.wisc.edu/iceberg.html⟩).

Long et al. (2002) carried out a comparative study of the number of Antarctic icebergs reported by the NIC and detected using a variety of satellite sensors, versus those observed in enhanced-resolution scatterometer data. This revealed an apparent increase in the number of icebergs apparent in both datasets and detected as a function of time over the period 1976 to 2001 (Figure 3.50, see color section), which the authors attributed to episodic calving events. For example, the significant increase in 1999 and 2000 in the number of icebergs tracked resulted from major calving events at that time from the Ross (1998, 2000) and Ronne (1999) Ice Shelves.

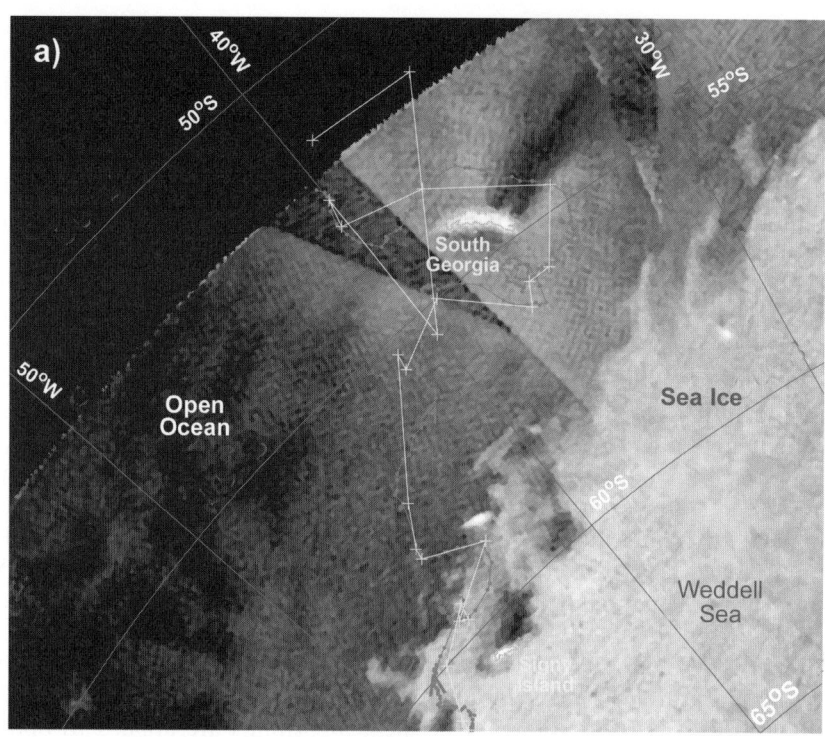

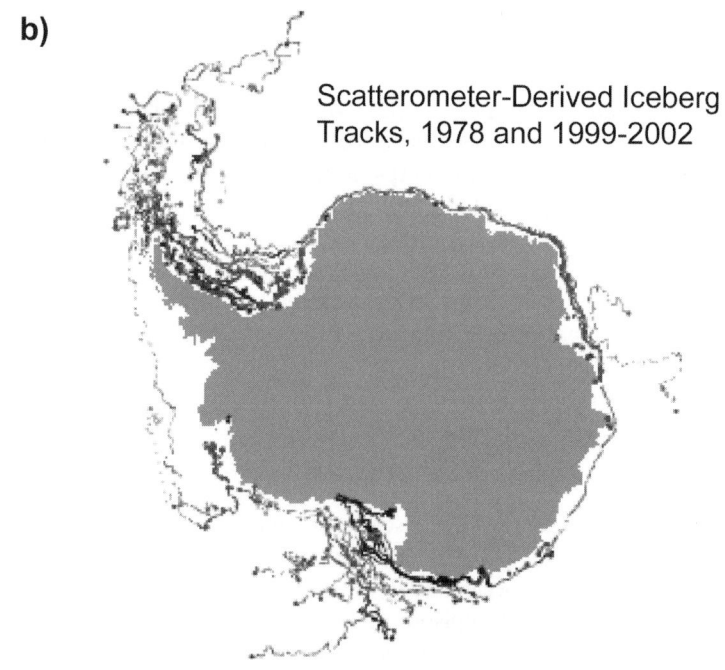

◄ **Figure 3.49.** (a) The scatterometer-derived drift track of iceberg A22B overlain on a QuikSCAT SeaWinds image (polar stereographic projection) from October 5, 2000. The continuous curved line denotes SCP positions, while the line marked by crosses shows NIC-reported locations and contains an erroneous position report (likely due to limited availability of imagery outside the sea-ice zone). The latter is the light gray region to the right, while the dark area is open ocean. Artifacts in the open ocean result from the combination of data from multiple passes. The bright arc in the upper center is South Georgia Island. (b) A compilation map of iceberg tracks derived from satellite radar scatterometer data for 1978 (Seasat) and 1992–2002 (ERS-1/-2, NSCAT, and QuikSCAT SeaWinds).

From (a) Long et al. (2002), reprinted with permission of the American Geophysical Union, copyright AGU 2002; and (b) the Antarctic Iceberg Tracking Database in the Scatterometer Climate Record Pathfinder at Brigham Young University (from ⟨http://www.scp.byu.edu/data/iceberg/database1.html⟩), image courtesy of David Long (Brigham Young University).

These bergs subsequently fragmented into a number of smaller bergs. In this case, the increase in bergs with time is also attributable to technological advances in iceberg detection and tracking techniques. The earlier NIC records were affected by data availability, with the jump in 1986 coinciding with the introduction of the DMSP OLS into tracking operations. Conclusive evidence has yet to emerge of a relationship between climate trends and the calving of large tabular bergs (Lazzara et al., 1999). In-depth information on the operational detection and tracking of icebergs is given for the Antarctic by Long et al. (2002) and the Arctic by Bertoia et al. (2004) and Gill (2001).

While SAR data provide a means of detecting small icebergs (see above), they are in general less readily available for operational iceberg tracking over large areas. An exception is the routine availability, in near-real time, of wide-swath (400 km) reduced (1 km) resolution data from the Envisat ASAR GMM. Such data are operationally used by the European *ICEMON* (ice monitoring) consortium (see Chapter 5 of Volume 1 of this book). Working in research mode, Gladstone and Bigg (2002) used conventional narrow-swath and high-resolution ERS SAR data to detect and track small- to medium-sized icebergs (200 m to 5 km across) in the Weddell Sea. Although manual, this technique is extended to include estimates of iceberg area flux, F, which is given by:

$$F = \frac{nA_m V_m}{D_m}$$

where n is the number of icebergs visible in a given image above a certain size threshold (determined by both SAR sensor and environmental parameters—e.g., sea state or sea-ice cover characteristics); A_m is the mean surface area of the icebergs; V_m is the mean iceberg velocity; and D_m is the mean distance traveled by the iceberg. Grounded icebergs are ignored in the analysis. Based on a limited dataset, Gladstone and Bigg (2002) estimated the westward mass flux of icebergs in the Antarctic Coastal Current (East Wind Drift) at the eastern entrance to the Weddell Sea to be 50–70 gigatonnes (Gt) per annum. Error sources, including those related to the 3-day time interval between ERS SAR acquisitions, and corrections are given in Gladstone and Bigg (2002).

Comparison of satellite data time series from various sources has shown that certain shallow-water regions act as major "grounding traps" for icebergs, both those produced locally and those drifting into the region (see Figure 3.40). Working in the region of George V Land (Antarctica) east of ~150°E, Frezzotti et al. (1998a) and Massom (2003), for example, have shown that, while many bergs drift far away from their point of origin soon after calving, others become grounded on shoals for periods of months to years (up to 10–20 years in the same region) (Figures 3.39 and 3.40). In this way, grounded icebergs have a profound impact on regional hydrographic conditions (Nøst and Österhus, 1998), ocean circulation (Lichey and Hellmer, 2001; Grosfeld et al., 2001), sea-ice conditions (Massom et al., 2001), polynyas (Massom et al., 1998), and ecology (Arrigo et al., 2002) (see Chapter 5 of Volume 1 of this book). The "stop–start" (i.e., ground–unground) drift behavior is illustrated in satellite data-based analyses by Frezzotti et al. (1998a), Keys et al. (1990), Young (1999), and Massom (2003), while MacAyeal et al. (2002) discussed kedging[5] as a means of detecting iceberg drift. A major current deficiency in studies of iceberg grounding is a lack of accurate ocean bathymetry data, particularly in regions perennially covered by sea ice. The potential exists to derive such information in a proxy fashion by combining information on grounded icebergs and thickness (the latter derived from satellite altimeter profiles).

Another line of research has been to place an AWS and a GPS on the massive B15A, a 4,000-km^2 iceberg that calved from the Ross Ice Shelf in March 2000 (MacAyeal et al., 2002). Combined with satellite image analysis, this approach provides very-high-resolution data on the drift behavior of individual large bergs, including collisions with the ice shelf. In this particular case, B15A grounded adjacent to major penguin breeding colonies close to Ross Island, where it had a major impact on penguin breeding success (Arrigo and van Dijken, 2003; Arrigo et al., 2002). The effect is captured in Figure 3.51 in a sequence of high-resolution visible images from the EOS Terra MISR images. See Chapter 5 of Volume 1 of this book for further discussion of this important effect. It also caused severe logistical problems for icebreakers resupplying McMurdo Station, and subsequently fractured to drift towards the Drygalski Glacier Tongue.

3.5.3 Measurement of iceberg thickness

Translating iceberg information into ice fluxes requires not only detailed measurement of the size and frequency distribution of the icebergs but also their thickness. This is a major limiting factor, given the difficulty in measuring iceberg thickness over large areas and for a wide range of iceberg sizes and shapes. Regarding the latter, tabular icebergs are common in Antarctica but rare in the Arctic. Understanding the migration and disintegration of large tabular icebergs also requires knowledge of their thickness as well as structural characteristics, while estimation of melt rate depends upon knowledge of change in shape and thickness

---

[5] Moving a ship by hauling in a hawser that is attached to an anchor dropped some distance away.

Figure 3.51. A sequence of cloud-free images (each $\sim$130 $\times$ 145 km in size) from the EOS Terra MISR showing iceberg movements and resultant changes in sea-ice distribution in the Ross Sea (Antarctica) from December 11, 2000 to December 9, 2001. Notable are the changing positions of icebergs C-16 and B-15A, and the increase in sea-ice extent in December 2001 compared with 2000. The icebergs initially calved from the Ross Ice Shelf in March 2000 (see Figure 3.38) and gradually drifted to a point northeast of McMurdo Sound where they created a barrier that dramatically altered regional wind and current patterns. Iceberg C-16, $\sim$50 km $\times$ 19 km in size, became grounded adjacent to Ross Island at the beginning of 2001, after rapid rotation during the latter part of 2000. The vast iceberg B-15A (only part of which is shown) also rotated and moved southwestwards to ground in the same area. These images were acquired by MISR's nadir (vertical-viewing) camera during Terra orbits 5235, 5497, 5599, 5759, 6192, and 10521. South is toward the top. The breeding success rate of several penguin colonies in the vicinity decreased drastically as a result of the iceberg groundings and their impact on the amount and concentration of sea ice in the region.

Imagery courtesy of NASA/GSFC/LaRC/JPL, MISR Team. Obtained from NASA's Visible Earth website $\langle$http://visibleearth.nasa.gov$\rangle$.

over time. One method for measuring structural, thickness, and basal characteristics is airborne ice-penetrating radar (Blankenship et al., 2002), but such measurements are spatio-temporally limited. Modern satellite altimeters can potentially be used to monitor the thickness of large tabular bergs by making systematic measurements of their elevation (draft) above sea level (Hawkins et al., 1991). An example is shown in

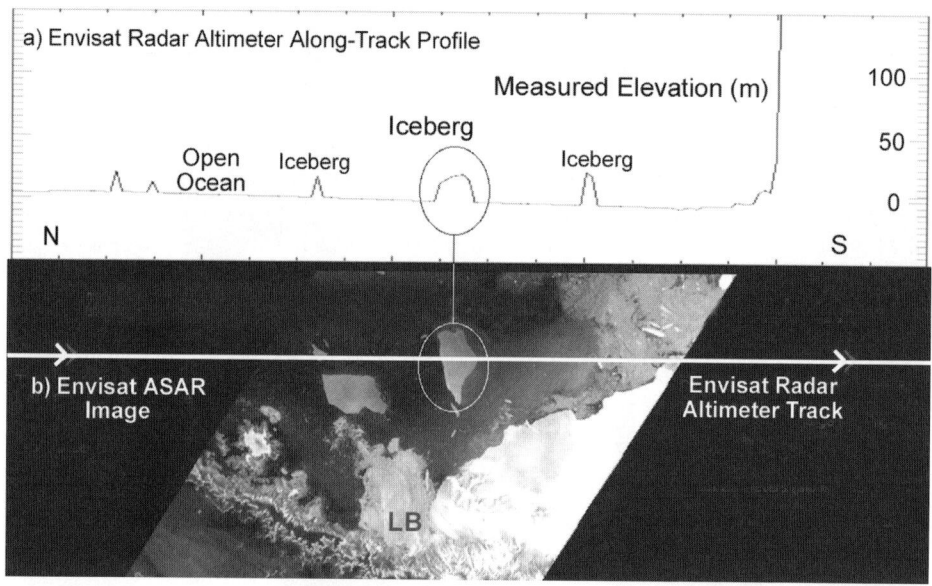

Figure 3.52. An along-track profile of the elevation (above sea level) measured by the Envisat Radar Altimeter (RA-2) of a large tabular iceberg in the NW Weddell Sea (shown below in the near-coincident Envisat ASAR image) on March 14, 2002 (orbit number 00192, descending). Icebergs such as this, which calved from the Ronne Ice Shelf, are seldom sampled by the altimeter at nadir, and may occur anywhere within the pulse-limited footprint. As such, the estimate of apparent height must be corrected for off-nadir distance. In this example, the iceberg "freeboard" elevation (above sea level) is $\sim$25 m, indicating a total thickness of $\sim$200 m. LB marks the disintegration of the Larsen B Ice Shelf—see Figure 3.27.
Image courtesy Mark Drinkwater (ESA) and ESA. Copyright ESA 2002, reproduced with permission.

Figure 3.52, of the along-track measurement profile of a large Antarctic tabular iceberg acquired by the Envisat advanced Radar Altimeter (RA-2). This instrument has a special tracking mode to enable measurement of abrupt topographic features such as ice-floe edge freeboards and ice sheet margins. In this example, the iceberg freeboard is measured at $\sim$25 m (using the surrounding ocean signature as a zero reference). With knowledge of the mean ice density, this quantity can be converted to total thickness. In this case, the iceberg thickness is estimated to be $\sim$200 m. Lack of observations of iceberg density is a major source of uncertainty. An assumption is also made of minimal penetration of the radar signal into the iceberg volume—i.e., that the reflection occurs from at or close to the ice surface. The effect of variable penetration of the altimeter signal into the iceberg is unknown for radar altimeters, and contributes to uncertainty in iceberg elevation and thickness estimates. It is, however, likely to approximate the penetration of the radar signal into the ice sheet proper (see Section 3.2.3). From the perspective of estimating the total freshwater flux from icebergs, and determining where and how it is redistributed, a major challenge is to acquire a sufficiently large number of observations on change in iceberg size and thickness to generate reliable and robust statistics.

Improved estimates of iceberg elevation and thickness are achievable from the innovative sensors onboard ESA's CryoSat and NASA's ICESat. On the former, the Ku-band (2.2-cm wavelength) SAR Interferometric Radar ALtimeter (SIRAL) is the first spaceborne sensor specifically designed to systematically measure sea-ice freeboard and thickness (see Chapter 5 of Volume 1 of this book for details). Note again that the CryoSat launch unfortunately failed in October 2005. Also promising is the GLAS onboard ICESat (Sergienko et al., 2004), with its relatively small surface footprint, high sampling rate along-track and high precision (Zwally et al., 2002a). Laser altimetry has the additional advantage of minimal penetration into the ice volume, but cannot penetrate cloud. Iceberg elevation is in fact a sub-product of the standard GLAS GLA13 sea-ice elevation/roughness product. For laser waveforms with more than one peak, iceberg elevation is calculated as the difference between the range offset of the maximum amplitude peak and the range offset of the first peak (the relevance of waveforms to altimetry is discussed in Section 3.6). These computations are made after tidal and atmospheric corrections have been applied, and the elevation computed is relative to the ellipsoid. It should be noted that this parameter is computed for all multiple-peak GLA13 records, even if the elevation is too high to be sea ice. For further details, see the NSIDC archive at ⟨http://nsidc.org/data/docs/daac/glas_altimetry/gla13_records.html⟩. A strong potential exists to combine information on the elevation above sea level of large icebergs derived from these new altimeters (GLAS and Envisat) with SAR data routinely collected by Envisat in GMM to measure the melt rate of large icebergs. The GMM data, at a spatial resolution of 1 km over a 400-km swath width, provide a means of identifying and routinely tracking large icebergs while monitoring their areal extent.

3.6 MEASUREMENT OF ICE SHEET TOPOGRAPHY/ELEVATION AND CHANGE IN ELEVATION

Without doubt, satellite remote sensing has had a dramatic impact on the large-scale measurement of ice sheet topography (elevation and slope characteristics). Precise topographic maps are of key importance as a means of constraining ice sheet dynamics in modeling. Recalling Section 1.2 (see p. 10), surface slope is an important determinant of the magnitude of the gravitational driving stress of the ice flow (Fahnestock and Bamber, 2001; Paterson, 1994). Moreover, accurate surface topography is required to estimate the flow direction and the divergence/convergence of flowlines (Rémy et al., 2001; Young et al., 1989), enabling computation of the so-called *balance velocity*[63] (see Section 3.8 for an explanation). Information on elevation is also required to enable estimations to be made of the rheological

---

[6] *Balance velocities* are the depth-averaged velocities necessary to maintain the steady-state shape of an ice sheet. They are estimated from data on surface slope, accumulation, and ice thickness (Paterson, 1994).

parameter values with respect to temperature, thereby enabling modification of the ice flow law (Rémy et al., 1993, 1996b). Precise maps of the direction and magnitude of surface slope also enable definition of ice divides and delineation of drainage basins, with this information being used for mass-balance calculations (Bamber et al., 2000a; Joughin and Tulaczyk, 2002; Testut et al., 2003). Moreover, surface topographic detail mirrors the interaction of the moving ice with the underlying bedrock (Paterson, 1994), and can be used to infer conditions at the bed–ice interface (Budd, 1970). Elevation also affects accumulation rates, through the atmospheric lapse rate and impacts on atmospheric dynamics, with topography also controlling the katabatic wind field (Parish and Bromwich, 1987). In addition, unusually flat regions are sometimes indicative of sub-glacial lakes (Ridley et al., 1993; Siegert et al., 2005), including Lake Vostok (Kapitsa et al., 1996). Located 4 km beneath the ice in East Antarctica, this extraordinary feature is 280×60 km in area and 500 m deep.

Recalling Section 1.2, the accurate measurement of elevation change (both regional and ice sheet-wide) is also of importance as a fundamental measure of ice volume, from which mass balance can be estimated (by the integrated approach) (Lingle and Covey, 1998). Amongst other things, this requires accurate initial baseline estimates of elevation, followed by consistent time series measurements (Madsen et al., 1999). This represents a major challenge, but again one that is well-suited to satellite techniques. The temporal scale of change is an important factor when addressing global change-related issues. Small seasonal to interannual variability in ice sheet elevation at a given location typically results from the complex interplay of a number of processes (DeConto and Pollard, 2002; Zwally et al., 2002a). These include (i) accumulation or ablation/sublimation, (ii) tectonic activity or isostatic rebound of the underlying bedrock, (iii) horizontal ice redistribution by ice creep and flow, (iv) the erosion and horizontal redistribution of unconsolidated snow by wind (particularly katabatic winds), and (v) compression ("densification") of the snow. The magnitude of this short-term variability in elevation is poorly known, but is estimated to be of the order of tens of centimeters to a meter over the accumulation zones of Greenland and Antarctica, increasing to a few meters in near-coastal regions in Greenland (Zwally et al., 2002a). The accurate monitoring of short-term changes is required to both infer the interannual variability of snow accumulation/melt rates and reveal longer term changes in ice-sheet volume. An additional factor for floating ice masses—e.g., glacier tongues and ice shelves—is the cyclical change in surface elevation resulting from vertical motion due to tidal flexure. These effects are discussed in Sections 3.8.2 and 3.9.

Profiles of surface elevation over limited areas can be obtained by kinematic GPS surveying, to an accuracy of $\sim \pm 5$ cm in the horizontal and ± 10 cm in the vertical (Eiken et al., 1997). This and other, traditional *in situ* methods of measuring ice-sheet surface elevation change, including surveying and cartographic techniques, are evaluated by Hagen and Reeh (2004). Other important data have emanated from airborne laser altimetry combined with kinematic GPS measurements (Csatho et al., 2005; Krabill et al., 1995, 2002; Spikes et al., 2003). Satellites alone can, however, offer large-scale coverage required to generate regional and ice-sheet-wide topographic maps. Aicraft and *in situ* measurements of course play an important role in constraining and validating these maps, and have immense value when combined with satellite data.

Another approach is to derive maps of elevation from satellite data using stereo-optical (photogrammetric) techniques. A modern example of a sensor designed to acquire stereo-graphic coverage of the surface is the PRISM (Panchromatic Remote-sensing Instrument for Stereo Mapping), to be launched on JAXA's (Japanese Aerospace eXploration Agency) ALOS satellite in late-2005. The objective of PRISM, which operates at 0.52–0.77 µm, is to obtain high-resolution stereo data (pixel size of 2.5 m) for the extraction of highly accurate DEMs. The instrument is a "three-line imager" with three independent catoptric systems for nadir-, fore-, and aft-viewing to achieve along-track stereoscopy. The nadir-looking telescope provides a swathwidth of 70 km (28,000 readout pixels), each of the fore- and aft-looking telescopes provide coverage over a swath of 35 km, while the nadir-looking telescope provides a swath of 70 km. The fore and aft telescopes are inclined by ±23.8° from nadir to realize a base to height/ratio of 1 at the satellite orbital altitude of 692 km. A campaign is underway to establish Ground Control Points (GCPs) with which to validate and assess the accuracy of PRISM-derived DEMs. See Tadono et al. (2004) for more details on PRISM. Other satellite sensors that have been used for stereoscopic mapping include SPOT HRS, the JERS-1 OPtical System (OPS), the Advanced Visible and Near Infrared Radiometer (AVNIR) on ADEOS and AVNIR-2 on ALOS, the GLobal Imager (GLI) on ADEOS-II (⟨http://shraku.eorc.nasda.go.jp/GLI/⟩), the LISS sensors onboard IRS satellites, and the EOS ASTER (Toutin and Cheng, 2001). Conventional stereo-optical techniques are, however, largely unsuccessful when applied to relatively featureless ice sheets (Pattyn, 1992). While they have been applied to mountainous near-coastal regions—e.g., Cheng et al. (2003)—these techniques generally cannot achieve the requisite vertical accuracies without adequate GCPs. Other limitations are the somewhat piecemeal coverage by narrow-swath sensors under a range of cloud cover and solar illumination conditions. In Greenland, for example, topographic maps of the steeper ice-sheet margins contain errors as large as 100s of meters (Ekholm, 1996). As we shall see, stereo-optical methods, while useful, have been largely superseded by other techniques. According to Bamber and Kwok (2004), the accuracy of DEMs constructed from satellite stereo photogrammetry is, at ~10 m, insufficient for mass-balance determination. Information on photogrammetric techniques using aircraft data is given by Hagen and Reeh (2004).

3.6.1 Surface elevation from satellite radar altimetry

Satellite altimetry, using radar and latterly laser sensors, is the primary method capable of providing the spatial coverage and density of surface elevation data, and at sufficient accuracy, needed to reduce current uncertainties in the mass balance assessment of the Antarctic and Greenland Ice Sheets and their contribution to change in eustatic sea level. In this section, we assess the contribution, strengths, and weaknesses of radar altimeters in this role. These are non-imaging, nadir-viewing sensors that exploit the ranging capability of radar to determine the topographic profile of the surface along the satellite ground track. They achieve this by transmitting very short pulses of electromagnetic (EM) radiation of known power towards the surface, then measuring the reflected energy within a number of narrow bands of time delays known as *range windows* or *gates* (Jensen, 1999; Rémy et al., 2001). The filtering of the return signal into range gates is termed *tracking*. Signal amplitudes in successive gates are then combined to create a *waveform*, as depicted schematically in #Figure 3.53a# (see color section). This is effectively a histogram of the received backscattered power sampled at the satellite, and results from the interaction of the altimeter's transmitted pulse with the scattering surface or volume. As a result, the waveform carries information on both (i) the one-way distance (*range*) between the satellite and surface and (ii) the nature of the surface within the range window. The following parameters can be extracted from this waveform: (i) the altimetric height, (ii) the backscattering coefficient σ^0 related to the waveform integral (in dB), (iii) the leading edge width (in m), and (iv) the trailing edge slope in a logarithmic scale. Representative waveforms from ice-sheet, sea-ice, and open-ocean surfaces are shown in Chapter 5 of Volume 1 of this book. The range is derived by the tracker device, which predicts the range by centering the waveform at a pre-designated tracking gate. Further discussion on the retrieval of surface roughness and sub-surface information from radar altimetry is covered in Sections 3.16.3 and 3.16.4.1, respectively.

Conventional satellite radar altimeters are based on *pulsewidth-limited geometry*, with pulse compression enabling high resolution in the range direction (Chelton et al., 1989; Fetterer et al., 1992; Robin et al., 1983). Observations are made along the ground track of the satellite at intervals of 660 m for Seasat and Geosat and $\sim$330 m for ERS, for example, with the separation between tracks depending on satellite orbital characteristics and latitude. As outlined below, careful processing of the data enables estimation of the minimum radar range from echo delays from within a *pulse-limited footprint* (the maximum surface area simultaneously illuminated) on the Earth's surface. The power in the range gates results from backscatter from range rings within the altimeter footprint, with each ring corresponding to one range gate in the digitally sampled waveform (Fetterer et al., 1992). As such, each range gate corresponds to a round-trip travel time for the EM energy of a few nanoseconds (depending on the satellite), which is equivalent to a fixed range resolution of the order of <1 m—e.g., 0.49 m for Seasat and Geosat. The pulse-limited footprint corresponds to that part of the waveform over which maximum power is measured, while the area mapped out by the pulse at the end of the last range gate is known as the *range window-limited footprint*.

The general principle of pulse-limited radar altimetry is illustrated schematically in Figure 3.54, which shows how an altimeter waveform would look when being scattered back from an "ideal" surface. The surface area illuminated by the pulse expands with time until the trailing edge of the pulse leaves the lowest reflecting points at nadir. As a result, the size of the pulse-limited footprint depends on both the compressed pulse length and the height distribution of the surface. Each scatterer echo outside the latter appears with relatively greater delay. The pulse-limited footprint diameter, associated with a corresponding area on the ground, is

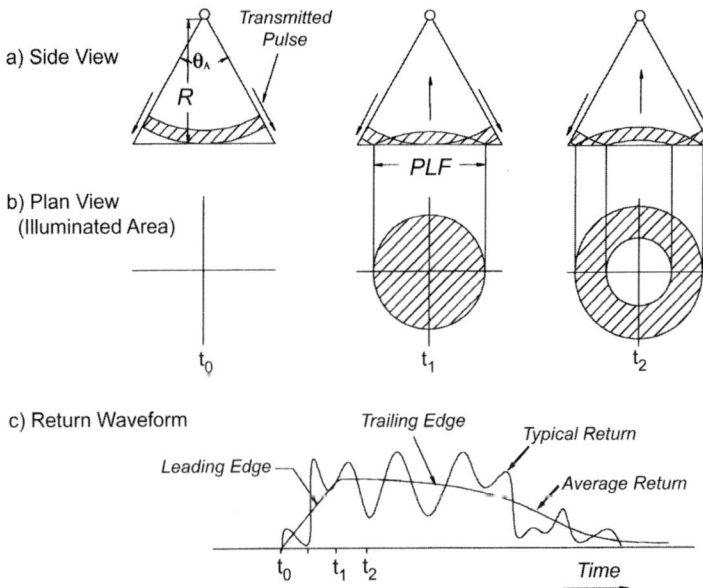

Figure 3.54. A schematic (not to scale) of the principle of pulse-limited altimetry. (a) A side view of the antenna beamwidth (θ_A), where R is the altimeter range. (b) The evolution with time, at t_0 through t_2, of three stages in the growth of the area illuminated on a surface by the Pulse-Limited Footprint (PLF). In this idealized case, the surface is uniformly rough (i.e., Lambertian), horizontally planar (i.e., no curvature or surface slope), and reflects all of the radiation incident on it—i.e., no absorption occurs. The radar pulse is transmitted at $t = 0$ (t_0). (c) The evolution with time of an altimeter waveform scattered from this surface. The average pulse waveform and a typical individual return are shown. For rough surfaces, the PLF is larger, given that the time required for the pulse to travel from crest to trough exceeds the effective pulse length. As the pulse moves outwards over the surface, the footprint becomes an annulus of increasing circumference yet constant area. Due to irregular surface slope, penetration of the radar signal into the ice and roughness, the illuminated area is typically irregular in shape or even discontinuous. At each point, the reflected energy is weighted both by the extinction of the radar wave in the snow/ice volume according to depth and by the instrument antenna gain (according to the distance to the antenna centre). Height measurement precision is determined both by the radar pulse length and the degree of averaging for each estimate, with height being defined as the minimum range between the radar and scatterer ensembles lying along the satellite ground track.

From Ridley and Partington (1988), after Rapley et al. (1983). Copyright Taylor & Francis 1988, reproduced with permission.

determined by the radar pulselength, with the leading edge corresponding to the rise with time of the power as the radar pulse illuminates an increasing area on the surface—i.e., the rise time of the leading edge of the return pulse corresponds to the spreading of the pulse illumination over the footprint (Bamber and Kwok, 2004). For conventional spaceborne pulse-limited radar altimeters, this footprint is of the order of a few kilometers in diameter for smooth surfaces—e.g., ~2 km for ERS and Geosat (1985–1988)—but expands markedly as large-scale surface roughness increases. The trailing edge of the waveform corresponds to the gradual decrease in the power in the waveform as the radar pulse expands over the surface to form an annulus (Figure 3.54b). With radar altimetry, the pulse return is used to determine the closest point on the surface to the satellite. It is apparent from Figure 3.54 that this may not necessarily correspond to the nadir point in the case of glaciers and ice sheets.

Early spaceborne radar altimeters onboard Seasat and Geosat (see Section 3.6.1.1) were designed primarily to measure ice-free ocean parameters—e.g., significant wave height, sea-level monitoring, and large-scale ocean circulation. For reasons noted in the previous paragraph and due to their relatively large surface footprint, satellite pulse-limited radar altimeters perform well over relatively smooth and normally distributed surfaces such as the open ocean and ice sheet inland regions, but poorly over regions of higher and more variable terrain, rough surfaces, and steeper or abrupt slopes such as those encountered around the periphery of ice sheets (Bamber, 1994a; Phillips, 1999; Rémy et al., 1999; Wingham, 1995). In such terrain, the radar pulse typically illuminates several different surface topographic features at the same time, resulting in waveforms which contain multiple maxima and exhibit unpredictable shape (see Figure 5.29 in Volume 1 of this book). This leads to frequent *loss of lock* (*track*) using open-ocean tracking techniques—i.e., the waveform leading edge can fluctuate back and forth within the range window or even disappear at times (Bamber and Kwok, 2004; Martin et al., 1983; Phillips, 1999). For this reason, modern altimeters are equipped with special onboard *ice modes* of operation to help maintain track over such surfaces—i.e., to retain the signal return within the timespan represented by the range window. Ice modes typically use a wider observation window (compared with ocean mode), which in the case of the ERS satellites was achieved by reducing the pulse bandwidth by a factor of 4 (from 28.8 to 115.2 m in width: Bamber and Kwok, 2004)—at the expense of a decrease in range resolution (by a factor of 4). In effect, the *onboard tracker* predicts the range of subsequent returns in order to ensure that the waveform remains and is "captured" within the timespan represented by the specified range window (Bamber and Kwok, 2004; Chelton et al., 2001). The switch between ocean and ice mode is automatically based upon a geographical mask. As shown in Figure 3.55a, tracking an echo of unpredictable shape can be achieved in ice mode by tracking the center of gravity of the return echo rather than its leading edge (⟨*http://envisat.esa.int/instruments/ra2/descr/tracking.html*⟩). Ambiguities are largely avoided in this fashion, as more than one leading edge may be present yet the location of the center of gravity remains unique. These data are of higher quality than those from earlier satellite radar altimeters that operated in open-ocean mode only. Further improvements have been made to the Envisat RA-2 (⟨*http://envisat.esa.int/*⟩), leading to improved geophysical products. These advances, which include more robust tracking over ice sheets, are outlined in Chapter 5 of Volume 1 of this book and in Benveniste et al. (2002), with a detailed account at ⟨*http://envisat.esa.int/instruments/ra2/descr/improvements.html*⟩.

The complications outlined above necessitate additional processing for radar altimeter data acquired by satellites over ice sheets (and sea-ice-covered oceans)

Sec. 3.6]   **Measurement of ice sheet topography/elevation and change in elevation**   215

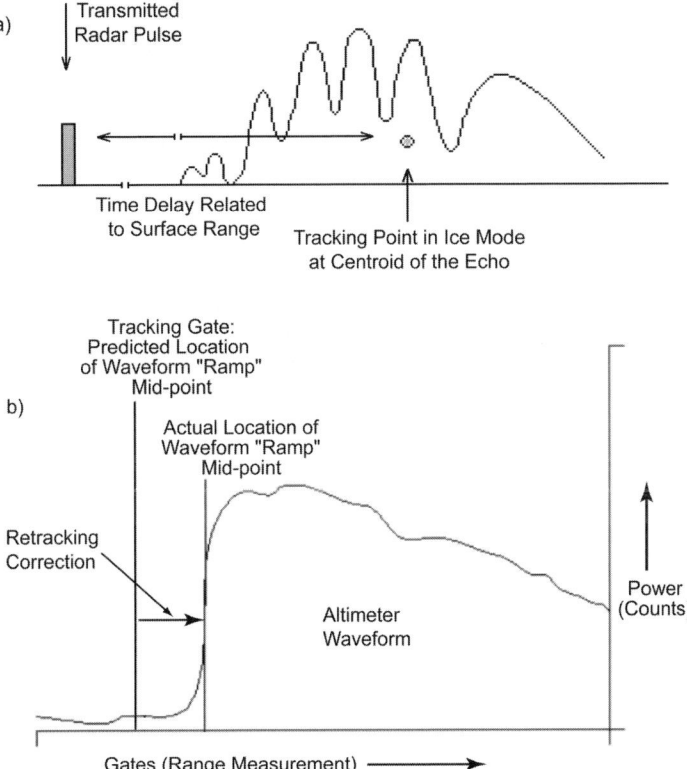

Figure 3.55. (a) A schematic of a typical radar altimeter waveform from a flat, uniformly rough (i.e., Lambertian) surface—e.g., an ice sheet interior—illustrating the concept of the ice mode of tracking used by ESA in their processing of Envisat radar altimeter data acquired over ice-covered surfaces. (b) Schematic illustration of the concept of retracking radar altimeter data. This step is necessary to correct for deviation of the waveform's leading edge from the specified altimeter tracking gate, and is carried out by computing the departure of the leading edge from the tracking gate and then correcting the satellite range measurement (and surface elevation) accordingly.

(a) After ESA (⟨ttp://envisat.esa.int/instruments/ra2/⟩). (b) From the NASA Ice Sheet Altimeter Pathfinder Program website at ⟨http://icesat4.gsfc.nasa.gov/ia_home/wiaas.html⟩. Reproduced with permission.

compared with the open ocean (Rémy et al., 1999). The height of the surface is derived from the leading edge of the waveform (Figure 3.53a), by determining which point along this edge corresponds to the average surface. For surfaces characterized by a Gaussian distribution of slopes—e.g., open-ocean areas—the location of the mean surface height within the antenna footprint is taken to be the half-power point (*retrack point*) of the waveform leading edge (Chelton et al., 2001; Féménias et al., 1993). For ice-free oceans, the retrack point corresponds to the mid-point of the waveform leading edge. The time interval between the transmitted pulse and this point is the classical radar altimeter measure of range between sensor and surface. Over ice-covered areas, however, a discrepancy often exists between the half-power point and the mid-point on the waveform leading edge. This is due to the interaction between the radar wave and a rough and generally irregular surface (Legrésy and Rémy, 1997; Wingham, 1995), slope-related errors, and the variable surface reflection and penetration of the radar wave into the snow/ice mass (Rémy et al., 2001). These factors result in an offset between the actual and recorded range to the surface (Bamber, 1994b).

In order to improve the accuracy of elevation estimates, it is therefore necessary to correct for this offset through post-processing by applying a range estimate refinement procedure known as waveform *retracking* (Figure 3.55b). A return waveform results from the interaction of the altimeter-transmitted pulse with the scattering surface or volume directly beneath the nadir-viewing sensor. Retracking of radar altimeter data involves determination of where the mean surface lies within the leading edge. Several different methods have been developed for retracking radar altimeter data over ice sheets—e.g., Bamber (1994a) for ERS data, and Martin et al. (1983) and Zwally et al. (1983, 1989) for earlier NASA altimeter missions. The NASA Goddard Space Flight Center (GSFC) waveform retracking algorithm (version 4), for example, derives range corrections by fitting a nine- or five-parameter function to each waveform and deletes ranges for which the waveforms are not representative of surface returns. This function describes a single- or double-surface return and calculates the range correction to the mid-point of the first return, which represents the average surface within the first pulse-limited footprint. The approach used to retrack ESA data, known as a threshold retracker, makes no assumptions about the properties of the return. For this technique, the retrack position is taken to be the first point along the waveform leading edge that exceeds a specified threshold of the waveform amplitude (Bamber and Kwok, 2004). Each retracking method has its own advantages and disadvantages, with no single all-purpose algorithm existing (Bamber and Kwok, 2004; Davis, 1996). For further information on current NASA tracking techniques, please see ⟨*http://icesat4.gsfc.nasa.gov/ia_home/retrack.html*⟩. Excellent additional background information on ice sheet radar altimetry is given by Bindschadler et al. (1989) and Zwally et al. (1990, 2002a), and by the NASA Goddard Space Flight Center ice altimetry website (⟨*http://icesat4.gsfc.nasa.gov/ia_home/AltDoc.html*⟩).

A *slope correction* is also necessary to provide the correct range for the corresponding ground location. The latter relates to the fact that ranges obtained by radar altimetry from a satellite to a sloping surface are affected by surface slopes and are not measurements of the surface elevation at the sub-satellite point. Instead, the reflecting point is shifted upslope from nadir to cause a slope-induced error between the true range to the sub-satellite point and the indicated range (Brenner et al., 1983). For a 1° slope, for example, the range distance between the nadir and nearest points is ~120 m (Bamber and Kwok, 2004). While the slope-induced error can be fairly large, it can typically be neglected for crossover analyses of data from different ground tracks as the slope is identical for both ascending and descending tracks at the crossover point. Other factors related to this error are assessed by Bamber and Kwok (2004).

Colour plates

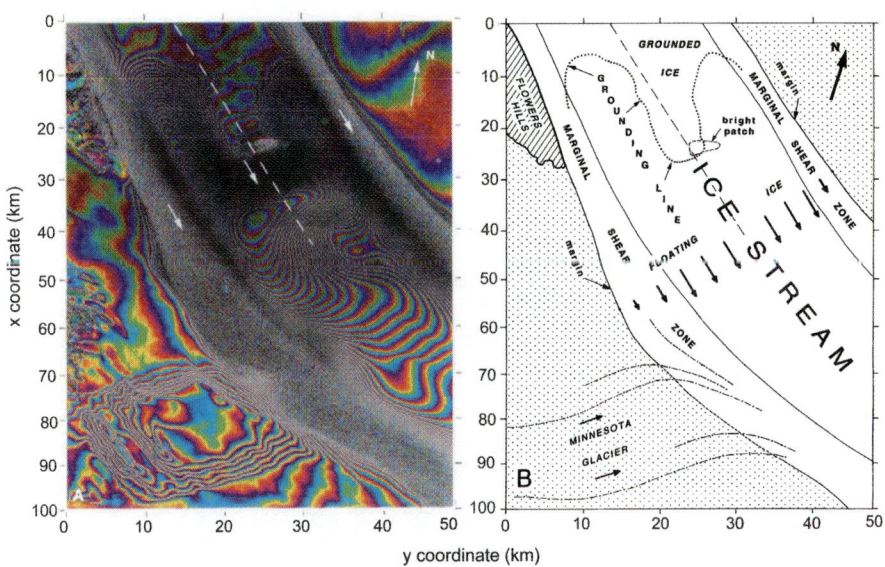

Figure 2.4. (a) Interferogram of ice motion in part of the Rutford Ice Stream, Antarctica, flowing at ~1 meter per day (m d^{-1}), with the direction of flow indicated schematically by arrows. As this section of the ice stream is floating, the fringe pattern contains the combined effects of horizontal ice flow motion and vertical tidal motion. The image center is at 78.43°S, 83.0°W. The color fringes represent the relative motion of the surface toward or away from the satellite—i.e., in the SAR line-of-sight (LOS) direction—over the 6-day period between the two radar observations comprising the interferogram. Each color cycle (from blue through yellow and red then back to blue) in this case represents a 2.8-cm change in range in the radar LOS—i.e., half of the wavelength of the ERS C-band SAR. In this case, the radar LOS azimuth is up (350.8° T). Superimposed on the interferogram is a conventional radar amplitude image in shades of gray. (b) Location map of the glaciological features shown in (a). The location of the grounding line (the limit of tidal flexure of the floating ice) is marked with a dotted line, and the ice flow direction is indicated by arrows.

From Goldstein et al. (1993). Courtesy of Herman Engelhardt (California Institute of Technology). Copyright AAAS 1993, reproduced with permission.

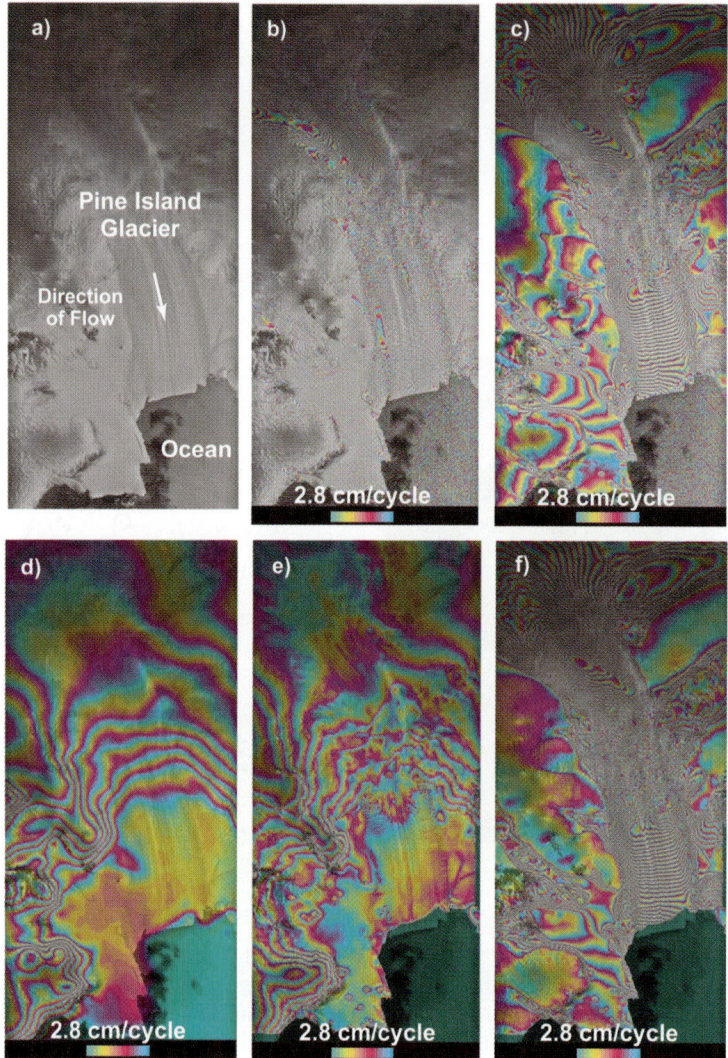

Figure 2.10. A series of images of Pine Island Glacier, West Antarctica, at ~75°S and 100°W, showing a sequence of processing steps to produce an InSAR-derived map of ice velocity. (a) A radar brightness (amplitude) image acquired by the right-looking ERS-1 C-band SAR on November 1995. The image swathwidth is 100 km, with a sample spacing of about 20 m on the ground. Pine Island Glacier, one of the largest dischargers of ice in Antarctica, is about 30 km wide and several hundred kilometers in length (see Chapter 3). (b) A raw interferogram of the glacier obtained by combining radar data acquired by ERS-1 and ERS-2 1 day apart (during the Tandem Mission). Each color cycle is equivalent to a relative displacement of the reflective surface of 28 mm over 1 day. All else being equal, the fringe rate in this case is governed by the baseline (distance) between the ERS-1 and ERS-2 orbits (the stereoscopic effect), the rate of glacier motion, and surface topography. (c) The baseline, calculated from precise orbit information, is removed to leave topography and glacier motion only. (d) A topographic map of the area derived from ERS satellite radar altimetry, at a 5-km resolution and with high precision on relatively flat areas and poor precision along steep slopes. (e) A differential interferogram constructed from ERS-1/2 data by combining two interferograms where ice motion is assumed to remain invariant. The resulting fringe pattern depends only on topography. Note that this assumption is violated on the floating part of the glacier which moves up and down with ocean tides, and is therefore exhibiting non-steady motion. The resulting differential interferogram, combined with control provided by radar altimetry, can be used to derive a precise topography of the grounded part of the glacier. In (f) the topography derived from (e) was removed to leave fringes associated with ice motion only (in the radar look direction).

Courtesy of Eric Rignot (NASA Jet Propulsion Laboratory). Imagery copyright ESA 1995.

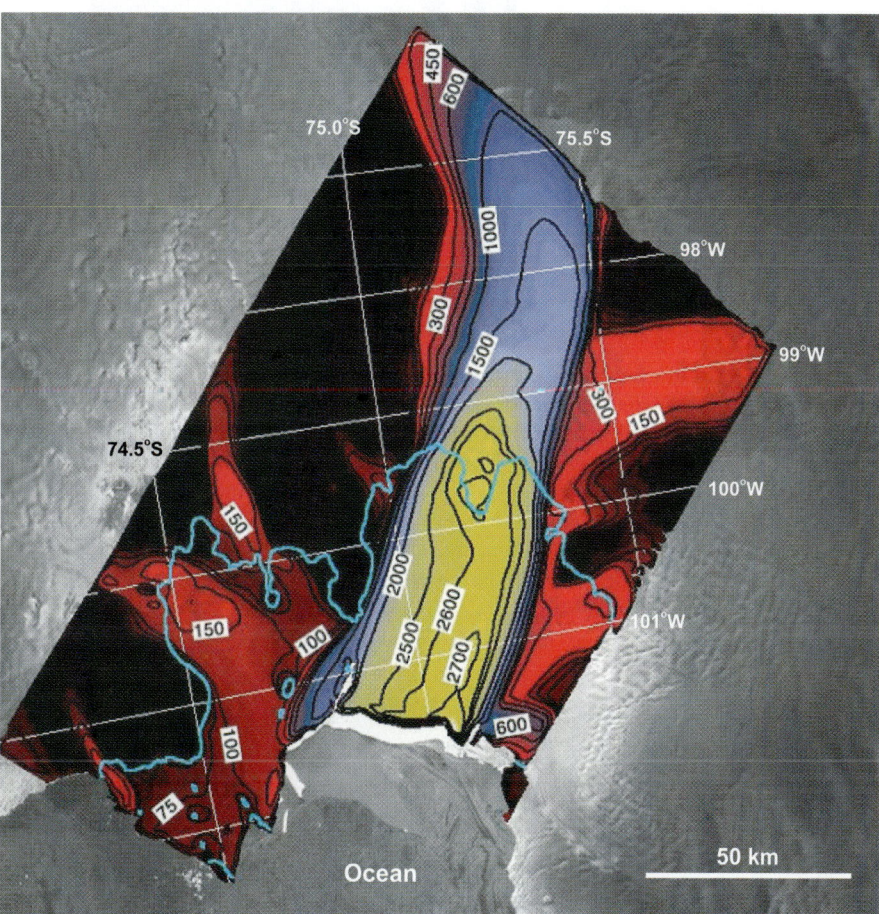

Figure 2.11. Map of ice motion of the lower reaches of Pine Island Glacier assembled from a mosaic of ERS-1/2 interferometry data from November 1995 (track 92, 482, 81), superimposed on a SAR amplitude image. It was constructed by combining images from ascending and descending passes, with the assumption that ice flows parallel to the ice surface are necessary to derive a vector of ice motion. Ice velocity is indicated in meters per annum. Precision is of a few meters per year, at 150-m posting. The blue line delineates the limit of grounded ice—i.e., the grounding line of the glacier. This in itself is another important application of InSAR (see Section 2.5.3 and Chapter 3). Tidal corrections are required over floating icemasses such as this.

Courtesy of Eric Rignot (NASA Jet Propulsion Laboratory). Imagery copyright ESA 1995.

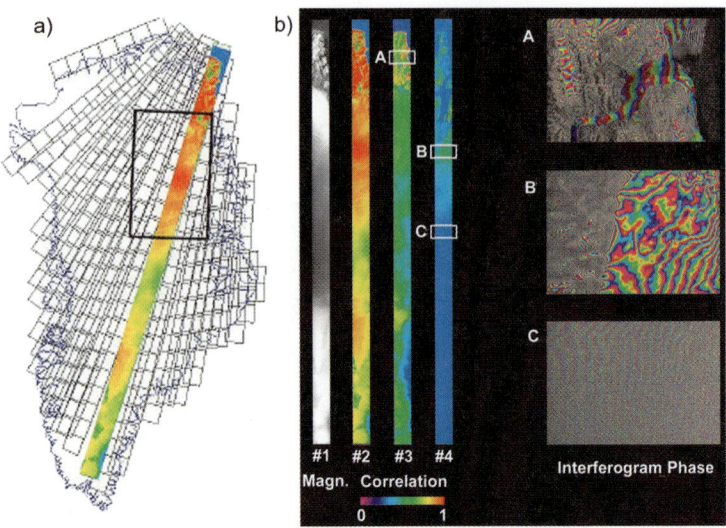

Figure 2.12. (a) Map of Greenland showing ERS coverage obtained through the Greenmap Project. The mapping of Greenland required approximately 3,000 standard ERS SAR frames. (b) Interferometric strip processing of ERS data (track 325) by the Greenmap Project (Danish Centre for Remote Sensing, Technical University of Denmark). SAR magnitude data are shown in strip #1, while strips #2 to #4 show maps of the interferometric correlation. The maps of the interferogram phase are shown in panels A, B, and C correspond to boxes A, B, and C in the strips. The acquisition dates are: strip #1, January 2–3, 1996; strip #2, February 6–7, 1996 (corresponding to the color strip in [a]); and strip #3, January 2 to February 7, 1996.

After ESA (2001). Courtesy of Johan Mohr (Technical University of Denmark). Images courtesy ESA 1996, reproduced with permission.

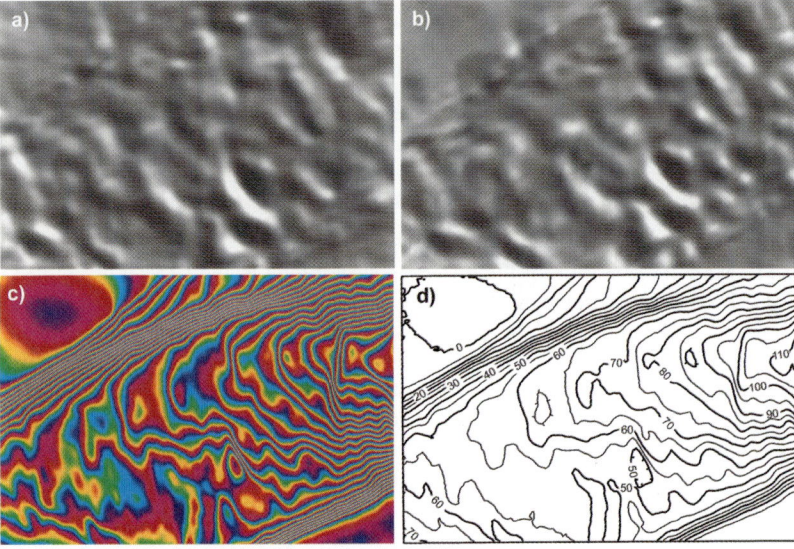

Figure 2.13. Results from the InSAR study of an ice stream in NE Greenland: (a) illuminated DEM derived from two interferograms by differential InSAR; (b) an enhanced resolution NOAA AVHRR band-2 (near-IR) image of the same region; (c) a 9-day interferogram of ice motion only; and (d) contour map of the velocity magnitude derived from (c), taking the velocity in the upper left-hand corner to be zero (note that in this case, the magnitude includes both the unresolved horizontal and vertical velocity components).

After Kwok and Fahnestock (1996). Copyright IEEE 2005, reproduced with permission.

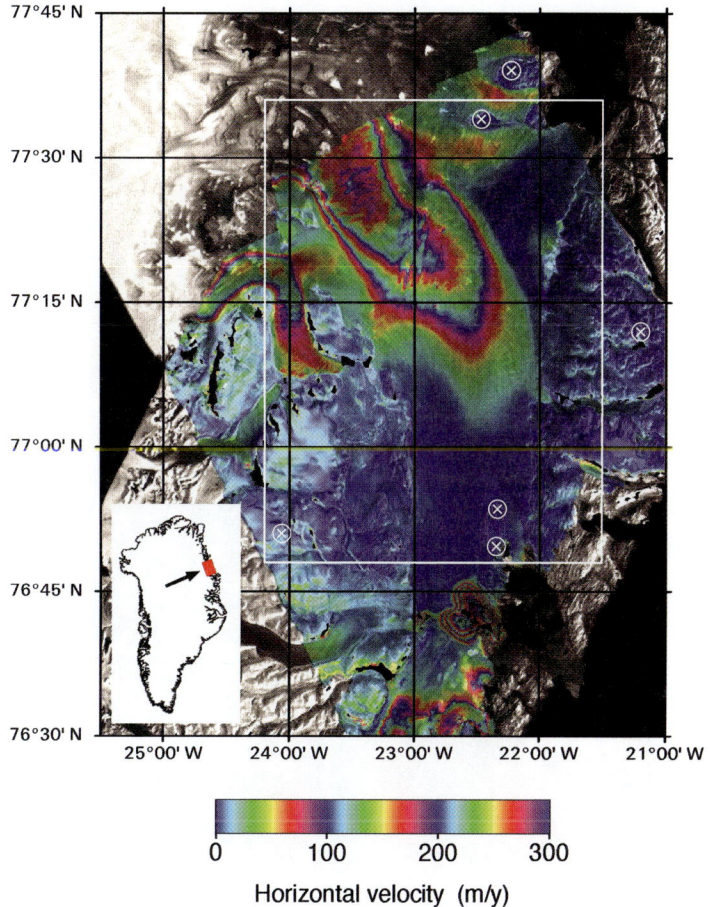

Figure 2.16. Color-coded map of the InSAR-derived annual horizontal ice displacement in the region around Storstrømmen Glacier (NE Greenland—see inset), and determined in the 90 × 90-km overlap region between ascending and descending orbit ERS SAR data. Ascending orbit ERS-1 data were acquired with a 70-day interval on January 31 and April 10, 1996, with ERS-2 repeats 1 day later. Descending orbit data were acquired with a 35-day interval on October 28 and December 2, 1995, with ERS-2 repeats 1 day later. In this case, the ambiguity heights (see Section 2.3.6) are approximately −70 m and +450 m for the ascending orbit data and −500 m and +5,000 m for the descending orbit data. The six symbols ⊗ mark the location of tiepoints on bedrock used to calibrate InSAR-derived velocities and heights, while the white box indicates the location of Figure 2.17. The heart-shaped feature towards the ice front (under the white box) represents an artifact caused by tidal motion, which is not constant whereas the processing assumes constant velocity. Tidal effects, and their removal, are covered in the next section.

From Mohr et al. (1998). Copyright Nature 1998, reproduced with permission.

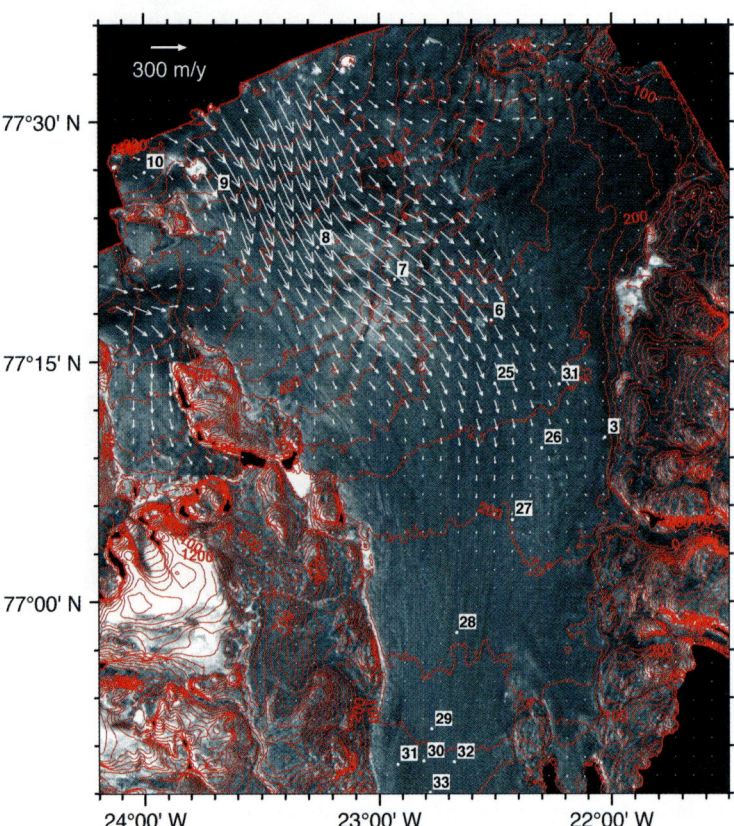

Figure 2.17. Map of InSAR-derived ice flow vectors and elevations from ERS data, from the region around Storstrømmen Glacier (NE Greenland) within the white box in Figure 2.16 (90 × 70 km) and displayed on a grayscale radar amplitude image. Data have been processed to a pixel spacing of ∼100 × 100 m, but only a small subset of the velocity vectors (white arrows) are displayed here. Surface elevation contours (in red) are given at 50-m intervals, with the elevations being derived from ascending orbit data only.

From Mohr et al. (1998). Copyright Nature 1998, reproduced with permission.

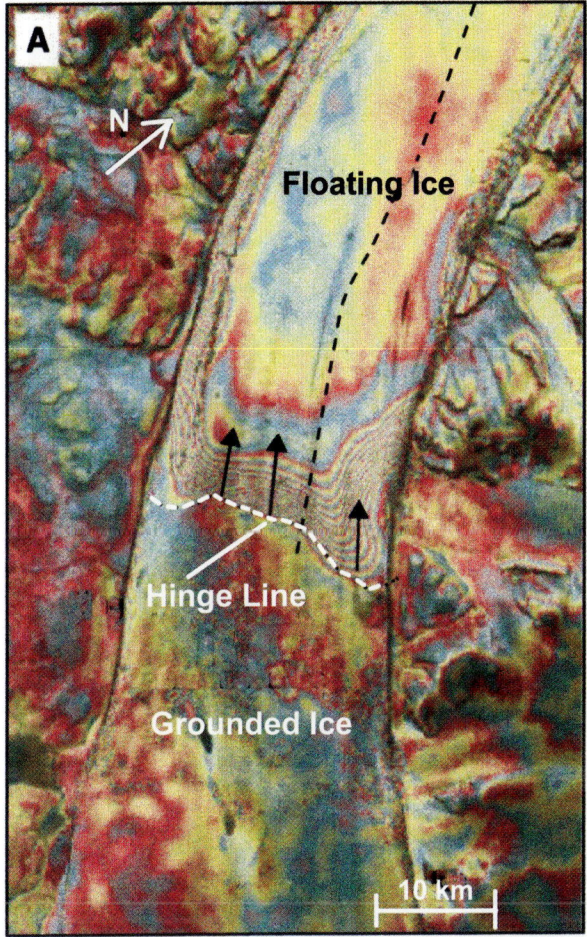

Figure 2.18. An important application of satellite radar interferometry is the systematic, precise, and detailed detection of the grounding line of ice sheet outlet glaciers. As the entire ice volume crossing the grounding line eventually melts into the ocean, it forms a natural boundary for calculating ice discharge. This example shows tidal displacements of the floating section of the Petermann Gletscher (N Greenland) obtained from quadruple-difference ERS radar interferometry. Each fringe, or 360° variation in phase, represents a 28-mm displacement of the glacier tongue toward the radar line of sight (23° from vertical in the case of ERS-1 and -2) due to forcing by the ocean tide. The SAR phase image is modulated by the radar brightness of the scene. The limit of tidal flexing of the floating glacier, or *hingeline*, is demarcated by the white dashed line. In such cases, the amplitude of tidal displacement increases rapidly from the hingeline (indicated by a high fringe rate) and subsequently decreases slowly toward the glacier front. Such a deformation pattern is in close agreement with model predictions from an elastic beam fixed at one end on bedrock (at the hingeline) and floating freely on the ocean (Holdsworth, 1969). The dashed line marks the profile of airborne laser altimeter and ice-sounding radar measurements (Krabill et al., 1995). Solid arrows indicate flow direction parallel to flowlines conspicuous in the radar amplitude images. Residual fringes on surrounding exposed rock are caused by inherent imperfections in the DEM used in regions of high topographic relief.

From Rignot et al. (1997). Copyright AAAS 1997, reproduced with permission.

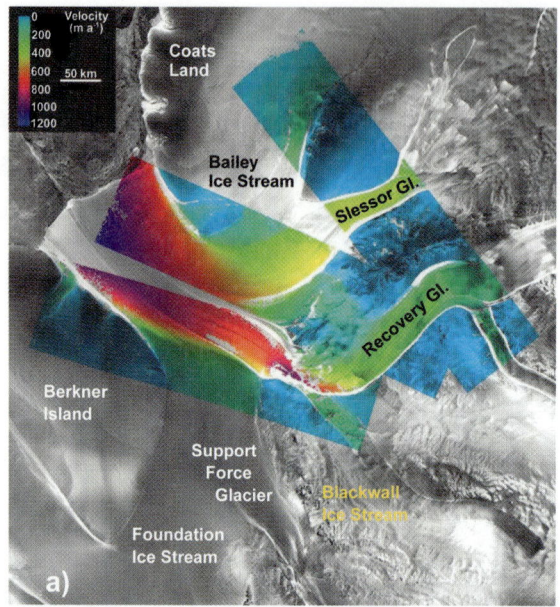

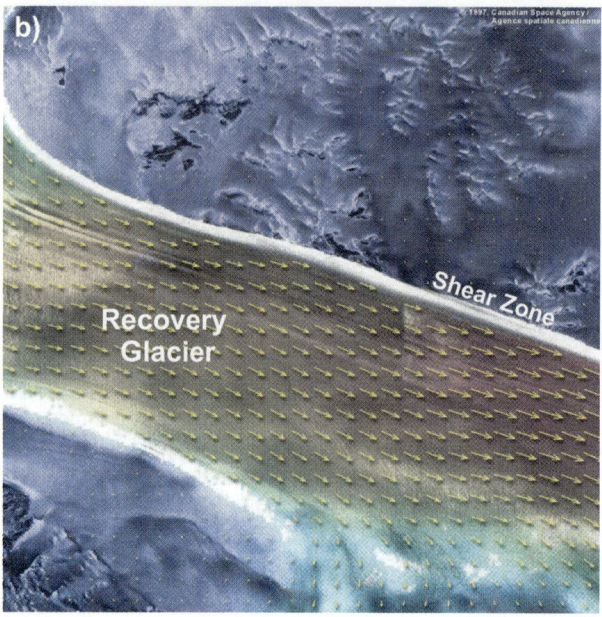

Figure 2.22. (a) Surface velocity derived by speckle tracking in three 1997 Radarsat-1 AMM repeat swaths on the Filchner Ice Shelf, Antarctica (location marked in inset). The background image is part of the AMM Antarctic mosaic. (b) A detail of Recovery Glacier, which enters the Filchner Ice Shelf, showing ice motion. This image (110 × 110 km) has been colored to highlight ice motion variability, with slow-moving ice in blue and faster moving ice in green through red tones. Ice towards the right-hand side of the image is moving at ∼340 m per annum. The glacier margins are characterized by shear margins, where heavy deformation and crevassing creates a strong backscatter and thus a bright target. It is not possible to retrieve useful ice motion data from such areas due to the high level of ice distortion.

(a) From Gray et al. (2001). Image obtained from ⟨http://www.ccrs.nrcan.gc.ca/ccrs/rd/apps/map/amm/amm1_e.html⟩. Radarsat-1 source images Canadian Space Agency 1997. The Antarctic Mosaic was produced by the Byrd Polar Research Center of Ohio State University from images processed by the Alaska SAR Facility. The velocity products were generated at the Canada Centre for Remote Sensing from data delivered by Radarsat International Inc. Copyright CSA 1997. Copyright Canadian Space Agency/Radarsat International 1997, reproduced with permission.

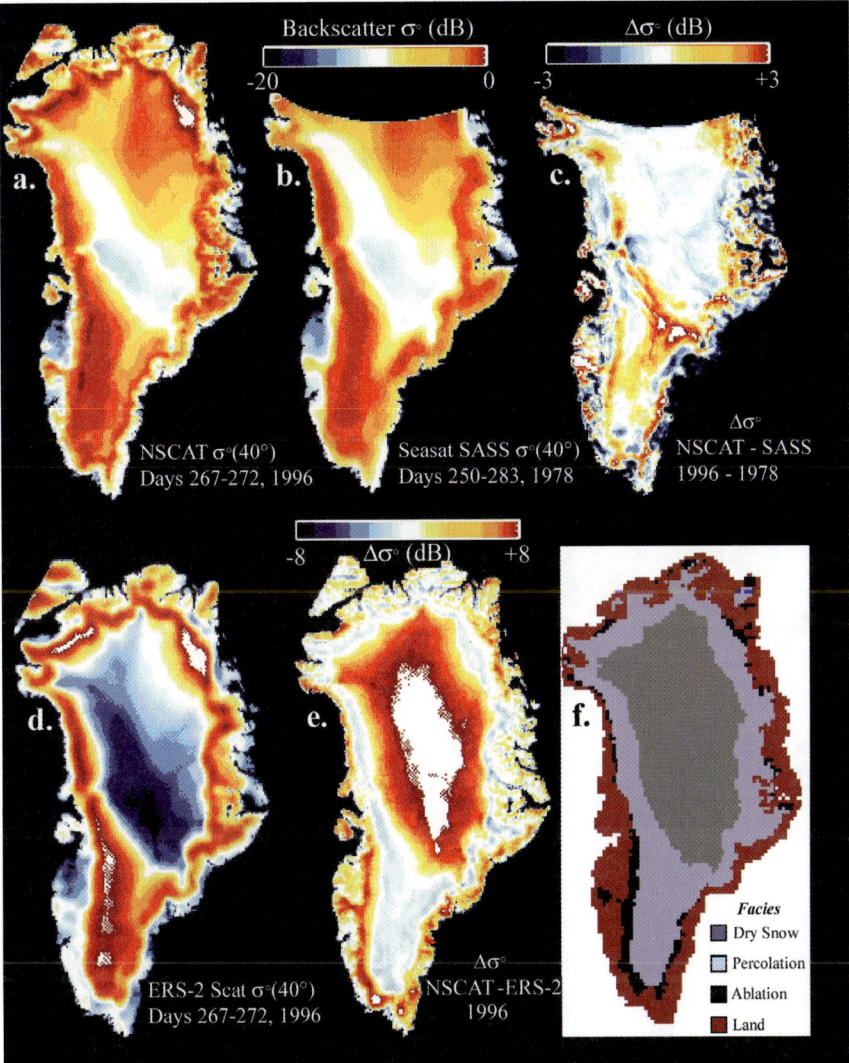

Figure 3.14. Comparison of A measurements (normalized to a 40° incidence angle) of the Greenland Ice Sheet made by the Seasat-A Scatterometer (SASS) with those of the NASA Scatterometer (NSCAT) and ERS-2 scatterometer made after an 18-year time interval—i.e., 1978 to 1996. The image in (a) was created from data acquired by NSCAT in September 1996, while the image in (b) was generated from Seasat data acquired during the same season in 1978. The image in (c) shows the difference ($\Delta\sigma^0$) between the first two images—i.e., 1996 minus 1978. In (a) and (b), the red areas are ice zones which normally experience some annual melting; the blue areas at the coastlines are exposed rock; the yellow and white areas in central Greenland are the dry-snow zone; and the light-blue areas are where annual snow accumulation is at its greatest. In (c), the red and white patches in the center indicate areas of significant change, with the pattern of changes being consistent with a 10-year warming trend. (d) A map of A data from the ERS-2 C-band scatterometer, for the time period corresponding to the NSCAT image in (a). (e) A map of the difference between the coincident NSCAT minus ERS-2 measurements. (f) A map of ice sheet facies derived from the NSCAT and ERS-2 data in (d) and (e). A temperature increase of more than 1°C took place between 1979 and the present day, except for the summer of 1992 when ash from the Mount Pinatubo volcanic eruption may have temporarily helped to cool the Northern Hemisphere.

Image courtesy of NASA, Mark Drinkwater (ESA), and Brigham Young University (Microwave Earth Remote Sensing, ⟨http://www.mers.byu.edu⟩). Copyright NASA.

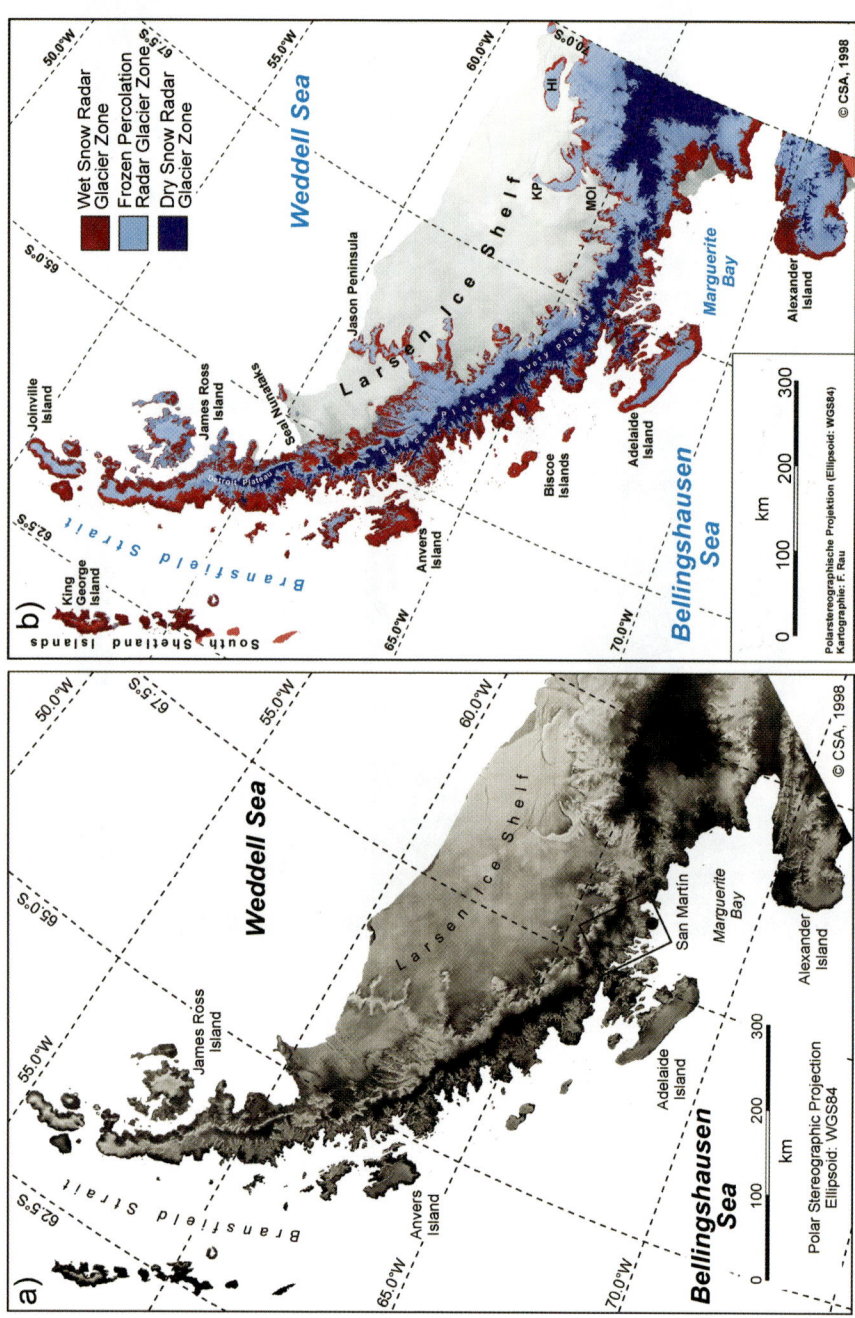

Figure 3.15. (a) Radarsat-1 ScanSAR mosaic of the northern Antarctic Peninsula, January 5, 1999. (b) "Radar facies" observed during December 12, 1998.

From Rau and Braun 2002. Copyright International Glaciological Society 2002, reproduced with permission.

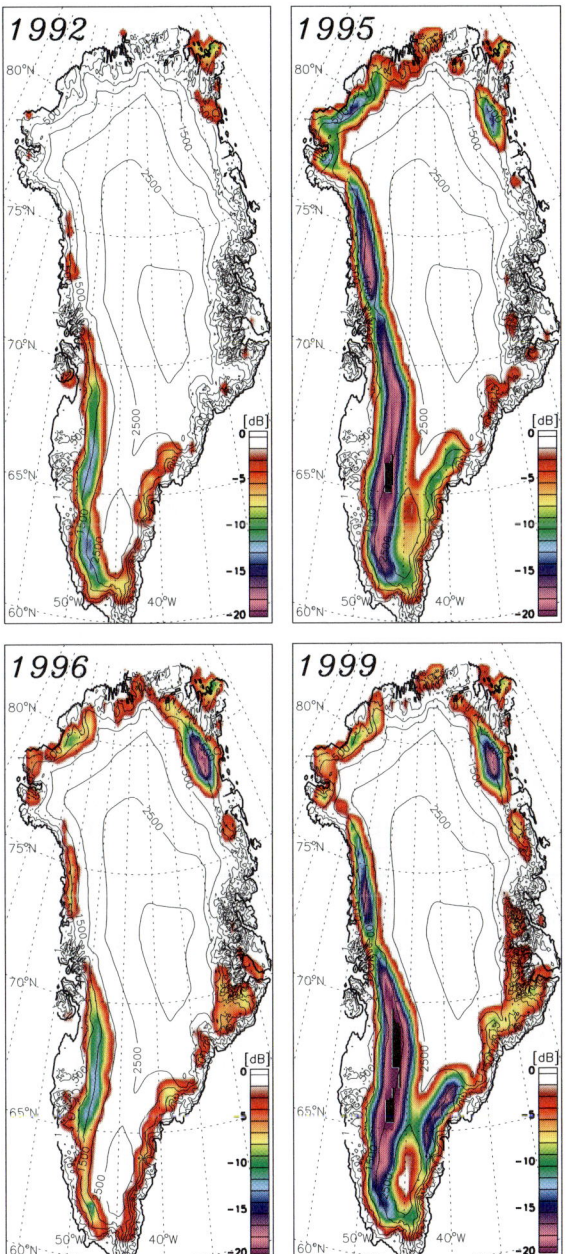

Figure 3.19. Maps of the areal extent of snowmelt in Greenland for the years 1992, 1995, 1996, and 1999, derived from ERS-1 and -2 radar scatterometer data. The color scale shows the maximum integrated reduction in σ_{40}° at the respective location, where summer dip is defined as the mean decrease in σ_{40}° within the melt area. In both cases, the data show the 3-day interval containing the maximum extent of the dip. The topographic contour interval is 500 m.

Imagery courtesy of Volkmar Wismann (IFARS, Germany). After Wismann (2000). Copyright IEEE 2000, reproduced with permission.

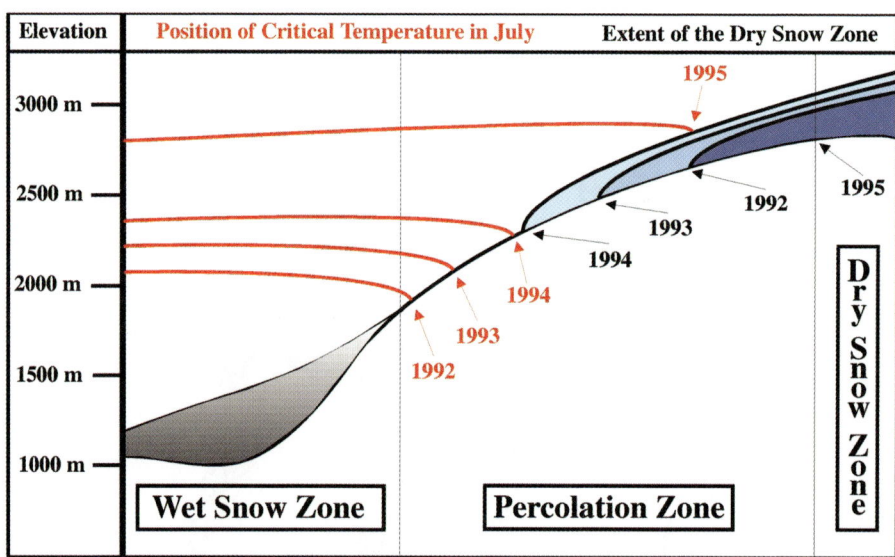

Figure 3.20. Schematic summarizing results from an analysis of ERS-1 scatterometer σ^o_{40} data along the western slope of the Greenland Ice Sheet from 1992 to 1995, showing the different facies. The 0°C isotherm, depicted as a red line, crept uphill over this time, to reach the dry-snow zone in the summer of 1995. This increase in the areal extent of melt coincided with snow accumulation at higher altitudes, with the addition of new snow on top of older firn resulting in a slight decrease in σ^o_{40}. This accumulation caused a gradual downhill progression of the dry-snow line (shown as a black line). A sudden interruption of this trend occurred in the summer of 1995, when high-altitude melt shifted the line back uphill.

From Wismann and Boehnke (1996). Copyright IEEE 1996, reproduced with permission.

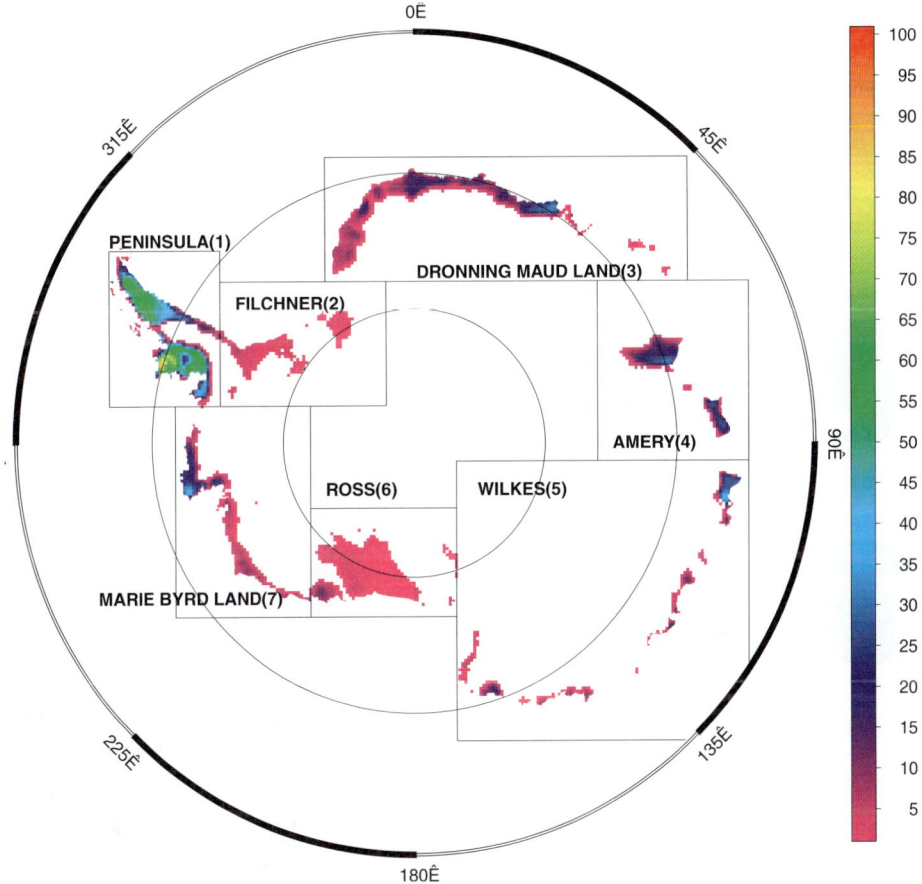

Figure 3.22. Map of the mean annual duration (in days) of circum-Antarctic melt derived from analysis of Nimbus-7 Scanning Multi-channel Microwave Radiometer (SMMR) and DMSP Special Sensor Microwave/Imager (SSM/I) data for the period 1980–1999.

From Torinesi et al. (2003). Copyright American Meteorological Society 2003, reproduced with permission.

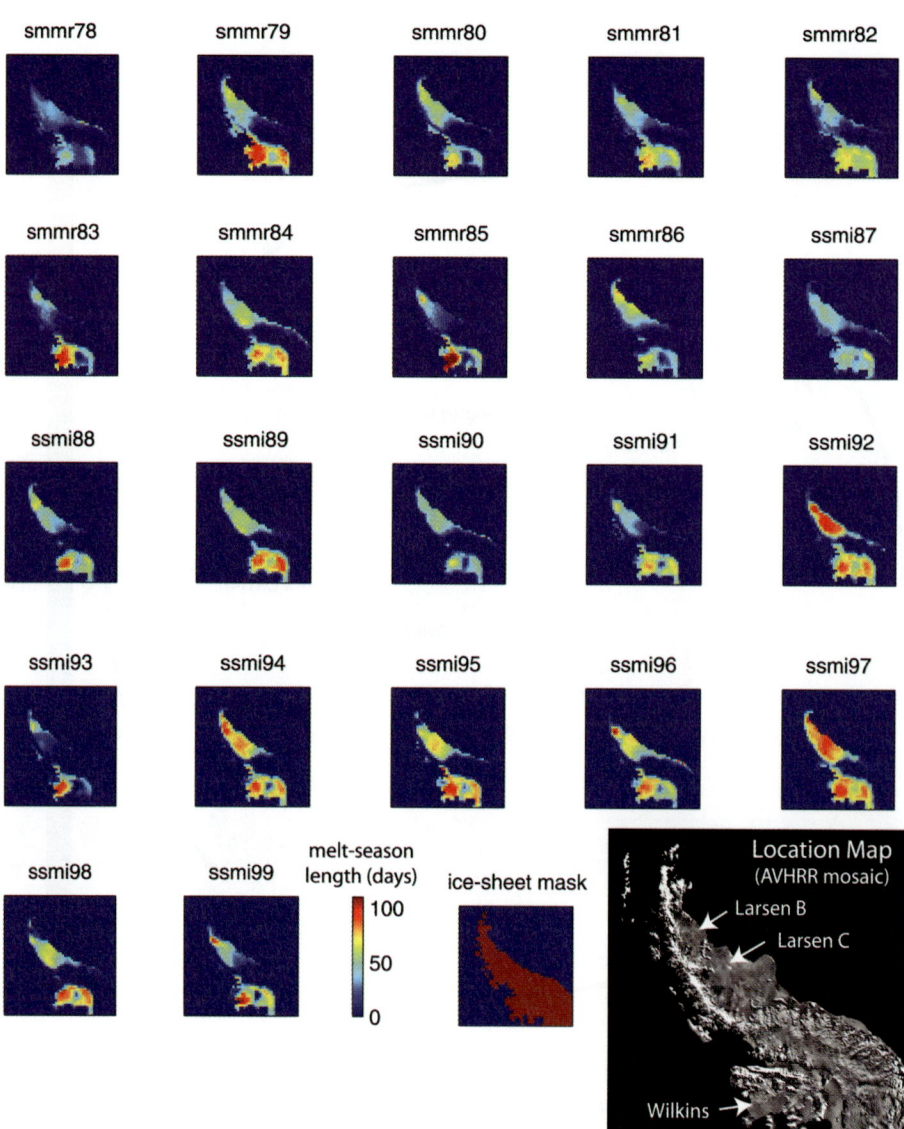

Figure 3.25. Maps of melt season lengths on the Antarctic Peninsula, using Nimbus-7 SMMR data from 1978 to 1986 and DMSP SSM/I data from 1988 to 1999, computed using the XPGR technique of Abdalati and Steffen (1997a). The location map is part of the U.S. Geological Survey AVHRR mosaic of Antarctica (see Figure 1.3).

After Fahnestock et al. (2002). Copyright International Glaciological Society 2002, reproduced with permission.

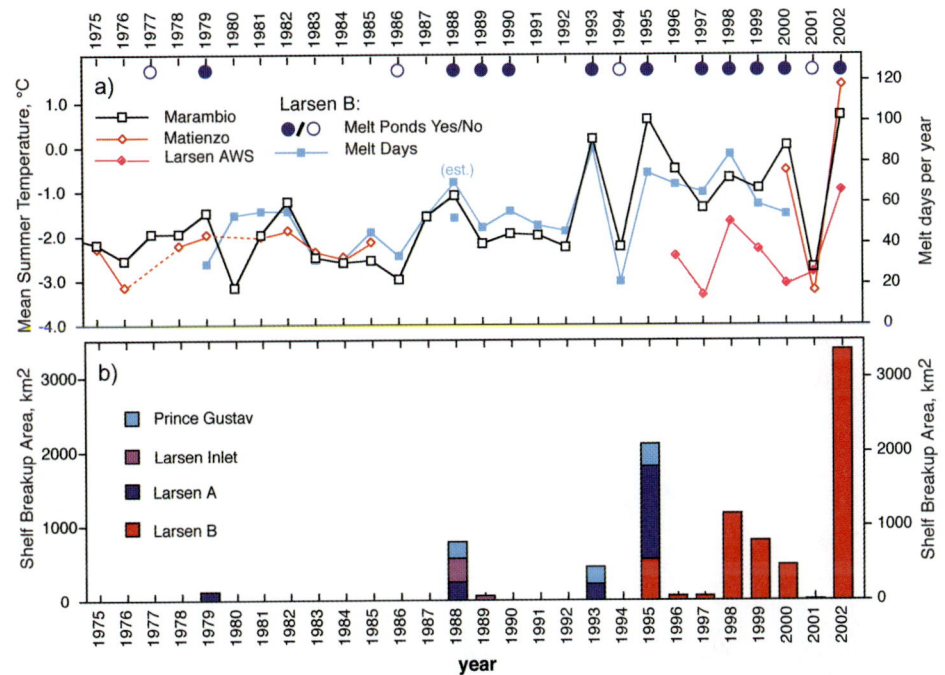

Figure 3.30. (a) Trends of mean surface air temperature for December–February from Marambio (64° 14′S, 56° 37′W) and Matenzio (64° 59′S, 60° 04′W) Stations and Larsen AWS (66° 57′S, 60° 54′W)—Skvarca and De Angelis, 2003), and melt days per annum for the central Larsen B Ice Shelf—derived from satellite passive-microwave data of 1 pixel of 25 × 25 km in area (Fahnestock et al., 2002). (b) The retreat of ice shelves in the NE Antarctic Pensinsula. The incidence of meltponds, determined from analysis of Landsat, AVHRR, and MODIS imagery, is also marked. Breakup events are shown when the timing is known, from analysis of satellite data, in a single year. In (a), the melt season length was estimated due to a lack of passive-microwave data between December 3, 1987 and January 12, 1988. Note that the January 1995 Larsen B breakup event is not included as it may not be related to climate-driven retreat, and that the several minor breakups in Larsen Inlet, Prince Gustav Channel, and Larsen A that took place in the 1980s cannot be precisely timed.

From Scambos et al. (2003). Copyright American Geophysical Union 2003, reproduced with permission.

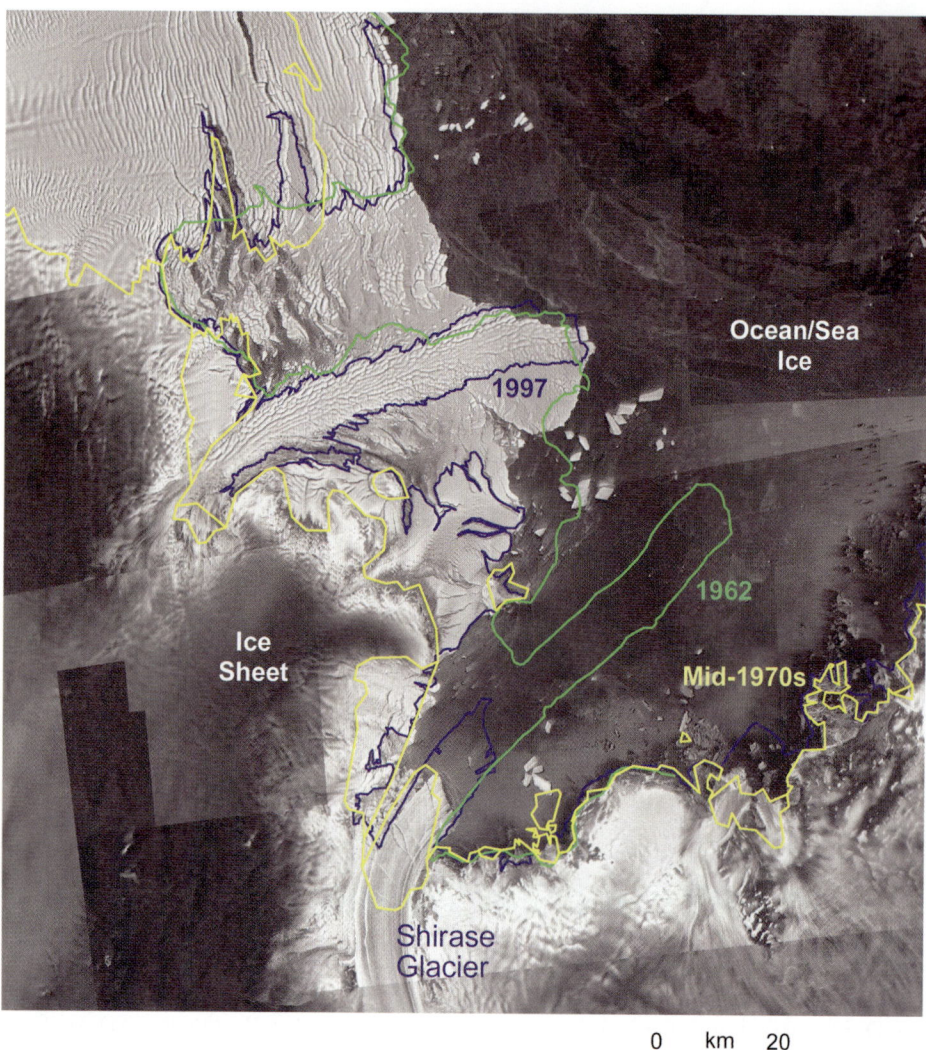

Figure 3.36. A Radarsat-1 image of the Shirase Glacier (East Antarctica) from the 2000 Antarctic Mapping Mission (AMM), with ice margin locations derived from other sources to show how the glacier retreated over the period 1962–2000. The 1997 coastline (obtained from the first Radarsat AMM) is given in blue, the 1962 coastline (from U.S. reconnaissance satellite imagery) in green, and the mid-1970s coastline (Landsat-derived) in yellow. This analysis indicates that the Shirase Glacier was at its farthest extent in 1962, then retreated ∼60 km in the mid-1970s to a position that almost corresponds to the 2000 position.

From ⟨http://www-bprc.mps.ohio-state.edu/rsl/radarsat/radarsat.html⟩. Courtesy Canadian Space Agency/NASA/Ohio State University, NASA Jet Propulsion Laboratory and Alaska Satellite (SAR) facility. Copyright Canadian Space Agency/Radarsat International 2005.

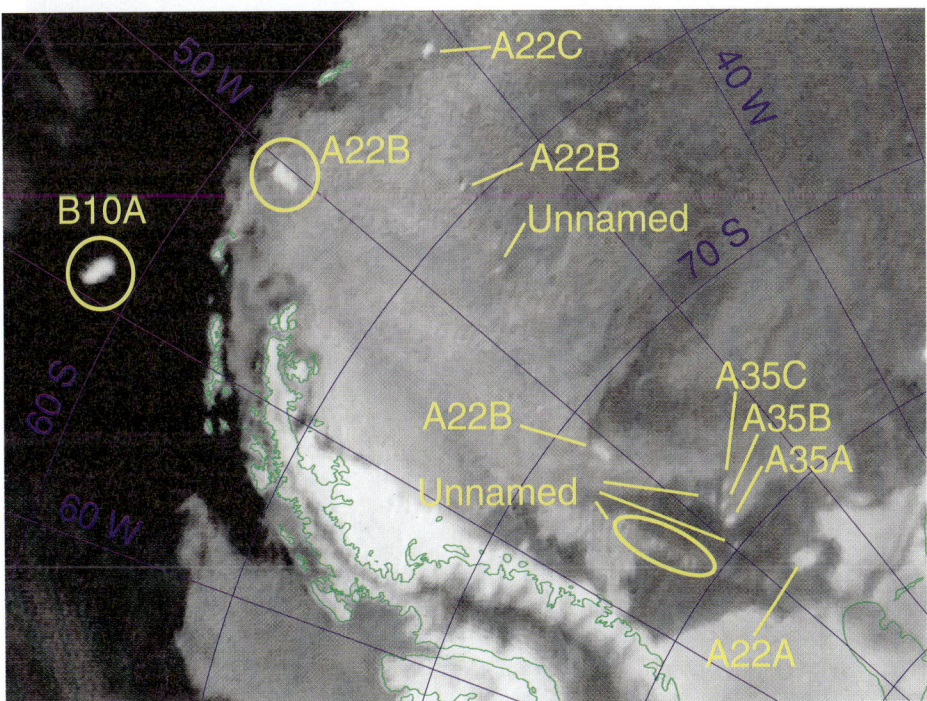

Figure 3.43. An enhanced resolution image from SeaWinds on QuikSCAT showing large named icebergs in the Weddell Sea (Antarctica) and adjacent to the Antarctic Peninsula on July 17, 1999. The gray-textured region is sea ice, while the dark area to the left is ice-free open ocean.

From Long et al. (2002). Copyright American Geophysical Union 2002, reproduced with permission.

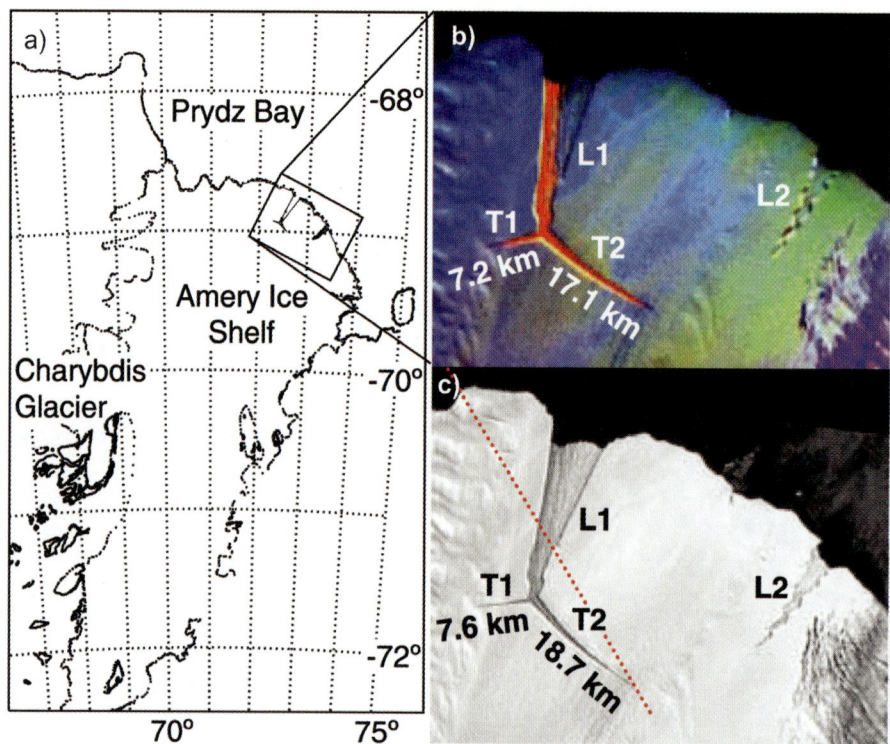

Figure 3.44. (a) Map showing the location of the Amery Ice Shelf rift system. (b) EOS Terra MISR false-color composite image of the red bands from the CF, AN, and CA cameras. Color here acts as a proxy for variations in angular reflectance which are related to surface texture, and which enhance the rifts. (c) A coincident Landsat-7 ETM band-4 image. Both satellite images were acquired on March 2, 2003. Data from the MISR and ETM are directly comparable as Terra and Landsat-7 follow the same groundtrack, but separated by 20 minutes. Differences in rift-length estimates (marked on the images) are due to differences in the pixel sizes of the two sensors. The red dashed line in (c) indicates the location of the profile in Figure 3.66.

From Fricker et al. (2005). Copyright American Geophysical Union 2005, reproduced with permission.

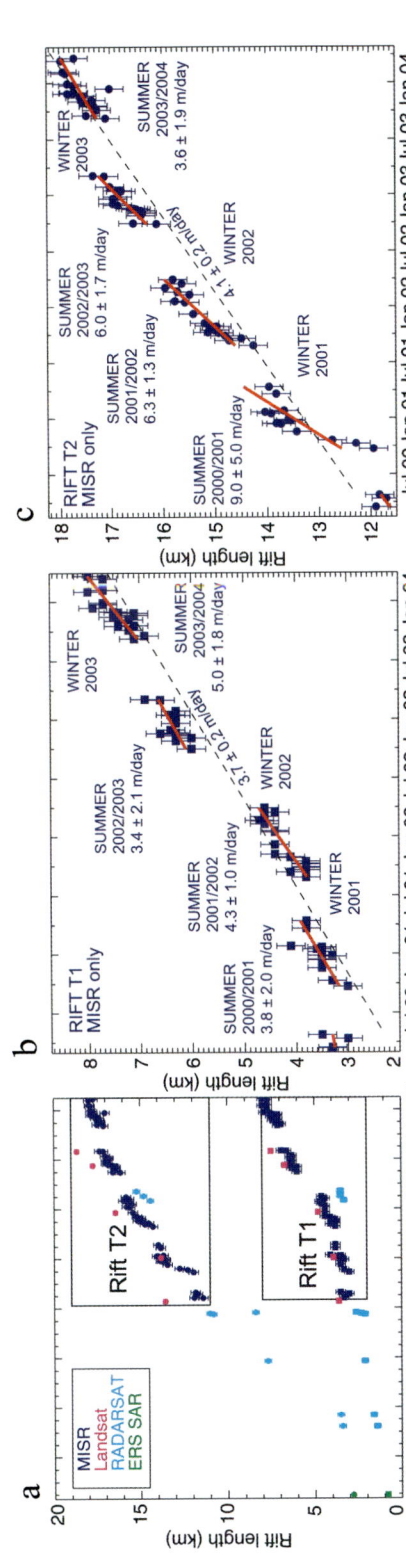

Figure 3.45. (a) Measured lengths of Amery Ice Shelf transverse rifts T1 and T2 (see Figure 3.44), derived from satellite imagery from 1996 to the present. For each image type, the error bars represent 1 pixel, where pixel sizes are: ETM = 12.5 m; Radarsat-1 SAR = 200 m; and MISR = 275 m. (b-c) MISR time series or T1 and T2, showing results of regression analysis of the change in riftlength with time.

From Fricker et al. (2005). Copyright American Geophysical Union 2005, reproduced with permission.

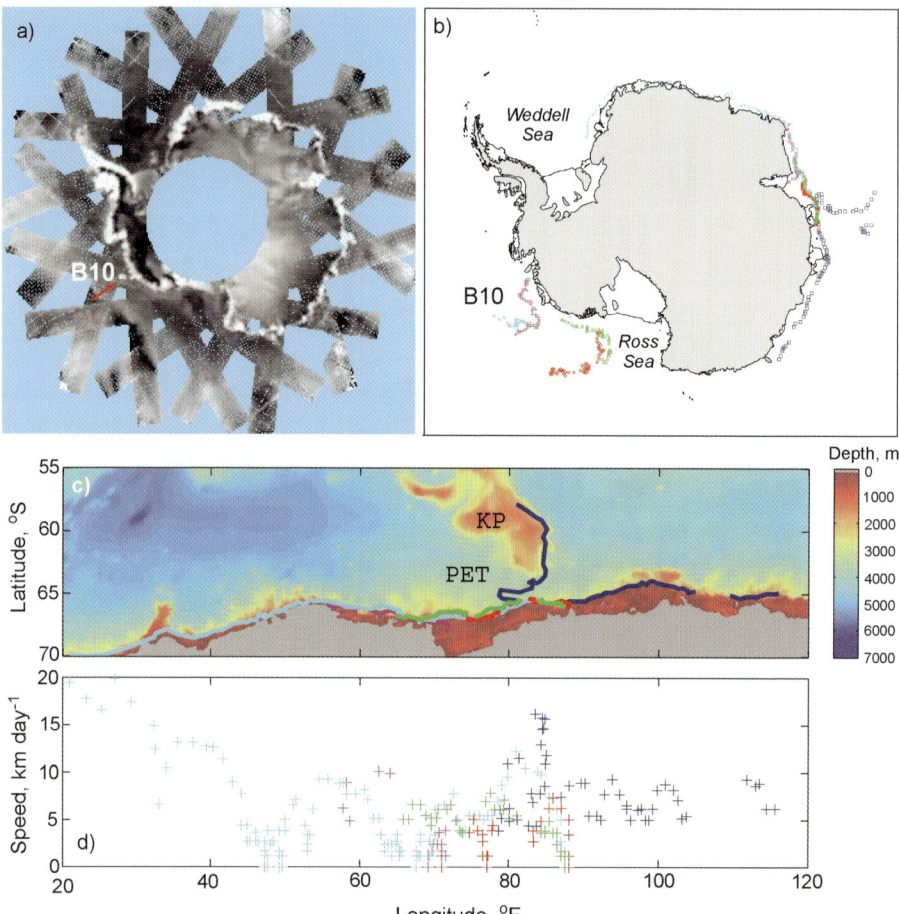

Figure 3.47. (a) Image showing the ERS wind scatterometer coverage of the Antarctic region south of latitude 55°S for a 1-day integration interval. Iceberg B10 (area ~8,000 km^2) is marked. (b) Drift tracks of eight large icebergs tracked in a time series of 5-day composite images of normalized backscatter. Close to the coast, drift is anti-clockwise within the Antarctic Coastal Current. Several tracks exhibit a retroflection, a stage with northward drift, then a drift to the east with the Antarctic Circumpolar Current (e.g., between ~80°E and 90°E (described by Tchernia and Jeannin (1980)) and in the western Bellingshausen Sea at about 130°W). (c) Map of seafloor bathymetry for the East Antarctic coastline between 20 and 120°E (ETOP05 data from NOAA, 1988). Drift tracks of five large icebergs are marked in different colours. PET is the Princess Elizabeth Trough and KP the Kerguelen Plateau. (d) Drift speeds of the icebergs in (c), with the color of crosses corresponding to the drift-track colours of individual icebergs in (c).

After Young (1999). ERS-1 wind scatterometer data copyright ESA 1994. Copyright CERSAT 1999, reproduced with permission.

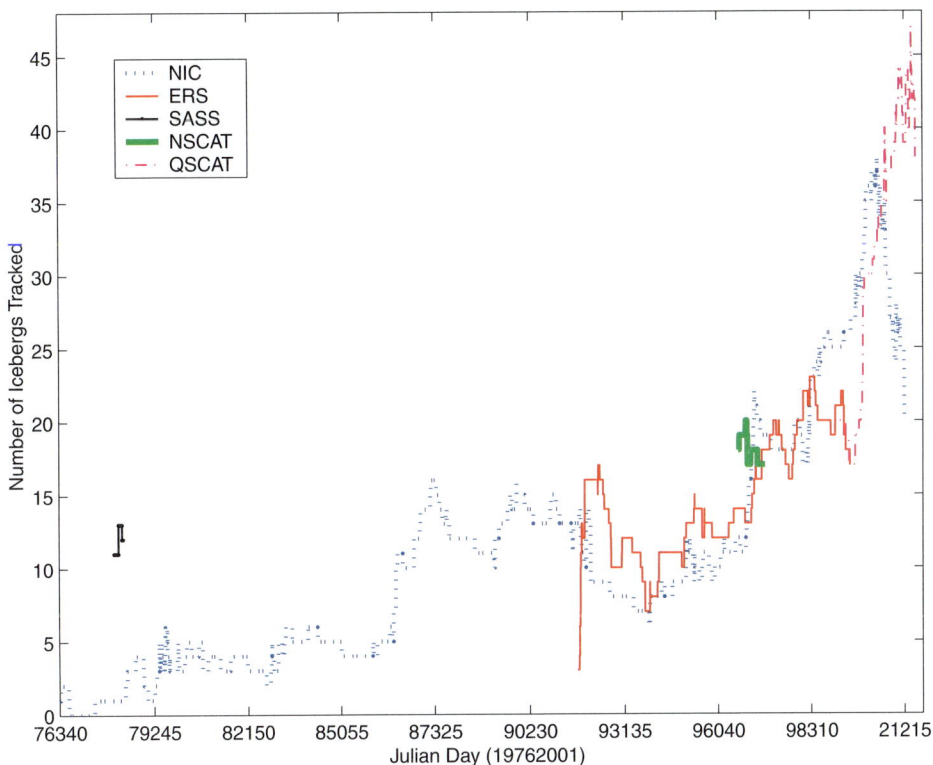

Figure 3.50. A plot of the number of icebergs detected and tracked versus time reported by the U.S. Navy/NOAA/Coastguard National Ice Center (NIC), and the Scatterometer Climate Record Pathfinder (SCP), the latter using enhanced resolution images from various scatterometers.

From Long et al. (2002). Copyright American Geophysical Union 2005, reproduced with permission.

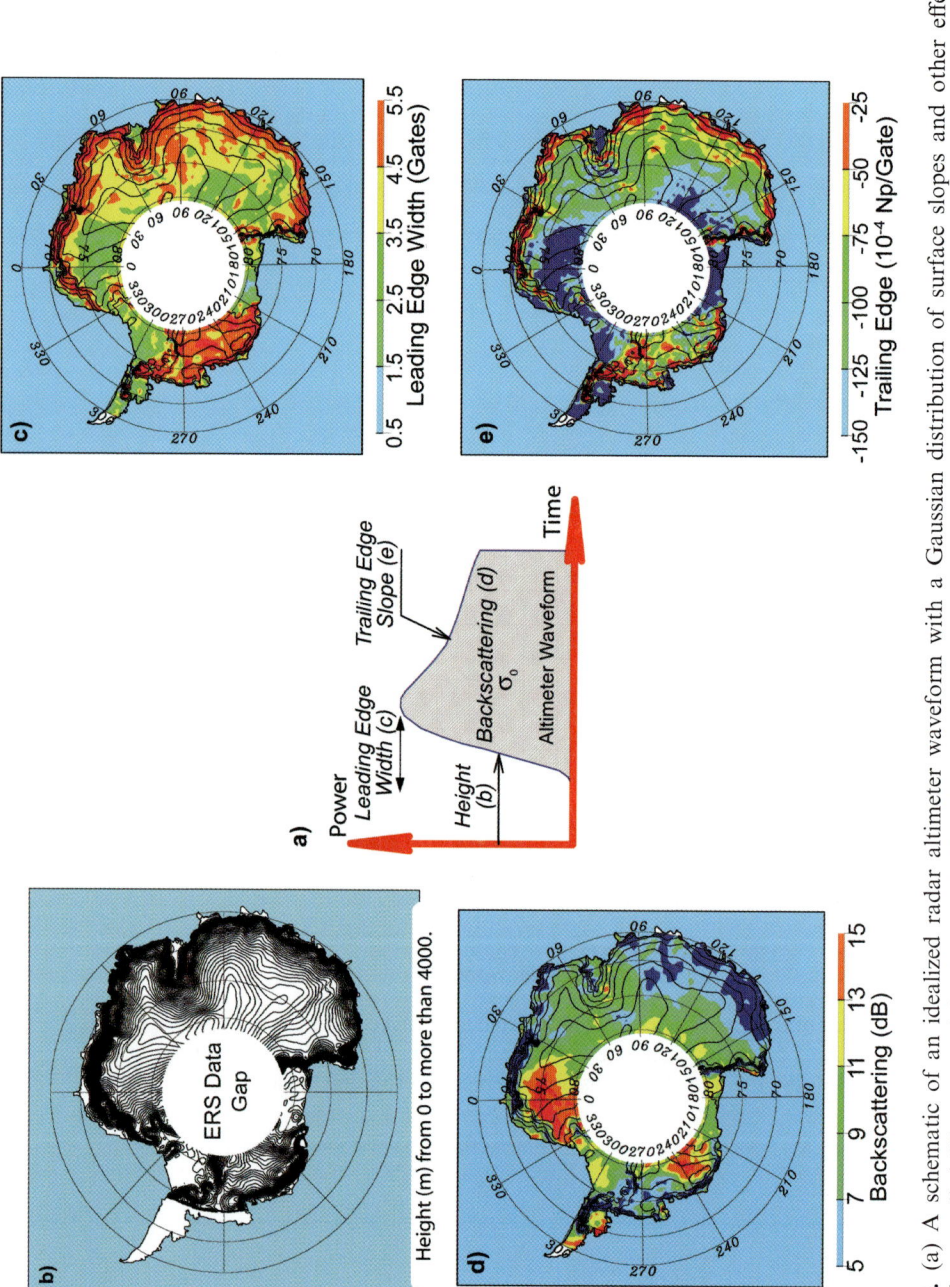

Figure 3.53. (a) A schematic of an idealized radar altimeter waveform with a Gaussian distribution of surface slopes and other effects following the so-called "Brown model" (Brown, 1977), illustrating the measurement principle of the altimeter and with key parameters marked. The corresponding scale (in km) of the temporal evolution is also given. Examples are also shown of the following parameters derived from the radar altimeter waveforms of ERS data and for the Antarctic Ice Sheet: (b) surface elevation (to >4,000 m), (c) the leading-edge width, (d) total backscattering, and (e) the trailing edge.

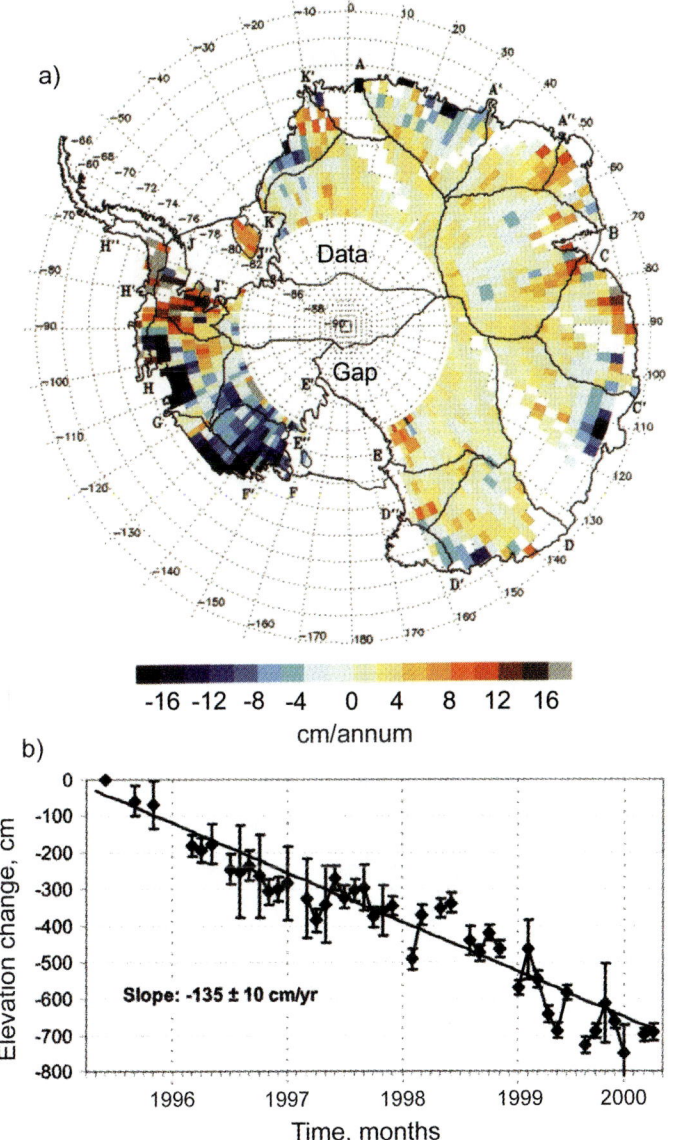

Figure 3.59. (a) A spatial plot of the trend in $\partial H/\partial t$ over the Antarctic Ice Sheet for the period June 1995 to April 2000, derived from ERS-2 radar altimeter data (123 million elevation change measurements) and after adjustment for variations in the backscattered power of the radar signal returns. Elevation change time series were computed for $1° \times 2°$ (latitude × longitude) regions north of 81°S (excluding ice shelves), and $0.6° \times 2.0°$ for the most southerly data between 81° and 81.6°S. Major drainage basin divides are marked. (b) The corresponding elevation change time series for a circular region (radius 20 km) centered around the outlet of the Pine Island Glacier at 75.35°S, 261.5°E.

From Davis and Ferguson (2004). Copyright IEEE 2004, reproduced with permission.

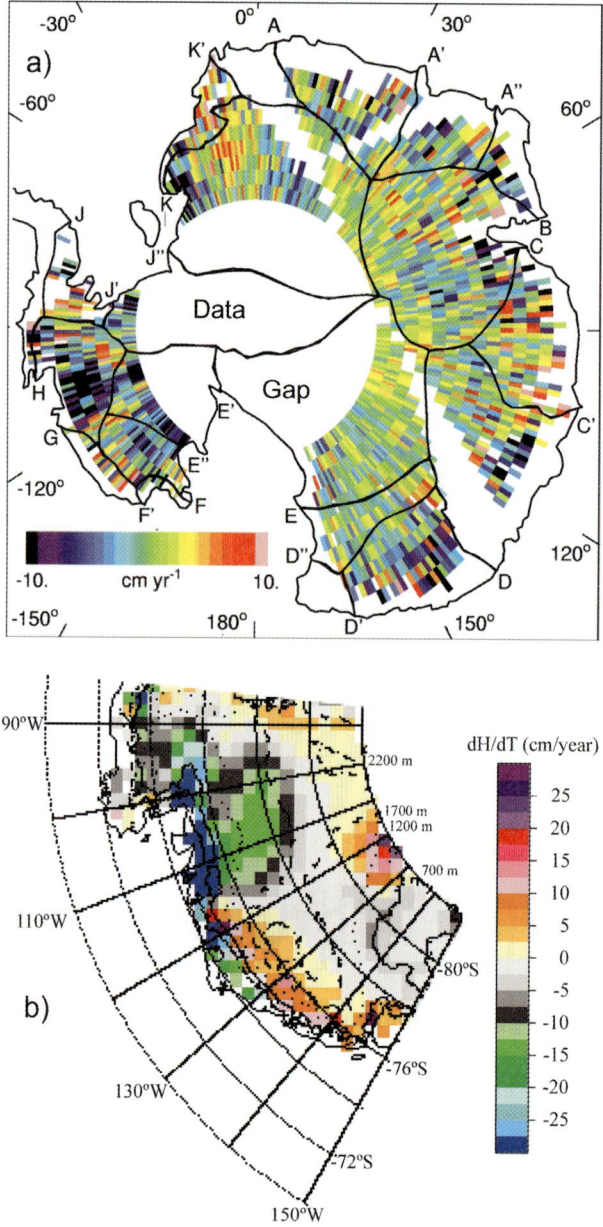

Figure 3.60. (a). Elevation change (in cm per annum) of the Antarctic Ice Sheet from 1992 to 1996 for 63% of the grounded Antarctic Ice Sheet, measured by the radar altimeters onboard ERS-1 and -2. The spatial resolution is about 1° by 1°. The datagap south of ∼82°S results from the orbital inclination of the ERS satellites. The dark lines are the boundaries of the major drainage basins, derived from ERS observations (Bamber and Huybrechts, 1996). (b) Map of elevation change of part of the West Antarctic Ice Sheet, derived from ERS-1 and -2 radar altimeter data (1992–1999). Values of $\partial H/\partial t$ are derived from multivariate linear and sinusoidal fits to $H(t)$ time series of elevation differences at crossover points within 100 km radius and ±250 m elevation of the grid center, and are mapped on a 50-km grid.

(a) From Wingham et al. (1998). (b) From Zwally et al. (2002a). (a) copyright AAAS 1998, reproduced with permission. (b) copyright Elsevier 2002, reproduced with permission.

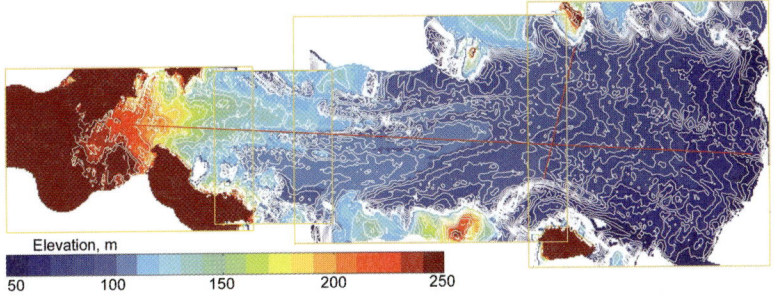

Figure 3.58. Surface elevation of the Amery Ice Shelf, determined from ERS radar altimeter data. Contour intervals are 2 m for elevations of 10–110 m, 10 m for 110–210 m, and 20 m for >210 m. The differences between the measured elevations at 73 *in situ* traverse GPS stations on the Lambert Glacier and the corresponding DEM have a mean of 12.2 m (DEM minus station elevation) and an r.m.s. of 14.8 m.

From Fricker et al. (2000a). Copyright International Glaciological Society 2000, reproduced with permission.

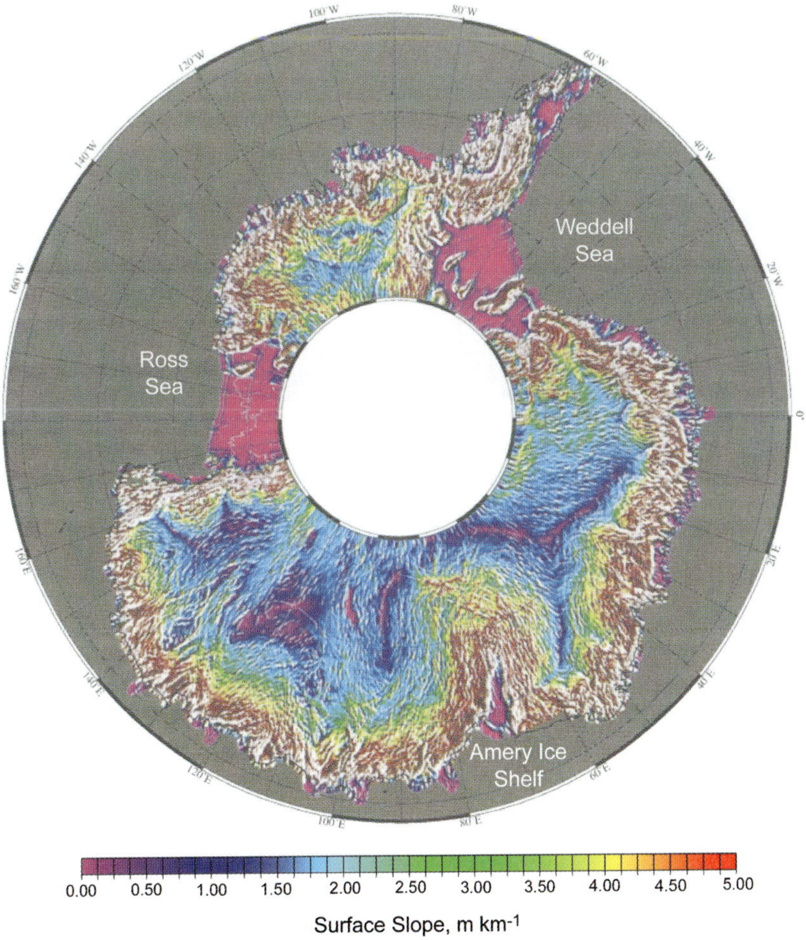

Figure 3.61. High-resolution map of the surface slope of the Antarctic Ice Sheet, derived from data from the ERS-1 Geodetic Mission (from April 1994 to March 1995), and using closely spaced data from the 168-day repeat cycle.

From Rémy et al. (1999). Copyright Blackwell Publishing 1999, reproduced with permission.

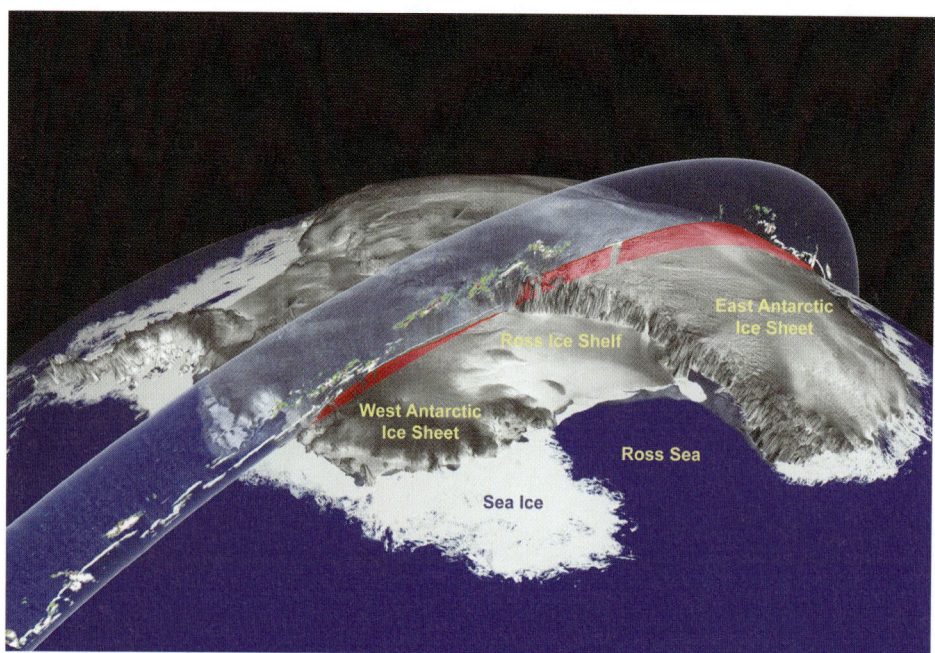

Figure 3.63. An illustration of ice sheet elevation and cloud data from the Geoscience Laser Altimeter System (GLAS) onboard NASA's ICESat on its first day of operation, February 20, 2003. On that day, the instrument collected a 1.064-μm-wavelength profile across Antarctica. The West and East Antarctic Ice Sheets are separated by the steep Transantarctic Mountains. Surface elevation profile (in red) is depicted relative to the Earth's standard ellipsoid, with a 50× vertical exaggeration. Data collected over sea ice and openwater in the adjacent Southern Ocean cannot be shown at this scale. Clouds of various thicknesses are indicated by colors changing progressively from light blue (depicting thin clouds) to white (opaque layers). Note that the laser cannot penetrate the thickest clouds, causing gaps in the elevation profile below. The three-dimensional background map of Antarctica is constructed of SAR imagery from Radarsat-1.

Image is courtesy of NASA (the Radarsat SAR base map is courtesy of the Canadian Space Agency).

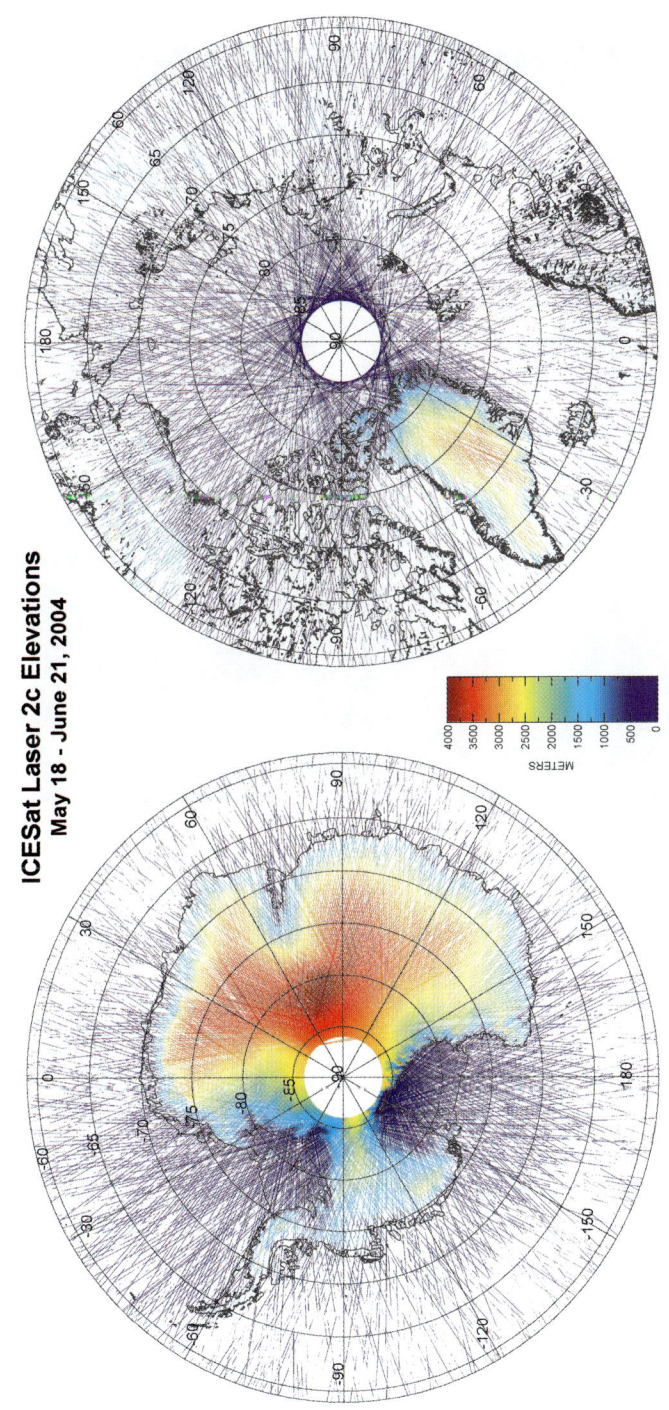

Figure 3.64. Surface elevation measurements of (a) Antarctica and (b) Greenland aquired by the ICESat GLAS during Mission 2c (May 18–June 21, 2004).

Courtesy of Christopher Shuman (NASA Goddard Space Flight Center, Code 614.1, Greenbelt, MD).

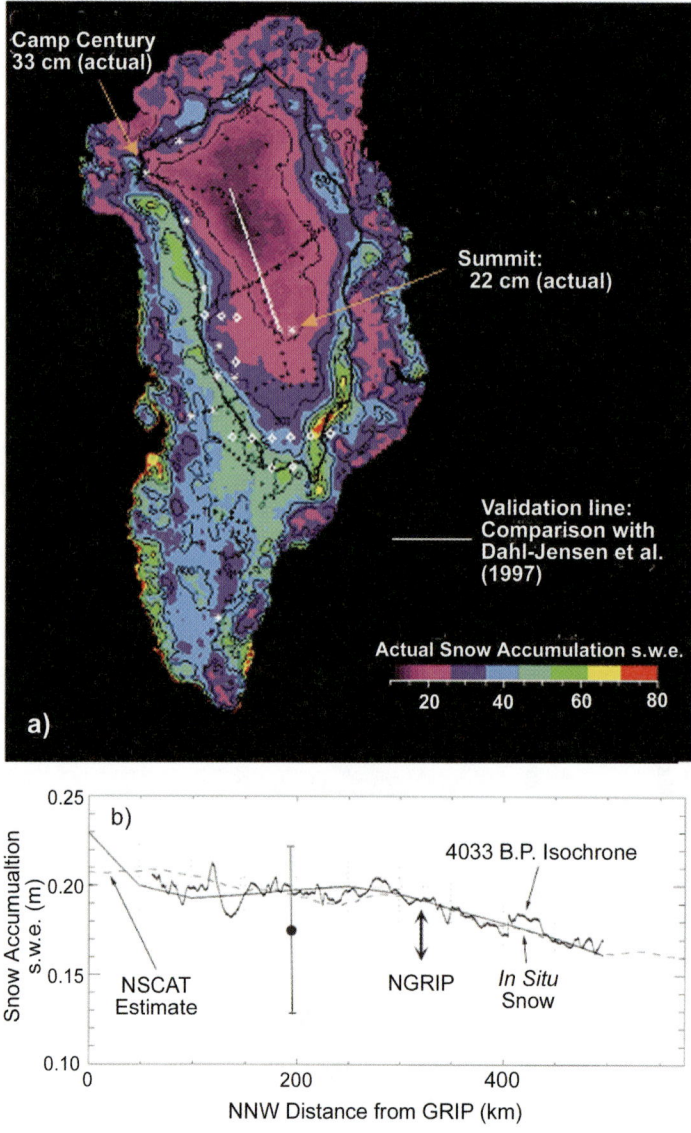

Figure 3.70. (a) Predicted snow accumulation in Greenland from NSCAT B values. Note that the accumulation estimates are valid only within the dry-snow zone, demarcated by the solid black line. This zone was determined from the ERS EScat–NSCAT (C band–Ku band) A difference, where $A = -1.8\,\text{dB}$ was used to define the lower dry-snow zone boundary. s.w.e. = snow-water equivalent. (b) Comparison of NSCAT accumulation retrieval (dashed line) from days 261–266, 1996 along the transect in (a). Dahl-Jensen et al. (1997) volcanic isochrone-inferred accumulation and *in situ* shallow core and snow pit data (after Bolzan and Stroebel, 1994; Clausen et al., 1988; Friedmann et al., 1995; and Ohmura and Reeh, 1991) are shown for comparison.

From Drinkwater et al. (2001). Copyright American Geophysical Union 2001, reproduced with permission.

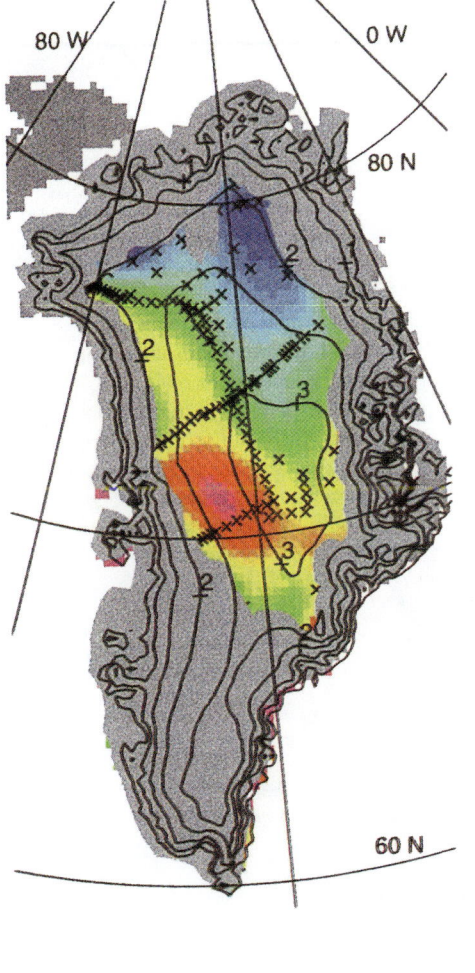
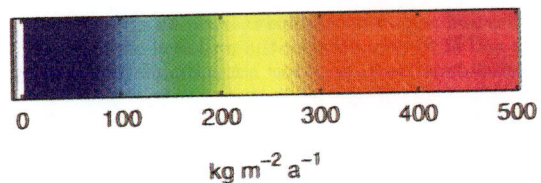

Figure 3.71. A map of the accumulation rate field in the dry-snow zone of Greenland, based on 4.5-cm wavelength Nimbus-7 SMMR observations from 1979 to 1985 and the relationship between the accumulation rate and the polarization of emission at this wavelength. Regions of seasonal melting are masked in gray (after Mote and Anderson, 1995). Elevation contours (in km) are from Ekholm (1996). Crosses mark locations of *in situ* observations of accumulation rate compiled by Ohmura and Reeh (1991).

From Winebrenner et al. (2001). Copyright American Geophysical Union 2001, reproduced with permission.

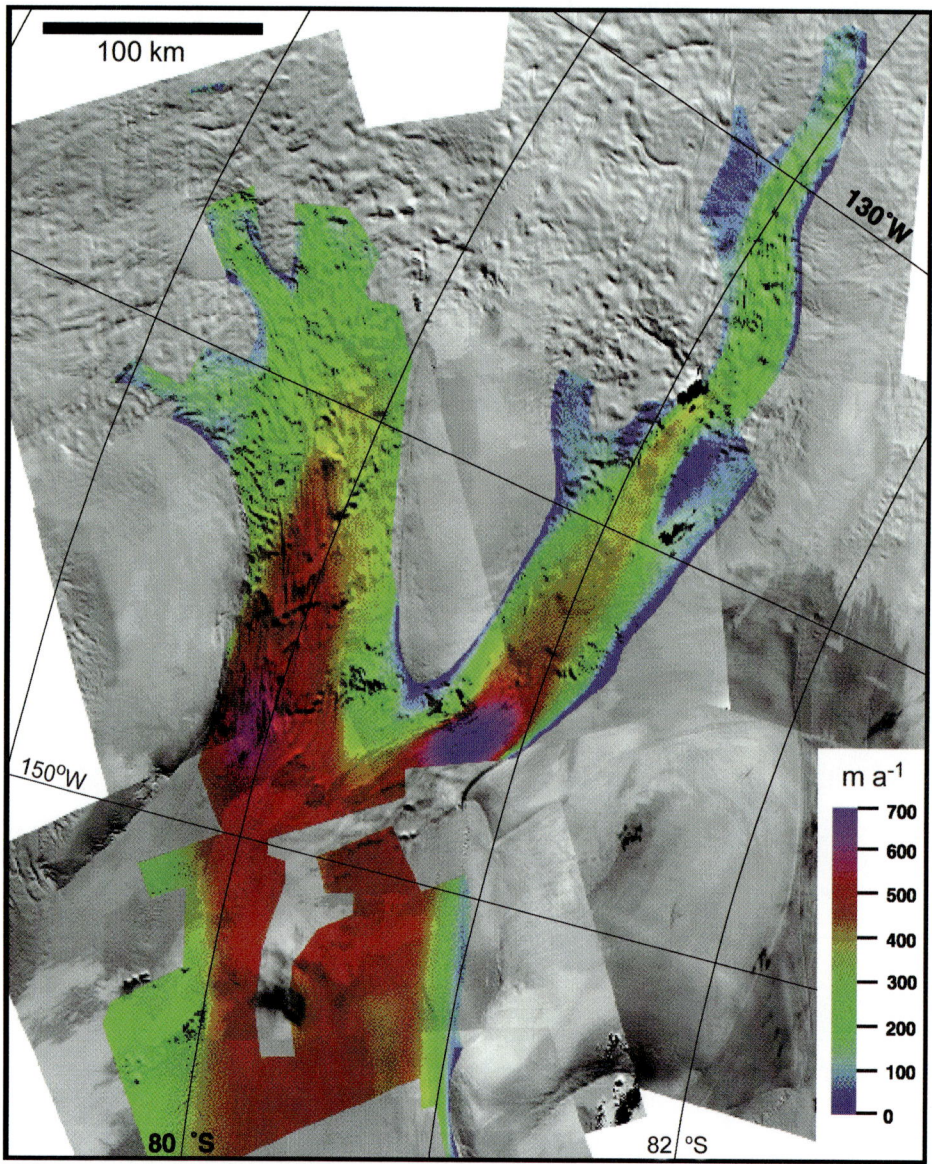

Figure 3.75. Color-coded map of the ice surface velocity magnitude (speed) of Ice Streams Bindschadler (D) and E (West Antarctica), derived from eight pairs of Landsat TM images using the technique outlined in Scambos et al. (1992), and superimposed on the Landsat image mosaic. Data are from the period January 1987 to January 1992.

From Bindschadler et al. (1996). Copyright International Glaciological Society 1996, reproduced with permission.

Figure 3.77. A shaded relief perspective view (looking from the east) of the ice flow field (white arrows) and elevation (blue contours) of Storstrømmen, a large outlet glacier draining the northeastern part of the Greenland Ice Sheet. While conventional InSAR techniques map only the radar line-of-sight component of the ice surface flow, the use of data from differing angles associated with both descending and ascending satellite orbits enables InSAR mapping of the full three-dimensional flow pattern. The deceleration in flow towards the glacier terminus— i.e., to the left of the figure—is contrary to normal outlet glacier flow but typical of a glacier that has recently surged. This map revealed for the first time that the northeastern branch of Storstrømmen, Kofoed-Hansen Brae, is currently a stagnant ice body.

From Mohr et al. (1998). Copyright Nature 1998, reproduced with permission.

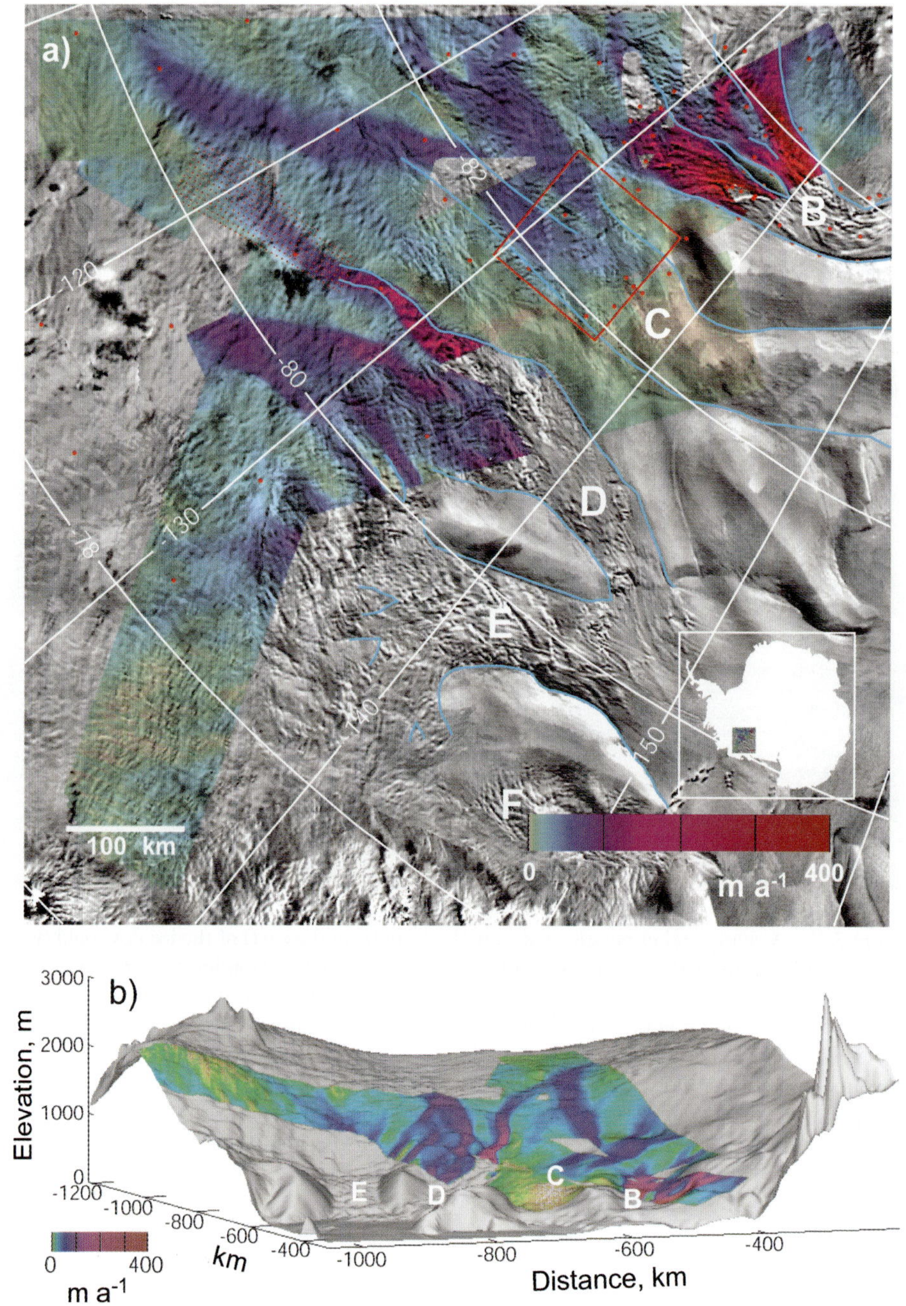

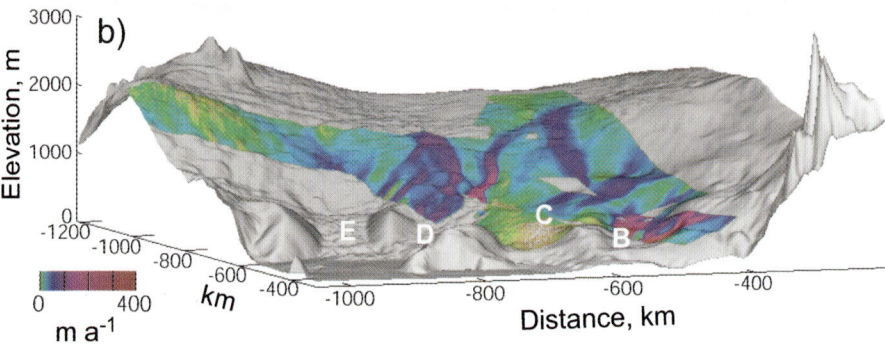

Figure 3.78. (a) Ice flow speed of West Antarctic ice streams (indicated by letter) determined from multiple swaths of Radarsat-1 AMM data co-registered with a mosaic of NOAA AVHRR data. In general, ice flow is from the top left to bottom right. Red dots show the position of *in situ* velocity measurements used for control. Blue lines show previously mapped ice stream margins. Surface velocity fields derived from InSAR and/or speckle tracking are particularly powerful when combined with DEM data. (b) Ice speed superimposed on surface elevation. View is upstream from the Ross Ice Shelf toward the inland ice divide. Part of the Transantarctic Mountains is on the right. The origin is at the South Pole.

From Joughin et al. (1999a). Copyright AAAS 1999, reproduced with permission.

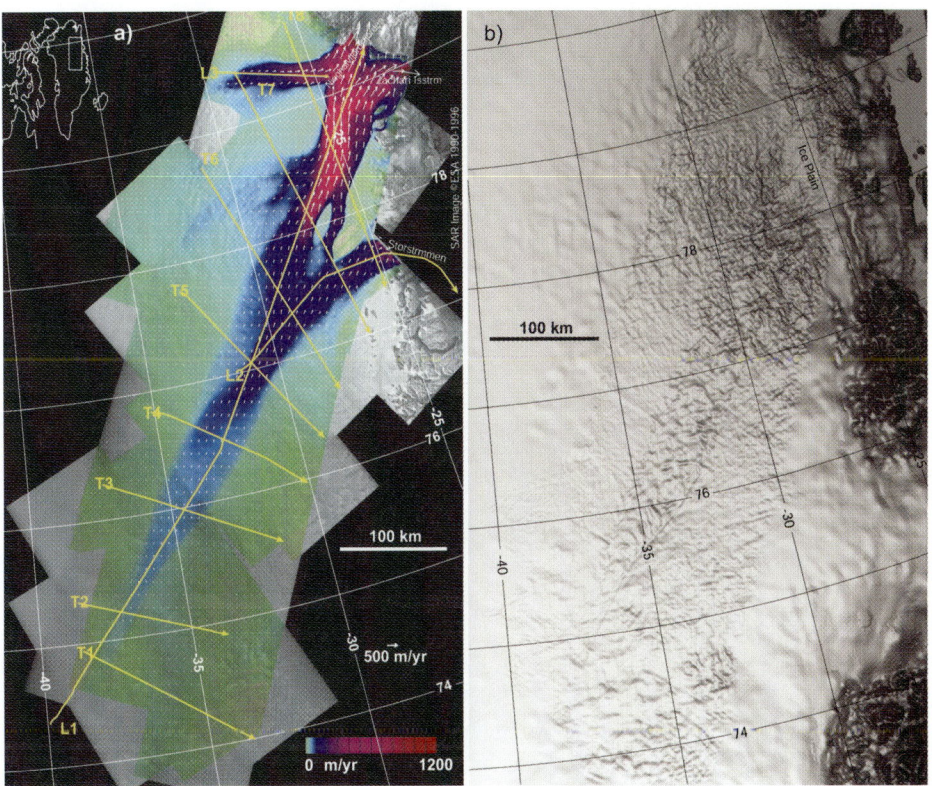

Figure 3.79. (a) An ice velocity map of the Northeast Greenland Ice Stream, derived from InSAR processing of ERS SAR data. These InSAR data were calibrated by balance velocity data, with ground control points providing an important means of validation. Speed is displayed in color over a mosaic of the SAR amplitude imagery, while subsampled velocity vectors are shown as white arrows. (b) A shaded surface Digital Elevation Model (DEM) of the same area. The light source and vantage point are directly overhead. The fine detail, with a horizontal resolution of better than 500 m, is the result of the blending of ERS InSAR (see Section 3.6.4) and radar altimetry data (Bamber et al., 2001). The data elsewhere are primarily from satellite radar altimetry where InSAR data are unavailable. The ice-free coastal data were photogrammetrically derived (Ekholm, 1996).

From Joughin et al. (2001a). SAR image copyright ESA 1990–1996. Copyright American Geophysical Union 2001, reproduced with permission.

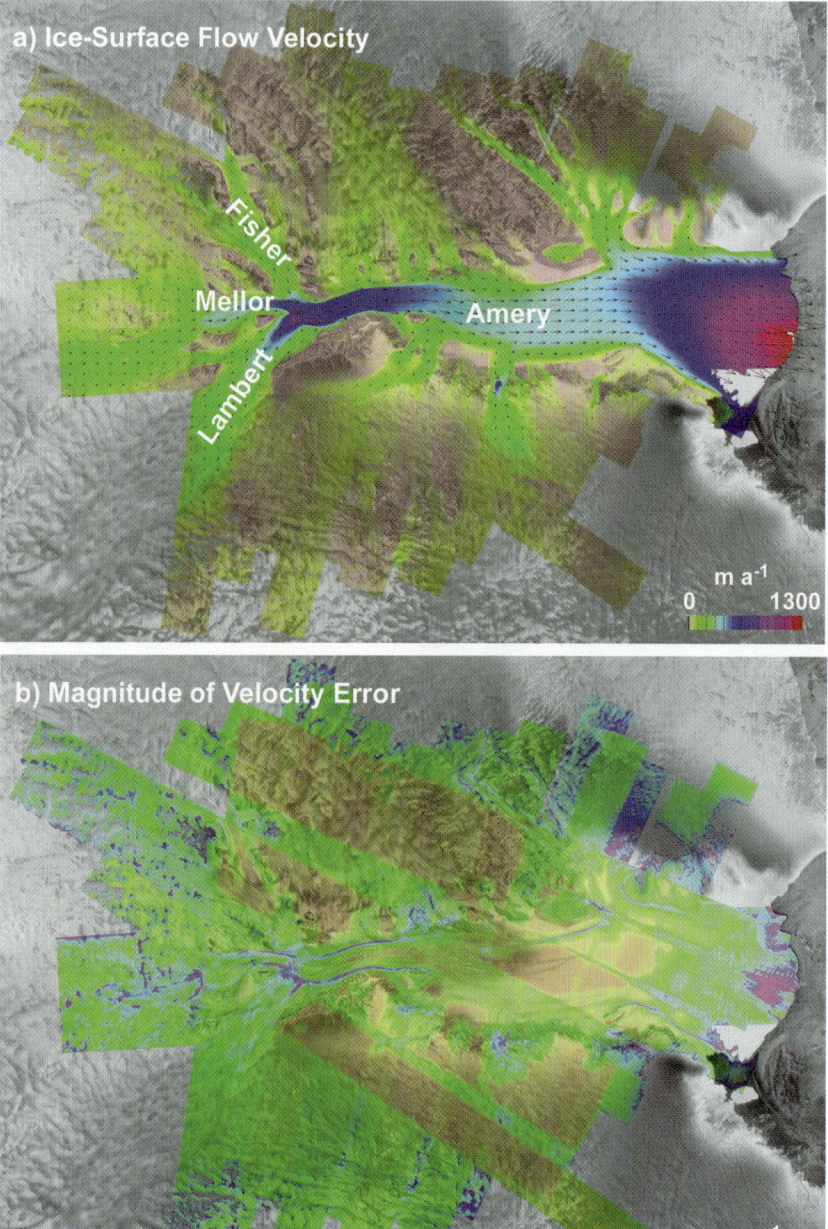

Figure 3.80. (a) Flow velocity of the Amery Ice Shelf, Lambert Glacier, and several other outlet glaciers. Speed is color-coded, and white arrows represent subsampled velocity vectors. The image dimensions are 911 × 687 km, and the projection is polar-stereographic with a standard latitude of 71°S. The map indicates speeds of nearly 750 m per annum just above the grounding zone at the confluence of Lambert, Fisher, and Mellor Glaciers. As the confined shelf widens downstream of the grounding zone, flow speed decreases to roughly 350 m per annum (at 200–300 km from the front). Closer to the front, the ice shelf begins to act more as an unconfined shelf, with speed increasing to over 1,200 m per annum at the front as longitudinal stresses begin to dominate over lateral drag. (b) Map of the magnitude of the velocity error for (a).

Imagery copyright Canadian Space Agency (1997, 2000). From Joughin (2002). Copyright International Glaciological Society 2002, reproduced with permission.

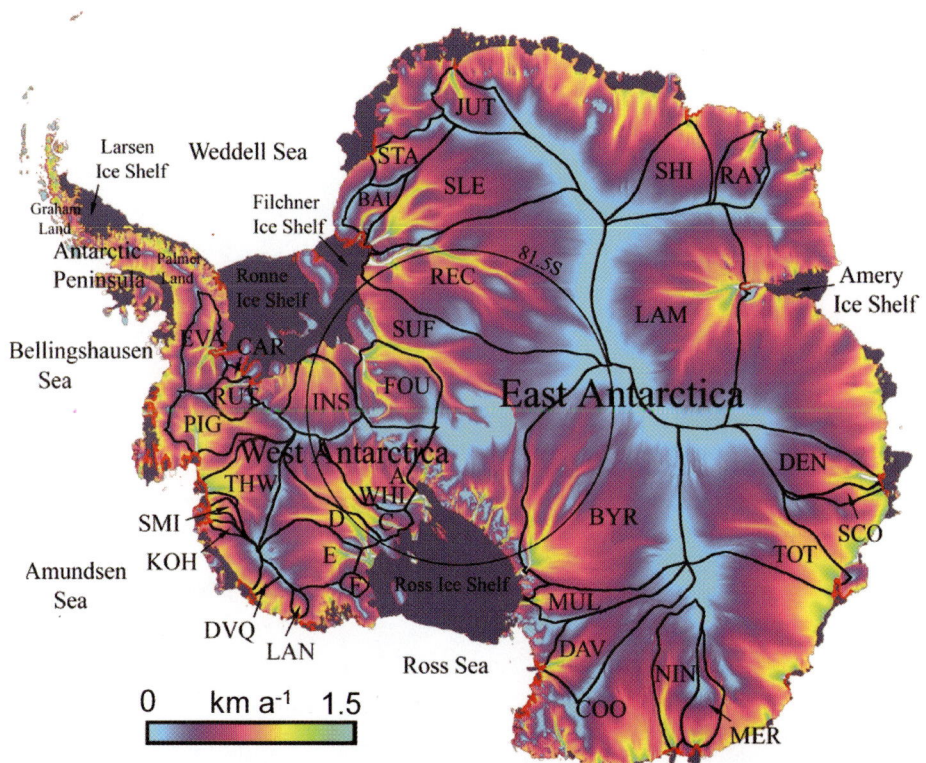

Figure 3.82. A computer model of ice balance velocities of the Antarctic Ice Sheet, based on a digital elevation model that incorporates ERS radar altimeter elevation measurements (between the steep coastal terrain and 81.4°S), and calculated by Bamber et al. (2000a). This invaluable tool indicates an intricate ice movement pattern. Coastal ice flows appear to be fed by complex systems of tributaries that penetrate hundreds of kilometers into the major drainage basins, and the major streams in East Antarctica dwarf the previously known West Antarctic ice streams. The locations of 33 Antarctic glaciers are overlain, with catchment areas marked by black lines. The mass budgets for each glacier are given in table 1 of Rignot and Thomas (2002). The glaciers are: A-F (A-F), Bailey (BAI), Byrd (BYR), Carlson (CAR), feeding eastern Cook Ice Shelf (COO), David (DAV), Denman (DEN), DeVicq (DVQ), Evans (EVA), Foundation (FOU), Institute (INS), Jutulstraumen (JUT), Kohler (KOH), Lambert/Amery/Fisher (LAM), Land (LAN), Mertz (MER), Mulock (MUL), Ninnis (NIN), Pine Island (PIG), Rayner (RAY), Recovery (REC), Rutford (RUT), Shirase (SHI), Slessor (SLE), Smith (SMI), Stanscomb-Wills (STA), Support-Force (SUF), Thwaites (THW), and Whillans (WHI).

From Rignot and Thomas (2002). Copyright AAAS 2002, reproduced with permission.

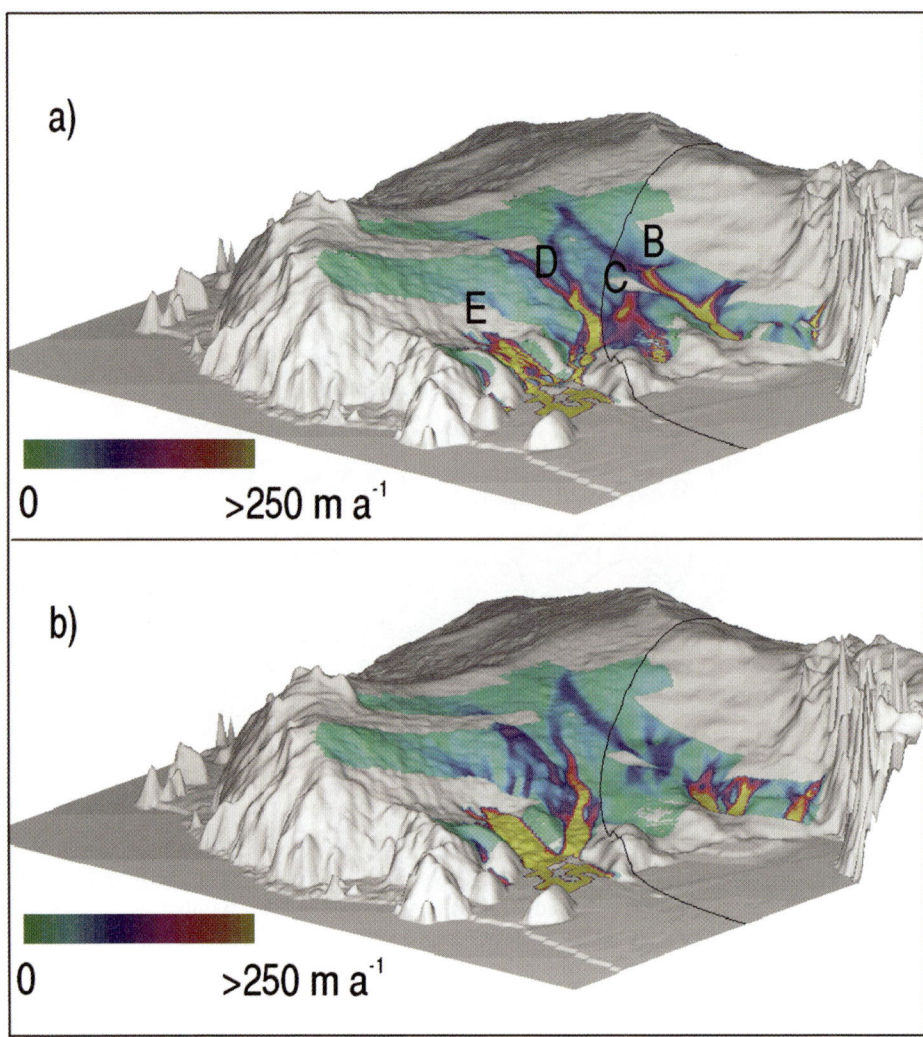

Figure 3.83. (a) Balance velocities for part of the Siple Coast, West Antarctica, showing parts of ice streams Whillans (B), C, Bindschadler (D), and E. (b) Surface velocities derived from Landsat feature tracking and from Radarsat-1 SAR data using speckle tracking and InSAR techniques combined (Joughin et al., 1999a). The agreement between (a) and (b) disintegrates to the south of 81.4°S (the black line)—i.e., the poleward limit of ERS-1 radar altimeter coverage. In both cases, the velocities are superimposed on a DEM derived from ERS-1 radar altimetry and terrestrial data, the latter poleward of the ERS-1 limit (Bamber and Bindschadler, 1997). The DEM was smoothed to remove short-wavelength undulations over a spatially variable distance of 20 times the ice thickness (Paterson, 1994).

From Bamber et al. (2000b). Copyright AAAS 2000, reproduced with permission.

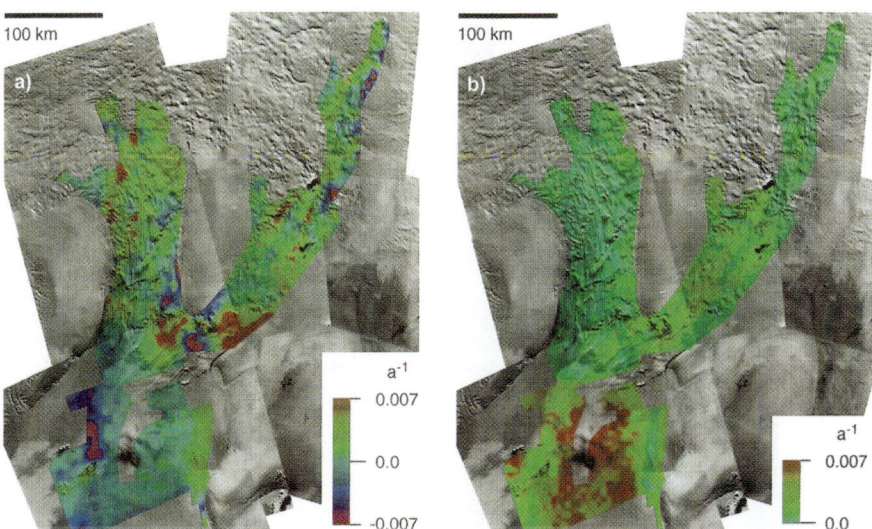

Figure 3.84. (a) Longitudinal strain rate (color-coded) for ice streams Bindschadler (D) and E (West Antarctica), calculated from Landsat-TM-derived ice surface velocity data collected over the period January 1987 to January 1992 (see Figure 3.75). For a glacier, the longitudinal strain rate is the rate of change in velocity, with respect to distance, in the direction of ice flow. (b) The spatial distribution of errors (1 standard deviation). Both are superimposed on the Landsat image mosaic.

From Bindschadler et al. (1996). Copyright International Glaciological Society 1996, reproduced with permission.

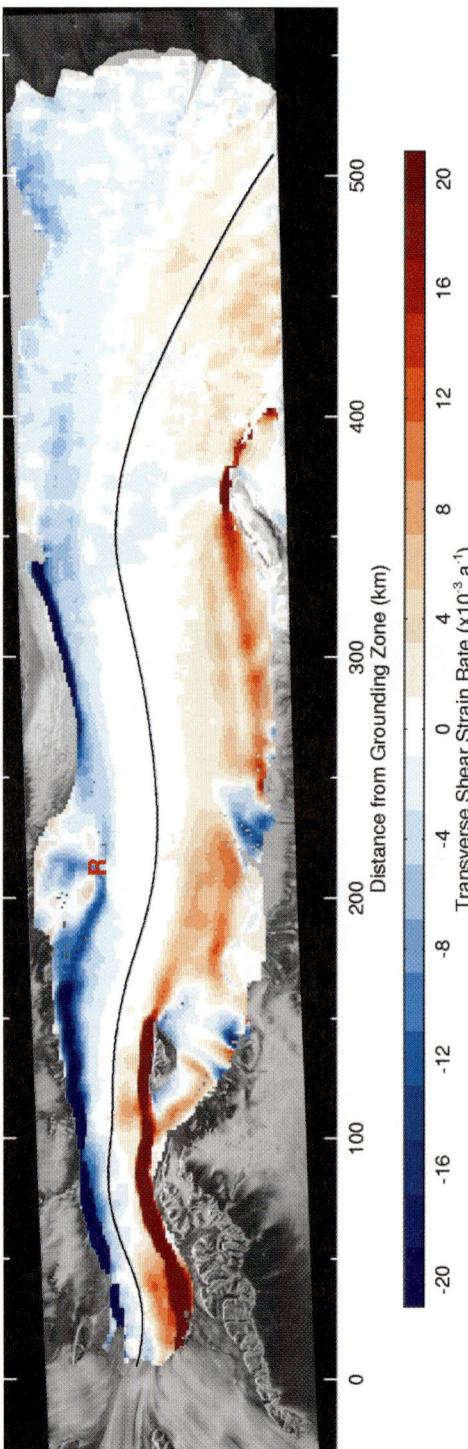

Figure 3.85. The spatial distribution of the transverse shear strain rate over the Amery Ice Shelf (East Antarctica), derived from the SAR maximum-coherence tracking-derived ice velocity data shown in Figure 3.81 and resolved relative to the local flow direction. Vertical strain rates were derived from the transverse and longitudinal values given in Young and Hyland (2002). In this example, strain rates vary systematically across and along the shelf, with strong shear bands at the ice stream margins' near-zero strain rates in the central part of the shelf, and significant rates towards the ice shelf front. The pattern also shows longitudinal bands of enhanced shear strain rate.

From Young and Hyland (2002). Copyright International Glaciological Society 2002, reproduced with permission.

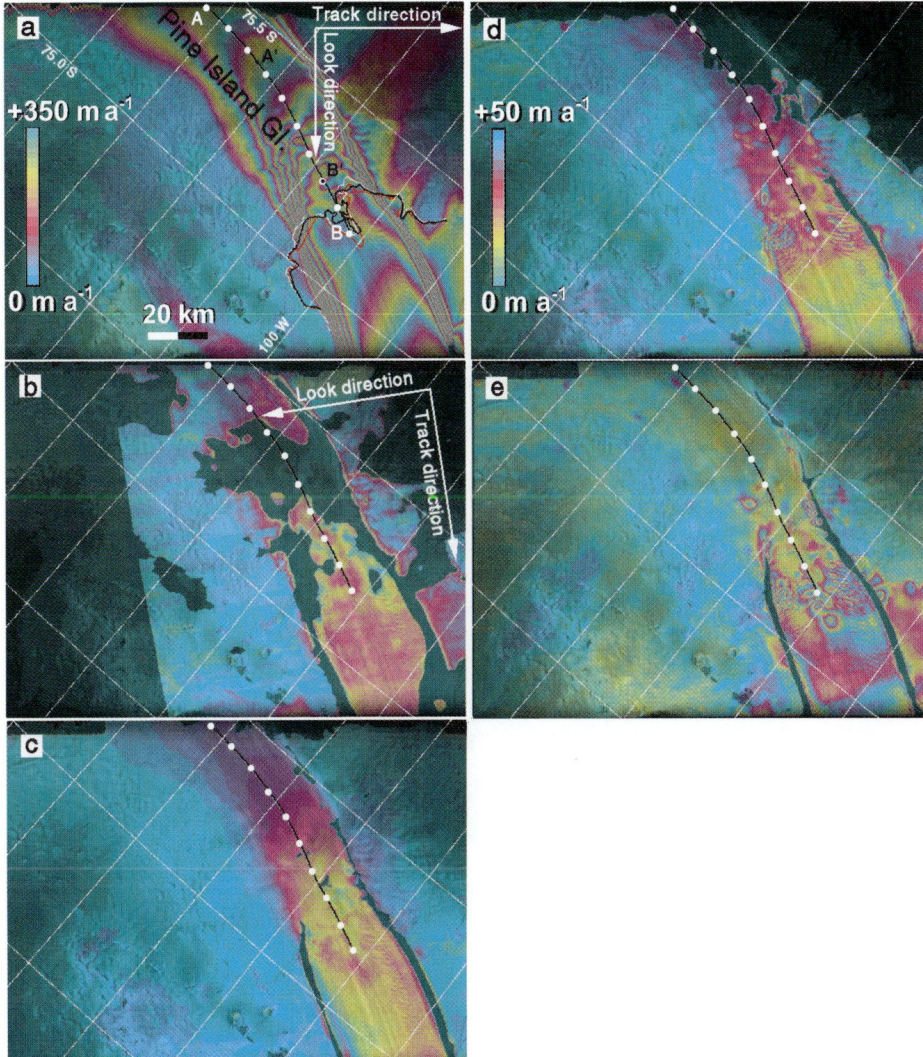

Figure 3.87. (a) Line-of-sight velocity of Pine Island Glacier (West Antarctica), November 11, 1995 (positive is downglacier) derived from InSAR analysis of ERS SAR data, with the grounding line position marked for 1992 (red line), 1996 (white), and 2000 (black). (b) Increase in alongtrack velocity between February 15, 1995 and November 11, 1995 (1,365 days). Increase in line-of-sight velocity measured between (c) November 11, 1995 and November 20, 1999 (1,470 days); (d) November 11, 1995 and February 24, 1996 (105 days); and (e) November 20, 1999 and March 4, 2000 (105 days). Each color cycle—i.e., from blue to red, yellow, and blue again—in (a) to (c) represents a 350 m per annum increment in velocity, and 50 m per annum in (d) to (e).

From Rignot et al. (2002). Copyright International Glaciological Society 2002, reproduced with permission.

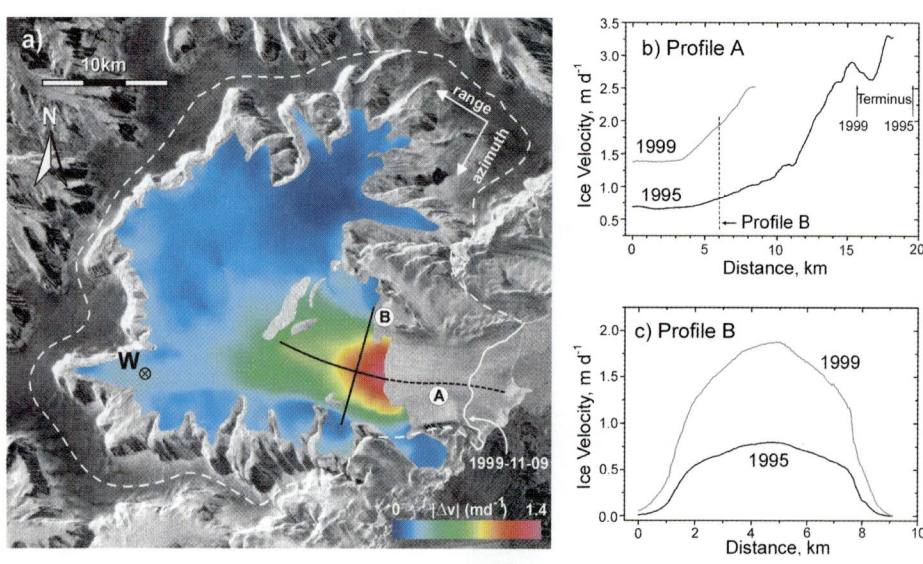

Figure 3.88. (a) Image of the ice velocity difference (in colors) of the Drygalski Glacier, Antarctic Peninsula, superimposed on an ERS SAR amplitude image of November 1, 1995. The velocity component in the acrosstrack direction at the surface is from ERS-derived interferograms from November 9–10, 1999 and October 31–November 1, 1995—i.e., from the ERS Tandem Mission. The white dashed line represents the approximate ice divide, while the continuous white line represents the terminus location on November 9, 1999. Changes in the magnitude of the velocity vector are given along (b) the longitudinal profile A (east-west) and (c) the transverse profile B (north-south), shown in (a).

From Rott et al. (2002). Copyright International Glaciological Society 2005, reproduced with permission.

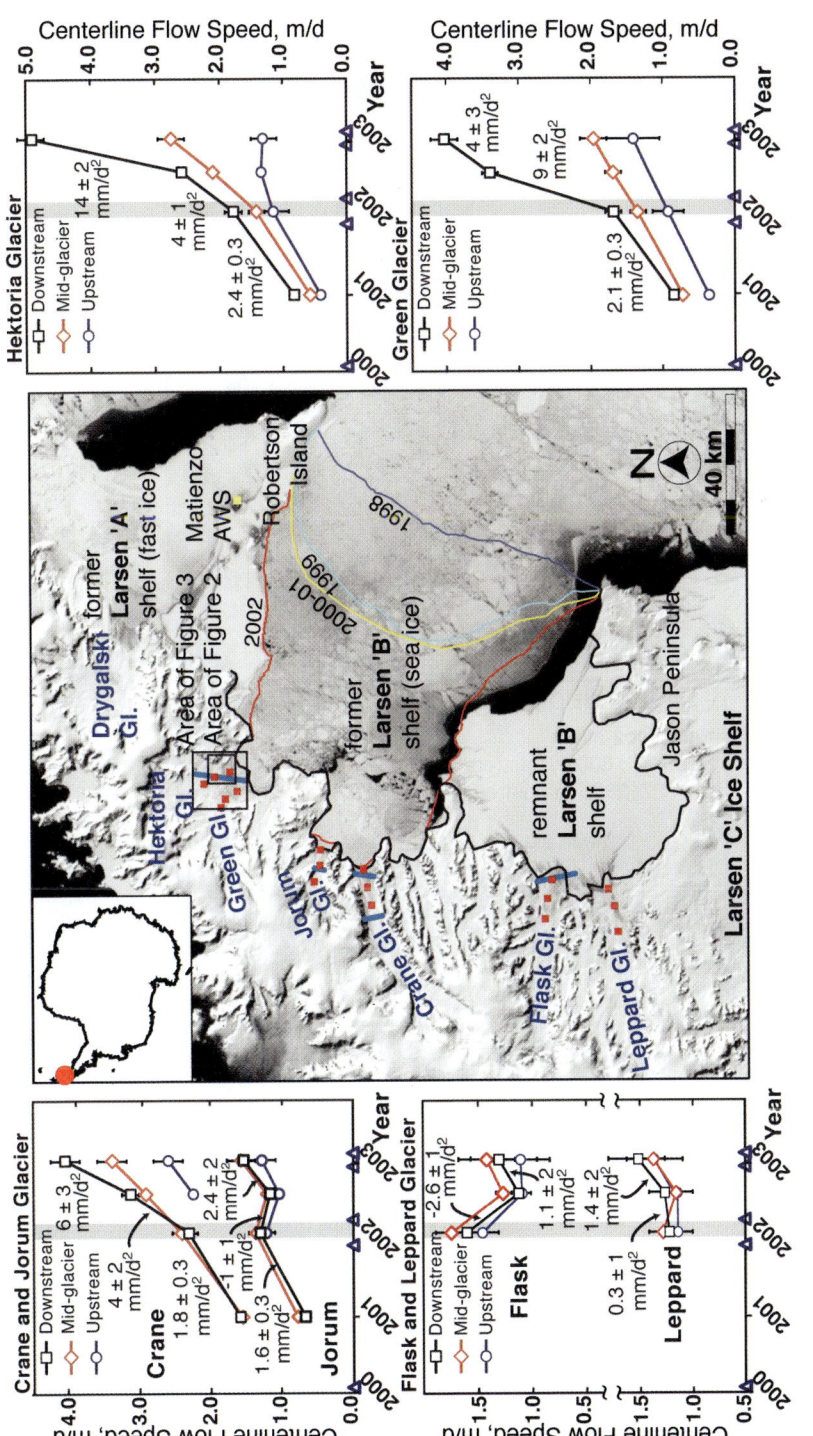

Figure 3.90. The central panel shows a cloud-free MODIS image from November 1, 2003 of the Larsen B Ice Shelf region, with April ice shelf extents and the grounding line marked (the latter in black, from Rack and Rott, 2004) between 1998 and 2002. The red boxes are speed measurement sites along lower-glacier centerlines. Short blue lines show locations of ICESat GLAS groundtracks over glacier trunks. The graphs show the centerline ice speeds and downstream site accelerations for six glaciers. The acquisition dates of Landsat 7 images are shown as triangles on the x-axes. The gray vertical bars represent the Larsen B Ice Shelf collapse event of February–March 2002.

From Scambos et al. (2004a). Copyright Cambridge Journals 2004, reproduced with permission.

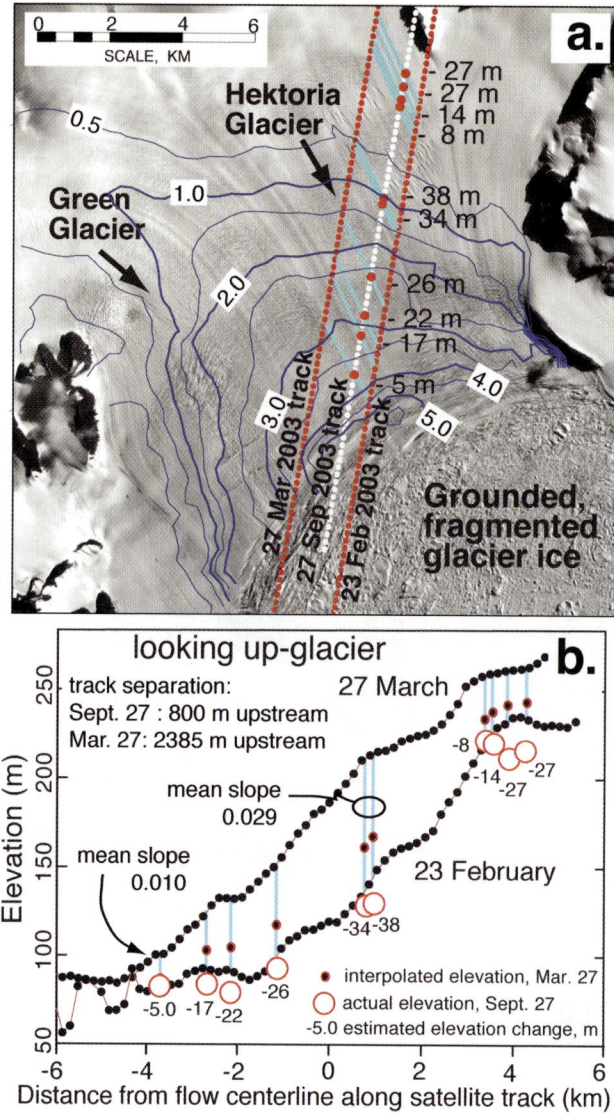

Figure 3.91. (a) Landsat 7 image of the Hektoria Glacier from February 20, 2003, with shotpoint locations from three ICESat tracks overlain. Contours are ice speed difference in meters per day between the periods December 6, 2001 to April 6, 2002 and December 18, 2002 to February 20, 2003. Valid elevation locations are shown as red dots. Blue bars connect three elevations along a flowline. Net elevation difference relative to our estimate is shown as text. (b) ICESat elevation profiles projected upflow. Red circles are sized to the estimated elevation error of ±6 m.

From Scambos et al. (2004a). Copyright Cambridge Journals 2004, reproduced with permission.

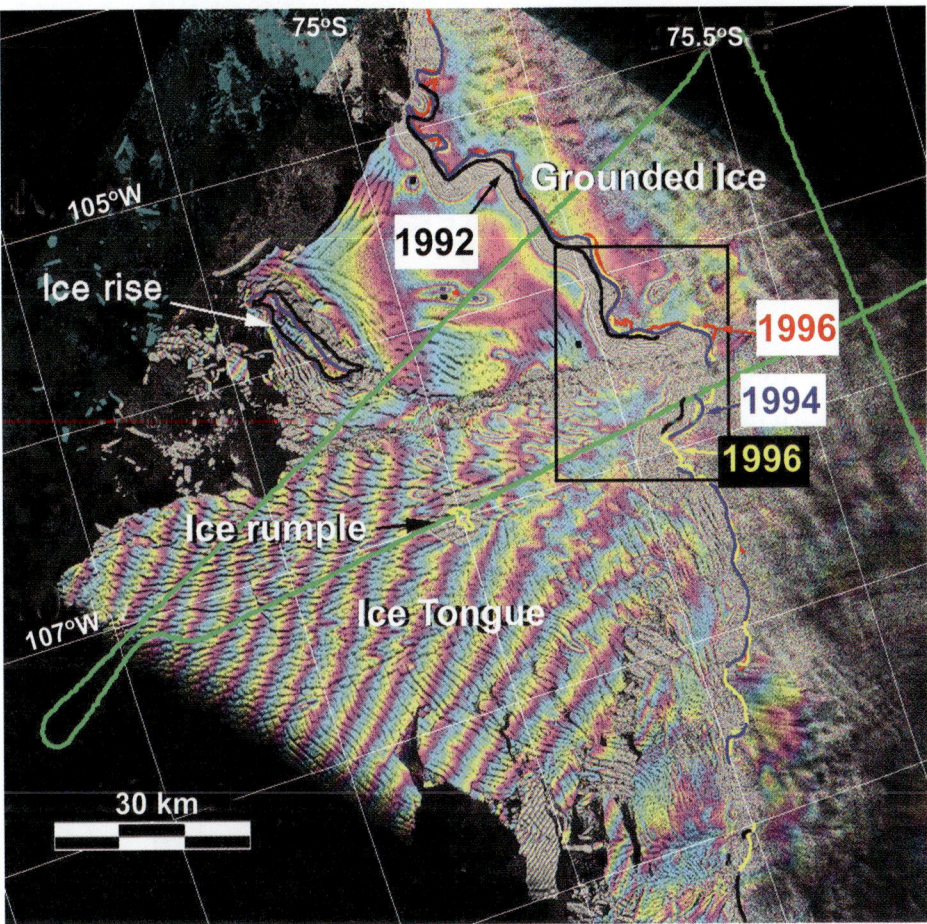

Figure 3.92. An InSAR image of the tidal motion of Thwaites Glacier (Antarctica) from 1996, derived from ERS C-band SAR data. Each color cycle (from blue through purple, yellow, and blue again) represents a 31-mm increment in vertical displacement of the glacier surface. The regular interferometric fringe pattern on the ice tongue is caused by the slow, solid block, horizontal rotation of the ice under the action of tidal currents. The tidal fringes are complex in this example, as a result of the ice flowing over and around series of topographic bumps and hollows. An example of an ice rumple, ~40 km downstream of the grounding line, is marked by an arrow. Grounding line locations inferred from InSAR are marked by lines in black (1992), blue (1994), red (1996a), and yellow (1996b).

From Rignot (2001). Copyright International Glaciological Society 2001, reproduced with permission.

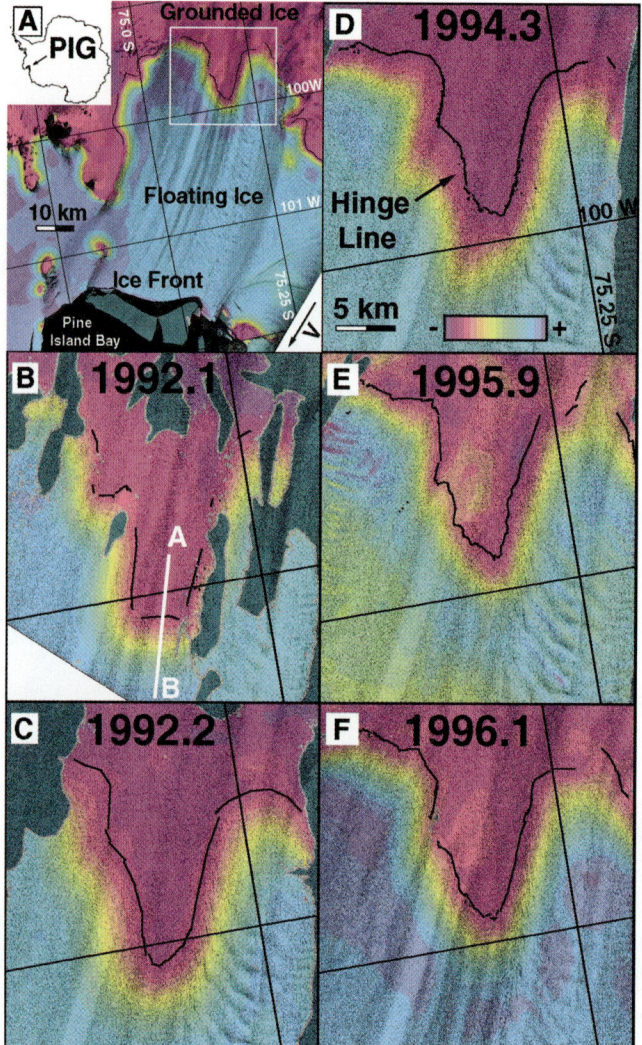

Figure 3.94. Normalized tidal displacements of Pine Island Glacier, West Antarctica, recorded with ERS differential interferometry, and color-coded (colour bar in D) from magenta (grounded ice) to yellow (glacier flexure zone) and blue (ice shelf ice in hydrostatic equilibrium with the ocean waters). The color tone is modulated by the radar brightness of the scene acquired by ERS-1 on January 21, 1996 (orbit 23627, frames 5589 and 5607). The fast-moving portion is revealed by flowline features conspicuous in the radar brightness image. Dark green signifies regions where no interferometric data are available. The hingeline position, retrieved from model fitting, is shown as a black, thin, continuous line separating grounded (magenta, minus sign) from floating ice (blue, positive sign). Its finger-shaped appearance in B to F indicates the presence of thicker ice at the glacier center than along its sides. In A, locally grounded areas or ice rises appear in magenta between the hingeline and the ice front, in areas where the ice shelf is virtually stagnant. Hingeline position and tidal displacements were recorded in: B, January 1992 (ERS-1 orbits 2970, 3056, and 3142; frame 5589); C, February 1992 (ERS-1 orbits 3260, 3346, and 3432; frame 5211); D, March 1994 (ERS-1 orbits 13826, 13869, and 13912; frame 5211); E November 1995 and January 1996 (ERS-1/2 orbit pairs 22614/2941 and 23616/3943; frame 5211); and F, January and February 1996 (ERS-1/2 orbit pairs 23627/3954 and 24128/4455; frame 5589). Between B and F, the hingeline retreated 5.0 ± 1.0 km in 3.78 years over a 275-m-wide region at the glacier center.

From Rignot (1998c). Copyright AAAS 1998, reproduced with permission.

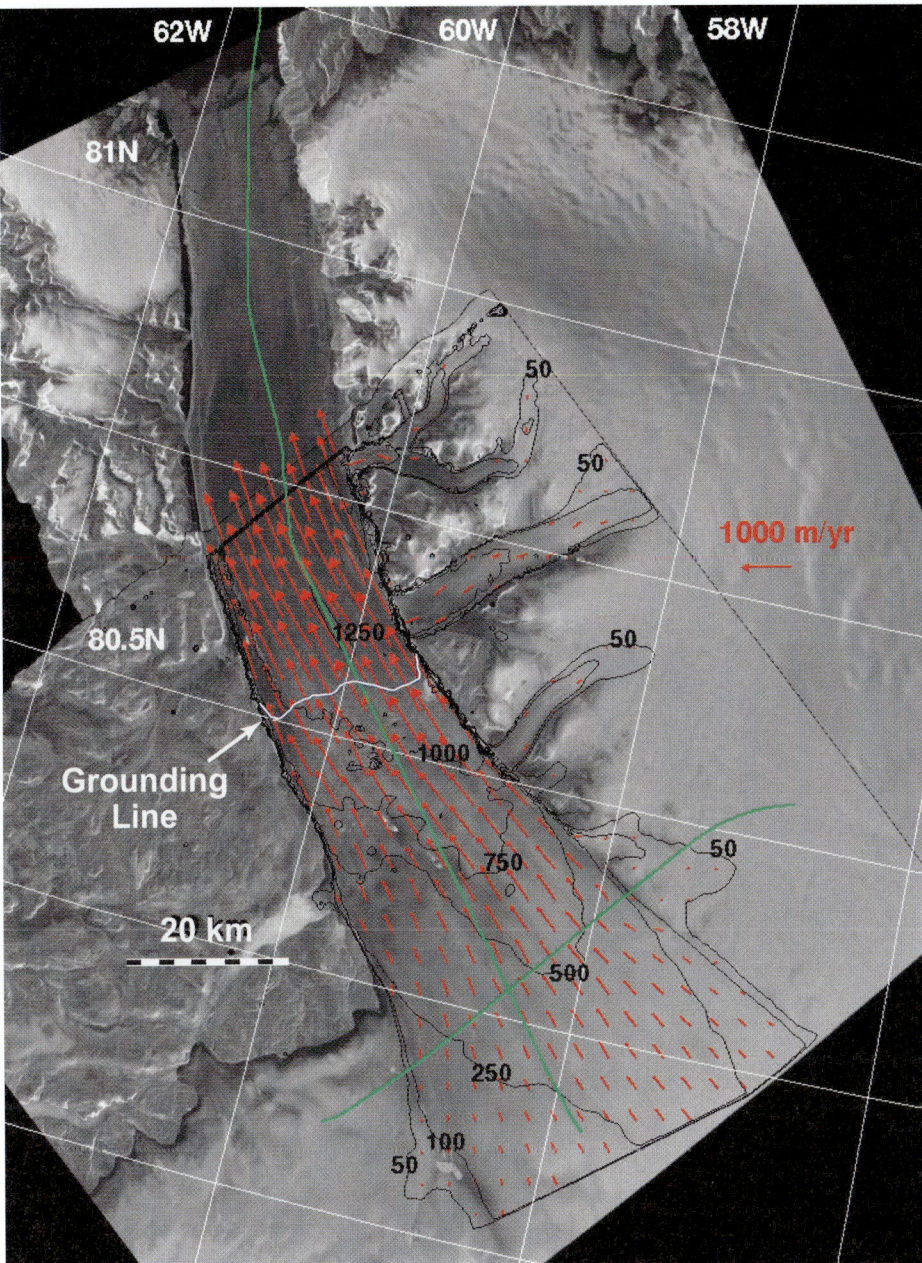

Figure 3.95. A vector velocity map of Petermann Gletscher (NW Greenland), derived from ERS InSAR ascending and descending track data, and superimposed on a SAR amplitude image. Flow vectors are in red, and velocity contours in blue. Ice-sounding radar thickness transects are marked as green lines. The grounding line is marked in black for 1992 and white for 1996. In this case, single-difference interferograms were used, assuming that ice flows parallel to the ice-sheet surface. For the floating part of the glacier—i.e., seaward of the grounding line—tidal motion was removed using the method described by Rignot et al. (2000b).

From Rignot et al. (2001). Copyright American Geophysical Union 2001, reproduced with permission.

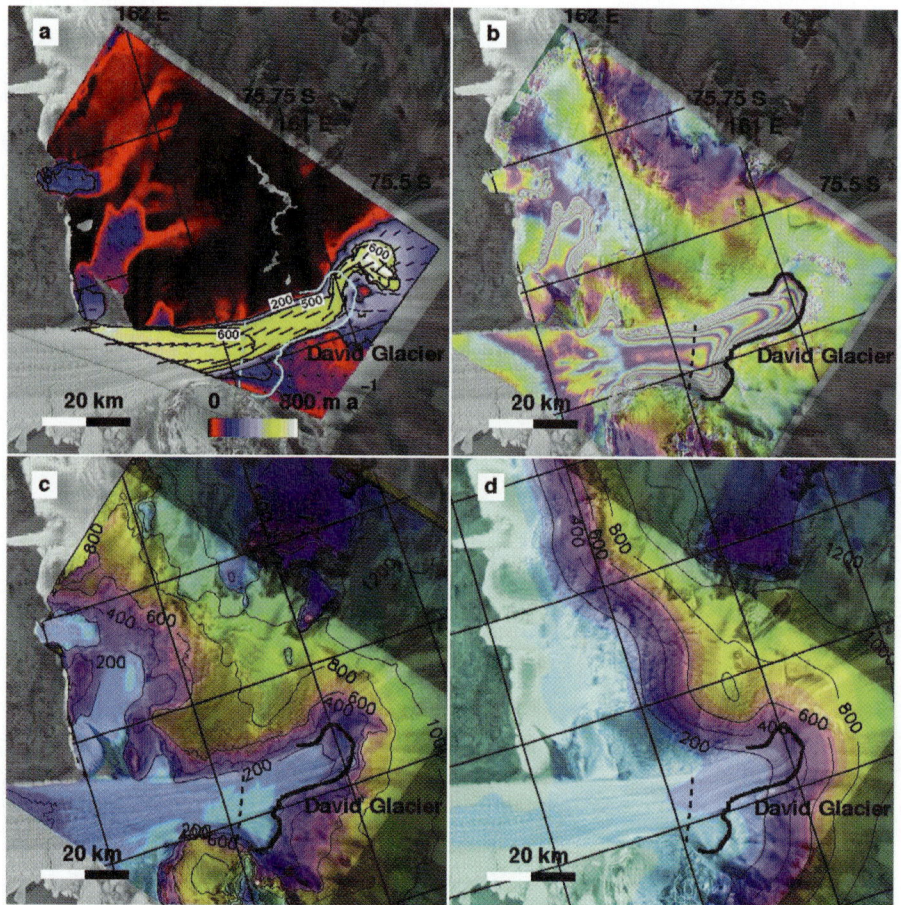

Figure 3.96. Satellite-derived products used in the computation of the mass balance of the David Glacier, Victoria Land (Antarctica): (a) velocity vector map (m per annum) derived from ERS InSAR data; (b) grounding line location and tidal motion derived from ERS InSAR, where each color cycle represents a 31-mm increment in surface vertical displacement due to changes in oceanic tide; and surface topography (200-m contour interval) derived from (c) ERS InSAR and (d) ERS radar altimetry. The grounding line and basal melt flux gates are shown in blue and dotted blue lines, respectively, in (a), and black and dotted black lines in (c) and (d). Latitude is plotted every 0.25°, and longitude every 1°.

From Rignot (2002a). Copyright International Glaciological Society 2002, reproduced with permission.

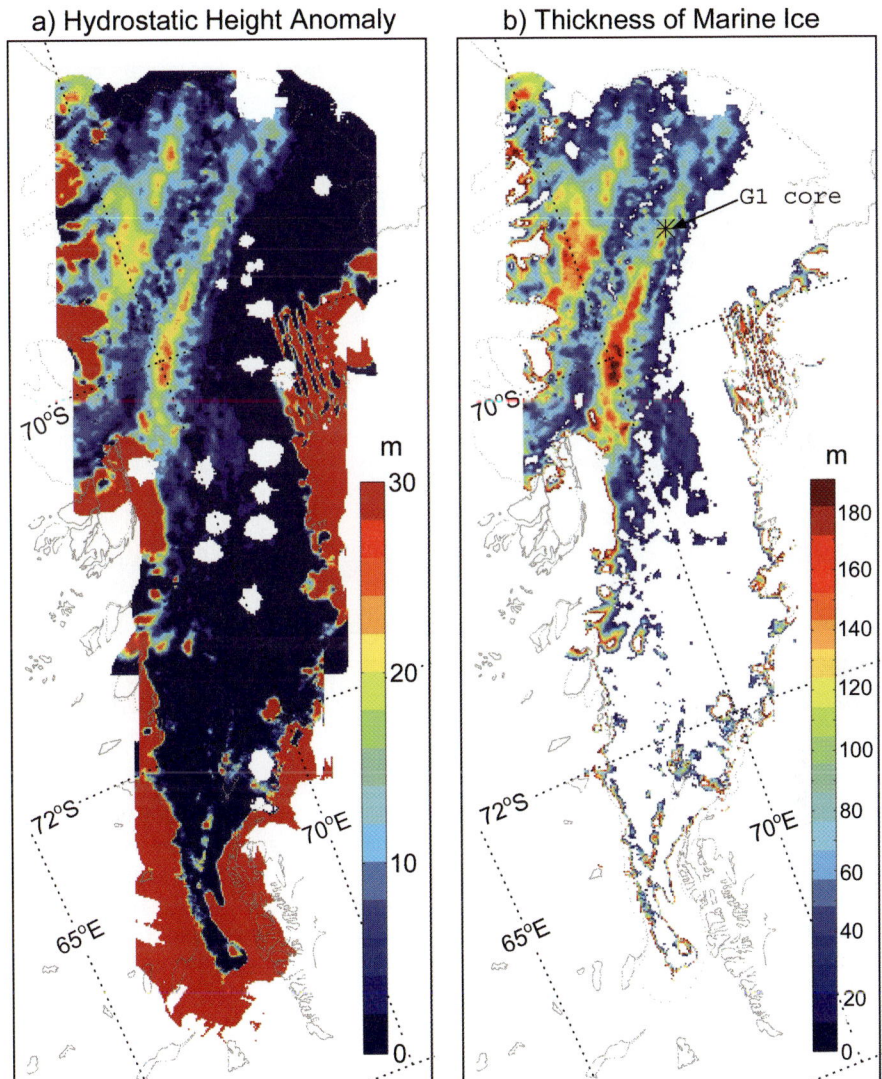

Figure 3.97. Distribution maps of (a) the hydrostatic height anomaly, and (b) the thickness of marine ice computed for the Amery Ice Shelf, Antarctica. Gray lines represent the base map of the region, and white regions in (a) correspond to null values.

From Fricker et al. (2001). Copyright American Geophysical Union 2001, reproduced with permission.

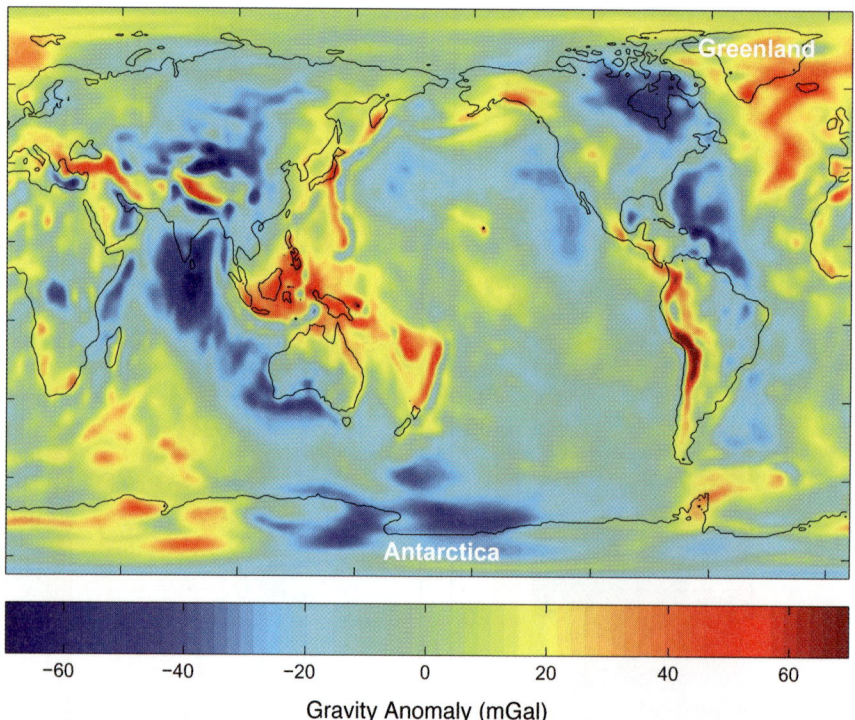

Figure 3.98. A map of the Earth's gravity anomalies based on 111 days of data collected during the commissioning phase of the GRACE data. This figure confirms that GRACE data are homogeneous, global, and accurate. Taken overall, the Earth's gravity field is determined by how the materials making up the Earth are distributed. As gravity changes over the surface of the Earth, so the weight of an object changes along with it. *Gravity anomaly* is a measure of how actual gravity deviates from *standard gravity*, which is the gravity value for a perfectly smooth and idealized Earth. A milligal (mGal) is a convenient unit for describing gravity variations over the Earth's surface, where $1\,\text{mGal} = 0.00001\,\text{m}\,\text{s}^{-2}$. This can be compared with the total gravity on the Earth's surface of $\sim 9.8\,\text{m}\,\text{s}^{-2}$. A key feature of a map of gravity anomalies such as this is that it tends to highlight shortwavelength features significantly better than is the case with a map of the geoid.

Image courtesy of NASA, NASA Jet Propulsion Laboratory, Center for Space Research (University of Texas), and GeoForschungsZentrum (GFZ) Potsdam.

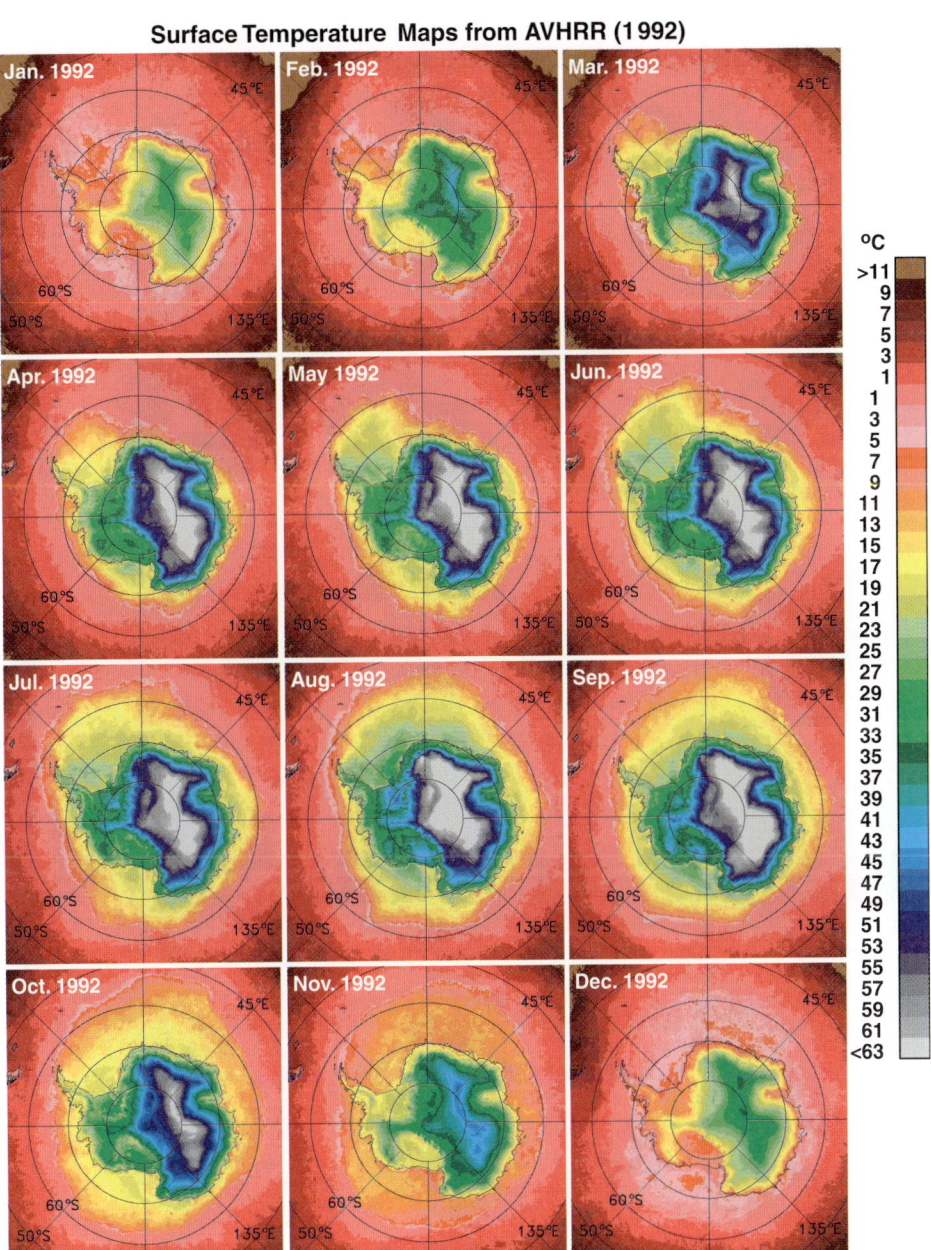

Figure 3.99. Color-coded maps of Antarctic mean monthly surface temperatures for 1992 derived from NOAA AVHRR data.

From Comiso (2000). Copyright American Meteorological Society 2000, reproduced with permission.

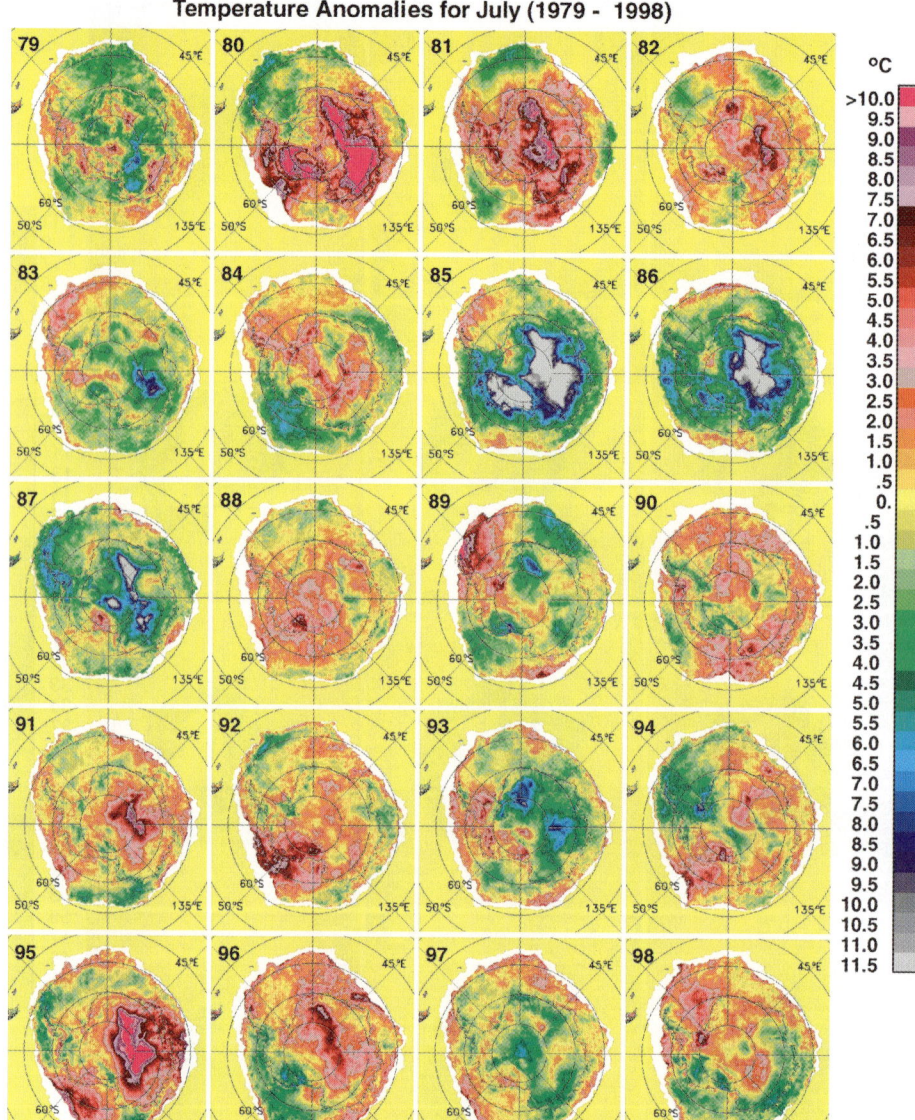

Figure 3.100. Color-coded monthly mean anomaly maps of Antarctic surface temperatures in July from 1979 to 1998, derived from NOAA AVHRR data by subtracting the long-term average (1979–1998) from each month.

From Comiso (2000). Copyright American Meteorological Society 2000, reproduced with permission.

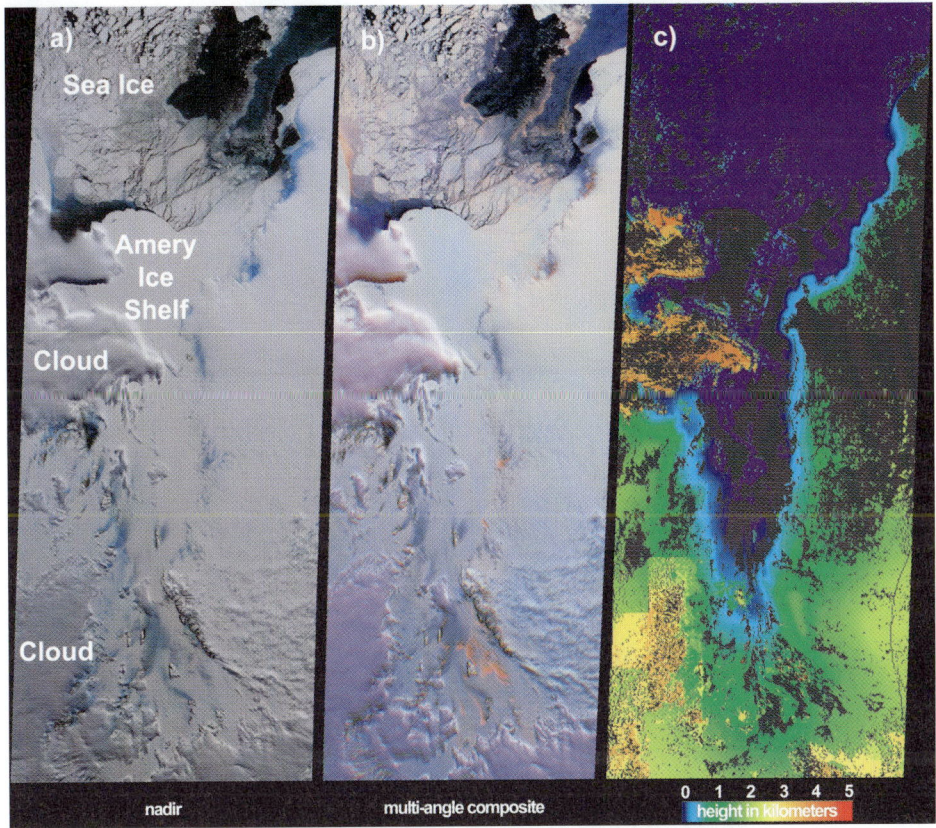

Figure 3.101. Views from the Multi-angle Imaging Spectro Radiometer (MISR) of the Amery Ice Shelf–Lambert Glacier system in East Antarctica on October 25, 2002, illustrating ice surface textures and cloud top heights. The left-hand panel (a) is a natural-color view from MISR's downward-looking (nadir) camera, while the central panel (b) is a multi-angular composite from three MISR cameras, in which color acts as a proxy for angular reflectance variations related to texture. Here, surfaces which predominantly exhibit forward-scattering (generally smooth surfaces) appear blue, while those mainly exhibiting backward-scattering (generally rough surfaces) appear red/orange. Textural variation for both the ice sheet and sea ice are also apparent. The red/orange pixels in the lower portion of the image correspond with a rough and crevassed region near the grounding zone. In the natural-color view, this rough ice is spectrally blue in color. Clouds exhibit both forward- and backward-scattering properties in the middle panel and thus appear purple, in distinct contrast with the underlying ice and snow. An additional multi-angular technique for differentiating clouds from ice is shown in the right-hand panel (c), which is a stereoscopically derived height field retrieved using automated pattern recognition involving data from multiple MISR cameras. Areas exhibiting insufficient spatial contrast for stereoscopic retrieval are shown in dark gray. Clouds are apparent as a result of their heights above the surface terrain. These data products were generated from a portion of the imagery acquired during EOS Terra orbit 15171. The panels cover an area of 380×984 km, and utilize data from blocks 145 to 151 within World Reference System-2 path 127.

Image courtesy of NASA/GSFC/LaRC/JPL, MISR Team ⟨http://eosweb.larc.nasa.gov/HPDOCS/misr/misr_html/lambert_amery_system.html⟩), Clare Averill, David J. Diner (NASA Jet Propulsion Laboratory), and Helen A. Fricker (Scripps Institution of Oceanography).

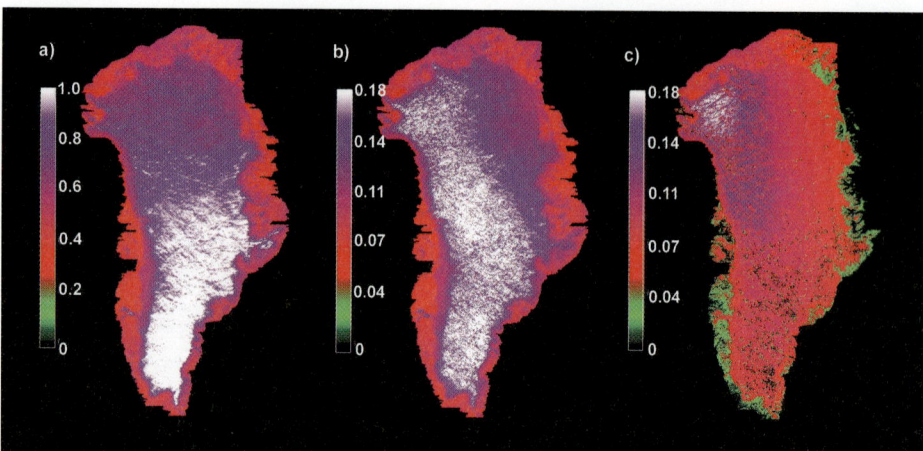

Figure 3.105. Monthly averaged surface albedo for August 1989 over the Greenland Ice Sheet and derived from cloud-free NOAA AVHRR visible and near-infrared radiances: (a) derived assuming that the snow surface is Lambertian; (b) retrieved from apparent surface reflectance using the DISORT model BRDF conversion technique (Stamnes et al., 1988); and (c) the difference between (a) and (b).

From Diner et al. (1999). Copyright American Meteorological Society 1999, reproduced with permission.

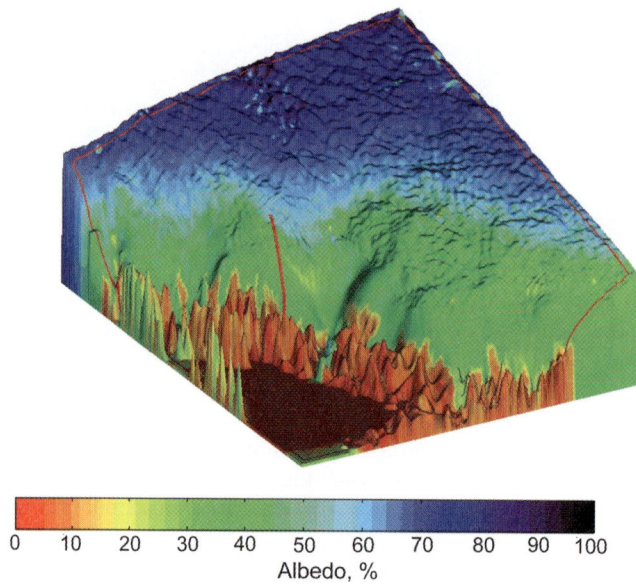

Figure 3.106. A three-dimensional map of the surface albedo for the ablation zone of the Jakobshavn Isbrae (Glacier) region of West Greenland on July 29, 1997 from the AVHRR Polar Pathfinder (APP) 1.25-km products (obtainable from ⟨http://nsidc.org/data/nsidc-0065.html⟩, Scambos et al., 2000]), and overlain on a DEM (data from ⟨http://nsidc.org/data/nsidc-0052.html⟩, Fahnestock et al., 1997). Large variations in the surface albedo are observed with distance from the dry-snow region towards the margin of the ice sheet. Several meltponds are evident in the image, with surface albedo near 20%. Also shown in the image is a transect (red line) extending from the margin of the ice sheet through three automatic weather stations towards the Swiss ETH/University of Colorado camp (at an elevation of 1,149 m above mean sea level). By extending from the ablation zone to the accumulation zone, these stations provide important data for comparison with satellite-derived albedo measurements. This in turn enables an assessment of the relative importance of the surface albedo in the ice sheet energy balance, and also provides key information on ablation rates.

Courtesy of Julienne Stroeve (NSIDC, University of Colorado).

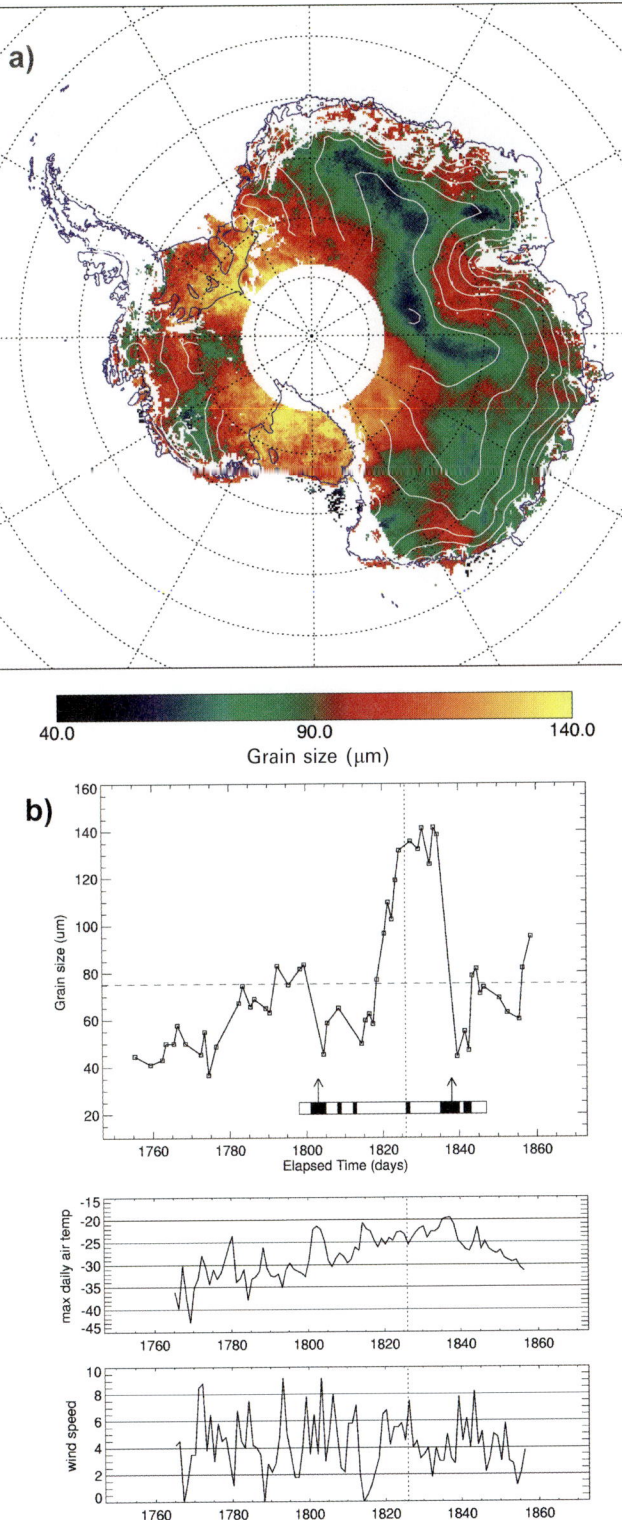

Figure 3.107. (a) The spatial variability of snow surface grain size on the Antarctic Ice Sheet derived from ERS-2 ATSR-2 radiance data, and (b) its temporal variability at Dome C in East Antarctica (74.5°S, 119.5°E, elevation 3,280 m), with plots of coincident wind speed and maximum daily air temperature at the near-surface.

From Young et al. (in prep.).

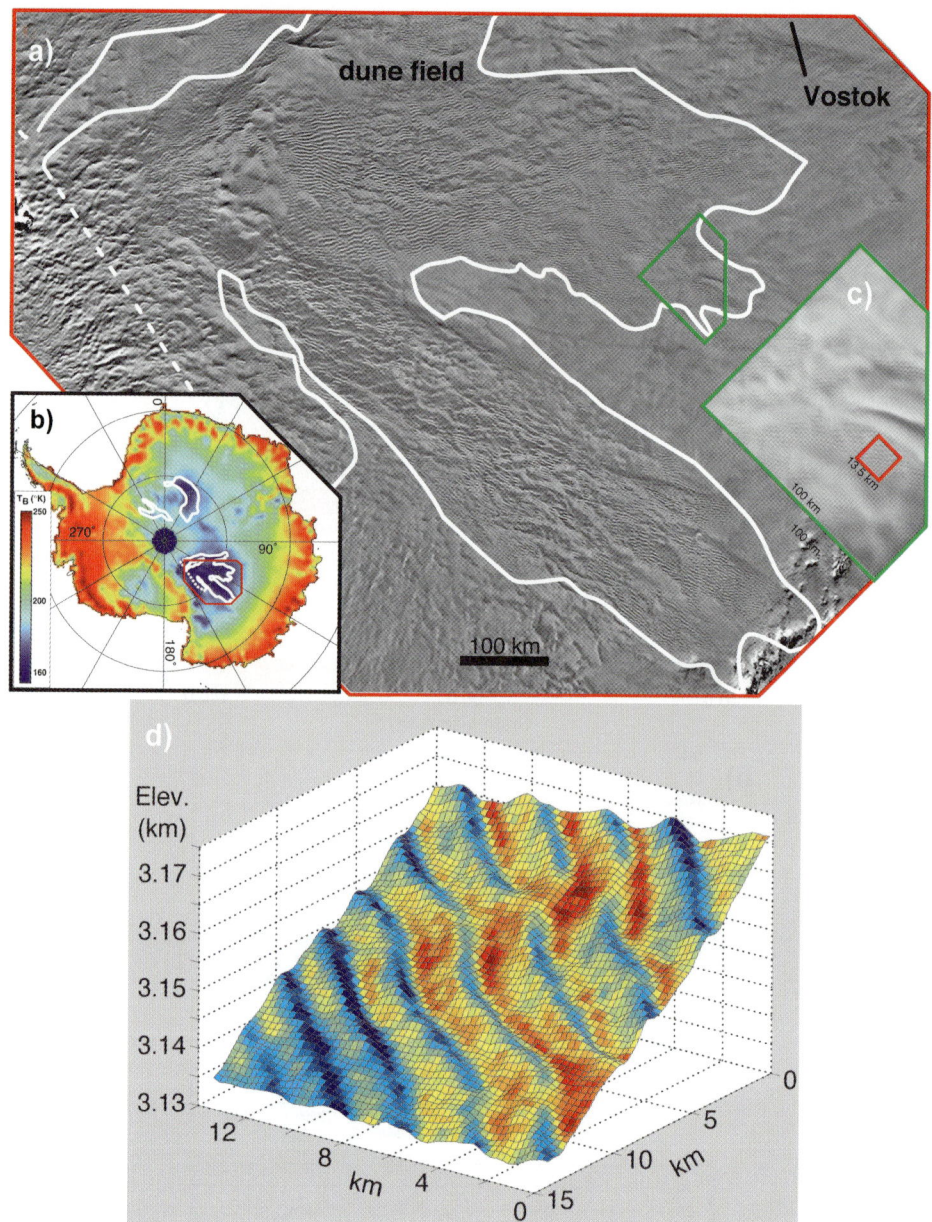

Figure 3.108. (a) A cloud-free AVHRR visible band image of a large Antarctic megadune field. (b) Locations of megadune fields identified in satellite imagery, plotted on an image of the annual average of thermally driven microwave emission from the snowpack at 37-GHz vertical polarization (DMSP SSM/I). (c) A 13.5×13.5-km ERS-1 SAR image of megadunes from the region outlined in red in (a). (d) A perspective view of the area outlined in red in (c). The topography was derived from interferometric processing of a tandem pair of coherent ERS-1 and-2 SAR images (acquired 1 day apart) (Kwok and Fahnestock, 1996). Color represents relative backscatter strength—i.e., blue is low and red high. In this area, megadune amplitudes are $\sim$4 m, and wavelengths of $\sim$2 km lead to a local change compared with regional slopes of $\sim$0.004.

From Fahnestock et al. (2000b). Copyright American Geophysical Union 2000, reproduced with permission.

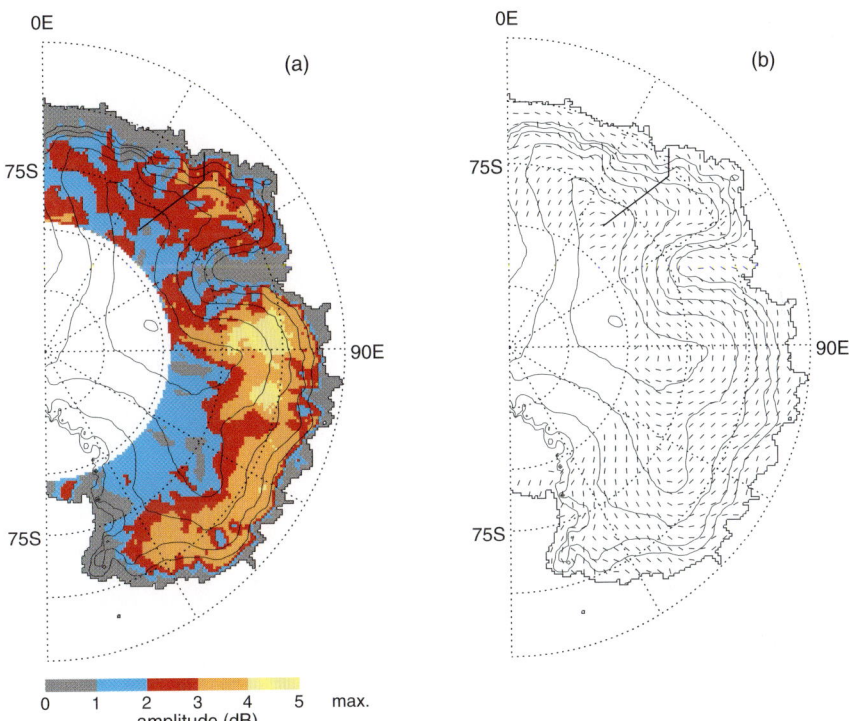

Figure 3.110. Maps of East Antarctica showing (a) the amplitude of the azimuthal variation of the mean backscatter of the snow/firn at C-band using ERS-1 wind scatterometer data, and (b) the orientations of the directional anisotropy of the backscatter coefficient of the snow cover, as defined by the antenna look direction for minimum backscatter coefficient. Values were calculated for 25 × 25-km cells, but only every fourth orientation value is plotted. Surface elevation contours are plotted at 500-m intervals starting at 1,500-m elevation. ERS-1 wind scatterometer data copyright ESA 1994.

From Furukawa and Young (1997), after Young et al. (1996). Copyright ESA 1997, reproduced with permission.

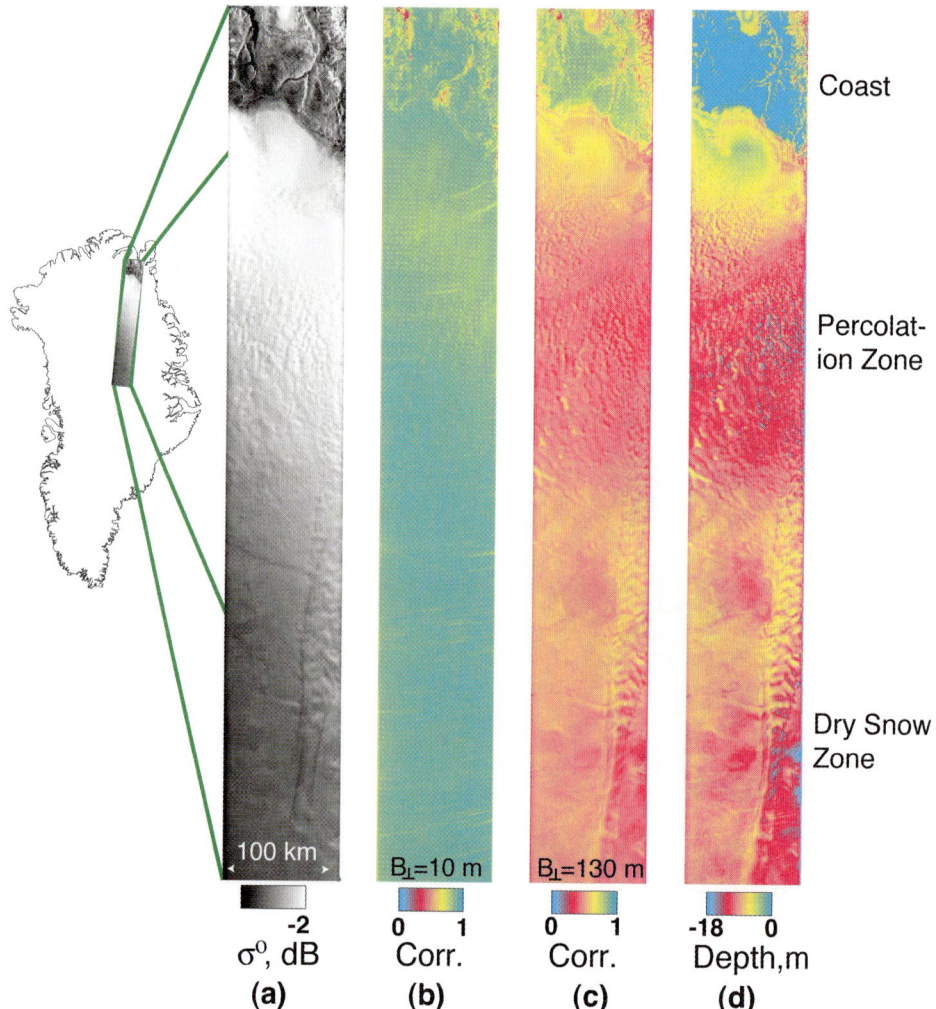

Figure 3.111. An 800-km-long strip of ERS C-band SAR images extending across NE Greenland from the ice sheet dry-snow zone (image bottom) through the percolation zone to the rocky coast (top). The ERS-1/-2 Tandem Mission images were acquired on January 11–12 and February 15–16, 1996: (a) a SAR amplitude (brightness) image; (b) a correlation image from a tandem image pair where the baseline is small (1 m in the interior to 24 m at the coast); (c) a correlation image with a much larger baseline of 124–134 m; and (d) a map of radar penetration depths inferred from (b) and (c) by assuming that (b) represents temporal decorrelation only by virtue of its small baseline. Little detail is apparent in the first correlation image (b), which is highly correlated due to its small baseline. Significant wave penetration and the increased baseline in the second correlation image (c) results in a significantly greater contrast between the different facies. In (d), the model was used to invert the correlation images into penetration depth.

From Hoen and Zebker (2000b). Copyright IEEE 2000, reproduced with permission.

A number of other corrections are required before useful geophysical parameters can be extracted from satellite radar altimeter measurements. Orbit errors and errors caused by variable atmospheric conditions should be accounted for in satellite measurements over both ocean and ice. Although radar altimeters can penetrate cloud cover, the atmosphere causes variations in the speed of the radar pulse and hence introduces errors into the derivation of surface topography. This effect depends primarily on the integrated water content of the atmosphere. Further corrections are required for ionospheric refraction (signal delay due to charged particles in the ionosphere) and the effects of dry atmospheric mass on the velocity of propagation of EM waves (Cudlip et al., 1994). Modern satellite missions employ dual-frequency altimeters to account for ionospheric effects—e.g., Topex-Poseidon and Envisat—and are equipped with PMW radiometers with frequencies centred on water vapor absorption bands—e.g., at 22 GHz—to correct for atmospheric effects. The dependencies on atmospheric pressure and water vapor are also accounted for with output from numerical models.

With these corrections and precise knowledge of the satellite orbit, then the time delay between pulse transmission and receipt can provide a measure of the elevation of the ice-sheet surface topography along the satellite ground track. Put simply, the altimeter transmits a radar pulse with a spherical wavefront at time T_t. This pulse reflects from the ice-sheet surface, and is received back at the satellite at time T_r. The altimeter range, or range to surface, R is then calculated as (NASA, 2003):

$$R = c(T_r - T_t)/2 \qquad (3.11)$$

where c is the speed of light. When subtracted from the satellite altitude, determined from precise knowledge of the satellite orbit, and referenced to a mathematical ellipsoid, the corrected altimeter range measurement yields a measure of the mean surface elevation of the ice sheet topography along the satellite ground track and above that ellipsoid. For modern altimeters, the height of the satellite above the reference ellipsoid is accurately determined by tracking the orbit from a globally distributed network of lasers and/or Doppler stations, with orbital parameters being further refined by incorporating orbital dynamics equations.

3.6.1.1 Satellite altimeter missions and ice sheets

The concept of using spaceborne radar altimeters to measure ice sheet topography was first proposed by Robin (1966). The satellite radar altimeter has been the "workhorse" sensor for the systematic monitoring of ice sheet elevation since the short Seasat mission of 1978. Together with the subsequent Geosat mission (1985–1990), Seasat provided ice sheet coverage to a maximum latitude of 72.1° only (Zwally et al., 1983, 1989). Although the Seasat and Geosat altimeters were designed only for tracking over open ocean, they acquired useful observations of the continental ice sheets, although much of the Antarctic continent remained uncovered. Gridded to a spatial resolution of 20 km, Seasat data resolved large-scale undulations only, and the mission lasted only 106 days. In spite of this, the Seasat radar altimeter acquired more than 600,000 useful range measurements over

Greenland and Antarctica and proved the concept. Grid resolution was improved to 10 km with the Ku-band (13.5 GHz) Geosat altimeter, which was also designed to track variable surfaces more precisely. This time series continued with the Geosat Follow-On mission (1998–present). Coverage was extended to 81.4° with the launch of ERS-1 in 1991, albeit over a less densely spaced grid over the nominal 35-day repeat orbit (April 1994–March 1995)—i.e., across-track sampling of 15 km at latitude 70°. This enabled observation of the entire Greenland Ice Sheet for the first time, and ~80% of the Antarctic Ice Sheet. The two ERS satellite altimeters were the first to have centimeter-scale orbit accuracy determination (achieved using sophisticated orbital-tracking devices). Subsequently, the combination of ERS-1 and -2 data in 168-day repeat cycles (termed the *Geodetic Mission*, from April 10, 1994 to March 21, 1995) created a denser gridding pattern for elevation data extraction. The ERS-1 satellite also flew in a 3-day repeat orbit, enabling study of variability over short timescales but at a low spatial resolution (Rémy et al., 2001).

Conventional ice sheet radar altimetry continues with the improved sensor (RA-2) onboard ESA's Envisat (⟨*http://envisat.esa.int/instruments/ra2/*⟩). This is a dual-frequency, pulse-limited radar operating at both Ku- and S-band (13.6 and 3.2 GHz, respectively), and is the first spaceborne altimeter to provide not only geophysical products but also average waveforms to the user at both frequencies plus "bursts" of up to 2,000 individual echo samples at Ku-band. Other satellite altimeters have been launched, but with orbital inclinations not optimized for high-latitude coverage. In operation from 1992 to 2001, the joint NASA/CNES (Centre National d'Études Spatiales) Topex-Poseidon mission carried a dual-frequency (C- and Ku-band, or 5.3 and 13.6 GHz, respectively) radar altimeter with fine and coarse tracking modes, but covered only the southern tip of the Greenland Ice Sheet (to a maximum latitude of 66°). Dual frequencies enable improved corrections for ionospheric effects. In spite of having neither an orbit inclination nor a tracking system that is optimized for ice sheet observation, the dual frequency of the Topex-Poseidon altimeter had the advantage of enabling better resolution of long-term trends in snowpack characteristics as they contribute to the altimeter signal (Rémy et al., 2001). This is possible due to the greater penetration depth of the lower frequency channel, and comparison of the two simultaneous signals. Note that the Envisat RA-2 is also a dual-frequency instrument. Single-channel radar atimeters tend to be more affected by this temporal snowpack artifact. As noted above, Topex-Poseidon is also equipped with a three-frequency PMW radiometer which, although designed to correct for radar wet atmospheric delay, provides useful simultaneous information on ice sheet microwave emissions (Rémy et al., 2001). The follow-on to Topex-Poseidon, Jason-1 (launched in December 2001), carries the Poseidon-2 altimeter (a dual-frequency sensor based on Topex but with improved sampling). Once again, its polar coverage is limited to relatively low latitudes only. By being optimized for ocean measurement, these systems are playing a leading role in the detection of regional and global sea-level change (Nerem and Mitchum, 2001a, b).

While the validity of early analyses (using Seasat and Geosat data) has since been questioned due to the fact that the instruments predated the era of accurate orbit determination and were unequipped with ice modes of operation (Davis et al.,

1998; Haines et al., 1994), they demonstrated the immense potential of satellite altimetry as a key ice sheet research tool. In an effort to produce a consistent time series of all data collected since the Seasat mission in 1978, the NASA Ice Sheet Altimeter Pathfinder Program (ISAPP) has reprocessed all available data using state-of-the-art algorithms and a consistent set of environmental corrections and orbit solutions. These data, and useful ancillary material, are available online at ⟨http://icesat4.gsfc.nasa.gov/ia_home/AltDoc.html⟩.

3.6.1.2 Satellite altimeter-derived digital elevation models

Data from the ERS altimeters in particular have revolutionized the quantitative representation of ice sheet topography, enabling the construction of high-resolution ice-sheet Digital Elevation Models (DEMs) of both Greenland (e.g., Bamber et al., 2000a, 2001a) and Antarctica (Bamber, 1994b; Bamber and Bindschadler, 1997; Jezek et al., 1999; Legrésy and Rémy, 1997; Liu et al., 1999). These have revealed numerous previously unknown details in surface roughness and topographic characteristics (Rémy et al., 1999).

A widely used altimeter-derived DEM of Antarctica—namely, the Radarsat-1 Antarctic Mapping Mission (AMM) DEM produced by Liu et al. (1999)—is shown in Figure 3.56. This product was initially constructed to aid the mosaicking and geometric rectification of Radarsat SAR data, but has since been adopted for ice-sheet-modeling purposes. While the foundation is ERS-1 radar altimeter data from the mid-1990s, it represents an improvement over similar products by also incorporating ancillary information together with sophisticated Geographic Information System (GIS) interpretation and error detection techniques (Liu et al., 1999). The satellite data gap poleward of ~81.4°S is typically filled with aircraft data (Bamber, 1994b), in this case from the Antarctic Digital Database or ADD (British Antarctic Survey, 1993). Similarly, Bamber et al. (2001a) constructed their Greenland DEM from a combination of satellite radar altimeter, airborne altimeter, aerial photography, and GPS data. The representation of the surface slope of the Antarctic Ice Sheet in Figure 3.56 gives a clear demonstration of the power of satellite observations, revealing as it does a wealth of new information on ice sheet glaciology and morphology. Surface slopes vary over several orders of magnitude—here brighter equates to flatter. Ice domes, ice divides separating the drainage basins, the merging of outlet glaciers into ice shelves and the surface expression of the subglacial Lake Vostok are all clearly visible. The image was computed from a 5-km-resolution gridded DEM (distributed by BEDMAP, based on the DEM of Liu et al., 1999). Altimeter signals are frequently lost in steep coastal regions, and the DEM was completed by Liu et al. (1999) using other available mapping information.

Hamilton and Spikes (2004) carried out an independent validation of this DEM using precision kinematic GPS profiling. They found that the DEM performs well overall at capturing the gross ice sheet morphology north of 81.4°S, but less so to the south. Their results further indicated that the horizontal resolution of the DEM is ~8 km at best (degrading to 10–50 km in the southernmost region unconstrained by ERS altimetry), meaning that significant topographic detail remains unresolved. Some errors are also apparent in absolute elevations, with differences between GPS

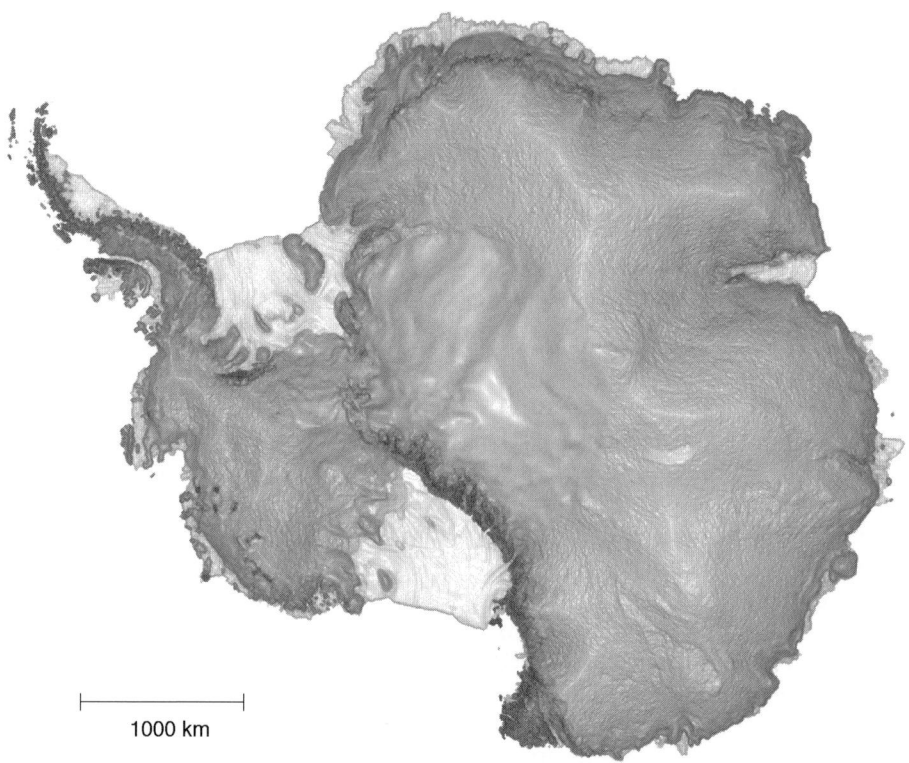

Figure 3.56. The high-resolution Radarsat-1 Antarctic Mapping Mission (AMM) Digital Elevation Model (DEM) of Antarctica, which combines topographic data from a variety of sources to provide consistent coverage. These sources are satellite radar altimetry, airborne radar surveys, the recently updated Antarctic Digital Database (version 2), and large-scale topographic maps from the U.S. Geological Survey (USGS) and the Australian Antarctic Division. Although the AMM DEM was created to facilitate the processing of Radarsat-1 AMM SAR data, it does not utilize any AMM radar data. The 1-km, 400-m, and 200-m resolution (horizontal) DEM data are provided in ARC/INFO and binary grid formats, and the 1-km and 400-m DEMs are also available in ASCII format (see ⟨*http://nsidc.org/data/docs/ daac/nsidc0082_ramp_dem_v2.gd.html*⟩).

Image generated by Roland Warner (Australian Antarctic Division and Antarctic Climate and Ecosystems Cooperative Research Centre), using data from Liu et al. (2001).

and DEM values of up to 50 m in certain relatively rugged inland regions. Hamilton and Spikes (2004) concluded that improvements could be made by incorporating ancillary data from AVHRR and MODIS (processed photoclinometrically—see Section 3.6.3) and the ICESat laser altimeter (next section).

A detailed look at the Pine Island Glacier region of West Antarctica is shown in Figure 3.57. The vertical accuracy of the Bamber and Bindschadler (1997) DEM is estimated to range from 1 to 10 m, depending on slope (Ekholm et al., 1995; Fahnestock and Bamber, 2001). This DEM is available online from the NSIDC at

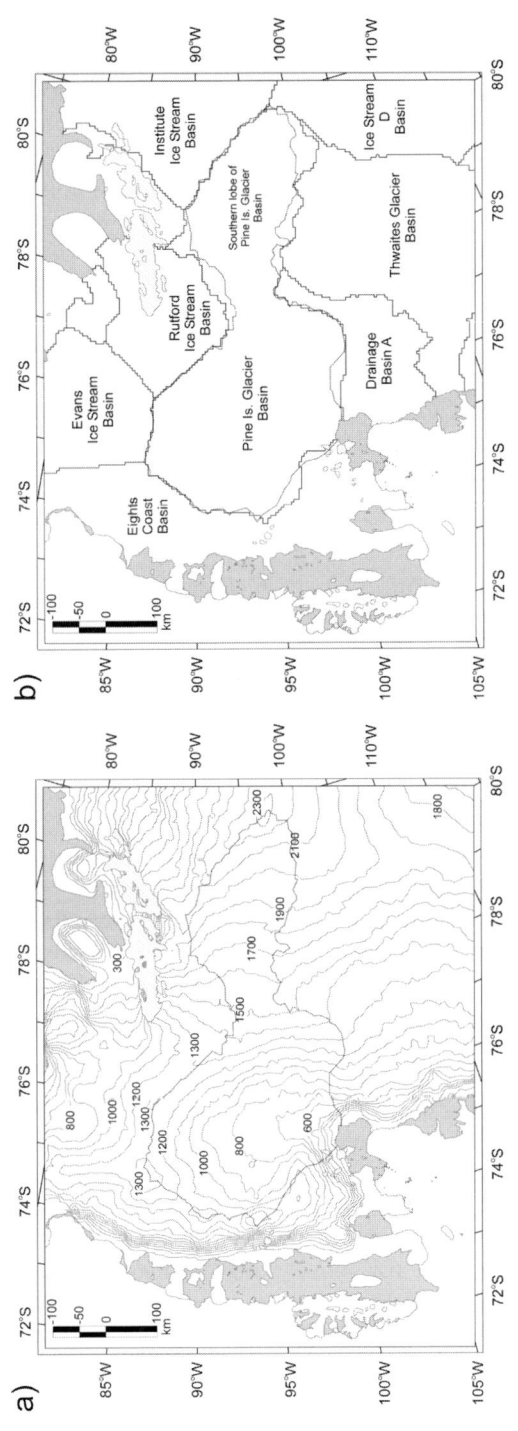

Figure 3.57. (a) Surface elevation map of the Pine Island Glacier (PIG) and surrounding ice sheet region derived from the Geodetic Mission of ERS-1 (which lasted from April 1994 until March 1995), and on a 5-km grid spacing (Bamber and Bindschadler, 1997). Contours are in meters above mean sea level. (b) Map of ice drainage basins in the same region, and derived from the same dataset. Solid lines denote drainage basin boundaries, while the dotted line is the boundary of the PIG drainage basin derived from Liu et al. (1999).

From Vaughan et al. (2001b). Copyright AGU 2001, reproduced with permission.

⟨http://nsidc.org/data/docs/daac/nsidc0076_antarctic_5km_dem.gd.html⟩. It has been possible to accurately delineate individual drainage basins and ice divides using these data, and determine flowlines and catchment areas (Vaughan et al., 2001b), as well as correct geometric and radiometric distortions in SAR images of the region (Noltimier et al., 1999).

Other workers have used satellite altimeter data to generate topographic maps of ice shelves. Fricker et al. (2000a), for example, constructed a 5-km-resolution DEM of the Lambert Glacier–Amery Ice Shelf system from ERS-1 altimeter data (Figure 3.58, see color section). This involved removal of the high-frequency vertical signal due to tidal motion. Satellite-derived DEMs have become an invaluable and integral part of ice-sheet numerical modelling, providing data for both input and validation (Hulbe and Payne, 2001). Moreover, DEMs have enabled the construction of ice-sheet *balance fluxes* (Bamber et al., 2000a, c; Budd and Warner, 1996). This powerful modeling tool is discussed in more detail in Section 3.8.3. The interested reader can access both DEM data and ancillary information from the excellent NSIDC website at ⟨http://nsidc.org⟩.

3.6.1.3 Radar altimeter measurement of elevation change over time

A key attribute of satellite radar altimetry is its ability to systematically acquire time series of surface elevations, from which changes in ice-sheet elevation over time ($\partial H/\partial t$) can be estimated. As noted above, the last decade has witnessed important technological improvements in spaceborne radar altimeters and data processing. This includes enhanced orbit error analysis and reduction (e.g., Davis et al., 2000; Kluever et al., 1997), orbit computation (Tapley et al., 1996), the identification of measurement system biases (e.g., Davis et al., 2000), and ice sheet retracking (Davis, 1996, 1997). As a result, ice-sheet satellite radar altimetry has evolved to a sufficient state of maturity that regional changes in surface elevation can be inferred with an accuracy on the order of a few centimeters per annum over periods of 5–10 years (Davis et al., 1998, 2000, 2001, 2005; Wingham et al., 1998; Zwally and Brenner, 2001; Zwally et al., 1998). Bamber et al. (1998) provide an assessment of accuracy based upon a comparison with airborne laser altimeter data in Greenland.

The concept of using altimeter-derived $\partial H/\partial t$ data to determine mass balance is based on the relationship between changes in elevation and changes in thickness and thus ice volume (Zwally, 1989; Zwally et al., 1989). Surface elevation changes are equivalent to ice thickness changes minus the vertical motion of the underlying bedrock (Rignot and Thomas, 2002), which is generally smaller and can be estimated separately (Wahr et al., 2001; Zwally et al., 2002a). As noted by Bamber and Kwok (2004), absolute elevation is not required for estimates of $\partial H/\partial t$, but rather the difference between two (or more) range estimates. This is generally obtained by comparing range estimates acquired at different epochs at the intersection of orbital tracks on the ground—i.e., at the crossover of ascending and descending orbits (*crossover analysis*) (Davis et al., 1998; Lingle and Covey, 1998).

Data processing includes correction of slope-induced errors (see Section 3.6.1.2), data filtering (quality checks), and adjustment of measurements at crossover points

(Bamber, 1994a, b; Lingle and Covey, 1998; Rémy et al., 1999; Zwally et al., 1989). As satellite altimeters view the surface along a narrow swath, closely spaced coverage of polar regions relies upon long orbital repeat intervals—the shorter the interval, the wider the ground-track spacing. Once again, this entails a trade-off between temporal resolution and coverage. In terms of ice-sheet research, dense topographic coverage is crucial for a variety of studies, from numerical modeling of ice dynamics to monitoring of key glaciological features, including ice divides and drainage basins, ice-shelf grounding lines, sub-glacial lakes, ice-shelf fronts, and ice rises. Both solid Earth and ocean tides also need to be taken into account when processing the data (Zwally et al., 1990). Time-variable vertical displacement due to tidal and atmospheric pressure effects is a key feature of floating ice shelves and glacier tongues, and is corrected for by incorporating data from *in situ* reference GPS programs or tide models.

An underlying assumption in using radar altimetry to measure and monitor $\partial H / \partial t$ is that the penetration depth of the radar signal into the snow/firn volume, and the relative contributions of surface and volume backscattering, remain invariant. As noted previously, altimetric estimates of ice sheet height can in fact be biased by the time-varying penetration of the radar pulse into the snow/firn mass (Davis, 1996; Davis and Moore, 1993; Davis and Zwally, 1993; Yi and Bentley, 1997), with errors of up to 8 m being reported by Ridley and Partington (1988). Accurate quantification of this effect is difficult because of the various interrelated factors affecting volume and surface backscattering (see Section 3.2.3). These include snow/firn wetness, morphology, and stratification (e.g., the presence/absence of buried icy layers), and snow-grain size and density, as they affect the electrical and geometrical properties of the snow/firn layer interacting with the radar pulse. Variations in the ratio of these two contributions lead to changes in both the amplitude of the waveform and the shape of its leading edge, which in turn affect the position of the retrack point (Bamber and Kwok, 2004). Davis (1993) and Davis and Moore (1993) developed a surface-/volume-scattering retracking algorithm to account for these effects on ice-sheet waveforms in order to minimize errors in elevation estimates. Other studies—e.g., Wingham et al. (1998)—have employed a correction for the surface backscatter coefficient, which is a measure of the waveform amplitude (see Figure 3.53).

A complication in using conventional single-frequency radar altimeters in particular to measure ice-sheet elevation change relates to the difficulty in separating the actual trend from the artificial trend due to long-term changes in snowpack characteristics (Rémy et al., 2001). A further challenge in determining long-term variability or trends is the need to also account for seasonal variations in elevation caused by variations in snowfall, firn compaction, or "densification" (Alley et al., 1982; Arthern and Wingham, 1998; Cuffey, 2001; Rémy and Legrésy, 2004; Rémy et al., 2002; Zwally and Li, 2002), and melting, in order to determine long-term variability. While short-term elevation changes occur both seasonally and interannually (Van der Veen and Bolzan, 1999), it is the longer term changes that are intimately linked to climate change and global sea level (Rignot and Thomas, 2002). Ferguson et al. (2004) address this issue and present an auto-regressive model designed to characterize seasonal and interannual variations in ice sheet elevation change time series from satellite altimeters, superimposed upon long-term linear trends.

In effect, the radar echo of pulse-limited altimeters depends on surface micro-roughness, the kilometer-scale roughness, the stratification of the snowpack, the radar extinction in the snow, and the antenna gain pattern (Brisset and Remy, 1996; Legrésy and Rémy, 1997; Rémy et al., 1995). As such, it has a frequency dependence. Dual-frequency altimeters such as those on Topex-Poseidon and Envisat provide a means of detecting the bias empirically (Rémy et al., 1996a), by increasing the volume of data relative to the number of unknowns. Working in central Greenland, Haardeng-Petersen et al. (1998) showed that satellite altimeter- and GPS-derived surface heights were in reasonably good agreement (difference <2 m), suggesting that the surface reflection signal of the altimetry dominates over volume scattering at this particular location.

New and improved estimates of $\partial H/\partial t$ are resulting from state-of-the-art processing of ERS radar altimeter data. An example is the production of a spatial plot of elevation change for a large part of the Antarctic Ice Sheet over the period from June 1995 to April 2000 by Davis and Ferguson (2004). Retracking of ice-mode return waveforms was carried out using the threshold algorithm of Davis (1997), with ice-sheet elevation differences (∂H) being computed at satellite crossover points for all possible combinations of monthly data subsets. Results are shown in Figure 3.59 (see color section). The time series were produced using the improved time series approach described in Davis and Segura (2001) and Ferguson et al. (2004), after adjustment for the backscattered power of the radar return signal. While the continent as a whole exhibited an apparent trend in $\partial H/\partial t$ of 0.4 ± 0.4 cm per annum over the 5-year period, there is again considerable regional variability. The trend over the East Antarctic Ice Sheet is 1.0 ± 0.6 cm per annum, while that of West Antarctica is -3.6 ± 1.0 cm per annum. Extreme changes are again noted over the Pine Island and Thwaites Glacier regions, with trends of -135 ± 10 cm per annum and -31.6 ± 5.2 cm per annum, respectively. These results are consistent with recent reports by Rignot and Jacobs (2002) of enhanced basal melting at the glacier's grounding lines from an increase in ocean temperatures.

The study by Davis and Ferguson (2004), and that of Davis et al. (2005), temporally extends that of Wingham et al. (1998), who produced a similar map covering the period 1992 to 1996. Such is the level of accuracy currently achievable over ice sheet interior regions that Wingham et al. (1998) were able to detect a -11.7 ± 1.0 cm per annum thinning trend averaged over the Pine Island Glacier and Thwaites Glacier drainage basin for the period 1992–1996 using ERS-1 and -2 data (Figure 3.60, see color section; see also Shepherd et al., 2001, 2002a). As noted earlier, this system is of particular concern, as it drains a large part of the WAIS and may be undergoing significant change (Vaughan et al., 2001b). In their satellite radar altimetry analysis, Shepherd et al. (2002b) showed that thinning of >25 m and 45 m occurred at the grounding zones of the Thwaites and Smith Glaciers, respectively, over the period 1991–2001. Wingham et al. (1998) further constructed a regional velocity map from ERS SAR interferometry to show that thinning is restricted to the fastest flowing sections of the glacier, and stress the paramount importance of extending the time series to determine whether the rate of glacier thinning is accelerating. Further analysis of the wider region from 85° to 150°W by Zwally et al. (2002a) revealed major regional variability in elevation change ranging from about −30 cm per annum (in the Thwaites/Pine Island drainage basin) to +20 cm per annum (on the ridge between Pine Island/Thwaites Glaciers and the Ross Ice Shelf drainage basin) (Figure 3.59b), and an average elevation change of the grounded ice of −4.3 cm per annum. The authors incorporated an estimate of 2.5 cm per annum for the average rate of bedrock uplift, after Huybrechts and Le Meur (1999), and also accounted for the peak-to-peak seasonal amplitude due to variations in firn compaction rates. Using ERS altimeter data, Shepherd et al. (2003) estimated a thinning of the Larsen Ice Shelf between 1992 and 2001.

In Greenland, satellite radar-altimeter measurements of the inland ice sheet have been supplemented by a multiyear campaign of aircraft laser altimeter measurements in the steeper sloping marginal regions and within PARCA (NASA's Program for Arctic Regional Climate Assessment) (Abdalati, 2001; Krabill et al., 1995, 1999; Thomas, 2001; Thomas et al., 2001a). For aircraft laser measurements, the surface spot size was only ~1 m, compared with several kilometers for satellite radar altimetry, leading to a high accuracy of the order of 10 cm in height estimation (Krabill et al., 1995). It appears from these new results that the higher elevation regions have been close to balance for the past few decades (Davis et al, 2000; Thomas et al., 2000b). This estimate is based on three independent analyses, using satellite radar altimeter data (areas south of 72°N), aircraft laser altimeter data, and comparing total surface accumulation with total ice discharge (i.e., volume balance estimates). The near-coastal region, on the other hand, appears to be undergoing significant thinning (Abdalati et al., 2001; Krabill et al., 2000), at rates up to 10 m per annum for one major outlet glacier (Thomas et al., 2000a). According to Krabill et al. (2000), this is contributing an estimated ~10% to the total observed global sea-level rise. Zwally et al. (2002a) urged caution in the interpretation of these results, as they are based on airborne laser altimeter measurements only, with coverage limited both spatially and temporally. Recent mass-balance observations in Greenland were evaluated by Thomas et al. (2001a). Improved estimates of changes in the surface elevation of Greenland were derived by Davis and Sun (2004) from Geosat and Geosat Follow-On data combined. Other studies of ice-sheet elevation change detected by satellite radar altimeter measurements include those of Bentley and Sheehan (1992), Lingle et al. (1994), Partington et al. (1991), and Yi et al. (1997).

3.6.1.4 Radar altimeter measurement of ice-sheet surface slope

Sophisticated processing has also enabled the derivation of detailed maps of ice-sheet surface slope. An example from Antarctica, and produced by Rémy et al. (1999) using data from the ERS-1 Geodetic Mission, is shown in Figure 3.61 (see color section). Fine detail is apparent, including 10-km-scale scars, undulations, and subglacial lakes. On the larger scale, the surface slope of Antarctica gradually increases from 0 to 3 m per kilometer inland, before abruptly increasing around the coastal margins. This break in slope, which occurs at an elevation of ~2,000 m, plays a central role both in the ice dynamics and the katabatic wind-flow regime (Pettré et al., 1986), and therefore local surface roughness and accumulation characteristics. Another striking feature apparent in these maps is the irregular slope intensity at small horizontal scales. As noted by Rémy et al. (1999), this raises the issue of what horizontal distance to use for averaging the slope parameter when applying the ice sheet flow law. Rémy et al. (1999) show that it is the small-scale behavior of the surface slope that yields information on key physical characteristics.

Another application has been the inference of regional ice-flow physical properties from altimeter-derived topography (Legrésy and Rémy, 1998; Rémy et al., 1999). Herzfeld (2004) produced an atlas of Antarctic topographic maps from satellite radar altimeter data. Conventional pulse-limited altimeters are, however, unable to resolve high-resolution features of the surface topography. The relationship between an individual echo profile and the surface elevation is not unique over topographic features with horizontal-wavelength scales which are similar to the altimeter antenna footprint (Wingham, 1995; Wingham et al., 1993). In addition, satellite radar altimeter data have been used to map ice-sheet surface properties (Legrésy and Rémy, 1998) and to study ice-sheet topographic detail related to bedrock and other characteristics—e.g., Brisset and Rémy (1996), Legrésy and Rémy (1997).

3.6.1.5 New-generation measurements: CryoSat

While first-generation satellite radar altimeters can provide sufficient measurement accuracy with which to constrain mass-balance calculations over intervals of 5–10 years, accuracies are insufficient over shorter time periods. Moreover, although conventional pulse-limited satellite radar altimeters acquire reasonably accurate elevation-change estimates over flatter interior portions of ice sheets (particularly when equipped with an ice sheet mode of operation), they perform much less successfully in marginal outer regions, where slopes are greater and undulations more prominent (Zwally et al., 2002a). As noted in Section 3.6.1, elevation errors in such regions are dominated by pulse spreading and loss of signal, with abrupt changes in slope often resulting in loss of track (Brenner et al., 1983). Accurate elevation, let alone change, measurements with conventional pulse-limited altimeters become difficult once the surface gradient exceeds the antenna beamwidth—e.g., $0.5°$ for ERS and Envisat. Van de Wal and Ekholm (1996) estimated the vertical accuracy of a satellite altimeter-derived DEM of Greenland in regions sloping more than $1°$ to be 75–100 m. Ice sheets typically consist of three distinct regions on the basis of surface slope: (i) a central plateau characterized by shallow slopes; (ii) steeper margins, often comprising outlet glaciers; and (iii) very low slopes on floating ice shelves and glaciers. Unfortunately, an estimated 23% and 17% of the surface areas of Greenland and Antarctica, respectively, exceed this limit (Phalippou and Wingham, 1999). Improved measurement of outer ice-sheet margins is critical, given that they are the most dynamic parts of ice sheets and are exposed to significant variability and perturbations in both atmospheric and oceanic forcing

(Alley and Whillans, 1991). They also exhibit the highest variability in accumulation, and may be responding rapidly to change (Krabill et al., 2000). Due to the scale and logistical difficulties involved, a circum-Antarctic aircraft laser altimeter campaign such as that described above for Greenland is impractical. Moreover, ~20% of the Antarctic interior remains beyond the reach of the pulse-limited ERS and Envisat radar altimeters due to their orbital configurations (ESA, 2003).

The need for higher resolution measurements to overcome these major limitations in the use of pulse-limited altimeters has led to the development of two innovative spaceborne systems—namely, NASA's ICESat (⟨http://icesat.gsfc.nasa.gov/⟩) and ESA's CryoSat. Given their mission requirements, they will make an immense contribution by addressing and possibly even closing out the uncertainty associated with current ice-sheet mass balance. Both are particularly exciting in that they carry emergent altimetry technologies for the first time in space and are specifically designed for the measurement of ice masses. Note again that the launch of CryoSat failed in October 2005. It is retained here in the hope that a follow-on satellite will be launched. CryoSat incorporates novel altimeter design features and radar-processing techniques (Drinkwater et al., 2004a; ESA, 2003; Wingham, 1999; Wingham et al., 2004; ⟨http://www.esa.int/livingplanet/cryosat⟩). Its payload, the Ku-band (frequency 13.6 GHz, 2.2 cm wavelength) SIRAL, will provide enhanced measurement not only of ice-sheet inland regions but also of the steeper ice-sheet margins (and glaciers and smaller ice caps) and ice shelves, as well as sea-ice thickness (see Chapter 5 of Volume 1 of this book). In effect, SIRAL combines conventional pulse-limited altimeter hardware with synthetic aperture and interferometric signal processing. The most revolutionary feature of SIRAL is its ability to locate target points in three-dimensional space, using a concept similar to the interferometric SAR technique described in Chapter 2. The CryoSat altimeter is designed with two receiving antennae forming an interferometer in the across-track direction—i.e., with a baseline of 1 m (when operating in interferometric SAR or "InSAR" mode), while the return signal in the along-track direction is processed using Doppler filtering to construct a synthetic aperture for an enhanced ground resolution of 250–300 m (Drinkwater et al., 2004a). The delay Doppler technique is described in detail by Raney (1998), and its advantages illustrated and evaluated in Figure 3.62.

When operating in InSAR mode, a second synthetic aperture system is utilized to form an across-track interferometer (ESA, 2003; Wingham et al., 2004), as noted above. The aim of this mode is to provide improved estimates of elevation over variable topography (Drinkwater et al., 2004a). The phase difference between echoes received at each antenna yields the angle between the off-nadir angle (the echo direction) and the interferometric baseline. The instrument concept is illustrated in Figure 5.85 of Volume 1 of this book. The across-track location and elevation of a given point on the surface is determined geometrically from the range and angle of the echo at each range. Phase multi-looking of the 64 echoes both reduces speckle and determines the coherence at each range. The effects of abrupt height variations are minimized in SIRAL by using narrowband-tracking pulses transmitted between successive wideband measurement bursts (ESA, 2003). CryoSat also includes state-of-the-art orbital-tracking technology, including the

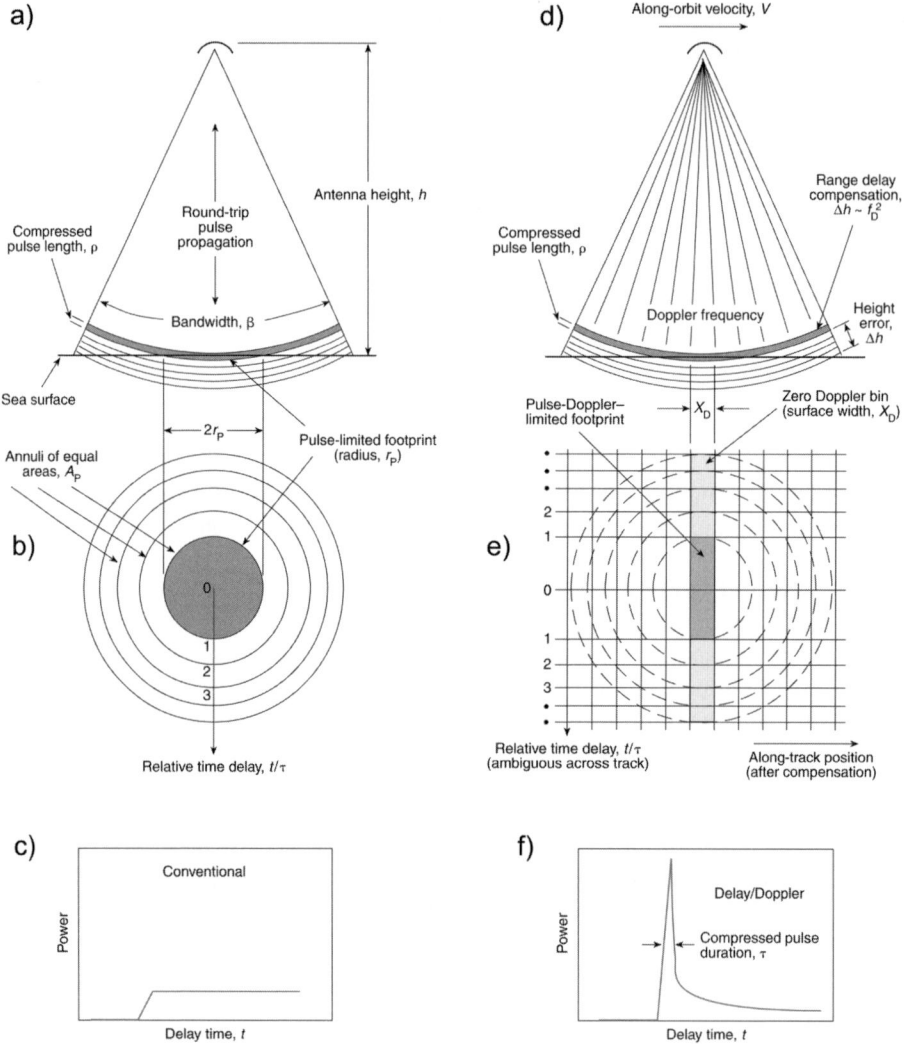

Figure 3.62. A schematic comparison of the illumination geometry (side view), footprint (plan view mapped to a flat-Earth surface), and impulse response for (a–c) a conventional pulse-limited radar altimeter, and (d–f) a delay-Doppler radar altimeter. Details of the latter are given in Raney (1998) and Gasparovic et al. (1999). The major innovative feature of the delay-Doppler technique is that returns from a given ensemble of transmissions along-track are coherently rather than incoherently processed (as is the case with pulse-limited systems), leading to a finer measurement precision that is better suited to ice-sheet (and sea-ice) applications. Doppler processing determines the location and size of the alongtrack surface footprint (e), which is substantially smaller than that of the pulse-limited system (b) and also relatively immune to surface topographic variations. Rather than being circular, as typically cited in the literature, footprints are elongated along-track, this effect being more pronounced for pulse-limited systems. This elongation results from the averaging of multiple pulses to create a multi-look waveform, over which time (e.g., 1 s) the satellite has advanced a considerable distance (typically approximately 6 km) in the along-track direction. By contrast, the relative location

splendidly named DORIS (Doppler Orbitography and Radio-positioning Integrated by Satellite) system to determine the orbit to an accuracy of 5 cm (⟨http://www.ign.fr/fr/PI/activites/geodesie/DORIS/index-en.html⟩). As a backup, a cluster of nine laser reflectors provide centimeter accuracy orbit determination using laser ground stations. The orientation of the SIRAL interferometric baseline is determined by a startracker system. Together, these innovative features enable CryoSat to collect accurate altimetric data over the steeper outer margins of ice sheets—i.e., within regions where conventional satellite radar altimeters typically suffer loss of track for the reasons outlined above. The SIRAL also has a conventional pulse-limited low-resolution mode for operation over ice-sheet inland regions. Moreover, coverage will be to a maximum latitude of 88°—significantly poleward of that afforded by ERS and Envisat. For the science mission, the nominal repeat cycle is 369 days (with a 30-day pseudo-subcycle). This is a compromise between a maximum number of crossover points and good spatio-temporal coverage—i.e., closely spaced ground tracks—at high latitudes (ESA, 2001). The issue remains that the mono-frequency Cryosat altimeter will be affected by a sub-surface signal that is difficult to accurately parameterize (Rémy et al., 2001). Further details of the CryoSat mission, which is designed to provide results which are compatible with those from conventional altimeter missions, are given in Chapter 5 of Volume 1 of this book.

3.6.2 Surface elevation from laser altimetry

The launch of the first spaceborne Earth-observing laser altimeter, the GLAS, on NASA's ICESat in January 2003 heralded another new and exciting era in the measurement of seasonal and long-term changes in ice sheet elevation from space. The technology behind GLAS is LIDAR, or LIght Detection And Ranging. LIDAR is a range-measuring device similar to radar, except that it transmits pulses of laser light instead of radar waves (Bufton et al., 1991). The measured surface elevation is determined from knowledge of the altitude of the satellite orbit above the Earth minus the range to the surface measured by GLAS, and corrected for delays caused by atmospheric refraction/delay (Mahesh et al., 2002) and instrument artifact effects. Variations in the Earth's surface elevation caused by ocean, solid, atmospheric load, and polar tides are again removed to obtain the surface signals of interest. The

of each delay-Doppler signal (derived height) is synchronized to coincide with the forward motion of the sensor, thereby eliminating footprint elongation ambiguities. A distinguishing feature of the CryoSat SIRAL is that it transmits radar pulses separated by intervals of only 50 μs, compared with ∼500 μs for pulse-limited systems (Drinkwater et al., 2004a). Satellite delay-Doppler footprints typically measure only 250 m along track (as is the case with SIRAL)—again a substantial improvement. As with pulse-limited altimeters, the cross-track footprint is determined by the pulse-limited condition (Fu and Cazenave, 2001; Raney, 1998). Note that this schematic is for a calm sea-surface only.

From Gasparovic et al. (1999). Reproduced with permission, copyright The Johns Hopkins University Applied Physics Laboratory.

GLAS instrument transmits 40 laser pulses (wavelength 1.064 μm) per second to the surface from an altitude of ~600 km. This produces a spot size of ~70 m in diameter with an along-track measurement spacing of ~172 m—both substantially smaller than those achievable with radar altimeters. The distance between spacecraft and ice sheet surface is measured to a predicted vertical accuracy of <15 cm under clear-sky conditions—again considerably better than is possible with conventional radar altimeters (Mitchell et al., 2003; Zwally et al., 2002a). The fact that surface height represents a mean value over the footprint area can introduce some surface-slope imprecision. The return pulsewidth, or time interval over which the signal is returned, increases with increasing surface roughness and/or slope (Brenner et al., 2000). Startrackers provide a precise laser-pointing reference to enable footprints to be located to 6 m horizontally, while an onboard GPS receiver enables radial orbit determination to better than 5 cm (Zwally et al., 2002a). Through careful control of spacecraft roll attitude, the laser footprint follows a reference ground track ("exact repeat mode") to ±50 m for each cycle. For further details on the theory of measurement, and individual error source contributions and their correction, see Zwally et al. (2002a, b, 2003a). Results from an ICESat track across Antarctica from the first day of operation are shown in Figure 3.63 (see color section).

For the first operational period beginning February 20, 2003, ICESat operations were in an orbit that repeated ground tracks every 8 days. This track pattern was selected for calibration and validation purposes, as rapid repeats allowed for the acquisition of refined pointing and orbit knowledge. During subsequent mission phases, GLAS has usually acquired data from a specific portion of a 91-day orbital pattern. The 33-day sub-cycle of the 91-day pattern was selected as the best compromise to achieve the mission's goals given that ICESat cannot be operated continuously as it was designed to do (⟨http://nsidc.org/data/icesat/orbit.html⟩). On the plus side, coverage extends to north and south latitudes of 86°. This covers all of the Greenland Ice Sheet and is again a marked improvement over previous systems (apart from CryoSat, had it been successfully launched), being key to extending coverage over most of the Antarctic Ice Sheet. Importantly, this provides adequate coverage of the marine-based WAIS and especially the dynamic Siple Coast ice streams and the highly active Amundsen Sea Embayment. The orbital inclination of 94° represents a compromise between the desire for a true polar orbit to observe the entire Antarctic Ice Sheet and the need for orbital crossovers for estimates of elevation change $\partial H/\partial t$ (Zwally et al., 2002a). Maps showing ground tracks and coverage over Antarctica and Greenland for similar latitude ranges for the 8-day repeat orbit are shown in Figure 3.64 (see color section). A color scale of the returned elevation values is shown from 0 to 4,000 m. Gaps in the track pattern are due to cloud cover that absorb/dissipate/scatter the emitted laser energy. The reader is also referred to Shuman et al. (in press).

The baseline results from GLAS are expected to greatly improve our knowledge of ice sheet topographic detail and its change over time (Figure 3.65). Improved tracking and the smaller footprint of GLAS data (~70 m compared with ~5 km for the ERS radar altimeter) results in substantially improved $\partial H/\partial t$ information in the regions of the ice surface that are not well-sampled by conventional radar

Sec. 3.6] Measurement of ice sheet topography/elevation and change in elevation   231

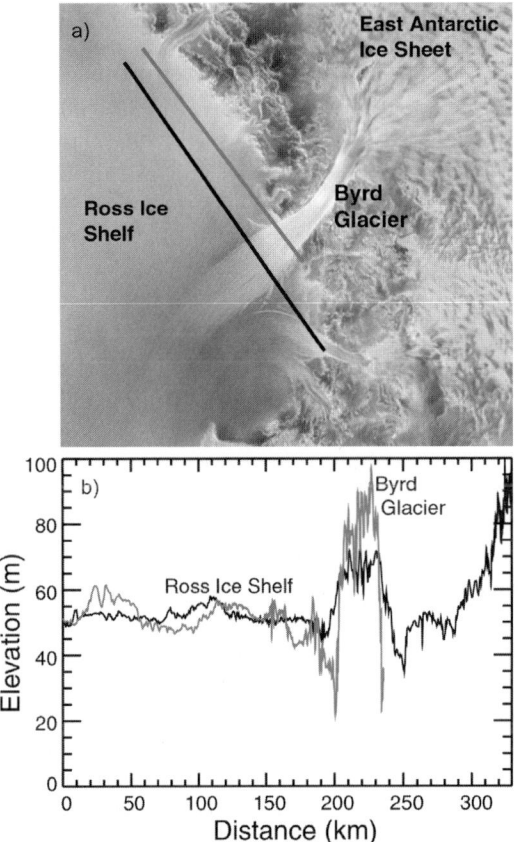

Figure 3.65. Two ICESat profiles across Byrd Glacier, which is the largest outlet glacier draining the East Antarctic Ice Sheet into the Ross Ice Shelf through the Transantarctic Mountains. In (a), the location of the profiles is marked on Radarsat-1 SAR imagery. The processed laser-altimeter-derived profiles of ice surface elevation in (b) show details of the troughs formed on the sides of the glacier as it enters the ice shelf. The differences between the elevations and widths of the glacier at the outer (black) and inner (gray) profile show how the glacier thins and spreads as it merges with the ice shelf. With time, ICESat's measurements of small changes in the elevations of the ice sheets, outlet glaciers, and ice shelves will yield key information on whether the rate of ice discharge into the ocean is increasing or decreasing and thus influencing sea level.
From Steitz et al. (2003). Reproduced with permission of NASA.

altimeters—i.e., steeper sloping ice margins and fast-flowing outlet glaciers and ice streams. These are regions where the largest elevation changes are likely to be found (Krabill et al., 2000; Scambos et al., 2004a, b; Yi et al., 2000). Moreover, the mission science objective is to measure profiles of ice-sheet elevation with sufficient temporal coverage and spatial density to determine interannual and longer term changes to an accuracy of <1.5 cm per annum per 100-km^2 area (Zwally et al., 2002a). Elevation change can again be determined both at orbital crossover points and along repeated ground tracks (Smith et al., 2005). The expected accuracy of a linear trend deduced from ICESat data over its designed 5-year mission lifetime and for all of Antarctica is $\sim$7 mm per annum of water equivalent (Zwally et al., 2002a). This includes errors in estimates of firn compaction and post-glacial rebound of the lithosphere and mantle (Peltier, 2001; Wahr et al., 2000). The elevation accuracy ranges from $\sim$80 cm over ice sheet regions with slopes of 1–2 degrees to $\sim$30 cm over low-slope areas—i.e., a distinct improvement over conventional radar altimeters. In their study showing thinning of West Antarctic glaciers with recent aircraft and satellite data, Thomas et al. (2004) determined the accuracy of ICESat measurements to be limited largely as a result of forward scattering in thin clouds and errors in laser pointing. Errors in elevation change were estimated to be ± 0.6 m at this point in the mission, which is a considerable improvement over radar altimeter measurements. Further improvements are expected with reprocessing efforts that are underway. An unprecedented level of detail is also obtainable over ice shelves (Fricker et al., 2004, 2005).

Preliminary analysis of GLAS data over ice shelves shows that their quality in terms of resolution of surface features is unprecedented (Fricker et al., 2005; Jay Zwally of NASA, pers. commun., 2004; Shuman, 2004; Shuman et al., in press). Indeed, GLAS is making a considerable contribution to the accurate measurement of the topography of ice shelf fronts. Recent analysis by Fricker et al. (2004, 2005) of data from the Amery Ice Shelf in East Antarctica shows that the new measurements are a significant improvement over previous measurements for defining ice shelf edge shapes and their changes with time. The retracking procedure employed enables not only the precise location of the ice front along the ICESat ground track, but also the resolution of detailed topographic signatures across the front of some ice shelves which are consistent with significant basal melting. Moreover, detailed measurements of rift depth are possible from space for the first time using GLAS data— e.g., from the Amery Ice Shelf (Fricker et al., 2005) (Figure 3.66)—a factor which will lead to improved understanding of the mechnisms of iceberg calving. As with radar altimeters, ICESat data acquired over floating ice masses need to be corrected for tidal variations to isolate non-tidal effects (Padman et al., 2002)—see Section 3.9. By the same token, ICESat will potentially provide data to constrain and thus enhance the performance of regional tide models.

As noted above, technical issues have prevented continuous operation of any of the GLAS lasers. Consequently, ICESat has operated since February 20, 2003 in bursts of $\sim$35–55 days (Shuman, 2004; Shuman et al., in press; Zwally, 2003, 2004a, b). Science data were acquired from February 20 to March 29, 2003 with GLAS Laser 1; from September 18 to November 19, 2004, February 17 to March 21, 2004, and May 18 to June 21, 2004 with Laser 2; and from October 3 to November 8, 2004 with Laser 3. Additional operation periods are anticipated through 2005 with Laser 3. Standard data products from ICESat/GLAS, as listed in Zwally et al. (2002a, b), are archived at the NSIDC ECS Distributed Active Archive Center (DAAC) at ⟨http://nsidc.org/data/icesat/⟩. The NSIDC archives and distributes 16 ICESat/GLAS products (Zwally et al., 2003b), including Level 1A, 1B, and 2 laser altimetry and atmospheric LIDAR data (see ⟨http://nsidc.org/daac/icesat/index.html⟩). A software tool designed to extract geoid and elevation

Sec. 3.6] Measurement of ice sheet topography/elevation and change in elevation 233

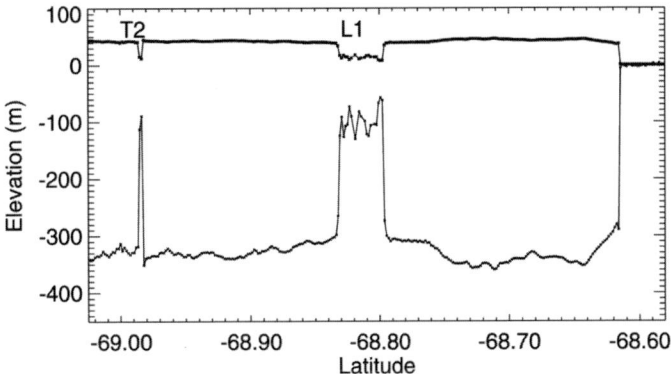

Figure 3.66. An elevation profile obtained by ICESat GLAS on October 18, 2004 across rifts T2 and L1 on the Amery Ice Shelf, for the ground track shown in Figure 3.44c. The ice bottom profile shown is derived from the hydrostatic relationship assuming the following column-averaged densities: 1,028 kg m^{-3} for seawater, 876 kg m^{-3} for ice-shelf ice, and 865 kg m^{-3} for interstitial "mélange" ice—e.g., in L1—(King, 1994b).
From Fricker et al. (2005). Copyright AGU 2005, reproduced with permission.

data from GLAS altimetry products—i.e., GLA06 and GLA12-15—is available from ⟨http://nsidc.org/data/icesat/tools.html⟩. Detailed descriptions of, and documentation for, GLAS products are available online at ⟨http://glas.wff.nasa.gov/⟩.

Laser altimetry has the unique capability of measuring the vertical distribution of surfaces within a single ground-resolution element (i.e., laser footprint). This result is achieved by digitizing the complex, time-varying return pulse energy (waveform) that comes from the reflection of a single laser pulse from the target surface. This waveform entails a measure of the vertical distribution of surface components weighted by illuminated area, reflectivity, and the spatial distribution of laser energy across the footprint. The use of visible radiation effectively removes another source of bias in conventional radar altimetry—namely, the variable penetration depth of the microwave signal into the snow/firn and the temporal and spatial variability in radar surface backscatter characteristics (Davis, 1993; Jezek and Alley, 1988; Ridley and Bamber, 1995; Ridley and Partington, 1988). As such, cross-mission crossover analysis provides a unique opportunity for calibration of the radar altimeters onboard ERS-2 (1995–2002), Envisat, and possibly a future CryoSat-type mission. The major disadvantage of spaceborne laser altimetry is its inability to penetrate significant cloud cover and to suffer reduced accuracies when penetrating thinner clouds. The GLAS LIDAR also directly measures cloud heights and aerosol structures (at a wavelength of 0.532 μm) (Spinhirne et al., 2005), for surface energy-balance calculations. It also obtains unique information such as heights and layering distributions on polar clouds, especially during the polar winter (see Chapter 4 of Volume 1 of this book). However, other technical issues have significantly reduced data return from the 0.532-μm channel.

ICESat data on seasonal and interannual changes in surface elevation may provide essential information for validation of atmospheric models of precipitation–evaporation in polar regions, as well as energy-balance models of surface melting and snow accumulation on the ice sheets. Combining the data with PMW-derived melt extent data also shows promise as a means of estimating meltwater volume from ice-sheet ablation zones. NASA's scientific strategy for monitoring glaciers and ice sheets also entails using GLAS in concert with an array of imaging sensors, including the Landsat-7 ETM+ (launched in 1999 and now partly functional), as well as Terra and Aqua MODIS, and Terra ASTER (launched December 1999). In fact, the ICESat orbit provides measurements coincident with those of EOS Terra and Aqua sensors quite frequently at high latitudes (Mahesh et al., 2004). An ICESat follow-on is under development and is anticipated before the end of 2010 (Koblinsky et al., 2001).

Continued measurements of ice sheet elevation change by satellites are the key to determining the variability in mass balance and sensitivity to climate change, as altimetry records are currently too short to enable confident distinctions between long-term ice sheet imbalance and short-term variation in surface mass balance (Houghton et al., 2001). Long (decadal-scale) time series of elevation changes will enable determination of the current ice-sheet mass balance, seasonal, interannual, and longer term changes in ice mass, causes of mass balance changes (e.g., by shifts in precipitation and ablation patterns, and ice-flow deceleration/acceleration), and estimation of the present and future contributions of changes in ice-sheet mass and volume to global sea-level rise (Zwally et al., 2002a). Long time series are required to better understand the "background" variability of processes determining elevation change and to enable assessment of the causes of observed change. These are essential prerequisites to developing a predictive understanding of ice-sheet variations, and a more accurate assessment of the present contribution of both major ice sheets to sea-level rise (Walsh et al., 2001). Another potentially important application of both ICESat and a future CryoSat-type mission is the measurement of changes in the surface elevation of icebergs over time (see Section 3.5.3). This could provide key information on the melt rates of icebergs, and thus improved estimates on their contribution to the ocean freshwater budget.

3.6.3 Improved digital elevation model construction using satellite image-based photoclinometry

While extremely useful, radar-altimeter-derived DEMs are typically limited by their coarse horizontal spatial resolution (of ~ 5 km). This in turn limits their inability to represent finer scale yet important ice-sheet topographic features—e.g., "undulation fields"—on the 1–10-km horizontal scale (Bindschadler and Vornberger, 1990; Scambos and Nereson, 1997; Seko et al., 1993). Ice-sheet topography at these scales is intimately related to bedrock topography and ice rheology (Budd, 1970; Budd and Jacka, 1989; Dowdeswell and McIntyre, 1987), and may be an indicator of variations in ice-sheet basal resistance (Balise and Raymond, 1985). Also of importance at this scale are relict ice-flow features, such as flowlines (Gudmundsson et al., 1998) and scars (Bindschadler and Vornberger, 1990; Scambos et al., 1999). Moreover, accumulation rates exhibit considerable variability across undulations with a horizontal scale of $\sim$5 km—e.g., in West Antarctica (Hamilton et al., 2000).

With these factors in mind, Scambos and Fahnestock (1998) developed an improved, enhanced-resolution DEM of NE Greenland by *photoclinometric processing* of cloud-free AVHRR data, by ingeniously combining existing altimeter-based DEMs (at a spatial resolution of 5 km) with satellite AVHRR images and exploiting the intimate relationship between ice-sheet surface slope and AVHRR radiometric sensitivity. By this method, digital image brightness in the visible to near-IR spectral region is quantitatively related to surface reflectivity and local slope orientation, the latter with respect to the direction of solar illumination. Filtered AVHRR images are compared with the coarse-resolution, independently derived DEM to determine the photometric relationship. This information is then used to convert unfiltered AVHRR data into physical surface slope measurements in the along-Sun direction in each image. This local detail is then combined with the regional elevation field to generate an enhanced DEM, thereby resolving features not present, or poorly resolved, in the original DEM, including possible paleo-drainage features in the Humboldt Glacier region of NE Greenland. The robustness of this approach as a means of "recovering" much of the topography missing from the original DEM was demonstrated by comparison with airborne laser-altimetry profiles. The improved topographic representation highlighted detail in the relief over ice streams, and resolved shallow troughs over the shear margins of ice streams (Scambos and Fahnestock, 1998). Information on governing relationships is given in Bindschadler and Vornberger (1994), Bindschadler et al. (2002), and Scambos and Fahnestock (1998). A critical assumption is that the surface behaves as a Lambertian reflector of constant reflectance—i.e., image brightness is directly related to the geometry of the surface topography. This approximation holds for near-nadir imagery acquired with moderate Sun elevations of 10–35° (Scambos and Haran, 2002), but not for large ranges of viewing and illumination angles (Nolin and Liang, 2000).

This technique of *data cumulation* was further refined by Scambos and Haran (2002), and extended to produce an image-enhanced DEM of the entire Greenland Ice Sheet, using a state-of-the-art DEM (Bamber et al., 2000a, 2001a) and 44 AVHRR images from the spring of 1997. Springtime was chosen due to the relative absence of clouds and uniform surface reflectance conditions. Channel-1 (0.58–0.68 µm) data are preferred due to their low sensitivity to snow-grain size variability compared with channel-2 data at 0.725–1.10 µm (Dozier et al., 1981). As shown in Figure 3.67, the new DEM has a much improved resolution, with a grid spacing of 625 m. By using multiple images rather than just two (as in Scambos and Fahnestock, 1998), slope information is derived over a wider range of solar illumination directions, thereby improving the photoclinometric procedure. Scambos and Haran (2002) have shown this to be a precise method as long as its inate potential for drift at large scales and noise at small scales can be constrained by combining with existing radar-altimetry-derived DEMs. The photoclinometric

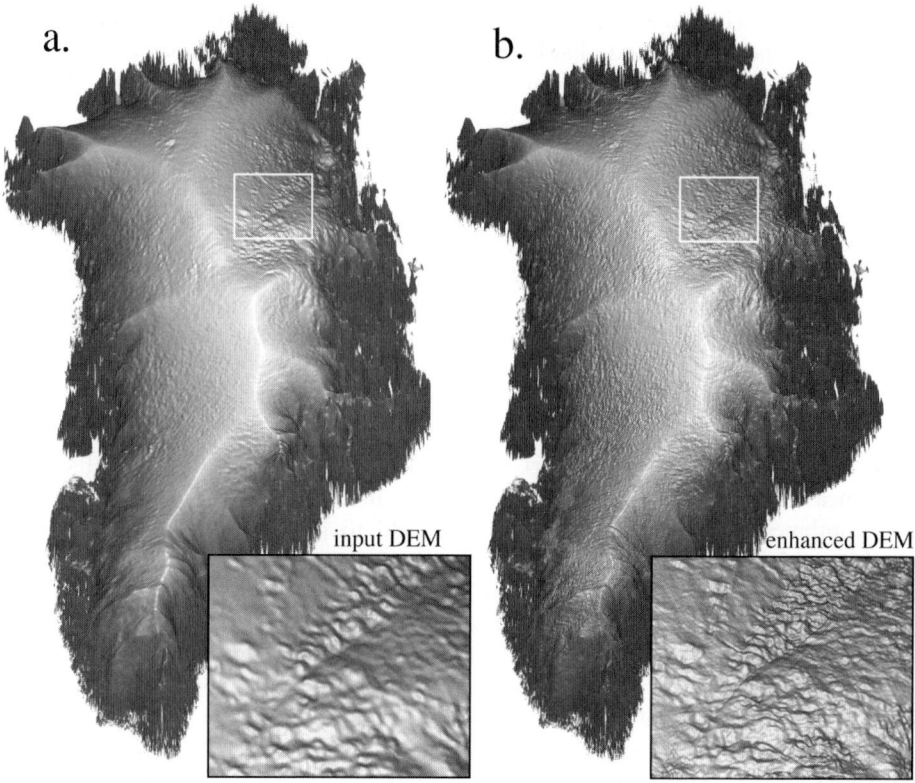

Figure 3.67. Shaded relief images of the Greenland Ice Sheet from (a) a satellite radar-altimeter-based DEM with a spatial resolution of ~5 km (Bamber et al., 2001a), and (b) the same DEM but photoclinometrically enhanced with data derived from NOAA AVHRR imagery and at a spatial resolution of ~1 km. Both images are grayscale representations of absolute slope. Insets (area = 187.5 × 156 km) to the lower right of each image show the "onset" of the newly identified ice stream in NE Greenland in more detail (this feature was discovered in ERS SAR imagery by Fahnestock et al., 1993).

From Scambos and Haran (2002). Copyright International Glaciological Society 2002, reproduced with permission.

technique adds detail on spatial scales of 3–20 km, thereby recovering a great deal of information missing in the original radar-altimeter-derived DEM.

Work is underway to improve this product even further using the enhanced radiometric and spatial resolution (250 m) offered by the MODIS onboard the EOS Terra and Aqua satellites. Scambos et al. (2002a) have constructed improved photoclinometric DEMs of Antarctica, using both AVHRR and MODIS data combined with satellite radar altimeter and airborne laser altimeter data. Figure 3.68 shows an example from the Ross Embayment ice streams, presented as a simulated three-dimensional perspective view. This region has experienced rapid recent changes in surface elevation as Whillans Ice Stream (B) draws down its

Sec. 3.6]   Measurement of ice sheet topography/elevation and change in elevation   237

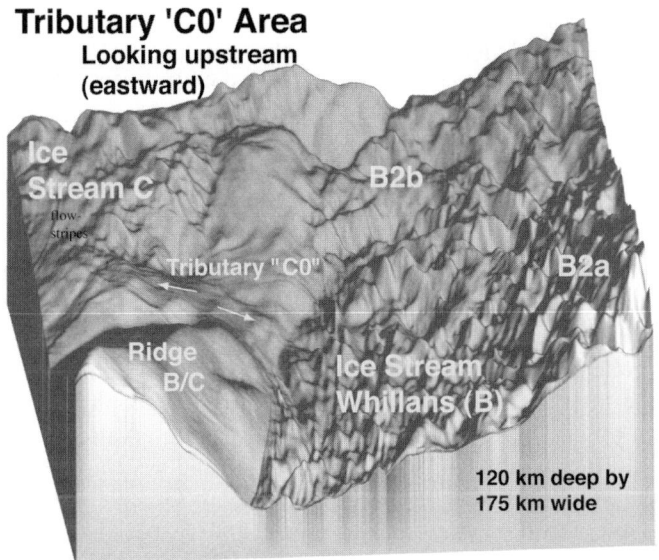

Figure 3.68. Sub-scene from the enhanced MODIS-derived photoclinometric DEM of the area near the onset regions of the Ross ice streams—i.e., part of the boundary between the Kamb Ice Stream (C) and the Whillans Ice Stream (B). The feature marked "C0", which was formerly a tributary to Kamb Ice Stream, has reversed slope and is now flowing back towards Whillans Ice Stream. The vertical exaggeration is $\sim$300.

From Scambos et al. (2002a). Copyright NSIDC 2002, reproduced with permission.

catchment to lower the surface relative to the Kamb Ice Stream (formerly known as Ice Stream C) (Conway et al., 2002; Scambos et al., 2002a). The new and improved DEM not only resolves these features but also provides important quantitative information on elevation at the flowlines and stream boundaries. The success of this technique is again dependent on the accurate detection and masking of clouds. Moreover, Haran and Scambos (2004) have produced a new DEM of central West Antarctica using MODIS and laser altimetry. With its improved spatial and radiometric resolution (12 bit), MODIS can map slopes as low as 0.0002, compared with 0.0007 for AVHRR (using photoclinometry). It should be noted that Terra MODIS level-1b 250-m data (MOD02QKM) are affected by a form of horizontal striping in contrast-enhanced ice sheet images, which must be removed to exploit their full radiometric sensitivity. Haran et al. (2002) address this issue.

3.6.4 Surface elevation from SAR interferometry (InSAR)

Covered in detail in Chapter 2, spaceborne repeat-pass InSAR is a unique and powerful method of determining both ice sheet height and/or motion in a simultaneous fashion. This is achieved from the phase information contained in highly correlated (coherent) and accurately co-registered pairs of single-look complex SAR images of the same scene but acquired a day to a few days apart and from slightly displaced locations along the satellite orbit track. Using the three-pass or double-differencing approach, two SAR interferograms with equal temporal baselines can be used to generate topographic maps of the ice sheet surface—i.e., by removing the contribution of ice motion from the measured phase signal. This approach works only where the ice flow remains invariant over the acquisition period of the constituent SAR images. Calibration using coincident *in situ* data and/or aircraft measurements—e.g., laser altimetry and ground-control points in the form of GPS measurements—improves the accuracy of the product and enables the derivation of absolute elevation values. The overall accuracy of satellite InSAR-derived ice sheet elevations is of the order of meters, depending on topography, coherence and availability of Ground Control Points (GCPs). Recalling Chapter 2, the interferometric coherence is affected by geophysical changes in the scene between observation, the time interval between observations, satellite-imaging geometry, the target brightness, signal bandwidth, and system noise.

Joughin et al. (1996a) determined topography in Greenland from ERS InSAR with an absolute elevation accuracy of $\sim$4 m, and a spatial resolution of $\sim$80 m (versus $\sim$10 km for satellite altimeter-derived DEMs). Forsberg et al. (2000) reported an accuracy of $\sim$15 m r.m.s. compared with independent GPS and aircraft laser data in the wet-snow and percolation facies of East Greenland. The study by Rignot et al. (2001) achieved a vertical precision of no better than $\pm$20 m in Greenland. Long-wavelength artifacts can at times undermine the accuracy of InSAR-derived DEMs. Joughin et al. (2001a) overcame this limitation by retaining only short-scale topographic interferometric information and combining it with satellite radar altimeter data (Bamber et al., 2001a) to form a DEM of the Northeast Greenland Ice Stream. The resultant improved horizontal grid resolution is typically better than 500 m (Ahlstrøm et al., 2002)—a significant enhancement on the altimetry-only DEM of $\sim$1 km at best (Ekholm, 1996). In other words, while the absolute accuracy of InSAR-derived ice sheet topographic measurements is generally poorer than that achievable by laser and radar altimeters (the latter for flat regions), the ice sheet is sampled in significantly greater horizontal detail (Joughin et al., 1996a; Kwok and Fahnestock, 1996; Mohr and Madsen, 1999; Mohr et al., 1998; Stenoien and Bentley, 2000).

An example, of a shaded surface three-dimensional representation DEM from the west coast of Greenland (Joughin et al., 1996a), was given in Chapter 2. Another example, from Joughin et al. (1998), is shown in Figure 3.69. The substantial improvement in detail afforded by InSAR—compared with satellite radar altimeter-generated DEMs—results in enhanced estimates of other important ice sheet parameters. Ahlstrøm et al. (2002), for example, found that the enhanced DEM resolution resulted in much-improved estimates of meltwater runoff in their modeling study of a drainage basin in West Greenland.

For floating ice masses, such as glacier tongues and ice shelves, better topographic maps are in general provided by satellite radar altimetry (Fricker et al., 2000a; Rignot et al., 2001), due to the relative difficulty of removing InSAR contamination by ocean tides. Where available, GPS measurements can aid the

Sec. 3.6] **Measurement of ice sheet topography/elevation and change in elevation** 239

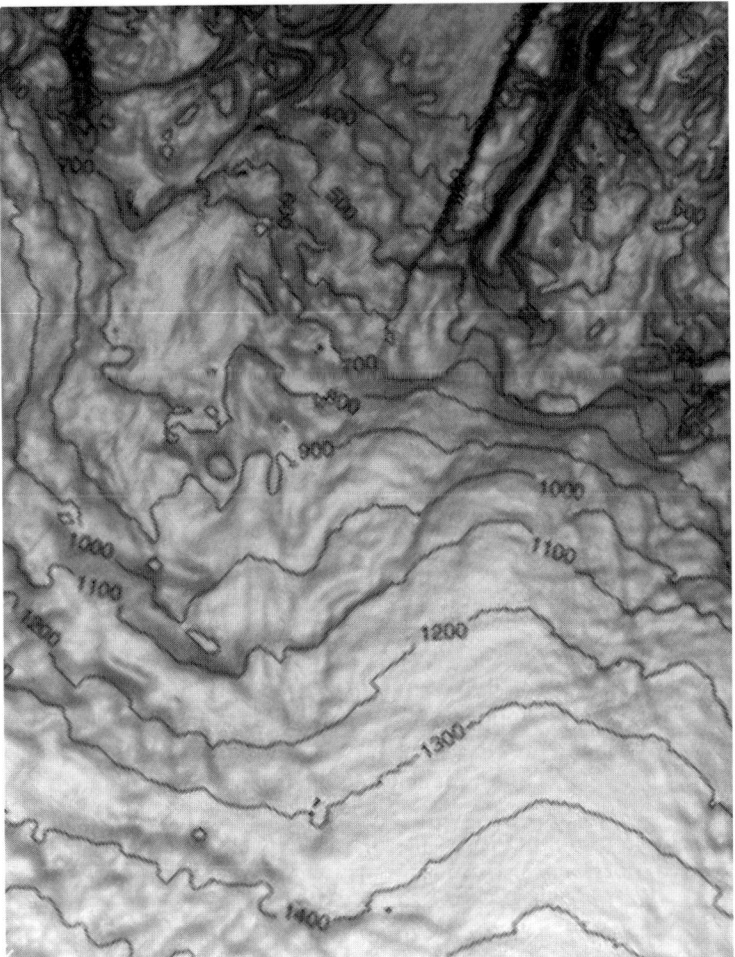

Figure 3.69. A high-resolution Digital Elevation Model (DEM) of Ryder Glacier, an outlet glacier at the northern edge of the Greenland Ice Sheet, derived from interferometric processing of ERS-1 SAR images. The DEM has a pixel spacing of 80 m. The effect of ice motion was canceled by double-differencing suitable pairs of interferograms (see Chapter 2). The DEM is shown as a shaded surface where the light is directed from above. Elevation contour intervals are 100 m. The SAR images used were acquired in March 1992, during the first ice phase of ERS-1. Approximately 1,600 tiepoints from the Danish KMS (National Survey and Cadastre) DEM (Ekholm (1996) were used to estimate the interferometric baseline. The relative short-scale accuracy of the InSAR-derived DEM is of the order of a few meters (Joughin et al., 1996b). Baseline and other errors may have introduced long-wavelength—i.e., >10 km—errors of up to a few decameters. Such errors have little impact on the accuracy of slope estimates. The DEM shows the topography for ice-covered areas only, as rugged topography makes it difficult to unwrap the phase in ice-free regions. The image width (x-direction) is ~75 km.

From Joughin et al. (1998). Copyright IEEE 1998, reproduced with permission.

removal of tidal contributions to the topographic signal (King et al., 2000), as can tidal models. Vaughan (1994) provides further information on the measurement of tidal flexure using kinematic GPS.

Interferometric data have also been used to fill in gaps where radar-altimeter-derived DEMs are most inaccurate—e.g., over steeply sloping terrain. Forsberg et al. (2000) reported that the application of InSAR techniques to the retrieval of heights and velocities in the outer margins of the East Greenland Ice Sheet is, however, complicated by the extreme topography present, which produces radar shadows and layovers and renders the unwrapping of interference fringes ambiguous. The assimilation of InSAR with ICESat GLAS data is enticing, particularly over steeper ice sheet margins where GLAS provides more accurate height retrievals than are possible with conventional radar altimetry (see Section 3.6.2). Error analysis remains a key current research area of satellite interferometry, as the latter becomes more extensively applied to mass-balance estimates which require a high degree of precision. Other inherent strengths and limitations of this extraordinary technique, including radar-penetration-depth-related biases, are evaluated in Chapter 2, and will not be repeated here. While data processing can be complex and the technique is some way from being routinely applicable, InSAR is without doubt an extremely exciting tool that is revolutionizing our ability to estimate ice-sheet mass balance as a result, especially when combined with other data. As we shall see in Section 3.11, InSAR-derived topographic information combined with InSAR velocity estimates and other data is making an extraordinary contribution to the derivation of ice fluxes and mass balance. Moreover, the emerging field of Pol-InSAR shows great promise as a means of eliminating the penetration bias that hampers ice-sheet elevation maps generated with single-channel interferometry (Dall et al., 2003, 2004)—see Chapter 2.

An alternative technique for constructing DEMs from SAR imagery is *radar-grammetry* or *stereo-radar imagery* (Leberl, 1990, 1998). This technique is similar to InSAR in that terrain elevation is derived from stereo-pairs of SAR images of the same area, but differs in that it uses amplitude rather than phase information from two images taken with a cross-track parallax (much larger than that used in InSAR). As a result, it is typically less sensitive and accurate than interferometry, and is generally not successfully applicable to featureless ice-sheet inland regions. Moreover, it is particularly sensitive to image speckle. Toutin and Gray (2000) provide an in-depth comparison of the general characteristics of DEM construction from remotely sensed data.

3.7 ACCUMULATION RATE

Accurate knowledge of patterns of accumulation and their variability across the ice-sheet surface is essential for the understanding of mass balance dynamics (Van de Wal, 2004; Van de Wal et al., 2001; Winebrenner et al., 2001), and on a variety of spatio-temporal scales. Current uncertainties in drainage-basin- and ice-sheet-wide accumulation rates, however, represent a primary error source in estimates of ice-sheet mass balance (Abdalati et al., 2004; Thomas et al., 2000b; Wahr et al., 2000). Net accumulation rates derived from recent estimates for Antarctica range, for example, from 2,020 Gt per annum (Giovinetto and Zwally, 2000) to 2,288 Gt per annum (Vaughan et al., 1999). As Richardson-Näslund (2004) points out, this difference corresponds to a global sea-level rise of ~0.7 mm per annum—a significant figure given that current estimates of overall global sea-level rise over the last century are 1.0–2.0 mm per annum (Church et al., 2001). Overall, greater knowledge of the interannual- and decadal-scale variability of accumulation is needed for mass-balance calculations using the direct-volume change method, while improved estimates of variability in accumulation rate (and the other important parameters) are required by the mass budget method as laid out in Section 1.2 (see p. 10) (Ohmura et al., 1999; Richardson-Näslund, 2004; Van der Veen and Bolzan, 1999). An improved knowledge of precipitation and redistribution processes over large regions is ideally required at scales of >10 km for ice-sheet- and atmospheric-modeling purposes (Abdalati et al., 2004).

Unfortunately, patterns of snow accumulation are notoriously difficult to measure *in situ*, and observations are sparse. It exhibits great spatial and temporal variability (Davis et al., 2001; McConnell et al., 2000a, b; Shuman et al., 2001; Van der Veen et al., 2001), related to large-scale weather patterns (Bromwich et al., 2000), to the prevailing wind regime on the regional scale, and on the local scale to surface topography (Goodwin, 1990). Precipitation falling on the ice-sheet surface is subject to reworking and downwind transport by synoptic and katabatic wind systems (Budd, 1970). At any given point on the ice sheet, net snow accumulation is the product not only of precipitation but also of snow lost and/or gained due to surface eolian redistribution, sublimation (Bintanja, 1998), and evaporation (Warren, 1996). This represents an error source in estimates of accumulation. Once again, *in situ* measurements are spatially very limited. Precipitation gauge undercatch at stations is well-documented. As such, a key issue in the interpretation of point measurements is their spatial representativeness, or lack thereof (Richardson-Näslund, 2004).

Abdalati et al. (2004) provide an assessment of current techniques used for measuring and estimating ice sheet accumulation rates. We shall briefly summarize these before moving on to satellite techniques. Local net accumulation rates, both over long periods and at a high temporal resolution, can be determined from ice/firn cores and snowpits with error estimates of ~5% for regional averages (Bales et al., 2001; Giovinetto and Zwally, 2000; Goodwin, 1991; McConnell et al., 2001; Morgan et al., 1991; Mosley-Thompson et al., 2001; Smith et al., 2002; Stenni et al., 2000). Multi-century data (from firn/ice core analyses) provide the backcloth against which to determine whether recently observed short-term changes reflect longer trends or natural variability (Alley and Bentley, 1988). These measurements are, however, logistically expensive and sparsely distributed, engendering considerable uncertainty in drainage-basin-scale accumulation estimates (Ohmura and Reeh, 1991). High-resolution ice-penetrating radar systems are being developed to extend such measurements spatially by detecting and mapping internal ice layers or isochrones in the upper 100–300 m, with individual layers being dated by a few ice-core analyses (Braaten et al., 2003; Fahnestock et al., 2001a; Kanagaratnam et al., 2001; Nereson et al., 2000; Richardson-Näslund, 2004; Welch and Jacobel, 2003). These systems are at present technologically limited to surface and airborne deployment, although work is underway to develop spaceborne systems (see Section 3.12). Another technique involves the use of sophisticated sonic devices attached to *in situ* AWSs and autonomous instrument packages. These provide satellite-transmitted point measurements of snow accumulation/erosion (Dahe et al., 2003), wind speed and direction, and firn compaction and firn temperature—e.g., as part of the 19-nation ITASE project. The latter aims to establish how atmospheric conditions over the last 200 years are reflected in the composition of the ice sheet's upper layers (Mayewski and Goodwin, 1997; ⟨http://www.ume.maine.edu/itase/⟩). Such configurations enable examination of the physical processes of snow deposition and erosion.

Modern ice sheet research uses output from General Circulation Models (GCMs) to prescribe climate and climate changes over the ice sheets in an effort to more realistically reproduce the spatio-temporal patterns of change in surface mass balance (Huybrechts, 2004; Huybrechts et al., 2004). An increasing emphasis is on forcing mass balance changes with GCM output—e.g., Fichefet et al. (2003); Huybrechts et al. (2004); O'Farrell et al. (1997); Van de Wal et al. (2001); and Wild and Ohmura (2000). Recent advances have taken place in the modeling of large-scale distributions of snowfall and net accumulation using atmospheric moisture convergence analyses based on improved meteorological analyses—e.g., from the U.S. National Centers for Environmental Prediction (NCEP)/National Centers for Atmospheric Research (NCAR) climate reanalysis dataset (Kalnay et al., 1996; Kistler et al., 2001) and European Centre for Medium-range Weather Forecasting (ECMWF) (Turner et al., 1999). Contemporary precipitation and accumulation fields have recently been produced both for Greenland (e.g., Bromwich et al., 1999, 2001; Hanna and Valdes, 2001; Hanna et al., 2001) and Antarctica (e.g., Connolley and King, 1996; Cullather et al., 1996; Smith et al., 1998; Vaughan et al., 1999). Such computations have benefited greatly from improvements in GCM performance over the polar regions in the past decade, with the incorporation of refined model physics and higher resolution (DeConto and Pollard, 2002; Wild et al., 2003). As a result, GCM distributions of annual precipitation over the Antarctic Ice Sheet, for example, agree with observations to within uncertainties in the observations themselves. Significant deficiencies remain, however, in the accurate simulation of seasonal cycles around coastal ice-sheet margins, where high accumulation rates occur. Fricker et al. (2000b) further concluded that current accumulation estimates are inadequate for the accurate estimation of mass balance, at least in the Amery Ice Shelf region of East Antarctica.

In general, models predict an increased melt contribution relative to accumulation in Greenland, but enhanced snowfall and accumulation in Antarctica (e.g., O'Farrell et al., 1997). The uncertainty in such predictions is again large, however. From an ice-sheet-modeling perspective, accurate accumulation-rate field information is required to construct maps of balance velocity (Fricker et al., 2000b), which form first-order estimates of ice sheet flow velocities—see Section 3.8 (Bamber et al., 2000a, c; Budd and Warner, 1996; Joughin et al., 1997). In addition, accurate maps of the spatio-temporal variability in snow accumulation are required to interpret ice-sheet surface elevation data from the new generation of high-precision satellite altimeters—namely, ICESat GLAS and a possible future CryoSat-type mission (see above). Seasonal and interannual variability must be removed in order to uncover long-term trends in surface elevation derived from time series of satellite altimetry data. With its minimal penetration depth, GLAS should in turn provide important accumulation information. Such information can also be inferred from the waveform characteristics of conventional radar altimeter data, with regions of high accumulation being characterized by low backscattering but a relatively high surface scattering due to enhanced surface-roughness features. The opposite is true for low-accumulation regions, due to a characteristic smoother surface (Rémy et al., 2001).

A promising approach is the synthesis of data from different yet complementary sources. The widely used map of accumulation rates for Antarctica compiled by Vaughan et al. (1999) used Nimbus-7 SMMR brightness–temperature data (1978–1987) as a background field in order to interpolate between the sparse *in situ* measurements available. This application requires low-frequency data, collected at $\sim 6\,GHz$. While such data were unavailable from the DMSP SSM/I series, low-frequency data have become available at an improved resolution with the launch of the AMSR sensors in 2002—i.e., a pixel size of $76 \times 44\,km$ versus $136 \times 89\,km$ for the SMMR.

While no all-encompassing technique to derive accumulation rate from satellite data currently exists, a number of avenues are being investigated in an effort to derive ice-sheet-wide accumulation patterns from satellite microwave remote sensing (Thomas et al., 2001a). The first comprises empirical methods. Using satellite radar scatterometer data from a variety of sources (i.e., ERS-1 and -2, NSCAT, and QuikSCAT), Drinkwater et al. (2001) estimated snow accumulation in dry-snow zones in Greenland from 1978 to 1996 by exploiting the strong relationship between the rate of decrease in backscatter coefficient at Ku-band frequencies and radar incidence angle for layered firn. The rationale is that this relationship is a function of the covariance of incidence angle, accumulation rate, snow-grain size, and density, and is therefore determined by pathlength-integrated volume-scattering effects in the upper firn layers. Results are shown in Figure 3.70 (see color section). Drinkwater et al. (2001) estimated retrieval errors to be $\sim 5\%$, by comparison of the satellite-derived accumulation with volcanic isochrone-inferred accumulation estimated by Dahl-Jensen et al. (1997) and snowpit and shallow core data (after Bolzan and Stroebel, 1994; Clausen et al., 1988; and Ohmura and Reeh, 1991) along the white line, and shown in Figure 3.70b. Once again, this technique exploits the frequent, high-incidence-angle coverage offered by radar scatterometers, with image enhancement (Early and Long, 2001; Long et al., 1993) resulting in a long-term dataset that is comparable with PMW in terms of its resolution. Indeed, Drinkwater et al. (2001) proposed that the acquisition of ice sheet daily coverage by combined active–passive sensors will result in an improved product. The launch of ADEOS-II in late 2002 provided such an opportunity, in that it carries both an AMSR and a SeaWinds-II Ku-band scatterometer. Unfortunately, however, the ADEOS-II AMSR failed in late 2003.

Wismann et al. (1996, 1997) and Wismann (1998) also derived snow accumulation over Greenland from satellite scatterometer data—i.e., from ERS-1 and -2, and

NSCAT. Their approach adopts a two-layer radar backscatter model to consider a firn layer buried by dry snow and to account for radar measurement parameters. By this method, they derived a snow accumulation rate from the estimated thickness of the dry-snow layer. In another study, Flach et al. (2005) demonstrated retrieval of dry-snow zone accumulation rate in the dry-snow zone of Greenland from *microwave-model inversion* of combined active (QuikSCAT) and passive (SSM/I) data.

Again working in Greenland, Forster et al. (1999) developed a coupled snow metamorphism–backscatter model, with observed accumulation rate and mean annual temperature as inputs, to predict dry-snow-zone accumulation as a function of C-band radar backscatter using ERS-1 SAR data. While promising, this technique encountered performance variability between swaths. A similar approach was adopted by Munk et al. (2003), who achieved an average difference between the satellite-derived accumulation-rate estimates and *in situ* measurements of $\sim$10% at best. Drinkwater et al. (2001) argued that low SAR incidence angles of 20–25° are sub-optimal for accumulation measurement, as surface roughness can significantly contribute to the measured backscatter (Jezek et al., 1993). This limitation, they suggest, can be largely overcome by the use of satellite radar scatterometers, which operate at higher incident angles (and longer wavelengths). Another empirical technique, developed by Bolzan and Jezek (1999), is based upon a strong correlation between annual (and multiyear) mean accumulation in a 150 × 150-km region of the Greenland Ice Sheet Summit and annual mean satellite PMW T_Bs at 19 and 37 GHz.

Other techniques adopt a more physical approach. Zwally (1977), for example, defined a relationship between dry-firn microwave emission, temperature, and accumulation rate by combining radiative transfer theory for scattering of ice grains in the uppermost few meters of firn with a model for isothermal grain growth. The rationale behind this approach is that low accumulation rates result in relatively large grain sizes within the penetration depth of the radiation and low T_B values, while smaller grains associated with high accumulation are responsible for much of the signal and result in higher observed T_Bs. As such, the T_B sensitivity varies inversely with the mean accumulation rate. The challenge is to derive absolute values of the accumulation rate from this relationship. This work was further developed by Zwally and Giovinetto (1995), who mapped large-scale accumulation rates in Greenland with an r.m.s. retrieval error of $\sim$10% using a technique based upon the covariance of *in situ* accumulation-rate observations and satellite microwave emission measurements at 19 GHz (1.55-cm wavelength), using Nimbus-5 Electrically Scanning Microwave Radiometer (ESMR) and coincident Nimbus-7 Temperature Humidity Infrared Radiometer (THIR) data. The physical basis of the covariance is the relationship between both microwave emissivity and accumulation rate and snow-grain size noted above—i.e., low-accumulation regions have larger grains near the surface, and therefore a lower microwave emissivity, than high-accumulation areas. Such an approach, which was also adopted by Giovinetto and Zwally (2000), requires regionally dependent empirical adjustments in emissivity/accumulation rate parameterizations. The relevant relationships at this shorter wavelength are, however, complicated by snow metamorphism and the development of large-grain depth hoar layers near the surface in response to high vertical temperature gradients.

The physical relationship between accumulation rate and microwave emission or backscatter depends upon a reliable and consistent relationship between accumulation rate and grain-size distributions at various relevant snow/firn depths (Alley, 1987; Comiso et al., 1982; Davis, 1995; Fily and Benoist, 1991; Sherjal and Fily, 1994; Zwally, 1977). As Winebrenner et al. (2001) pointed out, the firn–microwave interaction at 1–2-cm wavelengths is likely limited to the upper few meters (Zwally, 1977), within which the temperature is not isothermal but rather contains a strong seasonal signal. As a result, Winebrenner et al. (2001) developed a new technique to exploit microwave thermal emission data collected at longer wavelengths (4.5 cm, or 6.7 GHz frequency) using gridded 25-km-resolution T_B data from the Nimbus-7 SMMR (Gloersen et al., 1992) and snowpit data from Greenland (Ohmura and Reeh, 1991). The basis of this technique is that the emissivity at this wavelength is determined primarily by reflections related to density stratification (layering) within the firn, rather than by scattering from grains. Winebrenner et al. (2001) showed that a strong empirical relationship exists between firn density stratification (on scales up to a few centimeters) and accumulation rate, and also a strong theoretical link between density stratification and polarization of emission. The key layering characteristics here are mean layer thickness and the standard deviation of density fluctuations, which can be retrieved by spectral analysis of firn–pit density profiles (West et al., 1996). The new approach is also based on the fact that density layering in the snow has a greater impact on horizontally polarized emission than on vertically polarized. An important consideration is that the polarization ratio (see Section 3.3.3) is approximately independent of thermodynamic temperature for firn in the absence of melt (Arthern and Winebrenner, 1998). It provides a means of inferring the mean annual accumulation rate over large areas of the ice sheet which are unaffected by melt. The initial product is a 7-year average map of snow accumulation rates over central Greenland (Figure 3.71, see color section), accurate to an estimated 10%. Discrepancies occur in the presence of crusts, which strongly affect the polarization (Arthern and Winebrenner, 1998). Further work is essential to better understand the accumulation rate/firn-layering relationship. Similar accumulation maps were derived for Antarctica, using polarization of 6.7-GHz microwave emission, by Arthern and Winebrenner (1998) and Arthern et al. (2003). Unfortunately, a datagap exists from 1987 to 2002 in the availability of 6.7-GHz data (the DMSP SSM/I does not operate at this frequency). Improved estimates are likely, however, using the 6.925-GHz channel of the current AMSR(-E) data, given the improved spatial resolution compared with the SMMR. These data will be directly comparable with SMMR estimates for change detection. A drawback in using low-frequency data is the associated poor nominal spatial resolution—e.g., 136×89 km for SMMR and 76×44 km for AMSR.

It can be seen that the accurate measurement of absolute accumulation and accumulation rates using satellite techniques alone remains a major challenge, particularly equatorward of the dry-snow zone. The merging of satellite observations with modeling shows great potential, however. A key ongoing issue is the need to validate the satellite accumulation retrievals with detailed surface measurements. Meteorological measurements from AWS networks remain important as key means of estimating net water vapor flux to and from the surface (Box and Steffen, 2001). Sublimation/evaporation processes are highly sensitive to these fluxes, and represents a key yet poorly understood component of ice sheet mass balance, removing snow and ice and/or adding mass by condensation/deposition.

Massom et al. (2004) combined satellite PMW with modeled atmospheric analysis and AWS data to infer the unusual occurrence of snow accumulation events deep inland over the high-plateau desert zone of the East Antarctic Ice Sheet related to mid-latitude atmospheric-blocking anticyclone episodes. The latter, which have the effect of channeling moisture-laden cloud masses high up onto the Antarctic Plateau and into regions that are typically unaffected by synoptic weather systems, were detected in meteorological data and cloud pattern time series in satellite visible and TIR images. Following the work of Shuman and Alley (1993) and Mätzler et al. (1984), Massom et al. (2004) showed that the V/H ratio at 19 and 37 GHz is sensitive to short-term changes in ice sheet surface and near-surface properties related to snowfall events in low-accumulation regions (see also Section 3.16.1). The use of V/H ratios in this way minimizes the effect of changing physical temperatures on the observed signature of the radiating portion of the snow plus firn. By combining satellite microwave, AWS, modeled moisture fluxes, and *in situ* observations, Massom et al. (2004) found that a few such blocking high-related snowfall episodes, lasting only a matter of days, contributed a significant proportion of the mean (low) annual accumulation in inland sectors of the East Antarctic Ice Sheet—e.g., Dome C—where synoptic systems seldom penetrate to deliver snowfall. Earlier work—e.g., by Bromwich (1988)—suggested that, while snowfall occurs from clouds and is generated by orographic uplift and associated adiabatic cooling below an ice-sheet elevation of $\sim$3,000 m, clear-sky (ice crystal) precipitation in the form of "diamond dust" is the dominant mechanism above this level.

Rémy et al. (2001) highlighted the potential of dual-frequency radar altimeters as a means of inferring ice-sheet accumulation rates. This, they argue, could be achieved by inversion from the information on snow-grain size obtained at two different depths. The altimeter onboard Envisat shows promise in this respect, given its simultaneous operation at both 3.2 GHz (S-band) and 13.6 GHz (Ku-band). Further information on the extraction of information on near-surface characteristics of ice sheets from satellite radar altimeter data is given in Section 3.16.4.

An emerging field is the application of satellite SAR interferometry (see Chapter 2 and Section 3.8) to the measurement of spatial variations in ice-sheet accumulation. Working with ERS SAR data on the Greenland Ice Sheet, Hoen and Zebker (2000a) used all three products of interferometric processing—namely, phase, backscatter amplitude, and interferometric correlation—to estimate grain size and accumulation rate for a test site in the dry-snow zone. As the total interferometric correlation γ (see Chapter 2) decreases with increased volume scattering, it is in theory possible to infer snowpack-scattering characteristics (structure) from correlation observations if the thermal and temporal contributions to γ can be accounted for. Hoen and Zebker (2000a) observe greater backscatter but higher correlation on the leeward compared with the windward flank of small-scale ($\sim$10 km) topographic undulations, suggesting reduced-volume scattering in the snowpack there. Through simple modeling to relate (a) snow-grain size to backscatter and volume correlation and (b) the derived grain sizes to accumulation rate, they estimated variations in accumulation rate across the test area as a function of topography, finding a ~40% drop in accumulation between the windward and leeward sides. Further work is required to test the wider applicability of this technique and its accuracy.

An emerging technology that shows some promise for application to ice sheet research is Earth-reflected L-band signals from the Global Navigation Satellite System (GNSS). This system will be flown onboard TerraSAR-L, for example, from 2007 onwards (Zink, 2004). Wiehl et al. (2003) showed that the signal return from an ice sheet is complex but sensitive to both surface roughness and the internal structural characteristics of firn related to the accumulation rate, albeit averaged over a large scale. It follows that the accumulation rate may potentially be retrievable by inversion from GNSS data. The attractiveness of this technique is that GNSS signals are readily available free of charge, offer high spatial and temporal resolution, and require only passive instruments. Wiehl et al. (2003) suggested that the characteristic bistatic forward-scattering geometry makes this system complementary to other techniques—e.g., SAR and radar scatterometry. Given the large penetration depth of the L-band signal in dry firn—i.e., of the order of ~100 m (Mätzler, 2001; Rignot et al., 2000a)—the GNSS signals would contain information related to the millennial-scale accumulation rate only (Wiehl et al., 2003). This technique is in its infancy, and substantial work lies ahead before it becomes a widely used ice-sheet research tool. Wiehl et al. (2003) further suggested that it may be best used in conjunction with other L-band systems, including the 1.4-GHz (50-km resolution) Microwave Imaging Radiometer by Aperture Synthesis (MIRAS) sensor onboard ESA's Soil Moisture and Ocean Salinity (SMOS) mission (Kerr et al., 2000; Mätzler, 2001), to be launched in 2007 (⟨http://www.esa.int/livingplanet/smos⟩), the Hydrosphere State (HYDROS) mission to be launched in 2010 (Entekhabi et al., 2004), and the ALOS PALSAR (Hamazaki, 1999). A key feature of HYDROS is its unique ability to simultaneously acquire both passive and active measurements (the latter also at L-band), resulting in improved spatial reslution and accuracy (Long et al., 2005). Investigations are underway to assess the capability of L-band radiometry as a means of obtaining information on snow/firn layering at considerable depths of up to 150 m in dry polar firn. For technological information on GNSS, see Parkinson and Spilker (1996) and Seeber (2003). In spite of these advances and the potential of new technologies, the acquisition of a reliable and consistent large-scale picture of precipitation and accumulation over ice sheets (and sea ice) by satellite remote sensing remains elusive and a major source of uncertainty in mass-budget calculations.

3.8 ICE VELOCITY, STRAIN RATE, AND BALANCE VELOCITY/FLUX

Ice flow is intimately tied to ice sheet/glacier mass balance, geometry, and ice thickness (Paterson, 1994). As stated earlier, the gravitational driving stress downslope at the bed is directly proportional to the product of the surface slope and ice thickness (please see Paterson, 1994 for details). Ice-flow velocity controls the rate at which ice is transported from regions of accumulation to regions of ablation and/or loss by iceberg calving, and knowledge of the velocity and strain rate (related to the velocity gradient) is important in assessing the flow dynamics of ice sheets. Accurate ice-motion measurements are required to constrain models of ice-sheet and ice-shelf flow and dynamics. Such measurements are an essential prerequisite to the realistic estimation of ice-sheet mass budget, in concert with topography, ablation/accumulation, surface temperature, albedo, and ice thickness. Moreover, long-term and consistent satellite measurements of ice flow are required to interpret elevation change detected by ICESat and a possible future CryoSat-type mission (see Section 3.6). Information on the spatial variability of ice speed can also be used to infer spatial changes in basal resistance related, for example, to so-called *sticky spots* (Scambos et al., 2004a).

Understanding present day ice dynamics is also essential to any assessment of ice sheet stability. A high-profile example is the current debate about the possible causes and impact of apparent flow acceleration of outlet glaciers and ice streams (see Section 3.8.6) on the WAIS (Alley and Bindschadler, 2001; Bindschadler, 1998a; Bindschadler et al., 1998; Oppenheimer, 1998). All else being equal, such discharge-rate increases are predicted to lower surface elevation (Walsh et al., 2001), with some parties predicting a resultant collapse of the marine ice sheet (Mercer, 1968; Weertman, 1974, 1976; Thomas, 1979). Once again, our current lack of knowledge creates uncertainty and compounds the problem of forecasting future global sea-level rise (Gray et al., 1999). As we shall see, recent satellite-based studies are providing extraordinary new information with which to address these and other major questions, and are playing a major role in significantly reducing associated speculation.

Over the past decade, advances in satellite remote-sensing techniques have dramatically improved our ability to measure and monitor ice motion. *In situ* measurements using GPS technology enable high-precision point estimates of *three-dimensional ice velocity* and *ice kinematics*—e.g., Manson et al. (2000). Such data are, however, greatly limited in space and time and are logistically expensive. Once again, satellite remote sensing is eminently well-suited to large-scale measurement of ice-velocity fields and associated strain-rate distributions, enabling the collection of thousands of vectors over vast remote regions. Surface measurements remain, however, an important means of calibrating satellite retrievals (particularly where bedrock outcrops are lacking and on floating ice shelves, where tidal effects come into play: King et al., 2000; Padman et al., 2002; Rignot et al., 2000b). In this section, we provide a brief review of satellite techniques used to measure ice-sheet motion and the derivation of strain-rate information, and the estimation of discharge flux. Note that the *depth-averaged ice velocity* $\langle \vec{v} \rangle$ can be estimated from surface velocity measurements by applying a correction factor to account for the variation in horizontal ice velocity with ice depth. According to Wang and Warner (1998), this factor represents the major source of uncertainty in ice-flux calculations (see Section 3.11) if surface velocity and ice thickness are accurately known.

The following techniques are used to estimate ice sheet and glacier surface motion from satellite data:

- *feature tracking* using sequential time series of SAR amplitude images or cloud-free visible to TIR images; and
- *SAR interferometry* or *InSAR* (by *phase unwrapping*, *maximum-coherence tracking*, and *speckle tracking*—a form of intensity tracking).

By another technique—namely, *slant-range analysis* of satellite radar altimeter data— Zwally et al. (2002c) estimated the mean annual motion of Antarctic ice shelf outer margins (barriers) over the period 1978–1998. This technique is based on the short time (i.e., ~1 s) during which the radar altimeter that detects backscatter signals on a small surface elevation range or window fails to adjust to an abrupt elevation change (Martin et al., 1983; Thomas et al., 1984; Zwally and Brenner, 2001). Results are presented schematically in Figure 3.72, with drainage

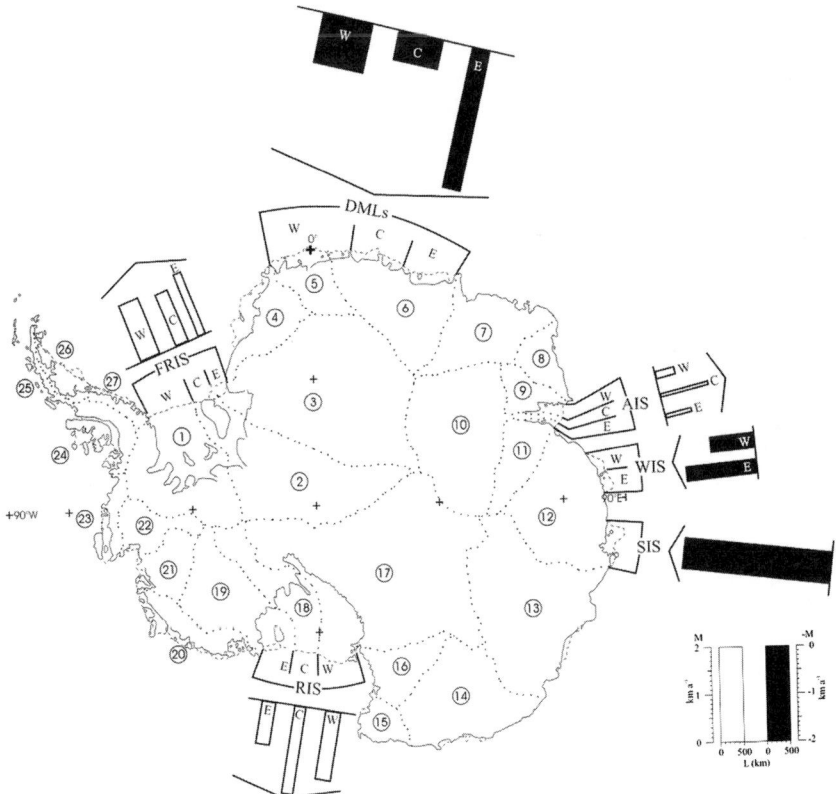

Figure 3.72. Barrier motion is estimated as the ratio between mean annual ice shelf area change for a particular interval, and the length of the discharge periphery. This value is positive if the barrier location progresses seaward, or negative if the barrier location regresses (break-back). Either positive or negative values are lower limit estimates because the method does not detect relatively small area changes due to calving or surge events. See table 1 in Zwally et al. (2002c) for a summary of minimum barrier-motion estimates for the ice shelf entities marked on this figure.

From Zwally et al. (2002c). Copyright International Glaciological Society 2002, reproduced with permission.

divides marked. The estimates represent the lower limit of regression or advance as the approach cannot account for either calving or surge events smaller than the barrier motion. This shortcoming is shared with all other methods that estimate barrier motion based on differences in barrier location over time (e.g., Keys et al., 1998).

3.8.1 Feature tracking

Feature tracking determines the ice-surface velocity field by detecting and tracking identifiable features that remain coherent in space and time—e.g., crevasses—in accurately ortho-rectified and co-registered sequential pairs of high-resolution images, using cross-correlation techniques. These features act as markers which persist in time and move with the ice. The immense potential of this technique as a means of greatly extending *in situ* measurements was realized with the launch of the Landsat program in 1972. Conventional image co-registration is by means of satellite orbital information and the matching of well-dispersed fixed points, such as nunataks and coastal features. The velocity field then represents a mean over the period between image acquisitions. The technique has been widely applied to the following satellite datasets:

- optical images—SPOT and Landsat, in particular (Bindschadler and Scambos, 1991; Bindschadler et al., 1994, 1996; Doake and Vaughan, 1991; Dwyer, 1995; Ferrigno et al., 1993; Frezzotti et al., 1998b, 2000; Lefauconnier et al., 1994; Lucchitta and Ferguson, 1986; Lucchitta et al., 1993a, 1989; MacDonald et al., 1989; Price and Whillans, 2001; Scambos et al., 1992; Stearns and Hamilton, in press; Vaughan et al., 1988; Whillans and Tseng, 1995; Whillans and Van der Veen, 1993);
- SAR amplitude images (Fahnestock et al., 1993; Jiabing et al., 2003; Lucchitta et al., 1994, 1995; and Lucchitta and Rosanova, 1997); or
- a combination of the two (e.g., Rosanova et al., 1998; Rosanova and Lucchitta, 2002).

Of these, optical feature tracking has been more widely used in ice-sheet research. The use of SAR amplitude image time series has been largely limited to a few key outlet glaciers, including the Pine Island Glacier in West Antarctica (Figure 3.73). Although this technique is unaffected by cloud cover, it suffers the same limitation as optical image feature tracking in that identifiable features and detectable displacements must be present between image acquisitions (Forster et al., 2003). By penetrating the surface, however, the SAR is likely to locate more tracking features related to internal inhomogeneities and crevasses. As a general rule, care must be taken when comparing data from sensors with different accuracies and attributes. For example, feature tracking is likely to be more accurate using ERS-1 SAR compared with earlier Landsat MSS data, due to the better location accuracy, internal geometry, and resolution of the former (Lucchitta and Rosanova, 1997). A variation on feature tracking—namely, SAR speckle tracking—is covered in Section 3.8.2.

Sec. 3.8] Ice velocity, strain rate, and balance velocity/flux 251

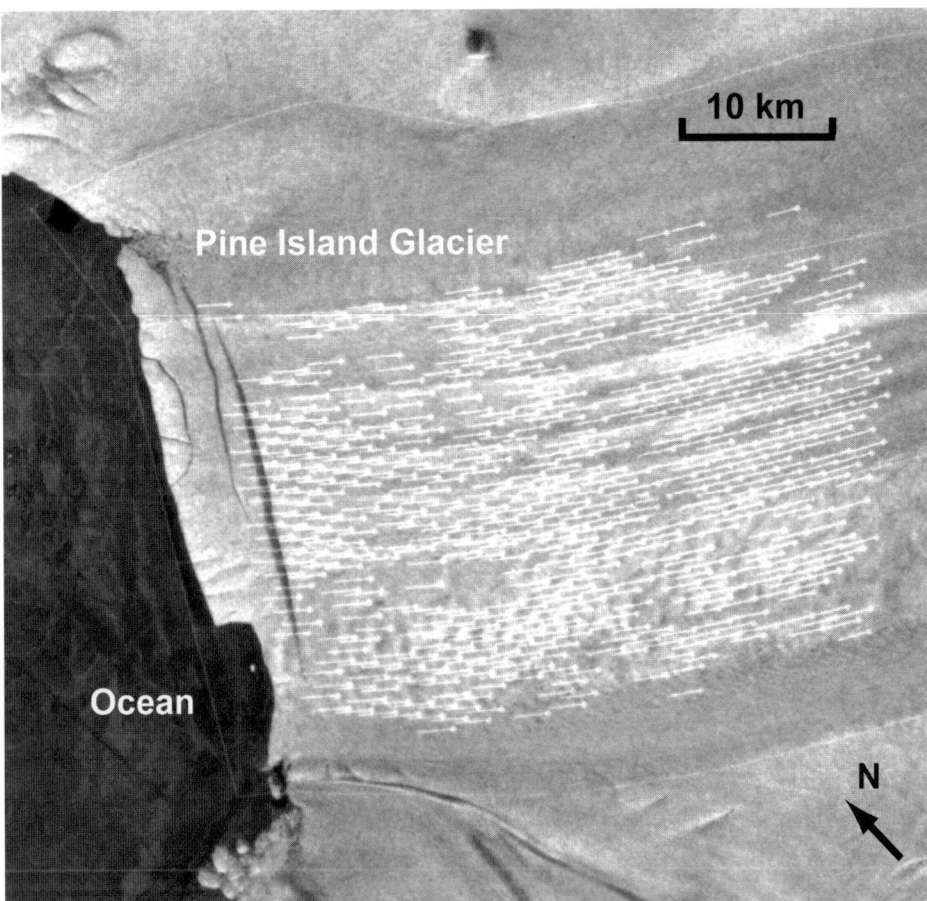

Figure 3.73. Surface ice-velocity vectors (shown as white lines, with motion to the left) superimposed on an ERS-1 image (February 9, 1992, images 2970–5607) of the lower part of Pine Island Glacier, Antarctica. Velocity vectors were computed by applying the automated cross-correlation method developed by Bindschadler and Scambos (1991) and Scambos et al. (1992) to the sequential images. The displacement vectors represent averages over the interval February 9, 1992 to December 4, 1992. The fast-moving central part of the glacier flows at $\sim$1.5 km per annum above and $\sim$2.6 km per annum below the grounding zone, with velocities as high as $\sim$2.8 km per annum at the terminus. Note the iceberg calving from the terminus.
From Lucchitta and Rosanova (1997). Copyright ESA 1992, reproduced with permission.

Conventional feature tracking is limited both by its relative lack of precision (based on the image pixel size), and its requirement that nunataks or other fixed points are present in the images to aid image co-registration. Unfortunately, such features are not typically present over large regions of both major ice sheets, yet these regions can be dynamically complex and are of central importance to the overall mass balance (Stephenson and Bindschadler, 1990). In response, Bindschadler and Scambos (1991) and Scambos et al. (1992) developed an ingenious means of accurately co-registering images over featureless ice-sheet inland regions. Using low pass-filtering over large image areas, this approach involves the correlation of subtle long-wavelength surface undulations (Dowdeswell and McIntyre, 1987; Swithinbank, 1988), related to underlying bedrock topography (Budd and Carter, 1971), to provide image co-registration to an accuracy of ~ 1 pixel. The rationale is that these large-scale features remain fixed relative to smaller scale surface features, which move with the ice. The second step involves applying an identical co-registration adjustment to two high pass-filtered images. The tracking stage is based on an image-to-image cross-correlation which applies the normalized cross-covariance correlation method (Bernstein, 1983) to the pattern of pixel brightness contained within sequential images, with small-scale features—e.g., snow dunes and crevasse scars— being used as markers. Surface feature displacement is mapped by (i) selecting small image areas centered on distinct features, or (ii) by dividing a large densely featured area into a grid and searching a subsequent image for matching areas using cross-correlation. As features are represented by small rectangular image areas or "chips" rather than single pixels, interpolation of peak correlation values enables the measurement of surface displacements to sub-pixel accuracy—i.e., 0.2 pixels, or $\sim 6\,\mathrm{m\,a}^{-1}$ for Landsat TM data (Scambos et al., 1992). The product is a detailed map of the velocity field comprising a high density of velocity measurements (Figure 3.74). Its precision depends on the time separation between images, and is a combination of systematic errors associated with chip correlation and image co-registration (Bindschadler, 1998b; Scambos et al., 1992). For Landsat TM data with a ground resolution of 28.5 m and a temporal separation of 2 years, Bindschadler and Scambos (1991) estimated that ice speeds of as low as 30 m per annum could be measured to a precision as great as ± 3 m per annum. As an indication of the extraordinary usefulness of this technique, Scambos et al. (2004a) were able to derive 7,380 vectors from a pair of Landsat TM images of Antarctica dating from 1986 and 1989.

Bindschadler et al. (1996) applied this technique to a surface velocity and mass balance study of Ice Streams Bindschadler (D) and E, two of the major streams feeding into the Ross Ice Shelf in Antarctica. The color-coded product in Figure 3.75 (see color section) entails over 75,000 measurements of velocity. As a result, it shows a level of detail and complexity that would have been impossible to determine from *in situ* data alone, but with an accuracy comparable with the latter. In this particular case, the new velocity field enabled a revision of the previously held belief notion that these ice streams were characterized by a "plug flow" pattern, with the central section of the streams discharging at a uniform speed (Bindschadler, 1998a). The use of high-resolution Landsat and SPOT data has greatly enhanced knowledge of the spatial extent of ice streams using both Landsat and SPOT data (e.g., Scambos and Bindschadler, 1991, 1993; Merry and Whillans, 1993), compared with previous studies which used moderate-resolution AVHRR data (e.g., Bindschadler and Vornberger, 1990). Subsequent studies have used high-resolution, narrower swath Landsat and SPOT imagery to focus in on regions of interest.

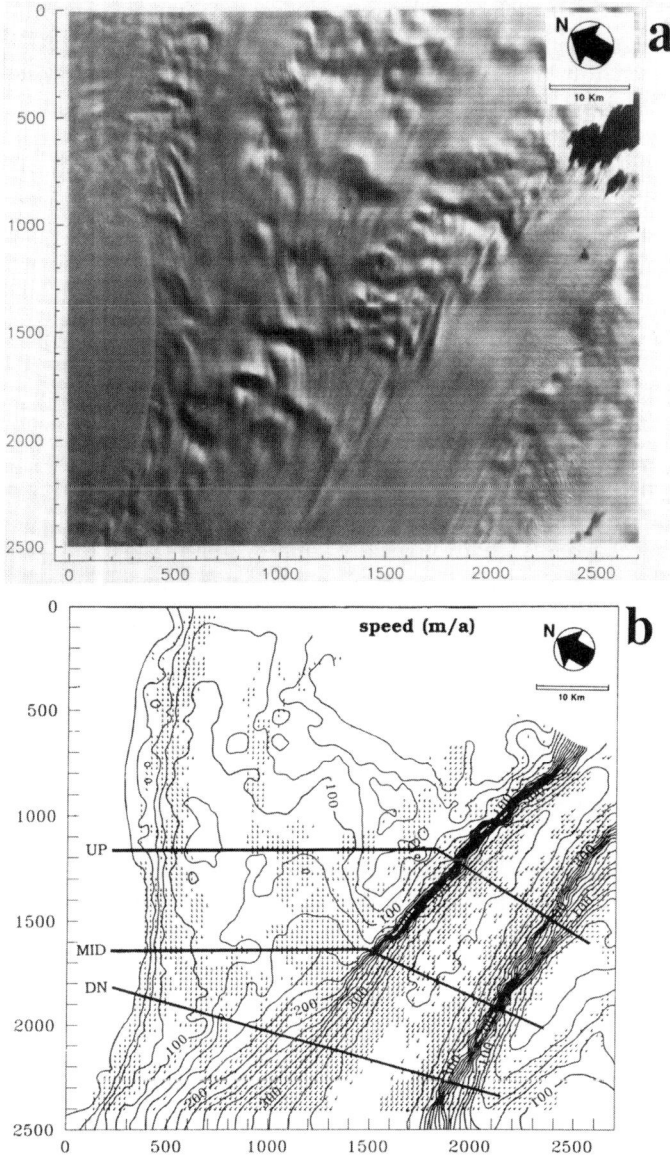

Figure 3.74. (a) A Landsat Thematic Mapper (TM) image of the upstream portion of Bindschadler Ice Stream (D) from January 16, 1987. Ice flow is from top to bottom in the image, with flowbands parallel to the flow direction. Axes are image coordinates for the study area in pixels. (b) Velocity field for the study region, with flow directions, locations, and magnitudes of the displacement measurements used for the mapping shown as arrows that scale with the speed. Contours are in meters per annum. In this case, the scene size is 77 km × 71 km.

From Scambos and Bindschadler (1993). Copyright International Glaciological Society 1993, reproduced with permission.

The availability of such high-density velocity data has also enabled the study of large-scale ice stream dynamics for the first time (MacAyeal et al., 1995). Recent ice dynamics research in West Antarctica has focused on ice stream onsets, defined as the location at which ice flow changes from inland-type flow (dominated by internal ice deformation) to streaming flow (dominated by basal processes) (Bindschadler et al., 2000, 2001b). Satellite remote sensing of ice flow is providing the opportunity to examine onset behavior and characteristics, which are centrally important to the long-term ice sheet balance state (Price et al., 2002). Bindschadler et al. (1997) used feature tracking to locate the onset of Ice Stream D (Bindschadler) and characterize the ice deformation and flow in its vicinity. Further work by Bindschadler (1997) and Price and Whillans (2001) suggests that certain onsets may indeed be migrating inland, thereby lengthening the influence of ice stream discharge and increasing the latter. The onset, discharge rate, and width of ice streams are controlled by the complex and poorly understood interplay of thermal, hydrological, topographic, and geological controls (Anandakrishnan et al., 1998; Bindschadler et al., 2001b; Jacobson and Raymond, 1998). Of importance is apparent switching between slow and fast modes of flow (Houghton et al., 2001), as observed in paleo-records. This may occur mechanically, hydrologically, and/or thermally, and is indicative of "surge-type" behavior, which is a form of cyclical instability (Kamb et al., 1985; Bindschadler, 1997; Reeh et al., 1994). The slow-flow process for ice sheets is internal creep, whereas fast-flow processes encountered in ice streams are sub-glacial sediment deformation, basal sliding, and enhanced creep (Alley, 1989; Engelhardt and Kamb, 1998; Kamb, 2001; MacAyeal et al., 1995). The improved understanding of these mechanisms involves a synthesis of modeling, *in situ* measurement, laboratory experiments, and satellite remote sensing.

Such studies have benefited from an extensive Landsat acquisition program over Antarctica in the early- to mid-1970s and another coastal-mapping program that has been in operation since 1984 under the auspices of the Scientific Committee on Antarctic Research (SCAR). Earlier MSS images are limited by their moderate resolution (~80 m), permitting tracking of only the larger patterns in the floating part of ice tongues or shelves. Subsequent TM images have a higher resolution (about 30 m), as do current Landsat-7 ETM+ images, but the data are expensive (although highly cost-effective compared with *in situ* measurements). Complementary ice-sheet coverage by SPOT tends to be limited to acquisitions during individual research campaigns. Current acquisition strategies for Landsat-7 ETM+ and Terra ASTER data (Kargel et al., 2005) are providing annual datasets (Raup et al., 2000)—e.g., for the Global Land Ice Measurement from Space (GLIMS) project[7] (⟨http://www.glims.org/⟩ and ⟨http://nsidc.org/data/glims/⟩), whereas earlier high-resolution studies are severely limited by their "spotty" coverage (Walsh et al., 2001). In Alaska alone, over 15,000 glaciers are being monitored in this fashion. A compilation of ice-velocity data from around Antarctica is available from the NSIDC at ⟨http://nsidc.org/data/velmap/⟩.

---

[7] The GLIMS project and results will be presented in detail in another Springer-Praxis book, due to be published in 2006.

In spite of its immense contribution, feature tracking using Landsat/SPOT or similar imagery has inherent weaknesses. Velocity retrieval errors, of up to a maximum of 15–20 m per annum, are insignificant for fast-moving ice streams and outlet glaciers (with typical velocities of 400–1,000 m per annum), but significant for slower moving small glaciers (with velocities of $\sim$100–150 m per annum) (Frezzotti et al., 1998b). The technique yields medium- to long-term average velocity only, with typically a year or more between sequential image pairs (e.g., Scambos et al., 1992). Moreover, the acquisition of suitable image pairs is severely hampered by persistent cloud cover (over ice-sheet margins, in particular) and, in the case of visible to near-IR data, polar darkness. Also, the density and distribution of velocity retrievals depends on the distribution and quality of features required for tracking. As a result, the highest density of observations is usually restricted to heavily crevassed ice-stream confluence regions, dropping off significantly inland where vast featureless terrains dominate (Young and Hyland, 2002). This results in a velocity field which is spatially irregular, even with a good dataset. As we shall see in the following section, the recent emergence of satellite SAR interferometry and coherence-tracking techniques fills a major gap in this respect, as well as providing additional key information. Having said this, an intriguing recent study by Bindschadler (2003) showed that sastrugi patterns detected in high radiometric resolution (12-bit) and spatial resolution (10 m) imagery from the EO-1 ALI persist over long-enough time periods that their movement with ice surface flow can be tracked in images separated by a number of months (using the cross-correlation technique of Scambos et al., 1992). This provides a potentially important source of ice-velocity data in relatively featureless uncrevassed regions of the central ice sheet (at least to the latitudinal limit of the sensor). Bindschadler (2003) demonstrated that the requisite correlation between images was largely unaffected by variations in Sun angle and azimuth or slight changes in surface reflectance. A current disadvantage is the narrow coverage of such sensors—i.e., 37 km for ALI.

Recent remote sensing and *in situ* studies (e.g., GPS and ice-penetrating radar programs) have identified significant changes in the configuration and flow speed of fast-moving ice streams on timescales ranging from decadal (e.g., Bindschadler and Vornberger, 1998) through century (Shabtaie and Bentley, 1987) to millennial (Bindschadler, 1998a; Casassa et al., 1991). These findings have necessitated an alteration of the paradigm of fixed ice-stream drainage channels, and show that understanding past discharge patterns holds a key to improving our understanding of present and possible future ice sheet behavior. "Scar-like" relict flow features, which indicate that dynamic ice behavior once occurred in regions which are now inactive, have been detected in AVHRR, SPOT, and Landsat imagery since 1990 (Bindschadler, 1998b; Bindschadler and Vornberger, 1990; Casassa et al., 1991; Jacobel et al., 1996, 2000; Scambos and Bindschadler, 1991; Stephenson and Bindschadler, 1990) (Figure 3.76). Such changes complicate the prediction of future ice-stream behavior, and image analysis of relict surface features—e.g., flow bands and scars—combined with ground-based radar studies of the stratigraphy (Jacobel and Gades, 1995; Jacobel et al., 2000) provide means of determining their timing and sequence. Fahnestock et al. (2000a) were able to infer a likely sequence of

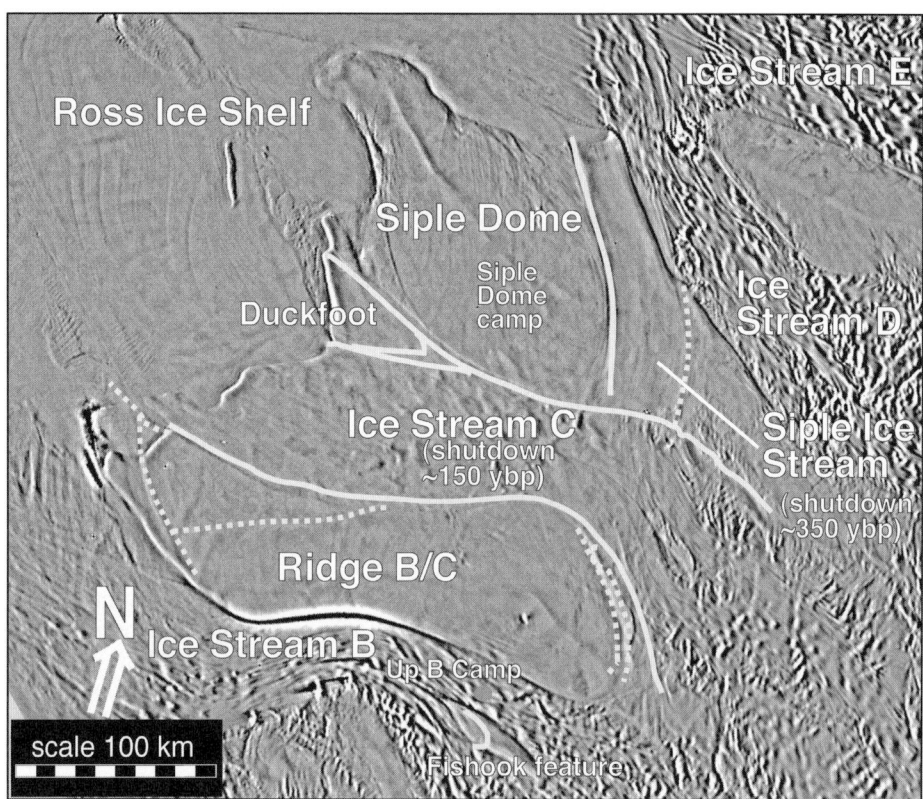

Figure 3.76. An overview of the central Siple Coast region of Antarctica in an AVHRR image mosaic, with subtle morphological features revealed by enhanced image processing—e.g., data cumulation. Confirmed former shear margins are marked as solid lines, while dashed lines represent suspected margins. Relict flow lines demarcate former ice stream trunk regions, while scar features mark former ice shear margins. Such is the large-scale and low elevation of such features that they are virtually impossible to detect and delineate at ground level.

Modified from Jacobel et al. (2000). Copyright International Glaciological Society 2000, reproduced with permission.

ice stream speed-ups, slowdowns, and catchment draw-downs on the Ross Ice Shelf over the last 1,000 years by combining satellite-derived flow-stripe information with radar profiles of inter-stream ridges. Such research has benefited greatly from image enhancement by data cumulation (see Section 3.6.3), with satellite data also playing a key role in guiding field activities. Intriguing patterns have emerged. For example, Kamb Ice Stream (C) appears to have stagnated from fast flow ~130 years ago to <10 m per annum at present (Retzlaff and Bentley, 1993), while the neighbouring Whillans Ice Stream (B) is currently fast-flowing (400–800 m per annum) (Anandakrishnan et al., 2001). Joughin and Tulaczyk (2002) presented evidence that the overall mass balance of the Ross Sea sector of the WAIS is now positive, largely as a result of a slowing down of Kamb Ice Stream (C) ~150 years ago.

3.8.2 SAR interferometry and ice velocity

In the decade since the ground-breaking study of Goldstein et al. (1993), *InSAR* has become established as a unique means of deriving not only ice-sheet topography (see Section 3.6.4) but also surface flow velocity, the latter to an unprecedented precision (Joughin et al., 2000). Measurements by InSAR of ice sheet and glacier flow dynamics (e.g., Joughin et al., 1995, 1996a, 2001a, b; Rignot et al., 1995, 2001, 2004; Winebrenner et al., 1997; Wunderle and Schmidt, 2000) are increasing the number of velocity data points by several orders of magnitude over large areas, and are dramatically improving our ability to constrain numerical modeling of key parameters controlling the dynamics of ice flow, including basal boundary conditions and ice rheology. Comparisons of InSAR-derived ice motion and coincident GPS measurements in regions of grounded ice suggest that ice velocity can be measured with a precision of $2-5\,m\,a^{-1}$ under favorable conditions (Joughin et al., 1999a, b, 2000; Mohr et al., 1998; Rignot et al., 1995), or in practice between 10 and $50\,m\,a^{-1}$ (Rignot, 2002a; Rignot et al., 2004). The principles, strengths, and limitations of SAR interferometry as they relate to ice-sheet research are dealt with in detail in Chapter 2, and will not be repeated here. Rather, in this section, we aim to illustrate the spectacular contribution that this technique is making to the study of ice-sheet dynamics.

By the satellite repeat-pass technique, two (or more) complex SAR images of the same surface area but acquired from slightly different orbit configurations and at different times are carefully co-registered and combined to exploit the subtle phase difference within each pixel and produce an interferogram. As noted previously, the interferometric phase contains contributions from both surface topography and coherent displacement along the look vector between acquisitions of the interferometric image pair (Bamler and Hartl, 1998). In effect, the sensitivity of the topographic term is determined by the spatial separation of the satellite sensor for the two measurements, termed the interferometric baseline (see Chapter 2 for details). The topography-related phase can be removed from the interferogram to produce a standalone surface displacement map by two methods: (i) using an independent and accurate DEM or (ii) multiple interferograms (Joughin et al., 1995, 1996a, b, 1998; Kwok and Fahnestock, 1996). The latter technique is a form of Differential InSAR, or DInSAR. By this method, topography is first isolated by double-differencing two interferograms of the same region with equal temporal baselines (see Section 3.6.4). The motion-only interferogram is then derived by additional differencing using the topography-only interferogram and either of the original interferograms. An example of ice-sheet velocity and topography obtained in this fashion is shown in Figure 3.77 (see color section).

An important prerequisite for accurate InSAR measurements of ice sheet velocity is knowledge of the interferometric baseline to an accuracy of a centimeter or better. Unfortunately, orbital-state vectors of spacecraft such as ERS-1 and -2 are insufficiently well-known to derive accurate baselines. Baseline errors, which result in a near-linear phase ramp in the interferogram (Bamber et al., 2000c), can be reduced by incorporating GCPs of known velocity and elevation—e.g., *in situ*

GPS measurements—to solve for the baseline. An alternative source of control, which can be used where sufficient GCPs are unavailable, is maps of balance velocities (see Section 3.8.3 for an explanation of the latter) (Joughin et al., 2001a). Another consideration is that an interferogram provides a measure of relative ice motion only, and GCPs are required within the interferogram to calibrate the data in order to provide an absolute reference (Mohr et al., 1998). Exposed bedrock outcrops—e.g., nunataks—provide tie-points of known zero-velocity suitable for this purpose (Mohr et al., 1998). In an ideal situation, control points would entail repeat GPS surveys of ice stakes (Joughin et al., 2001a). Where *in situ* tie-points are unavailable, which is often the case, balance velocities can be used as velocity tie-points in regions of slow flow (Bamber et al., 2000c; Joughin et al., 1997, 2001a). Regarding the latter, Bamber et al. (2000c) imposed a maximum threshold of 12 m per annum, beyond which the technique breaks down.

Importantly, InSAR is not dependent on solar illumination and cloud-free conditions, and can also provide measurements of ice discharge in featureless ice sheet interior regions where conventional feature-tracking techniques typically break down. Whereas feature tracking using Landsat/SPOT or SAR amplitude data produces velocity fields representing means over a period of months or even years, InSAR can uniquely provide a quasi-instantaneous velocity estimate—i.e., over a matter of days using current satellites. This is of key importance in that it enables resolution of short-term acceleration or deceleration in ice flow (Joughin et al., 2004a). Rignot et al. (1996) concluded that InSAR-derived velocities and those derived by conventional feature-tracking techniques overlap and complement each other. While individual InSAR-derived pixel velocities may not be as accurate as GPS measurements, satellite InSAR has the enormous advantage of acquiring a large number of measurements over a wide area, and over a denser spatial network than conventional feature-tracking techniques. This enables both the constraint of the ice dynamics (Legrésy et al., 2000) and the detection and monitoring of changes in ice-flow regimes. As a result, many subtle features in the fine-scale velocity structure and surface elevation are being revealed for the first time in many cases (Bindschadler, 1998b). Rignot (2001), for example, used ERS InSAR data to show that much of the floating tongue of the Thwaites Glacier (West Antarctica) is undergoing a solid-body rotation under the influence of ocean currents. Combining InSAR-derived velocity fields with accurate DEMs (either derived by InSAR or altimetry) is also yielding extraordinary results, and further insight into subtle flow regimes—e.g., Bamber et al. (2000b).

Given their high spatial density over large areas, InSAR velocity measurements are directly comparable with the output of ice-sheet models, and are enabling regional mass-budget estimates to be made, again often for the first time (see Section 3.11). This technique has broad application, and may be used to estimate the full vector motion field where crossing-orbit data—i.e., from ascending and descending passes—are available (Joughin et al., 1998; Mohr et al., 1998). An assumption is of surface-parallel flow (again see Chapter 2 for further details). Examples are shown in Figures 2.14 and 2.15. Where data are available from one pass only (which is unfortunately most often the case), then the InSAR-derived velocity is along the radar Line-Of-Sight (LOS) direction alone—i.e., it is one-dimensional. Interferometry is bounded by a number of strict constraints, such as temporal decorrelation of the phase, atmospheric artifacts, and long-wavelength errors, which limit the amount of suitable data currently available. Where suitable data are available, however, this extraordinary tool produces unique results, and is particularly powerful when supplemented by field and airborne measurements (again, please see Chapter 2).

On floating ice shelves and glacier tongues, the retrieval of ice velocity using InSAR (and SAR speckle tracking) techniques is complicated by a strong vertical tidal component. As a result, an incorrect compensation for vertical tidal motion will introduce a significant bias error for the range component of horizontal ice motion. King et al. (2000) estimated that vertical and horizontal movements of ice shelves can be as great as 0.3 m and 0.1 m, respectively, over a 1-hour period. As such, the retrieval of horizontal-creep flow (Hulbe et al., 1998) requires corrections for vertical tidal displacements, using differential interferometry (Rignot, 1996), data from regional tidal models (e.g., Padman et al., 2002), and/or coincident GPS measurements (Keller et al., 1997; King et al., 2000). Rignot et al. (2000b, 2001) removed tidal motion by employing a quadruple-difference interferogram with tidal predictions from the FES95.2 tidal model (Le Provost et al., 1998). Rack et al. (2000) further described tidal correction methods when deriving ice shelf motion fields from InSAR data, involving comparison of the interferometric analysis with field data and model calculations. Where suitable multiple interferograms and *in situ* tidal measurements are unavailable, the tidal component can be removed using tidal predictions alone, under the assumption that the floating ice mass deforms elastically (Rignot et al., 2004). Tidal model errors, due to imprecise bathymetric and grounding-line information and other errors, are estimated to be in the range of 10–50 cm (Padman et al., 2002). Care must be taken using satellite data due to possible tidal-aliasing effects, as stressed by Padman et al. (2002) see Section 3.9.

Another process that has a significant effect on height over floating ice, and its measurement by satellites, is the *Inverse Barometer Effect* or IBE (Gill, 1982; Padman et al., 2004; Rignot et al., 2000b). This entails a depression of the sea surface and floating ice of ∼1 cm per 1 millibar (mb) increase in surface atmospheric pressure. As such, a typical pressure anomaly for a polar low (cyclone) of ∼30 mb would result in a change in ice shelf height of ∼30 cm (Gray et al., 2001). According to Gray et al. (2001), a change in height of ∼20 cm due to a pressure change of ∼20 mb would lead to a cross-track InSAR velocity error of ∼5 m per annum. An underlying problem is that, as the IBEs associated with separate satellite overpasses are essentially decorrelated, this effect could result in significant height differences regardless of the tide-model accuracy (Gray et al., 2001). The IBE therefore represents a major error source in correcting satellite-derived ice shelf height estimates in order to derive mean flow rates from InSAR data and long-term trends in shelf elevation (and thickness) from satellite radar altimetry time series (Padman et al., 2002). In their speckle-tracking study of ice velocities on the Filchner Ice Shelf, Gray et al. (2001) incorporated Antarctic station barometric pressure data in an effort to understand and compensate for this effect. Another consideration is that satellite acquisitions may be separated by weeks or months, during which time basal melt/freeze or surface melt/accumulation/ablation will cause an ice shelf to adjust vertically. Systematic errors can result if these parameters remain unknown (King et al., 2000; Padman et al., 2002). Future consideration of more appropriate orbital periods for SAR satellites would minimize these limitations.

To date, InSAR applications have in practice been largely confined to regional rather than ice-sheet-wide studies due to a lack of suitable data. Notable exceptions are the ERS Tandem Mission (from October 1995 to June 1996, during which time ERS-2 followed ERS-1 in its orbit but separated by 24 hours), and the three Radarsat-1 AMMs (see pp. 75–76). More than 2,500 minutes of SAR data suitable for InSAR analysis were collected over Antarctica alone during the Tandem Mission. Covering a 30-day period in 1997, the first Radarsat-1 AMM (Jezek, 1998) enabled imaging of the important West Antarctic ice streams and 6 days of repeat-pass interferometric data. To date, these are the only InSAR data collected south of 80°S. Recall from Chapter 2 that the availability of both ascending and descending passes enables measurement of two components of the surface-velocity vector and also the removal of topographic effects—i.e., without requiring an independent DEM. Crossing Radarsat-1 InSAR swaths adjacent to the South Pole provided the first opportunity to derive surface velocities using multiple-look directions rather than the more conventional two-look directions—i.e., from ascending and descending orbits (Ford and Forster, 2002). This enables derivation of velocities for all three dimensions without the need for an assumption of surface-parallel flow (see Chapter 2). The follow-on MAMM in 2002 collected Radarsat-1 Fine-1 beam data to optimize the measurement of change detection and surface velocity (Jezek, 2002). Offering complete Antarctic coverage to the north of ∼80°S for the period September 3 to November 14, 2000, this represents the most comprehensive dataset for InSAR ice-motion studies yet, with three full cycles of data being acquired along both descending and ascending orbits (Joughin, 2002). Replication of MAMM occurred in September–December 2004, when MiniMAMM acquired repeat-pass data for three Radarsat-1 orbits along the same orbits (both descending and ascending) for the David and Pine Islands Glaciers and regions of the Filchner Ice Streams and Antarctic Peninsula.

While the 24-day repeat cycle is virtually ideal for measuring the relatively slow motion of inland ice—i.e., <100 m per annum (Geudtner et al., 1998; Gray et al., 1998; Joughin et al., 1999a)—interferometry tends to break down in faster flowing areas due to phase aliasing and decorrelation (this limitation also applies to shorter temporal baseline data) (Strozzi et al., 2002). Joughin et al. (2002) reported that phase decorrelation can also occur over a 24-day period where the accumulation rate exceeds ∼30 cm per annum—i.e., in the outer margins of the Antarctic Ice Sheet (Vaughan et al., 1999). In all cases, correlation levels for Radarsat-1 InSAR data are typically lower than for 1- or 3-day ERS-1/-2 data from the Tandem Mission (Joughin, 2002).

Fortunately, an alternative technique has been developed to derive useful ice velocity data from pairs of precisely co-registered SAR images in fast-flowing regions, such as ice streams (Gray et al., 1998; Joughin, 2002; Michel and Rignot,

1999; Rott et al., 1998b; Thiel et al., 1996). Known as *speckle tracking* (a form of *intensity tracking* or *cross-correlation optimization*), this technique exploits the cross-correlation function of coherent speckle patterns in pairs of real-valued SAR amplitude images. These functions, when sharply peaked, enable the derivation of surface displacements between coherent scenes in an interferometric pair (Gray et al., 2001; Joughin, 2002). Again, please see Chapter 2 for a fuller explanation. As such, image speckle is tracked in a manner similar to feature tracking with optical imagery. Speckle in SAR imagery is the pattern of strong radiometric variations produced by the coherent aspect of radar illumination. Whereas InSAR provides a measure of the ice surface displacement in the range direction only, speckle tracking yields both across-track and along-track information in an interferometric pair, albeit at a poorer resolution (comparable with the window size employed in the matching) and lower accuracy (Gray et al., 1998; Joughin et al., 2000; Michel and Rignot, 1999). Rignot (2002b) estimated the nominal precision of this technique to be 1/30 of a pixel (which is 4 m long along-track for ERS data), hence a precision in velocity of 50 m per annum for 1-day repeat-pass data—i.e., from the ERS Tandem Mission. Rignot (2002b) argued that this level of precision is acceptable for fast-moving outlet glaciers. The precision of speckle tracking improves to a few meters per annum with Radarsat data collected on a 24-day repeat-pass cycle with a 5-m pixel size (Joughin, 2002; Joughin et al., 1999a). As with conventional interferometry, speckle tracking is also limited by temporal and other decorrelation. Nonetheless, it can provide usable data in many cases where InSAR fails. Rignot et al. (2004) further showed that speckle tracking can be used to measure the velocity of ice shelves with little contamination from tides. Moreover, Strozzi et al. (2002) demonstrated that this technique is an invaluable source of information on ice motion associated with surging glaciers—e.g., Monacobreen in Svalbard. As noted earlier, tidal corrections are again required to account for tidal flexure and the inverse barometer effect (Gray et al., 2001).

Recent studies—e.g., Rignot (2002a) and Rignot et al. (2004)—have combined complementary InSAR and speckle-tracking techniques to improve regional coverage. Joughin et al. (1999a), for example, derived a surface-velocity map of much of the poorly-understood region flowing into the Whillans (B), Kamb (C), Bindschadler (D) Ice Streams and Ice Stream E, which drain the WAIS (see Figure 3.4). As the onset areas of tributary fast flow appear to have been migrating inland at hundreds of meters per annum (Bindschadler, 1997), understanding the flow in this region has become a major research priority (WAIS, 2003—see the *Amundsen Sea Embayment Science and Implementation Plan* at ⟨http://igloo.gsfc.nasa.gov/wais/ASEP-final.pdf⟩). The study by Joughin et al. (1999a) enabled the identification of previously unknown regional flow patterns, with a network of tributaries extending far into the ice sheet interior and providing the link from slow inland flow to rapid ice-stream flow. As such, it appears that the onset is not a point as previously thought, but rather a transition zone. The new results also suggest that tributaries feeding different ice streams emanate from the same source region, implying that the ice-stream flow is more interconnected than previously thought and should be modeled collectively. Another finding is that two tributaries feed into Kamb Ice Stream (C), which became stagnant ∼150 years ago (Retzlaff and Bentley, 1993), causing dynamic thickening of an extensive area at an estimated rate of 0.49 ± 0.02 m per annum (Joughin et al., 1999a). Figure 3.78 (see color section) is an example of ice motion data from the above study, again showing the extraordinary level of detail that can be achieved. Further work by Price et al. (2001a) examined flow features in SAR and visible satellite data to analyse post-stagnation behavior of the upstream areas of Kamb Ice Stream (C). This study enabled a reanalysis of hypotheses put forward to explain the stagnation of the ice stream.

Another example of the benefits of combining different techniques is provided by Joughin et al. (2001a). They produced the first large-scale map of the surface velocity of the Northeast Greenland Ice Stream at sub-kilometer resolution by combining SAR data from many crossing, ascending, and descending orbits of ERS-1 and -2 to derive vector measurements on the basis of the assumption of surface-parallel ice flow (Joughin et al., 1998; Mohr et al., 1998). In regions where data were only available with a single-track direction, speckle tracking was employed to derive vector measurements. The resultant map is shown in Figure 3.79 (see color section), alongside a satellite InSAR-derived DEM of the same region showing topographic variations associated with the flow features. This ice stream is a major 600-km-long feature that drains much of the northeast Greenland Ice Sheet (Joughin et al., 2001a), and which was first detected in SAR amplitude imagery (Fahnestock et al., 1993). Such is the level of detail that flow patterns in the onset area are revealed to be more complex than previously understood, with a second tributary entering from the south (Fahnestock et al., 2001b; Joughin et al., 2001a). The flow changes character downstream, apparently deviating from behavior expected from simple balance-flux estimates, which may be related to the surging behavior of Storstrømmen Glacier (Joughin et al., 2001a). Additional information from airborne Radio-Echo Sounding (RES) profiles and altimetry data is helping to define the character of this large feature, underlining the key importance of combining satellite with airborne and *in situ* techniques.

Further work by Joughin (2002) described an amalgamation of InSAR and speckle tracking to exploit the enhanced coverage of Radarsat-1 to build up a seamless composite mosaic map of ice velocity for the Lambert Glacier–Amery Ice Shelf system in Antarctica. This map, shown in Figure 3.80a (see color section), was derived using 26 interferometric pairs collected during AMM-2 and one pair collected during AMM-1. An estimate of the associated magnitude of the velocity error is given in Figure 3.80b. In areas where phase data are available from both ascending and descending orbits, the estimated errors are typically <1 m per annum, (e.g., brown regions in Figure 2.20). In other areas, the errors generally fall below about 5 m per annum. Near the shelf front, the errors approach 20 m per annum (due to the relatively poor correlation and the use of standard rather than fine-beam Radarsat-1 data). The Lambert Glacier drainage basin is one of six major Antarctic drainage basins (Giovinetto and Bentley, 1985), and drains an estimated ~16% of the area and 14% of the volume of the East Antarctic Ice Sheet (Fricker et al., 2000b) through only 1.7% of the coastline (Budd et al., 1967). As such, its state and behavior are major determinants of the mass balance of the ice sheet. This immensely important system has been the focus of intensive recent field activity, including the AMISOR project (Allison, 2003). Satellite remote sensing using complementary information from SAR, radar and laser altimeters, and a variety of visible to TIR sensors, is playing a major role in such projects by greatly extending detailed surface measurements both in space and time (e.g., Fricker et al., 2000a, b; Hyland and Young, 1998; Rignot, 2002a).

Young and Hyland (2002) also derived estimates of the surface motion of the Amery Ice Shelf with Radarsat-1 data—i.e., image pairs with a 24-day separation—but using maximum coherence as the measure of matching small samples taken from pairs of complex SAR images corresponding to the same part of the surface (Derauw, 1999). Results are shown in Figure 3.81. *Maximum-coherence tracking*, which is also referred to as the *coherence optimization* or *fringe visibility* algorithm (Strozzi et al., 2002), is comparable with speckle tracking but uses the phase information in complex-valued single-look SAR images rather than the speckle information in real-valued amplitude images (see Chapter 2). Young and Hyland (2002) removed the tidal component in the derived ice-shelf motion (Gray et al., 2001) by incorporating information from tide models outlined in Fricker et al. (2000a) and Padman et al. (2002). Please note that a summary of the comparative strengths and weaknesses of coherence tracking, differential InSAR, and intensity tracking is given in Section 2.8 (p. 109). While these new techniques have to some extent superseded feature tracking on visible satellite imagery, collection of ice velocities using the latter should continue for augmentation purposes to cover gaps in InSAR coverage due to lack of coverage or decorrelation.

Satellite-derived improvements in our understanding of patterns of ice flow unlock the potential of inverting topographic and surface motion data to obtain key information on basal topography and basal shear stress. This important application has recently been demonstrated by Joughin et al. (2001a, 2004b). These authors use control methods to invert a finite-element model of ice-sheet flow constrained by improved satellite-derived velocity data and other information to infer the basal shear stress under the Northeast Greenland and the Ross Ice Streams (Antarctica). Further details of the model used are given by MacAyeal (1989) and Hulbe and MacAyeal (1999). This is another example of the extraordinary advances that are taking place as a result of the availability of the new satellite datasets, providing important additional data with which to constrain ice-dynamics models and assess the possible impact of changes in ice flow characteristics. Future studies will also benefit greatly from anticipated continental-scale InSAR mapping of ice velocity by Radarsat-2 (Bamler and Holzner, 2004; Short and Gray, 2004) and other missions (see Chapter 2), provided the data are made available to the research community (Radarsat-2 is a commercial satellite). Another key factor is the availability of improved topographic information from ICESat GLAS (Joughin et al., 2004a; Shuman, 2004). Plans are afoot to launch a dedicated L-band InSAR satellite in the 2009/2010 timeframe (Smith, 2004). The main thrust of this project is to acquire measurements from multiple directions in order to determine actual surface displacement vectors. Further details are again provided in Chapter 2.

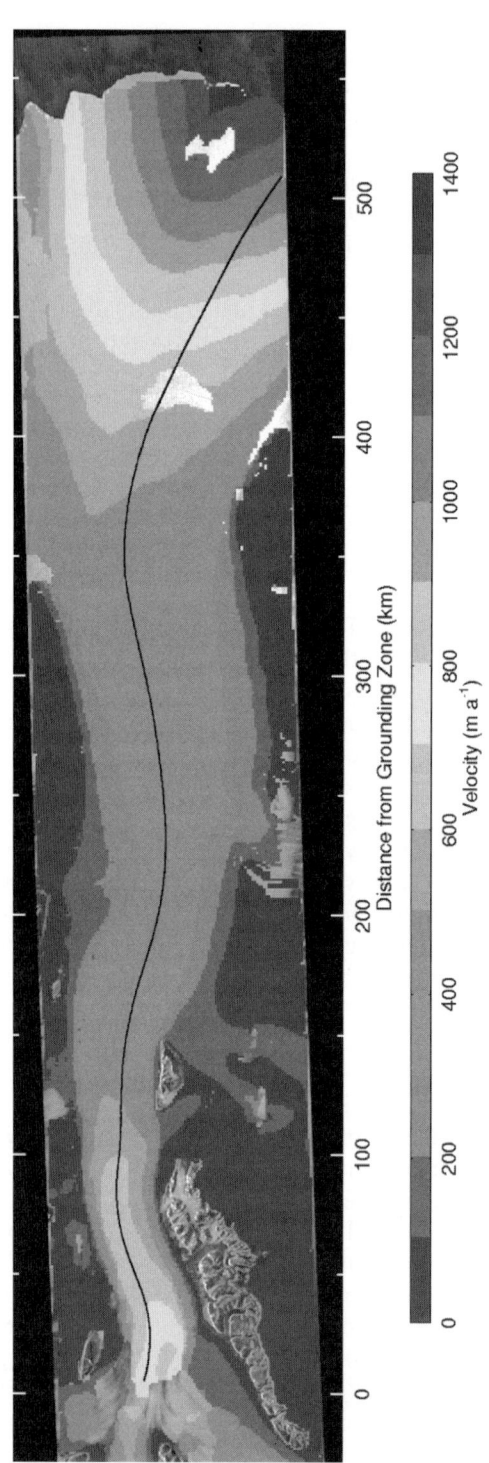

Figure 3.81. Spatial distribution of smoothed ice surface velocity magnitude over the Amery Ice Shelf and adjacent ice, produced by maximum-coherence tracking of coherent Radarsat-1 InSAR image pairs collected on September 24 and October 18, 1997. From the southern end of the shelf, velocity decreases from a high of about 800 m per annum to around 300 m per annum, and then increases to a maximum of about 1,350 m per annum at the center of the front.

From Young and Hyland (2002). Copyright International Glaciological Society 2002, reproduced with permission.

3.8.3 Balance velocities and fluxes

Recalling Section 1.1 (p. 1), an accurate assessment of the impact of an ice sheet on sea level requires accurate knowledge of its state of balance, which in turn necessitates observations of ice flow and estimates of net ice accumulation rates (Wang and Warner, 1998). The recent availability of accurate satellite radar-altimeter-derived DEMs has enabled the computation of another invaluable and powerful diagnostic tool—namely, maps of ice sheet *balance fluxes* and *balance velocities* (Budd and Warner, 1996). The concept refers to the conditions for steady state of a glacier/ ice sheet such that the outward flow distribution of ice exactly matches net accumulation—i.e., no change in thickness H occurs with time t at any point ($\partial H/\partial t = 0$)— and hence ice sheet geometry is maintained. More precisely, if over a horizontal-grid (x, y) domain, $A(x, y)$ is the net surface accumulation or ablation rate in ice-equivalent volume per unit area per unit time and $M(x, y)$ is the net basal freezing/melting rate, then the steady-state balance condition at any point with depth-averaged velocity $\bar{V}$ and ice thickness Z implies that:

$$A - M = \nabla \cdot (\bar{V} Z) \tag{3.12}$$

It is possible through integration at each point within a given drainage basin to determine the spatial pattern and vector values of balance flux (Budd and Warner, 1996). The *balance flux distribution* $\phi(x, y)$ of ice in an ice-stream–ice-shelf system is the hypothetical distribution of vector horizontal-flux density for unit width (in km^2 per annum) in the direction of ice flow, under steady-state conditions and for a given accumulation distribution, where:

$$\phi = \bar{V} Z \tag{3.13}$$

In other words, ϕ represents the spatial distribution of hypothetical ice (mass) fluxes required to locally balance a given accumulation/ablation distribution within a given drainage basin—i.e., the flow required to transport ice to counteract the effects of accumulation or loss to maintain a given ice-sheet thickness and shape (Bamber et al., 2000a, c; Budd and Warner, 1996; Vaughan et al., 2001b; Zwally and Giovinetto, 2001). The corresponding *balance velocity distribution* is given by:

$$\bar{V} = \frac{\phi}{Z} \tag{3.14}$$

Here, the divergence operator acting over the x, y domain is given as:

$$\nabla \cdot (\ ) = \left(\frac{\partial}{\partial x} i + \frac{\partial}{\partial y} j \right) \cdot (\ ) \tag{3.15}$$

where (i, j) are the unit vectors in the directions x and y, respectively. Please refer to Budd and Warner (1996) for further details, including discussion of basic assumptions and sources of uncertainty.

From the above, it can be seen that balance fluxes can be readily calculated with independent knowledge of ice accumulation and surface topography, while balance velocity estimation requires information on ice thickness. The requisite independent information on surface slope is typically obtained from the new and improved

DEMs derived from satellite altimeter data, with the balance-flux model making the assumption that the ice-flow direction is orthogonal to the elevation contours. While improved large-scale estimates of precipitation/snow accumulation distribution have recently become available (see Section 3.7), this remains a potentially large source of uncertainty in balance-flux calculations. Large-scale ice thickness data have recently been compiled—e.g., by the SCAR BEDMAP project in Antarctica (Lythe et al., 2001)—while more localized ice-thickness data typically come from aircraft radio-echo sounding campaigns. A map showing Antarctic coverage of the BEDMAP ice thickness datasets is available on the BEDMAP website at ⟨http://www.antarctica.ac.uk/aedc/bedmap/database/bedmap_coverage.html⟩. The resultant ice-thickness data are also available from this site.

An example of a balance-velocity map of Antarctica is shown in Figure 3.82 (see color section). This provides the most detailed picture yet of ice flow over the entire ice sheet. Balance velocities clearly delineate the flow patterns within each drainage basin, with major ice divides being visible as regions of low horizontal flow. Note again that the limit of ERS radar altimeter coverage of Antarctica is ~81.4°S and less accurate data are used to fill in the gap to the south of this (Bamber et al., 2000b; Huybrechts et al., 2000). This is also the case in the relatively steep outer margins of the ice sheet. An equivalent balance-velocity map of Greenland has been generated by Bamber et al. (2000a, c). Note that floating ice shelves are typically excluded due to the breakdown of assumptions regarding the influence of longitudinal stresses which break down in these regions (Bamber and Kwok, 2004).

The balance-flux model is a powerful diagnostic tool for assessing the ice flux and state of balance over either limited regions or entire ice masses, in order to interpret more spatially limited satellite measurements and field observations for the state of balance or dynamics. Determination of whether the ice sheet is in a state of imbalance or not (negative or positive mass budget) requires comparison of balance fluxes with actual (observed) ice fluxes. As stated by Wang and Warner (1998), accurate determination of the actual ice flux:

$$F_A = \eta V_S H \tag{3.16}$$

is based upon independent measurement of ice thickness H, and ice-surface velocity V_S, together with an estimate of the ratio (η) of depth-averaged velocity $\bar{V}$ to V_S. Again, satellite remote sensing involving feature tracking and InSAR techniques in particular has revolutionized the measurement of V_S over large areas of the ice sheets (although the entire ice sheets have yet to be covered). Few suitable measurements exist, however, of η and $\bar{V}$ (see Wang and Warner, 1998 for further information on ice-column-to-surface-velocity ratios).

This important tool provides a means of determining the net mass balance of an ice stream or glacier at its grounding line, by subtracting the measured flux estimate from the balance flux. As an example, Fricker et al. (2000b) used this technique with an ERS-1 radar-altimeter-derived DEM—5-km horizontal resolution (see Fricker et al., 2000a and Figure 3.60)—and various different accumulation distributions to compare balance and observed fluxes on the Lambert Glacier–Amery Ice Shelf drainage system, East Antarctica. Ice thicknesses were obtained from radio-echo sounding data and (on floating ice) by inversion of satellite-derived surface elevation data assuming that the ice is in hydrostatic equilibrium and using a density model. When considering an entire drainage basin, the balance flux through the "gate" of an outlet glacier (determined by integrating the accumulation of the entire catchment area in the direction of ice flow) can be compared with estimates of actual ice discharge (derived from ice thickness and velocity data) to give, for example, an estimate of the contribution to sea-level rise. Errors in the estimation of ice thickness by inverting surface elevation relate to uncertainty in the averaged ice-column density (Bamber and Kwok, 2004) and the effect of shear stresses in the vicinity of the margins of the floating ice mass (Reeh et al., 1997).

A major advantage of balance velocities is that they entail the only current means of providing an overview of ice-flow behavior over an entire ice sheet, and as such are extremely useful when examining basin-wide processes (Bamber et al., 2000a, b; Budd and Warner, 1996; Huybrechts et al., 2000; Joughin et al., 1997). The new balance-velocity datasets represent a major improvement over boundary conditions previously available for constraining ice-sheet numerical models, and provide an excellent snapshot of the motion of an ice sheet in steady state and new insights into its complex flow characteristics (Huybrechts, 2004; Huybrechts et al., 2000) and changes in flow regime. They also provide improved initial conditions for ice sheet dynamics models and are invaluable for model validation. Balance velocity maps further enable ice-dynamics models to more accurately resolve finer scale flow behavior. Bamber et al. (2000b), for example, presented new evidence based upon balance-velocity estimates to show that each major Antarctic drainage basin is fed by complex tributary systems that penetrate thousands of kilometers into the heart of the ice-sheet interior (Figure 3.83, see color section). These findings challenge the concept that the Antarctic plateau is a slow-moving and dynamically homogeneous region, and have important implications for modeling of dynamic response times of past and present ice-sheet geometries to climate forcing and its change. Moreover, uncertainties relating to the ice-sheet temperature regime, basal sliding, and the flow law can to some extent be constrained by comparison of balance velocities with model-derived dynamic velocities (Bamber et al., 2000a). While accurate where velocity gradients are small, balance velocities may lose accuracy in regions of fast flow—e.g., outlet glaciers (Layberry and Bamber, 2001)—and also in Antarctica to the south of ~81.4°S—i.e., where the DEM accuracy drops off (Bamber et al., 2000b). As Bamber et al. (2000c) pointed out, they are a derived quantity, and various assumptions are made in their derivation—e.g., that ablation and accumulation rates have remained relatively constant over the dynamic response time of a given drainage basin. It is important to note, for example, that balance velocities represent the mean depth-averaged flow over a period of the order of a millennium. Velocities measured by satellites, on the other hand, are short-term surface values. Relative strengths and weaknesses of balance velocities as a tool for investigating ice sheet velocity fields are evaluated by Bamber et al. (2000c).

Balance velocities have also been combined with new maps of *bedrock topography* to gain insight into how the latter controls ice flow patterns. Bedrock

DEMs are constructed by subtracting gridded ice-thickness data from an ice surface DEM (Vaughan et al., 2001b). In Antarctica, the *SCAR BEDMAP* project has compiled thickness data from various sources (including airborne radio-echo sounding) and collected over the past 50 years to produce a seamless suite of digital topographic models (Lythe et al., 2001; see p. 266). This includes grids of ice thickness over the grounded ice sheet and ice shelves, water-column thickness beneath the floating ice shelves, bed elevation beneath the grounded ice sheet, and bathymetry to 60°S, including sub-ice-shelf cavities. A similar ice thickness and bedrock topography dataset has also been produced for Greenland (Bamber et al., 2001b). In this case, the bedrock DEM was produced by subtracting ice thickness data obtained by airborne *Ice-Penetrating Radar* (IPR) from a DEM derived from satellite radar-altimeter data (mainly from Geosat and ERS) by Bamber et al. (2001b). Bamber et al. (2001b) show that comparing this new bed DEM with balance velocity estimates for Greenland leads to an improved understanding of bed-topographic controls over ice-flow patterns. Layberry and Bamber (2001) further combined these datasets to gain insight into the complex relationship between ice-sheet bed topography and ice flow, using additional information from the IPR signal to infer the presence of basal water. Basal melting is a prerequisite for basal sliding, which can lead to accelerated ice flow (Layberry and Bamber, 2001; Paterson, 1994). Once again, such investigations would have been impossible without recent advances in remote sensing.

3.8.4 Strain rates

The availability of satellite-derived high-density maps of ice sheet velocity on a regular grid has enabled determination of the *strain-rate tensor*, using the method formulated by Nye (1959). Surface *horizontal strain rates*, or *velocity gradients*, are related to the stresses experienced by the ice, through the flow law (Alley, 1992; Bindschadler, 1998b; Paterson, 1994)—please see the reviews by Van der Veen (1999) and Van der Veen and Payne (2004) for the governing equations. The derivation of ice strain rates does not require absolute velocity measurements—i.e., relative motions suffice. With ice surface velocities derived from Landsat TM data using the method of Scambos et al. (1992) and shown in Figure 3.75, Bindschadler et al. (1996) were able to derive horizontal strain-rate maps of Ice Streams D (Bindschadler) and E (Figure 3.84, see color section). Conversion of the ice displacements at each gridpoint to *longitudinal* ($\dot{\varepsilon}_L$), *transverse* ($\dot{\varepsilon}_T$), and *shear strain rates* was described in the appendix of Bindschadler et al. (1996). In addition, Bindschadler et al. (1996) calculated the vertical strain rate $\dot{\varepsilon}_Z$ (at the surface) from the horizontal strain rates, by applying the continuity condition of ice—i.e., $\dot{\varepsilon}_Z = -\dot{\varepsilon}_L - \dot{\varepsilon}_T$ (3.17). Information on the nature of *basal stress* can be inferred from the resultant maps, as can the important sub-glacial process of basal sliding (Bindschadler, 1998b; Bindschadler et al., 1996). MacAyeal et al. (1995) combined these same velocity data with ice-thickness and elevation data to compute the *basal resistance* (friction) pattern of Ice Stream E. In another example, Scambos et al. (2004a) investigated the relationship between *driving stress* and flow speed for the Institute Ice Stream (West Antarctica) by comparing Landsat-derived velocity estimates with elevation data from the Radarsat AMM-1 DEM (Jezek et al., 2000; Liu et al., 1999, 2001) and thickness information from the BEDMAP dataset (Lythe et al., 2001). This provides a straightforward means of investigating regions of probable changes in basal conditions (Price et al., 2001b; Scambos et al., 2004a).

Moving on to ice shelves, two major forces acting on an ice shelf are relevant to iceberg-calving mechanisms—namely, strain induced by the weight of the ice shelf causing it to thin and spread, and tidal action from underneath. Changes in these forces may influence the likelihood of ice-shelf breakup. Knowledge of the surface velocity and strain of ice shelves is important in determining their present kinematic state and detecting any change in that state. These parameters can be accurately measured by GPS (King et al., 2000). Larger scale measurements have recently become possible, however, with the emergence of InSAR and speckle-tracking techniques. Young and Hyland (2002), for example, derived the transverse component of the strain rate for the Amery Ice Shelf (Figure 3.85, see color section) from their SAR maximum-coherence tracking-derived velocity dataset (see Figure 3.81), and then used the velocity and transverse shear strain-rate distributions to describe key features of the flow regime and infer related deformation properties. Measurements from GPS campaigns remain immensely important, not least as a means of high-precision ground-truthing (validating) satellite techniques such as InSAR (King et al., 2000).

A similar study by Pattyn and Derauw (2002) used the surface velocity field from speckle tracking of ERS SAR data to calculate surface strain rates for the Shirase Glacier, a fast-flowing East Antarctic outlet glacier. This information was then used to estimate the vertically integrated force balance and to determine the major stress components resisting driving stress. From these results, the authors were able to infer that driving stress is largely balanced by basal drag, but with contributions from lateral drag of up to 15% of driving stress at the grounding zone. They further deduced that, upstream of the grounding zone, more than 90% of the total ice velocity is due to basal sliding. Comparison with a balance-flux distribution of the Antarctic ice sheet suggests that the glacier in the downstream part of the Shirase Glacier drainage basin is close to equilibrium, showing a slight negative imbalance.

Forster et al. (1998, 2003) presented a method for calculating longitudinal glacier strain rates directly from the wrapped phase of an InSAR interferogram.[8] This technique, which assumes that the ice flow path is known, enables calculation of strain rates from satellite SAR scenes lacking velocity GCPs or acquired in areas within which the phase is not continuously unwrappable from a velocity control point. As such, this approach foregoes the need to derive strain rates from either (a) the velocity field, which requires additional processing and also GCPs (Forster et al., 1999), or (b) the unwrapped phase (Fatland and Lingle, 2002), requiring phase unwrapping (see Chapter 2). Forster et al. (1998, 2003) combined InSAR-derived

---

[8] Recalling Chapter 2, the wrapped phase of a pixel in an unwrapped SAR interferogram corresponds to its height/displacement *modulo* an integer multiple of 2π.

relative ice displacements retrieved from a Radarsat-1 AMM image pair with flow direction information to calculate strain rates for the Recovery Glacier, a major East Antarctic outlet glacier draining into the Filchner Ice Shelf. They were able to apply an InSAR technique due to the high coherence in all regions apart from the lateral *shear margins* of the ice stream, in spite of the 24-day separation (see Raymond et al., 2001 for information on shear margins). Referring to the geometry for ice surface measurement from InSAR, presented schematically in Figure 2.10, Forster et al. (2003) calculated the longitudinal strain rate $\dot{\varepsilon}_L$ for the main glacier trunk along a flowline directly from the wrapped phase. Where a number of other assumptions are met, the longitudinal strain rate over distances greater than the ice thickness is given by:

$$\dot{\varepsilon}_L = \frac{\lambda}{4\pi T} \frac{\partial \phi}{\partial f} \frac{1}{\sin\theta \cos\xi} \qquad (3.18)$$

where λ is the radar wavelength; T is the time interval between the acquisition of the SAR images comprising the interferogram (the temporal baseline); ϕ is the wrapped phase; the phase gradient in the ice flow direction ($\partial\phi/\partial f$) is the fringe rate of the interferogram (again in the flow direction); θ is the radar look angle; and ξ is the angle on the surface between the radar look and ice-flow direction angles. Please see Forster et al. (2003) for a more in-depth explanation of the assumptions involved in calculating strain rates from InSAR data. In this case, the required information on ice flow direction can be estimated using several complementary approaches—i.e., from:

- local slope information derived from a DEM, assuming that ice flow is in the direction of maximum slope (Joughin et al., 1999a, b);
- flowlines observed in visible, near-IR and TIR (Bindschadler and Scambos, 1991) and SAR imagery (Rott, 1980); and
- an assumption that the ice flow is parallel to the glacier margins (Rignot et al., 1996).

Forster et al. (2003) further demonstrated that the tensile strength of the ice can be calculated from the InSAR-derived strain rate at the location of crevasse initiation (determined from SAR amplitude imagery by an increase in backscatter). This involves assumptions of a predictable temperature field and a simplified strain field. In the example from the Recovery Glacier Ice Stream using Radarsat-1 data, the strain rate at crevasse initiation was estimated to be ~0.002 per annum. The corresponding ice tensile strength ranged from 186–215 kPa (depending on the failure criteria used). An error analysis indicated errors of 17% for $\dot{\varepsilon}_L$ and 5.3% for the corresponding tensile strength. Forster et al. (2003) made the following recommendations based on reducing the relative error in $\dot{\varepsilon}_L$ calculated from wrapped phase:

- where $\dot{\varepsilon}_L$ is >0.02 per annum, 3-day repeat or ERS-1/-2 Tandem Mission data are required due to the large phase gradients;
- for strain rates of >0.005 per annum, steep look angles are required—e.g., of

ERS 35-day data or Radarsat-1 Extended Low Beam and Standard Beam S1 data;
- for strain rates of <0.005 per annum, errors are minimized by using the shallower look angles associated, for example, with the Radarsat-1 Extended High Beam and Standard Beam S7 modes (ERS 35-day data exhibit large errors).

A general recommendation is that perpendicular baselines of <300 m with errors of <1 m are preferred, while the relative DEM error should be <30 m. Further work using InSAR in West Antarctica has revealed significant variability in shear zone velocity (Frolich and Doake, 1998), while Rack et al. (2000) used InSAR analysis to study the deformation pattern of the Larsen Ice Shelf.

3.8.5 Detection of variability in ice flow and surge behavior

As we saw in Chapter 2, an exciting attribute of SAR interferometry is its unique ability to provide information on the short-term flow characteristics of outlet glaciers and ice streams (Goldstein et al., 1993; Joughin et al., 1995, 1996b; Kwok and Fahnestock, 1996; Mohr et al., 1998; Rignot et al., 1997a, b; Strozzi et al., 2002). Prior to the advent of InSAR, very little was known about the short-term variability in flow speed of these critically important features. Already alluded to in Section 3.8.2, the study of surging glaciers represents a particular challenge. During surge events, substantial volumes of ice can be transferred to lower altitudes, leading to a temporary acceleration of the melt rate (Luckman and Murray, 1999; Murray et al., 2001). Surge-type glaciers exhibit cyclical flow alternating between decades of slow flow punctuated by shorter periods of flow that is typically 10–1,000 times faster (Jiskoot et al., 2002; Murray et al., 2003). Such extreme velocities can undermine the ability of SAR interferometry to map the dynamics of surging glaciers, due to a loss of interferometric coherence by temporal decorrelation (Fatland and Lingle, 1998; Mohr et al., 1998; Murray et al., 2002). This is not always the case, however. For example, certain glaciers are characterized by less extreme surge episodes, which are amenable to measurement by satellite SAR interferometry (Dowdeswell et al., 1999; Luckman et al., 2002; Strozzi et al., 2000). Using ERS Tandem Mission data, Joughin et al. (1996c) detected a "mini" surge of Ryder Glacier in north Greenland, observing that the glacier speed increased by a factor of greater than 3 before returning to "normal", and all within the space of about 7 weeks. Working on Storstrømmen Glacier, also in Greenland, Mohr et al. (1998) observed an abrupt decrease in ice velocity after a surge event.

As with conventional interferometry, speckle tracking is also limited by temporal and other decorrelation. Nonetheless, it can provide usable data in many cases where InSAR fails, as seen previously. Moreover, Strozzi et al. (2002) demonstrated that this technique is an invaluable source of information on ice motion associated with surging glaciers—e.g., Monacobreen in Svalbard. In their study of the surge behavior of Monacobreen, Murray et al. (2003) further demonstrated the viability of using dual-azimuth differential InSAR combined with

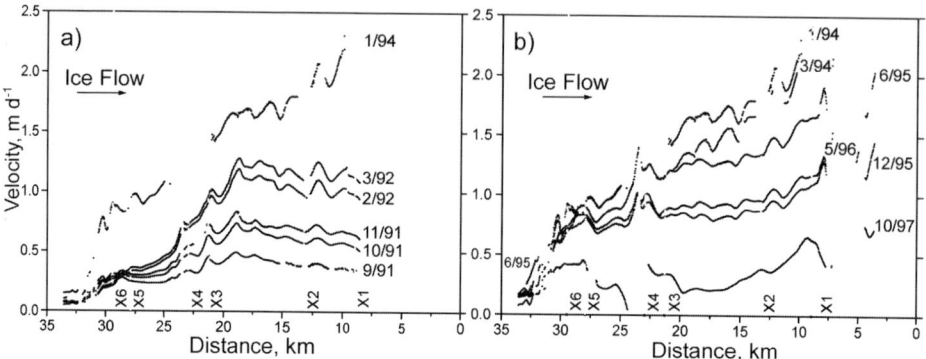

Figure 3.86. Time series of ice surface velocity derived from ERS SAR interferometry for a profile along Monacobreen (Svalbard), showing (a) a period of progressive velocity increase (September 1991 to January 1994) followed by (b) progressive velocity decrease (January 1994 to October 1997), punctuated by a temporary increase between December 1995 and May 1996. Data gaps indicate regions of incoherent data or where the flow direction approached the angle perpendicular to the SAR line of sight.

From Murray et al. (2003). Copyright AGU 2003, reproduced with permission.

intensity-correlation tracking (Strozzi et al., 2002)—using the latter where flow rates exceed those measurable with InSAR. Recall from Chapter 2 that data from three look directions are required to resolve the full three-dimensional flow (Joughin et al., 1998; Mohr et al., 1998). Ascending and descending pass data from ERS can be used to derive two-dimensional velocity fields, with the vertical flow component being estimated from a DEM by assuming flow that is parallel to the ice surface (Mohr et al., 1998; Murray et al., 2003; Reeh et al., 1999). See Section 3.8.2 and Chapter 2 for more detailed information on both techniques. The combination of InSAR and speckle-tracking techniques enables determination of glacier surface dynamics during surge episodes at an unprecedented spatio-temporal detail. Results from the ERS study by Murray et al. (2003) of Monacobreen, and covering periods of both an increase and decrease in flow velocity, are shown in Figure 3.86. Deceleration of flow is indicative that a surge event has recently occurred, and can be readily detected by InSAR techniques using both ascending and descending orbit data. Potential errors associated with the various techniques are discussed in Strozzi et al. (2002) and Murray et al. (2003), and are summarized in Chapter 2.

These new techniques promise to revolutionize our understanding of surge-type behavior and the mechanisms involved, which are of major interest yet poorly understood (Fowler and Johnson, 1996; Harrison and Post, 2002). Note that glacier surging has also been detected in digital analysis of high-resolution visible and near-IR data. Rolstad et al. (1997), for example, used feature-tracking techniques to detect the surge of a tidewater glacier in Svalbard, and similar studies have been carried out in the Russian High Arctic by Dowdeswell and Williams (1997). Jiskoot et al. (2002) examined controls on the surging behavior of East Greenland glaciers by constructing a glacier inventory from satellite images (ERS-1 and -2 SAR and Landsat 7 ETM+), aerial photographs, and maps, all assembled in a GIS. For an in-depth analysis of the current knowledge of glacier surge behavior, see the volume recently edited by Raymond and Van der Veen (2002).

Over the past 5 years, satellite interferometry has played an extraordinary role in the detection of both acceleration and deceleration of important outlet glaciers and ice streams. In their comparative study of past velocity measurements with those derived from the first Radarsat-1 AMM, Joughin et al. (2001b) revealed a pattern of slowdown extending from the upper reaches of Ice Streams A and Whillans (B) to well downstream of the grounding zone. Deceleration appears to have been fairly steady at a rate of 5.5 m per annum on the ice plain and sustained over intervals of up to 34 years. Further evidence for little or no contemporary flow acceleration or mass imbalance in the great ice streams draining the West Antarctic Ice Sheet (WAIS) is given by Whillans et al. (2001) and Doake et al. (2001). Joughin and Tulaczyk (2002) further reported an apparent thickening in the region, again based on satellite InSAR analysis. Further evidence of unexpected changes in West Antarctic ice stream velocities was given by Joughin et al. (2002). The oscillatory behavior of Siple Coast ice streams unearthed in the latter study may be linked to the possibility of unstable ice sheet behavior.

Illustrative of the regional complexity is the fact that the Pine Island Glacier (PIG) accelerated by fully 18 ±2% in 8 years (1992–2000) (Rignot et al., 2002). This acceleration, part of which is shown in Figure 3.87 (see color section) and was analyzed using force–balance techniques by Thomas and Rignot (in press), coincided with the significant glacier retreat and thinning outlined in Section 3.6.1 (Payne et al., 2004; Wingham et al., 1998; Zwally et al., 2002a), and as such differs from surge behavior, which entails simultaneous flow acceleration and thickening. Rignot et al. (2002) estimated that the PIG and Thwaites Glacier together contribute from $\sim$0.1–0.2 mm per annum to current global sea level rise, or up to 10% of the total. Joughin et al. (2003) compared InSAR data and sequential Landsat imagery to further identify and temporally constrain two PIG acceleration events separated by a period of at least 7 years (1987–1994). Analysis indicates that changes in driving stress consistent with observed thinning rates are of sufficient magnitude to explain much of the acceleration. Such changes are of great importance, given that the Pine Island–Thwaites Glacier system accounts for approximately 5% of the ice discharge of the entire Antarctic Ice Sheet (Vaughan and Bamber, 1998). Subsequent work has underlined that these fast-flowing Antarctic ice streams exhibit a range of dynamic behaviors (Bamber and Rignot, 2002), with evidence of changes in flow behavior.

Other studies have detected changes in ice flow regimes by feature tracking, taking advantage of the long though discontinuous time series of Landsat data— e.g., Stephenson and Bindschadler (1988). Rosanova et al. (1998) applied feature-tracking techniques to Landsat (and ERS-1 SAR) images to observe velocity changes in the Pine Island and Thwaites Glacier region over the period of 1973 to 1996. While studies in the 1980s estimated the ice front velocity to be 2.1–2.4 km per annum (Crabtree and Doake, 1982; Lindstrom and Tyler, 1984; Williams et al., 1982), recent studies have shown an increase to 2.6–2.8 km per annum (Lucchitta et al. 1995; Jenkins et al., 1997; Rignot et al., 2002). In a further analysis, Rosanova and Lucchitta (2002) co-registered earlier image sets from Landsat MSS (1973 and 1975), TM (1997 and 2000), and ERS-1 SAR (1994 and 1996) to a Landsat ETM+ image from 2000 to obtain the velocities of floating Pine Island Glacier ice. These data again supported a general increase in velocity, with most of the apparent acceleration occurring prior to the mid-1990s (see also Rosanova and Lucchitta, 2002). Rignot et al. (2002) suggested an increase in basal lubrication as a probable cause, while Jenkins et al. (1997) purported that a major calving event in the early 1970s may also have played a role.

3.8.6 Ice shelf buttressing, and the impact of ice shelf removal on outlet glaciers

While enhanced ice-shelf disintegration, such as that illustrated in Section 3.4.1, may not directly contribute to sea-level rise, it has been hypothesized that removal of the ice shelf "buttress" can have a substantial indirect effect via its impact on inflowing ice streams and glaciers (Mercer, 1978; Thomas et al., 1979). Frictional drag along the sides and at local grounded parts of an ice shelf creates "back stress", or longitudinal compressive force, which plays a role in the outflow of grounded ice (Thomas, 1973; Rist et al., 2002). The relationship between this back pressure and the stability of grounded ice upstream is a matter of considerable debate (Bindschadler, 1993; Bentley, 1998; Hindmarsh and Le Meur, 2001; Kenneally and Hughes, 2004). As outlined below, the recent demise of Antarctic Pensinsula ice shelves combined with advances in InSAR techniques has provided a unique opportunity to assess the rapid response of inflowing glaciers to the removal of an ice shelf "buttress". A number of modeling studies have challenged the speed increase/stress decrease impacts of buttress removal—e.g., Hindmarsh (1993). While Vaughan (1993b) observed no significant inflow change after the breakup of the Wordie Ice Shelf in Palmer Land, Rott et al. (2002) noted a threefold increase in flow velocity at the terminus of the Drygalski Glacier, the largest glacier in the region, in the 4 years after the major collapse of the Larsen A Ice Shelf in 1995 (using interferometric pairs of ERS SAR images) (Figure 3.88, see color section). Scambos et al. (2003) noted that this flow acceleration may also relate to an increase in meltwater percolation through the ice to facilitate basal sliding—e.g., Zwally et al. (2002b)—see Section 3.3. Further evidence of the impact of removal of the "buttressing" effect of the ice shelf was given by De Angelis and Skvarca (2003), who observed similar accelerations after 1995 following the collapse of the northern Larsen Ice Shelf, as well as glacier migration to several kilometers inland of their previous grounding zones (Figure 3.89).

In their analysis of five Landsat 7 ETM+ images from the period January 2000 to February 2003, Scambos et al. (2004a) showed that a twofold to sixfold increase occurred in the center-line speed of four glaciers flowing into the Larsen B Ice Shelf region following its unprecedented collapse in March 2002 (Figure 3.90, see color section). Over the same period, two glaciers farther to the south and still buttressed by an ice shelf—namely, the Flask and Leppard—showed no acceleration and little sign of acceleration, respectively. Ice velocities were derived by applying the feature-tracking algorithm of Scambos et al. (1992). Additional analysis of new satellite laser

Sec. 3.8]   Ice velocity, strain rate, and balance velocity/flux   275

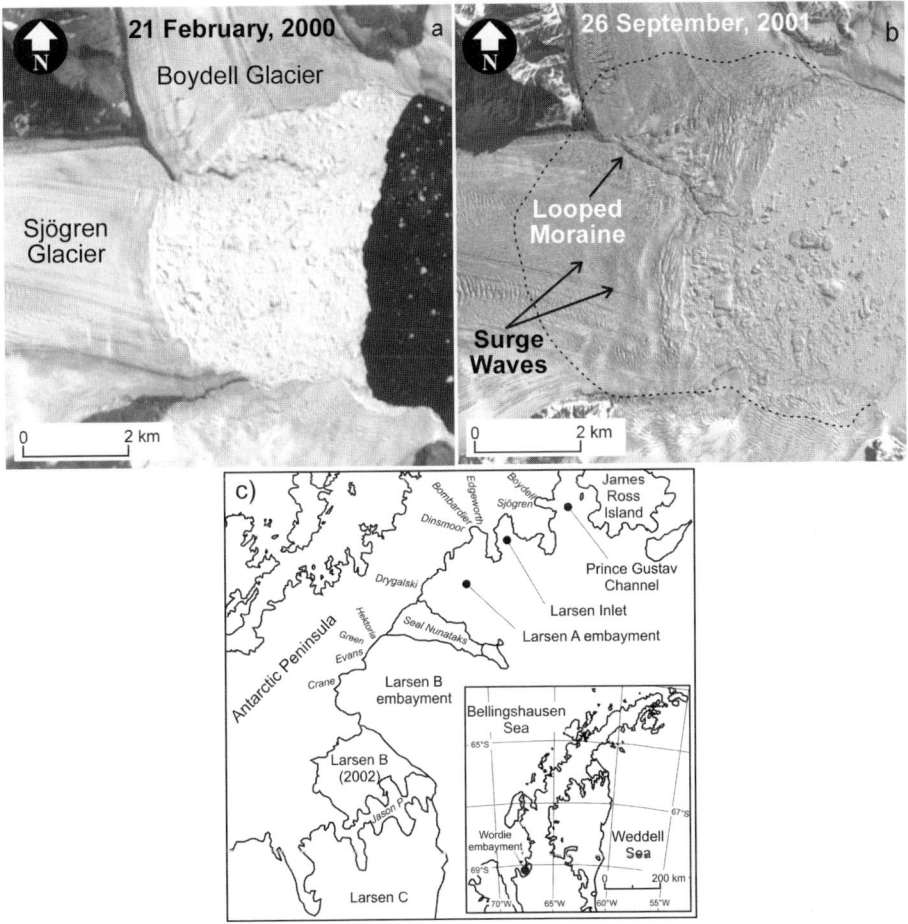

Figure 3.89. Surge of Boydell and Sjögren Glaciers, shown in a comparison of images from (a) Landsat 7 ETM+ band-8 (February 21, 2000), and (b) ASTER band-3 (September 26, 2001). The dashed line in (b) demarcates the location of already separated ice fronts as surveyed by airborne GPS on March 15, 2002. (c) Map of the Antarctic Peninsula, showing the locations of the glaciers. Enhanced surge behavior was also noted in the Edgeworth and Bombardier Glaciers.

From De Angelis and Skvarca (2003). Copyright AAAS 2003, reproduced with permission.

altimeter data from ICESat showed that the surface of one of the glaciers—namely, the Hektoria Glacier—lowered by as much as 38 ±6 m over a 6-month period 1 year after the ice-shelf breakup (Figure 3.91, see color section). Scambos et al. (2004a) further showed that seasonal variations in glacier flow speed occurred prior to the large post-collapse velocity increases, which they suggest provides evidence that both changes in the stress field and summer melt percolation (see Section 3.4.1) play a major role in glacier dynamics.

In a parallel and complementary study, Rignot et al. (2004) used interferometric SAR data collected by ERS-1/-2 and Radarsat-1 to also detect major glacier flow changes in response to the Larsen B ice shelf collapse in 2002. They observed accelerations after the collapse of up to a factor of 8 in glaciers flowing into the region. Another important outcome is the discovery of an estimated mass loss associated with flow acceleration that is in excess of $27\,km^3$ per annum, and an ice thinning rate of tens of meters per annum. Similarly, Rignot et al. (2005) combined ERS-1/-2 InSAR, Radarsat-1 speckle tracking, and airborne and ICESat laser altimetry to detect significant recent acceleration and thinning of glaciers flowing into the former Wordie Ice Shelf (West Antarctic Peninsula). This study concludes that additional observations of ice accumulation, thickness, and velocity are required to reduce uncertainties in mass-balance calculations.

Taken together, these results provide some of the first direct evidence for the stabilizing role of ice shelves, suggesting that their removal could have an important effect on ice-sheet elevation, mass balance, and eustatic sea level. The "health" of the Ross Ice Shelf is of obvious concern, given that major portions of both the East and West Antarctic Ice Sheets drain into it (each containing the equivalent of several meters of sea level rise) (Scambos et al., 2003). Once again, remote sensing is playing a key role in enabling monitoring of possible developments. Interestingly, similar contemporary acceleration appears to have taken place in glaciers unaffected by ice shelf collapse along the western flank of the Antarctic Peninsula, in feature-tracking analysis of satellite SAR data (Pritchard and Vaughan, 2004). The authors attribute this acceleration to enhanced basal lubrication as a result of increased summer melt, due to the observed southward retreat in recent decades of the $-9°C$ mean annual isotherm (Morris and Vaughan, 2003). In the Arctic, a recent acceleration has also been observed in the Jakobshavn Isbrae (Greenland) by Joughin et al. (2004b). This has coincided with substantial surface melting and dynamic thinning (Thomas et al., 2003). Again, such changes would have been difficult to quantify and even detect (over the larger area) without the benefit of remote sensing.

3.9 TIDAL DISPLACEMENT OF ICE SHELVES AND GLACIER TONGUES

As we saw in Section 3.8.2, accurate tidal predictions are essential to eliminate vertical tide motions when mapping the flow speeds and thickness changes of floating ice (Padman et al., 2002; Shepherd and Peacock, 2003). Tidal forcing itself plays an important yet poorly understood role in ice-shelf dynamics and thermodynamics around Antarctica and along the north coast of Greenland (Holdsworth, 1969, 1977; MacAyeal, 1985; Makinson and Nicholls, 1999). Most ocean-tide-induced differential motion occurs over a narrow band (typically a few kilometers wide) at the transition between grounded and floating ice and known as the tidal-flexure zone, with the flexure point being the limit of ice flexure from tidal movement (Schmeltz et al., 2002a). This is the site of important glaciological processes, which include cyclical bending and cracking at tidal frequencies, transmission of longitudinal stress gradients across the grounding zone, and a transition in flow style between ice-shelf dynamics and ice-sheet mechanics (Schmeltz et al.,

2002a; Vaughan, 1995). Tidal currents facilitate basal-ice melt and rift widening, and thus iceberg calving (Fricker et al., 2002a). Tides may also facilitate the flushing out of sediment and water from under the ice shelf by flexing the ice shelf/glacier tongue near the grounding zone. Moreover, Anandakrishnan et al. (2003) measured the flow velocity of Bindschadler Ice Stream (D) in West Antarctica to vary by a factor of 3 over the course of a day. This variability occurred both at and upstream of the grounding zone, with diurnal velocity fluctuations apparently driven by tidal processes beneath the Ross Ice Shelf. These results suggest that the ocean tide and ice shelf exert a significant yet poorly understood influence on the dynamics of ice stream flow. Large diurnal swings in velocity, characterized by *stick–slip motion*, have also been noted in the Whillans Ice Stream (Bindschadler et al., 2003).

Improved *in situ* tidal and bathymetric data are required to constrain tide models (Padman et al., 2002). The synthesis of remote sensing with modeling is enabling glaciologists to greatly extend *in situ* measurements of tidal flexing. The latter have been carried out using tiltmeters (e.g., Robin, 1958; Smith, 1991; Stephenson, 1984), gravimeters (Doake, 1992; Williams and Robinson, 1980), and GPS using both kinematic and static techniques (Bondesan et al., 1994; King et al., 2000; Reeh et al., 2000; Vaughan, 1995). Regarding remote sensing, recent research has exploited radar altimetry and InSAR to gain enhanced information on the tidal processes acting on ice shelves. Fricker and Padman (2002), for example, have shown that tidal signals can be detected using ERS-1 and -2 radar altimeter data (see also Shepherd and Peacock, 2003). Although the accuracy of a single measurement of ice-shelf elevation is only about 50 cm, it is possible with sufficient data to obtain the tidal coefficients for most major tidal frequencies to an accuracy of $\sim$3–5 cm, which is comparable with that achievable with regional tide models. Unfortunately, the maximum southward orbit of $\sim$81.4°S for the ERS and Envisat altimeters precludes coverage of the most southerly parts of the large Antarctic ice shelves—namely, the Filchner-Ronne and Ross. The new ICESat GLAS mission marks a significant improvement by extending coverage beyond all ice shelves to $\sim$86°S, and individual surface-elevation measurements are also more accurate (Zwally et al., 2002a).

Whereas satellite altimetry can monitor ice-shelf elevation changes over timescales of months to years (Fricker and Padman, 2002), differential InSAR alone can provide accurate snapshots of tide-driven vertical displacement over extensive areas ($\geq$100 km) and at a high spatial resolution (up to 20 m) and precision (e.g., Goldstein et al., 1993; Rignot, 1996; Rignot et al., 2000b), given the assumption of time-independent lateral ice flow. An example of the tidal displacements of the Thwaites Glacier tongue in West Antarctica is given in Figure 3.92 (see color section). To create this image, Rignot (2001) co-registered and double-differenced two Tandem Mission ERS-1 and -2 interferometric pairs, acquired 35 days apart in 1996. The phase signal associated with topography was removed from the double-difference interferogram using the Antarctic DEM of Bamber and Bindschadler (1997). The resultant *quadruple-difference interferogram* measures changes in the tidal displacement of the floating glacier tongue between four imaging epochs, plus noise (Rignot, 1996). As we shall see in the next section, the inner limit of the zone of tidal flexing demarcates the grounding line (as marked in Figure 3.92); in this case, the grounding line retreated by $\sim$14 km between 1992 and 1996. For a comparison of InSAR and GPS techniques, please see Rignot (1998a, b).

Once again, InSAR data are playing an important role in ice-shelf model development and improvement. Schmeltz et al. (2002a) used InSAR observations of tidal flexure on floating Antarctic and Greenland glaciers to validate and refine a finite element model simulation of tidal flexure on an elastic plate of ice. Comparison of the model results with InSAR data shows that tidal flexure is highly dependent on the ice thickness profile, and gives insight into the fracture mechanics of ice shelves. Furthermore, the inverse barometer effect—i.e., isostatic response of the ocean surface to changing atmospheric pressure—also contributes to vertical ice-shelf displacement (Rignot et al., 2000b)—see Section 3.8.2.

While InSAR techniques have recently been applied to the identification of tidal motion constituents (Hartl et al., 1994; Rignot et al., 2000b), it is again somewhat limited by the orbital periods of the current satellites, which are in phase with semi-diurnal tidal motion. This results in the need for long time periods to resolve the tidal components (Rignot et al., 2000b). The derivation of tidal and surface height change information from satellite altimeter data collected over ice shelves is complicated by aliasing (Padman et al., 2002). This arises from the complex relationship between tidal frequency and satellite-orbital repeat periods, which are relatively long— e.g., 183 days for ICESat (after an initial 3 months with an 8-day repeat interval) (Zwally et al., 2002a). With periods of ~ 0.5 and ~ 1 day, tides are under-sampled by satellites (Figure 3.93). Orbits can be designed to enable the removal of tides from a

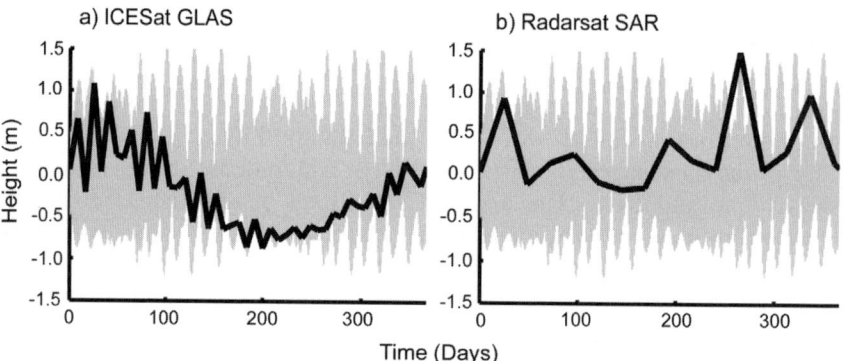

Figure 3.93. Examples of the *satellite-aliasing problem*, for (a) ICESat's Geoscience Laser Altimeter System (GLAS), and (b) Radarsat. The thin gray line in both is the predicted hourly tide height (m) at the point 71.5°S, 70°E on the Amery Ice Shelf for all of year 2004. In (a), the thick black line is the tidal time series that would be sampled by ICESat's GLAS with the "8-day" (actually 7.989-day) repeat cycle which was used during the ICESat Verification Phase. The satellite's sub-sampling produces a time series that contains a seasonal cycle, an artifact of the long aliasing periods for the S_2, K_2, K_1, and P_1 tidal harmonics. In (b), the thick black line is the tidal time series that would be sampled by Radarsat with its 24.0-day repeat cycle. The satellite's sub-sampling produces a time series whose mean value can be significantly different from the true mean height due to "freezing" of the S_2 (period = 12.00 h) semi-diurnal tidal harmonic. With the 24.0-day repeat cycle, the K_1 and P_1 tidal harmonics are indistinguishable from a true annual cycle.

After Padman et al. (2002). Copyright International Glaciological Society 2002, reproduced with permission.

long record of satellite altimeter data (Parke et al., 1987). As pointed out by Padman et al. (2002), Topex-Poseidon is currently the only altimetric satellite designed for this purpose (Parke et al., 1987; Smith, 1999), but provides no coverage poleward of 66.2°N and S. As a result, the most effective way of predicting tides over ice shelves and glacier tongues is through numerical modeling. Padman et al. (2002) used extensive, precise GPS measurements on the Amery Ice Shelf to constrain and develop an accurate tide model. Padman et al. (2002) further describe a new circum-Antarctic tide model for the seas surrounding Antarctica which uses data assimilation to improve its fit to available data. Other tide models are described by Le Provost et al. (1997), Makinson and Nicholls (1999), and Robertson et al. (1998). These improved models will in turn engender a better understanding of the complex interactions that occur between ice shelves and the ocean, including the dynamic effects of water circulation inside the ocean cavity underlying the ice. Such interactions have major implications for ocean water mass modification and the discharge of grounded ice. High-latitude tidal models will undoubtedly benefit from the availability of more precise grounding-line delineations by InSAR (see Section 3.10), although the lack of precise bathymetric information remains an issue in compromising their accuracy (Padman et al., 2002).

3.10 GROUNDING-LINE (ZONE) DETECTION AND MONITORING

The grounding line (zone), defined as the zone delineating the transition from a grounded to a floating regime, is highly sensitive to changes in ice thickness and is an important indicator of the dynamic state of the ice stream (Fricker et al., 2002b). It follows that changes in grounding-line position should entail early indicators of any thinning or thickening of the ice sheet caused by local or global climate change/shifts. Numerical models of ice flow require this information to realistically constrain basal ice conditions—i.e., basal stress conditions are significantly different for floating compared with grounded ice (Huybrechts et al., 1999). Moreover, grounding line location has an important impact on the shape and size of the sub-ice ocean cavity, which in turn affects sub-ice circulation, tidal processes, and basal melt and ice accretion distributions (Hughes, 1977; Mercer, 1978). These are key parameters in ice shelf–ocean interaction models—e.g., Beckmann and Goosse (2003); Hellmer and Jacobs (1992); Jenkins and Doake (1991); and Williams et al. (2002). As noted earlier, ice shelf melt in turn plays a key role in the formation of Antarctic Deep and Bottom Water and therefore ocean overturning and global ocean thermohaline circulation (Foldvik et al., 1985; Grosfeld et al., 1997; Hellmer, 2004; Meredith et al., 2000). Ice-sheet mass balance estimates are strongly dependent upon the accurate delineation of the grounding zone.

A number of *in situ* methods, including tiltmeters (e.g., Vaughan, 1995) and kinematic GPS (e.g., Smith, 1991), have been successfully used to detect and define ice-shelf grounding zones, but remain impractical for large-scale surveys. Prior satellite studies used high-resolution optical imagery—e.g., from Landsat (e.g., Hambrey and Dowdeswell, 1994; Swithinbank et al., 1988)—to infer grounding zones from changes in surface slope characteristics, but the approach tends to break down for fast-moving glaciers. A more recent approach, demonstrated by Fricker et al. (2002b), combined a DEM derived from ERS-1 radar altimeter data with ice thickness data measured by airborne ice-sounding radar, plus an ice-density estimate, in a hydrostatic (buoyancy) calculation to map the landward extent of floating ice on the Amery Ice Shelf, East Antarctica. These new results, confirmed by static GPS measurements and analysis of other satellite data, showed the grounding zone to be $\sim$240 km farther upstream than previously thought. This again illustrates the extraordinary advances that are resulting using modern satellite remote-sensing techniques.

Once again, satellite InSAR is making an immense contribution, with its unique ability to accurately map and monitor grounding line location on a large scale and to an unprecedented horizontal accuracy. This approach, which was first demonstrated by Goldstein et al. (1993), is based upon the detection of the limit of tidal flexure (vertical movement) of a floating ice shelf or glacier, to which InSAR is highly sensitive. This limit is often referred to as the *hinge line* position (Rignot, 1998a, b). When detected by InSAR, the latter is roughly equivalent to the grounding line location. Rignot (1996) automatically mapped the hinge line location in InSAR imagery by least-squares fitting an elastic-beam model of tidal flexure through profiles extracted from difference interferograms across the zone of flexure. This was carried out in a direction perpendicular to the iso-contours of glacier vertical displacement. The highest mapping precision is achieved in areas of large tidal motion, high SNR, and large hinge line radius of curvature; it degrades towards glacier margins where the signal is limited by SAR system resolution (Rignot, 1998b). Working on the Petermann Gletscher (NW Greenland) and using a three-pass InSAR approach, Rignot (1998a, b) estimated that grounding-line location can be detected and mapped with InSAR to a horizontal precision of 20–50 m at best. This is fully 1–2 orders of magnitude better than is possible using other techniques (Rignot, 1998b; Rignot et al., 1997a). The precision of detection/mapping also depends upon the character of the grounding zone, the amplitude of the tidal deformation signal, and the quality of the interferogram (i.e., the degree of phase coherence). Grounding-line locations migrate back and forth with ocean tide effects over a rough bed (Rignot et al., 2001). This acts to limit the horizontal precision of mean sea-level grounding line mapping to 100–200 m (Doake et al., 2001; Rignot, 1998b), although improved precision is possible if multiple interferograms are available (Rignot et al., 2001). For fast-flowing outlet glaciers, the use of InSAR data alone is complicated by the fact that the displacement signal contains a strong horizontal as well as vertical velocity component. Pötzsch et al. (2000) overcame this by using double-difference (i.e., four-pass) interferometry with ERS Tandem Mission data to recover the topography of the grounded ice. The routine application of such an approach is currently limited, however, by the lack of suitable data (a factor affecting the universal application of InSAR in general).

Satellite SAR interferometry is now a widely used tool for snapshot mapping of grounding-line location (Rignot, 1996, 2001; Rignot et al., 1997a, 2001), providing a baseline measurement against which to gauge grounding-line migration (Figure 3.94, see color section), which is thought to be an indicator of ice-sheet stability (particularly in West Antarctica) (Weertman, 1974). Indeed, recent measurements have enabled major revisions of grounding line location. In certain cases, they have revealed grounding line migration in response to changes in ice thickness (Gray et al., 2002; Rignot 1998a, b, c, 2002b), and also to detect changes in ephemeral grounding (Schmeltz et al., 2001). For example, recent analysis of ERS SAR interferometry showed that the position of the hinge line of the Pine Island Glacier had retreated at a rate of 1.2 ± 0.3 km per annum between 1992 and 1996 (Rignot, 1998c), which in turn implies that the ice thinned at a rate of 3.5 ± 0.9 m ice per annum—i.e., over the same period as radar altimeter measurements independently showed elevation change/thinning (Wingham et al., 1998). Moreover, and as seen above, Rott et al. (2002) used InSAR to investigate temporal changes in ice flow and grounding-zone behavior of glaciers feeding into the Larsen Ice Shelf after the major calving of the latter in 1995. Again, such changes have important implications for ice-sheet mass balance.

Similar studies in north Greenland have also measured a retreat in the grounding zones of major outlet glaciers, including the Petermann, Humboldt, and Ryder, at a rate of several hundred meters per annum over the period 1992 and 1996 (Rignot, 1998b; Rignot et al., 2000c, 2001)—see Figure 3.95 (color section). This rate of retreat implies localized ice thinning of the order of a meter per annum, which is unlikely to be caused by temporal changes in accumulation and/or ablation. This further implies that ice near the grounding zone must be thinning by anomalously high creep (Thomas et al., 2001a). Rignot et al. (2001) converted InSAR-derived glacier grounding-line migration into thinning rates using thickness slope measured *in situ* by ice-sounding radar (Chuah et al., 1996) and surface slope measured by airborne topographic mapping—i.e., by laser altimetry (Krabill et al., 1995, 1999). Recent studies using aircraft laser altimeters (Krabill et al., 1999, 2000; Thomas et al., 2000a) have shown that ice thinning is even more pronounced along the southeast coast of Greenland, possibly as a result of accelerated flow (Abdalati et al., 2001). In Section 3.11, we examine how recent developments have enabled the inference of ice flux across grounding zones, the net mass budget for glaciers, and the basal-melt rates of ice shelves.

3.11 ESTIMATES OF ICE DISCHARGE FLUX AND BASAL MELT/FREEZE RATES

Ice-discharge fluxes are typically calculated for gates across outlet glaciers and ice fronts using ice velocity and thickness data (Vaughan et al., 1995). The recent availability of InSAR data has greatly facilitated, and indeed revolutionized, such estimates. The accurate delineation of grounding zones by differential InSAR (Rignot, 1996; Rignot et al., 1997a) and the spatially dense mapping of ice velocity by InSAR (Rignot et al., 1997a) and SAR speckle tracking (Gray et al., 2001; Joughin et al., 1999a) has opened up the exciting possibility of estimating glacier and ice-stream mass balance more accurately than ever before by using the mass-budget approach (see Section 1.2, p. 10) (Mohr and Reeh, 2002; Mohr et al., 1998; Reeh et al., 2002). By the grounding-line flux approach, the mass budget of outlet glaciers is determined by subtracting ice discharge across the grounding zone (Rignot, 1998a, c, 2002a, b; Joughin and Tulaczyk, 2002; Rignot and Jacobs, 2002; Rignot et al., 2000a, c, 2002; Thomas and Rignot, 2002) from the estimated snow accumulation over the drainage basin upstream. The difference between these two large values determines whether the glacier is losing or gaining mass—i.e., whether it is in negative or positive mass balance, respectively. Studies using the grounding-line approach, which works best where there is a well-defined floating ice tongue (Bamber and Kwok, 2004), have also benefited greatly from advances in the estimation of ice thickness (by ice-sounding radar) and balance flux at the grounding line. These are the least well-constrained mass budget parameters. For Antarctica, the new input datasets include an accurate DEM from ERS-1 radar altimetry data combined with terrestrial data (Bamber and Bindschadler, 1997), digital maps of improved snow accumulation-rate estimates (e.g., Giovinetto and Zwally, 2000; Vaughan et al., 1999), and BEDMAP thickness data (Lythe et al., 2001). Individual drainage basin extent is delineated from the topographic information contained within the DEM (Hardy et al., 2000; Vaughan et al., 1999, 2001b).

With the above information and approach, the mass flux can be calculated for each glacier both at (across) the grounding zone and at a flux gate located some distance downstream. The glacier mass balance is then deduced by directly comparing ice discharge across the grounding line to a mass input calculated from snow accumulation. Following Rignot (1996), the ice flux Q from an outlet glacier at and below the hinge line is given by:

$$Q = \bar{h}\bar{V}_x W_y \qquad (3.19)$$

where $\bar{h}$ is the ice thickness; $\bar{V}_x$ the ice velocity in the x-direction averaged across the glacier width; and W_y is the glacier width in the y-direction (Paterson, 1994). An assumption is made that the basal and surface velocities of a floating glacier are equivalent, such that surface velocities measured by InSAR are equivalent to vertically integrated ice velocities. In the absence of direct measurements—e.g., by RES—the thickness of the floating ice is typically estimated from glacier surface elevations derived from a high-quality DEM, and assuming that the glacier/ice shelf is in hydrostatic equilibrium. This approach requires knowledge of the ice density profile with depth, and applies a conversion factor to the surface elevation data to account for ice and seawater densities (Bamber and Bentley, 1994; Rignot and Thomas, 2002; Vaughan et al., 1995). Rignot et al. (1997a), for example, assumed a mean glacier ice density, ρ_i, of $917 \, \text{kg m}^{-3}$ for the relatively warm ice of the coastal regions of Greenland (Hobbs, 1974), and a seawater density, ρ_w, of $1{,}030 \, \text{kg m}^{-3}$ (Foldvik et al., 1985). Ice thickness is then obtained by simply multiplying the surface elevation of the floating glacier by $\rho_w/\rho_w - \rho_i$, or 9.115 in this case. In another example, Rignot (2001, 2002b) inferred thicknesses of the floating ice of the Pine Island and Thwaites Glaciers from DEM data (referenced to sea level) by multiplying elevation by a constant derived from comparison of DEM and ice-sounding radar—i.e., BEDMAP.

Repeating the ice flux calculation at the flux gate downstream of the grounding zone enables an estimate to be made of the average basal melt rate of the ice shelf/floating glacier tongue under steady-state conditions, as follows. Using the conservation of mass relationship, Rignot (2002a), for example, then computed the steady-state basal melt of floating ice shelves/glaciers, $\dot{B}$, from the ice flux across the grounding zone ϕ_{GZ} and the flux estimated to pass through a gate downstream ϕ_{BM}—i.e., on the floating part, as:

$$\dot{B} = \frac{\phi_{BM} - \phi_{GZ} - \delta\phi_A}{\delta A} \qquad (3.20)$$

where $\delta\phi_A$ and δA are snow accumulation and ice-shelf area between the gates, respectively. An example of satellite-derived input, in this case for the David Glacier in Victoria Land (Antarctica), is given in Figure 3.96 (see color section). Results and errors are discussed in detail in Rignot (2002a). Using this combination of new information, Rignot (2001), for example, determined that the ice discharge of the Thwaites Glacier exceeded net accumulation over the drainage basin by ~30% over the period 1992–1996. Such estimates over short time periods would have been unthinkable before the advent of InSAR. In general, uncertainty in the estimate of outflow by this mass-budget technique is dominated by uncertainties in ice thickness estimates, of ~100 m when derived from hydrostatic equilibrium and tens of meters for RES data (Joughin and Tulaczyk, 2002; Joughin et al., 1999a; Lythe et al., 2001; Rignot and Thomas, 2002). Other factors again relate to uncertainty in depth-integrated ice density and estimates of snow accumulation (see Section 3.7). The new technique represents a major advance, however, by providing a means of more accurately deriving ice-basal melt rates. Prior to this, the difficulty in measuring this important parameter introduced a considerable error into estimates of the total mass balance using the component approach—i.e., determining calving flux and bottom mass balance.

It should be noted that other studies have used Landsat data to estimate the mass fluxes of outlet glaciers. Bindschadler et al. (1996), for example, used ice-surface velocity fields derived from feature tracking of Landsat time series (see Section 3.8.1), to calculate ice discharge fluxes through transverse gates on Ice Streams Bindschadler (D) and E in West Antarctica by specifying ice thickness based upon *in situ* measurements and assuming invariant ice velocity with depth. Based upon their results, the authors concluded that Ice Streams Bindschadler (D) and E were in a state of approximate equilibrium overall at the time of the study. In another example combining Landsat TM-derived velocity data (from 1986, 1989, and 1997) with ice thickness data from the BEDMAP 5-km-resolution grid (Lythe et al., 2001) and elevation data from the AMM-1 DEM (Liu et al., 1999, 2001), Scambos et al. (2004a) estimated the mass flux through a gate on the Institute Ice Stream (West Antarctica). For this, they used a seawater density of 1,028 kg m^{-3} and an estimated mean density of the ice column of 910 kg m^{-3}. The overall flux error was conservatively estimated to be ~10%, with one of the main error sources being uncertainty in ice thickness across the gate profile. Scambos et al. (2004a) extended this to an estimate of the mass balance of the ice stream catchment area by including accumulation estimates derived from the databases of Vaughan et al. (1999) and Giovinetto and Zwally (2000). The difference in the results obtained using these two inputs again underlines the current uncertainty in the accumulation term over ice sheets.

A comprehensive summary of recent satellite-derived results, entailing what is currently known about the mass balance of the Greenland and Antarctic Ice Sheets, is given by Rignot and Thomas (2002). This study includes a compilation of mass budget estimates for the 33 major Antarctic glaciers marked on Figure 3.82, determined using the approach outlined above. Such results represent a unique baseline against which to detect and measure change, and have enabled a major reappraisal of mass balance estimates over large regions. For example, substantial mass losses—i.e., negative mass balance—are observed for large areas which were previously considered to be gaining mass (Bentley and Giovinetto, 1991), including the Pine Island–Thwaites Glacier system (Rignot, 1998c, 2001, 2002b; Rignot et al., 2002), and the Totten and Lambert Glaciers (Rignot and Jacobs, 2002) (see table 1 in Rignot and Thomas, 2002 for details). On the other hand, the Ross ice streams in West Antarctica appear to be undergoing thickening (Joughin and Tulacyzk, 2002). The complex nature of the overall system is further illustrated by the fact that 12 glaciers surveyed in East Antarctica and 8 glaciers discharging into the Filchner-Ronne Ice Shelf are close to balance (Rignot, 2002a; Rignot and Thomas, 2002). According to Rignot and Thomas (2002), the current mass balance of the WAIS is -48 ± 14 Gt per annum), while that of East Antarctica is $+22 \pm 23$ Gt per annum. These studies once again emphasize the importance of ice–ocean interactions in the overall ice sheet mass balance, and the possible evolution of both grounded ice and ice shelves under a warmer climate regime. Although significant uncertainty remains in current measurements of balance and grounding line flux, there is a trend towards reduced uncertainty due to the extraordinary contribution of SAR interferometry.

An important outcome of these InSAR-driven studies has been the discovery that basal melt rates of floating outlet glaciers and ice shelves exceed the previous oceanographic estimates by Jacobs et al. (1996), Jenkins et al. (1997), and Hellmer et al., 1998. Rignot (2001), for example, estimated basal melt rates of 10–30 m per annum under two areas of the Thwaites Glacier in West Antarctica. Rignot and Jacobs (2002) calculated basal melt rates for 23 Antarctic outlet glaciers, directly seaward of their grounding zones and assuming steady-state conditions. Once again, this study reveals significant regional variability, with melt rates varying from <4 m per annum (for several glaciers flowing into the Filchner-Ronne Ice Shelves) to >40 m per annum for Pine Island Glacier. Such findings have major implications. For one thing, the basal melt mechanism in certain locations contributes to the formation of Antarctic Bottom Water (Foldvik et al., 1985; Meredith et al., 2000). As such, changes in melt may impact deep-ocean ventilation (Orsi et al., 2001) and global ocean circulation. Moreover, Rignot and Jacobs (2002) found that the basal melt rate has a strong positive correlation with ocean thermal forcing. Such findings are again of key importance, given the predictions of recent modeling studies.

Huybrechts and de Wolde (1999) and Warner and Budd (1998), for example, predict that rapid ice sheet collapse can result from enhanced basal melting. Rignot and Jacobs (2002) additionally emphasized the major issue of changing melt rates near the Antarctic grounding zone, underlining the vulnerability of floating glaciers and ice shelves to ocean temperature increases where deep water is in direct contact with the grounding zone. Their estimate is of a basal melt-rate increase of 1 m per annum per 0.1°C rise in ocean temperature—i.e., of Circumpolar Deep Water that already comes into contact with ice shelves (Schmeltz et al., 2002b). Description of the complex ocean–ice interaction processes involved, and the oceanic implications, is beyond the scope of this chapter, but is well-covered by Rignot and Jacobs (2002). Once again, such an analysis would not have been possible without satellite radar interferometry.

Similar studies have been carried out in north and northeast Greenland (Rignot et al., 1997a; Joughin et al., 1999b). In this work, InSAR-derived ice velocities were combined with measured and estimated ice thickness data to yield much-improved estimates of ice discharge from the major outlet glaciers controlling mass discharge from the Greenland Ice Sheet. For the Joughin et al. (1999b) study, ice thickness was measured by the University of Kansas airborne radar depth sounder. In a study by Rignot et al. (2001), measured thicknesses came from *in situ* ice-sounding radar profiles (Chuah et al., 1996), while larger scale estimates were derived by correcting single-difference ERS SAR interferograms for topography on floating ice using a DEM, with ice shelf thickness being inferred from the elevation estimates by assuming hydrostatic equilibrium. Rignot et al. (2001) then calculated ice discharge both at the grounding zones (again by assuming hydrostatic equilibrium of ice) and along cross-glacier ice-sounding radar profiles located upstream of the grounding zones. By comparing the results with estimates of iceberg-calving rates at the glacier front, Rignot et al. (2001) found that the ice output flux across the grounding line was on average 3.5 times larger than the ice front discharge. This implies that basal melting, estimated here to be of the order of 10–20 m per annum, is a dominant process of mass loss in northern Greenland (Rignot et al., 1997a). Rignot et al. (2001) estimated the error in mass flux to be of the order of 5% for a typical grounding-line velocity of 1,000 m per annum (with an uncertainty of 4 m per annum) and a thickness of 600 m (with an uncertainty of 30 m).

These new findings have overturned the previously held belief that the mass loss term in this sector is dominated by iceberg calving and surface melting. Moreover, comparison of the new ice discharge data with mass accumulation estimates suggests that more ice is being discharged from northern Greenland glaciers than accumulates in the interior. Once again, errors in current snow accumulation and runoff estimates remain large, however (Thomas et al., 2001a), as seen in Section 3.3.4. Drawing together available evidence, including data collected during the 1990s in NASA's PARCA initiative, Rignot and Thomas (2002) reported that the Greenland Ice Sheet is currently undergoing rapid thinning in its coastal margins, resulting in the loss of an estimated 50 km^3 of ice mass per year. This is equivalent to a global sea-level rise of ~0.13 mm per annum (assuming that the primary source is meltwater runoff). Further research on ice flow and melt processes is required, and is underway, to understand the origins and implications of this signal (Walsh et al., 2001). Rignot and Thomas (2002) believe that dynamic ice sheet response plays a key role, in the sense that an increase in meltwater percolating downwards via crevasses would lead to enhanced lubrication of the base moving across bedrock, as shown by Zwally et al. (2002b). For further information on these intriguing results, and how they are contributing to understanding the larger picture in Greenland, the reader is urged to consult Abdalati (2001). This volume gives excellent insight into the central role that satellite remote sensing now plays in large interdisciplinary polar research projects such as the PARCA initiative. Other work by Reeh et al. (2002) combined InSAR ice motion data with other data to determine ablation rates on Storstrømmen, the large outlet glacier in northeast Greenland. They concluded that current InSAR height estimates are insufficiently accurate to directly monitor ice sheet/glacier surface elevation changes due to ice buildup or ablation. On the other hand, the combination of detailed InSAR velocity measurements with ice ablation and thickness data, by means of the continuity equation, shows great promise in enabling the calculation of changes in surface elevation, and thus ice volume (Mohr et al., 1998; Reeh et al., 2002).

The complexity of ocean–ice shelf interaction processes is illustrated by the fact that melt rates vary substantially across individual ice shelves. It is thought that basal melt rates are particularly high near grounding zones as this is where floating ice attains its deepest draft and the pressure-dependent melting point related to the so-called "ice pump" mechanism (Lewis and Perkin, 1986) is at its lowest (Rignot, 1996). For example, Fricker et al. (2001) observed high basal melt rates of tens of meters per annum near the grounding zone of the Amery Ice Shelf in East Antarctica (Figure 3.97, see color section). While a few floating glaciers exhibit lower basal melt rates—e.g., 4 ±7 m per annum—this process appears to be a major source of mass loss in Antarctic ice shelves. The ice pump mechanism also results in freezing onto the ice shelf base at shallower depths further downstream (seaward), leading to a significant redistribution of ice under ice shelves (Jacobs et al., 1992; Jenkins and Bombosch, 1995). This was illustrated schematically in Figure 1.1 (see p. 3). Fricker et al. (2001) estimated the thickness distribution of this accretion layer of *marine ice* on the base of the Amery Ice Shelf from the hydrostatic (buoyancy) relationship using radar-altimeter-derived surface elevation data (see Figure 3.60) and ice thickness data from Russian airborne RES measurements plus estimates of ice/firn density and temperature. As the RES signal does not penetrate marine ice, but only the meteoric–marine ice boundary, the difference between the two maps is taken to represent the marine ice thickness. Such models again assume steady-state conditions. In this case, marine ice accretion was estimated in this fashion to be up to 190 m thick (Figure 3.97) and account for ∼9% of the total ice shelf volume. Underlying water mass changes resolved by oceanographic measurements confirm the net patterns of basal melt and freezing on the Amery Ice Shelf (Allison, 2003). Clearly, observations of ocean conditions under ice shelves and glacier tongues are required to validate and better understand the remote sensing and associated modeling results (Rignot et al., 2001). Submersible underwater vehicles—e.g., the unmanned Autonomous Underwater Vehicle (AUV) Autosub operated by the Southampton

Oceanography Centre (⟨http://www.soc.soton.ac.uk/PR/Autosub.html⟩)—play an important role in this respect (Jenkins et al., 2004), as do long-term mooring arrays. Such vehicles can provide information on the shape of the underwater cavity beneath ice shelves and floating glaciers, which is required for basal-melt-rate models (Hagen and Reeh, 2004). They also have the potential to collect profile measurements of oceanographic variables in the sub-ice water column—i.e., in regions that are inaccessible to conventional Conductivity Temperature Density (CTD) sensor measurements from ships. In addition, improved numerical modeling of the three-dimensional ocean and thermohaline circulation in the sub-ice-shelf cavity is essential to improving understanding of the exchange of ice and heat at the ice shelf base. This requires the coupling of an ice-shelf model to an ocean circulation model (Holland et al., 2003; Williams et al., 2002).

The studies outlined above illustrate the immense power and potential of these new satellite-based techniques, which provide accurate new baseline flux rates and inputs for ice shelf and ice sheet models. They provide systematic estimates of ice-shelf-basal melt/accretion rates over large areas that are inaccessible to direct observations from the sub-ice-shelf ocean cavity. They also underline the complementarity of satellite altimetry and interferometry for documenting change. Such estimates would have been impossible a few years ago, and are being refined as enhanced satellite data become available—e.g., from ICESat, a possible future CryoSat-type mission, and GRACE (see below). For example, improved definition of ice shelf topography and mass balance by ICESat should improve grounding-line flux estimates. To date, however, ice-sheet-wide estimates of mass balance have been restricted due to a lack of suitable satellite coverage poleward of 81.4°S, and the sign of mass balance of the East Antarctic Ice Sheet in particular remains unresolved (Rignot and Thomas, 2002). It is envisaged that ICESat and a future CryoSat-type satellite will play a key role in this respect by providing enhanced coverage of both the Antarctic Ice Sheet interior and the more steeply sloping coastal margins (Wingham et al., 2004; Zwally et al., 2002a).

A critical requirement is for improved coverage, both spatial and temporal, by satellite interferometric SAR systems. To date, no satellite SAR dedicated to interferometry has been launched, although plans are afoot to optimize coverage in future—e.g., a tandem Radarsat-2 and -3 mission (please see Chapter 2 for information on future systems). Note that wide-swath ScanSAR data have yet to be extensively used in InSAR studies, for reasons also laid out in Chapter 2. Other desirable and necessary improvements include better ice thickness estimates along all grounding zones, and their regular monitoring over time. While the new satellite missions will have the precision required for ice shelf elevation measurement, direct ice thickness measurements are preferable to improve the precision of mass fluxes.

Incidentally, a number of studies have utilized known Sun elevation and azimuth geometry to photoclinometrically convert shadow length in high-resolution visible images to ice cliff height for floating glacier tongues—e.g., the Pine Island Glacier (Bindschadler and Rignot, 2001), and the Ninnis Glacier in East Antarctica (Frezzotti et al., 1998a). In this way, the thickness of floating glaciers has been estimated, enabling change-in-thickness estimates through time series measurements.

3.12 THE MEASUREMENT OF CHANGES IN ICE MASS BY SATELLITE GRAVITY SENSORS

Of major importance has been the launch of the Gravity Recovery And Climate Experiment (GRACE) mission on July 21, 2003 (Tapley et al., 2004a, b), measurements from which provide an important constraint on total mass balance estimates derived from the other datasets and post-glacial rebound (Fukuda et al., 2002; Wahr et al., 2000). Gravity measurements are sensitive to changes in mass under the satellite, including mass redistribution due to crustal uplift and mantle inflow and changes in the mass of snow and ice contained in the ice sheet (Velicogna and Wahr, 2002; Wu et al., 2002). For a given mass change, surface height changes would be in the ratio of 1 : 3.5 : 10, respectively, for isostatic uplift, changes in solid ice, and changes in snow accumulation at the surface because of their different mass densities. Therefore, measurements of changes in both height (by altimetry) and gravity make it possible to improve spatio-temporal estimates of the magnitude of post-glacial rebound, with an accuracy that improves with the length of concurrent measurements (Wahr et al., 2000). The GRACE mission provides monthly estimates of global changes (anomalies) in mass at the Earth's surface, and averaged over several hundred kilometers. First results from the GRACE Gravity Model 01 (GGM01) are shown in Figure 3.98 (see color section). Prior to this mission, the Earth's gravity field was determined with poorer resolution data from different satellite missions and coverage was incomplete. The new GRACE model is estimated to be 10 to 50 times more accurate than all previous Earth gravity models at medium and long wavelengths (Tapley et al., 2004a). More information on the GGM01 model and GRACE is provided on the mission website at ⟨http://www.csr.utexas.edu/grace/gravity/ggm01/⟩.

The specified mission accuracy in the measurement of mean annual mass density over a 200-km radius disk is $1\,\text{gm cm}^{-2}$. Changes in the distribution of snow and ice on the polar ice sheets can in theory be determined to an estimated accuracy of better than 1 cm thickness (water equivalent). Such information is of direct benefit to the interpretation of ice sheet elevations measured by ICESat GLAS and averaged over similar scales (Velicogna and Wahr, 2002). Indeed, a more detailed geoid[9] map will improve the accuracy of satellite altimetry, SAR interferometry, and DEMs over ice sheets. The new data will also enable ice-sheet-related estimates of global sea-level change to an unprecedented accuracy of an estimated ∼0.2 mm per annum (Rignot and Thomas, 2002). As noted in Chapter 1, a strong complementarity exists between GRACE and IceSAT, in that the former measures changes in ice mass while the latter measures changes in volume (Bentley, 2004). When combined with radar/laser altimetry, the new and relatively high-spatial-resolution measurements of time-variable gravity also provide an important means of separating changes in ocean volume by thermal expansion versus mass, and their relative contributions to observed and projected sea-level change (Wahr et al., 2000; Wu et al., 2002).

---

[9] "Geoid" is the term given to the surface of the Earth that most closely approximates sea level in the absence of ocean currents, winds, and other "disturbing" forces.

Furthermore, these measurements potentially enable the separation of the current elastic response of the Earth's crust from its long-term viscous response (Bentley and Wahr, 1998; Velicogna and Wahr, 2002; Wahr et al., 2000). This will mark a significant improvement over current methods of correcting for post-glacial rebound, which typically use coupled models of ice sheet evolution and the visco-elastic response of the solid Earth (Huybrechts and Le Meur, 1999).

Recent work by Velicogna and Wahr (2005) and Velicogna et al. (2004) has confirmed that these fields could be used to recover mass imbalance and net snow accumulation for the Greenland and Antarctic Ice Sheets. Indeed, GRACE-derived estimates of Greenland mass change agree with ice mass imbalance estimates from radar altimeter elevations to within measurement errors. The same authors estimate net snow accumulation from the non-secular component of the GRACE signal. They conclude that it is possible to obtain monthly estimates of ice discharge through the grounding line from GRACE measurements if precipitation minus evaporation ($P - E$) and runoff are known. See ⟨*http://www.csr.utexas.edu/grace/*⟩ for further information on the GRACE mission, and NRC (1997) for information on gravity missions in general. Please also refer back to Section 1.2.2 (p. 16) for details of the similar GOCE mission.

3.13 THE RADAR SOUNDING OF ICE SHEETS FROM SPACE

Previous sections have stressed the key importance of measuring ice-sheet thickness, notably for the more accurate determination of mass balance—e.g., using satellite SAR interferometry—and contribution to sea level rise. Moreover, the present patterns and configuration of internal layers (isochrones) contain unique information on past conditions and changes in ice sheet accumulation and melt patterns and dynamics. In addition, the nature of the ice–bedrock interface has an immense impact on ice dynamics and therefore the response of an ice sheet to climate, and improved information on basal boundary conditions is required by models. Airborne and surface ice-sounding radar campaigns have collected invaluable data since the 1970s (Drewry et al., 1980; Robin, 1983), and significant technological advances have recently been made to yield higher resolution measurements of both ice-sheet thickness and internal layering (Gogineni et al., 1998, 2001, 2003). Such measurements are, however, greatly limited in their areal coverage. This deficiency has prompted the proposed development of a new class of scientific spaceborne radar—namely, an ice-sounding (ground-penetrating) system. The most notable system put forward to date is the MIMOSA (Mapping of Ice and MOnitoring of Sub Arctic areas) mission, which has recently been proposed to ESA as part of their Earth Explorer Opportunity Missions program. Details are given by Herique et al. (2000). The main objective of MIMOSA is the three-dimensional mapping of the Antarctic and Greenland Ice Sheets. The payload is an unfocused nadir-looking SAR operating at a low frequency (P-band, or ~300 MHz with a 10-MHz bandwidth) to enable ground penetration without excessive dielectric losses. The received signal is then a mixture of deep and surface echoes. Both instrument, with a large receive dynamic range, and processing are designed to maximize the coherent-power-to-clutter ratio. With these capabilities, it is hoped that MIMOSA will map the following key parameters over the entire ice sheet and at a 10-m vertical resolution:

- ice thickness up to 4,500 m;
- bedrock topography;
- internal ice layers from their reflective characteristics (up to a -60-dB reflection coefficient); and
- dielectric properties of ice.

When linked with coincident observations of surface parameters, these new measurements will potentially yield the following key information which is important as an input for numerical modeling and also a means of improving our knowledge of ice-sheet dynamics and history:

- total ice-sheet volume;
- derived ice-sheet stresses and stress gradients, which are components of the force–balance equation);
- history of ice flow related to layering characteristics;
- ice-flow velocity related to ice thickness, bedrock characteristics and topography, and internal-layer configuration—e.g., isochrone layers associated with volcanic eruption events; and
- physical evolution of ice (related to the estimated dielectric characteristics of ice).

Such measurements would play a key role in validating ice-dynamical models (Rommen et al., 2004).

The choice of sensor frequency is determined by the frequency-dependent dielectric characteristics of glacial ice. Referring to Section 5.1 of Volume 1 of this book, the propagation of EM waves in ice is determined by the relation $\varepsilon^{*} = \varepsilon_{0}(\varepsilon' - j\varepsilon'')$ (3.21) with the loss tangent $\tan \delta$ being given by $\tan \delta = \varepsilon''/\varepsilon'$ (3.22). The real part of the permittivity ε' is approximately constant and is taken to be equal to 3.15 for ice in the VHF/UHF frequency range (Bogorodskii et al., 1985; Herique et al., 2000). Dielectric absorption is calculated from the loss tangent, which depends upon the ice temperature and the EM wave frequency. The dielectric loss coefficient or absorption exhibits a low and approximately constant value up to a frequency of $\sim$300 MHz—i.e., the MIMOSA frequency. As such, the baseline frequency of the mission was chosen to be sufficiently high to minimize the physical size of the antenna required while remaining within the low radar attenuation band. For the Antarctic Ice Sheet, a mean attenuation of $\sim$3 dB km^{-1} is anticipated (Herique et al., 2000). Please see Herique et al. (2000) for further discussion of the physics of the measurement and potential sources of ambiguity and error (including ionospheric effects). Richardson and Holmlund (1999) and Richardson-Näslund (2004) provide information on the calculation of ice-sheet snow layer depths from ice-penetrating radar (in this case ground-based).

Operating on a small dedicated satellite in a true polar orbit at an altitude of $\sim$500 km, MIMOSA will carry out the first complete sounding coverage of the ice sheets over the planned 2-year mission. This new radar will complement existing spaceborne radar and optical systems in that it will observe the ice sheet to depths well beyond the surface and surficial layers, and at horizontal resolutions ranging from 1 km to tens of kilometers with monthly temporal coverage. Continuing aircraft and surface ice-sounding radar campaigns are required both to validate the new satellite measurements and to provide higher resolution observations in areas of interest.

3.14 ICE-SHEET SURFACE TEMPERATURE

Surface temperature is a fundamental parameter in the ice-sheet energy balance equation (1.6) (see p. 13). An obvious association is between surface (air) temperature and melt, as highlighted in Section 3.3. In Greenland, summer temperatures are high enough to cause widespread melting of the ice sheet surface. Recalling Section 3.4, new results from Zwally et al. (2002b) suggest that surface meltwater percolating through the ice mass to its base leads to accelerated ice sheet flow in Greenland. Given its warmer climate regime, Greenland is thought to be more susceptible to rapid melt than Antarctica in a present day global-warming scenario. Global climate models have predicted that warming over the Greenland Ice Sheet may set in motion a positive feedback cycle revolving around a decrease in surface albedo. This would lead to increased absorption of solar radiation which, the models suggest, will raise temperatures over the ice sheet, causing it to increasingly lose mass at its surface. The complementary measurement of albedo from space is covered in Section 3.14.1. Compared with Greenland, Antarctica is extremely cold. The annual mean temperature on the polar plateau is approximately $-50°C$, with mean austral-summer temperatures of about $-30°C$. Above-freezing temperatures do occasionally occur around the periphery, and more regularly in the Antarctica Peninsula region (for further information on Antarctic meteorology, please see King and Turner, 1997 and Bromwich and Stearns, 1993). In Antarctica, surface melting during summer is therefore mostly confined to peripheral ice shelves and some near-coastal parts of the ice sheet (Zwally and Fiegles, 1994). Of major concern, however, is the extraordinary warming trend in the Antarctic Peninsula region in recent decades (King and Comiso, 2003; Skvarca et al., 1999a), which has been manifesting itself in recent ice-shelf breakup events (see Sections 3.4.1 and 7.8.6). Once again, both Antarctic ice shelves (Scambos et al., 2000, 2003) and floating glacier tongues (Rignot et al., 2001) respond rapidly to changes in surface melt and basal heat fluxes, with the additional effects of meltwater in crevasses.

Accurate large-scale ice/snow physical temperature estimates are required as input into surface energy-balance models driven by turbulent heat fluxes and radiation—see Sections 1.2 (p. 10) and 1.6 (p. 13). These models comprise an important tool with which to predict the response of polar ice sheets to a changing climate, and to estimate parameters such as the total meltwater runoff within individual drainage basins (Ahlstrøm et al., 2002). In addition, accurate estimates of surface temperature are required, together with estimates of the sub-glacial heat flux, by ice sheet models to calculate the thermal condition of the ice–bed interface (Bindschadler, 1998b). Moreover, snow/ice temperature and its temporal variability determines the rate of surface and near-surface snow metamorphism, thereby affecting the size distribution of snow grains, which in turn impacts the optical and microwave properties of the snow/firn layer (Colbeck, 1987; Mätzler, 1998). Moreover, firn-compaction models (e.g., Zwally and Li, 2002) require improved monthly surface-temperature maps.

Satellite-interrogated Automatic Weather Stations (AWSs) and other autonomous *in situ* instrument packages play a major role in measuring near-surface ($\sim$2 m) air temperature and snow/ice temperature (Bromwich and Stearns, 1993; Steffen and Box, 2001). In Antarctica, AWSs have been in operation at some sites since 1980 (Shuman and Comiso, 2002; Shuman and Stearns, 2001). These data are available from the University of Wisconsin, Madison ($\langle ftp://ice.ssec.wisc.edu/\rangle$), with further information at $\langle http://uwamrc.ssec.wisc.edu/aws/\rangle$. The Greenland Climate Network, on the other hand, was established in 1994 (Box and Steffen, 2001; Steffen and Box, 2001). The coverage of AWS networks is, however, spatially and temporally limited, and insufficient to fully characterize large-scale decadal climate variability (Torinesi et al., 2003). This is also the case with meteorological data collected at Antarctic stations (Turner et al., 2002). A major aim of current research is to use satellite remote sensing to extrapolate these point temperature measurements over wider ice-sheet (and sea-ice) regions (Shuman and Comiso, 2002). Such data hold the key to characterizing regional to ice-sheet-wide climate variability and change.

Currently, two satellite datasets are used to calculate ice sheet (and sea ice) physical temperature—namely, thermal infrared (TIR) and passive microwave (PMW). Due to different wavelengths involved, and therefore different penetration depths into the snow/ice volume, techniques using these data in effect produce different quantities. TIR radiometry measures the "skin" surface temperature, whereas, for PMW radiometry, the derived physical temperature (T_C) emanates from some centimeters to meters below the surface. For the latter, the depth depends not only on the wavelength (frequency) and polarization of the radiation but also on the characteristics of the surface and near-surface layers. The following presents background information on these techniques, and gives recent applications. Please see Chapter 5 of Volume 1 of this book for additional information on the underlying principles of both techniques.

3.14.1 Thermal infrared techniques

Ice Sheet Surface Temperature (ISST), which is equivalent to the sea-ice product (*IST*—Ice Surface Temperature) in Chapter 5 of Volume 1 of this book, can be readily derived from cloud-free TIR satellite imagery, provided a correction for the intervening atmosphere is made—e.g., Bamber and Harris (1994). At the satellite wavelengths typically employed—i.e., 10.5–12.5 µm—the emissivity of snow and ice approximates unity (Dozier and Warren, 1982; Warren, 1982), but varies with viewing angle (Dozier and Warren, 1982; Key et al., 1997). For NOAA-7 AVHRR channel 4, for example, modeled snow emissivities vary from

0.988 to 0.9955 at 0° and 50° scan angles, respectively (Key et al., 1997). Ice sheet surface temperatures have been derived from cloud-free Landsat TM band-6 data by Orheim and Lucchitta (1988), Pattyn and Declair (1993), and Winther (1993b), but to relatively poor accuracy due, it is thought, to atmospheric effects and inaccurate pre-launch calibration effects (König et al., 2001). The NOAA AVHRR (Cracknell, 1997) is without doubt the "workhorse" sensor for measuring surface temperature over large areas (Steffen et al., 1993b), although Comiso (1994) also used single-channel data from the Nimbus-7 THIR (1978–1987) to derive ISST to an estimated accuracy of 2 K. Although AVHRR data have been available since the 1970s, continuous orbital records did not become digitally available until 1981 (Massom, 1991; Shuman and Comiso, 2002). Due to its excellent coverage (over a 2,240-km swathwidth) and short revisit interval (102 minutes) and ready data availability, the AVHRR is well-suited to operational measurement and the detection of change, at a medium resolution (of ∼1 km to 4 km, the latter in global coverage mode). The major limitation is cloud cover, necessitating both effective cloud removal/masking techniques and temporal averaging to ensure sufficient cloud-free coverage. As a result, the ISST data product is likely to be cold-biased, as they reflect cloud-free conditions (Fahnestock and Bamber, 2001).

Assuming that water vapor effects are minimal in the cold polar atmosphere, Comiso (1994, 2000, 2003) adopted a single-channel technique, deriving ISST directly from AVHRR channel-4 ($\lambda = 10.3$–$11.3\,\mu m$) data. This channel is chosen in preference to channel 5 ($\lambda = 11.5$–$12.5\,\mu m$) due to the lower sensitivity of the former to atmospheric water vapor. Results from Antarctica are shown in Figure 3.99 (see color section), with temperature anomaly maps in Figure 3.100 (see color section). King and Comiso (2003) used these data to show that a spatial coherence exists in interannual temperature trends in the Antarctic Peninsula region. In their linear regression analysis of AVHRR-derived ISSTs for the period 1981 through 2003 and north of latitude 60°N, Comiso and Parkinson (2004) noted a net warming trend of 0.85°C per decade over Greenland. Further analysis revealed a lengthening of the annual Greenland melt season of ∼4 days per decade.

Other algorithms typically employ the "split-window" approach to remove atmospheric water-vapor effects (Yamanouchi et al., 1987). This technique, which was originally developed for sea surface temperature retrieval (Minnett, 2001), involves the differencing of two spectrally adjacent channels with differing responses to atmospheric effects—e.g., channels 4 (10.3–$11.3\,\mu m$) and 5 (11.5–$12.5\,\mu m$) of the AVHRR (Kadosaki et al., 2002; Key and Haefliger, 1992; Yamanouchi et al., 2000). Haefliger et al. (1993) attained an agreement of 0.5 K between AVHRR ISST retrievals and *in situ* measurements in Greenland. Stroeve and Steffen (1998) also used this method to construct monthly mean maps of *ISST* from AVHRR data for the period 1989–1993, also for Greenland. Their results, with an estimated accuracy of better than 1 K for dry snow during summer (when cloud masking is optimal), confirm that ISST is strongly correlated with ice sheet topography. Maximum surface temperatures in summer occur during July and along the western coast and southern tip of Greenland, whereas large interannual variability in winter is associated with katabatic wind events. Although summer temperatures exhibit little variability, 1992 stands out as a particularly cold year, due, it is thought, to the volcanic eruption of Mount Pinatubo in 1991 (note that a similar anomalous signature showed up in the surface melt record, as discussed in Section 3.3).

As part of the NASA Mission to Planet Earth Pathfinder project, investigators at the U.S. NSIDC are currently using AVHRR data to calculate ice surface temperature data that will yield a 5-km grid-cell size for Antarctica on a twice-daily basis for the period 1982–present, with the last 4 years at 1.25-km resolution over most areas ($\langle http://nsidc.org/data/nsidc$-$0069.html\rangle$). The algorithm for this work is derived from Key and Haefliger (1992). Accuracy of the temperature measurement from the satellite data is expected to be ±1 K. A similar approach is followed by Key et al. (1997), who also incorporate atmospheric sounding data into their enhanced algorithm. Their AVHRR ISST algorithm is:

$$ISST = a + bT_{11} + c(T_{11} - T_{12}) + d[(T_{11} - T_{12})(\sec\theta - 1)] \quad (3.23)$$

where a, b, c, and d are empirically derived coefficients; T_{11} and T_{12} are the T_Bs (in K) for channels 4 and 5, respectively; and θ is the satellite scan angle. Data are also available for the Arctic. More research is required to validate these important data.

The measurement of surface temperature is continuing with the current MODIS sensors onboard the NASA Terra and Aqua satellites (Hall et al., 1998, 2004), offering improved spatial resolution and calibration. In future, the Visible Infrared Imaging Radiometer Suite (VIIRS) on the future U.S. National Polar-orbiting Operational Environmental Satellite System (NPOESS) missions will continue MODIS-like measurements (Crison and DeLuccia, 1998), and the Conical Microwave Imager/Sounder (CMIS) will measure surface thermal microwave emissions. The launch of the initial stage of NPOESS, termed the NPOESS Preparatory Project (NPP), is planned for 2006, with NPOESS proper being launched in 2009 (see Chapter 5 of Volume 1 of this book).

Stroeve et al. (1996) further examined the relationship between ERS-1 Along-Track Scanning Radiometer (ATSR) thermal radiances and snow surface temperature for the Greenland Ice Sheet through forward calculations of the LOWTRAN 7 radiative-transfer model, with inputs from radiosonde and other meteorological measurements. The ATSR is innovative in the sense that it is designed to provide much-improved atmospheric corrections and thus temperature retrieval accuracy by looking at the same point on the surface from two different angles. Stroeve et al. (1996) recommended that a dual-view algorithm using data from the 11- and 12-µm channels in both the nadir and forward views be used for applications in polar regions where atmospheric variability can be large. Both Stroeve et al. (1996) and Bamber and Harris (1994), who also used ATSR data, report an accuracy of better than 0.2 K from their algorithms. These techniques are also applicable to ERS ATSR/2 and Envisat AATSR data.

3.14.1.1 Cloud–ice discrimination

The derivation of surface information from satellite visible to TIR imagery is heavily dependent on the accurate discrimination of clouds from snow and ice, in order to mask the clouds. This represents a major challenge (see Chapter 5 of Volume 1 of this book for details of current techniques). New-generation satellite sensors have enhanced capabilities for distinguishing and even classifying clouds. Multi-angular measurements by the MISR, for example, show great promise as a means of distinguising thin cirrus cloud from snow/ice, and for improving cloud characterization (Di Girolamo and Wilson, 2003; Nolin et al., 2002a) (Figure 3.101, see color section). The MISR observes the daylit Earth continuously, and views the entire planet between 82°N and S every 9 days.

While computations of ISST from TIR data treat cloud as undesirable noise, it should again be remembered that cloud cover is in fact an important parameter which has a variable impact on the surface energy budget and the amount of energy available for snow/ice melt (Curry et al., 1996; Yamanouchi and Kawaguchi, 1984). Moreover, this impact depends on the type of cloud. This issue, and the important advances being made in cloud classification in satellite data, is evaluated by Bintanja and Van den Broeke (1996) and Cawkwell and Bamber (2002).

3.14.2 Passive-microwave techniques

Largely unaffected by cloud cover (particularly at lower frequencies), PMW brightness–temperature (T_B) data have the potential to provide alternative continuous and gap-free temperature distributions, extending from the 1970s into the future with improved AMSR(-E) data. Shuman and Comiso (2002), for example, used T_B data from the 37-GHz channel (V-pol) of the SSM/I (NSIDC, 2002). A disadvantage is that the temperature product is much coarser in resolution than that from the TIR-derived product—i.e., 25 km versus ∼1–4 km. For PMW sensors such as the Nimbus-7 SMMR, DMSP SSM/I, Aqua AMSR-E, and ADEOS-II AMSR, the measured T_B is the frequency- and polarization-dependent product of the emissivity and physical temperature of the radiating portion of the snow/ice volume (see Chapter 5 of Volume 1 of this book). Assuming that the snow is cold and dry and atmospheric effects are negligible, it is in principle possible to derive large-scale information on polar firn temperature and its variability. Referring back to Chapter 5 (Volume 1 of this book), the relationship between microwave thermal emission, or measured T_B, and snow/ice effective physical temperature T is described by the Rayleigh–Jeans approximation of Planck's equation and given by:

$$T = T_B \varepsilon^{-1} \qquad (3.24)$$

where T is the physical temperature integrated over a thickness approximating one penetration depth; and ε is microwave emissivity.

In practice, the retrieval of temperature is complicated by the complex nature and variability of wavelength-dependent microwave emission from polar firn and snow/ice. Microwave emissivity covers a significantly wider range of values than it does in the TIR, and depends on frequency-dependent radiative scattering by ice grains, controlled by density, grain size, layering/stratigraphy, and surface roughness

(Comiso et al., 1982; Foster et al., 1984; Rott et al., 1993; Steffen et al., 1993a; Zwally, 1977). It also has an angular and polarization dependence (see Chapter 5 of Volume 1 of this book). As such, T_B data values are a function of the snow and ice characteristics over the depth of emission ($\sim$1 m at 37 GHz for cold dry snow/firn: Shuman et al., 1995a; Surdyk, 2002a) rather than providing a pure surface temperature signal (Shuman and Stearns, 2001; Shuman et al., 1995b). At 37 GHz, the depth of emission corresponds largely to the penetration depth of the diurnal temperature cycle (Alley et al., 1997; Shuman et al., 1995a). At this wavelength, volume scattering on ice grains predominates, and temperature measurement uncertainties are due to variability in density stratification, surface roughness and conditions, and grain size (Shuman et al., 1995a). To account for approximately annual variations in snow emission characteristics, the T_B data can be calibrated to PMW temperatures (T_C) either empirically or by radiative-transfer modeling (Comiso et al., 1982; Fung and Chen, 1981). Shuman and Comiso (2002) used AWS air temperature data combined with an emissivity-modeling technique at specific ice sheet locations (Shuman et al., 1995a), while Das et al. (2002) combined daily satellite T_B values with modeled surface temperature data from the ECMWF dataset (Gibson et al., 1999). This produced mean emissivity values of 0.824 ± 0.02 and 0.800 ± 0.02 (corresponding to 37-V datasets from the SMMR for 1979–1987 and the SSM/I for 1988–1993, respectively).

The T_B measurement accuracy of the SSM/I at 37 GHz (V), for example, is ± 2 K (Hollinger et al., 1990). This channel is preferred due to its relatively shallow skin depth compared with the 19- and 22-GHz channels, resulting in a closer correlation with air temperature T_{air} (Shuman et al., 1995a; Van der Veen and Jezek, 1993) and a more rapid response to variability in T_{air}. Moreover, V-polarization remains relatively insensitive to surface variability compared with H-polarization (Shuman et al., 1993). Atmospheric effects are also assumed to be negligible at 37 V (Maslanik et al., 1989), due to the dry cold atmosphere above the central ice sheet (note that this may not always hold for ice-sheet margins and the sea-ice zone).

Working with 37-V T_B data from both the SMMR and SSM/I data, Das et al. (2002) showed that a high seasonal to interannual variability existed in temperature variance in the Siple Dome region of West Antarctica from 1979 to 1999. They further attributed an observed 5-year cycle in temperature variance to variations in the Southern Oscillation Index (SOI) (Cullather et al., 1996). Please see Shuman et al. (1995a) for further discussion on the empirical derivation of near-surface T_{air} from the satellite PMW record. Furthermore, Surdyk (2002a, b) showed that T_B data at 37 GHz can be used to detect and monitor short-term changes in T_{air} (Figure 3.102), such as those observed by Enomoto et al. (1998), over the cold, dry regions of an ice sheet. Surdyk (2002a) provided detailed discussion on the relationships between snow effective temperature, emissivity, and penetration depth. Note that, while higher frequency channels—i.e., 85 GHz—provide a closer approximation to T_{air}, due to their smaller penetration depth they are limited by their sensitivity to atmospheric effects (Surdyk, 2002a; Ulaby et al., 1981). In their analysis of seasonal and interannual trends in ERS scatterometer backscatter and DMSP SSM/I T_B data in Antarctica, Bingham and Drinkwater (2000) also noted

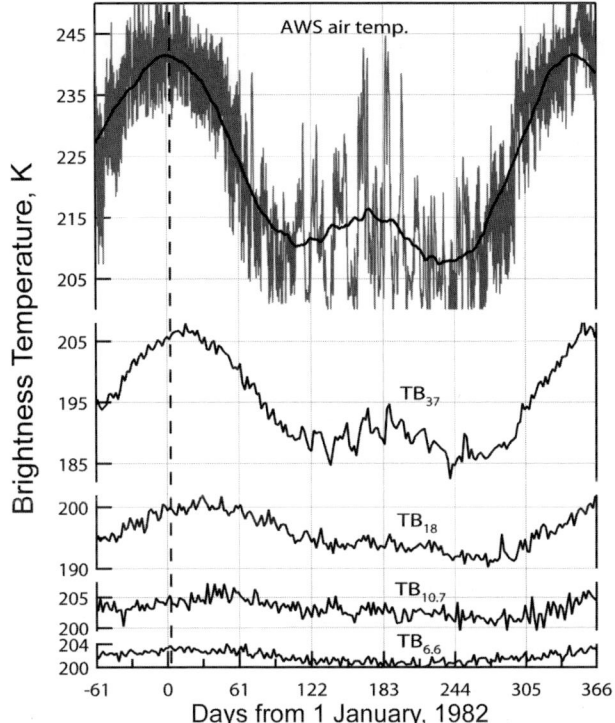

Figure 3.102. Comparison of the annual pattern of the air temperature measured by the AWS (eight times per day) at Dome C in East Antarctica (74.5°S, 119.5°E, elevation 3,280 m), and the four Nimbus-7 SMMR T_Bs for 1982. The y-axis scale is K. The thick line within the AWS record represents the 2-month smoothed pattern. The T_Bs were measured once every 2 days by SMMR.

From Surdyk (2002a). Copyright Elsevier 2002, reproduced with permission.

clear seasonal cycles related to T_{air} forcing via its effect on firn emissivity and absorption.

While the above techniques work on the premise that microwave emissions at 37 GHz track the overlying T_{air} with a minimal lag, longer wavelength emissions emanate from a greater penetration depth and represent a more stable temperature. Winebrenner et al. (1994), for example, used dual-polarized emissions measured at 6-cm wavelength. The penetration depth at 6.7 GHz (4.5-cm wavelength, and on the Nimbus-7 SMMR, ADEOS-II AMSR, and Aqua AMSR-E) is ~80 m in dry-snow zones, and is therefore affected by seasonal rather than daily temperature variations. Fine-scale density layering in the snow (of the order of 0.5 cm) affects both the polarization of emission (which is directly observable) and the effective emissivity (which is not) in such a way as to allow accurate estimation of emissivity from polarization (Winebrenner et al., 2001). The combination of emissivity and observed T_B thus yields an estimate of the 10–80-m firn temperature, which closely approximates the mean surface temperature. For comparison, Surdyk (2002a) computed the following penetration depths for the Antarctic Ice Sheet: 20–40 m at 6.6 GHz, 6–16 m at 10.7 GHz, 1–5 m at 18 GHz, and 0.1–1.4 m at 37 GHz.

Even longer wavelength, L-band systems are due for launch onboard SMOS in 2007 and HYDROS in 2010 (Long et al., 2005). The penetration depth at L-band (wavelength 23 cm) can be of the order of 100–150 m in cold, dry regions. Emission modeling by Mätzler (2001) also showed that emissivity approaches unity near 1 GHz, particularly at V-polarization close to the Brewster angle. Given these attributes, Mätzler (2001) suggested that SMOS has the potential to measure very stable thermal emissions from deep within the ice, and which may be related to geothermal heat sources. The spatial resolution is, however, limited to 50 km. Additional information on the applicability of this sensor to ice-sheet research is given by Drinkwater et al. (2004b). Further work is required to characterize ice sheet emissivity at L-band using theoretical and semi-empirical models. The overall aim is to develop algorithms to retrieve ice-sheet characteristics.

In general, ambiguous physical temperature retrievals can result from a number of factors when using PMW techniques. One relates to the difficulty of accounting for the complex spatial and significant temporal variability in ice-sheet microwave emissivity—e.g., due to surface melt (Das et al., 2002; Zwally and Fiegles, 1994) and snow metamorphism processes (e.g., depth hoar formation)—and changes in surface characteristics—e.g., crustal formation by radiation and wind processes and surface–hoar frost formation (Shuman et al., 1993). The former affect the degree of volume scattering, while surface changes impact the degree of surface scattering (Abdalati and Steffen, 1998; Shuman and Alley, 1993; Ulaby et al., 1981). Such events can bias temperature retrievals, and in a complex fashion. Another complication arises from the fact that the temperatures also represent an average over a large (25×25 km) pixel. The issue of temporal covariation of surface temperature and T_Bs is evaluated by Winebrenner et al. (2004).

Extending back to the 1970s, these datasets again provide key information with which to monitor change, and provide an important means of identifying seasonal to interannual temperature anomalies over vast regions of the Antarctic continent. Schneider et al. (2004), for example, combined analysis of T_Bs from the 37-GHz channels (V-pol) of the SMMR and SSM/I with AVHRR-derived ISSTs and NCEP/NCAR Reanalysis Project air pressure and temperature fields. Their study, based on principal component analysis, showed that much of the interannual variability in Antarctic surface temperature over the past 20 years can be explained by the influence of two large-scale modes of climate variability—namely, the El Niño–Southern Oscillation or ENSO (Kwok and Comiso, 2002) and the Southern Annular Mode (Schneider and Steig, 2002; Thompson and Solomon, 2002). See Chapter 5 of Volume 1 of this book for more details on these and other significant modes of variability, changes in which have contributed to observed cooling and warming trends in different Antarctic regions. This once again illustrates the power of the emerging trend of combining different but complementary datasets, both satellite- and model-derived.

Shuman and Comiso (2002) conclude that the different near-surface temperature datasets described above provide complementary information. Although individual daily differences between *in situ* and satellite temperatures can be quite large, the average errors are relatively small and well-constrained. They conclude that a combination of all methods may be necessary to determine a climate baseline which is sufficiently accurate, especially at high temporal resolution, to enable the evaluation of potential future temperature changes. For the TIR technique, monthly averages provide the most realistic representation of the large-scale ISST field. In addition, Shuman et al. (1997) have compared mean annual and seasonally averaged temperature derived from satellite PMW T_B data with δD and $\delta^{18}O$ isotopic variations measured in surface snow to better constrain the relationship of these isotopic climate indicators with average annual temperature (Shuman et al., 1997).

3.15 ICE SHEET SURFACE ALBEDO

Albedo is defined as the ratio of upwelling-to-downwelling radiative flux at the surface measured in a surface-parallel plane. *Broadband albedo* is albedo integrated over the complete solar spectrum—i.e., 0.3–3.0 μm (Stroeve, 2001; Stroeve et al., 1997).

While *reflectance* refers to this fraction for a given (single) incidence angle, albedo is the directional integration of reflectance over all Sun-view geometries. This is referred to as *diffuse* or *hemispherical* albedo if the surface is assumed to reflect equally at all angles—i,.e., *isotropically*. As we shall see, this is not the case with snow and ice. Recalling Section 1.2 (see p. 10), albedo is a key term in the surface energy balance—see (1.6)—and ice sheet mass balance (Greuell and Genthon, 2004; Oerlemans and Hoogendoorn, 1989), and is a fundamental parameter in climate modeling, also defining the lower boundary for radiative transfer (Hu et al., 1999). Importantly, it can vary greatly in space and time (König et al., 2001), and understanding the variation in the amount of solar radiation reflected and absorbed by the surface is a key factor in understanding climate processes and change (Steffen et al., 1993a). An example of this variation, as a function of snow age and state, is given in Figure 3.103. As the albedo of snow and ice is high, even a small change can dramatically modify the amount of incoming solar radiation absorbed, and thus the surface temperature and melt rate through the climate–albedo feedback process (Hall, 2003; Nolin and Stroeve, 1997). To reiterate, the latter is of major concern, given that climate models predict a regional warming over Greenland of 1 to 3 times the global mean, for example (Houghton et al., 2001). It follows that realistic and accurate parameterizations of surface albedo, and properties affecting it, are essential when modeling the polar ice sheets. Moreover, the penetration, or transmission, of shortwave radiation into the snow/ice volume drives (i) snow metamorphism and internal melt, and (ii) photo-chemical processes (Brandt and Warren, 1993). Knowledge of the optical properties of Antarctic snow can be applied to determine the vertical distribution of radiative heating in the snow during the Antarctic summer (Brandt and Warren, 1993).

Martonchik et al. (2000) provide an invaluable review of the reflectance nomenclature used in remote sensing. Detailed treatments of the theory of snow albedo and its calculation are given by Dozier (1989), Grenfell et al. (1994), Wiscombe and Warren (1980), and Warren (1982). These and other studies—

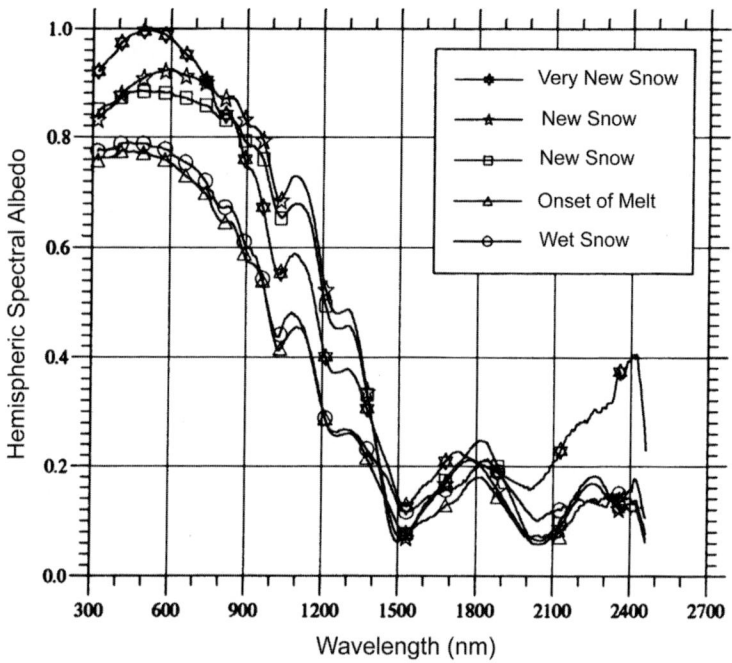

Figure 3.103. Hemispheric spectral albedo for very new snow (a few hours old), new snow (1–2 days old), the onset of melt, and wet snow in the spectral range of 300–2,500 nm, measured at one point on the surface of the Greenland Ice Sheet in 1991. All measurements were carried out at solar zenith angles between 58° and 68°, while the snow-grain diameter varied from 0.5 μm for very new snow, 0.1–0.2 μm for new snow, to 1–5 μm for damp and wet snow. These data were used to calibrate and validate AVHRR-derived albedo values.
From Steffen et al. (1993a). Copyright Springer-Verlag 1993, reproduced with permission.

e.g., Aoki et al. (2000); Green et al. (2002); Nolin and Dozier (2000); and Winther et al. (1999)—show that snow spectral reflectance is strongly determined by grain size (and shape), particularly in the near-IR, and can be modeled from the optical properties of ice. Other factors affecting snow/ice albedo are soot and dust (in the Arctic and in the visible: Warren and Clarke, 1986), solar incidence angle, and surface roughness (Leroux and Fily, 1998; Leroux et al., 1998a). For example, the specular, forward-scattering component of snow increases with grain size and low-Sun angle (Wiscombe and Warren, 1980). While surface grain size and roughness characteristics are poorly understood, progress is being made in their parameterization, both through *in situ* measurements—e.g., Gay et al. (2002)—and satellite remote sensing (see Section 3.16). See Chapter 5 of Volume 1 of this book for further discussion on the optical properties of snow.

Once again, satellite remote sensing offers the only practical means of measuring large-scale surface albedo and its spatio-temporal variability, using sensors operating at visible and near-IR wavelengths (Fily et al., 1997; Hall et al., 1989; Greuell and Knap, 2000; Knap and Oerlemans, 1996; Reijmer et al., 1999; Stroeve et al., 1997, 2000). This variability can occur rapidly by new snow deposition or melt, with changes in magnitude of 0.4–0.6 over periods of days to weeks (Winther, 1993a, c). Indeed, it is anticipated that satellite-derived albedos, and cloud amount and type/ optical properties, will in future serve as direct input for ice-sheet ablation models (Greuell and Genthon, 2004).

The retrieval of useful surface albedo measurements of ice sheets from satellite radiance observations involves a number of important steps, which are similar to those outlined for sea ice in Chapter 5 of Volume 1 of this book. Primarily, the data must be corrected for atmospheric effects (Vermote and Vermeulen, 1999). Second, as satellite sensors—such as the NOAA AVHRR (Toll et al., 1997), Envisat MERIS, Terra and Aqua MODIS, Terra MISR (Diner et al., 1998), and POLarization and Directionality of the Earth's Reflectances (POLDER) sensors onboard ADEOS-I, ADEOS-II, and PARASOL (Polarization and Anisotropy of Reflectances for Atmospheric Sciences coupled with Observations from a LIDAR)—measure in only narrow spectral bands, conversions must be applied to retrieve broadband information (Greuell et al., 2002; Greuell and Oerlemans, 2004; Liang, 2001; Strahler et al., 1999), which is required for energy-balance calculations. Albedo α is generally defined as the fraction of solar irradiance reflected from the surface (see Chapter 5 of Volume 1 of this book), and expressed as (Stroeve, 2001):

$$\alpha = \frac{\int \rho(\lambda) S(\lambda) \, d\lambda}{\int S(\lambda) \, d\lambda} \qquad (3.25)$$

where $\rho(\lambda)$ and $S(\lambda)$ represent spectral albedo and solar irradiance, respectively, at wavelength λ. Integration gives the broadband albedo over the solar spectrum, which is generally taken to be 0.3–3.5 µm; this characterizes the total shortwave energy reflected by the Earth's surface. Signals measured at the satellite are also affected by the Sun/sensor-viewing geometry and atmospheric scattering and absorption (Stroeve, 2001; again see Chapter 5 of Volume 1 of this book). Significant variability in reflectance can occur not only due to changes in wetness and grain size but also as a function of viewing and incident geometries, and at viewing geometries facing the Sun and large incidence angles in particular (Calvin et al., 2002). Moreover, albedo decreases from the visible into the near-IR (Grenfell et al., 1994).

The third step is required because satellite sensors such as the AVHRR, GLI, and MODIS are restricted to narrow fields-of-view and measure only a narrow angular portion of the total energy reflected from the surface. As such, the derived albedo is strictly a directional–hemispherical reflectance (Klein and Stroeve, 2002), whereas incident radiation is reflected into all angles in the entire hemisphere above the surface. Moreover, snow and ice surfaces reflect anisotropically (see below). As a result, there is a need to convert these angular measurements to hemispherical data. Recalling Chapter 5 of Volume 1, this "full-hemisphere" angular distribution of surface reflectance is termed the *Bidirectional Reflectance Distribution Function* (*BRDF*), or R, and is given in units of sr^{-1} (Warren, 1982). As illustrated geometrically in Figure 3.104, this describes how reflectance depends on solar and viewing angles (Nicodemus et al., 1977; Strahler et al., 1999) and is defined as the intensity of surface-leaving radiance divided by the irradiance incident from a single direction

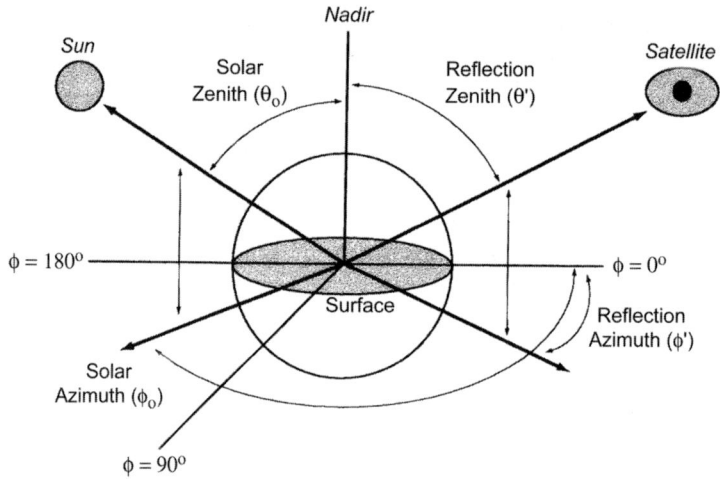

Figure 3.104. Schematic representation of Bidirectional Reflectance Distribution Function (BRDF) terminology.

From Klein and Hall (1999), after Steffen (1987). Copyright International Association of Hydrological Sciences 1987, reproduced with permission.

(Stroeve and Nolin, 2002a). After Klein and Hall (1999) and Tanikawa et al. (2002), the BRDF (R) at wavelength λ is defined by the upward-reflected radiance $I(\theta', \phi', \lambda)$ from the snow surface contributed by the direct solar beam and the incident flux $F_0(\theta_0, \phi_0, \lambda)$, on a surface normal to the beam (Warren, 1982):

$$R(\theta_0, \theta', \phi_0, \phi', \lambda) = \frac{dI(\theta', \phi', \lambda)}{\mu_0 \, dF_0(\theta_0, \phi_0, \lambda)} \qquad (3.26)$$

where (θ_0, ϕ_0) are the incident solar zenith and solar azimuth angles, respectively; (θ', ϕ') are the reflection (outgoing) zenith and azimuth angles, respectively; and $\mu_0 = \cos\theta_0$. For surfaces which are not flat, computation of BRDF depends upon accurate knowledge of the surface slope, through the use of accurate DEMs (Nolin and Dozier, 1993). The BRDF is also determined by the optical and structural properties of the ice/snow surface—e.g., transmission, reflection, multiple scattering, abosrption and emission, shadowing, and facet density and orientation density.

Knowledge of the BRDF is required to derive an albedo measurement from a satellite observation at specific view azimuth and zenith angles (Barnsley et al., 1994; Jin and Simpson, 1999, 2000; Klein and Hall, 1999). Moreover, it can be used to normalize satellite reflectance observations over a range of Sun-angle geometries to a common Sun-view geometry (Leroy and Roujean, 1994), thereby facilitating temporal comparisons and image mosaicking, for example. Another complicating factor that must be accounted for, however, is the fact that snow and ice are non-Lambertian, or anisotropic, reflectors—i.e., the distribution of reflectance is unevenly distributed among reflectance angles (Aoki et al., 2000; Grenfell et al., 1994; Klein and Stroeve, 2002; Tanikawa et al., 2002; Warren, 1982). Ice particles are strongly forward-scattering and transmit virtually all radiation at visible wavelengths, but are moderately to strongly absorbing in the near-IR (see Section 5.7.1 of Volume 1 of this book), with larger grain sizes leading to a decrease in reflectance (Grenfell et al., 1994; Warren, 1982; Warren and Wiscombe, 1980). As a result of the anisotropic behavior, the radiance measured at the satellite sensor is strongly dependent on the satellite and solar-viewing geometry at the time of data acquisition (Klein and Hall, 1999), increasing with an increase in the relative azimuth angle between Sun and sensor. Also, forward scattering becomes more pronounced as the solar zenith angle increases (Klein et al., 2000). Correction for anisotropic effects requires detailed knowledge of these geometrical relationships. As such, the BRDF provides a key means of explicitly prescribing surface reflectance in terms of its directional, spectral/optical, temporal, and spatial characteristics. This is critical, given the reflectance anisotropy of snow and ice and the resultant nonlinear relationship between the reflectance measured by a satellite radiometer from a given direction and the albedo. The angular shape and magnitude of the BRDF is determined by the geometric structure, density, and composition of the reflecting medium (Diner et al., 1999; Warren et al., 1998).

Correction for anisotropic effects has typically been accomplished by normalizing the BRDF by the albedo to get the anisotropic reflectance function, f (Klein and Hall, 1999; Steffen, 1987; Stroeve et al., 1997):

$$f(\theta_0, \theta', \phi_0 - \phi', \lambda) = \frac{\pi R(\theta_0, \theta', \phi_0 - \phi', \lambda)}{a_s(\theta_0, \lambda)} \qquad (3.27)$$

where a_s is the spectral albedo. Where $f > 1$, the reflectance observed at a given sensor zenith angle, solar zenith angle, and relative azimuth exceeds the true spectral albedo at that wavelength. Values of $f < 1$, on the other hand, indicate that the observed reflectance is less than the true spectral albedo (Klein and Hall, 1999). It is necessary to know the complete distribution of $f = (\theta_0, \theta_v, \phi_0 - \phi_v, \lambda)$ over all combinations of θ_0, θ', and $\phi_0 - \phi'$ in order to derive a spectral albedo from a satellite measurement of $R(\theta_0, \theta', \phi_0 - \phi', \lambda)$. Fortunately, advances in the modeling of BRDF have enabled the anisotropic scattering behavior of snow and ice to be taken into account. A recent approach, adopted by Fily et al. (1997), Nolin and Stroeve (1997), and Stroeve et al. (1997), has been to use the *Discrete-Ordinate Radiative Transfer model* (DISORT) of Stamnes et al. (1988) to model snow BRDF for the retrieval of albedo from satellite data. This enables the determination of shortwave reflectance at particular bidirectional-viewing geometries (Cabot and Dedieu, 1997; Nolin and Dozier, 1993). Nolin and Liang (2000) summarize recent advances in the modeling and measurement of snow BRDF. Difficulties arise in ablation zones, such as occur on the Greenland Ice Sheet, where the presence of large snow-free areas during seasonal melt periods undermines the use of anisotropy models based on snow. Another potential problem associated with the directionality of surface reflection may occur due to roughness at the scale of sastrugi. Warren et al. (1998) found f to vary by up to 8% for viewing angles of $<50°$ on the Antarctic Plateau due to sastrugi orientation, and recommended use of near-nadir data as a result.

Failure to accurately account for the bidirectional reflectance of snow, particularly at large viewing angles, can lead to large errors in albedo estimates from satellite optical data (Klein and Hall, 1999; Knap and Reijmer, 1998; König et al., 2001; Steffen, 1987; Stroeve et al., 1997; Tanikawa et al., 2002). This is illustrated in Figure 3.105 (see color section), which presents an albedo map of Greenland derived from AVHRR data, (a) assuming a Lambertian surface, and (b) applying a BRDF conversion to derive the albedo. The third map is the difference between (a) and (b), and shows differences not only in the magnitude of the albedo but also in the spatial distribution of areas of high and low albedo. Comparison with *in situ* data shows that the case incorporating BRDF more accurately represents the spatial distribution of albedo encountered over the ice sheet. The BRDF also enables improved image standardization/normalization by correcting for illumination and view angle effects.

In recent work, the approach outlined above has allowed climatologists to examine the large-scale seasonal variability of albedo on the Greenland Ice Sheet (Nolin and Stroeve, 1997; Stroeve et al., 1997). Strong variations result from increases in snow-grain size and exposure of the underlying ice cap as the seasonal snow cover ablates away. Impetus for this research comes from the recent findings, outlined in Section 3.3, of a significant increase in areal melt extent since 1979. As a result, increased attention is being paid to the coastal regions and ablation zones of Greenland, with emphasis on surface ablation, its sensitivity to boreal-summer warming, and possible albedo feedback effects (Stroeve, 2001; Stroeve and Nolin, 2003). As with skin surface temperature (Section 3.14.1), the most comprehensive dataset so far derives from the AVHRR, recently processed and compiled by the AVHRR Polar Pathfinder (APP) program (Barry, 1997; Maslanik et al., 1998; Scambos et al., 2002b; Stroeve, 2001; Stroeve et al., 2000; Wolfe et al., 2001). An example is given in Figure 3.106 (see color section) of an AVHRR-derived daily surface albedo field for the area of the west Greenland Ice Sheet and around the fast-flowing Jakobshavn Glacier (at ~69.2°N, 49.8°W). The draping of this dataset over a DEM (derived from ERS radar altimeter, field, and aerial data: Fahnestock et al., 1997) again demonstrates the additional benefits of combining different datasets. In this case, time series of such measurements can underline seasonal and interannual change and variability of albedo as a function of elevation, tying this information closely to changes in facies. Using these AVHRR data, Stroeve and Nolin (2003) found that years with anomalously low surface albedo corresponded to years in which the satellite PMW record indicates anomalously high (extensive) melt years. For further information on the APP dataset and data access, see: ⟨http://nsidc.org/data/pathfinders/index.html⟩.

The following sequence of processing steps was followed by Stroeve (2001) to produce consistent APP maps of the surface albedo of Greenland at a resolution of 5 km for the period 1981 to 1998 (please see Stroeve, 2001 for an assessment of accuracy). This approach is specific to the processing of AVHRR channel-1 and -2 data, although similar steps are required in the processing of similar satellite data—e.g., from MODIS. Satellite-specific calibration is carried out according to Rao and Chen (1994, 1999), with geo-referencing then utilizing the satellite orbit ephemeris model and clock corrections of Baldwin and Emery (1995). Next comes a conversion of the calibrated reflectances to Top-Of-Atmosphere (TOA) broadband reflectance ρ_{TOA}, which for snow and ice surfaces is:

$$\rho_{TOA} = 0.021\,577\,3 + 0.277\,479\rho_{1,TOA} + 0.506\,755\rho_{2,TOA} \qquad (3.28)$$

where $\rho_{1,TOA}$ and $\rho_{2,TOA}$ are the TOA reflectances in channels 1 and 2, respectively. As this is still a function of Sun and sensor geometry, it is not a broadband albedo. The next step involves using anisotropic correction factors (Suttles et al., 1988) to convert ρ_{TOA} to a broadband albedo α_{TOA}, where:

$$\alpha_{TOA} = \frac{\alpha_{TOA}}{f} \qquad (3.29)$$

where f is the anisotropic reflectance factor (3.24). Finally, surface albedo α_s is derived from its linear relationship with α_{TOA} (Koepke, 1989):

$$\alpha_s = \frac{\alpha_{TOA} - a}{b} \qquad (3.30)$$

where coefficients a and b were derived as a function of solar zenith angle, aerosol amount, water vapor (estimated from channel-4 and -5 data) and α_s using the Streamer radiative transfer model (Key and Schweiger, 1998). In this case, the aerosol optical depth is set by Stroeve (2001) at 0.06. Klein and Stroeve (2002) note that aerosols typically have a darkening effect over bright snow-/ice-covered surfaces, and can modify TOA reflectances at visible wavelengths by up to 20%, depending on aerosol type and optical depth (Nolin and Stroeve, 1997). As such, the correct parameterization of aerosol distribution and properties over snow and ice is an important factor, and remains a major challenge over snow and ice masses (Klein and Hall, 1999). Indeed, it remains a major source of error in satellite-derived albedo estimates, although one that can to some extent be corrected for by using data from new sensors with multi-angular measurement capability (Dozier and Painter, 2004).

Using this technique, Stroeve (2001) examined interannual variability in the surface albedo of the Greenland Ice Sheet, based on monthly mean data. Anomalously low albedos occurred during 1995 and 1998 in response to considerable surface melt in these years, while 1992 was characterized by a consistently high albedo, again probably due to lower temperatures and less surface melt as a result of the eruption of Mount Pinatubo. These results tie in closely with PMW and radar scatterometer assessments of interannual melt extent, as discussed in Section 3.3, and once again underline the benefits of gaining insight into processes and phenomena by using multiple datasets. Such work has benefited from comparison with *in situ* measurements from Greenland Climate Network AWSs (Steffen and Box, 2001), as shown in Stroeve et al. (2000).

The important albedo time series begun with the AVHRR is continuing with improved data from the MODIS sensor onboard NASA's Terra and Aqua spacecraft, the GLI onboard the Japanese ADEOS-II, and, in the future, VIIRS in the NPOESS program (see Chapter 5 of Volume 1 of this book). MODIS data provide complete spectral coverage for the first time. The MODIS is another current primary source of global data on snow/ice albedo, which is derived using a prototype clear-sky albedo algorithm (Klein and Hall, 1999; Klein and Stroeve, 2002; Klein et al., 2000). The standard NASA MODIS albedo/BRDF product combines surface reflectance data from both MODIS and MISR sensors to fit a BRDF in seven bands (0.4–2.5 µm) at a 1-km spatial resolution and on a 16-day cycle. The albedo estimates represent (i) a directional–hemispherical reflectance derived by integrating the BRDF over the exitance hemisphere for a single irradiance direction, and (ii) a bihemispherical reflectance derived by integrating the BRDF over all irradiance and viewing directions (Jin et al., 2002; Strahler et al., 1999). For further background information on albedo retrieval methodology, please see Stroeve and Nolin (2002a). The MODIS offers improved data products over the AVHRR by virtue of its accurate geolocation, routine characterization of atmospheric parameters, and onboard calibration (Barbieri et al., 1997). Daily and composite products are available from the NSIDC (⟨http://www.nsidc.org⟩). An assessment of the accuracy of this product, obtained through comparison with *in situ* measurements on the Greenland Ice Sheet, is given by Stroeve et al. (2005). Although improved accuracy is expected using MODIS data, more work is necessary to further reduce errors in satellite retrievals of surface albedo (and temperature). The ADEOS-II GLI has similar specifications and capabilities (again see Chapter 5 of Volume 1 of this book). Methods for deriving albedos from Landsat data are evaluated by König et al. (2001). Such data are most suitable in mountainous regions and over small ice masses, by virtue of their high spatial resolution.

Effective cloud detection and masking is again an essential prerequisite. As stated previously, the accurate detection of clouds over snow and ice is a challenging issue, and the success of the cloud detection and presence of sub-pixel-scale clouds remains a large source of uncertainty in albedo retrievals (Stroeve, 2001). Details of the MODIS cloud-masking algorithm are given in Ackerman et al. (1998), and of the atmospheric correction by Vermote et al. (1997). An assessment of current cloud-masking techniques for this purpose is given by Strahler et al. (1999), Stroeve (2001), and Klein and Stroeve (2002), and will not be expanded upon here. Please also refer back to Chapter 5 of Volume 1 of this book for additional information. Reflectance characteristics of snow and ice versus clouds in the 1–2.5-µm and 3–5-µm spectral regions are discussed by Calvin et al. (2002).

The approaches to albedo retrieval outlined above are valid for clear-sky conditions only. Other methods are, however, being developed to compute cloudy-sky surface albedos from satellite data—e.g., by Key et al. (2001). While albedo retrievals rely upon accurate cloud masking and removal, cloud cover is in itself an important component of the surface radiation budget (Arking, 1991; Bintanja and Van den Broeke, 1996; Cawkwell and Bamber, 2002; Shine et al., 1984; Chapter 4 of Volume 1 of this book)—with numerical models of ice sheet surface energy budget and mass balance depending on accurate parameterizations of cloud type and fraction. Unfortunately, the latter are often unrealistic, causing large uncertainties in both shortwave and longwave radiation fluxes. Cawkwell and Bamber (2002) address this issue by analysing ERS-2 ATSR/2 imagery in Greenland. They demonstrate the importance of reliable, quantitative cloud data in mass balance and other glaciological studies, and underline the sensitivity of the modeled ice-sheet radiation balance to cloud cover variability, with a nonlinear relationship between radiation balance and cloud fraction. The relationship between cloud cover and the radiation balance is highly complex, and further complicated by the complex effect of cloud type and fraction on surface albedo and temperature.

Three major factors contribute to errors in satellite-derived estimates of broadband albedo (Nolin et al., 2002a). These are (i) inaccurate atmospheric correction (particularly at visible wavelengths), (ii) errors associated with regression-based techniques for converting narrowband to broadband albedo, and (iii) errors from conversion of reflectance to albedo—i.e., in the process of accounting for the snow BRDF. The launch of advanced satellite instruments has enabled the more accurate derivation of large-scale albedo and BRDF (Leroy et al., 1997; Lucht et al., 2000; Strahler et al., 1999; Wanner et al., 1997). An exciting new development has been the advent of a new generation of spectro-radiometers capable of carrying out simultaneous multi-angular measurements at visible to near-IR wavelengths. Examples are the ERS-2 ATSR/2 (Stricker et al., 1995), Envisat Advanced ATSR (AATSR), EOS MISR (Diner et al., 1998, 2005), the ADEOS-I POLDER (August 1996 to July 1997) (Breon et al., 2002; Deschamps et al., 1994), and an identical instrument onboard ADEOS-II (2003), and the improved POLDER-P sensor launched on the European PARASOL satellite in December 2004. Information on these sensors is given in Chapter 5 of Volume 1 of this book. POLDER is a wide-FOV imaging radiometer that has provided the first systematic, global measurements of spectral, directional, and polarized characteristics of solar radiation reflected by the Earth–atmosphere system, and at a ground resolution of 5.3×6.2 km at nadir ($\langle http://smsc.cnes.fr/POLDER/\rangle$). It achieves this by observing targets from 12 directions. The PARASOL-P sensor acquires data in nine channels in the visible to near-IR (centered on 0.443–910 μm) over a scene of size $\sim$1,600 km (along-track) $\times$ 2,200 km (across-track) with a ground spatial resolution of 5.3×6.2 km at nadir. A key factor with these new sensors is that multi-angular observations provide a sampling of points at which bidirectional reflectance is measured, enabling enhanced use of physical models of snow BRDF such as DISORT (Stamnes et al. 1988). Moreover, multi-angular measurements yield improved estimates of aerosol optical depth and particle type, enabling more accurate atmospheric corrections to albedo estimates (Diner et al., 1999, 2005; Dozier and Painter, 2004).

Stroeve and Nolin (2002a) developed complementary algorithms to infer improved maps of ice sheet albedo from MISR data—one based on the spectral information, and the other exploiting angular information. Good agreement (to within 6%) was achieved in albedo retrieval compared with *in situ* measurements in Greenland. Stroeve and Nolin (2003) used albedos derived from MISR and MODIS data to focus in on the 2002 melt anomaly in West Greenland (see Section 3.3 and Figure 3.23). This analysis confirmed that melt began early in the Jakobshavn Isbrae drainage basin and was extensive in both that region and the northeastern sector of the ice sheet. See Jin et al. (2002) for "another angle" on BRDF/albedo retrieval with MISR data, and synergy with MODIS. Being sensitive to the geometric structure and composition of the reflecting surface, multi-angular measurements should provide additional information needed to characterize the surface and better determine albedo. Generally speaking, more work is necessary to take full advantage of the new technology, an example being the need to better incorporate surface roughness effects in snow/ice BRDF parameterizations (Warren et al., 1998). Excellent reviews of the new multi-angle technology, and its scientific applications, is given by Diner et al. (1999, 2005).

The benefits to be gained from using new, satellite hyperspectral data are discussed by Dozier and Painter (2004) and Nolin et al. (2002b). They used surface snow-grain-size information derived from high-resolution Hyperion data (see below) as input into a radiative-transfer model to compute spectral albedo over the broadband spectrum at 3-nm spectral resolution. As seen in Chapter 5 of Volume 1 of this book, this sensor, which was launched onboard the EO-1 platform in November 2000, was the first spaceborne hyperspectral sensor (Ungar et al., 2003; ⟨http://eo1.gsfrc.nasa.gov⟩). It collected simultaneous measurements in 220 bands over the spectral range of 0.4–2.5 μm and at a spatial resolution of 10–30 m, albeit over a narrow (7.5 km wide) swath with a scene length of 100 km. Calibration was carried out over the Greenland Ice Sheet by Bindschadler and Choi (2003). Nolin et al. (2002b) computed broadband albedo through integration after weighting the spectral albedo by the incoming solar irradiance. Comparison with *in situ* data from Greenland showed that this technique yields data that are accurate to <1%, which is significantly better than the 5–10% typically achievable using conventional satellite techniques—e.g., Stroeve and Nolin (2002b). Unfortunately, Hyperion coverage is limited to 7.75 × 100-km scenes. This approach, however, shows great promise, particularly given the future importance of hyperspectral remote sensing. For example, the VIIRS will be launched onboard the NPP satellite in 2006, then on the NPOESS series proper (starting in 2009). This sensor will collect data in numerous bands and over a wide swath (again see Chapter 5 of Volume 1 of this book).

Finally, satellite-derived albedo data can also be used to classify ice-sheet (and glacier) surface parameters related to mass balance. For example, the Snow Line Altitude (SLA) and Accumulation Area Ratio (AAR)—see Section 1.2 (p. 10)—can potentially be extracted from satellite visible data by virtue of the fact that firn and ice have a significantly lower albedo than new/seasonal snow—i.e., <0.5 versus ∼0.6–0.9, respectively. Recent studies have employed NOAA AVHRR (de Ruyter de Wildt et al., 2002) and high-resolution (15 m) Terra ASTER data (Khalsa et al., 2004) for this purpose, the latter work combining the satellite measurements—i.e., mapping of surface regimes and a stereoscopically derived DEM—with traditional glaciological measurements in a GIS framework. The ASTER is well-suited to this task. For one thing, it has improved spectral resolution compared with previous Landsat and SPOT sensors. Moreover, ASTER has adjustable gains, a key factor affecting the optimal imaging of highly reflective snow and ice masses which tend to saturate under normal gain settings (leading to loss of information). Indeed, ASTER has been programmed for the GLIMS project to acquire data from all regions with permanent "land" ice using optimally adjusted gain settings (Raup et al., 2000). Once again, however, its coverage is limited to a narrow swath.

Frezzotti et al. (2002a) showed that glazed (erosional and icy) surfaces can be mapped by virtue of their characteristic lower albedo than snow/firn or blue ice in

Landsat TM imagery. While ice and snow/firn albedos typically decrease from the visible to near-IR (Warren, 1982), glazed surfaces are characterized by spectral reflectances intermediate between snow and ice. As a result, they can be readily distinguished in false-color image composites using TM bands 2–4. Surface "glazing" occurs in winter by the formation of regelation ice films on the snow surface due to kinetic heating of wind-driven saltating drift snow, and in summer by radiational effects (Goodwin, 1990), combined with condensation–sublimation processes (Fujii and Kusunoki, 1982). Frezzotti et al. (2002a) further used the AVHRR band-2 : band-1 (near-IR : visible) ratio as an additional means of detecting glazed regions.

Areas of blue ice,[10] or areas of net ablation, have also been distinguished in AVHRR imagery by their distinctive visible to near-IR reflectance, also using band-2 : band-1 ratio data (Winther et al., 2001). These authors further distinguished melt- and wind-induced blue-ice areas. They also showed that the Landsat TM band-4 : band-3 ratio is a useful discriminator of snow and blue ice. Working in the Vestfold Hills (East Antarctica), Boresjo Bronge (1999) further showed TM band ratio 3 : 4 to be a useful discriminator of blue ice in Landsat imagery (see also Brown and Scambos, 2004). The improved detection of such regions, which are characterized by a mass balance which is zero or marginally negative (Bintanja et al., 1997, 2001; Frezzotti et al., 2002a; Winther et al., 2001), is required to better understand ice-sheet accumulation patterns and their retrieval from satellite microwave data (see Section 3.7). Moreover, such regions are of meteorological, glaciological, and climatological significance—e.g., for energy-budget calculations (Bintanja, 1999; Bintanja and Van den Broeke, 1995a)—and may act as a sensitive indicator of climate change (Bintanja and Van den Broeke, 1995b; Liston et al., 1999). Frezzotti et al. (2002a) also suggest that such information is of importance when choosing sites for firn/ice coring. Additional work by Nolin et al. (2002a) has underlined the strong potential of the new technology of spaceborne multi-angular measurement by Terra MISR as a means of detecting ablation-related blue-ice regions in Antarctica. Incidentally, blue-ice fields are unique regions on the Earth's surface for collecting meteorites (⟨*http://www-curator.jsc.nasa.gov/curator/antmet/program.htm*⟩).

3.16 GRAIN SIZE, IMPURITY CONTENT, AND SURFACE TO NEAR-SURFACE CHARACTERISTICS

3.16.1 Snow-grain size and impurity content

Accurate parameterizations of snow cover properties are required to improve climate and meteorological models (Bailey and Lynch, 2000a, b; Kärkäs et al., 2002; Smith

---

[10] While most of the Antarctica Ice Sheet has a snow/firn cover of ∼100 m overlying the ice, the ice is on the surface in blue-ice areas and there is no blanket of snow. This is due to the fact that wind and sublimation remove more snow than is accumulated by snowfall, causing a negative mass balance.

et al., 1998). Snow-grain size is another important yet poorly understood ice sheet parameter. Grain size is the primary snow parameter determining its spectral albedo at near-IR wavelengths (Aoki et al., 2003), whereas absorbing impurities (e.g., soot) affect its albedo in the visible spectrum (Dozier et al., 1981; Warren, 1982; Warren and Clarke, 1986; Warren and Wiscombe, 1980; Wiscombe and Warren, 1980). In fact, the average concentration of soot in the Arctic increased almost eight-fold over the period 1983–1998 (Hansen and Nazarenko, 2004). Grain size is also a proxy indicator of the thermodynamic state of the surface (Brandt and Warren, 1993). It also varies according to patterns of accumulation (i.e., potentially enabling the determination of fresh snowfall events), melt and snow metamorphism, and regional meteorological regimes—e.g., the surface wind field. Large-scale changes in snow-grain size provide further insight into changes in the snowpack energy balance (Nolin and Stroeve, 1997), and can help identify surface features, including blue-ice regions (see Section 3.15), snow-dune fields and melt regions. While surface measurements have been made—e.g., by Gay et al. (2002)—they are greatly limited in space and time. The only practical means of gaining large-scale information on spatial and temporal variability is again by satellite remote sensing.

Over the past two decades, an improved understanding of relationships between snowgrain size and reflectance (by enhanced radiative-transfer modeling) has combined with remote-sensing innovations to enable quantitative grain-size mapping from space (Nolin and Dozier, 2000). The general approach to retrieving grain-size information from satellite data is to employ radiative-transfer models to calculate snow reflectance in the near-IR region, which is then taken to be a nonlinear function of equivalent surface grain size at a given solar incidence angle (Nolin, 1993, 1998; Tanikawa et al., 2002). Spectral snow reflectance modeling work by Green et al. (2002) confirms that both snow-grain size and melt status information can be derived from cloud-free visible to near-IR measurements. Once again, anisotropic surface reflectance properties must be considered (Tanikawa et al., 2000), requiring knowledge of the BRDF (Leroux et al., 1998a, b; Sergent et al., 1998)—please refer to Section 3.15. The application of this approach to spaceborne imaging spectrometer datasets—e.g., from MISR—further requires atmospheric correction and the inclusion of terrain information—e.g., using an accurate DEM (Nolin and Dozier, 1993). Depth-dependent grain size retrieval from multi-spectral near-IR data is described by Li et al. (2001).

In addition to ice surface feature mapping, AVHRR data are also useful for grain size mapping (Warren et al., 1993; Winther et al., 2001). Reflectivity of snow in the near-IR (0.9–1.6 µm) is a strong function of grain size (Dozier et al., 1981; Grenfell et al., 1994; Wiscombe and Warren, 1980) as well as illumination, and a band ratio of AVHRR channel 1 ($\sim$0.8 µm) and channel 2 ($\sim$1.1 µm) gives a strong signal related to grain size while reducing illumination variation effects (Hall and Martinec, 1985). Data collected at $\sim$1.6 µm—e.g., by the AVHRR, MODIS, MERIS, ATSR, ATSR/2, AATSR, GLI (Stamnes and Li, 2001) and SPOT Vegetation—are particularly sensitive to surface snow-grain size. Snow-grain size has also been retrieved from Landsat TM data by Bourdelles and Fily (1993) and Fily et al. (1997), and from earlier Landsat data by Orheim and Lucchitta (1988, 1990). The Bourdelles and Fily (1993) Antarctic study showed that two large peaks occur in near-IR reflectance at 1.8 and 2.2 µm due to variations in the refractive index of ice. These peaks correspond to bands 5 (1.55–1.75 µm) and 7 (2.09–2.35 µm) of the Landsat TM and ETM+. Although the dependence on grain size only applies to wavelengths of ~0.8 µm, Bourdelles and Fily (1993) also highlighted the difficulty in using the region below 1 µm due to atmospheric effects, which can be significant but hard to characterize. Further work by Boresjo Bronge (1999) noted a relationship between grain size in East Antarctica and results from the band ratio 3:5 of the Landsat TM (see also Pattyn and Declair, 1993; Winther, 1993b). ASTER also has bands capable of mapping sub-pixel snow-grain size (Shi, 1999), with co-registered temperature data (Dozier and Painter, 2004).

Young et al. (2004) used ERS-2 ATSR/2 radiance measurements to determine the spatial distribution of surface snow-grain size and its temporal variability over an extensive area of the East Antarctic Ice Sheet. As noted above, scattering from a snow surface in the shortwave infrared part of the spectrum (0.9 to 3.5 µm) is strongly dependent on grain size and to a lesser extent on grain shape. Young et al. (2004) exploited this dependence together with measurements of snow surface reflectance obtained with the ATSR/2 instrument to map the spatial distribution of grain size over the Antarctic snow cover. The ratio of reflectance measurements at two different wavelengths—i.e., 0.87 and 1.6 µm—was used to account for local slope effects on the observed surface reflectance. A relation between this ratio of reflectance, grain size, and solar incidence angle was established from theoretical work, model calculations, and laboratory observations of reflectance from snow surfaces. The results presented in Figure 3.107 (see color section) show a large variation in grain size, from ~60 µm to >140 µm. In general, larger grain size is associated with warmer temperatures and lower elevations— e.g., over the major ice shelves—while the minimum grain size occurs along the high ridges of the ice-sheet interior.

Snow-surface grain size and the mass fraction of impurities were measured operationally, and globally every 4 to 16 days, for the first time from space using the GLI onboard ADEOS-II (2002–2003) (Nakajima et al., 1998). The algorithm used is described in Hori et al. (2001) and Stamnes (1999). See Section 5.9.14 of Volume 1 of this book for more details.

The new imaging spectrometry, or hyperspectral imaging, technology is particularly promising as a means of mapping surface grain size (Nolin and Dozier, 1993, 2000). This refers to acquisition of a complete reflectance spectrum of a scene by imaging over a large number of discrete, contiguous bands (see Chapter 5 of Volume 1 of this book). Early work by Hyvärinen and Lammasniemi (1987) showed the potential of using reflectance data collected at a high spectral resolution by relating changes in average ice-grain diameter to changes in the reflectance ratio of near-IR bands centered on 1.03, 1.26, and 1.37 µm. Nolin and Dozier (2000) used surface-particulate modeling combined with BRDF modeling for deriving snow-grain size from hyperspectral data. Such data have recently become available with the launch of the Hyperion sensor onboard EO-1 (see above). Specifically, they used a radiative transfer model to relate imaging spectrometer data collected at a strong ice absorption feature centered at $\lambda = 1.03$ µm (see Section 5.7.2 in Volume 1) to the optically equivalent grain size. Working in Greenland,

Nolin et al. (2002b) used cloud-free Hyperion hyperspectral data at 0.98–1.06 µm—i.e., spanning the same spectral absorption feature—within an atmospheric transmission window. As a result, surface reflectance measurements are largely unaffected by atmospheric water vapor, aerosols, and ozone. The above studies showed that the area of the absorption feature is directly related to the optically equivalent snow surface grain size. This enabled Nolin et al. (2002b) to use DISORT to create a lookup table to relate grain size to the scaled area, thereby computing snow-grain size in each Hyperion pixel. The resultant albedo estimated from the grain size information has been estimated to be accurate to within $\pm 2\%$.

With their ability to measure the polarization (as well as reflectance) properties of the surface, POLDER and POLDER-P have brought an additional dimension to the measurement of surface geophysical parameters such as snow-grain size. With polarimetric radiometers, radiative information is embedded in the polarization state of the transmitted, scattered, or reflected wave. In the case of POLDER(-P), the data allow for the construction of the Bidirectional Polarization Distribution Function (BPDF) as well as the BRDF. While promising, techniques using these combined datasets to derive information on the snow/ice surface require further work to develop a fuller understanding and realize their full potential. Please see Leroux et al. (1998b) for a discussion on the polarized bidirectional reflectance of snow in the near-IR. Mondet and Fily (1999) reported that POLDER measurements close to nadir, and with a viewing zenith angle of $<35°$, are preferred for grain-size measurements. It should be noted that none of the techniques outlined above can yield data during periods of polar darkness.

3.16.2 Changes in snow/firn characteristics inferred from passive-microwave data

Correlations between the ice-sheet surface and bedrock topography, and the accumulation environment, have been established in a few regions in Antarctica (Budd, 1970; Goodwin, 1988; Takahashi et al., 1997). Bedrock topography creates surface undulations which in turn constitute an inhomogeneous depositional and wind-erosional environment (Parish and Bromwich, 1987). Over cold, dry ice-sheet sectors—e.g., much of Antarctica and the Greenland interior—the behavior of time series of brightness–temperature (T_B) data measured by satellite PMW radiometers results from a complex interplay of snow temperature, layering, density, snow crystal size and shape, and surface roughness (Fily and Benoist, 1991; Foster et al., 1984, 1987; Jezek et al., 1993; Mätzler, 1987; Rott et al., 1993; Surdyk and Fily, 1995). Information on the crystal growth rate of Antarctic firn is given by Gow (1971).

A number of studies have attempted to infer information on the snow/firn characteristics of cold, dry ice sheets, and their variation with time/space, from satellite data. Surdyk (2002b), for example, combined Nimbus-7 SMMR and Radarsat SAR data with a DEM derived from ERS-1 radar altimeter data (Liu et al., 1999, 2001), the BEDMAP bedrock topography model (Lythe et al., 2001), and snowpit data in East Antarctica. Her study showed that snow characteristics inferred from variations in the satellite T_B data are a function of local wind conditions related to surface slope/topography and bedrock topography. Rather than corresponding to regions with the coldest air temperature at the highest elevations, the lowest T_Bs occurred in areas of wind shadow, where they were related to significant grain size enlargement by depth-hoar formation (Surdyk, 2002b). The presence of coarse grains acts to reduce ice sheet emissivity measured at 19–85 GHz ($\lambda = 1.55$–0.35 cm)—i.e., frequencies used by the SMMR, SSM/I, and AMSR PMW sensors—due to enhanced volume scattering (Abdalati and Steffen, 1998; Comiso et al., 1982; Foster et al., 1987; Surdyk and Fily, 1993, 1995; Zwally, 1977). Depth hoar results from large seasonal temperature gradients in regions associated with low accumulation rates, which in turn drive intense snow metamorphism (Giovinetto, 1963). Some refinement is required before depth-hoar formation, for example, can be unambiguously inferred from the satellite T_B record. Working in Greenland, Flach et al. (2005) demonstrated retrieval of dry-snow zone surface properties—i.e., *density*, grain size, *layer thickness*, and accumulation rate—from combined passive (SSM/I) and active (QuikScat) microwave data. Based on microwave-model inversion, this technique also provides an estimate of the geophysical error in radar-altimeter-derived measurements of surface elevation and thus mass change resulting from variable radar penetration of the surface (see Section 3.6.1.3). While such studies have been limited by the coarse resolution of SMMR and SSM/I data, they will benefit greatly from the use of improved-resolution Aqua AMSR-E, and ADEOS-II AMSR data. These studies also highlight the importance of *in situ* data as a means of interpreting the satellite signals. Such data are very limited, but are being collected during a number of international field operations—e.g., ITASE (⟨http://www.ume.maine.edu/itase/⟩).

Reflection of microwaves emitted from a snow/ice mass is a function of surface roughness (at the centimeter scale), snow density, incidence angle, and polarization (Shuman et al., 1993). At the 53.1° incidence angle of the SSM/I, which is close to the Brewster angle for snow, reflection of upwelling microwaves by density discontinuities/the surface is negligible at V-polarization (irrespective of other conditions), yet is significant for H-polarization (Fung and Chen, 1981; Rémy and Minster, 1991). Surface density generally increases with wind speed (Male, 1980). Under calm conditions (i.e., wind speeds of $<\sim 5\,\mathrm{m\,s^{-1}}$) such as those typically encountered over much of the inland ice-sheet regions, snowfall forms a soft surface layer with a bulk density as low as $200\,\mathrm{kg\,m^{-3}}$, whereas moderate to high winds during and after snowfall produce a fine-grained hard wind slab with a bulk density of $400\,\mathrm{kg\,m^{-3}}$ or more (Alley, 1988). Rémy and Minster (1991) further showed that winds can decrease the microwave reflection and increase H by their effect on ice sheet small-scale surface roughness. At SMMR, SSM/I, and AMSR wavelengths, centimeter-scale micro-roughness features are likely the dominant factor affecting microwave emission rather than 0.1–1.0-m-scale (vertical) sastrugi and snow dunes (Fung and Eom, 1982; Ulaby et al., 1981). Surface roughness on the latter scale range and greater is covered in the next section. The overall impact of an increase in surface roughness, and/or a decrease in surface density, is an increase in multiple scattering, which decreases H-reflection and increases H-emission while slightly increasing V-reflection and decreasing V-emission (Rémy and Minster, 1991).

Based upon these relationships, work by Shuman and Alley (1993) and Mätzler et al. (1984) revealed that the V/H ratio is sensitive to short-term changes in ice sheet surface and near-surface properties—e.g., related to snowfall or surface hoar-frost formation. Formed under clear, calm conditions and carpeting the surface in a low-density layer some millimeters to centimeters thick (Lang et al., 1984), surface hoar-frost reduces the near-surface density and increases surface roughness at a scale approximating the wavelength of the emitted radiation (Shuman and Alley, 1993;

Shuman et al., 1993). Together, these effects act to decrease the polarization and the value of V/H (Fung and Chen, 1981; Rémy and Minster, 1991). The use of a ratio minimizes the effect of changing the physical temperatures of the radiating portion (upper few meters) of the snow and firn. Working in central Greenland, Shuman et al. (1993) also found that abrupt changes in V/H resulted from changes in surface and near-surface microwave reflection rather than volume scattering from the upper few meters of snow (bulk). Working in the Dome C region of the high East Antarctic Plateau, Massom et al. (2004) again showed that abrupt changes in the microwave signature measured by the SSM/I at 19.35 GHz ($\lambda = 1.55$ cm), 37 GHz ($\lambda = 0.81$ cm), and 85.5 GHz ($\lambda = 0.35$ cm) resulted primarily from synoptic-scale changes in surface characteristics—e.g., due to surface hoar-frost formation/destruction and/or snowfall events (see Section 3.7). The basic assumption behind these studies is that the contribution of internal reflection and volume scattering remains constant and therefore negligible, given the abrupt nature of the observed changes in brightness–temperature. The study by Massom et al. (2004) benefited greatly from the availability of AWS meteorological data, which provided an important means of interpreting changes in microwave emission over time observed on a pixel-by-pixel basis. In general, the use of individual swath rather than daily average data is also recommended, as this enables the resolution of short-term changes in the microwave signature.

The presence of a high-density surface crust, on the other hand, acts to increase reflection at H- relative to V-polarization, leading to an increase in V/H. Due to the low net annual precipitation (Vaughan et al., 1999), icy surface crusts are a common feature of the central Antarctic Plateau (Palais et al., 1982), for example. According to a sensitivity study by Surdyk and Fily (1995), icy crusts/glazed surfaces produce polarization differences by inducing strong variations at interfaces.

Surdyk and Fily (1993, 1995) and Surdyk (2002b) applied a polarization ratio (*PR*) technique to satellite PMW data to infer information about ice-sheet surface and near-surface characteristics. Referring back to Section 3.3.3, the *PR* is defined as:

$$PR = \frac{TB_V - TB_H}{TB_V + TB_H} \quad (3.31)$$

where *TB* is the observed brightness temperature within a given pixel and subscripts *H* and *V* are the horizontal and vertical polarizations, respectively. At frequencies <10 GHz, *PR* is an indicator of the number of strata of varying grain size, density, and iciness over the uppermost 2 m or so (Surdyk and Fily, 1993, 1995), with higher *PR*s corresponding to more strata.

Surdyk and Fily (1993) further utilized the Gradient Ratio (*GR*) technique, whereby the frequency gradient between SMMR 18-GHz and 6.6-GHz vertical polarization data, $GR_{18,6.6}$, is defined as:

$$GR_{18,6.6} = \frac{\overline{TB_{18}} - \overline{TB_{6.6}}}{\overline{TB_{18}} + \overline{TB_{6.6}}} \quad (3.32)$$

where $\overline{TB_{6.6}}$ and $\overline{TB_{18}}$ are the annual means. Surdyk (2002b) observed that $GR_{18,6.6}$ is strongly correlated with the TB at 37V (TB_{37V}). Surdyk and Fily (1993) found a strong correlation between $GR_{18,6.6}$ and the mean snow-grain size within the uppermost 2 m; the lower the ratio, the larger the grain size. From this, Surdyk (2002b) deduced that an increase in $GR_{18,6.6}$ corresponds to highly stratified snow cover with a relatively small grain size. Such studies will benefit from the improved spatial resolution offered by ADEOS II AMSR and EOS Aqua AMSR-E.

3.16.3 Ice sheet roughness characteristics and proxy wind measurements

Ice sheet surface roughness can be characterized over a range of scales, from a few centimeters to ~100 kilometers. Roughness is determined by the bedrock topography, and the accumulation, erosional, and drift environment (Furukawa and Young, 1997; Long and Drinkwater, 2000). Due to persistent eolian redistribution of snow, snowdrift is a major process affecting the ice sheet surface on a variety of scales, from micro- to macro-scale (Goodwin, 1990; Watanabe, 1978). The creation of a range of wind-related surface features—e.g., glazed icy surfaces, sastrugi, and snow dunes—significantly alters the surface roughness (Inoue, 1989). This in turn has a complex feedback effect on the local to regional wind field (Fujii and Kusunoki, 1982; Van den Broeke and Bintanja, 1995) and *ipso facto* the snow accumulation process (Frezzotti et al., 2002b). Moreover, it was seen in Section 3.15 that surface roughness is a key determinant of ice-sheet albedo and BRDF (Warren et al., 1998). As such, both turbulent and radiative fluxes at the surface are strongly dependent on surface roughness (Bintanja and Van den Broeke, 1995a; Kärkäs et al., 2002). Large-scale surface roughness features, which are closely related to bedrock topography, can be measured by radar altimetry and SAR interferometry. Brisset and Rémy (1996), for example, used ERS-1 radar altimeter data to analyse kilometer-scale roughness characteristics of the Antarctic Ice Sheet surface. In a similar study of Greenland and Antarctica by Rémy et al. (1999), such features were found to be oriented at 45° to the ice flow direction. They discovered a number of features, including a 20-km wavelength undulation with an amplitude of 10 m in the region 74–81°S, 120–140°E, which they attributed to ice flowing over irregular bedrock topography.

Other recent studies have highlighted the ability of other sensors to measure important smaller scale roughness. Recent analysis of Landsat, SAR, and AVHRR imagery has even revealed the existence of fields of large "megadunes" (Fahnestock et al., 2000b; Frezzotti et al., 2002a, b) (Figure 3.108, see color section). Fahnestock et al. (2000b) determined the large-scale extent of Antarctic megadune fields from cloud-free AVHRR data, applying a band-pass spatial filter to highlight their subtle topographic manifestation. Possible mechanisms for their formation and maintenance include boundary-layer standing waves, firn ventilation, and eolian snow transport (Fahnestock et al., 2000b). In the Dome C region of East Antarctica, these dunes have a typical amplitude of 2–4 m, wavelength of 2–5 km, length of up to 100 km (Frezzotti et al., 2002b), and cover a vast area of $>500,000\,km^2$. Such features are virtually impossible to detect and map from the surface alone, given their horizontal scale and small topography (a few meters only). Megadune crests are aligned perpendicular to the prevailing katabatic wind direction (modeled by Parish and Bromwich, 1991), which is itself controlled by regional slope. As such, satellite mapping of their location, areal extent, and orientation can provide useful proxy wind-direction information from the vast data-sparse regions of the ice sheet between existing AWS stations (⟨http://uwamrc.ssec.wisc.edu/⟩), with implications for interpreting observed variability in large-scale snow accumulation patterns (Frezzotti et al., 2002a). This technique also enables monitoring of possible changes in surface-roughness patterns related to changes in atmospheric circulation patterns. Interestingly, megadune fields have remained largely unchanged in central Antarctica since at least 1963, as determined by comparison of modern imagery with DISP imagery by Fahnestock et al. (2000b).

Particularly promising is the new suite of spectrometers with multi-angle imaging capability. For example, POLDER data have been used to provide information on surface roughness on the scale of sastrugi (Leroux and Fily, 1998; Mondet and Fily, 1999). Moreover, the nine-channel EO-1 ALI has an enhanced ability to detect small-scale surface roughness features and undulations (Bindschadler, 2003). Nolin et al. (2002a) further demonstrated the very promising ability of the Terra MISR as a tool for characterizing ice surfaces on the basis of their surface roughness. Given the somewhat limited coverage offered by MISR—i.e., over a 380-km swath width—Nolin et al. (2002a) recommended the synergistic combination of MISR and MODIS data to optimize coverage while exploiting the multi-angular approach. They also show that MISR can effectively discriminate thin clouds—e.g., cirrus—from snow/ice. More work is in general required to relate angular ice/snow signatures, textural information, and key geophysical information in order to realize the exciting potential of these innovative and emerging technologies. Coordinated field programs such as ITASE (⟨http://www.ume.mai.e.edu/itase/⟩) again have a key role to play in this respect.

We noted in Section 3.6 how information on ice-sheet surface roughness can be derived from spaceborne radar-altimeter waveforms. Specifically, the shape of the waveform leading edge (Figure 3.53c) is predominantly determined by the height of sastrugi and snowdunes, as well as by kilometer-scale surface undulations and volume scattering (Legrésy and Rémy, 1997; Rémy et al., 2001). The waveform leading edge corresponds to the initial contact of the radar wave with the surface, and the ability of the radar wave to penetrate the surface as well as the surface roughness height can be derived from its characteristics of lengthening. The shape of the waveform trailing edge (Figure 3.53), on the other hand, is a function largely of the ratio between sub-surface and surface scattering as well as surface curvature and slope on the kilometer scale and by roughness distribution (Ulaby et al., 1982). It is typically obtained by linear regression on the logarithmic scale. An example of a map of this parameter, of Antarctica and using data from the ERS radar altimeter, is given in Figure 3.53e. A nominal value of -110×10^{-4} Np gates^{-1} is typical for a flat and largely non-penetrating surface such as an open ocean. Over ice sheets, lower values result from specular surfaces, while greater values result from enhanced volume scattering and/or surface slope (Rémy et al., 2001). These relationships are covered in more detail by Rémy et al. (2001). The derivation of proxy surface wind-field information is described by Legrésy et al. (1999).

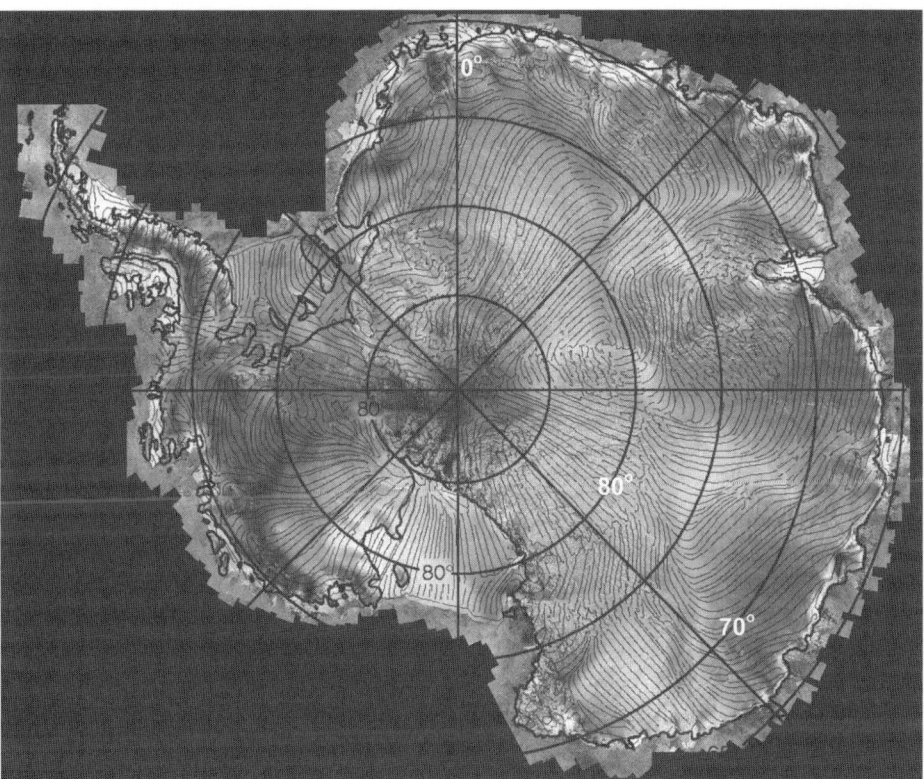

Figure 3.109. Streamlines of surface inversion winds on the Antarctic Ice Sheet inferred from NSCAT data, and superimposed onto the Radarsat 5-km-resolution Antarctic mosaic of SAR amplitude data from the Antarctic Mapping Mission (AMM).

From Long and Drinkwater (2000). AMM Mosaic courtesy Ken Jezek (Ohio State University). Radarsat data copyright Canadian Space Agency and Radarsat International, 1999. Copyright IEEE 2000, reproduced with permission.

Radar scatterometer backscatter and PMW emission are also highly sensitive to surface roughness characteristics, which combine to introduce azimuth angle dependencies in the data (Long and Drinkwater, 2000). In their analysis of radar scatterometer data in Antarctica, Long and Drinkwater (2000) noted a strong correlation between azimuthal variation and the orientation of the surface slope and small-scale roughness relative to the sensor look direction—i.e., the size, density, and orientation of dunes and sastrugi. The azimuth modulation pattern has an upslope across-wind maximum, and a downslope along-wind minimum, enabling mapping of the directionality in aligned, wind-induced surface roughness from space. Such information can be used to infer the direction of the mean (climatological) surface wind field, including the katabatic regime, over much of the ice-sheet interior (Ledroit et al., 1993; Long and Drinkwater, 2000) (Figure 3.109). Similar relationships between scatterometer (and radar altimeter) data and katabatic winds have been noted by Rémy et al. (1990, 1992) and Rémy and Minster (1991). Radar-altimeter characterization of ice-sheet surface and subsurface characteristics is covered in more detail in Section 3.16.4.1. Furukawa and Young (1997) and Young et al. (1996) also found a distinct directional anisotropy in measurements of the microwave backscatter coefficient acquired with the ERS-1 wind scatterometer (Figure 3.110, see color section). Analysis of the measurements obtained with different look directions for a given location shows there is a distinct directional anisotropy in the backscatter. The azimuthal modulation obtained by the suite of scatterometer antennae can be represented by a bisinusoidal function of the look direction of the beams (Ledroit et al., 1993). Furukawa and Young (1997) and Young et al. (1996) found a close correspondence between sastrugi alignment and wind direction measured at AWSs and the orientation of the anisotropy defined by the look direction for minimum backscatter, and that monthly mean values of the amplitude and orientation exhibit little variability through a year. They concluded that the character of the backscatter signal and associated anisotropy could, however, be affected by variable accumulation rates. Again more *in situ* measurements are required to validate and calibrate the satellite techniques and products, in order to better understand the mechanisms involved.

3.16.4 Active-microwave remote sensing of ice-sheet surface and near-surface characteristics

In this section, we briefly outline methods that have been developed to infer ice sheet near-surface characteristics from radar altimeter waveforms and from InSAR and Pol-InSAR processing of coherent SAR images. The inference of ice sheet near-surface characteristics using normalized SAR amplitude data and backscatter data from satellite wind scatterometers has been covered in earlier sections, and also in Section 2.3.4.

3.16.4.1 Near-surface characteristics from radar altimeter data

As noted in Section 3.6, the return-pulse waveform characteristics of radar altimeters provide information not only on the elevation of an ice sheet but also its surface and near-surface characteristics such as surface penetration/reflectivity and roughness (Legrésy and Rémy, 1997; Rémy et al., 2001), given the frequency-dependent penetration of the radar wave into the snow/ice volume to a depth of meters to tens of meters. Integration of the area under the waveform provides a measure of the total energy return from the surface footprint (related to the radar backscattering coefficient σ^0) (see Figure 3.53d). Due to penetration of the EM wave within the snowpack, this is the sum of both a volume and surface component (Rémy et al., 2001). The volume or sub-surface component relates to volume echo, which is attenuated by the snow extinction k_e. The latter is the sum of the scattering and absorption coefficients k_s and k, respectively. While k_s depends on the scatter (ice grain) volume (Fung and Eom, 1982), k depends on the snow temperature and radar wavelength (Mätzler, 1987). Typical values of k are $0.02\,\text{m}^{-1}$ to $0.03\,\text{m}^{-1}$ for snow temperatures from -20 to $-60°\text{C}$, respectively, while k_s varies from 0.03 to $0.12\,\text{m}^{-1}$

for ice-grain size of 0.2 to 1.0 mm, respectively (Fung and Eom, 1982; Rémy et al., 2001). To date, two types of volume-scattering models have been developed for application with satellite radar altimetry data: one based primarily upon scattering by ice grains of different size—e.g., Ridley and Partington (1988)—the other on reflection by internal layering—e.g., Jezek and Alley (1988) and Rémy et al. (1995).

Following Fung and Eom (1982), the surface-scattering coefficient σ_{surf} is given by:

$$\sigma_{surf} = \frac{R^2}{2S^2} \tag{3.33}$$

where R^2 is the Fresnel coefficient at the atmosphere–snow interface and S^2 is the r.m.s. of the surface slope. In this case, the Fresnel coefficient is given by:

$$R^2 = \frac{\left(\sqrt{\varepsilon'} - 1\right)^2}{\left(\sqrt{\varepsilon'} + 1\right)^2} \tag{3.34}$$

where ε' is the real part of the snow dielectric constant (see Chapter 5 of Volume 1 of this book) and depends on the snow density ρ_{snow} (Tiuri et al., 1984), such that:

$$\varepsilon' = 1 + 1.7\rho_{snow} + 0.7\rho_{snow}^2 \tag{3.35}$$

For snow densities of 350 and 400 kg m^{-3}, ε' ranges from 1.68 to 1.79, respectively, resulting in a range in R^2 of 1.08 dB (Rémy et al., 2001). Furthermore, the range in the surface roughness factor S results in a more important contribution to the range variation of surface scattering—i.e., of up to 20 dB—as reported by Rémy et al. (2001).

Figure 3.53d shows an example map of total backscatter from the Antarctic Ice Sheet, derived from ERS radar altimeter data. Apparent is a high degree of spatial variability e.g., >10 dB—in East Antarctica, where it appears to relate to katabatic wind intensity (Rémy et al., 1990). While research is aimed at inferring ice sheet surface and sub-surface characteristics and processes from radar-altimeter waveform characteristics (Davis and Zwally, 1993; Legrésy and Rémy, 1997, 1998; Partington et al., 1989; Ridley and Partington, 1988), this is a complex problem in that unambiguous separation of the different contributions is difficult. If accurate, such information entails another means of investigating regional, inter-decadal, and interannual variations in ice-sheet surface and near-surface characteristics (Davis and Zwally, 1993; Yi and Bentley, 1994). One approach is to analyze temporal variation, under the premise that short-term changes in the altimeter signature are most likely to result from changes in surface roughness (Rémy et al., 2001). Data from the 3-day repeat orbit phase of ERS-1, which lasted for 66 days in 1992, were used for this purpose by Legrésy and Rémy (1998). They studied synoptic-scale variability in surface-roughness parameters over the Antarctic Ice Sheet, then used a numerical radar-echo model to invert the data and retrieve information on snow penetration and stratigraphy. In this way, Legrésy and Rémy (1998) found that radar extinction at 13.6 GHz is dominated by snow-grain size. Particularly strong backscatter occurs from regions of low accumulation, where large ice grains occur within the upper few meters—i.e., within the radar penetration depth—due to high rates of snow thermodynamic metamorphism. As with many of the parameters described above, *in situ* measurements are required to verify satellite-derived variables (this is currently a major limitation).

3.16.4.2 *Near-surface characteristics from SAR interferometry (InSAR)*

We noted in Section 3.7 how interferometric processing of SAR data is a potential means of deriving information on spatial variations in accumulation patterns as they relate to topographic undulations. In this section, we extend the concept to include near-surface characteristics in general. This is a departure from the more conventional application of SAR amplitude or phase data to the study of firn characteristics (Section 3.2). Recent work by Hoen and Zebker (2000b) shows that maps of interferometric correlation—a by-product of the interferometric processing of SAR data (Chapter 2)—provide a novel means of mapping and monitoring the sub-surface characteristics of glaciers and ice sheets. This new approach is based on the fact that interferograms contain a decorrelating-phase noise component with statistics related to the scattering medium, with the degree of decorrelation increasing with increasing volume scatter. Radar penetration depths (one-way, $1/e$ point for power) within the upper layers of the ice mass can be estimated by modeling this effect. Hoen and Zebker (2000b) model the volume decorrelation by assuming a homogeneous scattering medium with exponential loss due to wave propagation, then invert the correlation observed in the ERS analysis to obtain maps of penetration depths.

Hoen and Zebker (2000b) applied this technique to ERS SAR data collected over Greenland. The results are shown in Figure 3.111 (see color section). From Figure 3.111d, it can be seen that penetration depths vary according to the ice sheet facies—from $\sim 10\,\mathrm{m}$ in the bare-ice zone to 15–38 m in the interior dry-snow zone (equivalent to results from Antarctica: Rott et al., 1993) and 15–35 m in the intervening percolation zone. The latter are unexpectedly large, with volume scatter apparently dominating rather than the scattering by icy structures in the uppermost few meters, as suggested by previous studies—e.g., Rignot (1995). Hoen and Zebker (2000b) suggest that more work is necessary to investigate this discrepancy, which may result from unmodeled scattering behavior and/or actual significant contributions from scatterers at substantial depth below the surface. Note also that phase decorrelation becomes much more pronounced with longer baselines—e.g., Figure 3.111c versus 3.111b. In general, information on penetration depth derived from the InSAR volume decorrelation term may prove to be useful for quantifying and compensating for the effect of variable penetration depth on ice surface elevation retrievals by satellite radar altimeters.

3.16.4.3 *Near-surface characteristics from polarimetric SAR interferometry (Pol-InSAR) data*

As described in Chapter 2, Pol-InSAR is a brand new field of research that represents the symbiosis of SAR polarimetry and InSAR. Recall from Chapter 5 of Volume 1 of this book that radar polarimetry comprises the science and methods involved in measuring and analyzing the complex scattering matrix of pixels in a radar image (Cloude et al., 2001). While much of the power of Pol-InSAR lies in other fields— e.g., terrestrial biomass mapping due to its ability to measure tree height and structure (Cloude and Treuhaft, 1999)—its application as a means of deriving information on near-surface ice sheet characteristics is currently being explored— e.g., Dall et al. (2003, 2004). Indeed, early results evaluated in Section 3.2.4 and Chapter 2 underline the exciting potential of this novel approach.

The basis of the Pol-InSAR technique is that the interferometric properties of polarimetric data can yield information on where the scattering signature is primarily emanating from. When two polarimetric images are obtained that satisfy the interferometric conditions of time separation and baseline and without a high degree of decorrelation, the complete polarimetric/interferometric information is stored in three 3×3 complex matrices, the coherence matrix of each image, and the analogous matrix formed from the products of the scattering vectors of the SAR image pair. A key factor has been the development of a phase-preserving polarimetric basis transformation, which enables the formation of interferograms between all possible elliptical polarization states (Cloude and Papathanassiou, 1998). An optimization procedure based on singular-value decomposition is then used to locate the polarizations of images 1 and 2 that maximize the interferometric coherence (i.e., reduce the effects of temporal and baseline decorrelation) between the images. Cloude and Papathanassiou (1998) further developed a modified coherent target decomposition method. The purpose of this procedure, when combined with the coherence optimization procedure, is to find the optimum scattering mechanisms that lead to the best differential-phase measurements— i.e., the highest coherence. The height difference between the physical scatterers possessing these mechanisms is then obtained from the interferometric phase difference. The physical interpretation of the results, and an assessment of their quality, can be determined by viewing the complex coherence of the interferogram as a function of polarization.

3.17 CONCLUSIONS

Our understanding of the Antarctic and Greenland Ice Sheets has advanced dramatically over the last decade in particular, due in large part to the availability of much-improved datasets from satellite remote sensing. This development has resulted in more accurate estimations of both ice sheet mass input and output, thereby enabling a significant refinement of mass-balance estimates and our understanding of the complex processes and feedbacks involved (Abdalati et al., 2004; Alley and Bindschadler, 2001; Thomas et al., 2001b). In Greenland, the coordination of new satellite and field observations within the NASA's PARCA program has effectively reduced uncertainty in elevation change above the equilibrium line from $\sim$10 cm per annum to $\sim$1–2 cm per annum (Abdalati, 2001; Thomas et al., 2001a). For the Greenland Ice Sheet overall, the reduction in uncertainty in the mass balance estimate has been by a factor of 2 (Payne and Bamber, 2004; Thomas et al., 2001a). The success of such projects is heavily reliant upon the effective fusion of satellite with complementary *in situ* and modeling data and techniques.

Importantly from an ice sheet numerical-modeling perspective, satellite data now provide greatly enhanced boundary conditions, improved data for validation, and an important means of constraining model physics, leading to enhanced model performance (Hulbe and Payne, 2001). The new observations have also forced a reconsideration of research priorities into the mass imbalance of the major ice sheets, and have brought to prominence many complex questions which remain unanswered (Abdalati et al., 2004).

As Payne and Bamber (2004) point out, a wealth of new information on the mass balance of the Antarctic Ice Sheet and evidence of recent and in places unprecedented change has come to light over the past decade, yet these results are constrained by a number of over-riding factors. From a satellite remote-sensing perspective, these include:

- the dearth of satellite observations in central East Antarctica—i.e., within the area currently beyond (poleward) satellite coverage (this is no longer an issue for Greenland);
- the short time period of large-scale, satellite-derived observations relative to the long response time of ice sheets; and
- major gaps in our knowledge of certain key parameters/variables, resulting from observational deficiencies.

The latter conspire to effectively limit overall error budget estimations of the great ice sheets. Poor knowledge of accumulation rates, for example, is currently undermining our ability to accurately establish the mean annual input to the ice sheets and how this is distributed over individual drainage basins. Along with poor knowledge of ice thickness and large errors in measured and modeled ablation rates, this remains a primary source of uncertainty in computations of ice-sheet mass balance using the component (flux) approach. As stated by Abdalati et al. (2004), knowledge of precipitation distribution and snow redistribution processes at horizontal scales of >10 km is required to construct the high-resolution continental-scale surface mass balance map required for atmospheric and ice-sheet modeling. Another important contributing weakness relates to large errors in measured and modeled large-scale ablation rates—a key factor for the Greenland Ice Sheet in particular. Clearly, the improved large-scale measurement of ice-sheet accumulation and ablation rates is a high priority—and a major challenge. Although the techniques outlined in this chapter for deriving accumulation and ablation rates from satellite data show great potential, they are in their infancy and substantial work lies ahead before routinely applicable and robust methods become available. As is typically the case, the combination of measurements from various different sources holds the greatest promise in this respect. Regarding ice thickness, measurements are required across the entire Antarctic Ice Sheet in addition to those contained within the BEDMAP dataset (Lythe et al.,

2001), and particularly along the coast and over specific drainage basins, including those of the major outlet glaciers (Abdalati et al., 2001).

Given their short length and fundamental importance, it is absolutely critical that key satellite time series are continued with no gaps, supplemented with data from new-generation sensors aimed at the measurement of ice-sheet parameters—e.g., ICESat GLAS and CryoSat. The launch of CryoSat unfortunately failed in October 2005—it is hoped that this important mission can be resurrected, given its key role in both ice-sheet and sea-ice monitoring and research. A major priority is to ensure continuation of InSAR missions beyond the lifetimes of ERS, Radarsat, and EnviSat in order to map changes in ice velocity and grounding-line location and fill in gaps in past InSAR coverage. To reiterate the point made in Chapter 2, it is strongly hoped that the major space agencies combine resources and expertise to launch dedicated and compatible systems in orbital configurations optimized for routine acquisition of interferometric data over ice sheets. Given the power and importance of InSAR, the improvement in coverage and the lengthening of the time series is an urgent priority. Also crucial is the continuation of coordinated field observations for validation purposes, to supply additional data that are currently unobtainable from satellites, and to acquire proxy measurements with which to extend satellite measurements back in time—e.g., through firn/ice core analysis. A PARCA-like experiment over Antarctica is highly desirable, yet is a more difficult proposition due to the larger distances and greater remoteness involved (compared with Greenland).

Particular attention needs to be paid to ice-sheet margins, where surface slopes are steeper and ice sheets are at their most dynamic and climate impacts are at their most variable. Unfortunately, these attributes create difficulties for certain important satellite methods—e.g., radar altimeters—and models also exhibit inadequacies due to their coarse resolution relative to the ice sheet features and processes occurring within the margins (Ritz et al., 1996). The outcome is that the present mass balance is particularly poorly known in these regions. In addition, since accumulation rates and air temperatures are substantially higher in coastal areas than in the interior, climate change has the potential to impact mass balance in these areas much more rapidly than inland (Abdalati et al., 2004). In general, current deficiencies relating to both observational and modeling schemes, and suggestions for improvements in ice sheet and atmospheric model physics and performance, for example, are further summarized by Abdalati et al. (2004) and Payne and Bamber (2004). In summary, major challenges clearly remain. The incentive is, however, great, given the wider, profound implications of changes in ice-sheet behavior and mass balance driven by global warming. The role that satellites have to play is immensely important.

3.18 REFERENCES

Abdalati, W. (ed.) (2001). PARCA: Mass balance of the Greenland Ice Sheet. *Journal of Geophysical Research*, **106**(D24), 33689–34058.

Abdalati, W. and K. Steffen (1995). Passive microwave-derived snow melt regions on the Greenland Ice Sheet. *Geophysical Research Letters*, **22**(7), 787–790.

Abdalati, W. and K. Steffen (1997a). Snowmelt on the Greenland Ice Sheet as derived from passive microwave satellite data. *Journal of Climate*, **10**(2), 165–175.

Abdalati, W. and K. Steffen (1997b). The apparent effects of the Mt. Pinatubo eruption on the Greenland Ice Sheet melt extent. *Geophysical Research Letters*, **24**, 1795–1797.

Abdalati, W., and K. Steffen (1998). Accumulation and hoar effects on microwave emission in the Greenland ice sheet dry snow zones. *Journal of Glaciology*, **148**, 523–531.

Abdalati, W. and K. Steffen (2001). Greenland Ice Sheet melt extent: 1979–1999. *Journal of Geophysical Research*, **106**(D24), 33983–33989.

Abdalati, W., W. Krabill, E. Frederick, S. Manizade, C. Martin, J. Sonntag, R. Swift, R. Thomas, W. Wright, and J. Yungel (2001). Outlet glacier and margin elevation changes: Near-coastal thinning of the Greenland Ice Sheet. *Journal of Geophysical Research*, **106**(D24), 33729–33742.

Abdalati, W., I. Allison, F. Carsey, G. Casassa, M. Fily, M. Frezzotti, H. A. Fricker, C. Genthon, I. Goodwin, Z. Guo et al. (2004). Recommendations for the collection and synthesis of Antarctic ice sheet mass balance data. *Global and Planetary Change*, **42**(1–4), 1–15.

Ackerman, S. A., K. I. Strabala, W. P. Menzel, R. A. Frey, C. C. Moeller, and L. E. Gumley (1998). Discriminating clear sky from clouds with MODIS. *Journal of Geophysical Research*, **103**(D24), 32141–32157.

Adam, S. A., A. Pietroniro, and M. M. Brugman (1997). Glacier snow-line mapping using ERS-1 SAR imagery. *Remote Sensing of Environment*, **61**, 46–54.

Ahlstrøm, A. P., C. Egede Boggild, J. J. Mohr, N. Reeh, E. Linz Christensen, O. B. Olesen, and K. Keller (2002). Mapping of the hydrological ice-sheet drainage basin on the West Greenland ice-sheet margin from ERS-1/-2 SAR interferometry, ice-radar measurement and modelling. *Annals of Glaciology*, **34**, 309–314.

Alley, R. B. (1987). Firn densification by grain-boundary sliding: A 1st model. *Journal de Physique*, **48**(C1), 249–256.

Alley, R. B. (1988). Concerning the deposition and diagenesis of strata in polar firn. *Journal of Glaciology*, **34**(118), 283–290.

Alley, R. B. (1989). Water-pressure coupling of sliding and bed deformation: I. Water system. *Journal of Glaciology*, **35**(119), 108–118.

Alley, R. B. (1992). Flow-law hypotheses for ice-sheet modeling. *Journal of Glaciology*, **38**, 245–256.

Alley, R. B. and C. R. Bentley (1988). Ice-core analysis on the Siple Coast of West Antarctica. *Annals of Glaciology*, **11**, 1–7.

Alley, R. B. and R. A. Bindschadler (eds.) (2001). *The West Antarctic Ice Sheet: Behavior and Environment* (AGU Antarctic Research Series No. 77). American Geophysical Union, Washington, DC.

Alley, R. B. and I. M. Whillans (1991). Changes in the West Antarctic Ice Sheet. *Science*, **254**, 959–963.

Alley, R. B., J. F. Bolzan, and I. M. Whillans (1982). Polar firn densification and grain growth. *Annals of Glaciology*, **3**, 7–11.

Alley, R. B., K. M. Cuffey, E. B. Evenson, J. C. Strasser, D. E. Lawson, and G. J. Larson (1997). How glaciers entrain and transport basal sediment: Physical constraints. *Quaternary Science Reviews*, **16**(9), 1017–1038.

Allison, I. (2001). Peephole through the ice: the AMISOR project. *Australian Antarctic Magazine*, **1**, 20–21.

Allison, I. (2003). *The AMISOR Project: Ice Shelf Dynamics and Ice–ocean Interaction of the Amery Ice Shelf* (FRISP Report No. 14). Geophysical Institute, University of Bergen, Norway, pp. 1–9.

Anandakrishnan, S., D. D. Blankenship, R. B. Alley, and P. L. Stoffa (1998). Influence of subglacial geology on the position of a West Antarctic ice stream from seismic observations. *Nature*, **394**, 62–65.

Anandakrishnan, S., R. B. Alley, R. W. Jacobel, and H. Conway (2001). The flow regime of Ice Stream C and hypotheses concerning its recent stagnation. In: R. B. Alley and R. A. Bindschadler (eds.), *The West Antarctic Ice Sheet: Behavior and Environment* (AGU Antarctic Research Series No. 77). American Geophysical Union, Washington, DC, pp. 283–294.

Anandakrishnan, S., D. E. Voigt, R. B. Alley, and M. A. King (2003). Ice stream D flow speed is strongly modulated by the tide beneath the Ross Ice Shelf. *Geophysical Research Letters*, **30**(7), 1361, DOI: 10.1029/2002GL016329.

Aoki, T., T. Aoki, M. Fukabori, A. Hachikubo, Y. Tachibana, and F. Nishio (2000). Effects of snow physical parameters on spectral albedo and bidirectional reflectance of snow surface. *Journal of Geophysical Research*, **105**, 10219–10236.

Aoki, T., A. Hachikubo, and M. Hori (2003). Effects of snow physical parameters on broadband albedos. *Journal of Geophysical Research*, **108**, 4616, doi:10.1029/2003JD003506.

Arigony, J., H. Saurer, R. Jaña, J. C. Simões, and H. Gossmann (2004). Monitoring snow parameters on the Antarctic Peninsula using satellite data: A new methodological approach. *Abstracts of the Interdisciplinary Workshop on Antarctic Peninsula Climate Variability: History, Causes and Impacts, Cambridge, UK, 16–18 September 2004*. British Antarctic Survey, Cambridge, UK, Vol. 1, p. 32.

Arking, A. (1991). The radiative effects of clouds and their impact on climate. *Bulletin of the American Meteorological Society*, **72**, 795–813.

Arrigo, K. R. and G. L. van Dijken (2003). Impact of iceberg C-19 on Ross Sea primary production. *Geophysical Research Letters*, **30**(16), 1836, DOI: 10.1029/2003GL017721.

Arrigo, K. R., G. L. van Dijken, D. G. Ainley, and M. A. Fahnestock (2002). Ecological impact of a large Antarctic iceberg. *Geophysical Research Letters*, **29**(7), DOI: 10.1029/2001GL014160.

Arthern, R. J. and D. P. Winebrenner (1998). Satellite observations of ice sheet accumulation rate. *Proceedings of IGARSS '98* (Vol. 4). Institute of Electrical and Electronic Engineers, Piscataway, NJ, pp. 2249–2251.

Arthern, R. A. and D. J. Wingham (1998). The natural fluctuations of firn densification and their effect on the geodetic determination of ice sheet mass balance. *Climate Change*, **40**, 605–624.

Arthern, R. J., D. P. Winebrenner, and D. G. Vaughan (2003). *Mapping Antarctic Accumulation Using Polarisation of Microwave Emission at 4.5 cm Wavelength* (Geophysical Research Abstracts, Vol. 5, 06616). European Geophysical Society. Copernicus Online Service and Information System (COSIS), Katlenburg-Lindau, Germany.

Ashcraft, I. S. and D. G. Long (2000). SeaWinds views Greenland. *Proceedings of International Geoscience and Remote Sensing Symposium IGARSS '00, 24–28 July 2000, Honolulu, HI*. Institute of Electrical and Electronic Engineers, Piscataway, NJ, Vol. 3, pp. 1131–1133. .

Bailey, D. A. and A. H. Lynch (2000a). Development of an Antarctic regional climate system model. Part I: Sea ice and large-scale circulation. *Journal of Climate*, **13**(8), 1337–1350.

Bailey, D. A. and A. H. Lynch (2000b). Development of an Antarctic regional climate system model. Part II: Station validation and surface energy balance. *Journal of Climate*, **13**(8), 1351–1361.

Baldwin, D. and W. J. Emery (1995). Spacecraft altitude variations in NOAA-11 infrared from AVHRR imagery. *International Journal of Remote Sensing*, **16**, 531–548.

Bales, R. C., J. R. McConnell, E. Mosley-Thompson, and B. Csatho (2001). Accumulation over the Greenland Ice Sheet from historical and recent records. *Journal of Geophysical Research*, **106**(D24), 33813–33826.

Balise, M. J. and C. F. Raymond (1985). Transfer of basal sliding variations to the surface of a linearly viscous glacier. *Journal of Glaciology*, **31**(109), 308–318.

Ballantyne, J. (2002). *A Multi-decadal Study of the Number of Antarctic Icebergs Using Scatterometer Data*. Available online at ⟨http://www.scp.byu.edu/data/iceberg/database1.html⟩.

Ballantyne, J. and D. G. Long (2003). *The Antarctic Iceberg Tracking Database*. Available online at ⟨http://www.scp.byu.edu/data/iceberg/database1.html⟩.

Bamber, J. L. (1994a). Ice sheet altimeter processing scheme. *International Journal of Remote Sensing*, **15**(4), 925–938.

Bamber, J. L. (1994b). A digital elevation model of the Antarctic Ice Sheet derived from ERS-1 altimeter data and comparison with terrestrial measurement. *Annals of Glaciology*, **20**, 48–64.

Bamber, J. L. and C. R. Bentley (1994). A comparison of satellite altimetry and ice-thickness measurements of the Ross Ice Shelf, Antarctica. *Annals of Glaciology*, **20**, 357–364.

Bamber, J. L. and R. A. Bindschadler (1997). An improved elevation dataset for climate and ice-sheet modelling: Validation with satellite imagery. *Annals of Glaciology*, **25**, 438–444.

Bamber, J. L. and A. R. Harris (1994). The atmospheric correction to satellite infrared radiometer data for polar regions. *Geophysical Research Letters*, **21**(19), 2111–2114.

Bamber, J. L. and P. Huybrechts (1996). Geometric boundary conditions for modelling the velocity field of the Antarctic ice sheet. *Annals of Glaciology*, **23**, 364–373.

Bamber, J. L. and R. Kwok (2004). Remote-sensing techniques. In: J. L. Bamber and A. J. Payne (eds.), *Mass Balance of the Cryosphere*. Cambridge University Press, Cambridge, UK, pp. 59–113.

Bamber, J. L. and E. Rignot (2002). Unsteady flow inferred for Thwaites Glacier and comparison with Pine Island Glacier, West Antarctica. *Journal of Glaciology*, **48**(161), 237–246.

Bamber, J. L., S. Ekholm, and W. Krabill (1998). The accuracy of satellite radar altimetry data over the Greenland Ice Sheet determined from airborne laser data. *Geophysical Research Letters*, **25**(16), 3177–3180.

Bamber, J. L., R. J. Hardy, P. Huybrechts, and I. Joughin (2000a). A comparison of balance velocities, measured velocities and thermo-mechanically modelled velocities for the Greenland Ice Sheet. *Annals of Glaciology*, **30**, 211–216.

Bamber, J. L., D. G. Vaughan, and I. Joughin (2000b). Widespread complex flow in the interior of the Antarctic Ice Sheet. *Science*, **287**(5456), 1248–1250.

Bamber, J. L., R. J. Hardy, and I. Joughin (2000c). An analysis of balance velocities over the Greenland ice sheet and comparison with synthetic aperture radar interferometry. *Journal of Glaciology*, **46**(152), 67–74.

Bamber, J. L., S. Ekholm, and W. B. Krabill (2001a). A new, high-resolution digital elevation model of Greenland fully validated with airborne laser altimeter data. *Journal of Geophysical Research–Solid Earth*, **106**(B4), 6733–6745.

Bamber, J. L., R. L. Layberry, and S. P. Gogineni (2001b). A new ice thickness and bedrock data set for the Greenland Ice Sheet: 1. Measurement, data reduction, and errors. *Journal of Geophysical Research*, **106**(D24), 33773–33780.

Bamber, J. L., D. J. Baldwin and S. P. Gogineni (2003). A new bedrock and surface elevation dataset for modelling the Greenland ice sheet. *Annals of Glaciology*, **37**, 351–356.

Bamler, R. and P. Hartl (1998). Synthetic aperture radar interferometry. *Inverse Problems*, **14**, R1–R54.

Bamler, R. and J. Holzner (2004). ScanSAR interferometry for Radarsat-2 and Radarsat-3. *Canadian Journal of Remote Sensing*, **30**(3), 437–447.

Bardel, P., A. G. Fountain, D. K. Hall, and R. Kwok (2002). Synthetic aperture radar detection of the snowline on Commonwealth and Howard Glaciers, Taylor Valley, Antarctica. *Annals of Glaciology*, **34**, 177–183.

Barnes, W. L., T. S. Pagano, and V. V. Salomonson (1998). Pre-launch characteristics of the Moderate Resolution Imaging Spectro-radiometer (MODIS) on EOS-AM1. *IEEE Transactions on Geoscience and Remote Sensing*, **36**, 1088–1100.

Barnsley, M. J., A. H. Strahler, K. P. Morris, and J.-P. Muller (1994). Sampling the surface Bidirectional Reflectance Distribution Function (BRDF): 1. Evaluation of current and future satellite sensors. *Remote Sensing Reviews*, **8**, 271–311.

Barry, R. G. (1997). Satellite-derived data products for the polar regions. *EOS, Transactions of the American Geophysical Union*, **78**(5), 52.

Baudoin, A., M. Shroeder, C. Valorge, M. Bernard, and V. Rudowski (2003). The HRS-SAP initiative: A scientific assessment of the High Resolution Stereoscopic instrument on board of SPOT 5 by ISPRS investigators. *Proceedings of ISPRS Workshop on High Resolution Mapping from Space, October 6–8, Hannover*. Institut für Photogrammetrie und GeoInformation, Universität Hannover, Germany.

Beckmann, A. and H. Goosse (2003). A parameterization of ice shelf–ocean interaction for climate models. *Ocean Modelling*, **5**, 157–170.

Benson, C. S. (1962). *Stratigraphic Studies in the Snow and Firn of the Greenland Ice Sheet* (SIPRE Research Report 70). U.S. Army Snow Ice and Permafrost Research Establishment, Corps of Engineers, Hanover, NH, 120 pp.

Bentley, C. R. (1998). Ice on the fast track. *Nature*, **394**, 21–22.

Bentley, C. R. (2004). Mass balance of the Antarctic Ice Sheet: Observational aspects. In: J. L. Bamber and A. J. Payne (eds.), *Mass Balance of the Cryosphere*. Cambridge University Press, Cambridge, UK, pp. 459–489.

Bentley, C. R. and M. B. Giovinetto (1991). Mass balance and sea-level change. *Proceedings of International Conference on the Role of the Polar Regions in Global Change, June 11–15, Fairbanks, Alaska*. University of Alaska, Fairbanks, pp. 481–488.

Bentley, C. R. and D. D. Sheehan (1992). Comparison of altimetry profiles over East Antarctic from Seasat and Geosat: An interim report. *Zeitschrift fur Gletscherkunde und Glazialgeologie*, **26**(1), 1–9.

Bentley, C. R. and J. M. Wahr (1998). Satellite gravity and the mass balance of the Antarctic Ice Sheet. *Journal of Glaciology*, **44**(147), 207–213.

Benveniste, J., M. Roca, S. Baker, D. Wingham, S. Laxon, O.-Z. Zanife, B. Legrésy, and F. Rémy (2002). Envisat radar altimetry products for cryospheric studies. *EGS XXVII General Assembly, April 21–26, Nice, France* (Abstract 06093).

Bernstein, R. (1983). Image geometry and rectification. In: R. N. Colwell (ed.), *Manual of Remote Sensing*. American Society of Photogrammetry, Falls Church, VA, pp. 881–884.

Bertoia, C., M. Manore, H. S. Andersen, C. O'Connors, K. Q. Hansen, and C. Evanego (2004). Synthetic aperture radar for operational ice observation and analysis at the

U.S., Canadian, and Danish National Ice Centers. In: C. R. Jackson and J. R. Apel (eds.), *Synthetic Aperture Radar Marine User's Manual*. National Oceanic and Atmospheric Administration, Washington, DC, pp. 417–442.

Bigg, G. R. (1999). An estimate of the flux of iceberg calving from Greenland. *Arctic Antarctic Alpine Research*, **31**, 174–178.

Bindschadler, R. (1993). Siple Coast Project research of Crary Ice Rise and the mouths of Ice Streams Band C, West Antarctica: Review and new perspectives. *Journal of Glaciology*, **39**(133), 538–552.

Bindschadler, R. A. (1997). Actively surging West Antarctic ice streams and their response characteristics. *Annals of Glaciology*, **24**, 409–414.

Bindschadler, R. (1998a). Future of the West Antarctic Ice Sheet. *Science*, **282**(5388), 428–429.

Bindschadler, R. A. (1998b). Monitoring ice-sheet behavior from space. *Reviews of Geophysics*, **36**, 79–104.

Bindschadler, R. (2001). History of Pine Island Glacier with Landsat imagery. *EOS, Transactions of the American Geophysical Union*, **82**(47), Fall Meeting Supplement, Abstract IP41A-08.

Bindschadler, R. (2003). Tracking subpixel-scale sastrugi with Advanced Land Imager. *IEEE Transactions on Geoscience and Remote Sensing*, **41**(6), 1373–1377.

Bindschadler, R. and H. Choi (2003). Characterizing and correcting Hyperion detectors using ice-sheet images. *IEEE Transactions on Geoscience and Remote Sensing*, **41**(6), 1189–1193.

Bindschadler, R. and E. Rignot (2001). Crack!: In the polar night. *EOS, Transactions of the American Geophysical Union*, **82**(43), 497–498 and 505.

Bindschadler, R. and T. A. Scambos (1991). Satellite-image-derived velocity field of an Antarctic ice stream. *Science*, **252**, 242–246.

Bindschadler, R. and W. Seider (1998). *Declassified Intelligence Satellite Photography (DISP) Coverage of Antarctica* (NASA TM-1998-206879). NASA Goddard Space Flight Center, Greenbelt, MD, 37 pp.

Bindschadler, R. A. and P. L. Vornberger (1990). AVHRR imagery reveals Antarctic ice dynamics. *EOS, Transactions of the American Geophysical Union*, **71**, 741–742.

Bindschadler, R. A. and P. Vornberger (1992). Interpretation of SAR imagery of the Greenland Ice Sheet using co-registered TM imagery. *Remote Sensing of Environment*, **42**, 167–175.

Bindschadler, R. A. and P. L. Vornberger (1994). Detailed elevation map of Ice Stream C Antarctica, using satellite imagery and airborne radar. *Annals of Glaciology*, **20**, 327–335.

Bindschadler, R. and P. Vornberger (1998). Changes in the West Antarctic Ice Sheet since 1963 from declassified satellite photography. *Science*, **279**, 689–692.

Bindschadler, R. and P. Vornberger (2000). Detecting ice-sheet topography with AVHRR, Resurs-01, and Landsat TM imagery. *Photogrammetric Engineering and Remote Sensing*, **66**(4), 417–422.

Bindschadler, R. A., H. J. Zwally, J. A. Major, and A. C. Brenner (1989). *Surface Topography of the Greenland Ice Sheet from Satellite Radar Altimetry* (NASA SP–503). NASA, Washington, DC, 105 pp.

Bindschadler, R. A., M. A. Fahnestock, P. Skvarca, and T. A. Scambos (1994). Surface-velocity field of the northern Larsen Ice Shelf, Antarctica. *Annals of Glaciology*, **20**, 319–326.

Bindschadler, R., P. Vornberger, D. Blankenship, T. Scambos, and R. Jacobel (1996). Surface velocity and mass balance of Ice Streams D and E, West Antarctica. *Journal of Glaciology*, **42**(142), 461–475.

Bindschadler, R. A., X. Chen, and P. L. Vornberger (1997). Surface velocity and strain rates at the onset of ice stream D, West Antarctica. *Antarctic Journal of the United States*, **32**(5), 41–43.

Bindschadler, R. A., R. B. Alley, J. Anderson, S. Shipp, H. Borns, J. Fastook, S. Jacobs, C. F. Raymond, and C. A. Shuman (1998). What is happening to the West Antarctic Ice Sheet. *EOS, Transactions of the American Geophysical Union*, **79**(22), 257, 264–265.

Bindschadler, R. A., X. Chen, and P. L. Vornberger (2000). The onset area of Ice Stream D, West Antarctica. *Journal of Glaciology*, **46**(152), 95–101.

Bindschadler, R., J. Dowdeswell, D. Hall, and J.-D. Winther (2001a). Glaciological applications with Landsat-7 imagery: Early assessments. *Remote Sensing of Environment*, **78**, 163–179.

Bindschadler, R. A., J. Bamber, and S. Anandakrishnan (2001b). Onset of streaming flow in the Siple Coast Region, Antarctica. In: R. Alley and R. Bindschadler (eds.), *The West Antarctic Ice Sheet: Behavior and Environment* (AGU Antarctic Research Series No. 77A). American Geophysical Union, Washington, DC, pp. 123–136.

Bindschadler, R. A., T. A. Scambos, H. Rott, P. Skvarca, and P. Vornberger (2002). Ice dolines on Larsen Ice Shelf, Antarctica. *Annals of Glaciology*, **34**, 283–290.

Bindschadler, R. A., M. A. King, R. B. Alley, S. Anandakrishnan, and L. Padman (2003). Tidally controlled stick–slip discharge of a West Antarctic ice stream. *Science*, **301**(5636), 1087–1089.

Bingham, A. W. and M. R. Drinkwater (2000). Recent changes in the microwave scattering properties of the Antarctic Ice Sheet. *IEEE Transactions on Geoscience and Remote Sensing*, **38**(4), 1810–1820.

Bingham, A. W. and W. G. Rees (1997). Satellite data synergies for monitoring Arctic ice. *Proceedings of 3rd ERS Scientific Symposium, March 17–21, Florence, Italy* (ESA SP-414, Vol. 2). ESA, Frascati, Italy, pp. 867–870.

Bintanja, R. (1998). The contribution of snowdrift sublimation to the surface mass balance of Antarctica. *Annals of Glaciology*, **27**, 251–259.

Bintanja, R. (1999) On the glaciological, meteorological and climatological significance of Antarctic blue ice areas. *Reviews of Geophysics*, **37**(3), 337–359.

Bintanja, R. and M. R. Van den Broeke (1995a). The surface energy balance of Antarctic snow and blue ice. *Journal of Applied Meteorology*, **34**(4), 902–926.

Bintanja, R. and M. R. Van den Broeke (1995b). The climatic sensitivity of Antarctic blue ice areas. *Annals of Glaciology*, **21**, 157–161.

Bintanja, R. and M. R. van den Broeke (1996). The influence of clouds on the radiation of ice and snow surfaces in Antarctica and Greenland in summer. *International Journal of Climatology*, **16**, 1281–1296.

Bintanja, R., S. Johnsson, and W. H. Knap (1997). The annual cycle of the surface energy balance of Antarctic blue ice. *Journal of Geophysical Research*, **102**(D2), 1867–1881.

Bintanja, R., C. H. Reijmer, and J. M. H. Hulscher (2001). Detailed observations of Antarctic blue-ice surfaces. *Journal of Glaciology*, **47**(158), 387–396.

Blankenship, D. D., D. L. Morse, and J. W. Holt (2002). An airborne radioglaciological survey of iceberg B15A on November 23, 2001 (AGU Fall Meeting, 6–10 December 2002, San Francisco). *EOS Trans. AGU*, **83**(47), Fall Meeting Suppl., Abstract C52B-01.

Bogorodskii, V. V., C. R. Bentley, and P. E. Gudmandsen (1985). *Radioglaciology: Glaciology and Quaternary Geology*. D. Reidel, Dordrecht, The Netherlands, 254 pp.

Bolzan, J. F. and K. C. Jezek (1999). Accumulation rate changes in central Greenland from passive microwave data. *Polar Geography*, **24**(2), 98–112.

Bolzan, J. F. and M. Stroebel (1994). Accumulation rate variations around Summit, Greenland. *Journal of Glaciology*, **40**, 56–66.

Bondesan, A., A. Capra, A. Gubellini, and J. Tison (1994). On the use of static GPS measurements to record the tidal response of a small Antarctic ice shelf (Hells Gate Ice Shelf, Victoria Land). *Geografia Fisica Dinamica Quaternaria*, **17**, 123–129.

Boresjo Bronge, L. (1999). Ice and snow classification in the Vestfold Hills, East Antarctica, using Landsat-TM data and ground radiometer measurements. *International Journal of Remote Sensing*, **20**(2), 225–240.

Bourdelles, B. and Fily, M. (1993), Snow grain-size determination from Landsat imagery over Adélie Land, Antarctica. *Annals of Glaciology*, **17**, 86–92.

Box, J. E. and K. Steffen (2001). Sublimation on the Greenland Ice Sheet from automated weather station observations. *Journal of Geophysical Research*, **106**(D24), 33965–33982.

Braaten, D., P. Kanagaratnam, T. Akins, and S. Gogineni (2003). *Measurement of Thickness of the Greenland Ice Sheet and High-resolution Mapping of Internal Layers* (Technical Report RSL 20780-2). Radar Systems and Remote Sensing Laboratory, University of Kansas, Lawrence, 12 pp.

Braithwaite, R. J. (1995). Positive degree-day factors for ablation on the Greenland ice sheet studied by energy-balance modelling. *Journal of Glacioogy*, **41**(137), 153–159.

Brandt, R. E. and S. G. Warren (1993). Solar-heating rates and temperature profiles in Antarctic snow and ice. *Journal of Glaciology*, **39**(131), 99–110.

Braun, M., F. Rau, H. Saurer, and H. Goßmann (2000). The development of radar glacier zones on the King George Island ice cap, Antarctica, during the austral summer 1996/97 as observed in ERS-2 SAR data. *Annals of Glaciology*, **31**, 357–363.

Brenner, A.C., R. A. Bindschadler, R. H. Thomas, and H. J. Zwally (1983). Slope induced errors in radar altimetry over continental ice sheets. *Journal of Geophysical Research*, **88**, 1617–1623.

Breon F. M., J. C. Buriez, P. Couvert, P. Y. Deschamps, J. L. Deuze, M. Herman, P. Goloub, M. Leroy, A. Lifermann, C. Moulin et al. (2002). Scientific results from the POLarization and Directionality of the Earth's Reflectances (POLDER). *Advances in Space Research*, **30**(11), 2383–2386.

Brisset, L. and F. Rémy (1996). Antarctic topography and kilometre-scale roughness derived from ERS-1 altimetry. *Annals of Glaciology*, **23**, 374–381.

Bromwich, D. H. (1988). Snowfall in high southern latitudes. *Reviews of Geophysics*, **26**(1), 149–168.

Bromwich, D. H. and C. R. Stearns (eds.) (1993). *Antarctic Meteorology and Climatology: Studies Based on Automatic Weather Stations*. American Geophysical Union, Washington, DC, 207 pp.

Bromwich, D. H., Q. S. Chen, Y. Li, and R. I. Cullather (1999). Precipitation over Greenland and its relation to the North Atlantic Oscillation. *Journal of Geophysical Research*, **104**, 22103–22115.

Bromwich, D. H., A. N. Rogers, P. Kallberg, R. I. Cullather, J. W. C. White, and K. J. Kreutz (2000). ECMWF analyses and reanalyses depiction of ENSO signal in Antarctic precipitation. *Journal of Climate*, **13**, 1406–1420.

Bromwich, D. H., Q.-S. Chen, L.-S. Bai, E. N. Cassano, and Y. Li (2001). Modeled precipitation variability over the Greenland Ice Sheet. *Journal of Geophysical Research*, **106**(D24), 33891–33898.

Brown, I. A., M. P. Kirkbride, and R. A. Vaughan (1999). Find the firn line!: The suitability of ERS-1 and ERS-2 SAR data for the analysis of glacier facies on Icelandic ice caps. *International Journal of Remote Sensing*, **20**(15), 3217–3230.

Brown, I. C. and T. A. Scambos (2004). Satellite monitoring blue ice extent near Byrd Glacier, Antarctica. *Annals of Glaciology*, **39**, 223–230.

Brown, G. S. (1977). The average impulse response of a rough surface and its application. *IEEE Transactions on Antennas and Propagation*, **AP-25**, 67–74.

Brunt, K. M. (2003). Using a GIS to monitor Ross Sea icebergs, Antarctica. *The Joint Antarctic Automatic Weather Station and Antarctic Meteorological Research Center Annual Meetings and the Antarctic Mesoscale Prediction System User's Workshop, Madison, Wisconsin, June 23–26* (preprint). University of Wisconsin, Madison, 4 pp.

Budd, W. F. (1970). Ice flow over bedrock perturbations. *Journal of Glaciology*, **9**(55), 29–48.

Budd, W. F. and D. B. Carter (1971). An analysis of the relationship between the surface and bedrock profiles of ice caps. *Journal of Glaciology*, **10**, 197–209.

Budd, W. F. and T. H. Jacka (1989). A review of ice rheology for ice sheet modelling. *Cold Regions Science and Technology*, **16**(2), 107–144.

Budd, W. F. and R. C. Warner (1996). A computer scheme for rapid calculations of balance-flux distributions. *Annals of Glaciology*, **23**, 21–27.

Budd, W., I. Landon-Smith, and E. Wishart (1967). The Amery Ice Shelf. In: H. Oura (ed.), *Physics of Snow and Ice*. Institute of Low Temperature Science, Sapporo, Japan, pp. 447–467.

Budd, W. F., T. H. Jacka, and V. I. Morgan (1980). Antarctic iceberg melt rates derived from size distributions and movement rates. *Annals of Glaciology*, **1**, 103–112.

Bufton, J. L., J. B. Garvin, J. F. Cavanaugh, L. Ramos-Izquierdo, T. D. Clem, and W. B. Krabill (1991). Airborne lidar for profiling of surface topography. *Optical Engineering*, **30**, 72–78.

Cabot, F. and G. Dedieu (1997). Surface albedo from space: Coupling bidirectional models and remotely sensed measurements. *Journal of Geophysical Research*, **102**, 19645–19663.

Calvin, W. M., M. Milman, and H. H. Kieffer (2002). Reflectance of Antarctica from 3 to 5 µm: Discrimination of surface snow and cloud properties. *Annals of Glaciology*, **34**, 121–126.

Casassa G. and I. M. Whillans (1994). Decay of surface topography on the Ross Ice Shelf, Antarctica. *Annals of Glaciology*, **20**, 249–253.

Casassa, G., K. C. Jezek, J. Turner, and I. M. Whillans (1991). Relict flow stripes on the Ross Ice Shelf. *Annals of Glaciology*, **15**, 132–138.

Cawkwell, F. G. L. and J. L. Bamber (2002). The impact of cloud cover on the net radiation budget of the Greenland Ice Sheet. *Annals of Glaciology*, **34**, 141–149.

Chelton, D. B., E. J. Walsh, and J. L. MacArthur (1989). Pulse compression and sea level tracking in satellite altimetry. *Journal of Atmospheric and Oceanic Technology*, **6**, 407–438.

Chelton, D. B., J. C. Ries, B. J. Haines, L.-L. Fu, and P. S. Callahan (2001). Satellite altimetry. In: L.-L. Fu and A. Cazenave (eds.), *Satellite Altimetry and the Earth Sciences: A Handbook for Techniques and Applications*. Academic Press, New York, pp. 1–131.

Chuah, T., S. Gogineni, C. Allen, and B. Wohletz (1996). *Radar Thickness Measurements over the Northern Part of the Greenland Ice Sheet* (Technical Report 10470-3). Radar Systems and Remote Sensing Laboratory, University of Kansas, Lawrence.

Church J. A., J. M. Gregory, P. Huybrechts, M. Kuhn, K. Lambeck, M. T. Nhuan, D. Qin, and P. L. Woodworth (2001). Changes in sea level. In: J. T. Houghton, Y. Ding, D. J.

Griggs, M. Noguer, P. Van der Linden, X. Dai, K. Maskell, and C. I. Johnson (eds.), *Climate Change 2001: The Scientific Basis* (contribution of Working Group 1 to the Third Assessment Report of the Intergovernmental Panel on Climate Change). Cambridge University Press, Cambridge, UK, pp. 639–694.

Clausen, H. B., N. S. Gundestrup, S. J. Johnsen, R. Bindschadler, and H. J. Zwally (1988). Glaciological investigations in the Crete area, central Greenland: A search for a new deep drilling site. *Annals of Glaciology*, **10**, 10–15.

Cloude, S. R. and K. P. Papathanassiou (1998). Polarimetric SAR interferometry. *IEEE Transactions on Geoscience and Remote Sensing*, **36**(5), 1551–1565.

Cloude, S. R. and R. N. Treuhaft (1999). The structure of oriented vegetation from polarimetric interferometry. *IEEE Transactions on Geoscience and Remote Sensing*, **37**(5), 2620–2624.

Cloude, S. R., K. P. Papathanassiou, and E. Pottier (2001). Radar polarimetry and polarimetric interferometry. *IEEE Transactions on Electronics*, **E84-C**(12), 1814–1822.

Colbeck, S. C. (1987). *A Review of the Metamorphism and Classification of Seasonal Snow Cover Crystals* (IAHS Publication 162). International Association of Hydrological Sciences, Christchurch, New Zealand, pp. 3–24.

Comiso, J. C. (1994). Surface temperatures in the polar regions using Nimbus-7 THIR. *Journal of Geophysical Research*, **99**(C3), 5181–5200.

Comiso, J. C. (2000). Variability and trends in Antarctic surface temperatures from *in situ* and satellite infrared measurements. *Journal of Climate*, **13**, 1674–1696.

Comiso, J. C. (2003). Warming trends in the Arctic. *Journal of Climate*, **16**(21), 3498–3510.

Comiso, J. C. and C. L. Parkinson (2004). Satellite-observed changes in the Arctic. *Physics Today*, **57**(8), 38–44.

Comiso, J. C., H. J. Zwally, and J. L. Saba (1982). Radiative transfer modeling of microwave emission and dependence on firn properties. *Annals of Glaciology*, **3**, 54–58.

Committee on Earth Observation Satellites (2000). *The Use of Earth Observing Satellites for Hazard Support: Assessments and Scenarios* (Final Report of the CEOS Disaster Management Support Group). National Oceanic and Atmospheric Administration, National Environmental Satellite, Data and Information Service, Silver Spring, MD, 214 pp.

Connolley, W. M. C. and J. C. King, (1996). A modelling and observational study of East Antarctic surface mass balance. *Journal of Geophysical Research*, **101**(D1), 1335–1343.

Conway, H., G. Catania, C. F. Raymond, A. M. Gades, T. A. Scambos, and H. Engelhardt (2002). Switch of flow direction in an Antarctic ice stream. *Letters to Nature*, **419**(3), 465–467.

Crabtree, R. D. and C. S. M. Doake (1982). Pine Island Glacier and its drainage basin: Results from radio echo-sounding. *Annals of Glaciology*, **3**, 65–70.

Cracknell, A. (1997). *The Advanced Very High Resolution Radiometer*. Taylor & Francis, London, 968 pp.

Crepon, M., M. N. Hoassais, and B. Saint Guily (1988). The drift of icebergs under wind action. *Journal of Geophysical Research*, **9**(C4), 3608–3612.

Crison, M. and F. DeLuccia (1998). Convergence of the civilian and military polar-orbiting environmental satellite systems: VIIRS (the Visible/Infrared Imager/Radiometer Suite) for NPOESS (National Polar-orbiting Operational Environmental Satellite System). *IGARSS '98*.

Csatho, B., T. Schenk, W. Krabill, T. Wilson, W. Lyons, G. McKenzie, C, Hallam, S. Manizade, and T. Paulsen (2005). Airborne laser scanning for high-resolution mapping of Antarctica. *EOS, Transactions of the American Geophysical Union*, **86**(25), 237–238.

Cudlip, W., D. R. Mantripp, C. L. Wrench, H. D. Griffiths, D. V. Sheehan, M. Lester, R. P. Leigh, and T. R. Robinson (1994). Corrections for altimeter low level processing at the Earth Observation Data Centre. *International Journal of Remote Sensing*, **15**(4), 889–914.

Cuffey, K. M. (2001). Interannual variability of elevation on the Greenland Ice Sheet: Effects of firn densification, and establishment of a multi-century benchmark. *Journal of Glaciology*, **47**(158), 369–377.

Cullather, R. I., D. H. Bromwich, and M. L. Van Woert (1996). Interannual variations in Antarctic precipitation related to El Niño–Southern Oscillation. *Journal of Geophysical Research*, **101**(D4), 19109–19118.

Curry, J. A., D. Randall, W. B. Rossow, and J. L. Schramm (1996). Overview of Arctic cloud and radiation characteristics. *Journal of Climate*, **9**, 1731–1764.

Dahe, Q., X. Cunde, I. Allison, B. Lingen, R. Stephenson, R. Jiawen, and Y. Ming (2003). Snow surface height variations on the Antarctic ice sheet in Princess Elizabeth Land, Antarctica: One year of data from an automatic weather station. *Annals of Glaciology*, **39**, 181–187.

Dahl-Jensen, D., N. S. Gundestrup, K. Keller, S. J. Johnsen, S. P. Gogineni, C. T. Allen, T. S. Chuah, H. Miller, S. Kipfstuhl, and E. D. Waddington (1997). A search in north Greenland for a new ice-core drill site. *Journal of Glaciology*, **43**, 300–306.

Dall, J., K. P. Papathanassiou, and H. Skriver (2003). Polarimetric SAR interferometry applied to land ice: First results, *Proceedings of IEEE 2003 International Geoscience and Remote Sensing Symposium IGARSS '03, Toulouse, July*. Institute of Electrical and Electronic Engineers, Piscataway, NJ, pp. 1432–1434.

Dall, J., K. P. Papathanassiou, and H. Skriver (2004). Polarimetric SAR interferometry applied to land ice: Modeling, *Proceedings of EUSAR 2004 Conference, Ulm, May*. VDE, Frankfurt, Germany, pp. 247–250.

Das, S. B., R. B. Alley, D. B. Reusch, and C. A. Shuman (2002). Temperature variability at Siple Dome, West Antarctica, derived from ECMWF re-analyses, SSM/I and SMMR brightness temperatures and AWS records. *Annals of Glaciology*, **34**, 106–112.

Davis, C. H. (1993). A surface and volume scattering retracking algorithm for ice sheet satellite altimetry. *IEEE Transactions on Geoscience and Remote Sensing*, **31**(4), 811–818.

Davis, C. H. (1995). Synthesis of passive microwave and radar altimeter data for estimating accumulation rates of dry polar snow. *International Journal of Remote Sensing*, **16**, 2055–2067.

Davis, C. H. (1996). Comparison of ice-sheet satellite altimeter retracking algorithms. *IEEE Transactions on Geoscience and Remote Sensing*, **34**(1), 229–236.

Davis, C. H. (1997). A robust threshold retracking algorithm for measuring ice-sheet surface elevation change from satellite radar altimeters. *IEEE Transactions on Geoscience and Remote Sensing*, **35**(4), 974–979.

Davis, C. H. and A. C. Ferguson (2004). Elevation change of the Antarctic Ice Sheet, 1995–2000, from ERS-2 satellite radar altimetry. *IEEE Transactions on Geoscience and Remote Sensing*, **42**(11), 2437–2445.

Davis, C. H. and R. K. Moore (1993). A combined surface- and volume-scattering model for ice-sheet radar altimetry. *Journal of Glaciology*, **39**(133), 675–686.

Davis, C. H. and D. M. Segura (2001). An algorithm for time-series analysis of ice-sheet surface elevations from satellite altimetry. *IEEE Transactions on Geoscience and Remote Sensing*, **39**, 202–206.

Davis, C. H. and S. Sun (2004). Long-term thinning of the southeast Greenland ice sheet from Seasat, Geosat, and GFO satellite radar altimetry. *IEEE Transactions on Geoscience and Remote Sensing Letters*, **1**(2), 47–50.

Davis, C. H. and H. J. Zwally (1993). Geographical and seasonal variations in the surface properties of the ice sheets by satellite radar altimetry. *Journal of Glaciology*, **39**, 687–697.

Davis, C. H., C. A. Kluever, and B. J. Haines (1998). Elevation change of the Southern Greenland Ice Sheet. *Science*, **279**, 2086–2088.

Davis, C. H., C. A. Kluever, B. J. Haines, C. Perez, and Y. Yoon (2000). Improved elevation change measurement of the southern Greenland ice sheet from satellite radar altimetry. *IEEE Transactions on Geoscience and Remote Sensing*, **3**(38), 1367–1378.

Davis, C. H., J. R., McConnell, J. Bolzan, J. L. Bamber, R. H. Thomas, and E. Mosley-Thompson (2001). Elevation change of the southern Greenland Ice Sheet from 1978 to 1988: Interpretation. *Journal of Geophysical Research*, **106**(D24), 33743–33754.

Davis, C. H., Y. Li, J. R. McConnell, M. M. Frey, and E. Hanna (2005). Snowfall-driven growth in East Antarctic Ice Sheet mitigates recent sea-level rise. *Science*, **308**(5730), 1898–1901.

De Angelis, H. and P. Skvarca (2003). Glacier surge after ice shelf collapse. *Science*, **299**, 1560–1562.

DeConto, R. and D. Pollard (2002). *The Antarctic Climate Evolution (ACE) Paleoclimate and Ice Sheet Modeling Workshop, Northampton, Massachusetts, May 30–June 2* (draft summary report).

Derauw, D. (1999). DInSAR and coherence tracking applied to glaciology: The example of Shirase Glacier. In: H. Saway-Lacoste (ed.), *"Fringe '99", Advancing ERS SAR Interferometry from Applications towards Operations, Liège, Belgium* (ESA SP-478). ESA, Noordwijk, The Netherlands.

de Ruyter de Wildt, M. S., J. Oerlemans, and H. Björnsson (2002): A method for monitoring glacier mass balance using satellite albedo measurements: Application to Vatnajökul, Iceland. *Journal of Glaciology*, **48**(161), 267–278.

Deschamps, P.-Y., F.-M. Bréon, M. Leroy, A. Podaire, A. Bricaud, J.-C. Buriez, and G. Sèze (1994). The POLDER mission: Instrument characteristics and scientific objectives. *IEEE Transactions on Geoscience and Remote Sensing*, **32**, 598–615.

Di Girolamo, L. and M. J. Wilson (2003). A first look at band-differenced angular signatures for cloud detection from MISR. *IEEE Transactions on Geoscience and Remote Sensing*, **41**(7), 1730–1734.

Diner, D. J., J. C. Beckert, T. H. Reilly, C. J. Bruegge, J. E. Conel, R. Kahn, J. V. Martonchik, T. P. Ackerman, R. Davies, S. A. W. Gerstl et al. (1998). Multi-angle Imaging SpectroRadiometer (MISR) Intrument Description and Experiment Overview. *IEEE Transactions on Geoscience and Remote Sensing*, **36**, 1072–1087.

Diner, D., G. P. Asner, R. Davies, Y. Knyazikhin, J.-P. Muller, A. Nolin, B. Pinty, C. B. Schaaf, and J. Stroeve (1999). New directions in Earth observing: Scientific applications of multi-angle remote sensing. *Bulletin of the American Meteorological Society*, **80**, 2209–2228.

Diner, D. J., B. H. Braswell, R. Davies, N. Gobron, J. Hu, Y. Jin, R. A. Kahn, Y. Knyazikhin, N. Loeb, J.-P. Muller et al. (2005). The value of multiangle measurements for retrieving structurally and radiatively consistent properties of clouds, aerosols, and surfaces. *Remote Sensing of Environment*, **97**(4), 495–518.

Doake, C. S. M. (1992). *Gravimetric Tidal Measurements on Filchner Ronne Ice Shelf* (Filchner-Ronne Ice Shelf Program, Report 6, pp. 34–39). Alfred-Wegener-Institute, Bremerhaven, Germany.

Doake, C. S. M. and D. G. Vaughan (1991). Rapid disintegration of the Wordie Ice Shelf in response to atmospheric warming. *Nature*, **350**, 328–330.

Doake, C. S. M., H. F. J. Corr, J. H. Rott, P. Skvarca, and N. W. Young (1998). Break-up and conditions for stability of the northern Larsen Ice Shelf, Antarctica. *Nature*, **391**, 778–780.

Doake, C. S. M., H. F. J. Corr, A. Jenkins, K. Makinson, K. W. Nicolls, C. Nath, A. M. Smith, and D. G. Vaughan (2001). Rutford Ice Stream, Antarctica. In: R. Alley and R. Bindschadler (eds.), *The West Antarctic Ice Sheet: Behavior and Environment* (AGU Antarctic Research Series No. 77). American Geophysical Union, Washington, DC, pp. 221–235.

Domack, E. W., D. Duran, K. McMullen, R. Gilbert, and A. Leventer (2002). Sediment lithofacies from beneath the Larsen B Ice Shelf: Can we detect ice shelf fluctuation? *EOS, Transactions of the American Geophysical Union*, **83**, F301.

Dowdeswell, J. A. and N. F. McIntyre (1987). The surface topography of large ice masses from Landsat imagery. *Journal of Glaciology*, **33**(113), 16–23.

Dowdeswell, J. A. and M. Williams (1997). Surge-type glaciers in the Russian High Arctic identified from satellite imagery. *Journal of Glaciology*, **43**, 489–494.

Dowdeswell, J. A., B. Unwin, A. M. Nuttall, and D. J. Wingham (1999). Velocity structure, flow instability and mass flux on a large Artic ice cap form satellite radar interferometry. *Earth and Planetary Science Letters*, **167**, 131–140.

Dozier, J. (1989). Spectral signature of alpine snow cover from the Landsat Thematic Mapper. *Remote Sensing of Environment*, **28**, 150–163.

Dozier, J. and T. Painter (2004). Multispectral and hyperspectral remote sensing of alpine snow properties. *Annual Review of Earth and Planetary Sciences*, **32**, 465–494.

Dozier, J. and S. G. Warren (1982). Effect of viewing angle on the infrared brightness temperature of snow. *Water Resources Research*, **18**(5), 1424–1434.

Dozier, J., S. R. Schneider, and D. F. McGinnis (1981). Effect of grain size and snowpack water equivalence on visible and near-infrared satellite observations of snow. *Water Resources Research*, **17**, 1213–1221.

Drewry, D. J. (ed.) (1983). *Antarctica: Glaciological and Geophysical Folio*. Scott Polar Research Institute, Cambridge University, UK.

Drewry, D. J. and E. M. Morris (1992). The response of large ice sheets to climatic change. *Philosophical Transactions of the Royal Society of London B*, **338**, 235–242.

Drewry, D. J., D. T. Meldrum, and E. Jankowski (1980). Radio echo and magnetic sounding of the Antarctic Ice Sheet, 1978–1979. *Polar Record*, **20**, 43–57.

Drinkwater, M. R. and C. C. Lin (2000). Introduction to the special section on emerging scatterometer applications. *IEEE Transactions on Geoscience and Remote Sensing*, **38**, 1763–1764.

Drinkwater, M. R. and D. G. Long (1998). Seasat, ERS-1/2 and NSCAT Scatterometer observed changes on the large ice sheets. *Proceedings of International Geoscience and Remote Sensing Symposium IGARSS '98, Seattle, Washington, 6–10 July 1998*. Institute of Electrical and Electronic Engineers, Piscataway, NJ, pp. 2252–2254.

Drinkwater, M. and D. Long (1999). Sea Winds imaging of polar ice sheets. *Proceedings of Quikscat Early Science Workshop, Embassy Suites Hotel, Arcadia, CA, November 2–5* (JPL project publication). NASA Jet Propulsion Laboratory, Pasadena, CA.

Drinkwater, M. R., D. G. Long, and A. W. Bingham (2001). Greenland snow accumulation estimates from satellite radar scatterometer data. *Journal of Geophysical Research*, **106**(D24), 33935–33950.

Drinkwater, M. R., R. Francis, G. Ratier, and D. J. Wingham (2004a). The European Space Agency's Earth Explorer Mission CryoSat: Measuring variability in the cryosphere. *Annals of Glaciology*, **39**, 313–320.

Drinkwater, M. R., N. Floury, and M. Tedesco (2004b). L-band Antarctic ice sheet brightness temperatures: Spectral emission modelling, temporal stability and impact of the ionosphere. *Annals of Glaciology*, **39**, 391–396.

Dwyer, J. L. (1995). Mapping tidewater glacier dynamics in East Greenland using Landsat data. *Journal of Glaciology*, **41**(139), 584–595.

Early, D. S. and D. G. Long (2001). Image reconstruction and enhanced resolution imaging from irregular samples. *IEEE Transactions on Geoscience and Remote Sensing*, **39**, 291–302.

Echelmeyer, K. A., W. D. Harrison, C. Larsen, and J. E. Mitchell (1994). The role of the margins in the dynamics of an active ice stream. *Journal of Glaciology*, **40**, 527–538.

Eiken, T., J. O. Hagen, and K. Melvold (1997). Kinematic GPS survey of geometry changes on Svalbard glaciers. *Annals of Glaciology*, **24**, 157–163.

Ekholm, S. (1996). A full coverage, high-resolution, topographic model of Greenland computed from a variety of digital elevation data. *Journal of Geophysical Research*, **101**(B10), 21961–21972.

Ekholm, S., R. Forsberg, and J. Brozena (1995). Accuracy of satellite elevations over the Greenland Ice Sheet. *Journal of Geophysical Research*, **100**(C2), 2687–2696.

Engelhardt, H. and B. Kamb (1998). Basal sliding of Ice Stream B, West Antarctica. *Journal of Glaciology*, **44**(147), 223–230.

Engeset, R. V., J. Kohler, K. Melvold, and B. Lundén (2002). Change detection and monitoring of glacier mass balance and facies using ERS SAR winter images over Svalbard. *International Journal of Remote Sensing*, **23**(10), 2023–2050.

Enomoto, H., H. Motoyama, T. Shiraiwa, T. Saito, T. Kameda, T. Furukawa, S. Takahashi, Y. Kodama, and O. Watanabe (1998). Winter warming over Dome Fuji, East Antarctica, and semi-annual oscillation in the atmospheric circulation. *Journal of Geophysical Research*, **D18**(103), 23103–23111.

Entekhabi, D., E. G. Njoku, P. Houser, M. Spencer, T. Doiron, Y. Kim, J. Smith, R. Girard, S. Belair, W. Crow et al. (2004). The Hydrosphere State (HYDROS) satellite mission: An Earth System Pathfinder for global mapping of soil moisture and land freeze/thaw. *IEEE Transactions on Geoscience and Remote Sensing*, **42**(10), 2184–2195.

ESA (2001). *CryoSat Mission and Data Description* (Report CS-RP-ESA-SY-0059). ESA, Noordwijk, The Netherlands.

ESA (2003). *CryoSat Science Report* (ESA SP-1272). ESA, Noordwijk, The Netherlands.

Fahnestock M. and J. Bamber (2001). Morphology and surface characteristics of the West Antarctic Ice Sheet. In: R. B. Alley and R. A. Bindschadler (eds.), *The West Antarctic Ice Sheet: Behavior and Environment* (AGU Antarctic Research Series No. 77). American Geophysical Union, Washington, DC, pp. 13–27.

Fahnestock, M. A. and R. A. Bindschadler (1993). Description of a program for SAR investigation of the Greenland Ice Sheet and an example of margin change detection using SAR. *Annals of Glaciology*, **17**, 332–336.

Fahnestock, M. A., R. A. Bindschadler, R. A. Kwok, and K. Jezek (1993). Greenland Ice Sheet surface properties and ice dynamics from ERS-1 SAR imagery. *Science*, **262**, 1530–1534.

Fahnestock, M., S. Ekholm, K. Knowles, R. Kwok, and T. Scambos (1997). *Digital SAR mosaic and elevation map of the Greenland Ice Sheet* (CD-ROM). National Snow and Ice Data Center, Boulder, CO.

Fahnestock, M. A., T. A. Scambos, R. A. Bindschadler, and G. Kvaran (2000a). A millennium of variable ice flow recorded by the Ross Ice Shelf, Antarctica. *Journal of Glaciology*, **46**(155), 652–664.

Fahnestock, M. A., T. A. Scambos, C. A. Shuman, R. J. Arthern, D. P. Winebrenner, and R. Kwok (2000b). Snow megadune fields on the East Antarctic plateau: Extreme atmosphere–ice interaction. *Geophysical Research Letters*, **27**(22), 3719–3722.

Fahnestock, M., W. Abdalati, S. Luo, and S. Gogineni (2001a). Internal layer tracing and age-depth-accumulation relationships for the northern Greenland Ice Sheet. *Journal of Geophysical Research*, **106**(D24), 33789–33798.

Fahnestock, M. A., I. Joughin, T. A. Scambos, R. Kwok, W. B. Krabill, and S. Gogineni (2001b). Ice-stream-related patterns of ice flow in the interior of northeast Greenland. *Journal of Geophysical Research*, **106**(D24), 34035–34046.

Fahnestock, M. A., W. Abdalati, and C. Shuman (2002). Long melt seasons on ice shelves of the Antarctic Peninsula: An analysis using satellite-based microwave emission measurements. *Annals of Glaciology*, **34**, 127–133.

Fatland, D. R. and C. S. Lingle (1998). Analysis of the 1993–1995 Bering Glacier (Alaska) surge using differential SAR interferometry. *Journal of Glaciology*, **44**(148), 532–546.

Fatland, D. R. and C. S. Lingle (2002). InSAR observations of the 1993–95 Bering Glacier (Alaska USA) surge and a surge hypothesis. *Journal of Glaciology*, **48**(162), 439–451.

Féménias, P., F. Rémy, P. Raizonville, and J. F. Minster (1993). Analysis of satellite altimeter height measurements above ice sheet. *Journal of Glaciology*, **133**, 591–600.

Ferguson, A. C., C. H. Davis, and J. E. Cavanaugh (2004). An autoregressive model for analysis of ice sheet elevation change time series. *IEEE Transactions on Geoscience and Remote Sensing*, **42**(11), 2426–2436.

Ferrigno, J. G. and W. G. Gould (1987). Substantial changes in the coastline of Antarctica revealed by satellite imagery. *Polar Record*, **23**(146), 577–583.

Ferrigno, J. G., B. K. Lucchitta, K. F. Mullins, A. L. Allison, R. J. Allen, and W. G. Gould (1993). Velocity measurements and changes in position of Thwaites Glacier/Iceberg Tongue from aerial photography, Landsat images and NOAA AVHRR data. *Annals of Glaciology*, **17**, 239–244.

Ferrigno, J. G., J. L. Mullins, J. Stapleton, R. A. Bindschadler, T. A. Scambos, L. B. Bellisime, J. Bowell, and A. V. Acosta (1994). Landsat TM image maps of the Shirase and Siple Coast ice streams, West Antarctica. *Annals of Glaciology*, **20**, 407–412.

Ferrigno, J. G., R. S. Williams Jr., E. Rosanova, B. K. Lucchitta, and C. Swithinbank (1998). Analysis of Coastal Change in Marie Byrd Land and Ellsworth Land, West Antarctica, using Landsat imagery. *Annals of Glaciology*, **27**, 33–40.

Ferrigno, J. G., J. L. Mullins, J. A. Stapleton, P. S. Chavez Jr., M. G. Velasco, R. S. Williams Jr., G. F. Delinski Jr., and D. A. Lear (2000). *Satellite Image Map of Antarctica* (revised and reprinted 2nd edn., 1996, Map I-2560—scale 1 : 5,000,000). U.S. Geological Survey/ National Science Foundation, Reston, VA.

Ferrigno, J. G., R. S. Williams, and K. M. Foley (2004). Coastal-change and glaciological map of the Saunders Coast area, Antarctica, 1972–1997. *Annals of Glaciology*, **39**, 245–250.

Ferrigno, J. G., R. S. Williams Jr., and J. L. Thomson (2002). *Coastal-change and Glaciological Maps of the Antarctic Peninsula* (BAS FS-017-02). U.S. Geological Survey, Reston, VA/ British Antarctic Survey, Cambridge, UK, 2 pp.

Fetterer, F., M. R. Drinkwater, K. Jezek, S. Laxon, R. Onstott, and L. Ulander (1992). Sea ice altimetry. In: F. D. Carsey (ed.), *Microwave Remote Sensing of Sea Ice* (AGU Geophysical Monograph 28), American Geophysical Union, Washington, DC, pp. 111-135.

Fichefet, T., C. Poncin, H. Goosse, P. Huybrechts, I. Janssens, and H. Le Treut (2003). Implications of changes in freshwater flux from the Greenland Ice Sheet for the climate of the 21st century. *Geophysical Research Letters*, **30**(17), 1911, DOI: 10.1029/2003GL017826.

Figa-Saldaña, J., J. J. W. Wilson, E. Attema, R. Gelsthorpe, M. R. Drinkwater, and A. Stoffelen (2002). The advanced scatterometer (ASCAT) on the Meteorological Operational (MetOp) platform: A follow on for European wind scatterometers. *Canadian Journal of Remote Sensing*, **28**(3), 404–412.

Fily, M. and J. P. Benoist (1991). Large scale statistical study of the Scanning Multichannel Microwave Radiometer. *Journal of Glaciology*, **37**, 129–139.

Fily, M., B. Bourdelles, J. P. Dedieu, and C. Sergent (1997). Comparison of *in situ* and Landsat Thematic Mapper derived snow grain characteristics in the Alps. *Remote Sensing of Environment*, **59**(3), 452–460.

Flach, J. D., K. C. Partington, C. Ruiz, E. Jeansou, and M. R. Drinkwater (2005). Inversion of the surface properties of ice sheets from satellite microwave data. *IEEE Transactions on Geoscience and Remote Sensing*, **43**(4), 743–752.

Flett, D. (2003). Operational use of SAR at the Canadian Ice Service: Present operations and a look into the future. *Proceedings of the 2nd Workshop on Coastal and Marine Applications of SAR, 8–12 September 2003, Svalbard, Norway*. ESA, Noordwijk, The Netherlands.

Foldvik. A. and T. Gammelsrød (1988). Notes on Southern Ocean hydrography, sea-ice and bottom water formation. *Palaeogeography, Palaeoclimatology, Palaeoecology*, **67**, 3–17.

Foldvik, A., T. Gammelsrød, N. Slotsvik, and T. Tørresen (1985). Oceanographic conditions on the Weddell Sea shelf during the German Antarctic Expedition 1979/80. *Polar Research*, **3**(2), 209–226.

Foldvik, A., T. Gammelsrød, S. Østerhus, E. Fahrbach, G. Rohardt, M. Schröder, K. W. Nicholls, L. Padman, and R. A. Woodgate (2004). Ice shelf water overflow and bottom water formation in the southern Weddell Sea. *Journal of Geophysical Research*, **109**, C0201, DOI: 10.1029/2003JC002008.

Ford, A. L. J. and R. R. Forster (2002). Three-dimensional ice velocities derived from multiple look direction InSAR: An example from Ice Stream A, West Antarctica. *AGU Fall Meeting* (Abstract C51A-0923). American Geophysical Union, Washington, DC.

Forsberg, R., K. Keller, C. S. Nielsen, N. Gundestrup, C. C. Tscherning, S. Nørvang Madsen, and J. Dall (2000). Elevation change measurements of the Greenland Ice Sheet. *Earth Planets Space*, **52**, 1049–1053.

Forster, R. R., K. C. Jezek, H. G. Sohn, A. L. Gray, and K. E. Matter (1998). Analysis of glacier flow dynamics from preliminary Radarsat InSAR data of the Antarctic Mapping Mission. *Proceedings of International Geoscience and Remote Sensing Symposium IGARSS '98, Seattle, WA, 6–10 July 1998*. Institute of Electrical and Electronic Engineers, Piscataway, NJ, pp. 2225–2227.

Forster, R. R., E. Rignot, B. L. Isacks, and K. C. Jezek (1999). Interferometric radar observations of glaciers Europa and Penguin, Hielo Patagónico Sur, Chile. *Journal of Glaciology*, **45**(150), 325–337.

Forster, R. R., K. C. Jezek, L. Koenig, and E. Deeb (2003). Measurement of glacier geophysical properties from InSAR wrapped phase. *IEEE Transactions on Geoscience and Remote Sensing*, **41**(11), 2595–2604.

Foster, J. L., D. K. Hall, A. T. C. Chang, and A. Rango (1984). An overview of passive microwave snow research and results. *Reviews of Geophysics and Space Physics*, **22**(2), 195–208.

Foster, J. L., D. K. Hall, and A. T. C. Chang (1987). Remote sensing of snow. *EOS, Transactions of the American Geophysical Union*, **68**, 681–684.

Fowler, A. C. and C. Johnson (1996). Ice sheet surging and ice stream formation. *Annals of Glaciology*, **23**, 68–73.

Fox, A. J. and A. P. R. Cooper (1998). Climate-change indicators from archival aerial photography on the Antarctic Peninsula. *Annals of Glaciology*, **27**, 626–642.

Frezzotti, M. (1997). Ice front fluctuation, iceberg calving flux and mass balance of Victoria Land glaciers (Antarctica). *Antarctic Science*, **9**(1), 61–73.

Frezzotti, M. and M. C. G. Mabin (1994). 20th century behaviour of the Drygalski Ice Tongue, Ross Sea, Antarctica. *Annals of Glaciology*, **20**, 397–400.

Frezzotti, M. and M. Polizzi (2002). 50 years of ice-front changes between the Adélie and Banzare Coasts, East Antarctica. *Annals of Glaciology*, **34**, 235–240.

Frezzotti, M., A. Cimbelli, and J. G. Ferrigno (1998a). Ice-front and iceberg behaviour along Oates and George V coasts, Antarctica, 1912–1996. *Annals of Glaciology*, **27**, 643–650.

Frezzotti, M., A. Capra, and L. Vittuari (1998b). Comparison between glacier ice velocities inferred from GPS and sequential satellite images. *Annals of Glaciology*, **27**, 54–60.

Frezzotti, M., I. E. Tabacco, and A. Zirizzotti (2000). Ice discharge of eastern Dome C drainage area, Antarctica, determined from airborne radar survey and satellite image analysis. *Journal of Glaciology*, **46**, 253–264.

Frezzotti, M., S. Gandolfi, F. La Marca, and S. Urbini (2002a). Snow dunes and glazed surfaces in Antarctica: New field and remote-sensing data. *Annals of Glaciology*, **34**, 81–88.

Frezzotti, M., S. Gandolfi, and S. Urbini (2002b). Snow megadune in Antarctica: Sedimentary structure and genesis. *Journal of Geophysical Research*, **107**(D18), 1–12.

Fricker, H. A. and L. Padman (2002). Tides on Filchner-Ronne Ice Shelf from ERS radar altimetry. *Geophysical Research Letters*, **29**(12), DOI: 10.1029/2001GL014175.

Fricker, H. A., G. Hyland, R. Coleman, and N. W. Young (2000a). Digital elevation models for the Lambert Glacier–Amery Ice Shelf system, East Antarctica, from ERS-1 satellite radar altimetry. *Journal of Glaciology*, **46**(155), 553–560.

Fricker, H. A., R. C. Warner, and I. Allison (2000b). Mass balance of the Lambert Glacier–Amery Ice Shelf system, East Antarctica: A comparison of computed balance fluxes and measured fluxes. *Journal of Glaciology*, **46**(155), 561–570.

Fricker, H. A., S. Popov, I. Allison, and N. Young (2001). Distribution of marine ice under the Amery Ice Shelf, East Antarctica. *Geophysical Research Letters*, **28**(11), 2241–2244.

Fricker, H. A., N. W. Young, I. Allison, and R. Coleman (2002a). Iceberg calving from the Amery Ice Shelf, East Antarctica. *Annals of Glaciology*, **34**, 241–246.

Fricker, H. A., I. Allison, M. Craven, G. Hyland, A. Ruddell, N. Young, R. Coleman, M. King, K. Krebs, and S. Popov (2002b). Redefinition of the grounding zone of the Lambert Glacier–Amery Ice Shelf system, East Antarctica. *Journal of Geophysical Research*, **107**(B5), DOI: 10.1029/2001JB000383.

Fricker, H. A., J. Bassis, and B. Minster (2004). GLAS observations of Antarctic ice shelf features: The vertical dimension. *EOS, Transactions of the American Geophysical Union*, **85**(47), Fall Meeting Supplement, Abstract C22A-01.

Fricker, H. A., N. W. Young, R. Coleman, J. N. Bassis, and J.-B. Minster (2005). Multi-year monitoring of rift propagation on the Amery Ice Shelf, East Antarctica. *Geophysical Research Letters*, **32**, L02502, DOI: 10.1029/2004GL021036.

Friedmann, A., J. C. Moore, T. Thorsteinsson, and J. Kipfstuhl (1995). A 1200-year record of accumulation from North Greenland ice cores. *Annals of Glaciology*, **21**, 19–25.

Frolich, R. M. and C. S. M. Doake (1998). Synthetic aperture radar interferometry over Rutford Ice Stream and Carlson Inlet, Antarctica. *Journal of Glaciology*, **44**(146), 77–92.

Fu, L.-L. and A. Cazenave (eds.) (2001). *Satellite Altimetry and Earth Sciences* (International Geophysics Series No. 69). Academic Press, San Diego, CA, 463 pp.

Fujii Y. and K. Kusunoki (1982). The role of sublimation and concentration in the formation of ice sheet surface at Mizuho Station, Antarctica. *Journal of Geophysical Research*, **87**, 4293–4300.

Fujii, Y., T. Yamanouchi, K. Suzuki, and S. Tanaka (1987). Comparison of surface conditions of the inland ice sheet, Dronning Maud Land, derived from NOAA AVHRR data with ground observation. *Annals of Glaciology*, **9**, 72–75.

Fukuda, Y., S. Aoki, and K. Doi (2002). Impact of satellite gravity missions on glaciology and Antarctic Earth sciences. *Polar Meteorology and Glaciology*, **16**, 32–41.

Fung, A. K. and M. F. Chen (1981). Emission from an inhomogeneous layer with irregular interfaces. *Radio Science*, **16**, 289–298.

Fung, A. K. and H. J. Eom (1982). Application of a combined rough surface and volume scattering theory to sea ice and snow backscatter. *IEEE Transactions on Geoscience and Remote Sensing*, **GE–20**, 4528–4536.

Furukawa, T. and N. W. Young (1997). Comparison of microwave backscatter measurements with observed roughness of the snow surface in East Queen Maud Land, Antarctica. *Proceedings of 3rd ERS Scientific Symposium, Florence, March 17–21* (ESA SP-414, Vol. 2). ESA, Noordwijk, The Netherlands, pp. 803–808.

Gaiser, P. W., K. M. St. Germain, E. M. Twarog, G. A. Poe, W. Purdy, D. Richardson, W. Grossman, W. Linwood Jones, D. Spencer, G. Golba et al. (2004). The WindSat spaceborne polarimetric microwave radiometer: Sensor description and early orbit performance. *IEEE Transactions on Geoscience and Remote Sensing*, **42**(11), 2347–2361.

Gasch, J., T. Arvidson, and S. N. Goward (2000). Fire and ice: An assessment of Landsat 7/ETM+ acquisitions over glaciers, volcanoes, Antarctica and sea ice. *Proceedings of International Geoscience and Remote Sensing Symposium IGARSS '00*. Institute of Electrical and Electronic Engineers, Piscataway, NJ, Vol. 2, pp. 457–459.

Gasparovich, R. F., R. K. Raney, and R. C. Beal (1999). Ocean remote sensing research and applications at APL. *Johns Hopkins APL Technical Digest*, **20**(4), 600–610.

Gay, M., M. Fily, C. Genthon, M. Frezzotti, H. Oerter, and J. G. Winther (2002). Snow grain-size measurements in Antarctica. *Journal of Glaciology*, **48**(163), 527–535.

Genthon, C. and G. Krinner (1998). Convergence and disposal of energy and moisture on the Antarctic Polar Cap from ECMWF reanalyses and forecasts. *Journal of Climate*, **11**, 1703–1716.

Geudtner, D., P. W. Vachon, K. E. Mattar, and A. L. Gray (1998). Radarsat repeat-pass SAR interferometry. *Proceedings of International Geoscience and Remote Sensing Symposium IGARSS '98, Seattle, WA, 6–10 July 1998*. Institute of Electrical and Electronic Engineers, Piscataway, NJ, pp. 1635–1636.

Gibson, J. K., M. Fiorino, A. Hernandez, P. Kållberg, X. Li, K. Onogo, S. Saarinen, and S. Uppala (1999). *The ECMWF 400-year Re-analysis (ERA-40) Project: Plans and Current Status* (10th Global Change Studies, pp. 369–372). American Meteorological Society, Boston.

Gilbert, R. and E. Domack (2003). Sedimentary record of disintegrating ice shelves in a warming climate, Antarctic Peninsula. *Geochemistry, Geophysics, and Geosystems*, **4**(4), 1038, DOI: 10.1029/2002GC00044.

Gill, A. E. (1982). *Atmosphere–Ocean Dynamics*. Academic Press, New York, 662 pp.

Gill, R. S. (2001). Operational detection of sea ice edges and icebergs using SAR. *Canadian Journal of Remote Sensing*, **27**(5), 411–432.

Giovinetto, M. B. (1963). *Glaciological Studies on the McMurdo-South Pole Traverse, 1960–1961* (Institute of Polar Studies Report, No. 7). Ohio State University, Columbus, OH.

Giovinetto, M. B. and C. R. Bentley (1985). Surface balance in ice drainage systems of Antarctica. *Antarctic Journal of the United States*, **20**, 6–13.

Giovinetto, M. and H. J. Zwally (2000). Spatial distribution of net surface accumulation of the Antarctic Ice Sheet. *Annals of Glaciology*, **31**, 171–176.

Gladstone, R. and G. R. Bigg (2002). Satellite tracking of icebergs in the Weddell Sea. *Antarctic Science*, **14**(3), 278–287.

Gladstone, R., G. R. Bigg, and K. Nicholls (2001). Iceberg trajectory modelling and meltwater injection into the Southern Ocean. *Journal of Geophysical Research*, **106**(9), 19903–19915.

Glen, J. W. and J. G. Paren (1975). The electrical properties of snow and ice. *Journal of Glaciology*, **15**(73), 15–38.

Gloersen, P., W. J. Campbell, D. J. Cavalieri, J. G. Comiso, C. L. Parkinson, and H. J. Zwally (1992). *Arctic and Antarctic Sea Ice, 1978–1987: Satellite Passive-microwave Observations and Analysis* (NASA Special Publication SP511, 290 pp.). NASA, Washington, DC.

Gogineni, S., T. Chuah, C. Allen, K. Jezek, and R. Moore (1998). An improved coherent radar depth sounder. *Journal of Glaciology*, **44**(148), 659–669.

Gogineni, S., D. Tammana, D. Braaten, C. Leuschen, T. Akins, J. Legarsky, P. Kanagaratnam, J. Stiles, C. Allen, and K. Jezek (2001). Coherent radar ice thickness measurements over the Greenland Ice Sheet. *Journal of Geophysical Research*, **106**(D24), 33761–33772.

Gogineni, S., G. Prescott, D. Braaten, C. Allen, K. Jezek, and the PRISM Research Team (2003). Polar radar for ice sheet measurements. *Proceedings of IGARSS '03, Toulouse, France, July*.

Goldstein, R. M., H. Engelhardt, B. Kamb, and R. M. Frolich (1993). Satellite radar interferometry for monitoring ice sheet motion: Application to an Antarctic ice stream. *Science*, **262**, 1525–1530.

Goodwin, I. D. (1988). *Ice Sheet Tpography and Surface Characteristics* (ANARE Research Notes No. 64). Australian National Antarctic Research Expedition, Australian Antarctic Division, Melbourne, 100 pp.

Goodwin, I. D. (1990). Snow accumulation and surface topography in the katabatic zone of eastern Wilkes Land, Antarctica. *Antarctic Science*, **2**(3), 235–242.

Goodwin, I. D. (1991). Snow-accumulation variability from seasonal observations and firn-core stratigraphy, eastern Wilkes Land, Antarctica. *Journal of Glaciology*, **37**(127), 383–387.

Gow, A. J. (1971). On the rates of growth of grains and crystals in south polar firn. *Journal of Glaciology*, **8**, 241–252.

Gray, A. L., K. E. Mattar, and P. W. Vachon (1998). InSAR results from the RADARSAT Antarctic Mapping Mission: Estimation of glacier motion using a simple registration procedure. *IGARSS '98, Proceedings of 18th International Geoscience and Remote Sensing Symposium, July 6–10, Seattle, Washington.* Institute of Electrical and Electronics Engineers, Piscataway, NJ, pp. 1638–1640.

Gray, A. L., K. E. Mattar, and N. Short (1999). Speckle tracking for 2-dimensional ice motion studies in polar regions. *Proceedings of 2nd International Workshop on ERS SAR Interferometry, "FRINGE '99", Liège, Belgium, November 10–12* (ESA SP-478, CD-ROM). ESA, Noordwijk, The Netherlands, pp. 1–8.

Gray, A. L., N. H. Short, K. E. Mattar, and K. C. Jezek (2001). Velocities and ice flux of the Filchner Ice Shelf and its tributaries determined from speckle tracking interferometry. *Canadian Journal of Remote Sensing*, **27**(3), 193–206.

Gray, A. L., N. Short, R. Bindschadler, I. Joughin, L. Padman, P. Vornberger, and A. Khananian (2002). RADARSAT interferometry for grounding-zone mapping. *Annals of Glaciology*, **34**, 269–276.

Green, R. O., J. Dozier, D. Roberts, and T. Painter (2002). Spectral snow-reflectance models for grain-size and liquid-water fraction in melting snow for the solar-reflected spectrum. *Annals of Glaciology*, **34**, 71–73.

Grenfell, T. C., S. G. Warren, and P. C. Mulen (1994). Reflection of solar radiation by the Antarctic snow surface at ultraviolet, visible and near-infrared wavelengths. *Journal of Geophysical Research*, **99**(D9), 18669–18684.

Greuell, W. and C. Genthon (2004). Modelling land ice surface mass balance. In: J. L. Bamber and A. J. Payne (eds.), *Mass Balance of the Cryosphere*. Cambridge University Press, Cambridge, UK, pp. 117–168.

Greuell, W. and W. H. Knap (2000). Remote sensing of the albedo and detection of the slush line on the Greenland ice sheet. *Journal of Geophysical Research*, **105**(D12), 15567–15576.

Greuell, W. and J. Oerlemans (2004). Narrowband-to-broadband albedo conversion for glacier ice and snow: Equations based on modeling and ranges of validity of the equations. *Remote Sensing of Environment*, **89**, 95–105.

Greuell, W., C. H. Reijmer, and J. Oerlemans (2002). Narrowband-to-broadband albedo conversion for glacier ice and snow based on aircraft and near-surface measurements. *Remote Sensing of Environment*, **82**, 48–63.

Grosfeld, K., R. Gerdes, and J. Determann (1997). Thermohaline circulation and interaction between ice shelf cavities and the adjacent ocean. *Journal of Geophysical Research*, **102**(C7), 15595–15610.

Grosfeld, K., M. Schröder, E. Fahrbach, R. Gerdes, and A. Mackensen (2001). How iceberg calving and grounding changes the circulation and hydrography in the Filchner Ice Shelf/Ocean System. *Journal of Geophysical Research*, **106**(C5), 9039–9056.

Gudmundsson, G. H., C. Raymond, and R. Bindschadler (1998). The origin and longevity of flow stripes on Antarctic ice streams. *Annals of Glaciology*, **27**, 145–152.

Haardeng-Petersen, G., K. Keller, C. C. Tscherning, and N. Gundestrup (1998). Modeling the signature of a transponder in altimeter return data and determination of the reflection surface of the ice cap near the GRIP camp, Greenland. *Journal of Glaciology*, **44**(148), 625–633.

Haefliger, M., K. Steffen, and C. Fowler (1993). AVHRR surface temperature and narrow-band albedo comparison with ground measurements for the Greenland Ice Sheet. *Annals of Glaciology*, **17**, 49–54.

Hagen, J. O. and N. Reeh (2004). *In situ* measurement techniques: Land ice. In: J. L. Bamber and A. J. Payne (eds.), *Mass Balance of the Cryosphere*. Cambridge University Press, Cambridge, UK, pp. 11–42.

Haines, B. J., G. H. Born, R. C. Williamson, and C. J. Koblinsky (1994). Application of the GEM-T2 gravity field to altimetric satellite orbit computation. *Journal of Geophysical Research*, **99**(C8), 16237–16254.

Hall, A. W. (2003). The role of surface albedo feedback in climate. *Journal of Climate*, **17**(7), 1550–1568.

Hall, D. K. and J. Martinec (1985). *Remote Sensing of Snow and Ice*. Chapman & Hall, New York, 189 pp.

Hall, D. K., J. P. Ormsby, R. A. Bindschadler, and H. Siddalingaiah (1987). Characterization of snow and ice reflectance zones on glaciers using Landsat Thematic Mapper data. *Annals of Glaciology*, **9**, 1–5.

Hall, D. K., A. T. G. Chang, I. L. Foster, C. S. Benson, and W M. Kovalick (1989). Comparison of *in situ* and Landsat derived reflectance of Alaskan glaciers. *Remote Sensing of Environment*, **28**(1), 23–31.

Hall, D. K., G. A. Riggs, and V. V. Salomonson (1995). Development of methods for mapping global snow cover using Moderate Resolution Imaging Spectroradiometer (MODIS) data. *Remote Sensing of Environment*, **54**(2), 127–140.

Hall, D. K., A. B. Tait, G. A. Riggs, V. V. Salomonson, J. Y. L. Chien, and A. G. Klein (1998). *Algorithm Theoretical Basis Document (ATBD) for the MODIS Snow-, Lake Ice- and Sea Ice-mapping Algorithms* (ATBD MOD10). NASA Goddard Space Flight Center, Greenbelt, MD. Available online at ⟨*http://modis. gsfc.nasa.gov/MODIS/Data/ ATBDs/atbd-mod10.pdf*⟩.

Hall, D. K., G. A. Riggs, V. V. Salomonson, N. E. DiGirolamo, and K. J. Bayr (2002). MODIS snow-cover products. *Remote Sensing of Environment*, **83**, 181–194.

Hall, D. K., J. Key, K. A. Casey, G. A. Riggs, and D. Cavalieri (2004). Sea ice surface temperature product from MODIS. *IEEE Transactions on Geoscience and Remote Sensing*, **42**(5), 1076–1087.

Hallikainen, M. and D. P. Winebrenner (1992). The physical basis of sea-ice remote sensing. In: F. D. Carsey (ed.), *Microwave Remote Sensing of Sea Ice* (AGU Monograph No. 68). American Geophysical Union, Washington, DC, pp. 29–46.

Hamazaki, T. (1999). Overview of the Advanced Land Observing Satellite (ALOS): Its mission requirements, sensors, and a satellite system. *Proceedings of ISPRS Joint Workshop Sensors and Mapping From Space, 27–30 September 1999*. International Society for Photogrammetry and Remote Sensing, Hannover, Germany.

Hambrey, M. J. and J. A. Dowdeswell (1994). Flow regime of the Lambert Glacier–Amery Ice Shelf system, Antarctica: Structural evidence from satellite imagery. *Annals of Glaciology*, **20**, 401–406.

Hamilton, G. S. and V. B. Spikes (2004). Evaluating a satellite altimeter-derived digital elevation model of Antarctica using precision kinematic GPS profiling. *Global and Planetary Change*, **42**, 17–30.

Hamilton, G., S. Arcone, N. Yankielun, and P. Mayewski (2000). Spatial variation in snow accumulation rates investigated using ground penetrating radar and GPS. *EOS, Transactions of the American Geophysical Union*, **81**(19), Spring Meeting Supplement, Abstract S21.

Hamley, T. C. and W. F. Budd (1986). Antarctic iceberg distribution and dissolution. *Journal of Glaciology*, **32**(111), 242–251.

Hanna, E. and P. Valdes (2001). Validation of ECMWF (re)analysis surface climate data, 1979–1998, for Greenland and implications for mass balance modelling of the ice sheet. *International Journal of Climatology*, **21**, 171–195.

Hanna, E., P. Valdes, and J. McConnell (2001). Patterns and variations of snow accumulation over Greenland, 1979–1998, from ECMWF analyses, and their verification. *Journal of Climate*, **14**, 3521–3535.

Hansen, J. and L. Nazarenko (2004). Soot–climate forcing via snow and ice albedos. *Proceedings of the National Academy of Sciences*, **101**, 423–428.

Haran, T. and T. A. Scambos (2004). Co-variations of ice sheet elevation and slope with accumulation, radar backscatter, and temperature in West Antarctica. *EOS Transactions of the American Geophysical Union*, **85**(47), Fall Meeting Suppl., Abstract C33C-0355.

Haran, T. M., M. A. Fahnestock, and T. A. Scambos (2002). De-striping of MODIS optical bands for ice sheet mapping and topography. *EOS, Transactions of the American Geophysical Union*, **83**(7), Fall Meeting Supplement, Abstract C12A-1003.

Harbin, M. (2005). Additional Antarctic acquisition opportunities. *Alaska Satellite Facility News and Notes*, **2**(1), 2. [Alaska Satellite Facility, Fairbanks, AK].

Hardy, R. J., J. L. Bamber, and S. Orford (2000). The delineation of major drainage basins on the Greenland Ice Sheet using a combined numerical modelling and GIS approach. *Hydrological Processes*, **14**(11–12), 1931–1941.

Hartl, P., K. H. Thiel, X. Wu, C. Doake, and J. Sievers (1994). Application of SAR interferometry with ERS-1 in the Antarctic. *Earth Observation Quarterly*, **43**, 1–4.

Haykin, S., E. O. Lewis, R. K. Raney, and J. R. Rossiter (1994). *Remote Sensing of Sea Ice and Icebergs*. John Wiley & Sons, New York, 686 pp.

Harrison, W. D. and A. S. Post (2002). How much do we know about glacier surging? *Annals of Glaciology*, **36**, 1–6.

Hawkins, J. D., S. W. Laxon, and H. A. Phillips (1991). Antarctic tabular iceberg multi-sensor mapping. *Proceedings of International Geoscience and Remote Sensing Symposium IGARSS '91, Helsinki, 3–6 June 1991*. Institute of Electrical and Electronic Engineers, Piscataway, NJ, Vol. 3, pp. 1605–1608.

Hellmer, H. H. (2004). Impact of Antarctic ice shelf melting on sea ice and deep ocean properties, *Geophysical Research Letters*, **31**(10), L10307, DOI :10.1029/2004GL019506.

Hellmer, H. H. and S. S. Jacobs (1992). Ocean interactions with the base of Amery Ice Shelf, Antarctica. *Journal of Geophysical Research-Oceans*, **97**(C12): 20305–20317.

Hellmer, H. H., S. S. Jacobs, and A. Jenkins (1998). Oceanic erosion of a floating Antarctic glacier in the Amundsen Sea. In: S. S. Jacobs and R. F. Weiss (eds.), *Ocean, Ice, and Atmosphere: Interactions at the Antarctic Continental Margin* (AGU Antarctic Research Series No. 75). American Geophysical Union, Washington, DC, pp. 83–99.

Herique, A., W. Kofman, P. Bauer, F. Rémy, and L. Phalippou (2000). A spaceborne ground penetrating radar: MIMOSA. *CEOS SAR Workshop, Toulouse, October 26–29, 1999* (ESA SP-450). ESA, Noordwijk, The Netherlands, pp. 645–650.

Herzfeld, U. C. (2004). *Atlas of Antarctica: Topographic Maps from Geostatistical Analysis of Satellite Radar Altimeter Data*. Springer-Verlag, Berlin, 364 pp.

Higgins, A. K. (1991). North Greenland glacier velocities and calf ice production. *Polarforschung*, **60**(1), 1–23.

Hindmarsh, R. C. A. (1993). Qualitative dynamics of marine ice sheets. In: W. R. Peltier (ed.), *Ice in the Climate System* (NATO ASI Series I, No. 12). Springer-Verlag, Berlin, pp. 67–99.

Hindmarsh, R. C. A. and E. Le Meur (2001). Dynamical processes involved in the retreat of marine ice sheets. *Journal of Glaciology*, **47**(157), 271–282.

Hobbs, P. V. (1974). *Ice Physics*. Clarendon Press, Oxford, UK, 837 pp.

Hoen, E. W. and H. A. Zebker (2000a). Topography-driven variations in backscatter strength and depth observed over the Greenland Ice Sheet with InSAR. *Proceedings of International Geoscience and Remote Sensing Symposium IGARSS '00*. Institute of Electrical and Electronic Engineers, Piscataway, NJ, Vol. 2, pp. 470–472.

Hoen, E. W. and H. A. Zebker (2000b). Penetration depths inferred from interferometric volume decorrelation observed over the Greenland Ice Sheet. *IEEE Transactions on Geoscience and Remote Sensing*, **38**(6), 2571–2583.

Holdsworth, G. (1969). Flexure of a floating ice tongue. *Journal of Glaciology*, **8**(54), 385–397.

Holdsworth, G. (1977). Tidal interaction with ice shelves. *Annales de Géophysique*, **33**(1), 133–146.

Holland, D. M., S. S. Jacobs, and A. Jenkins (2003). Modelling the ocean circulation beneath the Ross Ice Shelf. *Antarctic Science*, **15**(1), 13–23.

Hollinger, J. P., J. L. Pierce, and G. A. Poe (1990). SSM/I instrument evaluation. *IEEE Transactions on Geoscience and Remote Sensing*, **28**(5), 781–790.

Hori, M., T. Aoki, K. Stamnes, B. Chen, and W. Lei (2001). Preliminary validation of the GLI cryosphere algorithms with MODIS daytime data. *Polar Meteorology and Glaciology*, **15**, 1–20.

Houghton, J. T., Y. Ding, D. J. Griggs, M. Noguer, P. Van der Linden, X. Dai, K. Maskell, and C. I. Johnson (eds.) (2001). *Climate Change 2001: The Scientific Basis* (contribution of Working Group 1 to the Third Assessment Report of the Intergovernmental Panel on Climate Change). Cambridge University Press, Cambridge, UK, pp. 639–694.

Howell, C., J. Youden, K. Lane, D. Power, C. Randell, and D. Flett (2004). Iceberg and ship discrimination with Envisat multi-polarisation ASAR. *14th Annual Newfoundland Electrical and Computer Engineering Conference, October 12, St, John's, Newfoundland, Canada*. Available online at ⟨http://necec.engr.mun.ca/ocs/index.php?cf=1⟩.

Hu, B., W. Lucht, and A. H. Strahler (1999). The interrelationship of atmospheric correction of reflectances and surface BRDF retrieval: A sensitivity study. *IEEE Transactions on Geoscience and Remote Sensing*, **37**, 724–738.

Hughes, T. J. (1977). West Antarctic ice streams. *Reviews of Geophysics*, **15**, 1–46.

Hughes, T. (1983). On the disintegration of ice shelves: The role of fracture. *Journal of Glaciology*, **29**, 98–117.

Hughes, T. (1992). On the pulling power of ice streams. *Journal of Glaciology*, **38**(128), 125–151.

Hulbe, C. L. and M. A. Fahnestock (2004). West Antarctic ice stream discharge variability: Mechanism, controls, and pattern of grounding-line retreat. *Journal of Glaciology*, **50**(171), 471–484.

Hulbe, C. L. and D. R. MacAyeal (1999). A new thermodynamical numerical model of coupled ice sheet, ice stream, and ice shelf flow. *Journal of Geophysical Research*, **104**(B11), 25349–25366.

Hulbe, C. L. and A. J. Payne (2001). The contribution of numerical modelling to our understanding of the West Antarctic Ice Sheet. In: R. Alley and R. Bindschadler (eds.), *The West Antarctic Ice Sheet: Behavior and Environment* (AGU Antarctic Research Series No. 77). American Geophysical Union, Washington, DC, pp. 201–219.

Hulbe, C., E. Rignot, and D. MacAyeal (1998). Comparison of ice-shelf creep flow simulations with ice-front of Filchner-Ronne Ice Shelf, Antarctica, detected by SAR interferometry. *Annals of Glaciology*, **27**, 182–186.

Huybrechts, P. (2004). *Antarctica: Modelling*. In: J. L. Bamber and A. J. Payne (eds.), *Mass Balance of the Cryosphere: Observations and Modelling of Contemporary and Future Changes*. Cambridge University Press, Cambridge, UK, pp. 491–523.

Huybrechts, R. and J. de Wolde (1999). The dynamic response of the Greenland and Antarctic Ice Sheets to multiple-century climatic warming. *Journal of Climate*, **12**(8), 2169–2188.

Huybrechts, P. and E. Le Meur (1999). Predicted present-day evolution patterns of ice thickness and bedrock elevation over Greenland and Antarctica. *Polar Research*, **18**(2), 299–306.

Huybrechts, P., C. Mayer, F. Pattyn, and C. J. van der Veen (1999). A comparison of grounding-line treatments in numerical models under simplified conditions. *Geophysical Research Abstracts*, **1**(2), 590.

Huybrechts, P., D. Steinhage, F. Wilhelms, and J. Bamber (2000). Balance velocities and measured properties of the Antarctic Ice Sheet from a new compilation of gridded data for modelling. *Annals of Glaciology*, **30**, 52–60.

Huybrechts, P., J. Gregory, I. Janssens, and M. Wild (2004). Modelling Antarctic and Greenland volume changes during the 20th and 21st centuries forced by GCM time slice integrations. *Global and Planetary Change*, **42**, 83–105.

Hyland, G. and N. Young (1998). Wind-induced directional anisotropy of microwave backscatter and its impacts on imaging of the Antarctic continental snow cover. *Proceedings of the International Geoscience and Remote Sensing Symposium IGARSS '98, Seattle, WA*. Institute of Electrical and Electronic Engineers, Piscataway, NJ, pp. 1988–1990.

Hyvärinen, T. and J. Lammasniemi (1987). Infrared measurement of free-water content and grain size of snow. *Optical Engineering*, **26**, 342–348.

Inoue, J. (1989). Surface drag over the snow surface of the Antarctic plateau: 1. Factors controlling surface drag over the katabatic wind region. *Journal of Geophysical Research*, **94**, 2207–2217.

ITT (1982). *AVHRR/2 Advanced Very High Resolution Radiometer* (technical report). ITT Aerospace/Optical Division for NASA Goddard Space Flight Center, Greenbelt, MD.

Jacobel, R. W. and A. M. Gades (1995). Radar observations of a relict ice stream margin traversing Siple Dome, Antarctica. *Antarctic Journal of the United States*, **XXX**(5), 89–91.

Jacobel, R. W., T. A. Scambos, C. R. Raymond, and A. M. Gades (1996). Changes in the configuration of ice stream flow from the West Antarctic Ice Sheet. *Journal of Geophysical Research*, **101**(B3), 5499–5504.

Jacobel, R. W., T. A. Scambos, N. A. Nereson, and C. F. Raymond (2000). Changes in the margin of Ice Stream C, Antarctica. *Journal of Glaciology*, **46**(152), 102–110.

Jacobs, S. S. (1992). Is the Antarctic Ice Sheet growing? *Nature*, **360**(6399), 29–33.

Jacobs, S. S., D. R. MacAyeal, and J. L. Ardai Jr. (1986). The recent advance of the Ross Ice Shelf, Antarctica. *Journal of Glaciology*, **32**, 464–474.

Jacobs, S. S., H. H. Helmer, C. S. M. Doake, A. Jenkins, and R. M. Frolich (1992). Melting of ice shelves and the mass balance of Antarctica. *Journal of Glaciology*, **38**(130), 375–387.

Jacobs, S. S., H. H. Hellmer, and A. Jenkins (1996). Antarctic Ice Sheet melting in the southeast Pacific. *Geophysical Research Letters*, **23**(9), 957–960.

Jacobson, H. P. and C. F. Raymond (1998). Thermal effects on the location of ice stream margins. *Journal of Geophysical Research*, **103**(B6), 12111–12122.

Jenkins, A. and C. S. M. Doake (1991). Ice–ocean interaction on Ronne Ice Shelf. *Journal of Geophysical Research*, **96**(C1), 791–813.

Jenkins, A. and A. Bombosch (1995). Modeling the effects of frazil ice crystals on the dynamics and thermodynamics of ice shelf water plumes. *Journal of Geophysical Research*, **100**, 6967–6981.

Jenkins, A., D. G. Vaughan, S. S. Jacobs, H. H. Hellmer, and J. R. Keys (1997). Glaciological and oceanographic evidence of high melt rates beneath Pine Island Glacier, West Antarctica. *Journal of Glaciology*, **43**(143), 114–121.

Jenkins, A., H. Corr, K. Nicholls, C. Stewart, and C. Doake (2003). Interactions between ice and ocean near an ice shelf grounding line. *EOS, Transactions of the American Geophysical Union*, Fall Meeting Suppl., Abstract C21A-07.

Jenkins, A., D. Hayes, M. Brandon, Z. Pozzi-Walker, S. Hardy, and C. Banks (2004). *Oceanographic Observations at the Amundsen Sea Shelf Break* (FRISP Report 15, pp. 17–22). Forum for Research into Ice Shelf Processes, Geophysical Institute, University of Bergen, Norway.

Jensen, J. R. (1999). Radar altimeter gate tracking: Theory and extension. *IEEE Transactions on Geoscience and Remote Sensing*, **37**, 651–658.

Jezek, K. C. (1998). Flow variations of the Antarctic Ice Sheet from comparison of modern and historical satellite data. *Proceedings of International Geoscience and Remote Sensing Symposium IGARSS '98*. Institute of Electrical and Electronic Engineers, Piscataway, NJ, Vol. 4, pp. 2240–2242.

Jezek, K. C. (1999). Glaciological properties of the Antarctic ice sheet from Radarsat-1 synthetic aperture radar imagery. *Annals of Glaciology*, **27**, 33–40.

Jezek, K. C. (2002). RADARSAT-1 Antarctic Mapping Project: Change-detection and surface velocity campaign. *Annals of Glaciology*, **34**, 263–268.

Jezek, K. C. and R. B. Alley (1988). Effect of stratigraphy on radar-altimetry data collected over ice sheets. *Annals of Glaciology*, **11**, 60–62.

Jezek, K., J. P. Crawford, R. Bindschadler, M. R. Drinkwater, and R. Kwok (1991). Synthetic aperture radar observations of the Greenland Ice Sheet. *Proceedings of 2nd JPL AIRSAR Workshop, June 7–8, 1990* (JPL TP 90-56). NASA Jet Propulsion Laboratory, Pasadena, CA, pp. 21–28.

Jezek, K. C., M. R. Drinkwater, J. P. Crawford, R. Bindschadler, and R. Kwok (1993). Analysis of synthetic aperture radar data collected over the southwestern Greenland Ice Sheet. *Journal of Glaciology*, **39**(131), 119–132.

Jezek, K., P. Gogineni, and M. Shanableh (1994). Radar measurements of melt zones on the Greenland Ice Sheet. *Geophysical Research Letters*, **21**(1), 33–36.

Jezek, K. C., H. G. Sohn, and K. F. Noltimier (1998a). The Radarsat Antarctic Mapping Project. *Proceedings of International Geoscience and Remote Sensing Symposium IGARSS '98, Seattle, WA*. Institute of Electrical and Electronic Engineers, Piscataway, NJ, pp. 2462–2464.

Jezek, K. C., F. Carsey, J. Crawford, J. Curlander, B. Holt, V. Kaupp, K. Lord, N. Labelle-Hamer, A. Mahmood, P. Ondrus et al. (1998b). Snapshots of Antarctica from Radarsat-1. *Proceedings of International Geoscience and Remote Sensing Symposium IGARSS '98, Seattle, WA*. Institute of Electrical and Electronic Engineers, Piscataway, NJ, Vol. 3, pp. 1428–1430.

Jezek, K. C., H. Liu, Z. Zhao, and B. Li (1999). Improving a digital elevation model of Antarctica using radar remote sensing data and GIS techniques. *Polar Geography*, **23**, 185–200.

Jezek, K., H. Liu, Z. Zhao, and B. Li (2000). Improving a digital elevation model of Antarctica using radar remote sensing data and GIS techniques. *Polar Geography*, **23**(3), 185–200.

Jezek, K. C., R. Carande, N. Labelle-Hamer, K. Farness, and X. Wu (2003). RADARSAT 1 synthetic aperture radar observations of Antarctica: Modified Antarctic Mapping Mission, 2000. *Radio Science*, **38**(4), 8067, DOI: 10.1029/2002RS002643.

Jiabing, S., Z. Junqi, H. Dongmin, and S. Zhaohui (2003). Remote monitoring ice velocities of the Polar Record and the Dark Glaciers. *Chinese Journal of Polar Science*, **14**(2), 117–123.

Jin, Y., F. Gao, C. B. Schaaf, X. Li, A. H. Strahler, C. J. Bruegge, and J. Martonchik (2002). Improving MODIS surface BRDF/albedo retrieval with MISR multiangle observations. *IEEE Transactions on Geoscience and Remote Sensing*, **40**, 1593–1604.

Jin, Z. and J. J. Simpson (1999). Bidirectional anisotropic reflectance of snow and sea ice in AVHRR channel 1 and channel 2 spectral regions: Part I. Theoretical analysis. *IEEE Transactions on Geoscience and Remote Sensing*, **37**, 543–554.

Jin, Z. and J. J. Simpson (2000). Bidirectional anisotropic reflectance of snow and sea ice in AHRR channel 1 and 2 spectral regions: Part II. Correction applied to imagery of snow on sea ice. *IEEE Transactions on Geoscience and Remote Sensing*, **GE-38**(2), 999–1015.

Jiskoot, H., A. Luckman, and T. Murray (2002). *Controls on Surging in East Greenland Derived from a New Glacier Inventory* (Glaciological Data Report GD-30). National Snow and Ice Data Center, Boulder, CO, 62 pp.

Jonas, M. and D. G. Vaughan (1996). ERS-1 SAR mosaic of Filchner-Ronne Shelf ice. In: H. Oerter (ed.), *Filchner-Ronne Ice Shelf Programme (FRISP)* (Report 10). Alfred Wegener Institute, Bremerhaven, Germany, pp. 47–49. Available online at ⟨http://www.gfi.uib.no/forskning/frisp/FRISPRep10.pdf⟩.

Jonsson, S., N. Adam, and H. Bjornsson (1998). Effects of sub-glacial geothermal activity observed by satellite radar interferometry. *Geophysical Research Letters*, **25**, 1059–1062.

Joshi, M., C. J. Merry, K. C. Jezek, and J. F. Bolzan (2001). An edge detection technique to estimate melt duration, season and melt extent on the Greenland Ice Sheet using passive microwave data. *Geophysical Research Letters*, **28**(15), 3497–3500.

Joshi, M. D., J. F. Bolzan, K. C. Jezek, and C. J. Merry (1998). Classification of snow facies on the Greenland Ice Sheet using passive microwave and SAR imagery. *Proceedings of International Geoscience and Remote Sensing Symposium IGARSS '98, 7–10 July 1998, Seattle, WA*. Institute of Electrical and Electronic Engineers, Piscataway, NJ, pp. 1852–1854.

Joughin, I. (2002). Ice-sheet velocity mapping: A combined interferometric and speckle-tracking approach. *Annals of Glaciology*, **34**, 195–201.

Joughin, I. R. and D. MacAyeal (2002). Rift formation and growth on the Ross Ice Shelf. *EOS, Transactions of the American Geophysical Union*, **83**(47), Fall Meeting Supplement, Abstract C52A-02.

Joughin, I. and D. R. MacAyeal (2005). Calving of large tabular icebergs from ice shelf rift systems. *Geophysical Research Letters*, **32**(2), L02501, doi:10.1029/2004GL020978.

Joughin, I. and S. Tulaczyk (2002). Positive mass balance of Ross Ice Streams, West Antarctica. *Science*, **295**, 476–480.

Joughin, I. R., D. P. Winebrenner, and M. A. Fahnestock (1995). Observations of ice-sheet motion in Greenland using satellite radar interferometry. *Geophysical Research Letters*, **22**(5), 571–574.

Joughin, I., R. Kwok, and M. Fahnestock (1996a). Estimation of ice-sheet motion using satellite radar interferometry: Method and error analysis with application to Humboldt Glacier, Greenland. *Journal of Glaciology*, **42**(142), 564–575.

Joughin, I., D. Winebrenner, M. Fahnestock, R. Kwok, and W. Krabill (1996b). Measurement of ice-sheet topography using satellite-radar interferometry. *Journal of Glaciology*, **42**(140), 10–22.

Joughin I., S. Tulaczyk, M. Fahnestock, and R. Kwok (1996c). A mini-surge on the Ryder Glacier, Greenland observed via satellite radar interferometry. *Science*, **274**, 228–230.

Joughin, I., M. Fahnestock, S. Ekholm, and R. Kwok (1997). Balance velocities for the Greenland Ice Sheet. *Geophysical Research Letters*, **24**(23), 3045–3048.

Joughin, I. R., R. Kwok, and M. A. Fahnestock (1998). Interferometric estimation of three-dimensional ice-flow using ascending and descending passes. *IEEE Transactions on Geoscience and Remote Sensing*, **36**(1), 25–37.

Joughin, I., L. Gray, R. Bindschadler, S. Price, D. Morse, C. Hulbe, K. Mattar, and C. Werner (1999a). Tributaries of West Antarctic ice streams revealed by RADARSAT intererometry. *Science*, **286**, 283–286.

Joughin, I., M. Fahnestock, R. Kwok, P. Gogineni, and C. Allen (1999b). Ice flow of the Humboldt, Petermann and Ryder Gletscher, northern Greenland. *Journal of Glaciology*, **45**(150), 231–241.

Joughin, I. R., M. A. Fahnestock, and J. L. Bamber (2000). Ice flow in the northeast Greenland ice stream. *Annals of Glaciology*, **31**, 141–146.

Joughin, I., M. Fahnestock, D. MacAyeal, J. L. Bamber, and P. Gogineni (2001a). Observation and analysis of ice flow in the largest Greenland ice stream. *Journal of Geophysical Research*, **106**(D24), 34021–34034.

Joughin, I., S. Tulaczyk, R. Bindschadler, C. Hulbe, and S. Price (2001b). Variation in flow speed on Ice Stream B: Observation and analysis. *2001 WAIS Workshop, September 19–22*. Available online at ⟨ttp://igloo.gsfc.nasa.gov/wais/⟩.

Joughin, I., S. Tulaczyk, R. Bindschadler, and S. Price (2002). Changes in West Antarctic ice stream velocities: Observation and analysis. *Journal of Geophysical Research*, **107**(B11), 2289, DOI: 10.1029/2001/JB001029.

Joughin, I., E. Rignot, C. E. Rosanova, B. K. Lucchitta, and J. Bohlander (2003). Timing of recent accelerations of Pine Island Glacier, Antarctica. *Geophysical Research Letters*, **30**(13), 1706, doi:10.1029/2003GL017609.

Joughin, I., W. Abdalati, and M. Fahenstock (2004a). Large fluctuations in speed on Greenland's Jakobshavn Isbrae glacier. *Nature*, **432**, 608–610.

Joughin, I., D. G. MacAyeal, and S. Tulaczyk (2004b). Basal shear stress of the Ross ice streams from control method inversions. *Journal of Geophysical Research*, **109**, B09405, DOI: 10.1029/2003JB002960.

Kadosaki, G., T. Yamanouchi, and N. Hirasawa (2002). Temperature dependence of brightness temperature difference of AVHRR infrared split-window channels in the AVHRR. *Polar Meteorology and Glaciology*, **16**, 106–115.

Kaeaeb, A. (2001). Glacier monitoring from ASTER imagery: Accuracy and applications. *EOS, Transactions of the American Geophysical Union*, **82**(47), Fall Meeting Supplement, Abstract IP41A-06.

Kalnay, E., M. Kanamitsu, R. Kistler, W. Collins, D. Deaven, L. Gandin, M. Iredell, S. Saha, G. White, J. Woollen et al. (1996). The NCEP/NCAR 40-year reanalysis project. *Bulletin American Meteorological Society*, **77**, 437–471.

Kamb, B. (2001). Basal zone of the West Antarctic ice streams and its role in lubrication of their rapid motion. In: R. Alley and R. Bindschadler (eds.), *The West Antarctic Ice*

Sheet: Behavior and Environment (AGU Antarctic Research Series No. 77). American Geophysical Union, Washington, DC, pp. 157–199.

Kamb, B., C. Raymond, W. Harrison, H. Engelhardt, K. Echelmeyer, N. Humphrey, M. Brugman, and T. Pfeffer (1985). Glacier surge mechanism: 1982–1983 surge of Variegated Glacier, Alaska. *Science*, **227**, 469–479.

Kanagaratnam, P., S. P. Gogineni, N. Gundestrup, and L. Larsen (2001). High-resolution radar mapping of internal layers at the North Greenland Ice Core Project. *Journal of Geophysical Research*, **106**(D24), 33799–33812.

Kanagaratnam, P., S. P. Gogineni, V. Ramasami, and D. Braaten (2004). A wideband radar for high-resolution mapping of near-surface internal layers in glacial ice. *IEEE Transactions on Geoscience and Remote Sensing*, **42**(3), 433-440.

Kapitsa, A. P., J. K. Ridley, G. de Q. Robin, M. J. Siegert, and I. A. Zotikov (1996). A large deep freshwater lake beneath the ice of central East Antarctica. *Nature*, **381**, 684–686.

Kargel, J. S., M. J. Abrams, M. P. Bishop, A. Bush, G. Hamilton, H. Jiskoot, A. Kääb, H. H. Kieffer, E. M. Lee, F. Paul et al. (2005). Multispectral imaging contributions to Global Land Ice Measurements from Space. *Remote Sensing of Environment*, in press.

Kärkäs, E., H. B. Granberg, K. Kanto, K. Rasmus, C. Lavoie, and M. Leppäranta (2002). Physical properties of the seasonal snow cover in Dronning Maud Land, East Antarctica. *Annals of Glaciology*, **34**, 89–94.

Keller, K., R. Forsberg, and C. S. Nielsen (1997). Kinematic GPS for ice sheet monitoring and SAR interferometry in Greenland. *Proceedings of International Symposium on Kinematic Systems in Geodesy, Geomatics and Navigation, Banff, Canada, June 3–6* (Abstract KIS97). University of Calgary, Canada, pp. 525–528.

Kellogg, T. B. and D. E. Kellogg (1987). Recent glacial history and rapid ice stream retreat in the Amundsen Sea. *Journal of Geophysical Research*, **92**, 8859–8864.

Kenneally, J. P. and T. J. Hughes (2004). Fracture and back stress along the Byrd Glacier flowband on the Ross Ice Shelf. *Antarctic Science*, **16**(3), 345–354.

Kerr, Y. H., J. Font, P. Waldteufel, and M. Berger (2000). The Soil Moisture and Ocean Salinity Mission—SMOS. *ESA Earth Observation Quarterly*, **66**, 18–25.

Key, J. and M. Haefliger (1992). Arctic ice surface temperature retrieval from AVHRR thermal channels. *Journal of Geophysical Research*, **97**, 5885–5893.

Key, J. and A. J. Schweiger (1998). Tools for atmospheric radiative transfer: Streamer and Fluxnet. *Computers and Geosciences*, **24**, 443–451.

Key, J. R., J. B. Collins, C. Fowler, and R. S. Stone (1997). High-latitude surface temperature estimates from thermal satellite data. *Remote Sensing of Environment*, **61**, 302–309.

Key, J. R., X. Wang, J. C. Stroeve, and C. Fowler (2001). Estimating the cloudy-sky albedo of sea ice and snow from space. *Journal of Geophysical Research*, **106**, 12489–12497.

Keys, H. J. R. (1994). Ice giants a chip off the old B9. *Australian Geographic Magazine*, January–March, 22–23.

Keys, H. J. R., S. S. Jacobs, and D. Barnett (1990). The calving and drift of iceberg B-9 in the Ross Sea, Antarctica. *Antarctic Science*, **2**, 246–257.

Keys, H. J. R., S. S. Jacobs, and L. W. Brigham (1998). Continued northward expansion of the Ross Ice Shelf, Antarctica. *Annals of Glaciology*, **27**, 93–98.

Khalsa, S. J. S., M. B. Dyurgerov, T. Khromova, B. H. Raup, and R. G. Barry (2004). Space-based mapping of glacier changes using ASTER and GIS tools. *IEEE Transactions on Geoscience and Remote Sensing*, **42**(10), 2177–2183.

Kidwell, K. (2000). *NOAA KLM User's Guide* (revised edn). National Environmental Satellite Data Information Services (NESDIS), National Climatic Data Center, Boulder, CO. Available online at ⟨www2.ncdc.noaa.gov/docs/klm/index.htm⟩.

King, J. C. (1994a). Recent climate variability in the vicinity of the Antarctic Peninsula. *International Journal of Climatology*, **14**, 357–369.

King, E. C. (1994b). Observations of a rift in the Ronne Ice Shelf, Antarctica. *Journal of Glaciology*, **40**, 187–189.

King, J. C. and J. C. Comiso (2003). The spatial coherence of interannual temperature variations in the Antarctic Peninsula. *Geophysical Research Letters*, **30**, 1040, DOI: 10.129/2002GL015580.

King, J. C. and J. Turner (1997). *Antarctic Meteorology and Climatology*. Cambridge University Press, Cambridge, UK, 409 pp.

King, J. C., J. Turner, G. J. Marshall, W. M. Connolley, and T. A. Lachlan-Cope (2003). Antarctic Peninsula climate variability and its causes as revealed by analysis of instrument records. In: E. Domack, A. Leventer, A. Burnett, R. Bindschadler, P. Conley, and M. Kirby (eds.), *Antarctic Peninsula Climate Variability: Historical and Palaeoenvironmental Perspective* (AGU Antarctic Research Series No. 79). American Geophysical Union, Washington, DC, pp. 17–30.

King, M., L. Nguyen, R. Coleman, and P. J. Morgan (2000). Strategies for high precision processing of GPS measurements with application to the Amery Ice Shelf, East Antarctica. *GPS Solutions*, **4**(1), 2–12.

Kistler, R., E. Kalnay, W. Collins, S. Saha, G. White, J. Woollen, M. Chelliah, W. Ebisuzaki, M. Kanamitsu, V. Kousky et al. (2001). The NCEP-NCAR 50-year reanalysis: Monthly means CD-ROM and documentation. *Bulletin of the American Meteorological Society*, **82**, 247–267.

Klein, A. G. and D. K. Hall (1999). Snow albedo determination using the NASA MODIS instrument. *Proceedings of 56th Eastern Snow Conference, June 2–4, Fredericton, New Brunswick, Canada*, pp. 77–85.

Klein, A. G. and J. Stroeve (2002). Development and validation of a snow albedo algorithm for the MODIS instrument. *Annals of Glaciology*, **34**, 45–52.

Klein, A. G., D. K. Hall, and A. W. Nolin (2000). Development of a prototype snow albedo algorithm for the NASA MODIS instrument. *Proceedings of Eastern Snow Conference, 57th Annual Meeting, May 17–19, Syracuse, New York*, pp. 143–157.

Kluever, C. A., B. J. Haines, C. H. Davis, and Y. T. Yoon (1997). Orbit error analysis for ice-sheet growth studies using radar altimetry. *EOS, Transactions of the American Geophysical Union*, **78**(17), S103.

Knap, W. H. and J. Oerlemans (1996). The surface albedo of the Greenland Ice Sheet: satellite-derived and *in situ* measurements in the Sondre Stromfjord area during the 1991 melt season. *Journal of Glaciology*, **42**(141), 364–374.

Knap, W. H. and C. H. Reijmer (1998). Anisotropy of the reflected radiation field over melted glacier ice: Measurements in Landsat TM bands 2 and 4. *Remote Sensing of Environment*, **65**, 93–104.

Koblinsky, C., M. Rienecker, D. Adamec, W. Abdalati, and E. Lindstrom (2001). Oceans and ice: The slow dance of a complex system. *Proceedings of International Geoscience and Remote Sensing Symposium IGARSS '01, Sydney Australia, 9–13 July 2001*. Institute of Electrical and Electronic Engineers, Piscataway, NJ, Vol. 6, pp. 2859–2862.

Koepke, P. (1989). Removal of atmospheric effects from AVHRR albedos. *Journal of Applied Meteorology*, **28**, 1341–1348.

König, M., J.-G. Winther, and E. Isaksson (2001). Measuring snow and glacier ice properties from satellite. *Reviews of Geophysics*, **39**(1), 1–28.

König, M., J. Wadham, J.-G. Winther, J. Kohler, and A.-M. Nuttall (2002). Detection of superimposed ice on glaciers Kongsvegen and midre Lovenbreen, Svalbard, using SAR satellite imagery. *Annals of Glaciology*, **34**, 335–342.

Krabill, W. B., R. H. Thomas, C. F. Martin, R. N. Swift, and E. B. Frederick (1995). Accuracy of airborne laser altimetry over the Greenland Ice Sheet. *International Journal of Remote Sensing*, **16**, 1211–1222.

Krabill, W., E. Frederick, S. Manizade, C. Martin, J. Sonntag, R. Swift, R. Thomas, W. Wright, and J. Yungel (1999). Rapid thinning of parts of the southern Greenland Ice Sheet. *Science*, **283**, 1522–1524.

Krabill, W., W. Abdalati, E. Frederick, S. Manizade, C. Martin, J. Sonntag, R. Swift, R. Thomas, W. Wright, and J. Yungel (2000). Greenland Ice Sheet: High-elevation balance and peripheral thinning. *Science*, **289**, 428–430.

Krabill, W. B., W. Abdalati, E. B. Frederick, S. Manizade, C. F. Martin, J. G. Sonntag, R. N. Swift, R. H. Thomas, and J. G. Yungel (2002). Aircraft laser altimetry measurement of elevation changes of the Greenland Ice Sheet: Technique and accuracy assessment. *Journal of Geodynamics*, **34**, 357–376.

Kvaran, G., T. Scambos, and M. A. Fahnestock (1996). Improved AVHRR imagery of ice sheets via data cumulation: A theoretical and empirical evaluation. *EOS, Transactions of the American Geophysical Union*, **77**, F193.

Kwok, R. and J. C. Comiso (2002). Spatial patterns of variability in Antarctic surface temperature: Connections to the Southern Hemisphere Annular Mode and the Southern Oscillation. *Geophysical Research Letters*, **29**(14), DOI: 10.1029/2002GL015415.

Kwok, R. and M. A. Fahnestock (1996). Ice sheet motion and topography from radar interferometry. *IEEE Transactions on Geoscience and Remote Sensing*, **34**(1), 189–200.

Lane, K., D. Power, C. Randell, and D. Flett (2003). Validation of ENVISAT Synthetic Aperture Radar for Iceberg Detection. *13th Annual Newfoundland Electrical and Computer Engineering Conference, November 12, St. John's, Newfoundland, Canada*. Available online at ⟨http://www.ieee.nfld.net/NECEC03/abstracts/session2a.htm⟩.

Lang, R. M., B. R. Leo, and R. L. Brown (1984). Observations on the growth process and strength characteristics of surface hoar. *Proceedings of International Snow Science Workshop*. ISSW Workshop Committee, Aspen, CO, pp. 188–195.

Lange, M. and H. Kohnen (1985). Ice front fluctuations in the eastern and southern Weddell Sea. *Annals of Glaciology*, **6**, 187–191.

Larour, E., E. Rignot, and D. Aubry (2004). Modelling of rift propagation on Ronne Ice Shelf, Antarctica, and sensitivity to climate change. *Geophysical Research Letters*, **31**(16), L16404, DOI: 10.1029/2004GL020077.

Lazzara, M. A., K. C. Jezek, T. A. Scambos, D. R. MacAyeal, and C. J. Van der Veen (1999). On the recent calving of icebergs from the Ross Ice Shelf. *Polar Geography*, **23**(3), 201–212.

Layberry, R. L. and J. L. Bamber (2001). A new ice thickness and bed data set for the Greenland Ice Sheet: 2. Relationship between dynamics and basal topography. *Journal of Geophysical Research*, **106**(D24), 33781–33788.

Leberl, F. W. (1990). *Radargrammetric Image Processing*. Artech House, Norwood, MA, 595 pp.

Leberl, F. W. (1998). Radargrammetry. In: F. Henderson and A. J. Lewis (eds.), *Principles and Applications of Imaging Radar*. In: R. A. Ryerson (editor-in-chief), *Manual of Remote Sensing* (3rd edn., Vol. 2). John Wiley & Sons, New York, pp. 183–269.

Ledroit, M., F. Rémy, and J.-F. Minster (1993). Observations of the Antarctic ice sheet with the Seasat scatterometer: Relation to katabatic-wind intensity and direction. *Journal of Glaciology*, **39**(132), 385–396.

Lefauconnier, B., J. O. Hagen, and J. P. Rudant (1994). Flow speed and calving rate of Kongsbreen Glacier Svalbard, using SPOT images. *Polar Research*, **13**(1), 59–65.

Legrésy, B. and F. Rémy (1997). Altimetric observations of surface characteristics of the Antarctic Ice Sheet. *Journal of Glaciology*, **43**(144), 265–275.

Legrésy, B. and F. Rémy (1998). Using the temporal variability of the satellite radar altimetric observations to map surface properties of the Antarctic Ice Sheet. *Journal of Glaciology*, **44**, 197–206.

Legrésy, B., F. Rémy, and P. Schaeffer (1999). Different ERS altimeter measurements between ascending and descending tracks caused by wind induced features over ice sheets. *Geophysical Research Letters*, **26**, 2231–2234.

Legrésy, B., Rignot, E., and I. E. Tabacco (2000). Constraining ice dynamics at Dome C, Antarctica, using remotely sensed measurements. *Geophysical Research Letters*, **27**(21), 3493–3496.

Legrésy, B., F. Papa, F. Rémy, G. Vinay, M. van den Bosch, an O.-Z. Zanife (2005). Envisat radar altimeter measurements over continential surfaces and ice caps using the ICE-2 retracking algorithm. *Remote Sensing of Environment*, **95**, 150–163.

Le Provost, C., F. Lyard, J. M. Molines, M. L. Genco, and F. Rabilloud (1998). A hydrodynamic ocean tide model improved by assimilating a satellite altimeter-derived data set. *Journal of Geophysical Research*, **103**(C3), 5513–5529.

Leroux, C. and M. Fily (1998). Modelling the effect of sastrugi on snow reflectance. *Journal of Geophysical Research*, **103**(E11), 25779–25788.

Leroux, C., J. L. Deuze, P. Goloub, C. Sergent, and M. Fily (1998a). Ground measurements of the polarized bidirectional reflectance of snow in the middle infrared spectral domain: Comparisons with model results. *Journal of Geophysical Research*, **103**(D16), 19721–19731.

Leroux, C., J. Lenoble, G. Brogniez, J. W. Hovenier, and J. F. De Haan (1998b). A model for the bidirectional polarized reflectance of snow. *Journal of Quantitative Spectroscopy and Radiative Transfer*, **61**(3), 273–285.

Leroy, M. and J. L. Roujean (1994). Sun and view angle corrections on reflectances derived from NOAA/AVHRR data. *IEEE Transactions on Geoscience and Remote Sensing*, **32**, 684–697.

Leroy, M., J. L. Deuzé, F. M. Bréon, O. Hautecoeur, M. Herman, J. C. Buriez, D. Tanré, S. Bouffiès, P. Chazette, and J. L. Roujean (1997). Retrieval of atmospheric properties and surface bidirectional reflectances over the land from POLDER/ADEOS. *Journal of Geophysical Research*, **102**, 17023–17037.

Lewis, E. L. and R. G. Perkin (1986). Ice pumps and their rates. *Journal of Geophysical Research*, **91**, 11756–11762.

Li, W., K. Stamnes, B. Chen, and X. Xiong (2001). Retrieval of the depth dependence of snow grain size from near-infrared radiances at multiple wavelengths. *Geophysical Research Letters*, **28**(9), 1699–1702.

Liang, S. (2001). Narrowband to broadband conversion of land surface albedo. I: Algorithms. *Remote Sensing of Environment*, **76**, 213–238.

Lichey, C. and H. H. Hellmer (2001). Modeling giant iceberg drift under the influence of sea ice. *Journal of Glaciology*, **47**(158), 452–460.

Lindstrom, D. and T. J. Hughes (1984). Downdraw of the Pine Island Bay drainage basins of the West Antarctic Ice Sheet. *Antarctic Journal of the United States*, **19**(5), 56–58.

Lindstrom, D. and D. Tyler (1984). Preliminary results of Pine Island and Thwaites Glaciers study. *Antarctic Journal of the United States*, **19**(5), 53–55.

Lingle, C. S. and D. N. Covey (1998). Elevation changes on the East Antarctic Ice Sheet, 1978–1993 from satellite radar altimetry: A preliminary assessment. *Annals of Glaciology*, **27**, 7–18.

Lingle, C. S., L. Li-Her, H. J. Zwally, and T. C. Seiss (1994). Recent elevation increase on Lambert Glacier, Antarctica, from orbit cross-over analysis of satellite radar altimetry. *Annals of Glaciology*, **20**, 26–32.

Lisitzin, A. P. (2002). *Sea-Ice and Iceberg Sedimentation in the Ocean: Recent and Past*. Springer-Verlag, Berlin, 563 pp.

Liston, G. E. and J.-G. Winther (2004). *Antarctic Surface and Sub-surface Snow and Ice Melt Fluxes* (Terra Nostra, Schriften der Alfred-Wegener-Stiftung, 2004/4, Abstract Volume, Abstract S8/O14). Alfred Wegener Institute for Polar and Marine Research, Bremerhaven, Germany, pp. 218–219.

Liston, G. E., O. Bruland, J.-G. Winther, H. Elvehøy, and K. Sand (1999). Meltwater production in Antarctic blue-ice areas: Sensitivity to changes in atmospheric forcing. *Polar Research*, **18**(2), 283–290.

Liu, H., K. C. Jezek, and B. Li (1999). Development of an Antarctic digital elevation model by integrating cartographic and remotely sensed data: A geographic information system based approach. *Journal of Geophysical Research*, **104**(B10), 23199–23214.

Liu, H., K. Jezek, B. Li, and Z. Zhao (2001). *Radarsat Antarctic Mapping Project Digital Elevation Model Version 2* (digital media). National Snow and Ice Data Center, Boulder, CO.

Long, D. G. (2003). Reconstruction and resolution enhancement techniques for microwave sensors. In: C. H. Chen (ed.), *Frontiers of Remote Sensing Information Processing*. World Scientific, Hackensack, NJ, pp. 255–281.

Long, D. G. and M. R. Drinkwater (1994). Greenland ice sheet surface properties observed by the Seasat-A Scatterometer at enhanced resolution. *Journal of Glaciology*, **40**(135), 213–230.

Long, D. G. and M. R. Drinkwater (1999). Cryosphere applications of NSCAT data. *IEEE Transactions Geoscience and Remote Sensing*, **37**(3), 1671–1684.

Long, D. G. and M. R. Drinkwater (2000). Azimuth variation in microwave scatterometer and radiometer data over Antarctica. *IEEE Transactions on Geoscience and Remote Sensing*, **38**(4), 1857–1870.

Long, D. G., P. J. Hardin, and P. T. Whiting (1993). Resolution enhancement of spaceborne scatterometer data. *IEEE Transactions on Geoscience and Remote Sensing*, **31**(3), 700–715.

Long, D. G., M. R. Drinkwater, B. Holt, S. Saatchi, and C. Bertoia (2001). Global ice and land climate studies using scatterometer image data. *EOS, Transactions of the American Geophysical Union*, **82**(43), 503 (23 October 2001). Includes EOS Electronic Supplement at ⟨http://www.agu.org/eos_elec/010126e.html⟩.

Long, D. G., J. Ballantyne, and C. Bertoia (2002). Is the number of Antarctic icebergs really increasing? *EOS, Transactions of the American Geophysical Union*, **83**(42), 469–474.

Long, D. G., M. W. Spencer, and E. G. Njoku (2005). Spatial resolution and processing tradeoffs for HYDROS: Application of reconstruction and resolution enhancement techniques. *IEEE Transactions on Geoscience and Remote Sensing*, **43**(1), 3–12.

Löscher, B. M., H. J. W. de Baar, J. T. M. de Jong, C. Veth, and F. Dehairs (1997). The distribution of Fe in the Antarctic Circumpolar Current. *Deep-Sea Research II*, **44**(1/2), 143–187.

Loset, S. and T. Carstens (1993). Production of icebergs and observed extreme drift speeds in the Barents Sea. *12th Conference on Port and Ocean Engineering under Arctic Conditions—POAC 93, Hamburg, Germany, August 17–20*. HSVA, Hamburg, Germany, pp. 425–438.

Lubin, D. and R. A. Massom (2005). *Polar Remote Sensing: Volume 1. Atmosphere and Oceans*. Springer-Praxis, Chichester, UK.

Lucchitta, B. K. and H. M. Ferguson (1986). Antarctica: Measuring glacier velocity from satellite images. *Science*, **234**, 1105–1108.

Lucchitta, B. K. and C. E. Rosanova (1997). Velocities of Pine Island and Thwaites Glaciers, West Antarctica, from ERS-1 SAR images. *Proceedings of 3rd ERS Scientific Symposium, March 17–21, Florence, Italy* (ESA SP-414). ESA, Frascati, Italy, pp. 349–357.

Lucchitta, B. and C. Rosanova (1998). Retreat of northern margins of George VI and Wilkins ice shelves, Antarctic Peninsula. *Annals of Glaciology*, **27**, 41–46.

Lucchitta, B. K., H. M. Ferguson, F. J. Schafer, J. G. Ferrigno, and R. S. Williams Jr. (1989). Antarctic glacier velocities from Landsat images. *Antarctic Journal of the United States*, **24**, 106–107.

Lucchitta, B. K., L. M. Bertolini, J. G. Ferrigno, and R. S. Williams Jr. (1991). Monitoring the dynamics of the Antarctic coastline with Landsat images. *Antarctic Journal of the United States*, **26**, 316–317.

Lucchitta, B. K., K. F. Mullins, A. L. Allison, and J. G. Ferrigno (1993). Antarctic Glacier–Tongue velocities from Landsat images: First results. *Annals of Glaciology*, **17**, 356–366.

Lucchitta, B. K., C. E. Smith, J. A. Bowell, and K. F. Mullins (1994). Velocities and mass balance of Pine Island Glacier, West Antarctica, derived from ERS-1 SAR images. *Proceedings of 2nd ERS-1 Symposium: Space at the Service of Our Environment, Hamburg, Germany, October 11–14* (ESA SP-361). ESA, Noordwijk, The Netherlands, pp. 147–151.

Lucchitta, B. K., C. E. Rosanova, and K. F. Mullins (1995). Velocities of Pine Island Glacier, West Antarctica, from ERS-1 SAR images. *Annals of Glaciology*, **21**, 277–283.

Lucht, W., C. B. Schaaf, and A. H. Strahler (2000). An algorithm for the retrieval of albedo from space using semi-empirical BRDF models. *IEEE Transactions on Geoscience and Remote Sensing*, **38**, 977–998.

Luckman, A. J. and T. Murray (1999). Observations of surge-type glaciers in Svalbard using ERS SAR interferometry. *Proceedings of ESA Fringe '99 Meeting*.

Luckman, A., T. Murray, and T. Strozzi (2002). Satellite flow evolution throughout a glacier surge measured by satellite radar interferometry. *Geophysical Research Letters*, **29**(23), 2095, DOI: 10.1029/2001GL014570.

Lythe, M. B., D. G. Vaughan, and the BEDMAP Consortium (2001). BEDMAP: A new ice thickness and subglacial topographic model of Antarctica. *Journal of Geophysical Research*, **106**(B6), 11335–11351.

Lytle, V. I. and K. C. Jezek (1994). Dielectric permittivity and scattering measurements of Greenland firn at 26.5–40 GHz. *IEEE Transactions on Geoscience and Remote Sensing*, **32**(2), 290–295.

MacAyeal, D. R. (1985). Tidal rectification below the Ross Ice Shelf, Antarctica. In: S. Jacobs (ed.), *Oceanology of the Antarctic Continental Shelves* (AGU Antarctic Research Series No. 43). American Geophysical Union, Washington, DC, pp. 133–144.

MacAyeal, D. R. (1989). Large-scale flow over a viscous basal sediment: Theory and application to Ice Stream B, Antarctica. *Journal of Geophysical Research*, **94**(B4), 4071–4088.

MacAyeal, D. R., R. A. Bindschadler, and T. A. Scambos (1995). Basal friction of Ice Stream E, West Antarctica. *Journal of Glaciology*, **41**(138), 247–262.

MacAyeal, D., E. Rignot, and C. Hulbe (1998). Ice-shelf dynamics near the front of the Filchner-Ronne Ice Shelf, Antarctica, revealed by SAR interferometry: Model/interferogram comparison. *Journal of Glaciology*, **44**, 419–428.

MacAyeal, D. R., L. Padman, M. R. Drinkwater, M. Fahnestock, T. T. Gotis, A. L. Gray, B. Kerman, M. Lazzara, E. Rignot, T. Scambos, and C. Stearns (2002). *Effects of Rigid Body Collisions and Tide-forced Drift on Large Tabular Icebergs of the Antarctic*. Available online at ⟨http://geosci.uchicago.edu/~drm7/research/download.html⟩.

MacAyeal, D. R., T. A. Scambos, C. L. Hulbe, and M. A. Fahnestock (2003). Catastrophic ice shelf breakup by an ice shelf fragment capsize mechanism. *Journal of Glaciology*, **49**(164), 22–36.

MacDonald, T. R., J. G. Ferrigno, R. S. Williams Jr., and B. K. Lucchitta (1989). Velocities of Antarctic outlet glaciers determined from sequential Landsat images. *Antarctic Journal of the United States*, **24**, 105–106.

Madsen, S. N., J. J. Mohr, and N. Reeh (1999). Mapping Greenland by ERS-1/-2 InSAR for ice mass balance and dynamic studies. *Proceedings of Fringe '99, November 10–12, Liège, Belgium*. Available online at ⟨http://www.esa.int/fringe99⟩.

Mahesh, A., J. D. Spinhirne, D. P. Duda, and E. W. Eloranta (2002). Atmospheric multiple scattering effects on GLAS altimetry: Part II. Analysis of expected errors in Antarctic altitude measurements. *IEEE Transactions on Geoscience and Remote Sensing*, **40**(11), 2353–2362.

Mahesh, A., M. A. Gray, S. P. Palm, W. D. Hart, and J. D. Spinhirne (2004). Passive and active detection of clouds: Comparisons between MODIS and GLAS observations. *Geophysical Research Letters*, **31**, L04108, DOI: 10.1029/2003GL018859.

Makinson, K. and K. W. Nicholls (1999). Modeling tidal currents beneath Filchner-Ronne Ice Shelf and on the adjacent continental shelf: Their effect on mixing and transport. *Journal of Geophysical Research*, **104**, 13449–13465.

Male, D. H. (1980). Seasonal snow cover. In: S. C. Colbeck (ed.), *Dynamics of Snow and Ice Masses*. Academic Press, New York, pp. 305–395.

Manson, R., R. Coleman, P. Morgan, and M. King (2000). Ice velocities of the Lambert Glacier from static GPS. *Earth Planets and Space*, **52**(11), 1031–1056.

Martin, T. V., H. J. Zwally, A. C. Brenner, and R. A. Bindschadler (1983). Analysis and retracking of continental ice sheet radar altimeter waveforms. *Journal of Geophysical Research*, **88**, 1608–1616.

Martonchik, J. V., C. J. Bruegge, and A. H. Strahler (2000). A review of reflectance nomenclature used in remote sensing. *Remote Sensing Reviews*, **19**, 9–20.

Maslanik, J. A., J. Key, and R. G. Barry (1989). Merging AVHRR and SMMR data for remote sensing of ice and cloud in the polar region. *International Journal of Remote Sensing*, **10**(10), 1691–1696.

Maslanik, J., C. Fowler, J. Key, T. Scambos, T. Hutchinson, and W. Emery (1998). AVHRR-based Polar Pathfinder products for modeling applications. *Annals of Glaciology*, **25**, 388–392.

Massom, R. (1991). *Satellite Remote Sensing of Polar Regions: Applications, Limitations and Data Availability*. Belhaven Press, London/Lewis Publishers (CRC Press), Boca Raton, FL, pp. 307.

Massom, R. A. (2003). Recent iceberg calving events in the Ninnis Glacier region, East Antarctica. *Antarctic Science*, **15**(2), 303–313.

Massom, R. A., P. T. Harris, K. J. Michael, and M. J. Potter (1998). The distribution and formative processes of latent heat polynyas in East Antarctica. *Annals of Glaciology*, **27**, 420–426.

Massom, R. A., K. L. Hill, V. I. Lytle, A. P. Worby, M. J. Paget, and I. Allison (2001). Effects of regional fast-ice and iceberg distributions on the behaviour of the Mertz Glacier polynya, East Antarctica. *Annals of Glaciology*, **33**, 391–398.

Massom R. A., M. J. Pook, J. C. Comiso, N. Adams, J. Turner, T. Lachlan-Cope, and T. T. Gibson (2004). Precipitation over the Interior East Antarctic ice sheet related to mid-latitude blocking-high activity. *Journal of Climate*, **17**(10), 1914–1928.

Matsuoka, T., S. Mae, H. Fukazawa, S. Fujita, and O. Watanabe (1998). Microwave dielectric properties from Dome Fuji, Antarctica. *Geophysical Research Letters*, **25**, 1573–1576.

Mätzler, C. (1987). Applications of the interaction of microwaves with the natural snow cover. *Remote Sensing Reviews*, **2**, 259–392.

Mätzler, C. (1996). Microwave permittivity of dry snow. *IEEE Transactions on Geoscience and Remote Sensing*, **34**(2), 573–581.

Mätzler, C. (1998). Microwave properties of snow and ice. In: B. Schmitt, C. De Bergh, and M. Festou (eds.), *Solar System Ices*. Kluwer Academic, Dordrecht, The Netherlands, pp. 241–257.

Mätzler, C. (2001). Applications of SMOS over terrestrial ice and snow. *3rd SMOS Workshop, DLR, Oberpfaffenhofen, Germany, December*. ESA, Noordwijk, The Netherlands.

Mätzler, C. and E. Schanda (1984). Snow mapping with active microwave sensors. *International Journal of Remote Sensing*, **5**(2), 409–422.

Mätzler, C. and U. Wegmüller (1987). Dielectric properties of freshwater ice at microwave frequencies. *Journal of Physics D: Applied Physics*, **20**, 1623–1630. Errata, **21**, 1660 (1988).

Mätzler, C. and A. Wiesmann (1999). Extension of the microwave emission model of layered snowpacks to coarse-grained snow. *Remote Sensing of Environment*, **70**(3), 317–325.

Mätzler, C., R. O. Ramseier, and E. Svendsen (1984). Polarization efects in sea-ice signatures. *IEEE Journal of Oceanic Engineering*, **9**(5), 333–338.

Mayewski, P. A. and I. Goodwin (1997). *ITASE Science and Implementation Plan* (Joint PAGES/GLOCHANT Report. Available online at ⟨*http://www.ume.maine.edu/itase/*⟩.

McConnell, J. R., R. J. Arthern, E. Mosley-Thompson, C. H. Davis, R. C. Bales, R. Thomas, J. F. Burkhart, and J. D. Kyne (2000a). Changes in Greenland ice-sheet elevation attributed primarily to snow-accumulation variability. *Nature*, **406**, 877–879.

McConnell, J. R., E. Mosley-Thompson, D. H. Bromwich, R. C. Bales, and J. D. Kyne (2000b). Interannual variations of snow accumulation on the Greenland Ice Sheet (1985–1996): New observations versus model predictions. *Journal of Geophysical Research*, **105**(D3), 4039–4046.

McConnell, J. R., G. Lamorey, E. Hanna, E. Mosley-Thompson, R. C. Bales, D. Belle-Oudry, and J. D. Kyne (2001). Annual net snow accumulation over southern Greenland from 1975 to 1998. *Journal of Geophysical Research*, **106**(D24), 33827–33838.

McDonald, R. A. (1995a). Corona: Success for space reconnaissance—A look into the Cold War and a revolution in intelligence. *Photogrammetric Engineering and Remote Sensing*, **61**, 321–325.

McDonald, R. A. (1995b). Opening the Cold War sky to the public: Declassifying satellite reconnaissance imagery. *Photogrammetric Engineering and Remote Sensing*, **61**(4), 380–390.

McIntyre, N. F. (1986). The Antarctic ice sheet topography and surface bedrock relationship. *Annals of Glaciology*, **8**, 124–128.

Meier, M. F. (1994). Columbia Glacier during rapid retreat: Interactions between glacier flow and iceberg calving dynamics. In: N. Reeh (ed.), *Proceedings of Workshop on the Calving Rate of West Greenland Glaciers in Response to Climate Change*. Danish Polar Center, Copenhagen, 171 pp.

Mercer, J. H. (1968). Antarctic ice and Sangamon Sea level. *International Association of Scientific Hydrology Symposia*, **79**, 217–225.

Mercer, J. H. (1978). West Antarctic ice sheet and CO_2 greenhouse effect: A threat of disaster. *Nature*, **271**, 321–325.

Meredith, M. P., R. A. Locarnini, K. A. Van Scoy, A. J. Watson, K. J. Heywood, and B. A. King (2000). On the sources of Weddell Gyre Antarctic Bottom Water. *Journal of Geophysical Research*, **105**, 1093–1104.

Merry, C. J. and I. M. Whillans (1993). Ice-flow features on Ice Stream B, Antarctica, revealed by SPOT HRV imagery. *Journal of Glaciology*, **39**(133), 515–527.

Merson, R. H. (1989). An AVHRR mosaic image of Antarctica. *International Journal of Remote Sensing*, **10**(4/5), 669–674.

Michel, R. and E. Rignot (1999). Flow of Glaciar Moreno, Argentina, from repeat-pass Shuttle Imaging Radar images: Comparison of the phase correlation method with radar interferometry. *Journal of Glaciology*, **45**(149), 93–100.

Mika, A. M. (1997). Three decades of Landsat instruments. *Photogrammetric Engineering and Remote Sensing*, **LXIII**(7), 839–852.

Minnett, P. J. (2001). Satellite remote sensing: Sea surface temperatures. In: J. Steele, S. Thorpe and K. Turekian (eds.), *Encyclopaedia of Ocean Sciences*. Academic Press, London, pp. 2552–2563.

Mitchell, D., J. Sauber, D. Harding, C. Carabajal, W. Krabill, S. Manizade, and J. Bufton (2003). Expected ICESat measurement of glacier elevation change. *EGS-AGU-EUG Joint Assembly, Nice, France, 6–11 April 2003* (Abstract EAE03-A-04398). Copernicus Online Service and Information System (COSIS), Katlenburg-Lindau, Germany.

Mohr, J. J. and S. N. Madsen (1999). Error analysis for interferometric SAR measurements of ice sheet flow. *Proceedings of International Geoscience and Remote Sensing Symposium IGARSS '99, 28 June–2 July 1999, Hamburg, Germany*. Institute of Electrical and Electronic Engineers, Piscataway, NJ, Vol. 1, pp. 98–100.

Mohr, J. J. and N. Reeh (2002). Glacier surface velocity measurements from radar interferometry and the principle of mass conservation. *Proceedings of IGARSS 2002, June 22–28, Toronto*.

Mohr, J. J., N. Reeh, and S. N. Madsen (1998). Three-dimensional glacial flow and surface elevation measured with radar interferometry. *Nature*, **391**(6664), 273–276.

Mondet, J. and M. Fily (1999). The reflectance of rough snow surfaces in Antarctica from POLDER/ADEOS remote sensing data. *Geophysical Research Letters*, **26**(23), 3477–3480.

Moore J. C. and S. Fujita (1993). Dielectric properties of ice containing acid and salt impurity at microwave and low frequencies. *Journal of Geophysical Research*, **98**, 9769–9780.

Moreira, A. (2003). TerraSAR-X upgrade to a fully polarimetric imaging mode. *Proceedings of ESA Workshop, POLInSAR: Applications of SAR Polarimetry and Polarimetric Interferometry, January 14–16* (ESA SP-529). ESA European Space Research Institute (ESRIN), Frascati, Italy. Available online at ⟨http://www.earth.esa.int/polinsar⟩.

Morgan, V. I., I. D. Goodwin, D. M. Etheridge, and C. W. Wookey (1991). Evidence from Antarctic ice cores for recent increases in snow accumulation. *Nature*, **354**, 58–60.

Morris, E. M. (1999). Surface ablation rates on Moraine Corric Glacier, Antarctica. *Global and Planetary Change*, **22**(1/4), 221–231.

Morris, E. M. and D. G. Vaughan (2003). Spatial and temporal variation of surface temperature on the Antarctic Peninsula and the limit of viability of ice shelves. In: E. Domack, A. Leventer, A. Burnett, R. Bindschadler, P. Conley, and M. Kirby (eds.), *Antarctic Peninsula Climate Variability: Historical and Palaeoenvironmental Perspective* (AGU Antarctic Research Series No. 79). American Geophysical Union, Washington, DC, pp. 61–68.

Mosley-Thompson, E., J. R. McConnell, R. C. Bales, Z. Li, P.-N. Lin, K. Steffen, L. G. Thompson, R. Edwards, and D. Bathke (2001). Local to regional-scale variability of annual net accumulation on the Greenland ice sheet from PARCA cores. *Journal of Geophysical Research*, **106**(D24), 33839–33852.

Mote, T. L. (1998a). Mid-tropospheric circulation and surface melt on the Greenland Ice Sheet: Part II. Synoptic climatology. *International Journal of Climatology*, **18**(2), 131–145.

Mote, T. L. (1998b). Mid-tropospheric circulation and surface melt on the Greenland Ice Sheet: Part I. Atmospheric teleconnections. *International Journal of Climatology*, **18**(2), 111–129.

Mote, T. L. (2000). Ablation estimates for the Greenland Ice Sheet from passive microwave measurements. *Professional Geographer*, **52**, 322–331.

Mote, T. L. (2003). Estimation of runoff rates, mass balance, and elevation changes on the Greenland Ice Sheet from passive microwave observations. *Journal of Geophysical Research*, **108**(D2). 4056, DOI: 10.1029/2001JD002032.

Mote, T. and M. R. Anderson (1995). Evidence of an increase in snowpack melt on the Greenland ice sheet based on passive microwave measurements. *Journal of Glaciology*, **41**(137), 51–60.

Mote, T. L., M. R. Anderson, K. C. Kuivinen, and C. M. Rowe (1993). Passive microwave-derived spatial and temporal variations of summer melt on Greenland ice sheet. *Annals of Glaciology*, **17**, 233–238.

Mueller, D. R., W. F. Vincent, and M. O. Jeffries (2003). Break-up of the largest Arctic ice shelf and associated loss of an epishelf lake. *Geophysical Research Letters*, **30**(20), 2031, DOI: 10.1029/2003GL017931.

Munk, J., K. C. Jezek, R. R. Forster, and S. P. Gogineni (2003). An accumulation map for the Greenland dry-snow facies derived from spaceborne radar. *Journal of Geophysical Research*, **108**(D9), 4280, doi:10.1029/2002JD002481.

Murray, T., A. J. Luckman, and T. Strozzi (2001). Ice dynamics during Svalbard surges using satellite radar interferometry. *EOS, Transactions of the American Geophysical Union*, **82**(47), IP31A-11.

Murray, T., T. Strozzi, A. Luckman, H. Pritchard, and H. Jiskoot (2002). Ice dynamics during a surge of Sortebrae, East Greenland. *Annals of Glaciology*, **34**, 323–329.

Murray, T., T. Strozzi, A. Luckman, H. Jiskoot, and P. Christakos (2003). Is there a single surge mechanism?: Contrasts in dynamics between glacier surges in Svalbard and other regions. *Journal of Geophysical Research*, **108**(B5), 2237, doi:10.1029/2002JB001906.

Nakajima, T. Y., T. Nakajima, M. Nakajima, H. Fukushima, M. Kuji, A. Uchiyama, and M. Kishino (1998). Optimization of the Advanced Earth Observing Satellite II Global Imager channels by use of radiative transfer equations. *Applied Optics*, **37**, 3149–3163.

NASA (2003). GSFC Ice Altimetry home page: ⟨http://icesat4.gsfc.nasa.gov/⟩.

Neckel, H. and D. Labs (1984). The solar radiation between 3300 and 12,500 Å. *Solar Physics*, **90**, 205.

Nerem, R. S. and G. T. Mitchum (2001a). Observations of sea level change from satellite altimetry. In: B. C. Douglas, M. S. Kearney, and S. P. Leatherman (eds.), *Sea Level Rise: History and Consequences*. Academic Press, New York, pp. 121–163.

Nerem, R. S. and G. T. Mitchum (2001b). Sea level change. In: L. Fu and A. Cazenave (eds.), *Satellite Altimetry and Earth Sciences: A Handbook of Techniques and Applications*. Academic Press, New York, pp. 329–349.

Nereson, N. A., C. F. Raymond, R. W. Jacobel, and E. D. Waddington (2000). The accumulation pattern across Siple Dome, West Antarctica, inferred from radar-detected internal layers. *Journal of Glaciology*, **46**(152), 75–87.

Nghiem, S. V., K. Steffen, R. Kwok, and W.-Y. Tsai (2001). Detection of snowmelt regions on the Greenland Ice Sheet using diurnal backscatter. *Journal of Glaciology*, **47**(159), 539–547.

Nicodemus, F. E., J. C. Richmond, J. J. Hsia, I. W. Ginsberg, and T. Limperis (1977). *Geometrical Considerations and Nomenclature for Reflectance* (U.S. NBS Monograph No. 160). Institute for Basic Standards, Washington, DC.

NOAA (1988). *Digital Relief of the Surface of the Earth* (Data Announcement 88-MGG-020). NOAA NESDIS National Geophysical Data Center, Boulder, CO.

Nolin, A. (1993). Radiative heating in alpine snow. PhD thesis, University of California, Santa Barbara.

Nolin, A. W. (1998). Mapping the Martian polar caps: Applications of terrestrial optical remote sensing methods. *Journal of Geophysical Research*, **103**, 25851–25864.

Nolin, A. W. and J. Dozier (1993). Estimating snow grain size using AVIRIS data. *Remote Sensing of Environment*, **44**, 231–238.

Nolin, A. W. and J. Dozier (2000). A hyperspectral method for remotely sensing the grain size of snow. *Remote Sensing of Environment*, **74**, 207–216.

Nolin, A. W. and S. Liang (2000). Progress in bidirectional reflectance modeling and applications for surface particulate media: Snow and soils. *Remote Sensing Reviews*, **18**, 307–342.

Nolin, A. W. and J. Stroeve (1997). The changing albedo of the Greenland Ice Sheet: Implications for climate modelling. *Annals of Glaciology*, **25**, 51–57.

Nolin, A., F. Fetterer, and T. C. Scambos (2002a). Surface roughness characterizations of sea ice and ice sheets: Case studies with MISR data. *IEEE Transactions on Geoscience and Remote Sensing*, **40**, 1605–1615.

Nolin, A., B. Raup, J. Stroeve, and T. Scambos (2002b). *Mapping Snow Grain Size and Albedo on the Greenland Ice Sheet Using an Imaging Spectrometer* (Glaciological Data Report GD-30). National Snow and Ice Data Center, Boulder, CO, 73 pp.

Noltimier, K. F., K. C. Jezek, H. G. Sohn, B. Li, H. Liu, F. Baumgartner, V. Kaupp, J. C. Curlander, B. Wilson, and R. Onstott (1999). RADARSAT Antarctic Mapping Project; mosaic construction. *Proceedings of International Geoscience and Remote Sensing Symposium IGARSS '99*. Institute of Electrical and Electronic Engineers, Piscataway, NJ, Vol. 5, pp. 2349–2351.

Nøst, O. A. and S. Østerhus (1998). Impact of grounded icebergs on the hydrographic conditions near the Filchner Ice Shelf. In: S. S. Jacobs and R. F. Weiss (eds.), *Ocean, Ice, and Atmosphere: Interactions at the Antarctic Continental Margin* (AGU Antarctic Research Series No. 75). American Geophysical Union, Washington, DC, pp. 267–284.

NRC (1997). *Satellite Gravity and the Geosphere*. National Research Council, Washington, DC, 112 pp.

NSIDC (2002). *Antarctic Ice Shelf Collapses*. Available at ⟨http://www.nsidc.org/iceshelves/larsenb2002/⟩.

Nye, J. F. (1959). A method of determining the strain-rate tensor at the surface of a glacier. *Journal of Glaciology*, **3**(25), 409–419.

Oerlemans, J. and N. C. Hoogendoorn (1989). Mass balance gradients and climatic change. *Journal of Glaciology*, **35**(54), 399–405.

O'Farrell, S. P., J. L. McGregor, L. D. Rotstayn, W. F. Budd, C. Zweck, and R. C. Warner (1997). Impact of transient increases in atmospheric CO_2 on the accumulation and mass balance of the Antarctic Ice Sheet. *Annals of Glaciology*, **25**, 137–144.

Ohmura, A. and N. Reeh (1991). New precipitation and accumulation maps for Greenland. *Journal of Glaciology*, **37**, 140–148.

Ohmura, A., P. Calanca, M. Wild, and M. Anklin (1999). Precipitation, accumulation, and mass balance of the Greenland Ice Sheet. *Zeitschrift für Gletscherkunde und Glazialgeologie*, **35**(1), 1–20.

Oppenheimer, M., 1998. Global warming and the stability of the West Antarctic Ice Sheet. *Nature*, **393**, 325–332.

Orheim, O. (1984). Iceberg discharge and the mass balance of Antarctica. *Iceberg Research*, **8**, 3–7.

Orheim, O. (1985). Iceberg discharge and the mass balance of Antarctica. *Glaciers, Ice Sheets, and Sea Level: Effects of a CO_2-induced Climatic Change*. Committee on Glaciology, National Academy Press, Washington, DC, pp. 210–215.

Orheim, O. (1988) Antarctic icebergs: Production, distribution and disintegration. *Annals of Glaciology*, **11**, 205 (abstract).

Orheim, O. (1990). Extracting climatic information from observations of icebergs in the Southern Ocean. *Annals of Glaciology*, **14**, 352.

Orheim, O. (1993). Iceberg calving rates and the mass balance of Antarctica. *5th International Symposium on Antarctic Glaciology, Cambridge, UK, September*.

Orheim, O. and B. K. Lucchitta (1987). Snow and ice studies by Thematic Mapper and Multispectral Scanner Landsat images. *Annals of Glaciology*, **9**, 109–118.

Orheim, O. and B. K. Lucchitta (1988). Numerical analysis of Landsat Thematic Mapper images of Antarctica: Surface temperatures and physical properties. *Annals of Glaciology*, **11**, 109–120.

Orheim, O. and B. K. Lucchitta (1990). Investigating climate change by digital analysis of blue-ice extent on satellite images of Antarctica. *Annals of Glaciology*, **14**, 211–215.

Orsi, A. H., S. S. Jacobs, A. L. Gordon, and M. Visbeck (2001). Cooling and ventilating the abyssal ocean. *Geophysical Research Letters*, **28**, 2923–2926.

Ou, Z. (2004). An integrated spatial information system for ice service. *Proceedings of XXth ISPRS Congress 2004, Istanbul, Turkey, July 12–23* (Commission 1). Available online at ⟨thtp://www.isprs.org/istanbul2004/comm1/comm1.html⟩.

Padman, L., H. A. Fricker, R. Coleman, S. Howard, and L. Erofeeva (2002). A new tide model for the Antarctic ice shelves and seas. *Annals of Glaciology*, **34**, 247–254.

Padman, L., M. King, D. Goring, H. Corr, and R. Coleman (2004). Ice shelf elevation changes due to atmospheric pressure variations. *Journal of Glaciology*, **49**(167), 521–526.

Palais, J. M., I. M. Whillans, and C. Bull (1982). Snow stratigraphic studies at Dome C, East Antarctica: An investigation of depositional and diagenetic processes. *Annals of Glaciology*, **3**, 239–242.

Parish, T. R. and D. H. Bromwich (1987). The surface windfield over the Antarctic ice sheets. *Nature*, **328**, 51–54.

Parish, T. R. and D. H. Bromwich (1991). Continental-scale simulation of the Antarctic katabatic wind regime. *Journal of Climate*, **4**, 135–146.

Parke, M. E., R. H. Stewart, D. L. Farless, and D. E. Cartwright (1987). On the choice of orbits for an altimetric satellite to study ocean circulation and tides. *Journal of Geophysical Research*, **92**, 11693–11707.

Parkinson, B. W. and J. J. Spilker Jr. (1996). *Global Positioning System: Theory and Applications*. American Institute of Aeronautics and Astronautics, Boston, 643 pp.

Parrot, J. F., N. Lyberis, B. Lefauconnier, and B. G. Manby (1993). SPOT multispectral data and digital terrain model for the analysis of ice-snow fields on Antarctic glaciers. *International Journal of Remote Sensing*, **14**(3), 425–440.

Partington, K. C. (1998). Discrimination of glacier facies using multi-temporal SAR data. *Journal of Glaciology*, **44**(146), 480–488.

Partington, K. C., J. K. Ridley, C. G. Rapley, and H. J. Zwally (1989). Observations of the surface properties of the ice sheets by satellite altimetry. *Journal of Glaciology*, **35**(120), 267–275.

Partington, K. C., W. Cudlip, and C. G. Rapley (1991). An assessment of the capability of the satellite radar altimeter for measuring ice sheet topographic change. *International Journal of Remote Sensing*, **12**, 585–609.

Paterson, W. S. B. (1994). *The Physics of Glaciers* (3rd edn.). Butterworth-Heinemann, Oxford, UK, 480 pp.

Papathanassiou, K. P. and S. R. Cloude (2001). Single baseline polarimetric SAR interferometry. *IEEE Transactions on Geoscience and Remote Sensing*, **39**(11), 2352–2363.

Pattyn, F. (1992). Topographic mapping from SPOT in polar regions. *International Archives of Photogrammetry and Remote Sensing*, **29**(B4), 472–476.

Pattyn, F. (1996). Numerical modeling of a fast flowing outlet glacier: Experiments with different basal conditions. *Annals of Glaciology*, **23**, 237–246.

Pattyn, F. and H. Declair (1993). Satellite monitoring of ice and snow conditions in the Sør Rondane, Antarctica. *Annals of Glaciology*, **17**, 41–48.

Pattyn, F. and D. Derauw (2002). Ice-dynamic conditions of Shirase Glacier, Antarctica, inferred from ERS-SAR interferometry. *Journal of Glaciology*, **48**(163), 559–565.

Paulsen, T. and T. J. Wilson (1998). Declassified Intelligence Satellite Photographs (DISP) provide first high-resolution space-borne views of the Transantarctic Mountains in the Antarctic interior. *EOS, Transactions of the American Geophysical Union*, **79**(8), 97 and **79**(10), 125.

Payne, A. J., and J. L. Bamber (2004). Conclusions, summary and outlook. In: J. L. Bamber and A. J. Payne (eds.), *Mass Balance of the Cryosphere*. Cambridge University Press, Cambridge, UK, pp. 623–639.

Payne, A. J., A. Vieli, A. P. Shepherd, D. J. Wingham, and E. Rignot (2004). Recent dramatic thinning of largest West Antarctic ice stream triggered by oceans. *Geophysical Research Letters*, **31**, L23401, doi:10.1029/2004GL021284.

Peltier, W. (2001). Global glacial isostatic adjustment and modern instrumental records of relative sea level history. In: B. Douglas, M. Kearney, and S. Leatherman (eds.), *Sea Level Rise*. Academic Press, San Diego, CA, pp. 65–95.

Pettré, P., J. F. Pinglot, M. Pourchet, and L. Reynaud (1986). Accumulation distribution in Terre Adélie, Antarctica: Effect of meteorological parameters. *Journal of Glaciology*, **32**, 486–500.

Pfeffer, W. T., M. F. Meier, and T. H. Illangasekare (1991). Retention of Greenland runoff by refreezing: Implications for projected future sea level change. *Journal of Geophysical Research*, **96**(C12), 22117–22124.

Phalippou, L. and D. G. Wingham (1999). HSRRA: An advanced radar altimeter for ocean and cryosphere monitoring. *Proceedings of CEOS SAR Workshop, October 26–29*. ESA, Noordwijk, The Netherlands/Centre National d'Études Spatiale, Toulouse, France.

Phillips, H. A. (1998). Surface meltstreams on the Amery Ice Shelf, East Antarctica. *Annals of Glaciology*, **27**(1), 177–181.

Phillips, H. A. (1999). Applications of ERS satellite radar altimetry in the Lambert Glacier–Amery Ice Shelf system, East Antarctica. PhD thesis, Institute of Antarctic and Southern Ocean Studies and Antarctic CRC, Hobart, Tasmania.

Phillips, H. A. and S. W. Laxon (1995). Tracking of Antarctic tabular icebergs using passive microwave radiometry. *International Journal of Remote Sensing*, **16**(2), 399–405.

Pötzsch, A., B. Legresy, W. Korth, and R. Dietrich (2000). Glaciological investigation of Mertz Glacier, East Antarctica, using SAR interferometry and field observations. *Proceedings of the ERS–Envisat Symposium, October 16–20, Gothenburg, Sweden* (ESA SP-461, CD-ROM). Available online at ⟨http://envisat.esa.int/pub/ESA_DOC/gothenburg/500poetz.pdf⟩.

Power, D., J. Youden, K. Lane, C. Randell, and D. Flett (2001). Iceberg detection capabilities of Radarsat Synthetic Aperture Radar. *Canadian Journal of Remote Sensing*, **27**(5), 476–486.

Price, S. F. and I. M. Whillans (2001). Crevasse patterns at the onset to Ice Stream B, West Antarctica. *Journal of Glaciology*, **47**(156), 29–36.

Price, S. F., R. A. Bindschadler, C. L. Hulbe, and I. R. Joughin (2001a). Post-stagnation behaviour in the upstream regions of Ice Stream C, West Antarctica. *Journal of Glaciology*, **47**(157), 283–294.

Price, S. F., R. A. Bindschadler, C. L. Hulbe, and D. Blankenship (2001b). Force balance along a tributary and onset to Ice Stream D. *Journal of Glaciology*, **48**, 20–30.

Price, S. F., R. A. Bindschadler, C. L. Hulbe, and D. D. Blankenship. 2002. Force balance along an inland tributary and onset to Ice Stream D, West Antarctica. *Journal of Glaciology*, **48**(160), 20–30.

Pritchard, H. and D. Vaughan (2004). Accelerating glacier flow on the warming Antarctic Peninsula. *Proceedings of 4th WAIS Workshop*. Available online at ⟨igloo.gsfc.nasa.gov/wais/pastmeetings/abstracts04/Pritchard.htm⟩.

Pudsey, C. L. and J. Evans (2001). First survey of Antarctic sub-ice shelf sediments reveals mid-Holocene ice shelf retreat. *Geology*, **29**, 787–790.

Rack, W. and H. Rott (2003). Further retreat of the northern Larsen Ice Shelf and collapse of Larsen B. *16th International Workshop of the Forum for Research on Ice Shelf Processes (FRISP), June 25–26* (Report 14), Geophysical Institute, University of Bergen, Norway. Available online at ⟨http://gfi107.gfi.uib.no/forskning/frisp/⟩.

Rack, W. and H. Rott (2004). Pattern of retreat and disintegration of Larsen B Ice Shelf, Antarctic Peninsula. *Annals of Glaciology*, **39**, 505–510.

Rack, W., C. S. M. Doake, H. Rott, A. Siegel, and P. Skvarca (2000). Interferometric analysis of the deformation pattern of the northern Larsen Ice Shelf, Antarctic Peninsula, compared to field measurements and numerical modeling. *Annals of Glaciology*, **31**, 205–210.

Raney, R. K. (1998). The delay Doppler radar altimeter. *IEEE Transactions on Geoscience and Remote Sensing*, **36**, 1578–1588.

Rao, C. R. N. and J. Chen (1994). *Post-launch Calibration of the Visible and Near-IR Channels of AVHRR on NOAA-7, -9, and -11 Spacecraft* (NOAA Technical Report NESDIS 78). U.S. Department of Commerce, NOAA, Silver Spring, MD, 22 pp.

Rao, C. R. N. and J. Chen (1999). Revised post-launch calibration of the visible and near-infrared channels of the Advanced Very High Resolution Radiometer on the NOAA-14 spacecraft. *International Journal of Remote Sensing*, **20**(18), 3485–3491.

Rapley, C. G., H. D. Griffiths, V. A. Squire, M. Lefebvre, A. R. Birks, A. C. Brenner, C. Brossier, L. D. Clifford, A. P. R. Cooper, A. M. Cowan et al. (1983). *A Study of Satellite Radar Altimeter Operations over Ice-covered Surfaces*. ESA, Noordwijk, The Netherlands, pp. 1–224.

Rau, F. and M. Braun (2002). The regional distribution of the dry-snow zone on the Antarctic Peninsula north of 70°S. *Annals of Glaciology*, **34**, 95–100.

Rau, F., M. Braun, M. Friedrich, F. Weber, and H. Goßmann (2000a). Radar glacier zones and their boundaries as indicators of glacier mass balance and climatic variability. *Proceedings of EARSeL-SIG-Workshop on Land Ice and Snow, Dresden, Germany, June 16–17*.

Rau, F., M. Braun, H. Saurer, H. Goßmann, G. Kothe, F. Weber, M. Ebel, and D. Beppler (2000b). Monitoring multi-year snow cover dynamics on the Antarctic Peninsula using SAR imagery. *Polarforschung*, **67**(172), 27–40.

Raup, B., H. H. Kieffer, T. Hare, and J. Kargel (2000). Generation of data acquisition requests for the ASTER satellite instrument for monitoring a globally distributed target: Glaciers. *IEEE Transactions on Geoscience and Remote Sensing*, **38**, 1105–1112.

Raup, B. H., T. A. Scambos, and T. Haran (2005). Topography of streaklines on an Antarctic ice shelf from photoclinometry applied to a single Advanced Land Imager (ALI) image. *IEEE Transactions on Geoscience and Remote Sensing*, **43**(4), 736–742.

Raymond, C. F. and C. J. Van der Veen (2002). *The International Symposium of Fast Glacier Flow, Yakutat, Alaska, June, 10–14* (selected papers). Published in *Annals of Glaciology*, **36**, 2002.

Raymond, C. F., K. A. Echelmeyer, I. M. Whillans, and C. S. M. Doake (2001). Ice stream shear margins. In: R. Alley and R. Bindschadler (eds.), *The West Antarctic Ice Sheet: Behavior and Environment* (AGU Antarctic Research Series No. 77). American Geophysical Union, Washington, DC, pp. 137–155.

Reeh, N. (1994). Calving from Greenland glaciers: Observations, balance estimates of calving rates, calving laws. *The Workshop on the Calving Rate of West Greenland Glaciers in Response to Climate Change, September 13–15, 1993, Copenhagen, Denmark* (report). Danish Polar Center, Copenhagen, pp. 85–102.

Reeh, N., C. E. Bøggild, and H. Oerter (1994). *Surge of Storstrømmen: A Large Outlet Glacier from the Inland Ice of North-East Greenland* (Report Grønlands Geol. Unders. No. 162). GEUS, Copenhagen, pp. 201–209.

Reeh, N., H. H. Thomsen, O. B. Olesen, and W. Starzer (1997). Mass balance of North Greenland. *Science*, **278**(5336), 207–209.

Reeh, N., S. N. Madsen, and J. J. Mohr (1999). Combining SAR interferometry and the equation of continuity to estimate the three-dimensional glacier surface velocity. *Journal of Glaciology*, **45**(151), 533–538.

Reeh, N., C. Mayer, O. B. Olesen, E. L. Christensen, and H. H. Thomsen (2000). Tidal movement of Nioghalvfjerdsfjorden glacier, northeast Greenland: Observations and modelling. *Annals of Glaciology*, **31**, 111–117.

Reeh, N., J. J. Mohr, W. B. Krabill, R. Thomas, H. Oerter, N. Gundestrup, and C. E. Bøggild (2002). Glacier specific ablation derived by remote sensing measurements. *Geophysical Research Letters*, **29**(16), DOI: 10.1029/2002GL015307.

Rees, W. G., J. A. Dowdeswell, and A. D. Diament (1995). Analysis of ERS-1 Synthetic Aperture Radar from Nordaustlandet, Svalbard. *International Journal of Remote Sensing*, **16**(5), 905–924.

Reijmer, C. H., W. H. Knap, and J. Oerlemans (1999). The surface albedo of the Vatnajökull Ice Cap, Iceland: A comparison between satellite-derived and ground-based measurements. *Boundary-Layer Meteorology*, **92**(1), 125–144.

Reinartz, P., M. Lehner, R. Müller, and M. Shroeder (2004). Accuracy analysis for DEM and orthoimages derived from SPOT HRS stereo data without using GCP. *XXth ISPRS Congress, July 12–23, Istanbul* (Commission 1). International Archives of Photogrammetry, Remote Sensing and Spatial Information Services, p. 433ff.

Rémy, F. and B. Legrésy (2004). Antarctic ice sheet shape response to changes in outlet flow boundary conditions. *Global and Planetary Change*, **42**, 133–142.

Rémy, F. and J. F. Minster (1991). A comparison between active and passive microwave measurements of the Antarctic ice sheet and their association with surface katabatic winds. *Journal of Glaciology*, **37**(125), 3–10.

Rémy, F. and J. F. Minster (1997). Antarctica ice sheet curvature and its relation with ice flow and boundary conditions. *Geophysical Research Letters*, **24**, 1039–1042.

Rémy, F., C. Brossier, and J. F. Minster (1990). Intensity of a radar altimeter over continental ice sheets: A potential measurement of surface roughness and katabatic wind intensity. *Journal of Glaciology*, **36**, 133–142.

Rémy, F., M. Ledroit, and J. F. Minster (1992). Katabatic wind intensity and direction over Antarctica derived from scatterometer data. *Geophysical Research Letters*, **19**, 1021–1024.

Rémy, F., J. F. Minster, and P. Féménias (1993). Monitoring continental ice sheets by satellite altimetry. *Advances in Space Research*, **11**, 353–359.

Rémy, F., P. Féménias, M. Ledroit, and J. F. Minster (1995). Empirical microwave backscattering over Antarctica: Application to radar altimetry. *Journal of Electromagnetic Waves and Applications*, **9**(3), 463–474.

Rémy, F., B. Legrésy, S. Bleuzen, P. Vincent, and J. F. Minster (1996a). Dual-frequency TOPEX altimeter observation above Greenland. *Journal of Electromagnetic Waves and Application*, **10**, 1505–1523.

Rémy, F., C. Ritz, and L. Brisset (1996b). Ice sheet flow features and rheological parameters derived from precise altimetric topography. *Annals of Glaciology*, **23**, 277–283.

Rémy, F., P. Schaeffer, and B. Legrésy (1999). Ice flow physical processes derived from ERS-1 high resolution maps of Antarctica and Greenland Ice Sheets. *Geophysical Journal International*, **139**, 645–656.

Rémy, F., B. Legrésy, and L. Testut (2001). Ice sheet and satellite altimetry. *Surveys in Geophysics*, **22**(1), 1–29.

Rémy, F., L. Testut, and B. Legrésy (2002). Random fluctuations of snow accumulation over Antarctica and its relation with sea level change. *Climate Dynamics*, **19**, 267–276.

Retzlaff, R. and C. R. Bentley (1993). Timing of stagnation of Ice Stream C, West Antarctica, from short-pulse radar studies of buried surface crevasses. *Journal of Glaciology*, **39**(133), 553–561.

Richardson, C. and P. Holmlund (1999). Spatial variability in snow-layer depths in central Dronning Maud Land, East Antarctica. *Annals of Glaciology*, **29**, 10–16.

Richardson-Näslund, C. (2004). Spatial characteristics of snow accumulation in Dronning Maud Land, Antarctica. *Global and Planetary Change*, **42**, 31–43.

Ridley, J. (1993). Surface melting on Antarctic Pensinsula ice shelves detected by passive microwave sensors. *Geophysical Research Letters*, **20**(23), 2639–2642.

Ridley, J. K. and J. L. Bamber (1995). Antarctic field measurements of radar backscatter from snow and comparison with ERS-1 altimeter data. *Journal of Electromagnetic Waves and Applications*, **9**(3), 355–372.

Ridley, J. K. and K. C. Partington (1988). A model of satellite radar altimeter return from ice sheets, *International Journal of Remote Sensing*, **9**, 601–624.

Ridley, J. K., W. Cudlip, and S. W. Laxon (1993). Identification of sub-glacial lakes using the ERS-1 radar altimeter. *Journal of Glaciology*, **39**, 625–634.

Rignot, E. (1995). Backscatter model for the unusual radar properties of the Greenland Ice Sheet. *Journal of Geophysical Research*, **100**(E5), 9389–9400.

Rignot, E. (1996). Tidal flexure, ice velocities and ablation rates of Petermann Gletscher, Greenland. *Journal of Glaciology*, **42**(142), 476–485.

Rignot, E. (1998a). Radar interferometry detection of hinge-line migration on Rutford Ice Stream and Carlson Inlet, Antarctica. *Annals of Glaciology*, **27**, 25–32.

Rignot, E. (1998b). Hinge–line migration of Petermann Gletscher, north Greenland, detected using satellite radar interferometry. *Journal of Glaciology*, **44**, 469–476.

Rignot, E. (1998c). Fast recession of a West Antarctic Glacier. *Science*, **281**, 549–551.

Rignot, E. (2001). Evidence for rapid retreat and mass loss of Thwaites Glacier, West Antarctic. *Journal of Glaciology*, **47**(157), 213–222.

Rignot, E. (2002a). Mass balance of East Antarctic glaciers and ice shelves from satellite data. *Annals of Glaciology*, **34**, 217–227.

Rignot, E. (2002b). Ice-shelf changes in Pine Island Bay, Antarctica, 1947–2000. *Journal of Glaciology*, **48**, 247–256.

Rignot, E. and S. S. Jacobs (2002). Rapid bottom melting widespread near Antarctic Ice Sheet grounding lines. *Science*, **296**, 2020–2023.

Rignot, E. and D. MacAyeal (1998). Ice-shelf dynamics near the front of Filchner-Ronne Ice Shelf, Antarctica, revealed by SAR interferometry. *Journal of Glaciology*, **44**, 405–418.

Rignot, E. and R. H. Thomas (2002). Mass balance of polar ice sheets. *Science*, **297**(5586), 1502–1506.

Rignot, E. J., S. J. Ostro, J. J. van Zyl, and K. C. Jezek (1993). Unusual radar echoes from the Greenland Ice Sheet. *Science*, **261**(5129), 1710–1713.

Rignot, E., K. C. Jezek, and H. G. Sohn (1995). Ice flow dynamics of the Greenland Ice Sheet from SAR interferometry. *Geophysical Research Letters*, **22**(5), 575–578.

Rignot E., R. Forster, and B. Isacks (1996). Interferometric radar observations of Glaciar San Rafael, Chile. *Journal of Glaciology*, **42**(141), 279–291.

Rignot, E., S. Gogineni, W. Krabill, and S. Ekholm (1997a). Ice discharge from north and northeast Greenland as observed from satellite radar interferometry. *Science*, **276**, 934–937.

Rignot, E., S. Gogineni, W. Krabill, and S. Ekholm (1997b). Mass balance of north Greenland: Response, *Science*, Letter, **278**(5336), 209.

Rignot, E., K. Echelmeyer, and W. Krabill (2000a). Penetration depth of interferometric synthetic-aperture radar signals in snow and ice. *Geophysical Research Letters*, **28**(18), 3501–3504.

Rignot, E., L. Padman, D. R. MacAyeal, and M. Schmeltz (2000b). Analysis of sub-ice-shelf tides in the Weddell Sea using SAR interferometry. *Journal of Geophysical Research*, **105**(C8), 19615–19630.

Rignot, E., G. Buscarlet, B. Csatho, S. Gogineni, W. B. Krabill, and M. Schmeltz (2000c). Mass balance of the northeast sector of the Greenland Ice Sheet: A remote sensing perspective. *Journal of Glaciology*, **46**(153), 265–273.

Rignot, E., W. B. Krabill, S. P. Gogineni, and I. Joughin (2001). Contribution to the glaciology of northern Greenland from satellite radar interferometry. *Journal of Geophysical Research*, **106**(D24), 34007–34019.

Rignot, E., D. G. Vaughan, M. Schmeltz, T. Dupont, and D. MacAyeal (2002). Acceleration of Pine Island and Thwaites Glaciers, West Antarctica. *Annals of Glaciology*, **34**, 189–194.

Rignot, E., G. Casassa, P. Gogineni, W. Krabill, A. Rivera, and R. Thomas (2004). Accelerated ice discharge from the Antarctic Peninsula following the collapse of Larsen B Ice Shelf. *Geophysical Research Letters*, **31**, L18401, DOI: 10.1029/2004GL020697.

Rignot, E., G. Casassa, S. Gogineni, P. Kanagaratnam, W. Krabill, H. Pritchard, A. Rivera, R. Thomas, J. Turner, and D. Vaughan (2005). Recent ice loss from the Fleming and other glaciers, Wordie Bay, West Antarctic. *Geophysical Research Letters*, **32**, L07502, doi:10.1029/2004GL021947.

Rist, M. A., P. R. Sammonds, H. Oerter, and C. S. M. Doake (2002). Fracture of an Antarctic ice shelf. *Journal of Geophysical Research*, **107**, 1–13.

Ritz, C., A. Fabre, and C. Letreguilly (1996). Sensitivity of the Greenland Ice Sheet model to ice flow and ablation parameters: Consequences for the evolution through the last climatic cycle. *Climate Dynamics*, **13**(1), 11–24.

Robertson, R. A., L. Padman, and G. D. Egbert (1998). Tides in the Weddell Sea. In: S. S. Jacobs and R. F. Weiss (eds.), *Ocean, Ice, and Atmosphere: Interactions at the Antarctic Continental Margin* (AGU Antarctic Research Series No. 75). American Geophysical Union, Washington, DC, pp. 341–369.

Robin, G. de Q. (1958). Glaciology III: Seismic shooting and related investigations. *Norwegian–British–Swedish Antarctic Expedition, 1949–1952: Scientific Results* (Vol. V). Norsk Polarinstitutt, Oslo.

Robin, G. de Q. (1966). Mapping the Antarctic Ice Sheet by satellite radar altimetry. *Canadian Journal of Earth Science*, **3**, 893–901.

Robin, G. de Q. (ed.) (1983). Radio-echo studies of internal layering of polar ice sheets. *The Climatic Record in Polar Ice Sheets*. Cambridge University Press, Cambridge, UK, pp. 89–93.

Robin, G. de Q., D. J. Drewry, and V. A. Squire (1983). Satellite observations of polar ice fields. *Philosophical Transactions of the Royal Society of London*, **A309**, 447–461.

Rolstad, C., J. Amlien, J. O. Hagen, and B. Lundén (1997). Visible and near-infrared digital images for determination of ice velocities and surface elevation during a surge on Osbornebreen, a tidewater glacier in Svalbard. *Annals of Glaciology*, **24**, 255–261.

Rommen, B., C. Lin, J. Guijarro, and B. Ramirez Velado (2004). *Scientific Rationale for a Spaceborne P-band Ice-sounder* (Terra Nostra, Schriften der Alfred-Wegener-Stiftung, 2004/4, Abstract SS/P35). Alfred Wegener Institute for Polar and Marine Research, Bremerhaven, Germany, pp. 474.

Rosanova, C. and B. Lucchitta (2002). *Acceleration of Pine Island Glacier Ice Shelf* (Glaciological Data Report GD-30). National Snow and Ice Data Center, Boulder, CO, p. 76.

Rosanova, C. E., B. K. Lucchitta, and J. G. Ferrigno (1998). Velocities of Thwaites Glacier and smaller glaciers along the Marie Byrd Land coast, West Antarctica. *Annals of Glaciology*, **27**, 47–53.

Rott, H. (1980). Synthetic aperture radar capabilities for glacier monitoring demonstrated with Seasat SAR. *Zeitschrift für Gletscherkunde und Glazialgeologie*, **16**, 255–266.

Rott, H. (1989). Multispectral microwave signatures of the Antarctic ice sheet. In: P. Pampaloni (ed.), *Microwave Radiometry and Remote Sensing Applications*. VSP, Utrecht, The Netherlands, pp. 89–101.

Rott, H. and R. E. Davis (1993). Multifrequency and polarimetric SAR observations on alpine glaciers. *Annals of Glaciology*, **17**, 98–104.

Rott, H. and G. Markl (1989). *Improved Snow and Glacier Monitoring by the Landsat Thematic Mapper: Monitoring the Earth's Environment* (ESA SP-1102). ESA, Noordwijk, The Netherlands, pp. 3–12.

Rott, H. and T. Nagler (1993). Snow and glacier investigations by ERS-1 SAR: First results. *Proceedings of 1st ERS-1 Symposium, Space at the Service of Our Environment, Cannes, France, November 4–6, 1992* (ESA SP-359). ESA, Paris, pp. 577–582.

Rott, H., K. Sturm, and H. Miller (1993). Active and passive microwave signatures of Antarctic firn by means of field measurements and satellite data. *Annals of Glaciology*, **17**, 337–343.

Rott, H., T. Nagler, and D. Floricioiu (1995). Snow and glacier parameters derived from single channel and multiparameter SAR. *Proceedings of Symposium on Retrieval of Bio- and Geophysical Parameters from SAR Data for Land Applications, 10–13 October 1995*. CNES, Toulouse, France, pp. 479–488.

Rott, H., P. Skvarca, and T. Nagler (1996). Rapid collapse of northern Larsen Ice Shelf, Antarctica. *Science*, **271**, 788–792.

Rott, H., W. Rack, T. Nagler, and P. Skvarca (1998a). Climatically induced retreat and collapse of northern Larsen Ice Shelf, Antarctic Peninsula. *Annals of Glaciology*, **27**, 86–92.

Rott, H., M. Stuefer, A. Siegel, P. Skvarca, and A. Eckstaller (1998b). Mass fluxes and dynamics of Moreno Glacier, Southern Patagonia Icefield. *Geophysical Research Letters*, **25**(9), 1407–1410.

Rott, H., W. Rack, P. Skvarca, and H. D. Angelis (2002). Northern Larsen Ice Shelf Antarctica: Further retreat after collapse. *Annals of Glaciology*, **34**, 277–282.

Scambos, T. A. and R. A. Bindschadler (1991). Feature map of Ice Streams C, D, and E, West Antarctica. *Antarctic Journal of the United States*, **26**(5), 312–314.

Scambos, T. A. and R. A. Bindschadler (1993). Complex ice stream flow revealed by sequential satellite imagery. *Annals of Glaciology*, **17**, 177–182.

Scambos, T. A. and M. A. Fahnestock (1998). Improving digital elevation models over ice sheets using AVHRR-based photoclinometry. *Journal of Glaciology*, **44**, 97–103.

Scambos, T. A. and T. Haran (2002). An image-enhanced DEM of the Greenland Ice Sheet. *Annals of Glaciology*, **34**, 291–298.

Scambos, T. A. and K. Jezek (1999). New views of the West Antarctic: An overview of its glacio-morphology from Radarsat, AVHRR, and Landsat Data. *Proceedings of 6th Annual WAIS Workshop*. Available online at ⟨http://igloo.gsfc.nasa.gov/wais/pastmeetings/Agenda99.html⟩.

Scambos, T. A. and N. A. Nereson (1997). Satellite image and GPS study of the morphology of Siple Dome, Antarctica. *Antarctic Journal of the United States*, **30**, 91–93.

Scambos, T. A., M. J. Dutkiewitcz, J. C. Wilson, and R. A. Bindschadler (1992). Application of image cross-correlation software to the measurement of glacier velocity using satellite image data. *Remote Sensing of Environment*, **42**, 177–186.

Scambos, T. A., G. Kvaran, and M. A. Fahnestock (1999). Improving AVHRR resolution through data cumulation for mapping polar ice sheets. *Remote Sensing of Environment*, **69**, 56–66.

Scambos, T. A., C. Hulbe, M. Fahnestock, and J. Bohlander (2000). The link between climate warming and break-up of ice shelves in the Antarctic Peninsula. *Journal of Glaciology*, **46**(154), 516–530.

Scambos, T., T. Haran, D. D. Blankenship, and D. Morse (2002a). *Antarctic Photoclinometry Using AVHRR and MODIS* (Glaciological Data Report GD-30, Vol. 77). National Snow and Ice Data Center, Boulder, CO, pp. 130.

Scambos, T., T. Haran, C. Fowler, J. Maslanik, J. Key, and W. Emery (2002b). *AVHRR Polar Pathfinder Twice-daily 1.25 km EASE–grid Composites* (digital media). National Snow and Ice Data Center, Boulder, CO.

Scambos, T., C. Hulbe, and M. Fahnestock (2003). Climate-induced ice shelf disintegration in the Antarctic Peninsula. In: E. Domack, A. Leventer, A. Burnett, R. Bindschadler, P. Conley, and M. Kirby (eds.), *Antarctic Peninsula Climate Variability: Historical and Paleoenvironmental Perspective* (AGU Antarctic Research Series No. 79). American Geophysical Union, Washington, DC, pp. 335–347.

Scambos, T., J. Bohlander, B. Raup, and T. Haran (2004a). Glaciological characteristics of Institute Ice Stream using remote sensing. *Antarctic Science*, **16**(2), 205–213.

Scambos, T. A., J. A. Bohlander, C. A. Shuman, and P. Skvarca (2004b). Glacier acceleration and thinning after ice shelf collapse in the Larsen B embayment, Antarctica. *Geophysical Research Letters*, **31**, L18402, DOI: 10.1029/2004GL020670.

Schmeltz, M., E. Rignot, and D. MacAyeal (2001). Ephemeral grounding as a signal of ice-shelf change. *Journal of Glaciology*, **47**(156), 71–77.

Schmeltz, M., E. Rignot, and D. MacAyeal (2002a). Tidal flexure along ice-sheet margins: Comparison of InSAR with an elastic-plate model. *Annals of Glaciology*, **34**, 202–208.

Schmeltz, M., E. Rignot, T. K. Dupont, and D. MacAyeal (2002b). Sensitivity of Pine Island Glacier, West Antarctica, to changes in ice shelf and basal conditions: A model study. *Journal of Glaciology*, **48**, 552–558.

Schneider, D. P. and E. J. Steig (2002). Spatial and temporal variability of Antarctic ice sheet microwave brightness temperatures. *Geophysical Research Letters*, **29**(20), 1984, DOI: 10.1029/2002GL015490.

Schneider, D. P., E. J. Steig, and J. C. Comiso (2004). Recent climate variability in Antarctica from satellite-derived temperature data. *Journal of Climate*, **17**(7), 1569–1583.

Schowengerdt, R. (1987). *Modify University of Arizona Restoration, Software Module System, AVHRR Image Deconvolution/Resampling Capability* (Final Report to USGS Contract No. PO 060206-86). University of Arizona, 45 pp.

Seeber, G. (2003). *Satellite Geodesy: Foundations, Methods and Applications* (2nd edn.). Walter de Gruyter, Berlin, 589 pp.

Seko, K., T. Furukawa, F. Nishio, and O. Watanabe (1993). Undulating topography on the Antarctic Ice Sheet rvealed by NOAA AVHRR images. *Annals of Glaciology*, **17**, 55–62.

Sergent C., C. Leroux, E. Pougatch, and F. Guirado (1998). Hemispherical–directional reflectance measurements of natural snows in the 0.9–1.45 µm spectral range: Comparison with adding–doubling modelling. *Annals of Glaciology*, **26**, 59–63.

Sergienko, S. V., T. A. Scambos, D. R. MacAyeal, and J. L. Fastook (2004). Tabular icebergs in the South Atlantic: Melt ponding, melt pond geometry and margin evolution. *EOS, Transactions of the American Geophysical Union*, **85**(47), Fall Meeting Supplement, Abstract C22A-07.

Shabtaie, S. and C. R. Bentley (1987). West Antarctic ice streams draining into the Ross Ice Shelf: Configuration and mass balance. *Journal of Geophysical Research*, **92**(B2), 1311–1336. Erratum: *Journal of Geophysical Research*, **92**(B9), 9451 (1987).

Shepherd, A. and N. R. Peacock (2003). Ice shelf tidal motion derived from ERS altimetry. *Journal of Geophysical Research–Oceans*, **108**(C6), 3198, DOI: 10.1029/2001JC001152.

Shepherd, A., D. J. Wingham, J. A. D. Mansley, and H. F. J. Corr (2001). Inland thinning of Pine Island Glacier, West Antarctica. *Science*, **291**, 862–864.

Shepherd, A., D. J. Wingham, and J. A. D. Mansley (2002a). Inland thinning of the Amundsen Sea sector, West Antarctica. *Geophysical Research Letters*, **29**(10), 2-1–2-4.

Shepherd, A., D. J. Wingham, and J. A. Mansley (2002b). *Rapid Glacier Thinning along the Amundsen Coast, West Antarctica* (Glaciological Data Report GD-30). National Snow and Ice Data Center, Boulder, CO.

Shepherd, A., D. Wingham, A. Payne, and P. Skvarca (2003). Larsen Ice Shelf has progressively thinned. *Science*, **302**, 856–859.

Sherjal, I. and M. Fily (1994). Temporal variations of microwave brightness temperatures over Antarctica. *Annals of Glaciology*, **20**, 19–25.

Shi, J. (1999). An automatic estimation of snow fraction using ASTER simulated from AVIRIS in alpine regions. *Proceedings of 8th Annual JPL Airborne Geoscience Workshop, Pasadena, California*. NASA Jet Propulsion Laboratory, Pasadena, CA.

Shi, J. and J. Dozier (1993). Measurement of snow and glacier covered areas by single-polarization SAR. *Annals of Glaciology*, **17**, 72–76.

Shi, J. and J. Dozier (1995). Inferring snow wetness using C-band data from SIR-C's polarimetric synthetic aperture radar. *IEEE Transactions on Geoscience and Remote Sensing*, **33**(4), 905–914.

Shi, J. and J. Dozier (2000a). Estimation of snow–water equivalence using SIR-C/X-SAR: I. Inferring snow density and subsurface properties. *IEEE Transactions on Geoscience and Remote Sensing*, **38**(6), 2465–2474.

Shi, J. and J. Dozier (2000b). Estimation of snow–water equivalence using SIR-C/X-SAR: II. Inferring snow depth and particle size. *IEEE Transactions on Geoscience and Remote Sensing*, **38**(6), 2475–2488.

Shi, J., J. Dozier, and H. Rott (1994). Snow mapping in alpine regions with synthetic aperture radar. *IEEE Transactions on Geoscience and Remote Sensing*, **32**(1), 152–158.

Shi, J., J. Dozier, H. Rott, and R. E. Davis (1992). Snow and glacier mapping in alpine regions with polarimetric SAR. *Proceedings of IGARSS '92*, 2311–2314.

Shine, K. P., A. H. Henderson-Sellers, and R. G. Barry (1984). Albedo–climate feedback: The importance of cloud and cryosphere variability. In: A. L. Berger and C. Nicolis (eds.), *New Perspectives in Climate Modelling*. Elsevier, Amsterdam, pp. 135–155.

Short, N. H. and A. L. Gray (2004). Potential for Radarsat-2 interferometry: Glacier monitoring using speckle tracking. *Canadian Journal of Remote Sensing*, **30**(3), 504–509.

Shuman, C. A. (2004). ICESat's first year of measurements over the polar ice sheets. *EOS, Transactions of the American Geophysical Union*, **85**(17), Joint Assembly Supplement, Abstract C42A-01.

Shuman, C. A. and R. B. Alley (1993). Spatial and temporal characterization of hoar formation in central Greenland using SSM/I brightness temperatures. *Geophysical Research Letters*, **20**(23), 2643–2646.

Shuman, C. A. and J. C. Comiso (2002). *In situ* and satellite surface temperature records in Antarctica. *Annals of Glaciology*, **34**, 113–120.

Shuman, C. and C. Stearns (2001). Decadal-length composite inland West Antarctic temperature records. *Journal of Climate*, **14**(9), 1977–1988.

Shuman, C. A., R. B. Alley, and S. Anandakrishnan (1993). Characterization of a hoar-development episode using SSM/I brightness temperatures in the vicinity of the GISP2 site, Greenland. *Annals of Glaciology*, **17**, 183–188.

Shuman, C. A., R. B. Alley, S. Anandakrishnan, and C. R. Stearns (1995a). An empirical technique for estimating near-surface air temperatures in central Greenland from SSM/I brightness temperatures. *Remote Sensing of the Environment*, **51**, 245–252.

Shuman, C. A., R. B. Alley, S. Anandakrishnan, J. W. C. White, P. M. Grootes, and C. R. Stearns (1995b). Temperature and accumulation at the Greenland Summit: Comparison of high-resolution isotope profiles and satellite passive microwave brightness temperature trends. *Journal of Geophysical Research*, **100**(D5), 9165–9177.

Shuman, C. A., R. B. Alley, M. A. Fahnestock, P. J. Fawcett, R. A. Bindschadler, S. Anandakrishnan, and C. R. Stearns (1997). Detection and monitoring of annual indicators and temperature trends at GISP2 using passive microwave remote sensing data. *Journal of Geophysical Research*, **102**, 26877–26886.

Shuman, C. A., D. H. Bromwich, J. Kipfstuhl, and M. Schwager (2001). Multiyear accumulation and temperature history near the North Greenland Ice Core Project site, north central Greenland. *Journal of Geophysical Research*, **106**(D24), 33853–33866.

Shuman, C. A., H. J. Zwally, B. E. Schutz, A. C. Brenner, J. P. DiMarzio, W. Abdalati, and V. P. Suchdeo (in press). Ice sheets from ICESat: Summary. *Geophysical Research Letters*.

Siegert, M. J., S. Carter, I. Tabacco, S. Popov, and D. D. Blankenship (2005). A revised inventory of Antarctic subglacial lakes. *Antarctic Science*, **17**(3), 453–460.

Skvarca, E (1993). Fast recession of the northern Larsen Ice Shelf, Antarctic Peninsula monitored by space images. *Annals of Glaciology*, **17**, 317–321.

Skvarca, P. (1994). Changes and surface features of the Larsen Ice Shelf, Antarctica, derived from Landsat and Kosmos mosaics. *Annals of Glaciology*, **20**, 6–12.

Skvarca, P. and H. De Angelis (2003). Impact assessment of climatic warming on glaciers and ice shelves on northeastern Antarctic Peninsula. In: E. Domack, A. Burneet, A. Leventer, P. Convey, M. Kirby and R. Bindschadler (eds.), *Antarctic Peninsula Climate Variability: A Historical and Palaeoenvironmental Perspective* (AGU Antarctic Research Series No. 79). American Geophysical Union, Washington, DC, pp. 69–78.

Skvarca, P., W. Rack, H. Rott, and T. Ibarzábal y Donángelo (1998). Evidence of recent climatic warming on the eastern Antarctic Peninsula. *Annals of Glaciology*, **27**, 628–632.

Skvarca. P., W. Rack, and H. Rott (1999a). 34 year satellite time series to monitor characteristics, extent, and dynamics of Larsen B Ice Shelf, Antarctic Peninsula. *Annals of Glaciology*, **29**, 255–260.

Skvarca. P., W. Rack., H. Rott, and T. Ibarzábal y Donángelo (1999b). Climatic trend and the retreat and disintegration of ice shelves on the Antarctic Peninsula: An overview. *Polar Research*, **18**, 151–157.

Smith, A. J. E. (1999). *Application of Satellite Altimetry for Global Ocean Tide Modeling*. Delft University Press, Delft, The Netherlands, 200 pp.

Smith, A. M. (1991). The use of tiltmeters to study the dynamics of Antarctic ice-shelf grounding lines. *Journal of Glaciology*, **37**(125), 51–58.

Smith, B. E., C. R. Bentley, and C. F. Raymond (2005). Recent elevation changes on the ice streams and ridges of the Ross Embayment from ICESat crossovers. *Geophysical Research Letters*, **32**(21), L21S09, doi:10.1029/2005GL024365.

Smith, B. T., T. D. van Ommen, and V. I. Morgan (2002). Distribution of oxygen isotope ratios and snow accumulation rates in Wilhelm II Land, East Antarctica. *Annals of Glaciology*, **35**, 107–110.

Smith, I. N., W. F. Budd, and P. Reid (1998). Model estimates of Antarctic accumulation rates and their relationship to temperature changes. *Annals of Glaciology*, **27**, 246–250.

Smith, J. A. (ed.) (2004). *InSAR: Interferometric Synthetic Aperture Radar Concept Study Report*. NASA Jet Propulsion Laboratory, Pasadena, CA.

Smith, R. C., S. Stammerjohn, and K. S. Baker (1996). Surface air temperature in the Western Antarctic Peninsula region. In: R. M. Ross, E. E. Hofmann and L. B. Quetin (eds), *Foundations for Ecological Research West of Antarctic Peninsula* (Ant. Res. Series No. 70). American Geophysical Union, Washington, DC, pp. 104–121.

Sohn, H.-G. and K. C. Jezek (1999). Mapping ice sheet margins from ERS-1 SAR and SPOT imagery. *International Journal of Remote Sensing*, **20**(15/16), 3201–3216.

Sohn, H.-G., K. C. Jezek, and C. J. Van der Veen (1998). Jakobshavn Glacier, West Greenland: 30 years of spaceborne observations. *Geophysical Research Letters*, **25**(14), 2699–2702.

Spikes, V. B., B. Csatho, and I. M. Whillans (2003). Laser profiling of Antarctic ice streams: Methods and accuracy. *Journal of Glaciology*, **49**(165), 315–322.

Spinhirne, J. D., S. P. Palm, W. D. Hart, D. L. Hlavka, and E. L. Welton (2005). Cloud and aerosol measurements from GLAS: Overview and initial results. *Geophysical Research Letters*, **32**(22), L22S03, doi:10.1029/2005GL023507.

Stamnes, K. (1999). *Snow Grain Size/Impurities (CTSK2b1)* (draft algorithm description, NASDA Internal Document No. 3.4.2). National Space Development Agency of Japan, Tokyo, pp. 1–27.

Stamnes, K. and W. Li (2001). Retrieval of snow grain size and impurity from GLI: Atmospheric correction. *ADEOS–II GLI Workshop FY2001, November 14–16, Tokyo*. Available online at ⟨http://sharaku.eorc.nasda.go.jp/GLI/meet/2001/C_01.pdf⟩.

Stamnes, K., S.-C. Tsay, W. Wiscombe, and K. Jayaweera (1988). Numerically stable algorithm for discrete-ordinate-method radiative transfer in multiple scattering and emitting layered media. *Applied Optics*, **27**, 2502–2509.

Stearns, L. A. and G. S. Hamilton (2004). *Radarsat SAR Backscatter Variations and Ice Sheet Surface Conditions along US ITASE Traverse Routes in Antarctica* (Terra Nostra, Schriften der Alfred-Wegener-Stiftung, 2004/4, Abstract S21/P30). Alfred Wegener Institute for Polar and Marine Research, Bremerhaven, Germany, pp. 436.

Stearns, L. A. and G. S. Hamilton (in press). Velocities along Byrd Glacier, East Antarctica, derived from ASTER satellite imagery using automatic feature tracking. *Annals of Glaciology*, **41**.

Steffen, K. (1987). *Bidirectional Reflectance of Snow at 500–600 nm* (IAHS Publication No. 166). International Association of Hydrological Sciences, Christchurch, New Zealand, pp. 415–425.

Steffen, K. (1995). Surface energy exchange during the onset of melt at the equilibrium line altitude of the Greenland Ice Sheet. *Annals of Glaciology*, **21**, 13–18.

Steffen, K. and J. Box (2001). Surface climatology of the Greenland ice sheet: Greenland Climate Network 1995–1999. *Journal of Geophysical Research*, **106**(D24), 33951–33964.

Steffen, K. and R. Huff (2003). *A Record Maximum Melt Extent on the Greenland Ice Sheet in 2002* (digital media). Available online at ⟨http://cires.colorado.edu/steffen/melt/index.html⟩.

Steffen, K., W. Abdalati, and J. Stroeve (1993a). Climate sensitivity studies of the Greenland Ice Sheet using satellite AVHRR, SMMR, SSM/I and *in situ* data. *Meteorology and Atmospheric Physics*, **51**, 239–258.

Steffen, K., R. Bindschadler, C. Casassa, J. Comiso, D. Eppler, F. Fetterer, J. Hawkins, J. Key, D. Rothrock, R. Thomas et al. (1993b). Snow and ice applications of AVHRR in polar regions: report of a workshop held in Boulder, Colorado, May 20, 1992. *Annals of Glaciology*, **17**, 1–16.

Steffen, K., J. Box, and W. Abdalati (1996). Greenland surface climatology and GC–Net. *Program for Arctic and Regional Climate Assessment* (report). Cooperative Institute for Research in Environmental Sciences (CIRES), University of Colorado, pp. 25–30.

Steffen, K., S. V. Nghiem, R. Huff, and G. Neumann (2004). The melt anomaly of 2002 on the Greenland Ice Sheet from active and passive microwave satellite observations. *Geophysical Research Letters*, **31**(20), L20402, DOI: 10.1029/2004GL020444.

Steitz, D., E. Thompson, and C. M. O'Carroll (2003). *ICESat's Lasers Measure Ice, Clouds and Land Elevations* (digital media). Avilable online at ⟨http://icesat.gsfc.nasa.gov/⟩.

Stenni, B., F. Serra, M. Frezzotti, V. Maggi, R. Traversi, S. Becagli, and R. Udisti (2000). Snow accumulation rates in northern Victoria Land, Antarctica, by firn-core analysis. *Journal of Glaciology*, **46**(155), 541–552.

Stenoien, M. D. and C. R. Bentley (2000). Pine Island Glacier, Antarctica: A study of the catchment using interferometric synthetic aperture radar measurements and radar altimetry. *Journal of Geophysical Research-Solid Earth*, **105**(B9), 21761–21779.

Stephen, H. and D. G. Long (2000). Study of Iceberg B10A using scatterometer data. *Proceedings of International Geoscience and Remote Sensing Symposium IGARSS '00, Honolulu, HI, 24–28 July 2000*. Institute of Electrical and Electronic Engineers, Piscataway, NJ, Vol. 3, pp. 1340–1342.

Stephenson, S. N. (1984). Glacier flexure and the position of grounding lines on Rutford Ice Stream, Antarctica. *Annals of Glaciology*, **5**, 165–169.

Stephenson, S. N. and R. A. Bindschadler (1988). Observed velocity fluctuations on a major Antarctic ice stream *Nature*, **334**(6184), 695–697.

Stephenson, S. N. and R. A. Bindschadler (1990). Is ice-stream evolution revealed by satellite imagery? *Annals of Glaciology*, **14**, 273–277.

Stricker, N. C. M., A. Hahne, D. L. Smith, J. Delderfield, M. B. Oliver, and T. Edwards (1995). ATSR-2: The evolution in its design from ERS-1 to ERS-2. *ESA Bulletin*, **83**, 32–37.

Stroeve, J. (2001). Assessment of Greenland albedo variability from the Advanced Very High Resolution Radiometer Polar Pathfinder data set. *Journal of Geophysical Research*, **106**(D24), 33989–34006.

Stroeve, J. C. and A. W. Nolin (2002a). New methods to infer snow albedo from the MISR instrument with applications to the Greenland Ice Sheet. *IEEE Transactions on Geoscience and Remote Sensing*, **40**, 1616–1625.

Stroeve, J. C. and A. Nolin (2002b). *Comparison of Snow Albedo from MISR, MODIS and AVHRR with Ground-based Observations on the Greenland Ice Sheet* (Glaciological Data Report GD-30). National Snow and Ice Data Center, Boulder, CO, pp. 80.

Stroeve, J C. and A. W. Nolin (2003). Greenland albedo variability. *Proceedings of 7th Conference on Polar Meteorology and Oceanography and Joint Symposium on High-Latitude Climate Variations, May 12–16, Hyannis, MA* (Paper P12.2). American Meteorological Society, Boston.

Stroeve, J. and K. Steffen (1998). Variability of AVHRR-derived clear sky surface temperature over the Greenland Ice Sheet. *Journal of Applied Meteorology*, **37**(1), 23–31.

Stroeve, J., M. Haefliger, and K. Steffen (1996). Surface temperature from ERS-1 ATSR infrared thermal satellite data in polar regions. *Journal of Applied Meteorology*, **35**(8), 1231–1239.

Stroeve, J., A. Nolin, and K. Steffen (1997). Comparison of AVHRR-derived and *in situ* surface albedo over the Greenland Ice Sheet. *Remote Sensing of Environment*, **62**(3), 262–276.

Stroeve, J., J. Box, C. Fowler, T. Haran, J. Key, and J. Maslanik (2000). Intercomparison between *in situ* and AVHRR Polar Pathfinder-derived surface albedo over Greenland. *Remote Sensing of Environment*, **75**, 360–374.

Stroeve, J., J. Box, F. Gao, S. Liang, A. Nolin, and C. Schaaf (2005). Accuracy assessment of the MODIS 16-day albedo product for snow: Comparisons with Greenland *in situ* measurements. *Remote Sensing of Environment*, **94**(1), 46–60.

Strozzi T., A. Luckman, and T. Murray (2000). The evolution of a glacier surge observed with the ERS satellites. *Proceedings of ERS-ENVISAT Symposium, Gothenburg, Sweden, October 16–20*.

Strozzi, T., A. Luckman, T. Murray, U. Wegmüller, and C. L. Werner (2002). Glacier motion estimation using SAR offset-tracking procedures. *IEEE Transactions on Geoscience and Remote Sensing*, **40**(11), 2384–2391.

Surdyk, S. (2002a). Using microwave brightness temperature to detect short-term surface air temperature changes in Antarctica: An analytical approach. *Remote Sensing of Environment*, **80**(2), 256–271.

Surdyk, S. (2002b). Low microwave brightness temperatures in central Antarctica: Observed features and implications. *Annals of Glaciology*, **34**, 134–140.

Surdyk, S. and M. Fily (1993). Comparison of the passive microwave spectral signature of the Antarctic Ice Sheet with ground traverse data. *Annals of Glaciology*, **17**, 161–166.

Surdyk, S. and M. Fily (1995). Results of a stratified snow emissivity model based on the wave approach: Application to the Antarctic Ice Sheet. *Journal of Geophysical Research*, **100**(C5), 8837–8848.

Suttles, J. T., R. N. Green, P. Minnis, G. L. Smith, W. F. Staylor, B. A. Wielicki, I. J. Walker, D. F. Young, V. R. Taylor, and L. L. Stowe (1988). *Angular Radiation Models for Earth–Atmosphere System. Volume 1: Shortwave Radiation* (NASA Reference Publication 1184). NASA, Washington, DC, 144 pp.

Swithinbank, C. (1988). *Antarctica* (USGS Professional Paper 1386-B). U.S. Geological Survey, Washington, DC, 278 pp.

Swithinbank, C., R. S. Williams Jr., J. G. Ferrigno, B. A. Seekins, B. K. Lucchitta, and C. E. Rosanova (1997). *Coastal-change and Glaciological Map of the Bakutis Coast, Antarctica: 1972–1990* (Geological Investigations Map I-2600-F, scale 1:1,000,000). U.S. Geological Survey, Reston, VA.

Swithinbank, C., K. Brunk, and J. Sievers (1988). A glaciological map of Filchner-Ronne Ice Shelf, Antarctica. *Annals of Glaciology*, **11**, 150–155.

Swithinbank, C., R. S. Williams, J. G. Ferrigno, K. M. Foley, and C. E. Rosanova (2003). Coastal-change and glaciological map of the Bakutis Coast area. *Antarctica: 1972–2002* (text accompanying Map I-2600-F, 2nd edn., 10 pp.). U.S. Geological Survey, Reston, VA. Available online at ⟨http://pubs.usgs.gov/imap/2600/F/i2600f.pdf⟩.

Tadono, T., M. Shimada, M. Tatanabe, T. Hashimoto, and T. Iwata (2004). Calibration and validation of PRISM onboard ALOS. *Proceedings of XXth ISPRS Congress 2004, Istanbul, Turkey, July 12–23, 2004* (Commission 2). Available online at ⟨ttp://www.isprs.org/istanbul2004/comm2/comm2.html⟩.

Takahashi, S., R. Naruse, F. Nishio, T. Furukawa, H. Motoyama, S. Fujita, and S. Mae (1997). *East Queen Maud Land Enderby Land: Glaciological Folio, Ice Sheet Dynamics*. National Institute of Polar Research, Tokyo.

Tanikawa T., T. Aoki, and F. Nishio (2002). Remote sensing of snow grain-size and impurities from Airborne Multispectral Scanner data using a Snow Bidirectional Reflectance Distribution Function model. *Annals of Glaciology*, **34**, 74–80.

Tapley B. D., M. M. Watkins, J. C. Ries, G. W. Davis, R. J. Eanes, S. R. Poole, H. J. Rim, B. E. Schutz, C. K. Shum, R. S. Nerem et al. (1996). The JGM-3 geopotential model. *Journal of Geophysical Research*, **101**(B12), 28029–28049.

Tapley, B. D., S. Bettadpur, M. Watkins, and C. Reigber (2004a). The Gravity Recovery and Climate Experiment: Mission overview and early results. *Geophysical Research Letters*, **31**(9), L09607, DOI: 10.1029/2004GL019920.

Tapley, B. D., S. Bettadpur, J. C. Ries, P. F. Thompson, and M. Watkins (2004b). GRACE measurements of mass variability in the Earth system. *Science*, **305**(5683), 503–505.

Tchernia, P. and P. F. Jeannin (1980). Observations on the Antarctic East Wind Drift using tabular icebergs tracked by satellite Nimbus F (1975–1977). *Deep-sea Research*, **27**, 467–474.

Tchernia, T. and T. F. Jeannin (1984). Circulation in Antarctic waters as revealed by iceberg tracks 1972–1983. *Polar Record*, **22**(138), 263–269.

Teixidó, N., J. Garrabou, J. Gutt, and W. E. Arntz (2004). Recovery in Antarctic benthos after iceberg disturbance: Trends in benthic composition, abundance and growth forms. *Marine Ecology Progress Series*, **278**, 1–16.

Testut, L., R. Hurd, R. Coleman, F. Rémy, and B. Legrésy (2003). Precise drainage pattern of Antarctica derived from high resolution topography. *Annals of Glaciology*, **37**, 337–343.

Thiel K. H., P. H. Hartl, and X. Wu (1996). Monitoring the ice movements with ERS SAR interferometry in the Antarctic region. *Proceedings of the 2nd ERS Applications Workshop, London* (ESA SP-383, pp. 219–223). ESA, Noordwijk, The Netherlands.

Thomas, R. H. (1973). The creep of ice shelves: Interpretation of observed behaviour. *Journal of Glaciology*, **12**(64), 55–70.

Thomas, R. H. (1979). Effect of climatic warming on the West Antarctic Ice Sheet. *Nature*, **277**, 355–358.

Thomas, R. H. (2001). Remote sensing reveals shrinking Greenland Ice Sheet. *EOS, Transactions of the American Geophysical Union*, **82**(34), 369–373.

Thomas, R. H. (2004). Greenland: Recent mass-balance observations. In: J. L. Bamber and A. J. Payne (eds.), *Mass Balance of the Cryosphere*. Cambridge University Press, Cambridge, UK, pp. 393–436.

Thomas, R. H., E. Rignot, P. Kanagaratnam, W. Krabill, and G. Casassa (2004). Force-balance analysis of Pine Island Glacier, Antarctica, suggests cause for recent acceleration. *Annals of Glaciology*, **39**, 133–138.

Thomas, R. H., T. O. Sanders, and K. E. Rose (1979). Effect of warming on the West Antarctic Ice Sheet. *Nature*, **277**, 355–358.

Thomas, R. H., D. R. MacAyeal, D. H. Eilers, and D. R. Gaylord (1984). Glaciological studies on the Ross Ice Shelf, Antarctica, 1973–1978. In: D. Hayes and C. R. Bentley (eds.), *The Ross Ice Shelf: Glaciology and Geophysics* (AGU Antarctic Research Series No. 42). American Geophysical Union, Washington, DC, pp. 21–53.

Thomas, R. H., R. A. Bindschadler, R. L. Cameron, F. D. Carsey, B. Holt, T. J. Hughes, C. W. M. Swithinbank, I. M. Whillans, and H. J. Zwally (1985). *Satellite Remote Sensing for Ice Sheet Research* (NASA Technical Memorandum No. 86,233). NASA, Washington, DC, 32 pp.

Thomas, R. H., W. Abdalati, T. L. Akins, B. M. Csatho, E. B. Frederick, S. P. Gogineni, W. B. Krabill, S. S. Manizade, and E. J. Rignot (2000a). Substantial thinning of a major east Greenland outlet glacier. *Geophysical Research Letters*, **27**(9), 1291–1294.

Thomas, R., B. Csatho, M. Fahnestock, P. Gogineni, C. Kim, and J. Sonntag (2000b). Mass balance of the Greenland Ice Sheet a high elevations. *Science*, **289**(5478), 426–427.

Thomas, R. H. and PARCA Investigators (2001a). PARCA 2001, Program for Arctic Regional Climate Assessment (PARCA): Goals, key findings, and future directions. *Journal of Geophysical Research*, **106**(D24), 33691–33706.

Thomas, R., B. Csatho, C. Davis, C. Kim, W. Krabill, S. Manizade, J. McConnell, and J. Sonntag (2001b). Mass balance of higher-elevation parts of the Greenland Ice Sheet. *Journal of Geophysical Research*, **106**(D24), 33707–33716.

Thomas, R. H., W. Abdalati, W. B. Krabill, S. Manizade, and K. Steffen (2003). Investigation of surface melting and dynamic thinning of Jakobshavn Isbrae, Greenland. *Journal of Glaciology*, **49**(165), 231–239.

Thomas, R., E. Rignot, G. Casassa, P. Kanagaratnam, C. Acuña, T. Akins, H. Brecher, E. Frederick, P. Gogineni, W. Krabill et al. (2004). Accelerated sea-level rise from West Antarctica. *Science*, **306**(5694), 255–258.

Thompson, D. W. J. and S. Solomon (2002). Interpretation of recent Southern Hemisphere climate change. *Science*, **296**(5569), 895–899.

Tiuri, M. E., A. H. Sihvola, E. G. Nyfors, and M. T. Hallikainen (1984). The complex dielectric constant of snow at microwave frequencies. *IEEE Journal of Oceanic Engineering*, **OE-9**, 377–382.

Toll, D. L., D. Shirey, and D. S. Kimes (1997). NOAA AVHRR land surface albedo algorithm development. *International Journal of Remote Sensing*, **18**(18), 3761–3796.

Torinesi, O., M. Fily, and C. Genthon (2003). Variability and trends of the summer melt period of Antarctic ice margins since 1980 from microwave sensors. *Journal of Climate*, **16**(7), 1047–1060.

Toutin T. (2002). Three-dimensional topographic mapping with ASTER stereo data in rugged topography. *IEEE Transactions on Geoscience and Remote Sensing*, **40**(10), 2241–2247.

Toutin, T. (2004). Comparison of stereo-extracted DTM from different high-resolution sensors: SPOT-5, EROS-A, IKONOS-II, and QuickBird. *IEEE Transactions on Geosciene and Remote Sensing*, **42**(10), 2121–2129.

Toutin, T. and P. Cheng (2001). DEM generation with ASTER stereo data. *Earth Observation Magazine*, **10**(6), 10–13.

Toutin, T. and L. Gray (2000). State-of-the-art of elevation extraction from satellite SAR data. *ISPRS Journal of Photogrammetry and Remote Sensing*, **55**, 13–33.

Turner, J., W. M. Connolley, S. Leonard, G. R. Marshall, and D. G. Vaughan (1999). Spatial and temporal variability of net snow accumulation over the Antarctic from ECMWF Re–analysis Project data. *International Journal of Climatology*, **19**, 697–724.

Turner, J., J. C. King, T. A. Lachlan-Cope, and P. D. Jones (2002). Recent temperature trends in the Antarctic. *Nature*, **418**, 291–292.

Ulaby, F. T. (1996). *Fundamentals of Applied Electromagnetics*. Prentice Hall, Englewood Cliffs, NJ, 407 pp.

Ulaby, F. T., R. K. Moore, and A. K. Fung (1981). *Microwave Remote Sensing, Active and Passive: Vol. 1. Microwave Remote Sensing Fundamentals and Radiometry*. Addison-Wesley, Reading, MA, 456 pp.

Ulaby, F. T., R. K. Moore, and A. K. Fung (1982). *Microwave Remote Sensing, Active and Passive: Vol. 2. Radar Remote Sensing and Surface Scattering and Emission Theory*. Addison-Wesley, Reading, MA, 609 pp.

Ungar, S. G., J. S. Pearlman, J. A. Meldenhall, and D. Reuter (2003). Overview of the Earth Observing One (EO-1) mission. *IEEE Transactions on Geoscience and Remote Sensing*, **41**(6), 1149–1159.

Van de Wal, R. S. W. (2004). Greenland: Modelling. In: J. L. Bamber and A. J. Payne (eds.), *Mass Balance of the Cryosphere*. Cambridge University Press, Cambridge, UK, pp. 437–458.

Van de Wal, R. S. W. and S. Ekholm (1996). On elevation as input for mass-balance calculations of the Greenland ice sheet. *Annals of Glaciology*, **23**, 181–186.

Van de Wal, R. S. W., M. Wild, and J. de Wolde (2001). Short-term volume changes of the Greenland Ice Sheet in response to doubled CO_2 conditions. *Tellus*, **53B**, 94–102.

Van den Broeke, M. and R. Bintanja (1995). The interaction of katabatic winds and the formation of the blue-ice areas in East Antarctica. *Journal of Glaciology*, **41**, 395–408.

Van den Broeke, M. R., D. van As, C. H. Reijmer, and R. S. W. van de Wal (2004). The surface radiation balance in Antarctica using automatic weather stations. *Journal of Geophysical Research*, **109**, doi:10.1029/2003JD004394.

Van der Sanden, J. J. and S. G. Ross (eds.) (2001). *Applications Potential of Radarsat-2: A Preview* (report prepared for the Canadian Space Agency). Natural Resources Canada and Canada Centre for Remote Sensing, Ottawa, 117 pp.

Van der Veen, C. J. (1991). State of balance of the cryosphere. *Reviews of Geophysics*, **29**, 433–455.

Van der Veen, C. L. (1998). Fracture mechanics approach to penetration of surface crevasses on glaciers. *Cold Regions Science and Technology*, **27**, 31–47.

Van der Veen, C. J. (1999). *Fundamentals of Glacier Dynamics*. A. A. Balkema, Rotterdam, The Netherlands, 462 pp.

Van der Veen, C. J. and J. F. Bolzan (1999). Interannual variability in net accumulation on the Greenland Ice Sheet: Observations and implications for mass balance measurements. *Journal of Geophysical Research*, **104**(D2), 2009–2014.

Van der Veen, C. J. and K. C. Jezek (1993). Seasonal variations in brightness temperature for central Antarctica. *Annals of Glaciology*, **17**, 300–306.

Van der Veen, C. J. and A. Payne (2004). Modelling land ice dynamics. In: J. L. Bamber and A. J. Payne (eds.), *Mass Balance of the Cryosphere*. Cambridge University Press, Cambridge, UK, pp. 167–226.

Van der Veen, C. J., D. H. Bromwich, B. M. Csatho, and C. Kim (2001). Trend surface analysis of Greenland accumulation. *Journal of Geophysical Research*, **106**(D24), 33909–33918.

Vaughan, D. G. (1993a). Relating the occurrence of crevasses to surface strain rates. *Journal of Glaciology*, **39**(132), 255–266.

Vaughan, D. G. (1993b). Implications of the breakup of the Wordie Ice Shelf, Antarctica, for sea level. *Antarctic Science*, **5**, 403–408.

Vaughan, D. G. (1994). Investigating tidal flexure on an ice shelf using kinematic GPS. *Annals of Glaciology*, **20**, 372–376.

Vaughan, D. G. (1995). Tidal flexure of ice shelf margins. *Journal of Geophysical Research*, **100**(B4), 6213–6224.

Vaughan, D. and J. Bamber (1998). Identifying areas of low-profile ice sheet and outcrop damping in the Antarctic ice sheet by ERS-1 satellite altimetry. *Annals of Glaciology*, **27**, 1–6.

Vaughan, D. G and C. S. M. Doake (1996). Recent atmospheric warming and retreat of ice shelves on the Antarctic Peninsula. *Nature*, **379**, 328–331.

Vaughan, D. G. and M. Jonas (1996b). Measurements of velocity of Filchner-Ronne Ice Shelf. *Filchner-Ronne Ice Shelf Programme (FRISP)*, **10**, 111–116.

Vaughan, D. G., C. S. M. Doake, and D. R. Mantripp (1988). *Topography of an Antarctic Ice Stream* (SPOT 1 image utilization, assessment, results, report). Centre National d'Études Spatiales, Toulouse, France, pp. 167–174.

Vaughan, D. G., R. M. Frolich, and C. S. M. Doake (1994). ERS-1 SAR: Stress indicator for Antarctic ice streams. *Proceedings of 2nd ERS-1 Symposium* (ESA SP-361). ESA, Noordwijk, The Netherlands, pp. 183–186.

Vaughan, D. G., J. Sievers, C. S. M. Doake, H. Hinze, D. R. Mantripp, H. Sandhager, H. W. Schenke, A. Solheim, and F. Thyssen (1995). Subglacial and seabed topography, ice thickness and water column thickness in the vicinity of Filchner-Ronne-Schelfeis, Antarctica. *Polarforschung*, **64**(2), 75–88.

Vaughan, D. G., J. L. Bamber, M. Giovinetto, J. Russell, and A. P. R. Cooper (1999). Reassessment of surface mass balance in Antarctica. *Journal of Climate*, **12**, 933–946.

Vaughan, D. G., G. J. Marshall, W. M. Connolley, J. C. King, and R. Mulvaney (2001a). Devil in the detail. *Science*, **293**, 1777–1779.

Vaughan, D. G., A. M. Smith, H. F. J. Corr, A. Jenkins, C. R. Bentley, M. D. Stenoien, S. S. Jacobs, T. B. Kellogg, E. Rignot, and B. Lucchitta (2001b). A review of Pine Island Glacier, West Antarctica: Hypotheses of instability vs. observations of change. In: R. Alley and R. Bindschadler (eds.), *The West Antarctic Ice Sheet: Behavior and Environment* (AGU Antarctic Research Series No. 77). American Geophysical Union, Washington, DC, pp. 237–256.

Vaughan, D. G., G. J. Marshall, W. M. Connolley, C. Parkinson, R. Mulvaney, D. A. Hodgson, J. C. King, C. J. Pudsey, and J. Turner (2003). Recent rapid regional climate warming on the Antarctic Peninsula. *Climate Change*, **60**, 243–274.

Velicogna, I. and J. Wahr (2002). A method for separating Antarctic postglacial rebound and ice mass balance using future ICESat Geoscience Laser Altimeter System, Gravity Recovery and Climate Experiment, and GPS satellite data. *Journal of Geophysical Research*, **107**(B10), 2263, DOI: 10.1029/2001JB000708.

Velicogna, I. and J. Wahr (2005). Greenland mass balance from GRACE. *Geophysical Research Letters*, **32**(18), L18505, doi:10.1029/2005GL023955.

Velicogna, I., J. Wahr, E. Hanna, and P. Huybrechts (2005). Short-term mass variability in Greenland, from GRACE. *Geophysical Research Letters*, **32**, L05501, doi:10.1029/2004GL021948.

Vermote, E. F. and A. Vermeulen (1999). *Atmospheric Correction Algorithm: Spectral Reflectances (MOD09)* (Algorithm Theoretical Basis Document). NASA Goddard Space Flight Center, Greenbelt, MD, 107 pp.

Vermote, E. F., D. Tanré, J. L. Deuze, M. Herman, and J. J. Morcrette (1997). Second simulation of the satellite signal in the solar spectrum: An overview. *IEEE Transactions on Geoscience and Remote Sensing*, **35**(3), 675–686.

Viehoff, T. and A. Li (1995). Iceberg observations and estimation of submarine ridges in the western Weddell Sea. *International Journal of Remote Sensing*, **16**(17), 3391–3408.

Vincent, W. F., D. R. Mueller, and P. Van Hove (2004). Break-up and climate change at Canada's northern coast, Quttinirpaaq National Park. *Meridian*, Spring/Summer, 1–6.

Vinje, T. E. (1980). Some satellite-tracked iceberg drifts in the Antarctic. *Annals of Glaciology*, **1**, 83–87.

Vinje, T. E. (1989). Icebergs in the Barents Sea. *Proceedings of 8th International Offshore Mechanics and Arctic Engineering Conference (OMAE)*, The Hague, Netherlands, March 19–23. American Society of Mechanical Engineers, New York, pp. 139–145.

Vornberger, P. L. and R. A. Bindschadler (1992). Multi-spectral analysis of ice sheets using co-registered SAR and TM imagery. *International Journal of Remote Sensing*, **13**(4), 637–645.

Wahr, J., D. Wingham, and C. Bentley (2000). A method of combining ICESat and GRACE satellite data to constrain Antarctic mass balance. *Journal of Geophysical Research*, **105**(B7), 16279–16294.

Wahr, J., T. van Dam, K. Larson, Kristine, and F. Olivier (2001). GPS measurements of vertical crustal motion in Greenland. *Journal of Geophysical Research*, **106**(D24), 33755–33760.

WAIS (2003). *The West Antarctic Ice Sheet Initiative: A Multi-disciplinary Study of Rapid Climate Change and Future Sea Level*. Avilable online at ⟨http://igloo.gsfc.nasa.gov/wais/⟩.

Walsh, J. E., J. Curry, M. Fahnestock, M. C. Kennicutt II, A. D. McGuire, W. B. Rossow, M. Steele, C. J. Vorosmarty, R. Wharton, C. Elfring et al. (2001). *Enhancing NASA's Contributions to Polar Science*. National Academy Press, Washington, DC, 138 pp.

Wang, W. L. and R. C. Warner (1998). Simulation of the influence of ice rheology on velocity profiles and ice-sheet mass balance. *Annals of Glaciology*, **27**, 194–200.

Wanner, W., A. H. Strahler, B. Hu, P. Lewis, J.-P. Muller, X. Li, C. L. Barker Schaaf, and M. J. Barnsley (1997). Global retrieval of bidirectional reflectance and albedo over land from EOS MODIS and MISR data: Theory and algorithm. *Journal of Geophysical Research*, **102**(D14), 17143–17161.

Warner, R. C. and W. E. Budd (1998). Modelling the long-term response of the Antarctic ice sheet to global warming. *Annals of Glaciology*, **27**, 161–168.

Warren, S. G. (1982). Optical properties of snow. *Reviews of Geophysics Space Phys.*, **20**(1), 67–89.

Warren, S. G. (1984). Optical constants of ice from the ultraviolet to the micro-wave. *Applied Optics*, **23**(8), 1206–1225.

Warren, S. G. (1996). Antarctica. *Encyclopedia of Climate and Weather*. New York, Oxford University Press, pp. 32–39.

Warren, S. G. and A. D. Clarke (1986). *Soot from Arctic Haze: Radiative Effects on the Arctic Snowpack* (Glaciological Data 18). World Data Center for Glaciology (Snow and Ice), Boulder, CO, pp. 73–77.

Warren, S. G. and W J. Wiscombe (1980). A model for the spectral albedo of snow: II. Snow containing atmospheric aerosols. *Journal of Atmospherical Science*, **37**(12), 2734–2745.

Warren, S. G., R. E. Brandt, and R. D. Boime (1993). Blue ice and green ice. *Antarctic Journal of the United States*, **28**, 255–256.

Warren, S. G., R. E. Brandt, and P. O'Rawe Hinton (1998). Effects of surface roughness on bidirectional reflectance of Antarctic snow. *Journal of Geophysical Research*, **103**, 25789–25807.

Watanabe, O. (1978). Distribution of surface features of snow cover in Mizuho Plateau. *National Institute of Polar Research*, Special Issue 7, 44–62.

Weertman, J. (1973). *Can a Water-filled Crevasse Reach the Bottom Surface of a Glacier?* (IAHS Publication No. 95). International Association of Hydrological Sciences, Christchurch, New Zealand, pp. 139–145.

Weertman, J. (1974). Stability of the junction of an ice sheet and an ice shelf. *Journal of Glaciology*, **13**(67), 3–11.

Weertman, J. (1976). Glaciology's great unsolved problem. *Nature*, **260**, 284–286.

Weidick, A. (1994). Fluctuations of West Greenland calving glaciers. In: N. Reeh (ed.), *Workshop on the Calving Rate of West Greenland Glaciers in Response to Climate Change* (report). Danish Polar Institute, Copenhagen, pp. 141–168.

Weidick, A., C. Andreasen, H. Oerter, and N. Reeh (1996). Neoglacial glacier changes around Storstrømmen, North-East Greenland. *Polarforschung*, **64**(3), 95–108.

Welch, B. C. and R. W. Jacobel (2003). Analysis of deep-penetrating radar surveys of West Antarctica, US-ITASE 2001. *Geophysical Research Letters*, **30**(8), 1444, doi:10.1029/2003GL017210.

Wendler, G., K. Ahlnas, and C. S. Lingle (1996). On Mertz and Ninnis Glaciers, East Antarctica. *Journal of Glaciology*, **42**, 447–453.

West, R. D., D. P. Winebrenner, L. Tsang, and H. Rott (1996). Microwave emission from density-stratified Antarctic firn at 6 cm wavelength. *Journal of Glaciology*, **42**(140), 63–76.

Whillans, I. M. and Y. Tseng (1995). Automatic tracking of crevasses on satellite images. *Cold Regions Science and Technology*, **23**, 201–214.

Whillans, I. M. and C. J. Van der Veen (1993). New and improved determinations of velocity of Ice Streams B and C, West Antarctica. *Journal of Glaciology*, **39**, 483–490.

Whillans, I. M., C. R. Bentley, and C. J. van der Veen (2001). Ice Streams B and C. In: R. B. Alley and R. A. Bindschadler (eds.), *The West Antarctic Ice Sheet: Behavior and Environment* (AGU Antarctic Research Series No. 77). American Geophysical Union, Washington, DC, pp. 257–281.

Wiehl, M., B. Legrésy, and R. Dietrich (2003). Potential of reflected GNSS signals for ice sheet remote sensing. *Progress in Electromagnetics Research*, **PIER 40**, 177–205.

Wiesmann, A. and C. Mätzler (1999). Microwave emission model of layered snowpacks. *Remote Sensing of Environment*, **70**(3), 307–316.

Wild, M. and A. Ohmura (2000). Changes in mass balance of the polar ice sheets and sea level under greenhouse warming as projected in high resolution GCM simulations. *Annals of Glaciology*, **30**, 197–2003.

Wild, M., P. Calanca, S. C. Scherrer, and A. Ohmura (2003). Effects of polar ice sheets on global sea level in high-resolution greenhouse scenarios. *Journal of Geophysical Research*, **108**(D5), 4165, DOI: 10.1029/2002JD002451.

Williams, M. J. M., A. Jenkins, and J. Determann (1998a). Physical controls on ocean circulation beneath ice shelves revealed by numerical models. In: S. S. Jacobs and R. F. Weiss (eds.), *Ocean, Ice, and Atmosphere: Interactions at the Antarctic Continental Margin* (AGU Antarctic Research Series No. 75). American Geophysical Union, Washington, DC, pp. 285–299.

Williams, M. J. M., R. C. Warner, and W. F. Budd (1998b). The effects of ocean warming on melting and ocean circulation under the Amery Ice Shelf, East Antarctica. *Annals of Glaciology*, **27**, 75–80.

Williams, M. J. M., R. C. Warner, and W. F. Budd (2002). Sensitivity of the Amery Ice Shelf, Antarctica, to changes in the climate of the Southern Ocean. *Journal of Climate*, **15**, 2740–2757.

Williams, R. N., W. G. Rees, and N. W. Young (1999). A technique for the identification and analysis of icebergs in synthetic aperture radar images of Antarctica. *International Journal of Remote Sensing*, **20**(15/16), 3183–3199.

Williams, R. S. Jr. and J. G. Ferrigno (1995). Landsat images of Greenland. In: A. Weidick (ed.), *Greenland: Satellite Image Atlas of Glaciers of the World* (USGS Professional Paper 1386-C). U.S. Geological Survey, Reston, VA, pp. C116–C141.

Williams, R. S. Jr. and J. G. Ferrigno (1998). *Coastal-change and Glaciological Maps of Antarctica* (USGS Fact Sheet, FS-050-98). U.S. Geological Survey, Reston, VA, 2 pp.

Williams, R. T. and E. S. Robinson (1980). The ocean tide in the Southern Ross Sea. *Journal of Geophysical Research*, **85**(C11), 6689–6696.

Williams, R. S. Jr., J. G. Ferrigno, T. M. Kent, and J. W. Schoonmaker, Jr. (1982). Landsat images and mosaics of Antarctica for mapping and glaciological studies. *Annals of Glaciology*, **3**, 321–326.

Williams, R. S. Jr., D. K. Hall, and C. S. Benson (1991). Analysis of glacier facies using satellite techniques. *Journal of Glaciology*, **37**(125), 120–128.

Williams, R. S. Jr., J. G. Ferrigno, C. Swithinbank, B. K. Lucchitta, and B. A. Seekins (1995). Coastal change and glaciological maps of Antarctica. *Annals of Glaciology*, **21**, 284–290.

Williams, R. S. Jr., J. G. Ferrigno, C. Swithinbank, B. K. Lucchitta, B. A. Seekins, and C. E. Rosanova (1997). Glaciological delineation of the dynamic coastline of Antarctica. *Antarctic Journal of the United States*, **32**(2), 7 pp. Available online at ⟨http://www.nsf.gov/pubs/1998/nsf9824/ch10.htm⟩.

Willis, C. J., J. T. Macklin, K. C. Partington, K. A. Teleki, W. G. Rees, and R. G. Williams (1996). Iceberg detection using ERS-1 synthetic aperture radar. *International Journal of Remote Sensing*, **26**, 1777–1795.

Winebrenner, D. P., E. D. Nelson, R. Colony, and R. D. West (1994). Observation of melt onset on multiyear Arctic sea ice using the ERS-1 synthetic aperture radar. *Journal of Geophysical Research*, **99**(C11), 22425–22441.

Winebrenner, D. P., I. R. Joughin, and M. A. Fahnestock (1997). Interferometric SAR for observation of glacier motion and firn penetration. *Proceedings of 3rd ERS Scientific Symposium, March 17–21, Florence, Italy* (ESA SP-414). ESA, Frascati, Italy.

Winebrenner, D. P., R. J. Arthern, and C. A. Shuman (2001). Mapping Greenland accumulation rates using observations of thermal emission at 4.5-cm wavelength. *Journal of Geophysical Research*, **106**(D24), 33919–33934.

Winebrenner, D. P., E. J. Steig, and D. P. Schneider (2004). Temporal co-variation of surface and microwave brightness temperatures in Antarctica, with implications for the observation of surface temperature variability using satellite data. *Annals of Glaciology*, **39**, 346–350.

Wingham, D. J. (1995). The limiting resolution of ice-sheet elevations derived from pulse-limited satellite altimetry. *Journal of Glaciology*, **41**(138), 414–422.

Wingham, D. (1999). The first of the European Space Agency's opportunity missions: CryoSat. *Earth Observation Quarterly*, **63**, 21–24.

Wingham, D. J., C. G. Rapley, and J. G. Morley (1993). Improved resolution ice sheet mapping with satellite radar altimeters. *EOS, Transactions of the American Geophysical Union*, **74**(10), 113–166.

Wingham, D. J., A. J. Ridout, R. Scharroo, R. J. Arthern, and C. K. Shum (1998). Antarctic elevation change from 1992 to 1996. *Science*, **282**, 456–458.

Wingham, D. J., L. Phalippou, C. Mavrocordatos, and D. Wallis (2004). The mean echo and echo cross–product from a beamforming interferometric altimeter and their application to elevation measurements. *IEEE Transactions on Geoscience and Remote Sensing*, **42**(10), 2305–2323.

Winther, J.-G. (1993a). Landsat TM derived and *in situ* summer reflectance of glaciers in Svalbard. *Polar Research*, **12**(1), 37–55.

Winther, J.-G. (1993b). Studies of snow surface characteristics by Landsat TM in Dronning Maud Land, Antarctica. *Annals of Glaciology*, **17**, 27–34.

Winther, J.-G. (1993c). Short- and long-term variability of snow albedo. *Nordic Hydrology*, **24**(2–3), 199–212.

Winther, J.-G., S. Gerland, J. B. Ørback, B. Ivanov, A. Blanco, and J. Boike (1999). Spectral reflectance of melting snow in a high Arctic watershed on Svalbard: Some implications for optical remote sensing studies. *Hydrological Processes*, **13**(12/13), 2033–2049.

Winther, J.-G., M. N. Jespersen, and G. E. Liston (2001). Blue-ice areas in Antarctica derived from NOAA AVHRR satellite data. *Journal of Glaciology*, **47**(157), 325–334.

Wiscombe, W. J. and S. G. Warren (1980). A model for the spectral albedo of snow: I. Pure snow. *Journal of Atmospherical Science*, **37**(12), 2712–2733.

Wismann, V. (1998). Snow accumulation on Greenland derived from NSCAT and ERS scatterometers data. *Proceedings of Workshop on Emerging Scatterometer Applications, 5–7 October* (ESA SP-424). ESTEC, Noordwijk, The Netherlands, pp. 87–90.

Wismann, V. (2000). Monitoring of seasonal snowmelt on Greenland with ERS scatterometer data. *IEEE Transactions on Geoscience and Remote Sensing*, **38**(4), 1821–1826.

Wismann, V. and K. Boehnke (1996). Dramatic decrease in radar cross-section over Greenland observed by the ERS-1 scatterometer between 1991 and 1995. *Proceedings of International Geoscience and Remote Sensing Symposium IGARSS '96, Lincoln, NE, May*. Institute of Electrical and Electronic Engineers, Piscataway, NJ, Vol. 4, pp. 2014–2016.

Wismann, V. and K. Boehnke (1997). *Monitoring Snow Properties on Greenland with ERS Scatterometer and SAR* (ESA SP-414, Vol. 2). ESA, Noordwijk, The Netherlands, pp. 857–861.

Wismann, V. A., K. Boehnke, A. Cavanié, R. Ezraty, F. Gohin, D. H. Hoekman, and I. H. Woodhouse (1996). *Land Surface Observations Using the ERS-1 Windscatterometer: Part II* (report for the ESA, ESTEC Contract 11103/94/NL/CN). ESA, Noordwijk, The Netherlands.

Wismann, V. A., D. P. Winebrenner, K. Boehnke, and R. J. Arthern (1997). Snow accumulation on Greenland estimated from ERS Scatterometer data. *Proceedings of International Geoscience and Remote Sensing Symposium IGARSS '97*. Institute of Electrical and Electronic Engineers, Piscataway, NJ, Vol. 4, pp. 1823–1825.

Wolfe, J., C. Fowler, and T. Scambos (2001). Monitoring long-term regional changes in the Arctic and Antarctic with AVHRR 5 km Polar Pathfinder data. *EOS, Transactions of the American Geophysical Union*, **82**(47), Fall Meeting Suppl., Abstract IP22B-0694.

Wong, A. P. S., N. L. Bindoff, and A. Forbes (1998). Ocean ice-shelf interaction and possible bottom water formation in Prydz Bay. In: S. S. Jacobs and R. F. Weiss (eds.), *Ocean, Ice, and Atmosphere: Interactions at the Antarctic Continental Margin* (AGU Antarctic Research Series No. 75). American Geophysical Union, Washington, DC, pp. 173–187.

Wu, X., M. Watkins, R. Kwok, E. Ivins, and J. Wahr (2002). Measuring present-day secular change in Greenland ice mass with future GRACE gravity data. *Journal of Geophysical Research*, **107**(B11), 2291, DOI: 10.1029/2001JB000543.

Wunderle, S. and J. Schmidt (2000). Fluctuations and ice-flow velocity of the Northeast and McClary Glaciers on the Antarctic Peninsula derived from remote sensing data and SAR interferometry. *Polarforschung*, **67**(1/2), 41–52.

Yamaguchi, Y., A. B. Kahle, H. Tsu, T. Kawakami, and M. Pniel (1998). Overview of the Advanced Spaceborne Thermal Emission and Reflection Radiometer (ASTER). *IEEE Transactions on Geoscience and Remote Sensing*, **36**(4), 1062–1071.

Yamaguchi, Y., H. Fujisada, A. B. Kahle, H. Tsu, M. Kato, H. Watanabe, I. Sato, and M. Kudoh (2001). ASTER instrument performance, operational status, and application to Earth sciences. *Proceedings of International Geoscience and Remote Sensing Symposium IGARSS '01*. Institute of Electrical and Electronic Engineers, Piscataway, NJ, pp. 1215–1216.

Yamanouchi, T. and S. Kawaguchi (1984). Longwave radiation balance under a strong surface inversion in the katabatic wind zone, Antarctica. *Journal of Geophysical Research*, **89**, 11771–11778.

Yamanouchi, T., K. Suziki, and S. Kawaguchi (1987). Detection of clouds in Antarctica from infrared multispectral data of AVHRR. *Journal of Meteorological Society of Japan*, **65**(6), 949–961.

Yamanouchi, T., N. Hirasawa, G. Kadosaki, and M. Hayashi (2000). Evaluation of AVHRR cloud detection at Dome Fuji Station, Antarctica. *Polar Meteorology and Glaciology*, 110–116.

Yi, D. and C. R. Bentley (1994). Analysis of satellite radar altimeter return waveforms over the East Antarctic Ice Sheet. *Annals of Glaciology*, **19**, 137–142.

Yi, D. and C. R. Bentley (1997). A retracking algorithm for satellite radar altimetry over an ice sheet and its applications. In: S. C. Colbeck (ed.), *Glaciers, Ice Sheets, and Volcanoes: A Tribute to Mark F. Meier* (U.S. Army CRREL Special Report 96, Vol. 27, pp. 112–120). U.S. Army Cold Regions Research and Engineering Laboratory, Hanover, NH.

Yi, D., C. R. Bentley, and M. D. Stenoien (1997). Seasonal variation in the apparent height of the East Antarctic Ice Sheet. *Annals of Glaciology*, **24**, 191–198.

Yi, D., J. B. Minster, and C. R. Bentley (2000). The effect of ocean tidal loading on satellite altimetry over Antarctica. *Antarctic Science*, **12**(1), 119–124.

Young, N. W. (1998). Antarctic iceberg drift and ocean currents derived from scatterometer image series. *Proceedings of a joint ESA–Eumetsat Workshop on Emerging Scatterometer Applications: From Research to Operations* (ESA SP-424). ESA, Noordwijk, The Netherlands, pp. 125–132.

Young, N. W. (1999). ERS scatterometer observations of Antarctic iceberg drift. *Centre ERS d'Archivage et de Traitement (CERSAT) News*, **10**, 1–2.

Young, N. W. (2001). An iceberg the size of Jamaica! *Australian Antarctic Magazine*, **1**, 24 25.

Young, N. W. (2002). More of those big bergs: Where are they now? *Australian Antarctic Magazine*, **3**, 36.

Young, N. W. and G. Hyland (1997). Applications of time series of microwave backscatter over the Antarctic region. *Proceedings of 3rd ERS Scientific Symposium, March 17–21, Florence, Italy* (ESA SP-414). ESA, Frascati, Italy, pp. 1007–1014.

Young, N. W. and G. Hyland (1998). Interannual variability of Antarctic snow melt events derived from scatterometer data. *Proceedings of 1998 IEEE International Geoscience and Remote Sensing Symposium (IGARSS '98), July 6–10, Seattle, Washington*. Institute of Electrical and Electronics Engineers, Piscataway, NJ, pp. 2261–2263.

Young, N. W. and G. Hyland (2002). Velocity and strain rates derived from InSAR analysis over the Amery Ice Shelf, East Antarctica. *Annals of Glaciology*, **34**, 228–234.

Young, N. W., I. D. Goodwin, N. W. J. Hazelton, and R. J. Thwaites (1989). Measured velocities and ice flow in Wilkes Land, Antarctica. *Annals of Glaciology*, **12**, 192–197.

Young, N. W., D. Hall, and G. Hyland (1996). Directional anisotropy of C-band backscatter and orientation of surface microrelief in East Antarctica. In: J. Kingwell (ed.), *Proceedings of 1st Australian ERS Symposium, February 6, Canberra* (COSSA Publication 037). University of Tasmania, Hobart, pp. 117–126.

Young, N. W., D. Turner, G. Hyland, and R. N. Williams (1998). Near-coastal iceberg distributions in East Antarctica, 50°E–145°E. *Annals of Glaciology*, **27**, 68–74.

Young, N., G. Hyland, J. Anderson, and M. Fily (2004). Surface wind field and snow grain size over Antarctica derived from ATSR data. *Proceedings of Envisat Symposium, September 6–10, Salzburg, Austria.* Available online at ⟨http://www.congrex.nl/04a06/Programme04.html⟩.

Young, N. W., M. Frezzotti, M. Mancini, and J. Anderson (in press). Mapping surface wind field from satellite and field surveys (East Antarctica). *Annals of Glaciology*, **39**.

Young, N. W., M. Fily, and G. Hyland (in prep.). Spatial distribution of surface grain size of the Antarctic snow cover.

Zahnen, N., F. Jung-Rothenhausler, H. Oerter, F. Wilhems, and H. Miller (2002). Correlation between Antarctic dry snow properties and backscattering characteristics in Radarsat SAR imagery. *Proceedings of ERSel-LISSIG Workshop: Observing Our Cryosphere from Space, Bern, Switzerland, March 11–13.* Available online at ⟨http://las.physik.uni-oldenburg.de/eProceedings/vol02_1/02_1_zahnen1.pdf⟩.

Zhou, G. and K. Jezek (2002). Satellite photograph mosaics of Greenland from the 1960a era. *International Journal of Remote Sensing*, **23**(6), 1125–1142.

Zhou, G., K. Jezek, W. Wright, and J. Granger (2002). Orthorectification of 1960s satellite photographs covering Greenland. *IEEE Transactions on Geoscience and Remote Sensing*, **40**(6), 1247–1259.

Zink, M. (2004). The TerraSAR-L interferometric mission objectives. *Proceedings of Fringe '03 Workshop, December 1–5, 2003, Frascati, Italy* (ESA SP-550). ESA, Noordwijk, The Netherlands.

Zink, M., C. Buck, J.-L. Suchail, R. Torres, A. Bellini, J. Closa, Y.-L. Desnos, and B. Rosich (2001). The radar imaging instrument and its applications: ASAR. *ESA Bulletin*, **106**, 46–55.

Zwally, H. J. (1977). Microwave emissivity and accumulation rate of polar firn. *Journal of Glaciology*, **18**(79), 195–215.

Zwally, H. J. (1989). Growth of the Greenland Ice Sheet: Interpretation. *Science*, **246**, 1589–1591.

Zwally, J. (2003). *ICESat's First Measurements of Polar Ice: EGS-AGU-EUG Joint Assembly, Nice, France, 6–11 April 2003* (Geophysical Research Abstracts, Vol. 5, 08000). European Geophysical Society, Katlenburg-Lindau, Germany.

Zwally, H. J. (2004a). Advances in measuring Antarctic sea-ice thickness and ice-sheet elevations with ICESat laser altimetry. *XXVIII SCAR and COMNAP XVI Meeting, Bremen, Germany, July 25–31* (abstract volume). Alfred Wegener Institute for Polar and Marine Research, Bremerhaven, Germany, 86 pp.

Zwally, H. J. (2004b). ICESat's measurements of ice sheet elevation changes and sea ice thickness distributions. *Meeting of European Geosciences Union, Nice, France, April 25–30* (abstract). European Geosciences Union, Nice, France.

Zwally, H. J. and A. C. Brenner (2001). Ice sheet dynamics and mass balance. In: L.-L. Fu and A. Cazenave (eds.), *Satellite Altimetry and Earth Sciences*. Academic Press, San Diego, CA, pp. 351–369.

Zwally, H. J. and S. L. Fiegles (1994). Extent and duration of Antarctic surface melting. *Journal of Glaciology*, **40**(136), 463–476.

Zwally, H. J. and M. B. Giovinetto (1995). Accumulation in Antarctica and Greenland derived from passive-microwave data: A comparison with contoured compilations. *Annals of Glaciology*, **21**, 123–130.

Zwally, H. J. and M. B. Giovinetto (2001). Balance mass flux and ice velocity across the equilibrium line in drainage systems of Greenland. *Journal of Geophysical Research*, **106**(D24), 33717–33728.

Zwally, H. J. and J. Li (2002). Seasonal and interannual variations of firn densification and ice-sheet surface elevation at the Greenland summit. *Journal of Glaciology*, **48**(161), 199–207.

Zwally, H. J., R. A. Bindschadler, A. C. Brenner, T. V. Martin, and R. H. Thomas (1983). Surface elevation contours of Greenland and Antarctic ice sheets. *Journal of Geophysical Research*, **88**, 1589–1596.

Zwally, H. J., A. C. Brenner, J. C. Major, R. A. Bindschadler, and J. G. Marsh (1989). Growth of the Greenland Ice Sheet: Measurement. *Science*, **246**, 1587–1589.

Zwally, H. J., A. C. Brenner, J. C. Major, T. V. Martin, and R. A. Bindschadler (1990). *Satellite Radar Altimetry over Ice: Vols. 1, 2, and 4. Processing and Corrections of Seasat Data over Greenland* (NASA Reference Publication 1233). NASA, Washington, DC.

Zwally, H. J., A. C. Brenner, and J. P. DiMarzio (1998). Technical comment: Growth of the southern Greenland ice sheet. *Science*, **281**, 1251.

Zwally, H. J., B. Schutz, W. Abdalati, J. Abshire, C. Bentley, A. Brenner, J. Bufton, J. Dezio, D. Hancock, D. Harding et al. (2002a). ICESat's laser measurements of polar ice, atmosphere, ocean, and land. *Journal of Geodynamics*, **34**, 405–445.

Zwally, H. J., W. Abdalati, T. Herring, K. Larson, J. Saba, and K. Steffen (2002b). Surface melt–induced acceleration of Greenland Ice Sheet: Flow. *Science*, **297**, 218–222.

Zwally, H. J., M. A. Beckley, A. C. Brenner, and M. B. Giovinetto (2002c). Motion of major ice-shelf fronts in Antarctica from slant-range analysis of radar altimeter data, 1978–1998. *Annals of Glaciology*, **34**, 255–262.

Zwally, H. J., A. C. Brenner, H. Cornejo, M. Giovinetto, J. L. Saba, and D. Yi (2003a). Antarctic ice-sheet mass balance from satellite radar altimetry 1992 to 2001. *Proceedings of the 23rd IUGG General Assembly* (Abstract SM11/10A/B18-001). International Union of Geodesy and Geophysics, University of Colorado, Boulder, CO.

Zwally, H. J., R. Schutz, C. Bentley, J. Bufton, T. Herring, J. Minster, J. Spinhirne, and R. Thomas (2003b). *GLAS/ICESat L2 Antarctic and Greenland Ice Sheet Altimetry Data V001* (digital media). National Snow and Ice Data Center, Boulder, CO.

Appendix

Parameters of Synthetic Aperture Radar Missions

As discussed in Volume 1, Synthetic Aperture Radar (SAR) is perhaps the single most powerful tool for high-latitude remote sensing, by virtue of its ability to map surface features at high spatial resolution in all weather. This appendix tabulates the instrumental parameters of the major polar orbiting SAR sensors since 1978, and illustrates the steady increase in capabilities.

Table A.1. Frequency and wavelength of IEEE (Institute of Electrical and Electronics Engineers) radar band designation.

Band	Frequency (GHz)	Wavelength, λ (cm)
P-Band	0.3–1	30–100
L-Band	1–2	15.0–30.0
S-Band	2–4	7.5–15.0
C-Band	4–8	3.75–7.5
X-Band	8–12	2.5–3.75
Ku-Band	12–18	1.67–2.5
K-Band	18–27	1.11–1.67
Ka-Band	27–40	0.75–1.11

Table A.2. The NASA Seasat SAR.

Parameters	Seasat SAR
Mission month	July–October
Mission year	1978
Altitude (km)	800
Orbital inclination	108°
Frequency (GHz)	1.28
Polarization	HH
Swath width (km)	100
Azimuth (AZ) resolution (m)	25 (four-look)
Range (RG) resolution (m)	25
Peak power (kW)	1
Bandwidth (MHz)	19

Table A.3. The Japan Aerospace Exploration Agency (JAXA) Japanese Earth Resources Satellite-1 (JERS-1) SAR.

Parameters	JERS-1 SAR
Launch date	11 February 1992
Operation end date	12 October 1998
Altitude (km)	570
Orbital inclination	98°
Orbital period (minutes)	96
Repeat cycle (days)	44
Local time at descending node	10:30–11:00 AM
SAR observation frequency (GHz)	1.3 (L-band)
Bandwidth (MHz)	15
Polarization	HH
Resolution (m)	18 (three-look)
Nadir angle	35°
Swath width (km)	75

Table A.4. The European Space Agency (ESA) Advanced Microwave Instrument (AMI) aboard the Earth Remote Sensing satellites (ERS-1, ERS-2).

Parameters	AMI aboard ERS-1 and ERS-2
Launch dates (ERS-1, ERS-2)	17 July 1991, 21 April 1995
Operation end date (ERS-1)	10 March 2000
Altitude (km)	785
Orbital inclination	98.5°
Orbital period (minutes)	100
Repeat cycle (days)	35
Local time at descending node	10:30 AM
SAR observation frequency (GHz)	5.3 (C-band)
Bandwidth (MHz)	15.55
Polarization	VV
Resolution, along track (m)	6–30
Resolution, cross track (m)	26.3
Nadir angle	23°
Swath width (km)	102.5

Table A.5. The Canadian Space Agency RADARSAT-1 SAR (Luscombe et al., 1993) [Reference: Luscombe, A. P., I. Ferguson, N. Shepherd, D. G. Zimcik, and P. Naraine, (1993). The RADARSAT synthetic aperture radar development. *Canadian Journal of Remote Sensing*, **19**(4), 298–310.]

Parameters	RADARSAT-1 SAR
Launch date	04 November 1995
Altitude (km)	798 km
Orbital inclination	98.6°
Orbital period (minutes)	100.7
Repeat cycle (days)	24
Local time at ascending node	18:00–00:15
SAR observation frequency (GHz)	5.3 (C-band)
Bandwidth (MHz)	11.6, 17.3, or 30.0
Polarization	HH

RADARSAT-1 SAR Imaging Modes

Mode	Nominal resolution (m)	Number of beams	Swath width (km)	Nadir angle (deg)
Fine	8	15	45	37–47
Standard	30	7	100	20–49
Wide	30	3	150	20–45
ScanSAR Narrow	50	2	300	20–49
ScanSAR Wide	100	2	500	20–49
Extended High	18–27	3	75	52–58
Extended Low	30	1	170	10–22

Table A.6. The ESA Environmental Satellite (ENVISAT) Advanced Synthetic Aperture Radar (ASAR).

Parameters	ENVISAT ASAR
Launch date	01 March 2002
Altitude (km)	800 km
Orbital inclination	98.55°
Orbital period (minutes)	100.6
Repeat cycle (days)	35
Local time at descending node	10:00 AM
SAR observation frequency (GHz)	5.33 (C-band)
Polarization	VV, HH, or cross (HH&HV or VV&VH)

ENVISAT ASAR imaging modes

Mode	Polarization	Resolution (m)	Swath width (km)	Characteristics
Image Mode (IM)	VV or HH	30	58–109	Seven selectable swaths
Alternating Polarization Mode (AP)	HH/VV, HH/HV, or VV/VH	30	58–109	Two co-registered images per acquisition in one of seven selectable swaths; radiometric resolution reduced as compared with IM
Wide Swath (WS)	VV or HH	150	405 × 405	ScanSAR mode
Global Monitoring (GM)	VV or HH	1000	405 × 405	ScanSAR mode
Wave Mode (WM)	VV or HH	30	5 × 5 or 5 × 10 vignettes	Vignettes are acquired at regular intervals along track, and are converted to wave spectra for ocean monitoring

Table A.7. Canadian Space Agency RADARSAT-2 SAR

Parameters	RADARSAT-2 SAR
Launch date	Mid-2005
Altitude (km)	798 km
Orbital inclination	98.6°
Orbital period (minutes)	100.7
Repeat cycle (days)	24
Local time at ascending node	18:00
SAR observation frequency (GHz)	5.405 (C-band)
Polarization	HH, VV, HV, VH

RADARSAT-2 imaging modes (courtesy Canadian Space Agency)

Polarization	Beam mode	Resolution RG × AZI (m)	Swath width (km)
Selective single	Ultra-Fine	3 × 3	20
Transmit H or V, Receive H or V	Multi-Look Fine	11 × 9	50
Polarimetric	Fine Quad-Pol	11 × 9	25
Transmit H and V on alternate pulses, Receive H and V on every pulse	Standard Quad-Pol	25 × 28	25
Selective polarization	Standard	25 × 28	100
Transmit H or V,	Wide	25 × 28	150
Receive H or V or (H and V)	Low Incidence	40 × 28	170
	High Incidence	20 × 28	70
	Fine	10 × 9	50
	ScanSAR Wide	100 × 100	500
	ScanSAR Narrow	50 × 50	300

Table A.8. JAXA Advanced Land-Observing Satellite (ALOS) Phased-Array-type L-band Synthetic Aperture Radar (PALSAR). Courtesy A. Rosenqvist, JAXA.

Parameters	PALSAR
Launch date	Mid-2005
Altitude (km)	691.65
Orbital inclination	98.16°
Repeat cycle (days)	46 days (2 days for side-cycle instruments)
Local time at descending node	10:30 AM
SAR observation frequency (GHz)	1.27 (L-band)
Polarization	Single, Dual, Quad-polarization

PALSAR imaging modes

Mode:	High resolution			ScanSAR	Polarimetry
	Single polarization	Dual polarization	Direct downlink		
Polarization	HH or VV	HH/HV or VV/VH	HH or VV	HH or VV	HH/HV + VV/VH
Nadir angle (deg)	8–60 (39 typical)	8–60 (39 typical)	8–60 (39 typical)	18–43	8–30 (24 typical)
Resolution (m)	7–44 (10 at 39°)	14–88 (20 at 39°)	14–88 (20 at 39°)	100 (multi-look)	24–89 (30 at 24°)
Swath width	40–70	40–70	40–70	250–350	20–65

Table A.9. The ESA TerraSAR-L. Courtesy M. Zink, ESA-ESTEC.

Parameters	TerraSAR-L
Launch date	2008
Altitude (km)	630
Repeat cycle (days)	14
Local time at ascending node	18:00
SAR observation frequency (GHz)	1.258 (L-band)
Polarization	Single, Dual, Quad-polarization

TerraSAR-L imaging modes

Mode	Nadir angle (deg)	Resolution RG x AZI (m)	Swath width (km)
Quad-polarization	20–36	9 × 5	40
Dual-polarization	20–45	9 × 5	70
Single-polarization	20–45	5 × 5	70
ScanSAR Dual Pol	20–45	50 × 50	>200
ScanSAR Single Pol	20–45	20 × 5	>200
Wave Mode	20–45	9 × 5	20

Table A.10. The DLR TerraSAR-X. Courtesy S. Lehner, DLR.

Parameters	TerraSAR-X
Launch date	April 2006
Altitude (km)	514
Orbital inclination	97.44°
Repeat cycle (days)	11
Local time at ascending node	18:00
SAR observation frequency (GHz)	9.65 (X-band)
Polarization	Single, Dual, Quad-polarization

TerraSAR-X Imaging Modes

Mode	Coverage RG × AZI (km)	Resolution RG × AZI (m)	Polarization	Nadir angle range (deg)
HR Spotlight	10 × 5	(1.5–3.5) × 1.0	Single, Dual, Quad	20–55
Spotlight	10 × 10	(1.5–2.3) × 2.0	Single, Dual, Quad	20–55
StripMap	30 × ≤1,650	(1.7–3.5) × 3.0	Single	20–45
Polarimetric StripMap	15 × ≤1,650	(1.7–3.5) × 6.0	Dual, Quad	20–45
ScanSAR	100 × ≤1,650	(1.7–3.5) × 16.0	Single, Dual, Quad	20–45
Experimental 300-MHz Spotlight	10 × 5	(0.6–1.5) × 1.0	Single, Dual, Quad	20–55
Experimental Dual Receive StripMap	30 × ≤1,650	(1.7–3.5) × 1.5	Single, Dual, Quad	20–45

Index

AATSR 164, 166, 294, 307, 310
Ablation, past 23
Ablation rates 174, 267, 322
  uncertainty in current estimates 322
Ablation zone 159, 161, 176, 180, 248
  of Greenland Ice Sheet 303–304
Absorbing impurities 310
Absorption
  bands 217
  coefficient 318–319
  microwave 41, 161, 166, 190, 290, 318–319
    coefficient 318–319
    relationship to snow-grain size 319
    shortwave 299, 311–312
Accumulation, relationship with surface/bedrock topography 312
Accumulation Area Ratio (AAR) 19, 308
Accumulation rate
  see also Ice sheet, accumulation 240–247, 267, 282
  effect of mid-latitude blocking anticyclones on 246
  in situ measurement techniques 7, 241–242
  information from radar altimetry 243
  InSAR techniques 246–247
  modeling of 242, 282, 284
  laser altimeter techniques 243
  passive-microwave techniques 243, 244–246, 309
  radar-altimeter techniques 246
  radar-scatterometer techniques 243–244, 318
  past 7, 23
  SAR techniques 244
  uncertainties in 240–242, 284, 322
Accumulation zone 159, 248, 314, 319–320
ADEOS, JAXA spacecraft 160, 167, 203, 205, 211, 243, 301, 307
ADEOS-II, JAXA spacecraft 60, 74, 158, 167, 174, 211, 243, 295, 297, 301, 305–307, 311, 313, 315
Adiabatic cooling 246
Aerial photography 144, 185, 219, 272
Aerosol 13
  amount 305
  effects on surface albedo measurement 305, 307
  optical depth 305
    satellite measurement techniques 307
  particle type 305
    satellite measurement techniques 307
  structure, satellite laser measurement techniques 233
Air pressure 22, 217
Aircraft measurements 23–24, 41, 44, 49, 82, 100, 116, 207, 210–211, 220, 225, 262
Alaska Satellite Facility (ASF) 73, 75–76

Albedo (shortwave) 13, 299
  broadband 299, 301, 305, 308
  factors affecting 300–301
  definition 299, 301
  diffuse 299
  facies 304
  Greenland Ice Sheet 300, 304–305
  hemispheric spectral 299, 300
  impact of clouds upon 13, 301
  impact of Mount Pinatubo eruption on 305
  impact of soot and dust on 300, 310
  relationship with surface temperature and melt 299
  satellite measurement techniques 299–309
    atmospheric effects on 301, 305
    clear-sky algorithms 304–306
    cloudy-sky algorithms 306–307
    correction of aerosol effects 307
    correction for bidirectional reflectance distribution function 301–307
    effects of aerosols on 305
    error sources 306
    hyperspectral techniques 308
    in mountainous regions 306
    multi-angular techniques 305, 307–308
    narrow-to-broadband conversion 301, 304–305, 307
    processing steps
      AVHRR 304–305
      MODIS 306
    validation of 305–306, 308
  snow surface see Snow, albedo 250–251, 254–255
  spectral 301, 303
  surface 13, 164, 174
    surface classification using 308–309
    temporal variability of 301, 304–305
    wet snow 300
ALI 140, 143–144, 151, 255, 316
ALOS, JAXA spacecraft
  see also PALSAR 78–80, 102, 111, 211
  orbital parameters 79–80, 102
Amery Ice Shelf 108–109, 151, 153, 157, 161, 183, 186, 188, 196, 198, 232–233, 242, 262–264, 266–267, 269, 278–279, 286
Amery Ice Shelf Ocean Research (AMISOR) project 151, 157, 263

AMSR 174, 243, 245, 295, 297, 313, 315
AMSR-E 174, 245, 295, 297, 313, 315
Amundsen Sea Embayment 230
  Science and Implementation Plan 261
Anisotropic reflectance factor 303, 305, 310
Anisotropy 302–303, 305
  directional, in radar backscatter 318
Antarctic Bottom Water formation 2–3, 175, 279, 284
Antarctic Circumpolar Wave (ACW) 14
Antarctic Coastal Current see East Wind Drift
Antarctic coastal mapping program, Landsat 254
Antarctic Convergence 189
Antarctic Deep Water 175, 279
Antarctic Digital Database (ADD) 219–220
Antarctic Dipole 14
Antarctic Ice Sheet 1–2, 4–5, 8–10, 16, 20, 22, 57, 65, 75–76, 95, 98, 101–102, 107, 109, 137, 140, 145, 151–153, 155–156, 158–159, 161, 166, 174–175, 182, 186, 198, 210–211, 218–220, 225, 230, 232–233, 260–261, 262, 266, 269, 278–279, 283, 286, 291, 315–317, 319
  acceleration of grounded-ice discharge from 9
  air temperature over 16
  areal extent 4
  contribution to global sea-level rise 4–5
  drainage basin size 16
  mass budget 10, 186
  mass loss from 8
  surface roughness characteristics of 315–318
  thickness 4
  total backscatter (satellite radar altimeter), map of 319
  volume 175
Antarctic Mapping Mission (AMM), Radarsat-1 61, 75–76, 98, 101, 107, 109, 137, 155–156, 158, 182, 186, 219–220, 260, 262, 317
  digital elevation model 219–220, 269, 283
  strain rate measurements with 269–271
Antarctic Meteorological Research Center (AMRC) 203
Antarctic Oscillation 170

Antarctic Peninsula 7, 9, 57, 159–160, 166, 172, 177–178, 183, 274–275, 291
  ice-shelf collapse on 7, 9, 274–276, 291
  MiniMAMM coverage of 76, 158, 260
  warming trend in 166, 172, 180, 275–276, 291, 293
Antarctic plateau 314
Antenna 290
Aqua, NASA spacecraft 174, 245, 295, 297, 313, 315
ARC/INFO 220
Arctic 9, 111, 300, 310
Arctic marine system 316
Arctic Ocean 6
Arctic Oscillation (AO) 14
Areal extent of glaciers, changes in 186
Argon, US spy satellite series 181, 200
Argos telemetry 21
ASAR 78–79, 114, 151, 161, 192–193, 196, 199, 208–209
ASCAT 166
Ascending node 258–259
ASTER 139, 140–141, 183–184, 211, 234, 254, 275, 308, 311
Atlantic Ocean 203
Atmospheric circulation 2, 13–14, 20
Atmospheric modeling 241–242, 323
Atmospheric moisture convergence modeling 58, 242
Atmospheric sounding 294
Atmospheric temperature 2
Atmospheric circulation variability, modes of 8, 14
ATSR(/2) 166, 294, 306–307, 310–311
Australian Antarctic Division 220
Automatic weather stations (AWSs) 13, 22, 169, 172, 174, 242, 246, 293
  Antarctic 22, 293, 297
  Greenland 22, 293
  measurement of accumulation rate with 22, 242
  measurement of air/surface temperature with 22, 292
  measurement of firn temperature with 242
  measurement of meteorological variables with 22, 292, 315–316
  measurement of water-vapor flux with 245–246
  on icebergs 206

validation of satellite albedo data using 305, 315, 318
Autonomous underwater vehicle (AUV) 286–287
Autosub 286–287
AVHRR 22, 93, 143, 145–151, 158, 166, 190, 202, 220, 235–237, 252, 255–256, 292–293, 298, 300–301, 304–306, 308–310, 315
  calibration 304
  geo-referencing 304
AVNIR(-2) 211
Azimuth angle 47
Azimuthal modulation, of radar backscatter data 318

Backscatter, radar see Microwave
Baffin Bay 196–197
Baffin Bay Low 172
Balance velocity see Ice sheet, balance velocity
Band ratioing 139
Band-pass filter 315
Bandwidth, detector 63, 79
Bare ice zone 320
Basal melt/freeze rates of floating ice, estimates of 283–287
Bathymetry, ocean 101, 195, 201, 206, 259, 268, 277
BEDMAP 219, 266, 268–269, 282–283, 322
Bedrock 176, 226
  DEMs 267–268
  elevation 12, 17
  outcrops 82, 93–94, 104, 109, 140, 248, 250–251, 258
  surface topography, relationship with 226, 312, 315
  topography 234, 252, 267–268, 312
  topography, relationship with accumulation 312, 315
  topography, relationship with ice motion 268, 315
  vertical motion 222
Berg Analysis and Prediction System (BAPS) 193
Bidirectional Polarization Distribution Function (BPDF) 312

Bidirectional Reflectance Distribution
    Function (BRDF)
  see also Reflectance 301–308, 310–311
  angular shape and magnitude of 303
  dependence on surface optical and
    structural properties 302–303
  dependence on surface slope 302
  effect of surface melt on 303
  effect of surface roughness on 303, 308, 315
  satellite measurement of 307–308
  uncertainties in 303
Bimodal histograms 172–173
Biogeochemical cycling 2, 6
Bindschadler Ice Stream, West Antarctica
    144, 146, 252–253, 261, 283
  mass budget of 283
  velocity variations 277
Blocking anticyclones 246
Blue ice 65, 140, 174, 309–310
  detection/monitoring in satellite imagery
    309
  role in meteorite collection 309
Bombardier Glacier, Antarctica 275
Bottom Water see also Antarctic Bottom
    Water 2
Boundary-layer standing wave, atmospheric
    315
Boydell Glacier, Antarctica 275
Brewster angle 298, 313
Brigham Young University (USA) 203
Brightness temperature 295–299, 312, 314
Brine rejection 2
Buttressing effects of ice shelves see Ice
    shelves, buttressing effect
Byrd Glacier, Antarctica 231
Byrd Polar Research Center 76

Calibration, onboard sensor 150
Canadian Eastern Seaboard 193
Canadian Ice Service (CIS) 193, 195
Canadian Space Agency (CSA) 75, 155
Cape Adare, Antarctica 191
Cape Crozier, Antarctica 191
Casey Station, Antarctica 4
CBERS-2, Brazilian–Chinese satellite 141
CHAMP, German satellite mission 18
Circumpolar Deep Water (CDW) 2, 285
Clear-sky precipitation see Diamond dust
Climate change, Antarctic 8

Climate, past 6–7, 19
  reconstruction of, factors affecting 164
Climate modeling 299, 309
Climate variability 1, 14, 292
Climate warming 3–4, 6–9, 20, 23, 40, 101,
    139, 159, 174–175, 186, 189, 210,
    223, 234, 291, 299, 309
  abrupt 7–8
  amplification at high latitudes 9
  past 6–7
Clock corrections 304
Cloud
  amount (fraction) 13, 301, 306–307
  contamination of satellite optical imagery
    139, 141, 145, 164, 202, 211, 230,
    233, 255, 293
  detection and classification 295
  height measurement, by satellite laser
    altimetry 233
  impact on surface albedo 301, 306–307
  impact on surface temperature 307
  masking 237, 293–295, 306
  optical properties 301
  reflectance characteristics 306
  role in surface energy budget 13, 295,
    306–307
  sub-pixel 306
  susceptibility 280
  tropospheric cirrus 295, 316
  type 301, 306
CMIS 294
CNES, Centre National d'Études Spatiales
    115, 218
Coastal (change) mapping see Ice sheet
    margins
Coherence optimization see SAR maximum-
    coherence tracking
Coherence tracking of ice motion
  accuracy of 108–109
  technique 108–109
  tie-points 109
  relative strengths and weaknesses 109–111
Coherent imaging system 42, 47
Cold halocline layer (CHL) 334
Communication bandwidth 143
Constant False Alarm Rate (CFAR)
    algorithm (iceberg detection)
    195–197
Corona, U.S. spy satellite series 145

COSMO-SkyMed, European spacecraft program 79
Crary Ice Rise (Antarctica) 145, 152
Crevasses 39, 106, 137, 150, 153, 176, 180, 189–190, 250, 252, 270, 291
Cross-Polarization Gradient Ratio (XPGR) 170–172
Crusts, surface 245
  effect on microwave emission 298, 314
CryoSat, ESA spacecraft 21, 116, 175, 209, 227, 229–230, 234, 243, 248, 287
Crystal orientation fabrics, ice 15, 23
CTD measurements 287

Danish Meteorological Institute 193, 195–197
Data cumulation 147–149, 152, 235–236, 256
David Glacier, Antarctica 76, 158, 260
Decadal variability/change 175, 188
Declassification, image 144–145
Deep Water 2
Defense Intelligence Satellite Program, U.S. 144–145, 180, 184, 316
Deficiencies in current knowledge 322
Deglaciation, past epochs of 4
Degree-day models 12, 165–166, 174
Delft Institute for Earth-oriented Space Research (The Netherlands) 75
DEM *see* Digital Elevation Model
Density, ice 14, 17, 174, 208, 280, 282–283, 295
  satellite measurement techniques 313
Density, sea water 233, 282–283
Depolarization effects 170
Depth hoar 58, 154, 244, 298, 313
  impact on microwave emission and volume scattering 313–314
Descending node 258–259
DESCW 74–75
Diamond dust 246
Dielectric constant 41, 153–154, 290
Dielectric discontinuities, in glacial ice 190
Dielectric loss 23, 289–290
Dielectric properties of ice 290
Dielectric properties of snow 166, 223, 290, 319

Digital Elevation Model 42, 49, 52–53, 58, 60, 62, 64, 72, 82, 89, 92–93, 95, 102, 105, 156, 211, 219–222, 226, 238, 257–258, 268, 271, 302, 304, 312
  improved using satellite image-based photoclinometry 234–237
  of Antarctica 277
  stereoscopic 308
Directional hemispherical reflectance 301
Discharge rate *see* Ice sheet, discharge rate
DISORT radiative transfer model 303, 307, 312
Distributed Active Archive Center (DAAC), NASA 75, 232
DLR, German aerospace agency 18, 44, 115
DMSP, U.S. satellite series *see also* SSM/I 145, 170–173, 190, 201–202, 243, 245, 296, 298, 313
Dolines 142
Dome C 24, 246
Doppler tracking of satellites 217, 229
DORIS 62, 79, 229
Dronning Maud Land (East Antarctica) 154
Dry-snow line 166
Dry-snow zone 65, 158, 160, 166, 168, 171, 243–246, 313, 320
Drygalski Glacier, Antarctica 274
Drygalski Glacier Tongue, Antarctica 206
Dynamic range, detector 56, 290
Dynamic response time 6–7, 267, 322
Dynamic thresholding 177

East Antarctic Ice Sheet 8, 107, 153, 224, 231, 262, 276
  mass balance of 284
East Wind Drift 195, 202, 205
ECMWF 174, 242
Ecology, high latitude 6, 206
Edge-detection algorithms 172, 177
Edge enhancement 177
Edge following 177
Edgeworth Glacier, Antarctica 275
EISMINT 23
El Niño–Southern Oscillation (ENSO) 14, 298
Electromagnetic radiation (waves) 42–43, 47, 212, 217

Elevation (surface), satellite measurement techniques
  InSAR 46–54
  photoclinometry 47, 93, 142, 220, 234–237, 287
  Radar altimetry 93–95, 213, 215–229, 233–235, 238, 248, 266–268, 277, 282, 312
  stereoscopy 47, 308
Elevation angle 47
Ellesmere Island 177
EMISAR 64, 112
Emission, passive microwave 295–296, 313–314
  factors affecting 313–314, 317
Emission modeling 298
Emissivity 169
  modeling 296
  passive microwave 295–296
    physical controls on 295–296, 298, 312–314
  thermal infrared 292–293
    scan-angle dependency 293
Energy balance/budget, surface 2, 12–13, 165–166, 174, 233–234, 291, 301, 309–310
  models 12–13, 291
  role of clouds in 295, 306–307
Energy flux equation 12–13
Envisat, ESA spacecraft see also ASAR 62, 74, 78–80, 111, 114, 151, 155, 161, 162, 164, 208–209, 217–218, 226–227, 301, 333
EO-1, NASA spacecraft 140, 143, 151, 308, 311
Equator-to-pole temperature gradient 2
Equatorial coordinate system 64
Equilibrium line 10, 19, 159–160, 165, 175–176, 321
Equilibrium line altitude 19
EROS Data Center, U.S. Geological Survey 145
EROS-A 140
ERS Tandem Mission 45, 59, 61, 71, 73–75, 99–100, 102, 114, 260–261, 270–271, 277
ERS-1, ESA spacecraft 45, 52–54, 57, 59–63, 71, 73–77, 79–80, 83–84, 94, 98–99, 102, 107–108, 114, 167–168, 177, 193, 201, 203, 205, 243, 257, 260–262, 270–271, 277, 280–281, 294, 296, 312, 318
  ice orbit phase 76
  geodetic orbit phase 76, 218
  maximum latitudinal coverage 75
  orbit-state vectors 75
  orbital cycles/parameters 76, 81
  radar altimeter 212–213, 216, 218–219, 221–222, 225–227, 266, 268, 277, 286, 304, 316, 319
  SAR 45, 52–53, 57, 59–63, 70–71, 73–77, 79–80, 83–84, 94, 98–99, 107, 142, 151, 155, 177, 179–181, 183, 185–186, 193, 198, 200, 205, 224, 238–239, 244, 246, 251, 270–274, 276–277, 320
ERS-2, ESA spacecraft 45, 52–54, 57, 59–63, 71, 73–77, 79–80, 83–84, 98–99, 107, 167–168, 201, 203, 205, 243, 257, 260–262, 280–281, 296, 306, 311, 318
  maximum latitudinal coverage 75
  orbital parameters 76, 81
  orbit-state vectors 75
  radar altimeter 212–213, 216, 218, 224–227, 230, 233, 268
  SAR 45, 52–54, 57, 59–63, 70–71, 73–77, 79, 80, 83–84, 98–100, 102, 108, 114, 177, 179–180, 183, 186, 193, 198, 205, 224, 238, 246, 260–261, 270272, 276–277, 320
ERS wind scatterometer (EScat) 296
ERTS-1, NASA spacecraft see also Landsat 139
ESA Earth Observation User Service Portal 58
ESCAT 167–168, 201, 203, 205, 243, 296, 318
ESMR 244
ETH/CU Camp, Greenland 169
ETM+ see Landsat ETM+
European Space Agency (ESA) 74–75, 84, 116, 193, 215
Exo-atmosphere solar irradiance 143

Facet density 302
Facies, ice sheet 64, 112–113
  changes in 304

microwave penetration depth of 320
satellite mapping and monitoring of 158–164, 170–171, 304
False-color imagery 309
Fast ice *see* Sea ice
Feedback mechanism, ice–albedo–temperature 9, 13, 164, 291, 299, 304
FES95.2 tide model 259
Field operations, role of satellites in planning 137, 141, 151, 256
Field-of-view (FOV) 142, 301
Filchner Ice Shelf, Antarctica 105–108, 259, 270
Filchner Ice Streams 76, 158, 260
Filchner–Ronne Ice Shelf 155, 186, 277, 284
Fimbul Ice Shelf 140, 183
Firn 17, 20, 58, 64–65, 154–155, 162, 165, 170, 176, 182, 223, 243, 312, 320
  backscatter effects 243
  characteristics inferred from passive-microwave data 312–315
  compaction models 292
  core measurements 241, 243, 323
  densification/compaction 17, 223, 225, 232, 242
  grain growth processes 312
  impermeability, and its role in ice-shelf breakup 180, 182
  layering/stratigraphy 65, 295–296
  microwave properties 292, 312–313
  optical properties 292, 308
  stratification/layering 224, 245, 247, 289, 312
    satellite measurement techniques 247, 289, 318–321
  temperature 245, 297–298, 312
  temperature, AWS measurement techniques 242
  ventilation 315
Firn cores 241, 243
Firn line 159
Firn–ice boundary 165
Flask Glacier, Antarctica 274
Flow bands *see* Flow stripes
Flow law, ice-sheet 14–15, 23
Flow stripes 147, 150–152, 157
Flux gate 282–283

Flux (irradiance)
  downwelling 308
  extraterrestrial solar 301, 302
Force balance equation, ice sheet 273, 290
Foreshortening, radar image 67, 69, 89, 155
Freeboard *see* Sea ice
Freshwater budget *see* Ocean, freshwater budget
Fresnel coefficient 319
Fringe visibility tracking *see* SAR maximum-coherence tracking

Gain, detector 308
Geikie Ice Cap, Greenland 64
Geoid 50
  satellite measurement techniques 18–19, 288
Geopotential height 172
George V Ice Shelf (Antarctica) 183, 191, 194, 202
Geosat, NASA satellite mission 217–218
Geosat-Follow On 225
Geothermal heat flux 15
GINSAR 110–111
Glaciers
  see also Ice streams
  Arctic 6
  climate change, impact of 6
  dynamic response time 6
  geometry 19
  Greenland 189, 272
  inventory of Greenland 272
  mass balance of 6, 19–20, 40
    annual 20
  mass balance, *in situ* measurement techniques 20
  mechanics 20
  modeling 5, 15
  mountain 6, 66
    contribution to global sea-level rise 6
    impact on freshwater budget of the Arctic Ocean 6
    recent retreat of 6
    response to climate warming 6
  outlet *see also* Ice streams 219, 226, 231, 248, 255, 260–261, 267, 269, 280, 323
    accelerated flow of 9, 248, 274–276
    katabatic-wind channeling in 2

Glaciers (cont.)
    strain rate of, calculated from InSAR wrapped phase 269–271
    surge behaviour, satellite detection of 271–276
    temperate 1
    thermodynamics 20
    thinning of 273, 276, 281
    tidewater 59, 110–111, 272
    valley 141
    velocity 189, 255
Glacier front 59
Glacier tongue 2, 96, 175, 184, 189, 198, 210
    advance/retreat, satellite monitoring of 190
    elevation of 210
    melt of 291
    motion, satellite measurement techniques 254, 258, 261
        tidal effects on 259, 277
        inverse barometer effects on 259–260, 278
        tidal effects on satellite measurement techniques 2
    tidal motion of 96, 101, 210, 223
GLAS see also ICESat 65, 95, 112–113, 153, 175, 198, 207, 209, 220, 227, 229–234, 240, 243, 263, 277–278, 287–288
Glazing, surface 57–58, 309
    satellite detection of 308–309
GLI 190, 211, 301, 305–306, 310–311
Global climate change see Climate warming
Global Climate/Circulation Model (GCM) 9, 12, 15, 242
    uncertainties in predictions from 9
Global Land Ice Management from Space (GLIMS) project 254, 308
Global mean temperature 4
Global Monitoring Mode (GMM, Envisat ASAR) 193, 196, 205, 209
Global Positioning System (GPS) 21–22, 41, 62, 78, 82, 89, 93–95, 99, 101, 112–113, 141, 175–176, 186, 206, 219–220, 223–224, 238, 240, 248
    airborne surveying with 275
    ice kinematics measurement with 210, 219, 240, 248

    ice-sheet velocity measurement with 255, 258, 269
    grounding line detection with 279–280
    kinematic techniques 277, 279
    onboard satellites 230
    surface elevation measurement by 210
    three-dimensional velocity measurement with 248
    tidal flexure measurements with 269, 277, 279
    validation of satellite DEMs using 219–220
    validation of satellite ice-motion data with 248, 257–258, 269
Global sea-level rise see Sea-level rise, global
Global warming see climate warming
GNSS 247
GOCE, European satellite 18–19, 289
GRACE, U.S.–German satellite 18, 287–289
GRACE-FO, U.S.–German satellite 18
Gradient ratio (GR) 314–315
    relationship with snow grain size 315
Grain size (snow/firn), depth-dependent 310
Grain size, surface
    see also Snow, grain size 309–310
    factors affecting 310
    impact on surface albedo 300
    satellite measurement of 308, 310–313, 315
Gravity field, satellite measurement techniques 18–19, 288–289
    benefits to global sea-level change estimates 288
    benefits to surface elevation measurements 288
    measurement of geoid by 288
    measurements of ice-mass change/redistribution by 288–289
    measurement of isostatic rebound by 289
Greenhouse gases and related warming 5, 20
Greenland Climate Network (GC-Net) 174, 305
Greenland glaciers 189, 272
Greenland Ice Sheet 1–2, 4–6, 9–10, 16, 20–22, 41, 44, 53, 57, 59, 63–65, 71, 76, 82, 84, 90, 94–95, 100, 112, 147, 155, 158, 160–161, 165, 167,

171–172, 176–177, 210–211, 218, 222, 224–226, 235, 281, 285, 291, 293, 300, 305, 311–313
  acceleration of grounded-ice discharge from 9
  albedo 300, 305
  areal extent 4
  contribution to global sea-level rise 4–5
  deglaciation, possible irreversibility of 4
  drainage basin size 16
  dynamic behaviour, changes in 9
  dynamic response time of 6
  flow rates 16
  grain size, surface 311–313
  mass budget of 10, 16
  mass loss of, recent 5, 9
  melt extent, surface 305
    recent changes in 9
  surface roughness 315
  surface temperature 305
  thickness 4, 285
  thinning of 9, 281, 285
  warming trend over 9, 293
Greenland Mapping Project (Greenmap) 84
Greenland Summit 64, 95
Ground-Control Points (GCPs) 45, 82, 89, 93–97, 139, 211, 238, 257–258
Grounded ice
  loss of 8–9, 11
  thickness of 23
Grounding-line/zone 3, 16, 101, 106, 223–224, 259, 266, 273, 277, 285, 289
  definition 2
  dependence on tidal effects 280
  detection and monitoring 279–282
    in situ measurement techniques 279
    satellite InSAR measurement techniques 40, 60, 113, 280–281, 323
      precision of 280
    satellite optical measurement techniques 279–280
    satellite radar altimeter measurement techniques 280
  migration of 3, 9, 101, 274, 277, 281

Hazard detection and monitoring 116
Heat flux, basal 291–292

Heat sink, ice sheets/ice shelves/icebergs as 2–3
Hektoria Glacier (Antarctica), surface elevation change of 275
High Salinity Shelf Water (HSSW) 2–3
Hinge line 101
  detection by satellite techniques 280–281
Historical records, of ice-front change 186–187
Hoar frost, surface 58, 298, 313–314
  effect on microwave emission 313–314
  effect on surface micro-scale roughness 313–314
  formation processes 58, 313
Holocene 7
Humboldt Glacier, Greenland 235
  grounding-line migration of 281
  thinning of 281
Humidity 166
HYDROS, NASA satellite mission 248, 298
Hydrostatic relationship 233, 280, 282
Hyperion 308, 311–312
Hyperspectral remote sensing 308, 311–312

Ice–bedrock interface, observation techniques 23
Ice caps, small 1
Ice cores 7, 164, 241, 243, 323
  accumulation-rate records from 7
  information on atmospheric temperature and circulation from 7
  factors affecting interpretation of 20–21
  use of satellite data in choosing sites 309
Ice deformation experiments, laboratory 23
Ice discharge flux, estimates of 281–287
Ice dome 146, 219
Ice dynamics *see* Ice sheet, dynamics
  models 15, 279
Ice front 3, 175, 177, 184, 186–188, 198, 201
  velocity 273–275
Ice grain deformation 15
Ice lenses 153, 160, 190
Ice mass flux 186
Ice pipes 153, 160, 190
Ice pump mechanism 286
Ice rises 15, 223

Ice sheet
  ablation (rate) 6, 11, 14, 16–17, 19, 23, 58, 100, 210, 234, 248, 265, 301
  accumulation (rate) 3, 7, 10–13, 16–17, 19, 22–23, 58, 95, 100–101, 156, 186, 209–210, 225–226, 234–235, 240, 244, 248, 265–266, 276, 288–289
  accumulation zones 210
  advance/retreat, rate of 3
  albedo 2, 174–175, 180, 248
  as topographic barrier to atmospheric circulation 2
  balance flux 222, 262, 265–268
    distribution 265
  balance velocity 95, 209, 242, 265–268
    as control for InSAR 95, 258
    distribution 265
  barrier motion 249–250
  basal boundary conditions 15, 23–24, 257, 269, 289
  basal flow 14, 176
    global importance of 1–9
    hydrology 15
  basal lubrication 15, 276, 286
  basal processes 3, 6, 8, 138, 164, 210, 254
  basal resistance variations 234, 248, 268
  basal shear stress 14, 263
  basal sliding 14–15, 175–176, 254, 267–268, 286
  basal slip 14
  basal topography 141, 263
  basal water, detection of 268
  basal water pressure 175
  bedrock topography 234, 252, 267–268, 312
  borehole measurements 176
  catchment areas 222
  collapse, debate on 8–9
  coastal configuration 21
  coastal margins see Ice sheet margins
  continuity equation 11, 14
  deformation 14–15, 23, 254
  discharge flux 248, 279
  discharge rate 6, 9, 225, 231, 267, 289
  drainage basins 16–17, 113, 210, 219, 221–224, 262, 265–267, 282, 322–323
  drainage divides 156, 210, 219, 222–223, 249–250, 266

  dynamics 6–7, 9–10, 14, 17, 40, 95, 99–100, 155, 186, 209, 226, 248, 257, 263, 267, 290
  elevation, surface 2, 9, 16, 17, 47, 50, 62, 174, 209–220, 242–243
    radargrammetric techniques 240
    satellite InSAR measurement techniques 237–240
    satellite laser altimeter measurement techniques 229–234
    satellite radar altimeter measurement techniques 211–220, 226–229
    stereoscopic techniques 211
  elevation change 11–12, 17, 164, 175, 210, 248, 276, 286
    airborne laser altimeter measurement techniques 210, 227
    factors affecting 210
    GPS measurement techniques 210
    satellite laser altimeter measurement techniques 230–234
    satellite radar altimeter measurement techniques 222–229
    surveying of 210
  emergence velocity 12
  equilibrium state 10
  feedback mechanisms 15
  flow
    controls on 14
    finite-element model of 263
  flow bands 255
  flow law 14, 23, 209, 267–268
  flow, creep see internal creep
  flow, meltwater lubrication of 164
  flow, patterns of 3, 23, 96, 97–100, 138, 141, 145, 152, 156, 209–210, 234, 261, 266
  flow, slow 254
  flow acceleration see also Surge behaviour 9, 15, 175, 234, 268
  flow direction 97, 99, 209, 253, 270
  flow features, relict see also Flow stripes and streak lines 235, 255
  flow history 290
  flow stability 147
  flow bands 253
  flow lines 97–98, 140, 222, 235, 237, 270
    convergence/divergence of 209
  flux divergence 98

flux calculations 248, 265–266
force balance 8
geometry 247, 265, 267
glaciology 20
global climate system, role in 2, 3, 6, 20
gravitational driving stress 14–15, 209, 247, 268–269, 290
grounding *see also* Grounding line 8
hydrology 254
ice flux 101, 240
indicators of global warming/variability 3, 20
internal creep 210, 254, 259
internal deformation 254
internal layers, remote sensing of 23, 241, 290–291, 313
marine, possible collapse of 248
mass balance 4–6, 10–15, 40, 100, 159, 164, 174, 177, 210–211, 234, 240–242, 246–247, 252, 262, 273, 276, 279, 281, 289, 306, 308, 321–322
   current uncertainties in estimates of 5, 21, 276
   in situ measurement techniques 20
   local 11–12
   satellite measurement techniques 9, 40, 113, 116, 222, 225, 240, 266–267, 276, 286–287, 289, 321–322
   specific 11
   total 12
mass budget 12, 16–17, 153, 186, 198, 247, 248, 282–287
mass conservation 11, 100
mechanics 15, 20, 23, 254
melt
   see also Melt, surface 4, 6, 8, 11, 14, 21, 102, 147, 156, 159, 160–161, 242
   impact on palaeo record interpretation 164
   duration 168–169
   extent, interannual variability 168
   extent, seasonal 168, 234
   rates 210
meltwater runoff 12, 161, 174, 234, 238, 285, 289, 291
microwave penetration of 63–66, 77, 80, 112, 208, 223, 240, 247, 320

modeling 7–8, 15, 20–21, 23, 41, 113, 209, 219, 222–223, 241, 254, 257, 263, 267, 290, 322
   performance of 20, 23
morphological features 141, 147, 219
motion
   see also Ice sheet, velocity, surface 15, 50, 59, 62
   satellite measurement techniques 50–54, 247–264
      feature tracking 39–40, 91, 103, 113, 249, 250–256, 266, 274
      limitations of 39, 255, 258
      maximum-coherence tracking 108–110, 249, 255
      SAR interferometry 39, 42, 45, 50–54, 70–71, 73–74, 76, 80, 83–84, 89–100, 224–225, 249, 255, 257–264, 266, 323
      sastrugi tracking 255
      speckle tracking 101–108, 110, 249
orography 2
paleo-drainage features 235, 254
past history/conditions 6, 23, 139
pressure melt point 176
relief, surface 96
response to climate change 20
rheology 234, 257
role in global climate system 2, 20
scars 225, 234, 252, 255
schematic 10
sea-level rise, contribution to 4, 8–9, 18–19
sensitivity to ocean warming 175
shape 8, 209
shear deformation 176
shear margins 95
shear stress, basal 14–15
slope, surface 14, 95–96, 98–99, 143, 209, 219, 247, 270, 316, 319
stability/instability 8, 40, 113, 248, 273–274, 276, 281
stratigraphy 255, 319
steady state 95, 100, 209, 265, 267
"sticky spots" 248
strain rate 11, 59, 96, 155, 248
   longitudinal 11, 59, 268, 270
   satellite measurement techniques 268–271

408 Index

Ice sheet (*cont.*)
   shear 268
   tensor 268
   transverse 11, 268, 269
  sublimation 10, 12, 210, 241, 246
  submergence/emergence velocity 100
  surface boundary conditions 15
  surface geometry 14
  surface roughness 93, 212, 219, 226, 243, 315–318, 319
  surface slope 225–226, 265
  surface temperature *see* Temperature, surface
  tectonic activity and 210
  temperature 7, 154, 210, 267
  thermodynamics 10, 14–15, 20, 40, 254
  thickness 6, 23, 95, 100–101, 209, 222, 247–248, 265–266, 270, 276
   in situ/aircraft measurement techniques 23, 285, 289
   satellite measurement techniques 289–291
    uncertainties in 322–323
  topography
   see also Ice sheet, elevation 2, 7, 64, 95, 112, 150, 226, 234–237, 248, 254
   impact on atmospheric dynamics 210
   impact on atmospheric lapse rate 210
   impact on katabatic winds 210, 226, 316
  troughs 231, 235
  undulations 147, 225–226, 234–235, 252
  velocity, basal 14
  velocity, depth-averaged 95, 209, 248, 265–266
  velocity gradients 94, 267–268
  velocity, seasonal dependence 90
  velocity, surface
   see also Ice sheet, motion 11–12, 14, 16–17, 145, 157, 224–225, 234, 242
   satellite measurement ofmeasurement techniques 247–264, 276
  velocity, vertical 14, 23
  velocity, vertically integrated 282
  velocity gradient 248
  volume changes, measurement techniques 7, 210, 222, 225, 234, 290
  wind erosion/redistribution 210, 309

Ice sheet margins, satellite detection and mapping of 76, 95, 141, 151, 158, 175–186, 323
  Landsat mapping project of Antarctica 177
Ice sheet–ocean interactions *see also* Ice shelf–ocean interaction 2–3, 6
Ice shelf 2, 15–16, 53, 175
  ablation rates 286
  advance/retreat, satellite measurement techniques 142
  atmospheric pressure effects on 223
  back stress 274, 279
  basal ice accretion 2, 3, 11–12, 260, 279, 286
  basal melt *see* Melt
  bottom profile 233
  buttressing effect 9, 177
   satellite measurement techniques 274–276
   impact of removal of 274–276
  calving *see* Iceberg
  collapse/breakup 7, 175–176, 182, 274–276, 291
  deformation 271
  DEMs 222, 266, 280
   tidal effects on 222
  elevation and elevation change of 210, 222, 231–232, 238, 240, 277
  effect of tidal aliasing on satellite measurement of 278–279
  flux 189
   gates 267
  flow 3, 248
  fracture mechanics 278
  freeze rates 3, 11
  fronts 223, 232, 262
  geometry, changes in 8
  ice-column density 233, 267, 280
  kinematics 269
  mass budget 153
  mass change 9
  mélange 198, 233
  melt 2, 3, 5–6, 8–9, 11–12, 16, 21, 172, 180, 224–225, 260, 268, 279, 281, 284
   impact on sea ice 21
   InSAR measurements of 284–286
   oceanographic measurements of 284
   rates 3

relationship to ocean temperature rise 285
role in oceanic heat loss 3
models 2–5, 278, 287
ocean-swell flexing of 198
schematic 3
shear stresses 15, 267
slopes 226
stability 183
strain rate 269
  measurement from InSAR wrapped phase data 269–271
  transverse 269
stress field 3, 275
sub-ice cavity 3, 268, 279, 287
temperature 3, 189
thickness 139, 266–267, 285
  airborne ice-penetrating radar measurement of 285
thinning 224–225, 269
tidal motion (flexing) of 53, 96, 101, 108, 189, 198, 210, 222–223, 240, 248, 269, 277–279
tidal processes 277
velocity 189, 259–261, 269
  inverse barometer effects on satellite measurement techniques 259–260, 278
  tidal effects on satellite measurement techniques 259
vulnerability to ocean warming 3
Ice shelf–ocean interaction 2–3, 8–9, 175–176, 279, 284–285
  models 3, 279, 287
Ice Shelf Water (ISW) 2–3
Ice stream 16, 107, 113, 146, 155, 157, 177, 186, 235
  confluence regions 255
  discharge conditions, past 255
  discharge rate 254, 273–274
  drainage channels 255
  draw-downs 256
  dynamics 254
  flow 8, 15–16, 102, 147, 149, 252
    acceleration of 248, 256, 272
    deceleration of 256, 272
    plug 252
    variability, satellite detection of 271–276
  flow features, relict 255–256

flow lines 256
grounding-line migration of *see also* Grounding line/zone 281
inter-stream ridges 256
margin scars 147
onset detection, satellite techniques 254, 261, 262
shear margins 89, 147, 155, 157, 235, 256, 270
stability/instability 8, 273
  factors affecting 8
stick–slip motion of 277
surge behaviour, satellite detection of 271–276
thickness 282–287
thinning of 281
tributaries 261–262, 267
trunk regions 256
velocity *see also* Ice sheet, velocity 255
width 254
Ice Stream A 273
Ice Stream B *see* Whillans Ice Stream
Ice Stream C *see* Kamb Ice Stream
Ice Stream D *see* Bindschadler Ice Stream
Ice Stream E 252, 261, 268, 283
Ice viscosity 6
Iceberg
  aircraft measurements of 196, 207
  area flux 205
  A22B 204–205
  B9B 192, 194–195
  B15 191–192, 194, 196, 206–207
  B16 192, 194
  B17 192, 194
  B20 191
  C08 192, 194–195
  C09 194–195
  C16 207
  C19 191
  D14 196
  Antarctic Meteorological Research Center database 203
  calving 3, 10, 12, 16, 177, 183, 186, 189, 194–196, 248, 285
    Ninnis Glacier 194–195, 198, 200–201
    Pine Island Glacier 183–184, 186, 274
    seasonal dependence of 189, 198
  calving events, satellite detection of 190, 194–195, 198–199, 251

Iceberg (*cont.*)
  calving flux, definition of 189
  calving front *see* Terminus, glacier; Ice shelf, front
  calving mechanisms 189, 198–199, 232, 269
  calving rate 186, 189
  detection and size statistics 186, 188–199
    passive-microwave techniques 201
    radar scatterometer techniques 198, 202–204
    Polarimetric SAR techniques 199
    SAR techniques 190, 192–198, 209
    sensor icidence-angle dependence 193–194
    visible–thermal infrared techniques 190–193, 202–203
  distinction from sea ice in satellite imagery 190, 202–203
  draft 202
  drift
    bathymetry and 201–202, 206
    ocean-currents and 199, 201–202
    wind and 199
  drift (rates) 188, 199, 201–206
    satellite measurement techniques
      passive-microwave 201
      radar scatterometer 201, 202
  drift trajectories 188, 190, 192, 195–196, 204–206, 209
    modeling of 201
    satellite-tracked beacon tracking of 201
  elevation and elevation change 234
  fracture 189
  fragmentation 203
  freeboard 189
  freshwater source, as a 188, 199, 208–209, 234
  grounded/grounding 21, 175, 192, 195, 199–200, 202, 205, 207
  grounding traps 206
  icebergs as a hazard 188, 199, 206
  impact of sea ice on 198
  impact on benthic faunal communities 188
  impact on ecology 191, 206
  impact on fast ice 21, 188, 199, 201
  impact on ocean circulation 206–207
  impact on ocean primary production 189
  impact on ocean sedimentation 188
  impact on pack ice 21, 191, 199, 201, 206–207
  impact on penguin colonies 191, 206–207
  impact on polynya formation 21, 191
  impact on rate of carbon sequestration 189
  kedging 206
  manual analysis 197
  mass flux 205
    calculation of 17
  melt 3, 21, 189–190, 199, 206–207
    role in oceanic heat loss 3
  melt rate 209, 234
  microwave penetration of 208
  microwave signature of 201
  microwave (radar) target, as 190–191
  naming convention, Antarctic 203
  National Ice Center operational database 202–203
  ocean-surface roughening, impact on iceberg detection 193
  operational detection and tracking of 202–203, 205
  proxy measure of ocean currents/circulation, as 196, 201
  residence time 206
  Scatterometer Climate record Pathfinder (SCP) database 203–205
  shape 206–207
  ship-borne observations 189–190, 194
  size and frequency distributions 206, 209
  source of iron and nutrient input into the ocean 189, 199
  structural characteristics 206–207
  thickness, airborne measurement techniques 207, 209
    thickness, satellite measurement techniques 206–209
      laser altimeter techniques 209
      radar altimeter techniques 206–208
  tongue 175
  visible–thermal infrared techniques 193
ICEMON 205
ICESat, NASA spacecraft
  see also GLAS 21, 65, 95, 112–113, 153, 175, 207, 209, 229–234, 248, 275
  atmospheric products 232–233
  data archiving 232–233
  ice-sheet accumulation, measurement techniques 243

maximum latitudinal coverage 230, 277
mission phases 230, 232
penetration depth 243
software 232–233
surface elevation measurements 229–234, 276
ICESat Follow-On 234
Ikonos 140, 142, 144
Ikonos II 140
Image segmentation 193
In situ measurements 7, 21–23, 41, 62, 78, 82, 89, 93–95, 99–101, 112–113, 116, 141, 164, 175–176, 186, 206, 210–211, 219–220, 223–224, 230, 238, 240–241, 243, 248, 255, 258, 262, 269, 275, 277, 279, 280, 308, 309, 312–313, 316, 320–323
Indian Ocean 203
InSAR *see* SAR interferometry
InSAR satellite, NASA JPL (proposed) 114–115
Institute Ice Stream, West Antarctica 269
  mass budget of 283–284
Intensity, radar backscatter 154, 182
  factors affecting 154–155, 156, 158, 223
Intensity, specific (of radiation; radiance) 302
Intensity tracking *see* Speckle tracking
Interferogram *see* SAR interferometry, interferogram
Interferometric baseline *see* SAR interferometry, interferometric baseline
Interferometric cartwheel 114
Interferometry *see* SAR interferometry
Inter-glacial 6
Inter Governmental Panel on Climate Change (IPCC) 4, 5
International Arctic Science Committee (IASC) 6
International Trans-Antarctic Scientific Expedition (ITASE) 137
Inverse barometer effect 101, 259–260, 278
Irradiance *see* Flux
IRIS 126
IRS, Indian satellite series 140–141, 211
IRS-1A 141
IRS-1D 140–141
ISMASS 9

Isochrones 241, 243, 289–290
  volcanic eruption events in ice cores 243, 290
Isostatic (post-glacial) rebound 6, 17, 210, 232, 289
  satellite measurement of 288–289
Isotope fractionation 20–21, 299
Isotropic reflectanceintensity 299
ITASE 242, 313, 316

Jakobshavn Isbrae 9, 16, 59, 165, 177, 304
  recent flow acceleration of 9, 276
  recent melt and dynamic thinning of 9, 276
Jason-1 218
JAXA, Japan Aerospace Exploration Agency 75
JERS-1, JAXA spacecraft 60, 75, 77, 79–80, 155, 198, 211
Jetty Peninsula, Antarctica 153
Jutulstraumen Ice Stream, Antarctica 140

Kamb Ice Stream 143, 146, 152, 237, 256, 261–262
  stagnation/deceleration 256, 261–262
Katabatic winds 2, 13, 58, 210, 226, 241, 293
  definition 2
  direction of, inferred from satellite data 317
  effect on surface roughness 317–318
  effects of surface slope on 316
  effects on ice-surface temperature 293
KATE-200 181, 184, 201
KMS (Denmark) 239

Laboratory experiments 254
Lake Vostok 210, 219
Lakes, sub-glacial 210, 219, 223, 225
Lakes, surface 176, 178
Lambert Glacier, East Antarctica 107, 157, 262–263, 266–267, 284
Lambertian reflectance 302–303
Lambertian surface 302–304
Landsat, NASA spacecraft 141–142, 145, 147–148, 159, 177–178, 180–181, 184, 188, 200, 250, 252–253, 255, 258, 273, 279, 283, 293, 308–309, 310–311, 315

Landsat 1, NASA spacecraft 139
Landsat 7, NASA spacecraft 254, 272, 274–275
Landsat 7 ETM+ 139–140, 143–144, 164, 178, 183, 185, 198, 234, 254, 272, 274–275, 311
Landsat Antarctic coastal mapping project 141
Landsat image atlas of Antarctica 141
Landsat image atlas of Greenland 141
Landsat Multispectral Scanner (MSS) 139–140, 181, 184, 188, 200, 250, 253
Landsat Return Beam Vidicon (RBV) 140
Landsat Thematic Mapper (TM) 139–140, 143, 148, 164–165, 184, 188, 200, 252–253, 268, 283, 293, 309–311
  geolocation error of 139–140
  measurement of surface temperature 293
Larsen A Ice Shelf, collapse of 274, 281
Larsen-B Ice Shelf 7, 142, 172, 178–180
  buttressing effect of 274–276
  collapse of 7, 177–183, 274–276
  thinning of 180, 276
Larsen C Ice Shelf 183
Larsen Ice Shelves 9, 225, 271
Larsen Inlet 180
Laser altimetry 7, 18, 21, 65, 95, 112–113, 153, 175, 198, 209, 229–235, 263, 275, 278, 287, 323
  aerosols, satellite measurement techniques 233
  aircraft 82, 89, 93, 95, 112, 210–211, 222, 225, 227, 235–236, 281
  accuracy of 225
  cloud contamination of 230, 233
  cloud heights, satellite measurement techniques 233
  ice-sheet elevation and elevation change, satellite measurement techniques 229–234, 275
    accuracy of 229–232
    crossover analysis 230, 233
    error sources, correction 229–230, 232
    precise orbit determination 230
  ice-shelf elevation and elevation change, satellite measurement techniques 232
    tidal effects 232

  iceberg elevation and elevation change, satellite measurement techniques 234
  retracking techniques 232
  rift depths, satellite measurement techniques 232–233
  surface slope, measurement of 281
  tidal aliasing of 278
Laser tracking of satellites 217, 229
Last-Glacial Maximum 7, 183
Latent heat 13, 166
Layover, radar image 67, 69, 83, 88–89, 155, 240
Leppard Glacier 274
Lidar 229
LISS-III 141, 211
Longwave radiation 13
LOWTRAN radiative transfer model 294

Mantle inflow 288
Maps 272
Marine ice 2–3, 286
Mass balance
  see also Ice sheet, mass balance 10–21
  component (flux) approach 17, 189, 283
  definition of 11
  integrated approach 17
  local 11–12
  measurement techniques 16–21
  net 12
  specific 11–12
  total 12
Mass budget
  see also Ice sheet, mass budget
  definition of 12
  satellite measurement of 282–287
Maximum-likelihood classification 171
McMurdo Station, Antarctica 75, 191, 206, 207
Mega-dune fields 315–316
Melt, surface
  see also Ice sheet, melt 16, 21, 57, 159, 164–175, 291, 304
  effects of cloud on 295
  impact on surface albedo 300–301
  impact on passive-microwave emissivity 298
  impact on radar backscatter 160–161, 166

onset, duration and extent 167, 169, 170–171, 180, 182–183
  rates 165–166, 174–175, 209, 291
  satellite detection of 164–175
    active microwave techniques 57, 166–169, 305
    InSAR techniques 57
    passive microwave techniques 169–173, 174, 234, 304–305
    visible-thermal IR techniques 164–166
  season length 182
visible–thermal IR techniques 164–166
Melt lakes 165
Meltponds, surface 142, 178, 180
Meltwater
  see also Ocean, freshwater budget
  downward percolation 175–176, 180, 274–275, 286, 291
  impact on ocean thermohaline circulation 6
  in-flow channels 176
  role in ice-shelf break-up 275
  runoff 21, 161, 164, 174–176, 285, 291
MERIS 114, 151, 301, 310
Mertz Glacier 191–192, 194–195, 200–201, 202
Mertz Glacier Polynya 195
Meteorites on ice sheets, satellite detection of 309
Meteorology, Antarctic 291–292
Meteorological forcing 17
MetOp, ESA spacecraft series 168
Micro-satellites 115
Microwave
  absorption see Absorption
  active see also Radar altimetry, Radar scatterometry, SAR interferometry, Synthetic aperture radar 23
  backscatter 48, 54, 151–152, 154, 167, 212, 223, 233, 243–245, 249, 319–320
  backscatter coefficient 155, 212, 243, 318
  backscatter modeling 112
  emissivity 244–246, 295–298
  extinction 224, 318
    relationship to snow-grain size 319
  loss tangent 290
  model inversion techniques 244, 313
  passive see Passive microwave

penetration depth see Penetration depth
polarization see Polarization
radar azimuth 47, 317–318
radar line-of-sight (LOS) 51–52, 91, 96–97, 100, 259
radar look angle 96–97, 270, 272
radar slant range 47, 51, 96
reflection
  factors affecting 313–314
  internal 314
  surface 224
reflection coefficient 290
reflectivity, radar 41, 154, 166, 216
scattering 295
  multiple 313
  surface 57, 112, 161, 190, 243, 316, 319
  volume 57, 64–65, 112, 160, 166, 190, 223–224, 243, 313–314, 319–320
transmissivity 166
volume echo 318
volume scattering models 319
Middle-infrared (mid-IR) 311
MIMOSA, ESA spacecraft 289–291
Mini-Modified Antarctic Mapping Mission (MiniMAMM, Radarsat-1) 76, 158, 260
MIRAS 247
Mission to Planet Earth (NASA) 294
MISR 183–185, 198, 206–207, 295, 301, 305–307, 309–310, 316
  cloud classification/detection techniques 295, 315
  maximum latitudinal coverage of 295
MODIS 142, 149–153, 157, 164, 166, 177–178, 190–192, 220, 234, 236–237, 301, 304–307, 310, 316
  geo-location of 306
  horizontal striping artifact 237
  onboard calibration 306
Moisture flux, atmospheric 58
Moisture pathways, atmospheric 13
Monacobreen, Svalbard 261
  surge behaviour of 271–272
Motion, ice sheet see Ice sheet, motion
Moraine, looped 275
Moulin 176

Mount Pinatubo eruption (1991)
  impact on Greenland Ice Sheet temperature and melt extent 305
  impact on satellite measurements 171–172, 294
MSS see Landsat MSS
Multi-angle optical remote sensing 183–185, 198, 206–207, 295, 301, 305–310, 315, 316
  surface roughness information from 316
Multi-polarized area ratio (iceberg detection algorithm) 199

Narrow-to-broadband conversion 301
NASA 75, 155, 218, 225, 227, 285, 294, 321
NASA Ice Sheet Altimeter Pathfinder Program 215, 219
NASA Goddard Space Flight Center 216
  Ice Altimetry website 216
NASA Jet Propulsion Laboratory 114
NASDA see JAXA
National Ice Center (U.S. Navy/NOAA/U.S. Coast Guard) 202–203, 205
National Snow and Ice Data Center (NSIDC, USA) 149, 151, 170, 177, 183, 209, 220, 222, 232, 254, 294, 306
NCEP/NCAR Reanalysis dataset 242, 298
Near infrared (near-IR) 139–193, 202–203, 235, 263, 270, 309
Nimbus 5, NASA spacecraft 244
Nimbus 7, NASA spacecraft 170–172, 174, 244–245, 293, 295, 312–314
Ninnis Glacier, Antarctica 192, 194–195, 198, 200–201, 287
NOAA 22, 143
NOAA polar orbiter spacecraft see also AVHRR 21
Normalised radar cross section 168
North American Trough (NAT) 172
North Atlantic Ooscillation (NOA) 14, 172
North East Greenland Ice Stream 235–236, 262, 263
NPOESS, U.S. satellite series 294, 305, 308
NPOESS Preparatory Project (NPP) 294, 308
NSCAT 160, 167, 203, 205, 243, 317
NSF 23

Nunatak see Bedrock outcrops
Nyquist sampling rate 85

Oates Coast, Antarctica 140
Ocean
  bathymetry 201
  circulation 3, 175, 188, 206, 284
    impact of icebergs on 206
  currents 3, 187–188, 191, 195–196, 201, 258
    measurements by InSAR 115–116
  freshwater budget 3, 6, 9, 11, 21, 164, 188, 208, 234
  heat loss 3
  ice sheet interactions see Ice sheet–ocean interaction
  ice shelf interactions see Ice shelf–ocean interaction
  modeling 287
  overturning 279
  salinity anomalies 188
  sediment cores 183
  sedimentation 188
  stratification 188
  temperature 3, 8, 139, 285
  thermal forcing 284
  thermohaline circulation 6, 164, 279
  warming 3, 8, 101, 175
    impact on ice-shelf/glacier basal melt 8
  water-mass formation 2–3
  water-mass mixing/modification 2–3, 21
  waves and swell 189
Offset tracking 110
OLS 145, 190, 202, 205
OPS 211
Optical remote sensing 39
  image co-registration 250, 252
Orbit
  ephemeris 62, 139, 304
  precision information 75, 78, 217, 227, 229–230
  state vectors 75
  Sun-synchronous 100
Orbital
  dynamics 217
  state vectors 257
  tracking 217, 227, 229–230

Pacific Ocean 203
Pack ice *see* Sea ice
Palmer Land, Antarctica 274
PALSAR 78–80, 151, 155, 192
  polarimetric capabilities of 111
PAN 141
Parallax, range 48–50, 52, 240
PARASOL 307
PARCA 23, 95, 159, 225, 285–286, 321, 323
Passive microwave 169–175, 180, 201, 217–218, 234, 243–247, 285, 292, 295–299, 312–315, 317
  atmospheric effects on 296
  low-frequency 243, 247, 298
  techniques, changes in snow/firn characteristics 312–315
  techniques, ice-sheet accumulation 243–244, 313
Penetration depth 63–66, 77, 80, 112, 154–155, 160, 162, 190, 208, 216, 223, 240, 244, 247, 295–298, 319, 320
  diurnal temperature cycle effect on 296
  frequency-dependence of 64–65
  different ice-sheet facies, of 320
  maps of, from InSAR 320
Penguins
  Emperor 191
  icebergs and 191, 206, 207
Percolation zone 65, 158–160
Petermann Gletscher (Glacier), Greenland 53, 280
  grounding-line migration of 281
  thinning of 281
Phase
  see also SAR interferometry 42–43, 47–48, 228, 229
  difference 43–44, 51
  flat-Earth 51
  shift 42–44, 48
Phasor, interferometric 48
Photoclinometry 47, 93, 142, 151, 220, 234–237, 287
Photogrammetry 211
Photometric relationship 235
Pine Island Glacier, West Antarctica 73, 220–221, 224–225
  basal melt rate of 284
  calving of 183–184, 186, 274

  contribution to global sea-level rise 273
  DEM of 220–223
  elevation change of 281
  flow acceleration of 273
  grounding-line migration of 281
  mass budget of 284
  MiniMAMM coverage of 76, 158, 260
  motion of 250–251
  thickness measurement of 282–283, 287
  thinning of 273, 281
Planck function 295
Pléiades, French satellite series 140
Polar darkness 2, 255, 312
Polar Pathfinder Program 294
  AVHRR surface albedo dataset 304
  AVHRR surface temperature dataset 294
Polarimateric InSAR (Pol-InSAR) 65, 111–112, 116, 163–164, 240
  aircraft measurements 112
  coherence matrix 111
  ice-sheet phase detection 163–164
  measurement of ice-sheet near-surface properties 320–321
  optimization procedures 111, 321
  penetration depth and 112, 240, 321
  phase-preserving transformations 111, 321
  pre-processing of 112
  processing of 112, 321
  scattering mechanisms 111–112
  target classification using 111
  target decomposition methods 111, 321
  temporal decorrelation effects 164
Polarimetric microwave radiometers, passive 158
Polarimetric radiometers, optical 312
Polarimetric SAR *see* Synthetic Aperture Radar polarimetry
Polarization
  alternating mode (Envisat ASAR) 199
  impact on InSAR coherence 65, 321
  multi-polarization 151
  of passive-microwave sensor frequencies 58, 245–246, 292, 296, 313–315
  factors affecting 313–315
  radar 111, 163
  ratio (PR) 160, 170, 245–246, 314
  factors affecting 313–314
  of visible-near infrared wavelengths 307, 312

Polarized bidirectional reflectance *see*
    Bidirectional Polarization
    Distribution Function
POLDER(-P) 301, 307, 312, 316
PolInSAR *see* Polarimetric InSAR
Polynyas 21
  impact of icebergs on 188
Positive degree-day temperature models *see*
    Degree-day models
Post-glacial rebound *see* Isostatic rebound
Power-to-Mean-Ratio algorithm, iceberg
    detection 195–197
PRARE 77
Precipitation 58, 322
Precipitation rates 2, 13, 19, 174
Precipitation minus evaporation 12–13, 234, 241, 289
Prince Gustav Channel Ice Shelf 9, 180
Principal component analysis (PCA) 139, 147
PRISM 211
Pulse-limited footprint 208

Quantization *see also* Radiometric
    resolution 143
Queen Maud Land, East Antarctica 140
QuickBird 140
QuickScat, NASA spacecraft *see also*
    SeaWinds 167–169, 172, 202–205, 243–244, 313

Radar Altimeter-2 (RA-2) 162, 208–209, 217–218, 226–227, 233, 277
Radar altimetry 4, 7, 17–18, 93, 155, 161–163, 198, 206–209, 211–230, 233–235, 243, 246, 249, 263, 266, 268, 277–280, 282, 286–287, 304, 312, 315–320, 323
  accumulation information from 243, 246
  airborne 219, 262
  antenna beamwidth 213, 226
  antenna footprint 226
  antenna gain 213, 224
  atmospheric effects 217
    passive-microwave correction of 217–218
  backscattering coefficient 211
  barrier motion, measurement techniques 249–250
  corrections 217
  crossover point/analysis 216, 222–224, 229
  data filtering 222
  delay-Doppler techniques 227–229
  DEMs 93–95, 219–222, 234–235, 238, 266–268, 282, 312
  dual-frequency 224, 246
  grounding-line detection and mapping 280
  ice modes 214
  ice-sheet stratification, satellite measurement of 247, 319
  interferometric techniques 227–228
  ionospheric effects 217–218
  loss of track (lock) 214, 219
  maximum latitudinal coverage 227, 229, 266, 277
  measurement techniques, ocean parameters 214
    ocean circulation 214
    sea-level monitoring 214
    significant wave height 214
  measurement system biases 222
  meltwater runoff detection 161
  measurement of near-surface ice-sheet characteristics 212, 313, 318–319
  ocean mode 214
  orbit error analysis 222
  orbital-tracking techniques 227, 229
  penetration of signal 318
    correction of 313, 320
  pulse-limited techniques 94–95, 212–213, 224, 226–229
  pulse-limited footprint 212–214, 216
  range, measurement techniques 212, 215–217, 227, 229
  range gates 212, 215
  range resolution 212
  range rings 212
  range windows 212
  range window-limited footprint 212
  retrack point 215, 223
  single-frequency 223–224
  slope correction 216, 222
  surface elevation, measurement techniques 213, 215–229, 277
    error sources, and correction of 216, 223–224, 229, 233, 313, 318, 320
  surface elevation change, measurement techniques 222–225, 277

accuracy of 224, 226, 248
    effect of tidal aliasing on 278–279
surface properties, measurement
    techniques 226
surface roughness, measurement
    techniques 212, 316, 319
surface roughness effects 224, 316
surface slope, measurement techniques 225
synthetic aperture processing of 227
tidal coefficients 277
total backscatter 318–319
tracker device 212, 214
tracking gate 212, 215
volume scattering models 319
waveforms 212–213, 243, 318–319
    amplitude 223
    ice sheet 212, 214
    ice-sheet accumulation effects 243
    integration of area under 318
    leading edge of 212–213, 215–216, 223
        ice-sheet 316
        ocean 316
        relationship to surface roughness 316
        shape 316
    multi-look 227, 229
    ocean 212, 215
    ramp 215
    retracking 215–216, 223–224
    shape 214–215, 316
    trailing edge of 214
Radar-echo sounding, airborne 7, 153
Radar, ice-penetrating 23–24, 139, 207,
    241–242, 255, 268
    measurement of internal layers with 23
Radar, ice-sounding 280–281, 285, 289–290
    airborne 7, 285, 289–290
    in situ 7
    measurement of ice thickness 285
    measurement of internal layering 289
    path-length 42, 48
    satellite 289–291
Radar scatterometry 58, 159–162, 166–169,
    172, 201–205, 243–244, 247,
    296–297, 317–318
    dataset A 160
    image enhancement techniques 167–168,
        202–203
    measurement of ice-sheet accumulation
        243–244

measurement of ice-sheet facies 159–162
measurement of ice-sheet melt 166–169
measurement of ice-sheet surface
    roughness 317–318
measurement of ice-sheet surface wind-
    field, proxy 317–318
measurement of icebergs 198, 201
Radargrammetry 240
Radarsat-1, CSA spacecraft 54, 60–61, 72,
    75–76, 78, 80, 98, 101, 104, 107,
    138, 145, 155–158, 166, 177, 180,
    182, 186, 188, 192–195, 197–199,
    202, 219–220, 231, 260–262,
    269–271, 273, 276, 283, 312, 317
    Antarctic Mapping Project (RAMP) 76,
        155
    Antarctic Mapping Mission *see* Antarctic
        Mapping Mission, Radarsat-1
    beam modes 76
    orbital parameters 78–79
Radarsat-2, CSA spacecraft 60, 74, 78, 80,
    98, 109, 114, 137, 151, 158, 199,
    263
    orbital parameters 79
    polarimetric capabilities of 111, 158
    scanning modes 109
Radarsat-3, CSA spacecraft 114
    polarimetric capabilities of 114
Radiance *see* Intensity
Radiation balance, high-latitude 2
Radiation budget, surface 2, 306
Radiative flux
    incoming longwave 13, 299
    incoming shortwave 13, 299
    outgoing longwave 13
Radiative heating 299
Radiative transfer 299, 305, 308, 310
Radiative transfer equation 296
Radio-echo sounding, aircraft 23–24, 100,
    262, 266–268, 286
Radiometric resolution 140, 142–143,
    145–147, 149–150, 236
Radiometric sensitivity 143, 235, 237
Radiosonde data 294
Range offset, of laser waveforms 209
Range parallax 48–49
Rayleigh–Jeans approximation 295
Recovery Glacier, Antarctica 76, 270

Refractive index
  ice 310–311
Reflectance 143, 149, 235, 299, 302–303, 306
  shortwave *see also* Albedo, Bidirectional Reflectance Distribution Function 299, 301, 309, 311
  top-of-atmosphere 304–305
Reflection *see* Microwave, reflection
Reflection, specular 300, 316
Reflection azimuth 302
Reflection zenith 302
Reflectivity 139, 143, 233, 235, 310
Regelation ice films 58, 309
Region growing algorithms 177
Rennick Glacier, Antarctica 140
Repeat interval (cycle), orbital 77–79, 143, 278–279
Reservoir compounds 137, 141–143
Resolution cell 42, 46–47
Resourcesat 419
Resurs, Russian spacecraft series 143
Rift formation and propagation 40, 147, 150, 152, 183–184, 186, 191, 198, 232–233, 276
Ronne Ice Shelf 204
Ross Ice Shelf 108, 145, 147–150, 152, 186–187, 190–192, 195–196, 202–203, 206–207, 225, 231, 252, 256, 276–277
Ross Island 191, 207
Ross Sea 191, 207
Roughness, surface 13, 41, 93, 154, 212, 219, 224, 243, 295–296, 315–319
  factors affecting 154–155, 315–318
  impact on BRDF 303, 308, 315
  impact on directionality of surface reflection 303
  impact on microwave emissionvity 312–314, 317
  impact on snow accumulation 315
  impact on surface albedo 300, 315
  impact on surface wind field 315
  satellite measurement techniques 315–318
    InSAR techniques 315
    passive-microwave techniques 317
    radar altimeter techniques 315–316
    radar scatterometer techniques 317–318
    visible and near-infrared techniques 315–316
    multi-angular techniques 316
  scales of 315
  small-scale 313–314
  wind effects on 313–314
Rumples, ice 139, 153
Rutford Ice Stream, Antarctica 52
Ryder Glacier, N. Greenland 239
  grounding-line migration of 281
  thinning of 281
  surge behaviour of 271

SAR *see* Synthetic Aperture Radar
SAR interferometry 7, 9, 18, 24, 39–135, 186, 198, 224, 227, 237–240, 246–247, 249, 255, 257–264, 266, 274, 276, 279–288, 315, 318, 320
  across-track techniques 45–54, 79
  advantages/disadvantages of 113, 258, 263
  aircraft techniques 44, 49, 95, 100, 115
  along-track techniques 44–45, 79, 98, 107, 115–116
    ocean-current measurement by 115–116
  altitude of ambiguity *see* Ambiguity height
  ambiguity height 50
  ambiguity velocity 50
  atmospheric (tropospheric) propagation-delay effects 46, 48, 51, 56, 60–61, 113, 259
  azimuth streaks 60–61
  basal melt rates, measurement of 282–287
  baseline attitude 55
  baseline, critical 63, 81
  baseline, effective 51
  baseline errors 62, 257–258
  baseline information 74–75
  baseline, interferometric (spatial) 43, 45, 47, 48–51, 54, 62, 65, 70–71, 76–77, 78, 83, 89, 91–92, 99, 101, 111, 113, 227, 257
  baseline, optimal 70–71, 74
  baseline, parallel 104
  baseline, perpendicular 47, 50, 63, 271
  baseline scaling 92, 94
  baseline, temporal 44, 50, 52, 57, 59, 66, 70–71, 83, 89, 91, 96
  baseline, zero 51
  baseline length 54, 70–71, 74

Index

baseline-related trade-offs 49, 70–71
calibration and validation 41, 95
clock drift, impact of 56–57, 100
coherence *see* SAR interferometry, phase coherence
coherence image (map) 65, 69, 73, 76, 83
constraints, and sources of ambiguity and error 46, 49, 55–71, 80–81, 84–85, 99–102, 259
co-registration, image 99, 103
correlation coefficient 56
crossing-orbit data 258–259
data requirements 74–80
decorrelation, baseline *see also* SAR interferometry, phase decorrelation 56, 62–63, 70–71, 320–321
decorrelation, processing *see also* SAR interferometry, phase decorrelation 56
decorrelation, target motion/rotation *see also* SAR interferometry, phase decorrelation 59
decorrelation, temporal *see also* SAR interferometry, phase decorrelation 48–49, 55–60, 76–78, 80, 102, 113–114, 259–260, 271, 321
decorrelation, thermal *see also* SAR interferometry, phase decorrelation 55–57
decorrelation, volume 63–66
DEM, InSAR-derived 98–99, 113, 238–240, 262
developments and outstanding issues 112–116
differential InSAR 45–46, 52–54, 90–101, 257, 259, 271, 277
  relative strengths and weaknesses 109–111
discharge flux, measurement of 281–287
displacement, measurement techniques 90–92, 96–97
Doppler information, use of 95
double-differencing 83, 90, 238–239, 257, 280
dual-antenna systems 44-45, 49, 115
error analysis 113
error sources 52–53, 94–96, 99–100, 113, 240
fast ice, measurement techniques 41

filtering techniques 73, 83
four pass 280
freeware packages *see also* SAR interferometry, software packages 73
frequency, optimal 80
fringe aliasing effects 107
fringe patterns, interferometric 50, 54, 59, 83, 85, 90–91, 94, 102, 270
geometric co-registration 81–82
geometric distortion effects 67
ground-control points *see also* SAR interferometry, tie-points 45, 82, 89, 93, 94–97, 257–258
ground range 89
grounding line, detection and monitoring of 40, 60, 113, 280–281, 323
hazard detection and monitoring with 116
hydrological applications 116
ice deformation, measurement techniques 45
ice discharge rates, measurement techniques 40
ice-sheet accumulation patterns, measurement techniques 246–247, 320
ice-sheet motion, measurement techniques 39, 42, 45, 50–54, 70–71, 73–74, 76, 89–100
ice-sheet near-surface characteristics, measurement of 313, 320
ice-sheet surface roughness, measurement of 315
ice-sheet topography/elevation, measurement techniques 39, 42, 46–54, 70–71, 76, 80, 83–84, 89–93, 237–240
ice velocity, measurement techniques 53–54, 71, 76, 80, 83–84, 89–90, 94, 96, 224–225, 257–264, 266, 276, 323
  error sources 99–101
  one-dimensional 258–259
  two-dimensional 95–99, 272
  three-dimensional 96–99, 260, 263, 272
image co-registration 75, 81–82
image geocoding 52, 73, 75, 89
image mosaic processing 84, 113

SAR interferometry (cont.)
  interferogram 43, 45, 50, 52, 55, 82, 89–92, 94–95, 238, 270, 280, 320–321
  interferogram correlation, maps of 320
  interferogram, definition of 54–55
  interferogram, generation 73, 82–84
  interferogram, topography only 257
  interferogram geo-coding 89
  ionospheric propagation-delay effects 46, 48, 51, 56, 60–61, 113
  L-band mission, future 263
  long-wavelength residuals (artifacts) 93, 99–100, 238–239
  look directions 97–100, 272
  mass budget estimates 40, 240, 281–287
  master image 81–82
  maximum latitudinal coverage 98, 260
  melt effects 57, 80
  multi-look processing 83, 108, 193
  multi-satellite 79, 114–115
  need for data continuity 323
  orbits, ascending and descending 99, 258–260, 262
  path length 42, 48, 58, 91
  penetration depth effects 63–66, 77, 80, 112, 240
  penetration depth maps 64–65, 320
  phase aliasing 260
  phase coherence (correlation) 46, 48–49, 55, 58–59, 62, 67, 69, 75, 80, 83, 88–89, 95, 102, 108–109, 111, 238, 246, 270–271
  phase decorrelation 51, 55–70, 88, 91, 271, 320
    surface-type discrimination using 65
    wavelength-dependence 58, 77
  phase decorrelation, estimation of 67, 320
  phase difference see SAR interferometry, phase, interferometric
  phase discontinuities 83, 85
  phase gradient 85, 89, 110, 270
  phase, interferometric 42–44, 46, 48, 51–52, 54, 83, 91, 96, 257
  phase noise 48–49, 51, 59, 79, 85, 99
  phase shift 44, 51, 83
  phase uncertainty 49, 53
  phase, unwrapped 51
  phase unwrapping 54–55, 59, 63, 66–67, 70, 73, 83–89, 92–95, 99, 106–107, 109, 113
    branch cuts techniques 86
    error sources 88–89
    fast-Fourier transforms 87
    fringe-line detection technique 86–87
    Goldstein's branch cut algorithm 86
    least-squares methods 87
    minimum cost flow algorithms 87
    minimum discontinuity algorithms 86
    minimum L^p-norm methods 88
    minimum-norm methods 85, 87–89
    minimum spanning tree algorithm 87
    path-dependent methods 87
    path-following methods 85–87
    quality-guided path methods 87
    residue compensation methods 87
    single minimum Steiner tree algorithm 87
    unweighted multi-grid technique 87
  phase, wrapped 269–271
    glacier strain rate calculation with 269–271
  phase-to-height conversion techniques 89
    baseline rotation method 89
    integrated incidence angle method 89
    normal baseline method 89
  phase-to-height sensitivity 49
  polarization effects 65, 321
  precision 51, 257
  pre-processing steps 82–84
  principles and terminology 41–55
  processing errors 70
  processing steps 71–89
  quadruple differencing 259, 277
  range, measurement techniques 42, 83
  repeat cycles, satellite 102
  repeat-pass (multi-pass) techniques 40–41, 43, 45–54, 55, 57, 76, 79, 91, 97, 110, 115, 237–238, 257
  resolution, horizontal 94
  resolution, vertical 94
  rift propagation, measurement of 40
  satellite missions 287
  satellite orbital parameters 78–79, 81, 102
  ScanSAR imagery and 78–81, 84, 109, 115
  sea-ice motion, measurement techniques 116
  selection of suitable images 74–80

separation of topographic and motion/
    displacement terms 51–54
single-pass techniques 43–45, 48–51
slant-range geometry 89, 91, 96
slave image 81–82
slope, impact of 63, 97–99
software packages 73–75
speckle, impact on phase coherence 66, 83
steep terrain effects 63, 67, 89
strain rates, measurement techniques
    269–271
  tensile strength of ice, measurement
    from 270
surface-parallel flow, assumption of 97,
    99–100, 104, 258–260, 262, 272
surge behaviour, measurement techniques
    40, 91, 110, 113, 258, 262, 271–273
  errors in 272
synergy with other satellite sensors
    113–114
system noise effects 51
tandem satellite missions, future 114–115
three-pass 280
tidal aliasing effects 259
tidal effects, correction of 101, 238, 240,
    259
tidal flexure/motion, effects of 96, 101, 105,
    259
tidal flexure/motion, measurement
    techniques 40, 53, 60, 277–278
tie-points see also ground-control points
    62, 76, 82, 93–95, 239
tranferential approach 110–111
trade-offs 49, 58, 70–71
tutorial 75
volcanic/tectonic applications 116
volume-decorrelation modeling 320
wavelength-dependence 58, 77, 80, 96,
    114–115
SAR Maximum-Coherence Tracking 40,
    113, 263–264
  advantages/disadvantages of 263
  ice-shelf velocity measurement 263–264
    tidal-effect removal 263
  strain rates, measurement techniques 269
SAR Speckle Tracking 40, 78–79, 99,
    101–109, 113, 260–262
  advantages of 106, 263
  amalgamation with InSAR 262

atmospheric propagation delays 106
azimuth streaking 107
baseline 103
calibration of 103–104
constraints on 261
definition 102
disadvantages of 106–107, 263
ground-control points 103
ice velocity measurements 260–262, 276,
    281
  uncertainties in 107, 261
image co-registration 103, 107, 250
image mosaicking 107
inverse barometer effect on 261
ionospheric propagation delays 107
maximum latitudinal coverage of 260–261
precision of 107, 261
processing steps 103, 105
relative strengths and weaknesses of
    109–111
surge behaviour, measurement techniques
    261, 271–272
  errors in 272
tidal effects on 105–108, 259, 261
Sastrugi 57, 143–144, 255, 315–316
  alignment, relationship with surface wind-
    field 317–318
  effect on BRDF 303
  effect on microwave emission 313
  effect on radar backscatter 316–318
Satellite orbital tracking 62
Satellite remote sensing
  general advantages of 21, 23, 175
  ice sheet applications of, summary 138
  planning of field campaigns, importance in
    138, 153
Satellite sensor classes, summary 1, 138
Satellites, role in transmitting data from in
    situ instruments 21
Saturation, detector 143, 308
SCAR 9, 254, 266
Scatterers, point 54
Scatterers, surface 42, 59
Scattering
  centers 51
  matrix, complex 321
  phase, invariant 48
Scatterometer Climate record Pathfinder
    (SCP) 167–168, 203, 205

Scatterometer Image Reconstruction with Filtering (SIRF) algorithm 161
Sea ice, impact on precipitation/accumulation rates 20
Sea ice 2–3, 14, 20, 41, 201, 296
  albedo 191
  concentration 20, 207
  dynamics 21, 191
  edge location 195
  elevation 209
  extent 20, 189–190, 207
  fast ice 21, 41, 111
    impact of grounded icebergs on 21
    impact of ice-sheet coastal configuration on 21
    impact on icebergs 195, 198
  formation 2–3
  freeboard 208–209
    impact of ice-shelf melt on 21
    impact on iceberg calving 189
    melt-back 190
    motion 116
    operational observation of 195
    polynya see Polynya
    precipitation over 247
    salt rejection 2
    surface roughness, remote sensing of 209
    surface temperature 292
    thickness 209
      laser altimeter measurement techniques 209
    thickness distribution 21
      impact of icebergs on 21
Sea-level rise, global 4–9, 218, 223, 231, 234, 265, 273
  ice sheets and 4–10, 20, 40, 164, 175, 276, 285, 289
  ice shelves and 177, 276
  impacts of 5
  impact of past ice-sheet conditions on 6–7
  model predictions of 4, 5, 9
    current uncertainties in 4–6
  radar-altimeter measurement techniques 4, 218
  rates of 4
  role of satellites in reducing uncertainty in 248
  thermal expansion (ocean) contribution 4
  tide gauge measurement techniques 4

Seasat, NASA spacecraft 45, 161–162, 166, 203, 205, 212, 214, 217–219
Seasat radar altimeter 212, 214, 217–219
Seasat-A scatterometer (SASS) 161–162, 166, 203, 205
Sea surface temperature (SST) 293
SeaWinds see also QuikScat 168, 202–205
SeaWinds II 243
Search array 209
Sediment see Sub-glacial, sediments
Sensible heat 66
Sensor zenith angle 303
Service Argos see Argos telemetry
Shackleton Ice Shelf (East Antarctica) 183
Shadowing, radar image 67, 69, 155, 240
Shallow-ice approximation 14
Ship detection, in SAR imagery 193, 199
Shirase Glacier, East Antarctica 186
Shortwave radiation 299
  absorption 299
  driving snow metamorphism 299
  driving snow photo-chemical processes 299
  penetration 299
  reflection 302
  scattering, multiple 302
  transmission 299, 302
Shuttle Imaging Radar (SIR-C), NASA 455
Shuttle Radar Topography Mission (SRTM), NASA 44
Signal-to-Clutter Ratio (iceberg detection algorithm) 199
Signal-to-noise ratio (SNR) 56, 67, 69, 77, 80, 82–83, 143
Single-look complex (SLC) imagery, SAR 48, 72–74, 82
Siple Coast 108, 256
  ice streams 230, 273
Siple Dome 256
SIRAL see also CryoSat 116, 209, 227, 229
Sjögren Glacier, Antarctica 275
Skin surface temperature see also Temperature, surface 4
Slessor Glacier, Antarctica 102
Slope see Ice sheet, slope
Smith Glacier 224
SMMR 170–172, 174, 243, 245, 295–298, 312
SMOS, ESA satellite mission 247, 298

Index   423

Snow
  absorption coefficient 318–319
  accumulation patterns, ice-sheet factors affecting 310, 313, 315–316
  age 299
  albedo *see* Albedo
  anisotropic reflectance characteristics of 302–303
  characteristics inferred from passive-microwave data 312–315
  densification 210
  density 154, 223, 243, 295, 303, 312–313, 319
  deposition 143
  dielectric constant 153–154
  dielectric properties 166, 223, 290, 319
  drift 309, 315
  dunes 252, 313, 315, 317
  emissivity 244–245, 292–293, 312–314
  erosion 143, 315
  extinction coefficient 224, 318
  grain growth, model 244
  grain shape 300, 311–312
  grain size *see also* Grain size 147, 154, 174, 223, 235, 243–244, 246, 292, 295–296, 303, 309–312, 319–320
    effect on microwave absorption coefficient 319
    effect on microwave extinction coefficient 319–320
    effect on surface albedo 299–300, 304
    relationship with accumulation rate 243–247, 310
    relationship with surface temperature 292, 310
    remote sensing of 308, 310–313, 315, 319–320
  granularity 154
  impurity content 310
    satellite measurement techniques 311
  kinetic heating of 58
  layering/stratification 223–224, 247, 289, 295–296, 312–313
  liquid-water (free water) content 57, 154, 160–161, 166
  melt 57, 310
  melt and refreezing 154, 164–173
  metamorphism 58, 154, 244, 292, 298–299, 310, 313

  microwave backscatter 151–152, 243–244
  microwave backscatter coefficient 155, 318
  microwave penetration depth 222, 318
  microwave properties 155, 292, 313–314, 318–320
  moisture content, volumetric 167
  optical properties 292, 299–300, 302–303
  penetration depth *see* Penetration depth
  pit measurements 241, 243, 312
  properties, effect of surface winds upon 313–314
  radiative transfer modeling 244
  redistribution, aeolian 10, 57, 143, 210, 241, 315
  reflectance 147, 235
    optical spectral 299–300, 306, 310–311
  reflectivity 154, 235, 310, 318
  saltation 58, 309
  temperature *see also* Temperature, surface 154, 295–299, 312
  temperature gradients 244
  wetness 57, 154, 223
  wetness, effect on backscatter 167
Snow line 14, 19, 160
Snow Line Altitude (SLA) 19, 160, 308
Snow/firn near-surface characteristics
  changes inferred from satellite passive-microwave data 312–315
  satellite InSAR techniques 320
  satellite radar altimeter techniques 318–320
  satellite polarimetric interferometry techniques 320–321
  temporal variability 319
Solar azimuth angle 302
Solar incidence angle 300
Solar spectrum 299
Solar zenith angle 143, 302–303, 305
Soot 310
  impact on surface albedo 300
South Georgia 204–205
South Pole 155, 260
Southern Aannular Mmode (SAM) 8, 14, 298
  positive index state of 8
Southern Oceann 189
Southern Oscillation 170
Southern Ooscillation Iindex (SOI) 296

424 Index

Soyuz Kosmos, Russian spacecraft series 181, 184, 201
Spatial resolution 236
Speckle, in radar imagery 66, 83–190, 193, 240, 261
Spectral resolution 311
Spectral shift filtering 63
SpectralonTM 110
Spectroscopy, imaging *see* Hyperspectral imaging
Specular reflector 300, 316
SPOT, French spacecraft/sensor program 139, 141, 143, 145, 147, 177, 211, 250, 252, 254–255, 258, 308, 310
SPOT 1-4 141
SPOT 5 140
SPOT HRG 139
SPOT HRS 139–140, 211
SPOT HRV 143
SPOT Vegetation 310
SSM/I 170–174, 201, 243–245, 296, 298, 313
Startracker system 229–230
Stefan–Boltzmann law 13
Stereoscopy 47, 140–141, 211
Stereo-radar techniques 240
Storm tracks 13
Storstrømmen Glacier, NE Greenland 100, 262
   surge behaviour of 271
Streamer, radiative transfer model 305
Sub-glacial
   conditions 15
   fabric (till) 14–15
   hydrological conditions 8, 15
   lakes *see* Lakes, sub-glacial
   processes 15
   sediments 8, 14, 23
      deformation 14, 254
Superimposed ice 19, 159
Surface boundary conditions 15
Surface features, detection and mapping of 137–164
   high-resolution visible-thermal IR methods 137–145
   moderate-resolution visible–thermal IR methods 145–151
   passive-microwave radiometers 158
   radar scatterometers 158
   SAR techniques 151–158

Surface roughness *see* Roughness, surface
Surge behaviour and flow variability 15, 40, 90–91, 110, 254, 261
   detection of 271–276
   high-resolution visible–near IR 272–275
   InSAR techniques 40, 91, 110, 113, 258, 262, 271–274, 276
   Speckle-tracking techniques 261, 271–272
Surge waves, on a glacier 275
Svalbard 261, 271–272
Swiss Camp, Greenland 165
Synthetic aperture radar (SAR) 7, 18, 24, 39, 41, 43–52, 54–57, 60–67, 70–74, 76–84, 90–95, 98–116, 138, 142, 145, 151, 154–161, 163–164, 170, 174, 177, 179–180, 182–183, 185–186, 188, 190, 192–200, 202, 205–206, 208–209, 219, 231, 237–240, 244, 246–247, 249–251, 255, 257–258, 264, 266, 268–274, 276–278, 280–287, 289–291, 315, 317, 320–321, 323
   aircraft 44–45
   amplitude 41–43, 47, 54, 58, 98, 102, 190, 246, 262, 270, 318, 320
   antenna size 290
   C-band 47, 54, 58, 60, 64–65, 77, 80, 154–155
   complex images 42
   distortion, correction of 222
   dual-antenna systems 44–45, 49
   incidence angle 100, 201
   Ku-band 155
   L-band 45, 47, 58, 60, 64–65, 77, 80, 155
   looks 59, 69
   multi-look processing 59, 79, 83
   P-band 289
   phase *see* Phase
   polarimetricy 43, 111, 151, 158, 161–164, 199, 318, 320–321
   polarimetry, utility for ice-sheet facies detection 161–163
   polarimetry, utility for ice-sheet melt detection 166
   polarimetry, utility for sea-ice classification 111
   principles of 41–42
   resolution cell 42

ScanSAR imaging 78–81, 155, 166, 182, 192–195, 198, 287
single-antenna systems 44–45
split-antenna systems 78
techniques, ice-sheet facies 160–164
unfocussed 289
wavelength 43
X-band 47, 77, 154

Tandem Mission, Radarsat-1 and 2 114
Tape recorders, onboard 77
Temperature, air 13, 19, 166, 169, 291, 296, 313
Temperature gradients, seasonal variability 313
Temperature, ice/firn
   see also Temperature, surface
   comparison with isotopic fractionation 299
   diurnal cycle 296
   relationship with air temperature 296–297
   satellite passive-microwave techniques 295–299
      atmospheric effects on 296
      emissivity effects 295–298
      polarization effects 296
Temperature, vertical profile 14–15
Temperature, surface 13, 166, 248, 291–292
   Antarctic 291
   AVHRR Polar Pathfinder dataset 294
   effect on snow metamorphism 291
   effect on surface melt 291
   Greenland 291, 293–294
   satellite thermal IR measurement techniques 292–295
      atmospheric effects on 293
      cloud masking of 293–295
      correction of atmospheric effects 294
      single-channel techniques 293
      split-window techniques 293–294
   trends 298
Terminus, glacier 184–186
Terra, NASA spacecraft 139–140, 149, 152, 164, 183, 185, 198, 206–207, 234, 254, 294–295, 301, 305, 308–309, 316
TerraSAR-L 77, 80, 164, 247
   orbital parameters 79
   polarimetric capabilities 111, 114–115

TerraSAR-X 77–78, 164
   orbital parameters 79
   polarimetric capabilities 116
Texture analysis, image 190
Thermal infrared (TIR) 139–142, 145–147, 150, 159, 166, 177, 180, 183, 190, 193, 202–203, 205, 244, 249–250, 252–255, 263, 268, 270, 272, 274–275, 283, 292–295, 298–299, 315
THIR 244, 293
Thwaites Glacier, West Antarctica 224–225, 258, 273
   basal melt rate of 284
   mass budget of 283–284
   tidal displacements of 277
   thickness measurement of 282–283
Tidal currents 276
Tidal displacement of ice shelves and glacier tongues 276–279
   GPS measurement techniques 277
   in situ measurement techniques 277
   satellite measurements techniques 277
      InSAR techniques 280–281
Tidal effects on
   aliasing of satellite data 259, 278-279
   cracking 276
   flushing of sub ice-shelf sediments and water 276–277
   ice-sheet mechanics 276
   ice-shelf dynamics 276–277
   ice-stream flow velocity 277
   iceberg calving 276
   rift widening 276
   water-mass modification in sub ice-shelf cavity 279
Tidal flexing 277–278, 280-281
Tidal-flexure zone 276
Tidal harmonics 278
Tide models 100, 105–106, 108, 223, 232, 240, 259, 263, 277, 279
Titanic, RMS 193
TOPEX–Poseidon 4, 217–218, 224, 279
   maximum latitudinal coverage of 279
Totten Glacier, East Antarctica 284
Transantarctic mountains 152, 231
Transferential algorithm 110–111
Transition zone 160

United States Geological Survey (USGS) 141, 220
University of Colorado 183
University of Kansas 285

Validation of satellite data 21–23, 41, 116, 143, 211, 245, 313, 320
VECTRA 84, 114
Ventilation, deep-ocean 284
VIIRS 294, 305, 308
Victoria Land, Antarctica 140
Visible–thermal IR imagery
  high-resolution 98, 137–145
  moderate resolution 145–151
Visible–near infrared techniques 47, 139–152, 159, 164–165, 177–178, 180–185, 188, 190–192, 200, 202–203, 205–207, 211, 234–237, 249–250, 252–255, 263, 268, 270, 272, 274–275, 279–280, 283, 287, 299–312, 315–316
Visible-thermal IR imagery
  high-resolution 98, 137–145
  moderate resolution 145–151
Volume extinction coefficient 166, 213, 224
Vostok Station, Antarctica 254

WAIS *see* West Antarctic Ice Sheet
Ward Hunt Ice Shelf (Ellesmere Island) 177
Water vapor 114, 217
  absorption bands 217
Weddell Sea 192–193, 204, 205

West Antarctic Ice Sheet (WAIS) 7–8, 95, 107, 145–146, 158, 224, 230, 248, 256, 273, 276
  mass balance of 284
  recent thinning of 7
  volume of 8
West Ice Shelf (East Antarctica) 183
Wet-snow zone 64, 159–161
Whillans Ice Stream 145–146, 152, 236–237, 256, 261
  velocity variations 277
Wide Field Sensor (WiFS) 141
Wilkes Sub-glacial Basin 4
Wind conditions, local surface
  dependence on surface slope 312
  direction, proxy measurement techniques 316–318
  effect on accumulation 313
  effect on surface roughness 313
  effect on surface snow conditions/characteristics 313
Wind processes on ice-sheet surface *see also* Roughness, surface 309, 312–313
Wind slab 313
Wind speed and direction (velocity) 13, 22, 58, 166, 242
WindSat, NASA spacecraft 158
Wordie Ice Shelf, Antarctica 274, 276
World Glacier Monitoring Service 1

X-band satellite data reception 151

Printing: Mercedes-Druck, Berlin
Binding: Stein+Lehmann, Berlin